The Economics of Global Warming

The International Library of Critical Writings in Economics

Series Editor: Mark Blaug

Professor Emeritus, University of London
Professor Emeritus, University of Buckingham
Visiting Professor, University of Exeter

This series is an essential reference source for students, researchers and lecturers in economics. It presents by theme an authoritative selection of the most important articles across the entire spectrum of economics. Each volume has been prepared by a leading specialist who has written an authoritative introduction to the literature included.

A full list of published and future titles in this series is printed at the end of this volume.

The Economics of Global Warming

Edited by

Tom Tietenberg

Mitchell Family Professor of Economics
Colby College
Waterville, Maine

THE INTERNATIONAL LIBRARY OF CRITICAL WRITINGS IN ECONOMICS

An Elgar Reference Collection
Cheltenham, UK • Brookfield, US

Published by
Edward Elgar Publishing Limited
8 Lansdown Place
Cheltenham
Glos GL50 2HU
UK

Edward Elgar Publishing Company
Old Post Road
Brookfield
Vermont 05036
US

A catalogue record for this book is available from the British Library

Library of Congress Cataloging in Publication Data
The economics of global warming / edited by Tom Tietenberg.
(The international library of critical writings in economics ; 74) (An Elgar reference collection)
Includes bibliographical references and index.
1. Air—Pollution—Economic aspects 2. Global warming—Economic aspects. 3. Greenhouse effect, Atmospheric—Economic aspects. 4. Air—Pollution—Taxation. 5. Emissions trading. 6. Carbon taxes. I. Tietenberg, Thomas H. II. Series. III. Series : An Elgar reference collection.
HC79.A4E254 1997
363.73'92—dc20 96–32489
CIP

ISBN 1 85898 356 8

Printed in Great Britain by Galliard (Printers) Ltd, Great Yarmouth

Contents

C Deriving Optimal Solutions

PART III THE CHOICE OF INSTRUMENTS

A Carbon Taxes

B Tradeable Permits

Acknowledgements

The editor and publishers wish to thank the authors and the following publishers who have kindly given permission for the use of copyright material.

Academic Press, Inc. for article: Michael Hoel (1994), 'Efficient Climate Policy in the Presence of Free Riders', *Journal of Environmental Economics and Management*, **27** (3), November, 259–74.

American Economic Association for articles: Ralph C. d'Arge, William D. Schulze and David S. Brookshire (1982), 'Carbon Dioxide and Intergenerational Choice', *American Economic Review*, **72** (2), 251–6; Thomas C. Schelling (1992), 'Some Economics of Global Warming', *American Economic Review*, **82** (1), March, 1–14; Robert Mendelsohn, William D. Nordhaus, and Daigee Shaw (1994), 'The Impact of Global Warming on Agriculture: A Ricardian Analysis', *American Economic Review*, **84** (4), September, 753–71; John P. Weyant (1993), 'Costs of Reducing Global Carbon Emissions', *Journal of Economic Perspectives*, **7** (4), Fall, 27–46; Graciela Chichilnisky and Geoffrey Heal (1993), 'Global Environmental Risks', *Journal of Economic Perspectives*, **7** (4), Fall, 65–86.

Frank Cass & Co. Ltd for article: Geoffrey Bertram (1992), 'Tradeable Emission Permits and the Control of Greenhouse Gases', *Journal of Development Studies*, **28** (3), April, 423–46.

Elsevier Science B.V. for articles and excerpt: Michael Hoel (1992), 'Carbon Taxes: An International Tax or Harmonized Domestic Taxes?', *European Economic Review*, **36**, 400–406; Dale W. Jorgenson and Peter J. Wilcoxen (1993), 'Reducing US Carbon Emissions: An Econometric General Equilibrium Assessment', *Resource and Energy Economics*, **15** (1), 7–25; William D. Nordhaus (1993), 'Rolling the "DICE": An Optimal Transition Path for Controlling Greenhouse Gases', *Resource and Energy Economics*, **15** (1), 27–50; Stephen C. Peck and Thomas J. Teisberg (1993), 'Global Warming Uncertainties and the Value of Information: An Analysis Using CETA', *Resource and Energy Economics*, **15** (1), 71–97; Adam Rose and Brandt Stevens (1993), 'The Efficiency and Equity of Marketable Permits for CO_2 Emissions', *Resource and Energy Economics*, **15** (1), 117–46; Charles D. Kolstad (1994), 'The Timing of CO_2 Control in the Face of Uncertainty and Learning', in E.C.V. Ierland (ed.) *International Environmental Economics: Theories, Models and Applications to Climate Change, International Trade and Acidification*, 75–96.

Elsevier Science Ltd (Butterworth Heinemann) for articles: Alan S. Manne and Richard G. Richels (1991), 'Buying Greenhouse Insurance', *Energy Policy*, **19** (6), July/August, 543–52; Alan Manne, Robert Mendelsohn and Richard Richels (1995), 'MERGE: A Model for Evaluating Regional and Global Effects of GHG Reduction Policies', *Energy Policy*, **23** (1),

17–34; Richard Richels and Jae Edmonds (1995), 'The Economics of Stabilizing Atmospheric CO_2 Concentrations', *Energy Policy*, **23** (4/5), April/May, 373–8; Robert C. Lind (1995), 'Intergenerational Equity, Discounting, and the Role of Cost–Benefit Analysis in Evaluating Global Climate Policy', *Energy Policy*, **23** (4/5), April/May, 379–89; Thomas C. Schelling (1995), 'Intergenerational Discounting', *Energy Policy*, **23** (4/5), April/May, 395–401.

Institute for International Economics for excerpts: William R. Cline (1992), 'The Economic Benefits of Limiting Global Warming', in *The Economics of Global Warming*, Chapter 3, 81–138 and references; William R. Cline (1992), 'Benefit–Cost Synthesis', in *The Economics of Global Warming*, Chapter 7, 277–320 and references.

International Association for Energy Economics for articles: William D. Nordhaus (1991), 'The Cost of Slowing Climate Change: A Survey', *Energy Journal*, **12** (1), 37–65; John Whalley and Randall Wigle (1991), 'Cutting CO_2 Emissions: The Effects of Alternative Policy Approaches', *The Energy Journal*, **12** (1), 109–24.

International Monetary Fund for articles: William R. Cline (1993), 'Give Greenhouse Abatement a Fair Chance', *Finance & Development*, **30** (1), March, 3–5; Nancy Birdsall and Andrew Steer (1993), 'Act Now on Global Warming – But Don't Cook the Books', *Finance & Development*, **30** (1), March, 6–8.

Kluwer Academic Publishers for article: David G. Victor (1991), 'Limits of Market-Based Strategies for Slowing Global Warming: The Case of Tradeable Permits', *Policy Sciences*, **24** (2), May, 199–222.

Macmillan Magazines Ltd for article: Cynthia Rosenzweig and Martin L. Parry (1994), 'Potential Impact of Climate Change on World Food Supply', *Nature*, **367**, January, 133–8.

MIT Press for excerpt: John Whalley and Randall Wigle (1991), 'The International Incidence of Carbon Taxes', in Rudiger Dornbusch and James M. Poterba (eds), *Global Warming: Economic Policy Responses*, Chapter 7, 233–63.

Oxford University Press for articles: Scott Barrett (1990), 'The Problem of Global Environmental Protection', *Oxford Review of Economic Policy*, **6** (1), 68–79; Alan Ingham, James Maw and Alistair Ulph (1991), 'Empirical Measures of Carbon Taxes', *Oxford Review of Economic Policy*, **7** (2), 99–122; Bjorn Larsen and Anwar Shah (1994), 'Global Tradeable Carbon Permits, Participation Incentives, and Transfers', *Oxford Economic Papers*, **46** (Suppl.), October, 841–56.

United Nations Publications Board for article: Tom Tietenberg (1992), 'Relevant Experience with Tradeable Entitlements', in *Combating Global Warming: Study on a Global System of Tradeable Carbon Emission Entitlements*, Chapter IV, United Nations Conference on Trade and Development, 37–54.

Every effort has been made to trace all the copyright holders but if any have been inadvertently overlooked the publishers will be pleased to make the necessary arrangement at the first opportunity.

In addition the publishers wish to thank the Library of the London School of Economics and Political Science, the Marshall Library of Economics, Cambridge University, and B & N Microfilm, London, for their assistance in obtaining these articles.

Introduction

Tom Tietenberg

Few challenges pose a greater threat to our economic and political institutions than global warming. A number of characteristics conspire to create this complexity:

- Although the science behind the greenhouse effect is well established, the scientific evidence behind the specific contributions made by human-generated pollution is less so. As a practical matter this not only complicates the search for solutions, but it undermines the political will to take action (Schelling, 1992).
- Global warming is a phenomenon which evolves slowly; the emissions and damages are widely separated in time. Not only does this force a longer planning horizon than usual, but it raises serious ethical questions about the appropriateness of standard optimality techniques such as cost–benefit analysis (Cline, 1992; D'Arge, Schulze and Brookshire, 1982; Lind, 1995; Schelling, 1995).
- Several insights flow from the recognition that the control of human generated greenhouse gases is a global public good. International solutions must be sought as unilateral approaches are unlikely to be sufficient. Yet international cooperation is difficult since nonparticipants cannot be excluded from the benefits of successful control. Nations can become 'free riders' by leaving control to others, an action which undermines international cooperative efforts (Barrett, 1990; Hoel, 1994).
- The benefits and costs of global warming will not be uniformly distributed across nations. Experts believe that the developing countries, especially island nations, would tend to suffer most because of the vulnerability of their economies and their relative lack of adaptive strategies (Schelling, 1992; Rosenzweig and Parry, 1994). Yet these are not the nations which are currently emitting most of the greenhouse gases. Furthermore, depending upon the chosen approach, the burdens of controlling greenhouse gases will not fall equally either (Whalley and Wigle, 1991).
- The cost of being wrong is enormous. Most strategies for controlling global warming focus to no small extent on reduction in emissions of gases from the combustion of fossil fuels. Fossil fuels are currently the fuels of choice and the modern economy has become heavily dependent on them. Forcing a rapid transition away from these fuels raises the spectre of considerable economic disruption. On the other hand, failing to respond in a timely manner to the challenge of global warming could subject the planet to unprecedented ecological stresses. The consequences of these stresses can only be guessed at in part because many of the dynamic feedback effects which would be set in motion are simply not completely understood (Cline, 1992).
- The risks posed by global warming are not the kinds of risks normally captured effectively by traditional economic analysis. In particular the risks associated with climate change are poorly understood (even the likelihoods of various outcomes are difficult to discern), endogenous (in the sense that human actions affect the level of risk), collective (in the sense that climate change will affect large numbers of people in the same way) and irreversible (Chichilnisky and Heal, 1993).

This, then, is the challenge to policy and to those who have something to contribute to the evolution of that policy. While many disciplines can contribute in important ways to that process, economics is, and should be, a full partner in that debate. In this book I have attempted to draw together some of the most important studies on the economics of global warming to make them easily available in one place. They run the gamut from methodological concerns over discounting and the appropriate treatment of uncertainty to implementation issues raised about the two most commonly discussed policy instruments for controlling global warming – carbon taxes and tradeable carbon emission permits.

Background

One class of global pollutants, popularly called greenhouse gases, absorb the long wavelength (infrared) radiation from the earth's surface and atmosphere, trapping heat that would otherwise radiate back into space. The existing mix and distribution of these gases within the atmosphere is in no small part responsible for both our hospitable climate on the earth and the inhospitable climate on other planets. Changing the mix of these gases could modify the climate and the ecosystems which have adapted to the current climate.

Although carbon dioxide is the most abundant and the most studied of these greenhouse gases, many others have similar thermal radiation properties. These include the chlorofluorocarbons, nitrous oxide and methane.

The current concern over the effect of this class of pollutants on climate arises because emissions of these gases are increasing over time, changing their mix in the atmosphere. Evidence is mounting that by burning fossil fuels, levelling tropical forests, and injecting more of the other greenhouse gases into the atmosphere, humans are creating a thermal blanket capable of trapping enough heat to raise the temperature of the earth's surface. The 1995 Intergovernmental Panel on Climate Change (IPCC) report anticipates a surface air warming of between 1.0 and 3.5° C (1.8 and 6.3° F) by the year 2100. A 1.5° C increase in the temperature within a century would produce the warmest climate in 6000 years.

Temperature increases of this magnitude could also trigger a rise in sea level. The Intergovernmental Panel on Climate Change projects that global sea level will rise from 20 to 86 centimetres by 2100, with 49 centimetres the best guess for the year 2100. This compares with a rise of only 15 centimetres over the last century. Much larger increases could be expected if the warming caused the West Antarctic ice sheet to disintegrate, a low-probability event.

Damages, Costs and the Search for an Optimal Response

Damage Estimates

What are the likely consequences of temperature increases of this magnitude? While a number of studies exist (see the survey in Cline, 1992), most effort has gone into predicting the effects on agriculture. Mendelsohn, Nordhaus and Shaw (1994) have estimated that the damages in the United States would be small because higher valued crops could be substituted for the crops which are more vulnerable to the climate-induced increases in temperature. While these estimates are much more optimistic than others, Mendelsohn and his co-authors attribute this to the fact that other studies fail to include all the possible adaptation strategies farmers could employ.

Others, with a more global focus, are less optimistic. Rosenzweig and Parry (1994), for example, conclude that while adaptability may be high in the industrialized countries, limited opportunities for farmers in developing countries means that they are likely to bear the brunt of the problem. Since many of those countries are already producing too little food, global warming would only exacerbate the problem.

The implication of this analysis, that nations have different stakes in the international search for solutions to global warming, goes well beyond their differential ability to adapt. Regions with a characteristically cold climate, such as large portions of the Soviet Union, may actually benefit from this warming trend, while others that are naturally somewhat arid may see marginal agricultural land become unproductive desert, triggering a diminished capacity for raising food. These differences (and the rather divergent senses of urgency they promote) could prove quite divisive in the search for cooperative solutions to the global warming problem.

Control Costs

How expensive is greenhouse gas control? In this collection of essays four answers to that question are provided. The analysis of Jorgenson and Wilcoxen (1993) features a methodologically unique model. It is distinguished by the considerable sectoral detail it embodies (35 industrial sectors) in a general equilibrium model, by the extensive database (1947–85) used to estimate its parameters, and by treating productivity growth endogenously.

Whereas the Jorgenson–Wilcoxen model considers only the USA, Whalley and Wigle (1991) present a one-period (1990–2030) general equilibrium model which has a global focus. Dividing the world into six regions this highly aggregated model is capable of explicitly considering international trade effects.

The Nordhaus (1991) survey combines the various estimates of the costs of controlling emissions to construct marginal and total global cost functions. His estimates suggest that a modest reduction of greenhouse gas emissions can be obtained at low cost.

The final essay in this section (Weyant, 1993) presents the results of a comparative exercise involving subjecting some models to over two dozen common control scenarios. It discusses the resulting projections and concludes: 'The cost of a long-run global carbon emissions control program designed to limit emissions to their 1990 level (about 6 billion metric tons) is projected to be about 2.5% of world GDP by the year 2043, plus or minus 1%.'

The Search for Optimality

How much should be invested in greenhouse gas control now? That is a terribly important, if difficult, question to answer. With looming social needs for funds (health care, education, crime, etc.), how large should current commitments to the control of global warming be?

The three essays in this section of the book take three rather different approaches to that question and, not surprisingly, come up with different answers. The DICE (Dynamic

Integrated Climate-Economy) model developed by Nordhaus (1993) is the only explicitly optimizing model. Based on the Ramsey model of economic growth, its estimated trajectories can be interpreted as the most efficient path for slowing climate change, given initial endowments. The optimal policy involves modest current investments and a continual increase in global temperatures, at least until 2105.

Cline (1992) in contrast constructs a simple model to forecast benefits and costs of global warming control as a function of 36 global warming strategies. Benefit–cost ratios are calculated for each. Summarizing these results he concludes: 'using the central assumptions of this study, if there is weighting of outcomes to take account of risk aversion, the benefit–cost analysis finds that aggressive abatement action is warranted'.

What explains the difference between these two studies? One factor, the choice of discount rates, seems to explain most of the difference. Cline uses a lower discount rate and both he and Nordhaus agree this is a central element of the difference. (This controversy over the appropriate discount rate to use is covered in subsequent essays in this book.)

Another difference results from the way the possibility of high damage outcomes are treated. In his choice of which scenarios to emphasize Cline tends to favour those which weigh the possibility of high damage outcomes more heavily. Nordhaus tends to recognize these possibilities qualitatively in the final caveat section of this paper rather than give them prominence in this calcuation of optimal strategies.

Manne, Mendelsohn and Richels (1995) present another nonoptimizing model which is designed to trace the implications of various strategies for controlling global warming in an internally consistent way. Dividing the world into five geopolitical regions (the USA, other OECD countries, the former Soviet Union, China and the Rest of the World), this model is used to calculate the costs and benefits that result from four alternative policies, as well as to characterize the sensitivity of these outcomes to changes in some key underlying parameters (the willingness to pay to avoid ecological damages, the cost of alternative emission control policies and the discount rate). Summarizing these results the authors conclude: 'Although stabilizing concentrations buys the most – in terms of reduced market and non-market damages – it also carries the highest price tag. Costs outweigh benefits by almost eight to one. From a purely benefit–cost perspective the only policy that appears attractive is the delayed tax.' (The delayed tax scenario involves imposing a five dollars per ton tax on carbon emissions after a delay of 30 years.)

The final essay (Richels and Edmonds, 1995) uses a cost-effectiveness framework rather than a benefit–cost framework. Starting with predefined atmospheric CO_2 concentration targets (as opposed to attempting to derive the efficient targets endogenously), this study examines both the total cost of reaching various targets as well as the optimal paths for reaching them. The significant conclusion is that the time path for reaching the target is as important a policy choice as the target itself.

The Choice of Instruments

Carbon Taxes

What policy approaches are available to deal efficiently with the global warming problem?

Although global warming and ozone depletion impose an environmental cost, currently that cost is not being borne, or even recognized, by those who ultimately control the magnitude of the problem. Furthermore, those choosing unilaterally to reduce their emissions expose themselves to the higher costs associated with prevention strategies.

A 'carbon tax', which has been implemented unilaterally in a few countries and is currently being widely discussed in Europe and the United States, could be one component of this package. The initial focus on carbon dioxide, rather than on all greenhouse gases, is due not only to its current importance in the global warming problem, but also due to the relative ease with which carbon dioxide emissions can be monitored (Victor, 1991). Ultimately taxes would necessarily be imposed on other gases as well.

The transition to a more sustainable economic system in atmospheric terms will depend upon the development of new technologies and upon much greater levels of energy efficiency than are currently being achieved. Those transitions will not occur unless the prevailing economic incentives support and encourage them. Once the appropriate taxes were in effect the incentives would be changed; greater energy efficiency and the development of new technologies would become top priority objectives.

As desirable as carbon charges or taxes are in principle, formidable barriers stand in the way of their implementation in practice. Three issues must be resolved before a carbon tax becomes an international reality. First, some fair means of distributing the burden of control must be identified and implemented. Appropriate mechanisms for distributing the revenue can be used to moderate tax outcomes which are perceived as unfair. Secondly, the specific form of the tax must be agreed upon. Should the system consist of harmonized domestic taxes or an international tax collected by some international agency? Finally, the system must be designed in such a way as to encourage nations to participate in the agreement as well as to minimize the adverse consequences of those nations which chose to be nonparticipating free riders (Hoel, 1994).

Hoel (1992) examines three of the most commonly discussed tax designs. Domestic carbon taxes designed to achieve domestic goals would be a cost-effective means of achieving those goals, but the free rider problem suggests that a collection of domestic policies may not add up to efficient global policy. The second and third options envision: (1) an international carbon tax collected by an international agency with the revenue reimbursed to the governments of the participating countries on the basis of fixed reimbursement shares and (2) internationally harmonized domestic taxes, involving the same tax rate applied to all countries. The analysis concludes that while these last two options may look very similar in principle, in practice they produce very different distributional outcomes. Because nations keep the revenue they collect from harmonized domestic taxes, this approach represents the special case of the international tax approach in which the reimbursement shares are designed to produce no net outflow of revenue.

Whalley and Wigle (1991) provide a detailed empirical look into the international incidence of these two different approaches to carbon taxation. Their model investigates controls stringent enough to achieve a global carbon reduction of 50 per cent, a larger reduction than considered by most analysts. Their results indicate not only that the revenues from such a tax would be very large, but that the distributional differences between domestic taxes and an international tax are enormous. In particular the global tax allows the developing countries and the centrally planned economies to benefit financially from global warming control, while they experience substantial losses from harmonized domestic taxes. As a consequence of the resulting large transfers to the developing countries which are a part of the global tax package, financial losses to the industrialized nations are substantially greater with the global tax than with harmonized domestic taxes.

While the availability of revenue from carbon taxes offers a significant opportunity for encouraging developing countries to participate in agreements, that opportunity can only be exploited by agreements which envision some flow of funds from the industrialized to the developing countries. Whether those flows are politically feasible remains to be seen.

Tradeable Permits

Tradeable permits offer an alternative approach for controlling global warming which approaches the distributional issue from a slightly different vantage point. The process would be initiated by establishing a cap on the amounts each nation could emit on an annual basis. Nations achieving reductions of greenhouse gases sufficient to keep them under their cap could sell excess allowances to other nations. By purchasing these allowances, the acquiring nation could increase its emissions of greenhouse gases by an equivalent amount. The total emission limit negotiated by the agreement would be binding; no trade leading to a net increase in aggregate emissions above this limit would be approved.

As long as the sum of emissions was equal to the total permitted annual emissions, cost-effectiveness could be achieved regardless of how these emission allowances were initially allocated across countries. An international market in emission allowances would facilitate their movement from those countries with the capacity to create them most cheaply to those countries faced by very high costs of additional control. The capacity to achieve cost-effectiveness regardless of how the initial emission allowances are allocated, is a significant attribute of a tradeable permits approach that can be exploited for developing a market approach to cost-sharing. The potential cost savings from introducing trading could be large (Manne and Rutherford, 1994).

The tradeable permits concept has been applied to air pollution control since the mid-1970s. A review of the lessons to be derived from that experience suggests that several characteristics of the global warming problem are compatible with the circumstances when tradeable permits have worked best (Tietenberg, 1992). In particular greenhouse gases are uniformly mixed (which implies that only the amount emitted is a policy concern; the location does not matter); the global nature of the market offers the maximum potential for avoiding problems such as the rise of market power; emission trading allows the issue of who pays for control to be separated from who undertakes control, and it fits rather naturally into a convention which is based upon meeting specific emission targets (as the Climate Change Convention is).

The key discussion required to implement a tradeable permits system is how to allocate the permits. Appealing to the ethical principle that all of the planet's inhabitants should have an equal right to emit carbon, Bertram (1992) proposes and analyses a system initiated by allocating permit on the basis of population. For most of the industrialized countries of the world current emissions far exceed emissions allowed by a population-based initial allocation. The desire to meet the allowed emissions goal at least cost motivates an industrialized country's demand for acquiring permits from developing countries. The resulting financial transfers are substantial ($50 billion per year with permit prices of $20/ton).

Population-based initial allocations are, of course, not the only possible strategies rooted in ethical principles. Rose and Stevens (1993) identify ten possible allocation rules which flow from a variety of ethical foundations; two of these are analysed in some detail: (1) cutting back on emissions proportionally across all nations and (2) distributing large proportions of the permits to the poorest nations. This paper suggests that in either case trading would reduce costs substantially compared with the pre-trade allocation and, given the initial distribution of permits, would make all countries better off.

Larson and Shah (1994), however, point out that designing international agreements solely on the basis of general principles of fairness ignores a substantial impediment to solving the global warming problem – participation in any international agreement is voluntary. Countries which end up being injured by those principles can refuse to participate. Furthermore, since international agreements are difficult to enforce, it is important to construct agreements which provide all parties with incentives to live up to the agreement.

Larson and Shah (1994) argue that none of the principles examined by Bertram (1992) or Rose and Stevens (1993) satisfy these conditions because they place too high a burden either on the OECD countries or the developing countries. The allocation scheme they propose (and analyse) involves allocating all non-OECD countries permits equal to their projected emissions, while the OECD countries would be allocated permits equal to the world emissions target minus the allocations to the non-OECD countries. This, they argue, would provide incentives to participate to the OECD countries (they could lower their costs by one half compared with unilateral reductions), while the non-OECD countries would receive sufficient permits to implement current development plans.

A step toward a tradeable permits approach was established by the Climate Change Convention which went into force on 21 March, 1994. Reducing carbon dioxide emissions to 1990 levels by the end of the decade was specified as the cap. Limiting carbon emissions to specific levels now appears to be an internationally recognized objective.

The convention identifies a set of nations (listed in Annex I) committed to achieving a stabilization of carbon dioxide emissions by 2000. Since the *laissez-faire* outcome would involve increasing emissions, emission reductions will be required to meet the target. Other nations have subscribed to the general principles of the convention, but are not on the list of nations committed to meeting the stabilization target.

As the Annex I nations seek to reduce emissions sufficiently to meet the target, they have two choices: they can either find the reductions within their own borders or they can seek them in other nations. An official process, known as Joint Implementation, has been established to explore and to develop a set of procedures for transborder transfers of reduction offsets. Pilot projects, funded by the GEF, have been initiated in China, Jamaica, Mexico and the Philippines.

Hedging Strategies

Cutting emissions is, of course, not the only possible response to the challenge of global warming. Another involves investments in reducing the uncertainty concerning the ultimate impacts of global warming. How should these strategies be balanced with emissions reduction and how important is additional research and development?

A relatively recent literature, based upon decision tree analysis, provides both a systematic framework for thinking about these questions and some initial crude answers. This literature recognizes that policies chosen in the face of uncertainty are likely to be wrong. Acquiring more information and spreading out the reductions allows the policies to be more carefully tailored and, therefore, reduces those costs. It also delays action which intensifies the ultimate damages. Balancing off the cost reductions from purposeful delays with the potentially higher damage levels provides a basis for the simultaneous calculation of optimal emissions reduction strategies and strategies to improve information. It also allows the value of additional information to be calculated.

After providing an excellent introduction to the decision tree approach, Manne and Richels (1991) use a model of the USA to evaluate the differences between a 'learn, then act' strategy and an 'act, then learn' strategy. While their model holds the cumulative amount of emissions constant over the 60-year planning horizon for this comparison, the 'act, then learn' scenario requires policies to be determined now for the next 20 years without knowing how serious a problem global warming will turn out to be. It therefore selects a 'middle of the road' policy which is not perfectly appropriate for either the high damage or low damage outcomes. By contrast, the 'learn, then act' scenario delays action, but then bases the policies on better information over this 20-year period. After the 20-year period all uncertainty is considered resolved in both scenarios.

Manne and Richels conclude that the value of better information is potentially substantial, but that value is quite sensitive not only to its accuracy, but also to the prospects for new supply and conservation technologies. Furthermore, they find that programmes to reduce climate uncertainty and to develop new supply and conservation options lower the need for immediate precautionary cutbacks.

The Manne–Richels model is for the United States. Would these results hold on a global basis and over a longer time horizon? According to Peck and Teisberg (1993) they do. Using a global model with a 200-year planning horizon (which was shortened for some of the specific calculations in the article), they conclude that the value of information is certainly large enough to justify current research and development efforts and perhaps large enough to justify higher funding.

Earlier policy choices are more likely to be suboptimal due to the uncertainty associated with identifying the appropriate approach. According to the Peck–Teisberg model, as long as the policy choices are optimal at each point in time, given the prevailing uncertainty, ample time is available to resolve uncertainties. However, if the political process chooses suboptimal policies (as the authors in footnote 26 seem to suggest is a likely possibility) and this choice can be prevented by early resolution of uncertainty, the urgency of resolving uncertainty is dramatically increased.

Kolstad (1994) focuses on how the irreversibility of emission control investments affects conclusions about the value of information and the desirability of undertaking immediate reductions. In his stochastic growth model Kolstad focuses on the consequences of embedding autonomous learning (where the mere passage of time reduces uncertainty). He finds that if emissions control is perfectly reversible (no sunk capital), then, due to the long time lag between emission and damages, autonomous learning does not appreciably affect today's optimal emissions control policies. On the other hand, with irreversible control investments, the possibility of rapid learning reduces current optimal control investments.

Ethics and Intergenerational Discounting

A more fundamental concern involves defining the appropriate procedures for incorporating intergenerational fairness considerations into the analysis. Noneconomists are nearly universal in their condemnation of the use of discounting in deriving optimal solutions for very long-term problems such as global warming. The charge is that this technique short-changes future generations.

Defining Appropriate Discount Rates

Since the present value component of cost–benefit analysis emphasizes near-term over long-term consequences, the application of discounting in cost–benefit analysis will weigh the current costs of controlling emissions more heavily than the distant future damages. The classic debate is between those who prefer a high discount rate based on the high opportunity cost of capital (Birdsall and Steer, 1993), implying less current control, and those who believe it should be very low based upon a low rate of time preference (Cline, 1993), implying more.

As the Lind (1995) review of the theoretical foundations for discounting reveals, the typical formula for the rate of time preference (which in equilibrium is equal to the marginal rate of return on investment) involves the sum of two variables: (1) the utility rate of discount (also called the 'pure rate of time preference') and (2) the product of the elasticity of the marginal utility of consumption and the rate of growth of consumption. The implication is that as long as consumption is growing, the discount rate is positive, even with a zero rate of time preference. Furthermore, using a discount rate lower than the marginal return on investment, regardless of how derived, runs the risk that inferior investments will be justified.

Schelling (1995) calls into question two of the ethical foundations of discounting using time preference as the basis. Arguing that discounting for time preference *is* certainly appropriate for decisions delaying consumption streams later in one's own life, Schelling concludes that it *is not* appropriate when applied to problems (such as global warming) where the beneficiaries of deferred consumption are future generations. On the other hand, discounting to take account of economic growth assumes (mistakenly in Schelling's view) that future generations will be richer than current generations, even after accounting for the future damages. Even after economic growth the prime beneficiaries of global warming control (the developing countries) can be expected to be 'significantly poorer than the abatement-financing countries are now'. According to this provocative analysis, relying on a rate of time preference foundation for discounting implies a negative rather than a positive discount rate. Using a negative discount rate obviously conflicts with the desire to assure that only projects with a high rate of return are undertaken.

The Suitability of Intergenerational Discounting Framework

At the most abstract level the issue is not about the appropriate level of the discount rate, but rather whether the present value criterion is a sufficient framework for incorporating ethical considerations into global warming policy choices. Based upon their analysis of the premises behind discounting, both Lind (1995) and Schelling (1995) conclude that it is not.

D'Arge, Schulze and Brookshire (1982) take a different approach. They develop a formal analysis of some ethical propositions and show that the discount rate which satisfies specific ethical norms and the market rate of return on investments are equal only under some special conditions.

What is the appropriate means of taking ethical considerations into account? The conclusion which emerges from this analysis is that discounting may be perfectly appropriate, even for a problem such as global warming, but *not* in the conventional manner. From a sustainability perspective the issue should not be (as it has been) how best to incorporate discounting into the *objective function*. Rather, the issue should be how best to design the *constraints* in order to assure sustainability, using the present value criterion to select from among the infinite number of sustainable outcomes.

Economic optimality involves maximizing an objective function subject to certain constraints. The constraints define the feasible set. Once the constraints are defined so as to include only sustainable outcomes, cost–benefit analysis can be used to find the most efficient sustainable outcome. For global warming this principle can be applied in three steps. First, an acceptable upper limit on temperature increase or greenhouse gas concentration can be defined. Secondly, cost–benefit analysis using the present value criterion can be used to select the optimal sustainable path to each limit. Thirdly, the analysis can be used to facilitate both the choice of the most appropriate limit and the best path for reaching it. This is precisely the kind of analysis which is presented in Richels and Edmonds (1995).

References

Manne, A.S. and T.F. Rutherford (1994), 'International Trade in Oil, Gas and Carbon Emission Rights: An Intertemporal General Equilibrium Model', *Energy Journal*, **15** (1), 57–76.

Part I
General

[1]

OXFORD REVIEW OF ECONOMIC POLICY, VOL. 6, NO. 1

THE PROBLEM OF GLOBAL ENVIRONMENTAL PROTECTION

SCOTT BARRETT
London Business School[1]

Suppose that land is communally owned. Every person has the right to hunt, till, or mine the land. This form of ownership fails to concentrate the cost associated with any person's exercise of his communal right on that person. If a person seeks to maximize the value of his communal rights, he will tend to overhunt and overwork the land because some of the costs of his doing so are borne by others. The stock of game and the richness of the soil will be diminished too quickly. It is conceivable that those who own these rights, i.e. every member of the community, can agree to curtail the rate at which they work the lands if negotiating and policing costs are zero ... [However,] negotiating costs will be large because it is difficult for many persons to reach a mutually satisfactory agreement, especially when each hold-out has the right to work the land as fast as he pleases. [Furthermore,] even if an agreement among all can be reached, we must yet take account of the costs of policing the agreement, and these may be large, also.

Demsetz (1967, pp. 354–5)

I. INTRODUCTION

Demsetz's influential paper on the development of private property rights makes depressing reading for anyone concerned about global common property resources such as the oceans and atmosphere. Demsetz's view—and it is one that is shared by many others—is that users of a communally owned resource will fail to come to an agreement on managing the resource even though it is in the interest of all users to co-operate and reduce their rates of use of the resource. The reason is that if this improved situation is attained, every user will earn even higher returns by free-riding on the virtuous behaviour of the remaining co-operators. As a consequence, united action on the part of users can be expected to be unstable; co-operative agreements, even if they are reached, will not persist. The only

[1] I have benefited greatly from comments made by David Pearce on an earlier draft of this paper.

way out of the common property dilemma, as Demsetz makes clear, is intervention by '*the* state, *the* courts, or the leaders of *the* community' (emphasis added). In Demsetz's example, the intervention manifests itself in the development of private property rights to the resource, but the intervention could just as easily involve regulation.

The reason this view is disquieting is that for global common property resources there is no World Government empowered to intervene for the good of all. To be sure, there do exist international institutions—most notably the United Nations Environment Programme—which have been given the mandate to co-ordinate international environmental protection efforts. But none of these institutions can dictate what is to be done; that requires agreement by the parties concerned. The problem is perhaps best exemplified by the International Whaling Commission (IWC), which was established to conserve whale stocks, but whose best efforts in this regard have been repeatedly foiled. IWC membership is open to any country, and this leaves open the possibility that the whales could be protected for the global good. But any member can object to a majority decision, and hence render that decision meaningless. For example, a 1954 proposal to prohibit the taking of blue whales in the North Pacific was rejected by the only members who hunted blue whales in this ocean—Canada, Japan, the US, and the USSR—and hence did nothing to protect this species. In 1981 the IWC sought to ban the use of the non-explosive harpoon for killing minke whales. The ban was objected to by Brazil, Iceland, Japan, Norway, and the USSR. Since these were the only countries that hunted minke whales, the ban had no effect.[2]

Because national sovereignty must be respected, the problem of conserving global common property resources is no different from that described by Demsetz. The only way out of the global common property dilemma is agreement. Yet, just as in the situation Demsetz describes, there are strong incentives for governments not to co-operate, or to defect from an agreement should one be reached. This is the crux of the problem of managing global common property, and what distinguishes this problem from the long studied one of common property management under the jurisdictional control of a central authority.

Attempts to correct global, unidirectional externalities will encounter similar difficulties. Consider the problem where certain activities by one country harm all others. A good example is deforestation of Amazonia by Brazil. The rain forests play a crucial role in the protection of biological diversity and in the functioning of the carbon cycle. When standing, the rain forests serve as habitat to about a half of all wildife species and absorb carbon dioxide, one of the so-called greenhouse gases. When the forests are burned, masses of species can become extinct and substantial quantities of greenhouse gases are emitted. If the rights to generate these externalities are vested in the one country, as indeed they are in the case of Brazilian deforestation, then the others will have to pay this nation to cease its destructive activities. If the externality affected only one other country, then bargaining might be possible; the externalities might be internalized without outside intervention.[3] But in the case of global externalities, all countries except the generator suffer. All sufferers might be willing to bribe the generator to cease its harmful activities. But a contribution by any one country would confer benefits on all others and not just the one making the compensating payment. The others could therefore do better by free-riding. But then so too could the one that contemplated making the payment. Co-operation would again be foiled.[4] Mechanisms exist that can lead countries to reveal their preferences for global public goods truthfully (see, e.g., Groves and Ledyard, 1977), and hence for correct-

[2] See Lyster (1985).

[3] See Coase (1960). Bargaining has in fact taken place at the bilateral level. A famous example is the Trail Smelter case. The Canadian smelter emitted pollutants that crossed the US border. The case was arbitrated by an international tribunal comprised of an American, a Canadian, and a Belgian. The tribunal found that Canada was liable for damages, and also established emission regulations for the plant. The judicial decisions on this case make fascinating reading. See Trail Smelter Arbitral Tribunal (1939, 1941).

[4] Demsetz (1967, p. 357) argues that in the large numbers case, 'it may be too costly to internalize effects through the market-place'. Elsewhere, Demsetz (1964) argues that it might in fact be optimal for the externality not to be internalized since the costs of internalization should include the costs of transacting the agreement. But even then the free-rider problem would prevail; intervention, were it possible, might still be desired.

OXFORD REVIEW OF ECONOMIC POLICY, VOL. 6, NO. 1

ing global externalities. But in the absence of a World Government these mechanisms cannot be employed without the consent of the sovereign nations themselves. Every country would be better off if it agreed to participate in the revelation exercise. But each would do even better if others participated and it did not. All will therefore choose not to participate. The crux of the problem of correcting global externalities, like that of managing global common property, is that global optimality demands global co-operation, and yet the incentives facing individual countries work in the opposite direction.

The theoretical arguments for supposing that co-operation will not develop are compelling. But they can hardly be complete. Co-operation *does* take place and is often codified in international agreements. Some of these are woefully ineffective—a famous example being the International Convention for the Regulation of Whaling (1946) which established the IWC. Others do appear to have achieved a great deal. Of these last, the Montreal Protocol on Substances that Deplete the Ozone Layer (1987) seems the most impressive, because it demands that its many signatories undertake substantial reductions in their emissions of ozone-depleting chlorofluorocarbons (CFCs) and halons. Though agreements dealing with unidirectional externalities are rare and almost invariably toothless, there is one—the World Heritage Convention—that at least holds some promise. This agreement places responsibility for safeguarding natural environments like the Serengeti and the Galapagos Islands on a community of nations, and could be invoked to protect the remaining tropical rain forests. There is clearly a need to explore why international co-operation might develop, and what the significance of particular forms of co-operation might be.

To make any progress we will need a basis from which to assess whether co-operation can in fact be expected to achieve much. Contrary to Hardin's (1968) famous allegory of the commons, the absence of co-operation need not lead to tragedy. Section II discusses some of the parameters that are important in determining the potential gains to co-operation. Having drawn the boundaries, we then consider how we might move closer to the full co-operative solution, the global optimum. Non-co-operation may sometimes wear the disguise of co-operation, and section III shows that an outcome better than the purely nationalistic one can emerge even where binding agreements are absent. Effective management of global environmental resources does however seem to rely on the more formal institution of international law. Section IV discusses the rudiments of a model that explains why countries would co-operate when the free-rider problem must surely bite, and what international agreements mean for global social welfare and the welfare of citizens of individual countries. Just as failure to co-operate may not lead to tragedy, so co-operation may not buy us very much. Indeed, combining the analysis of the potential gains to co-operation with this model of formal agreements, it can be shown that co-operation is sometimes hardest to obtain when it is most needed.

II. THE POTENTIAL GAINS TO CO-OPERATION

Where a global externality is unidirectional, the country causing the externality will, in the absence of a negotiated settlement, ignore the damages its activities impose on other countries. This is the full *non-co-operative* outcome. The full *co-operative* outcome is found by internalizing the externality. In this case the country inflicting the externality chooses its actions so as to maximize the net benefits of all countries, including itself. Global net benefits will of course be higher in this case. The difference between the global net benefits for the co-operative and non-co-operative outcomes defines the potential gains to co-operation.

Where the externalities of concern are reciprocal in nature, every country has some incentive to take unilateral action even in the absence of a binding agreement. Furthermore, the strength of this incentive will depend on the actions taken by all other countries. An example of a reciprocal externality is the emission of a global pollutant. If one country reduces its emissions, it will benefit from the improved environmental quality, provided other countries do not increase their emissions so as to fully offset the one country's action. The other countries will benefit partly by being able to increase their emissions somewhat and partly by enjoying a cleaner environment (again, provided their increase in emissions does not entirely offset the one country's extra abatement). The extent of the benefit enjoyed by the conserving country will clearly depend on the actions taken by the other countries, and of course all countries are subject to a similar calculus. It is this interdependence which makes calculation of the potential gains to co-operative management of global common property more difficult. It is better, then, that we work with a specific model.

To fix ideas, reconsider the problem of global pollution. Suppose that the relevant number of countries is N. One might think that N would include all the world's countries, but that need not be so, a point we return to later. Let us however suppose for simplicity that N does include all countries and that each is identical. Each, therefore, emits the same quantity of a pollutant *ex ante*—that is, before the game is played—and each faces the same abatement cost and benefit functions. The problem is then perfectly symmetric. To simplify the analysis further, assume that the *marginal* abatement cost and benefit functions are linear. Clearly, the marginal abatement cost schedule for each country must depend on its own abatement level, while each country's marginal abatement benefit function must depend on *world-wide* abatement.

In the absence of any co-operation, each country will maximize its own net benefits of abatement and in so doing will choose a level of abatement at which its own marginal abatement cost equals its own marginal abatement benefit.[5] This is the non-co-operative (Nash equilibrium) solution to this game, and it is shown as abatement level Q^* in Figure 1. Were countries to co-operate fully, they would seek to maximize the global net benefits of abatement. Since we have assumed that all countries are identical, the global net benefits of abatement can be defined as the sum of every country's net benefits of abatement. In maximizing the global net benefits of abatement, each country will choose a level of abatement at which its own marginal costs of abatement equal the global marginal benefits of abatement, or the sum of the marginal abatement benefits enjoyed by all countries.[6] The full co-operative solution to this game is shown as abatement level Q^{**} in Figure 1.

Figure 1
Graphical Illustration of the Potential Gains to Co-operation

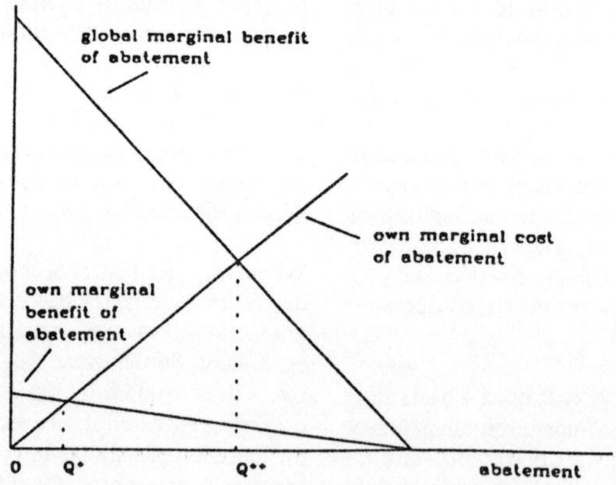

[5] I am assuming here that every country believes that its choice of an abatement level will not alter the choices of the other countries; that is, I am assuming zero conjectural variations. One can impose positive or negative conjectures, but these assumptions would be *ad hoc*. Alternatively, we could determine a consistent conjectures equilibrium—that is, one in whose neighbourhood every country's conjectures are confirmed by the responses of the other countries. Cornes and Sandler (1983) find that consistent conjectures can lead to even greater overuse of the resource compared with the Nash equilibrium.

[6] This is of course nothing but a restatement of Samuelson's (1954) rule for the optimal provision of public goods. For an alternative presentation of these principles, see Dasgupta's (1982) model of a global fishery.

OXFORD REVIEW OF ECONOMIC POLICY, VOL. 6, NO. 1

One sees immediately that the full co-operative solution demands greater abatement but, equally, gives to every country a greater net benefit. For a given size of N, the difference between Q^* and Q^{**} can be shown to depend on the slopes of the marginal abatement benefit and cost curves. Denote the (absolute value of the) slope of each country's marginal abatement benefit curve by the letter b, and the slope of each country's marginal abatement cost curve by the letter c. Then it can be shown that the discrepancy will tend to be small whenever c/b is either 'large' or 'small' (see Barrett, 1989). The approximate implication of this result is that fairly innocuous pollutants (that is, pollutants for which b is small) that are very costly to control (that is, that have a large c) do not cause too great a problem. Nor do extremely hazardous pollutants (that is, pollutants associated with a large b) that are cheap to control (that is, pollutants that have a small c). In the former case, even collective action will not call for large abatement levels. In the latter, countries will want to abate substantial quantities of emissions unilaterally. The real problem is with pollutants whose marginal abatement benefit and cost curves are both either steep or flat—that is, hazardous pollutants that are costly to control, and mildly offensive pollutants that can be controlled at little cost. Of these, the former type naturally causes the greatest concern, for the cost of failing to co-operate in this case is very, very high.

Unidirectional and reciprocal externalities are plainly different in their effects. A country has no incentive to abate its emissions if the externality is unidirectional (provided side payments are ruled out), even if the pollutant is highly toxic. Not so if the externality is reciprocal in nature. For then the emitting country will have strong private incentives to control its emissions.

III. NON-CO-OPERATIVE ENVIRONMENTAL PROTECTION

The politics of global environmental protection are not as sterile as the above models would imply. Once we permit alternative strategies to be chosen, widen the choice sets themselves, and allow motivations other than self-interest (narrowly defined) to guide decision-making, global environmental protection can be enhanced.

(i) Supergames

In the above games, strategies are chosen once, and the games are never repeated. But in the common property game countries are unavoidably locked in a continuing relationship, and this leaves open the possibility that they may retaliate and hence that co-operative strategies may be countenanced. Suppose then that all countries choose one of two strategies: they either choose the full co-operative abatement level Q^{**} initially, and continue to choose Q^{**} in every future period provided all other countries chose Q^{**} in every previous period; or they choose not to co-operate in any period.[7] Then the co-operative trigger strategy will constitute an equilibrium to this supergame provided the rate of discount is sufficiently low (for then the gains to choosing the non-co-operative strategy will be low, too; see Friedman, 1986). Although this supergame equilibrium entails co-operation, the co-operation is tacit and is enforced by means of a non-co-operative mechanism, retaliation; there is no explicit agreement, no open negotiation.[8]

One variant of this game involves countries adopting a convention that says that a subset of countries should co-operate in the face of free-riding by the others (see Sugden, 1986). It may, for example, be believed that the industrialized countries should co-operate to reduce emissions of CFCs or greenhouse gases (because it was their emissions that caused the environmental reservoirs to be filled in the first place), and that the poor countries should be allowed to free ride. In this game, each of the members of the subset could adopt the trigger strategy with respect to the subset, and each of the others could simply choose their optimal non-co-operative responses. This latter possibility may be fragile, because different countries may have different views about which should co-operate and

[7] If countries choose not to co-operate in the initial period, they need not choose Q^* in this period. The reason for this is that the optimal non-co-operative abatement level is contingent on the abatement levels chosen by the other countries.

[8] Axelrod's (1984) tournaments of the repeated prisoners' dilemma game suggest that behaviour in the disguise of co-operation may well emerge. Whether his findings carry over to the common property game with many players, however, is as yet unknown.

which should not. This ambiguity may help explain why unilateral action to reduce CFC emissions was limited (US Environmental Protection Agency, 1988, p. 30576):

> In 1978 the United States restricted the use of CFCs in aerosols. While several nations adopted similar restrictions (e.g. Sweden, Canada, Norway) and others partially cut back their use (European nations, Japan), there was no widespread movement to follow the United States' lead. Concerns existed then that other nations had failed to act because the United States and a few other nations were making the reductions thought necessary to protect the ozone layer. Similar concerns exist today that unilateral action could result in 'free riding' by some other nations.

It is not obvious how sufferers of a unidirectional externality could punish the offending nation. But so long as the countries are engaged in some form of exchange, potent weapons may be at the sufferers' disposal. The Packwood–Magnuson Amendment to the US Fishery Conservation and Management Act (1976) requires the US government to retaliate whenever foreign nationals compromise the effectiveness of the IWC. An offending nation automatically loses half its allocation of fish products taken from US waters, and if the country refuses to improve its behaviour within a year, its right to fish in US waters is revoked.

(ii) Matching

In the one-shot common property game, each country chooses an abatement level and nothing else. There is no reason why a country's choices need be limited in this way. We could, for example, allow countries to choose a 'base' abatement level—that is, a level of abatement which is not explicitly contingent on the abatement levels chosen by other countries—and a 'matching rate'—in our example, a fraction of the sum of all countries' base abatement levels. In effect, countries would then voluntarily subsidize each other's abatement levels. In the original game, one unit of abatement by country i buys country i one unit less of global emission (assuming that the emissions of all other countries are held fixed). In the matching game, one unit of abatement by country i may yield a much greater reduction in global emission. Matching might improve matters—Guttman (1978) shows that under certain conditions matching can sustain the full co-operative solution. But in the management of global common property, matching is rarely invoked. Recently, Norway announced that it would allocate one-tenth of one per cent of its GDP each year to a fund on climate change if other industrial countries matched its contribution (on a GDP percentage basis). The offer has yet to be taken up. Environmental groups in the United States have argued that the US should unilaterally surpass the reductions specified in the Montreal Protocol and at the same time impose restrictions on imports of products containing or made with CFCs from countries that fail to agree to make the same reductions. Their plea was rejected by the authorities (US Environmental Protection Agency, 1988, p. 30574).

(iii) Morality

In the models discussed thus far, the welfare of every country is assumed to depend solely on its *own* net benefits. An alternative way of looking at the problem is to assume that countries act according to some moral principle which requires that they take stock of the effect their actions have on the welfare of other countries. For example, suppose every country but one reduces its emissions of some global pollutant by at least x tonnes. Let us further suppose that the recalcitrant nation would like all others to reduce their emissions by y tonnes each. Then the leaders of the recalcitrant nation might feel compelled to obey the rule: if $y > x$ then we are morally obligated to reduce our emissions by at least x tonnes. If countries obey this rule, then the free-rider problem can be mitigated (see Sugden, 1984). That moral principles may guide non-co-operative abatement is suggested by the following remarks made by the House of Commons Environment Committee (1984, p. lxxi) in its report recommending that the UK join the Thirty Per Cent Club:[9]

> As our inquiry has progressed the stance of the United Kingdom has become increasingly isolated by its refusal to legislate to reduce SO_2 and NO_x emissions. Since our work began three West European countries

[9] The Thirty Per Cent Club consists of the countries that have signed the Protocol to the 1979 Convention on Long-Range Transboundary Air Pollution on the Reduction of Sulphur Emissions or their Transboundary Fluxes by at least 30 Per Cent (1985). The UK has not joined this 'club' but it has committed itself to substantial reductions in sulphur dioxide and nitrogen oxides emissions by agreeing to comply with the European Community Large Combustion Plant Directive.

have joined those already in the 30 per cent club, and several Eastern European countries have committed themselves to reduce transfrontier emissions by 30 per cent. SO_2 emissions in the United Kingdom have indeed fallen by 37 per cent since 1970, but the levels of high-stack emissions which affect remote areas have not fallen. In 1970, when the 37 per cent fall began, we were the largest emitter in Western Europe. In 1984, we are still the largest emitter. NO_x emissions have not fallen. In Western Europe only West Germany deposits more SO_2 in other countries than does the United Kingdom, and further significant reductions cannot be achieved by either without controls.

The Committee's concern in this passage lies less with the net benefits to the UK of reducing emissions than with how UK abatement has lagged behind the rest of Europe. Indeed, the Committee's evaluation did not even consider whether the other European countries were abating more simply because it was in their own self interest to do so. The argument seems to be: 'The other European nations are reducing their emissions, so we should, too.' Compared with the supergame problem, co-operation in this case is not instrumentally important—the Environment Committee did not seek to reduce UK emissions so that others would reduce theirs even further—but intrinsically important. Concerns about fairness have been shown to militate against the free-rider problem in experimental tests (Marwell and Ames, 1981). However, in the next section we shall see that in a bargaining situation, obeying moral principles may serve only to undermine the cause of environmental protection.

IV. CO-OPERATIVE ENVIRONMENTAL PROTECTION

In the absence of co-operation, outcomes better than the full non-co-operative one can sometimes develop, at least in principle. But such instances seem to be rare. Even two close neighbours with strong trading ties can fail to arrive at a preferred solution, as the disagreement between the United States and Canada over the exploitation of the North Pacific fur seal illustrates. Following an initial conflict between the two nations over the pelagic seal hunt, a Tribunal Arbitration was convened at the request of the two parties (with Great Britain acting for Canada). In late 1893 the Tribunal decided that the United States did not have territorial jurisdiction over the Bering Sea, and hence could not keep Canadian sealing vessels out of these waters. This effectively sanctioned open-access harvesting of the species, and co-operation proved impossible to secure (Paterson and Wilen, 1977, p. 94):

Following the decision of the Tribunal, the diplomatic efforts of both Great Britain and the United States had been directed to convincing the other to reduce its sealing in order to allow the herd to recover from earlier depradations. No agreement could be reached and in 1897 the United States unilaterally forbade its citizens to engage in pelagic sealing in the North Pacific. At the same time quota adjustments were made as the herd diminished in size. So strong was the reaction to the declining herd size and the continued Canadian pelagic hunt that a bill reached the [United States] Senate which called for the complete destruction of the herd. It did not pass . . .

Better management of the population had to await the signing of the North Pacific Fur Seal Treaty by Great Britain, the United States, Japan, and Russia in 1911—a remarkable agreement that remains in force today.[10] Effective management of global environmental resources seems to demand that countries co-operate openly and put their signatures on international agreements, treaties, and conventions. Explanations for why co-operation of this kind might emerge are offered below.

(i) International Environmental Agreements

Consider the following modification to the common property game described earlier. Suppose a subset of the N identical countries 'collude' by signing an international environmental agreement and that the remaining countries continue to act non-co-operatively. Suppose further that the signatories to the agreement choose their collective abatement level while taking as given the abatement decision functions of the non-signatories, while the latter countries continue to behave atomistically and choose their abatement levels on the assumption that the abatement levels of all other countries are fixed. That is, the signatories act as 'abatement leaders', and the non-signatories as 'abatement followers'. Quite clearly, we would

[10] For legal background on this treaty and its successors, see Lyster (1985).

like the number of signatories, the terms of the agreement, and the abatement levels of all non-signatories to be determined jointly. We also require that the agreement itself is stable. A stable agreement is one where non-signatories do not wish to sign the agreement and signatories do not wish to renege on their commitment. Then it can be shown that for identical countries with linear marginal abatement benefit and cost functions a stable international environmental agreement always exists (Barrett, 1989).

The solution to this problem exhibits many of the features of actual agreements. The net benefits realized by both signatories and non-signatories are higher than in the earlier problem where negotiation was ruled out. What is more, the signatories would like the non-co-operators to sign the agreement. However, non-signatories do better by free-riding.

It is important to emphasize that the agreement is *self-enforcing*. Any signatory that renounces its commitment can reduce its abatement level and hence its costs. However, in pulling out of the agreement, the number of co-operators is reduced and the agreement itself is weakened; the remaining co-operators reduce their abatement levels, too. A signatory will want to pull out of an agreement only if the saving in abatement costs exceeds the resulting loss in benefits. Similarly, a country that joins an agreement will have to abate more and hence incur higher costs. But the very act of joining will strengthen the agreement; the other co-operators will also increase their abatement levels. Joining appears attractive if the resulting increase in benefits realized by the new signatory exceeds the increase in costs that this country must incur in committing itself to the terms of the agreement.

Real treaties are not rewritten with every defection or accession, but mechanisms are at work that have a similar effect. It is common for treaties to come into force only after being ratified by a minimum number of signatories. The Montreal Protocol did not come into force until it had been ratified by at least eleven countries representing at least two-thirds of global consumption of the controlled substances. It is also common for treaties to be reviewed and altered when necessary and often at regular intervals. The Montreal Protocol is being reviewed in 1990, and is then expected to be strengthened. That this agreement is self-enforcing is suggested by a comment made by the US Environmental Protection Agency (EPA) (1988, p. 30573):

> EPA judged that the obvious need for broad international adherence to the Protocol counseled against the United States' deviating from the Protocol, because any significant deviation could lessen other countries' motivation to participate.

Self-enforcement is essential in any model of international environmental agreements because nation states cannot be forced to perform their legal obligations. A country can be taken to the International Court of Justice for failing to comply with the terms of a treaty, but only with the defendant's permission. Even then, the disputing countries cannot be forced to comply with the Court's decision.[11]

What are the gains to having international environmental agreements? The answer depends partly on the number of relevant and potential signatories. When N is large, international environmental agreements can achieve very little no matter the number of signatories. The reason, quite simply, is that when N is large, defection or accession by any country has only a negligible effect on the abatement of the other co-operators.

Determination of N is not always a trivial matter. Some treaties do not restrict participation, but in these cases many of the signatories may have no effective say in environmental protection. Over 100 countries have signed the 1963 Partial Nuclear Test Ban Treaty, but only a few signatories possess nuclear weapons technology. The 1967 United Nations Treaty on Principles Governing the Activities of States in the Exploration and Use of Outer Space including the Moon and Other Celestial Bodies has been ratified by scores of countries but not by the two with space technology capabilities—the US and USSR. Other treaties explicitly restrict participation. The Agreement on the Conservation of Polar Bears can only be signed by five circumpolar countries (Canada, Denmark (including Greenland), Norway, the US, and the

[11] See Lyster (1985) for a discussion of other compliance mechanisms.

USSR). To become a signatory to the Antarctic Treaty of 1959 a country must maintain a scientific research station in the Antarctic and be unanimously accepted by existing parties to the agreement. In these cases non-signatories may quite clearly be affected by how the signatories manage the resource. Signatories to the Antarctic treaty voted recently to allow mineral exploration, despite appeals by non-signatories to designate Antarctica a nature reserve.

In the above model, *N* was assumed to represent both the number of countries that emit a (uniformly mixed) pollutant into the environment and the number harmed as a consequence. However, for some problems the number of emitters may be less than the number of sufferers (for global pollutants, all countries). When the number of emitters is small, an international environmental agreement signed by a subset of emitters may well have a significant effect on the welfare of these countries. However, the effect on global welfare may still be small because the emitters have no incentive to take into account the welfare losses suffered by non-emitting nations. The appropriate way to account for countries that do not emit the pollutant but nevertheless suffer the consequences of others' emissions is to admit side payments—payments which induce emitting nations to undertake greater abatement but which leave all parties no worse off compared to the situation where side payments are forbidden. We return to the side payments issue later.

The gains to international co-operation can also be shown to depend on *c*, the slope of each country's marginal abatement cost curve; and *b*, the (absolute value of the) slope of each country's marginal abatement benefit curve. For a given size of *N*, the number of signatories to a treaty increases as *c/b* falls. This suggests that we should expect to observe a large number of signatories (in absolute terms) when *N* is 'large', the marginal abatement cost curve is flat, and the marginal abatement benefit curve steep. However, we already know that when *c/b* is 'small' the benefit of having an agreement is diminished. It is commonly asserted that treaties signed by a large number of countries accomplish little of substance: 'The greater the number of participants in the formulation of a treaty, the weaker or more ambiguous its provisions are likely to be since they have to reflect compromises making them acceptable to every State involved' (Lyster, 1985, p. 4). This analysis suggests that the reason treaties signed by a large number of countries appear to effect little additional abatement is not that the signatories are heterogeneous—although that may be a contributing factor. Nor is the reason solely that in these cases *N* is also large. A major insight of the model is that a large subset of *N* will sign an agreement only when the non-co-operative and full co-operative outcomes are already close.

This latter observation may not seem consistent with all the evidence. The Montreal Protocol, for example, demands of its signatories significant reductions in the production and consumption of the hard CFCs and halons, and about 46 countries have already signed the agreement—a fairly large number by any standard. As Table 1 shows, the effect of the agreement on ozone depletion is estimated to be very significant. Percentage ozone depletion is estimated to be reduced from 50 to 2 per cent in 2075 as a result of the agreement. But of course each country has some incentive to take unilateral action in reducing emissions; in doing so all other countries will benefit, but so too will the country taking the action. Furthermore, non-signatories to the agreement may well face an incentive to abate less than they would otherwise for the simple reason that greater abatement on the part of signatories improves the environment for non-signatories as well. Hence it is by no means clear that the agreement necessarily means that the environ-

Table 1
Estimates of the Reduction in Percentage Ozone Depletion Effected by the Montreal Protocol

Case	2000	2025	2050	2075
No controls	1.0	4.6	15.7	50.0
Montreal Protocol	0.8	1.5	1.9	1.9

Source: US EPA (1988), Table 3, p. 30575.

ment and global welfare will be significantly better off, contrary to what the figures in Table 1 imply. What the model does suggest is that so many countries would not have committed themselves to the agreement in the first place unless they already intended to take substantial unilateral action. In other words, although the agreement itself may effect only little additional abatement, the very fact that so many countries have signed the agreement suggests that the potential gains to co-operation were in this instance not very great.

What does the model predict about the prospects of an agreement being reached on global warming? *N* will again be large, and this will militate against significant united action. However, in this case *c/b* will be large, too; the marginal costs of abating carbon dioxide emissions will rise very steeply as fossil fuels must be substituted for and energy is conserved. This suggests that the number of signatories to an agreement would be small, and that little additional abatement could be effected by co-operation. The tragedy is that in the case of global warming fuller co-operation would probably result in huge gains in global welfare.[12]

(ii) Leadership

It is sometimes asserted that countries should, on their own, do more than the non-co-operative solution demands of them. US environmental groups, for example, have argued that the US should have taken greater unilateral action before the Montreal Protocol was drafted, that it should now comply with the terms of the Protocol in advance of the deadlines, and that it should exceed the agreed emission reductions and phase out production and consumption of these chemicals entirely. The House of Commons Energy Committee (1989, p. xvii), in its recent investigation on the greenhouse effect, recommended '. . . that the UK should . . . consider setting an example to the world by seriously tackling its own emission problems in advance of international action, especially where it is economically prudent to do so'.

We have already seen that such 'unselfish' unilateral actions need not be matched by other countries. The United States, Canada, Sweden, and Norway banned the use of CFCs in non-essential aerosols in the late 1970s, and yet other countries did not reciprocate.[13] Unilateral restrictions on pelagic sealing in the North Pacific by the US were not duplicated by Canada. We have also seen that countries may wish to give in to their moral beliefs and embrace a less insular view of their responsibilities. An important question is whether 'unselfish' unilateral action can be expected to have a positive influence on international negotiations. If one country (or group of countries) abates more than the Nash non-co-operative solution demands, and all others choose the abatement levels that are optimal for them in a non-co-operative setting, will the environment be any better protected when international treaties are later negotiated?

In a two-country analysis, Hoel (1989) shows that the answer depends on whether the unilateral action is taken before agreement is reached and is not contingent on that agreement or whether the action is a commitment to abate more than the negotiated agreement requires. Hoel shows that in the former case, 'unselfish' unilateral action may compromise negotiations and lead, ultimately, to *greater* emissions than would have occurred had both countries behaved 'selfishly'. In the latter case, however, the country's announced commitment to overfulfil its negotiated abatement level can be expected to reduce total emissions.

There is an obvious incentive compatability problem with this tactic, for the 'unselfish' country could do better by reneging on its commitment (the agreement is therefore *not* self-enforcing). Nevertheless, the analysis shows that the desire by environmentalists and others to reduce total emissions may not be well served by their calls for 'unselfish' unilateral action, a point that the EPA stressed in defending its ozone depletion policy (1988, p. 30574): 'Unilateral action by the United States would not significantly add to efforts to protect the ozone layer and could even be counter productive by undermining other nations' incentive to participate in the Protocol.'

It is important to note that the US and the European Community announced their intentions to phase out production and consumption of the ozone-

[12] The analysis given here is by necessity over-simplified. For a more detailed analysis, see Barrett (1990).

[13] Reciprocity was certainly not full. The European Community, for example, passed two decisions limiting production capacity of the so-called hard CFCs (CFC-11 and -12) and reducing their use in aerosols by 30 per cent.

depleting chemicals by the end of the century *after* the Montreal Protocol came into force but before re-negotiation talks had started. It would be wrong, however, to ascribe these developments simply to 'unselfish' behaviour. After all, the world's largest manufacturer of CFCs, US-based Du Pont, announced its intention to phase out production of CFCs by the end of the century *before* the phase-out decisions were taken by the US and EC. Three days after the EC decided to phase out CFCs, the chairman of the leading European producer of CFCs, ICI, declared that production of CFCs should cease 'as soon after 1998 as is practicable'. Much more is at work here.[14]

(iii) Efficient Co-operation

Signatories to an international environmental agreement are assumed to maximize the net benefits accruing to the *group*. This means, among other things, that the marginal abatement costs of every signatory must be equal; the abatement undertaken by the group must be achieved at minimum total cost.

How realistic is this assumption? In the case of the Montreal Protocol, the assumption is not very wide of the mark. The Protocol imposes on every industrial country signatory an obligation to reduce its production and consumption of CFCs by an equal percentage. This requirement on its own is inefficient because at the margin the costs of complying with the Protocol will surely vary. For example, the UK can apparently meet its obligations by simply prohibiting the use of CFCs in aerosols—an action that is nearly costless. The US banned the use of CFCs in aerosols many years ago, and hence can meet the terms of the Protocol only by instituting more costly measures. However, the Protocol allows limited international *trading* in emission reductions. For any signatory, CFC production through mid-1998 can be 10 per cent, and from mid-1998 onwards 15 per cent, higher than it would have been without trading provided the increase in production by this signatory is offset by a decrease in production by another signatory. Furthermore, trades of consumption (but, strangely, not production) quotas are permitted by the Protocol within the European Community. These provisions will help increase the efficiency of attaining the total emission reduction implicit in the agreement, although they almost certainly do not go far enough.

(iv) Side Payments

The equilibrium in the model of international environmental agreements is determined by a concept of stability that prohibits side payments. An important question is whether side payments might effect a Pareto improvement. To investigate this issue, reconsider the concept of stability employed in the model. In equilibrium, non-signatories do better than signatories, but no country can do better by changing its status. Signatories want non-co-operators to sign the agreement, because their net benefits would then increase. But non-signatories do worse by signing. Hence, without compensating payments, non-signatories will not want to sign the agreement. It is in this sense that the agreement is stable.

However, the very fact that signatories do better if non-co-operators sign the agreement suggests that trade might be possible. In particular, it might be possible for signatories to make side payments to a subset of non-co-operators to encourage them to sign the agreement. All might be made better off. It is in fact very easy to show that this can happen, that an international environmental agreement that specifies abatement levels *and* side payments can manage the global common property resource better than one that prohibits side payments.

An important feature of the World Heritage Convention is that it does admit side payments. The Convention established a World Heritage Fund that is used to help protect natural environments of 'outstanding universal value'. Each party to the Convention (there are over 90 signatories) is required to provide the Fund every two years with at least one per cent of its contribution to the regular budget of UNESCO.[15] In practice this means that the Fund is almost entirely financed by the industrial countries. Clearly, both the industrial and poor countries benefit from the Convention—otherwise

[14] See Barrett (1990) for an analysis of these developments.

[15] The United States and the UK continued to contribute to the Fund even after withdrawing their funding from UNESCO.

they would not have signed it—but the poor countries may not have signed the Convention were it not for the Fund. Though the Fund is small, the mechanism could prove instrumental in protecting many of the world's remaining natural environments, including the tropical rain forests.

The World Heritage Convention is unique among international environmental agreements for incorporating side payments. But the need for side payments is not unique to the conservation of natural environments and wildlife. It has, for example, become increasingly clear that the success of the Montreal Protocol will ultimately hinge on the accession of non-signatories to the Protocol. Concern has specially been voiced about the need to get the poor countries—and, in particular, China and India—to sign the Protocol. These countries have declared their need for financial assistance, and the issue of setting up a global fund for protecting the ozone layer is likely to dominate discussion at the next meeting of the Protocol's signatories. It would seem to be something that cannot be avoided.

REFERENCES

Axelrod, R. (1984), *The Evolution of Cooperation*, New York, Basic Books.
Barrett, S. (1989), 'On the Nature and Significance of International Environmental Agreements', mimeo, London Business School.
— (1990), *Ozone Holes, Greenhouse Gases, and Economic Policy*, London, Institute of Economic Affairs (forthcoming).
Coase, R. H. (1960), 'The Problem of Social Cost', *Journal of Law and Economics*, 3, 1–44.
Cornes, R. and Sandler, T. (1983), 'On Commons and Tragedies', *American Economic Review*, 73, 787–92.
Dasgupta, P. (1982), *The Control of Resources*, Cambridge, MA., Harvard University.
Demsetz, H. (1964), 'The Exchange and Enforcement of Property Rights', *Journal of Law and Economics*, 7, 11–26.
— (1967), 'Toward a Theory of Property Rights', *American Economic Review*, 57, 347–59.
Friedman, J. W. (1986), *Game Theory with Applications to Economics*, Oxford, Oxford University Press.
Groves, T. and Ledyard, J. (1977), 'Optimal Allocation of Public Goods: A Solution to the "Free Rider" Problem', *Econometrica*, 45, 783–809.
Guttman, J. M. (1978), 'Understanding Collective Action: Matching Behavior', *American Economic Review Papers and Proceedings*, 68, 251–5.
Hardin, G. (1968), 'The Tragedy of the Commons', *Science*, 162, 1243–8.
Hoel, M. (1989), 'Global Environmental Problems: The Effects of Unilateral Actions Taken by One Country', Working Paper No. 11, Department of Economics, University of Oslo.
House of Commons Energy Committee (1989), *Energy Policy Implications of the Greenhouse Effect*, vol. 1, London, HMSO.
House of Commons Environment Committee (1984), *Acid Rain*, vol. 1, London, HMSO.
Lyster, S. (1985), *International Wildlife Law*, Cambridge, Grotius.
Marwell, G. E. and Ames, R. E. (1981), 'Economists Free Ride, Does Anyone Else?', *Journal of Public Economics*, 15, 295–310.
Paterson, D. G. and Wilen, J. (1977), 'Depletion and Diplomacy: The North Pacific Seal Hunt, 1886–1910', *Research in Economic History*, 2, 81–139.
Samuelson, P. (1954), 'The Pure Theory of Public Expenditure', *Review of Economics and Statistics*, 36, 387–9.
Sugden, R. (1984), 'Reciprocity: The Supply of Public Goods through Voluntary Contributions', *Economic Journal*, 94, 772–87.
— (1986), *The Economics of Rights, Co-operation and Welfare*, Oxford, Basil Blackwell.
Trail Smelter Arbitral Tribunal (1939), 'Decision', *American Journal of International Law*, 33, 182–212.
— (1941), 'Decision', *American Journal of International Law*, 35, 684–736.
US Environmental Protection Agency (1988), 'Protection of Stratospheric Ozone; Final Rule', *Federal Register*, 53, 30566–602.

[2]

Some Economics of Global Warming

By Thomas C. Schelling *

Global warming from carbon dioxide was an esoteric topic 15 years ago, unknown to most of us. But in a few years, helped along by some hot summers, it has climbed to the top of the international agenda. Cabinets, Parliaments, and heads of government have issued pronouncements on reducing carbon emissions, and in June of this year more than a hundred governments will be represented by ministers or heads of government at a great United Nations Conference on Environment and Development to be held in Rio de Janeiro. Together with nongovernmental organizations representing labor, business, students, environmentalists, scientists, and groups concerned with health and child development and family planning, these representatives are expected to need 25,000 hotel rooms. A "framework agreement" is widely expected, together with some institutional arrangements that will keep global environmental issues permanently on every government's agenda. And at the center of these issues will be the phenomenon that has come to be known as the "greenhouse effect."

The greenhouse effect itself is simple enough to understand and is not in any real dispute. What is in dispute is its magnitude over the coming century, its translation into changes in climates around the globe, and the impacts of those climate changes on human welfare and the natural environment. These are beyond the professional understanding of any single person. The sciences involved are too numerous and diverse. Demography, economics, biology, and the technology sciences are needed to project emissions; atmospheric chemistry, oceanography, biology, and meteorology are needed to translate emissions into climates; biology, agronomy, health sciences, economics, sociology, and glaciology are needed to identify and assess impacts on human societies and natural ecosystems. And those are not all.

There are expert judgments on large pieces of the subject, but no single person clothed in this panoply of disciplines has shown up or is likely to. So, I venture to offer my judgment.

I

First on the principle. The metaphor of the greenhouse is not quite appropriate, but the basic idea is not in dispute. The earth is bathed in sunlight, some reflected and some absorbed. If the absorption is not matched by radiation back into space, the earth gets warmer until the intensity of that thermal radiation matches the absorbed incoming sunlight. Some gases in the atmosphere that are transparent to sunlight absorb radiation in the infrared spectrum, blocking that outward radiation and warming the atmosphere. When the atmosphere has warmed enough to intensify the thermal radiation so that it matches the absorbed incoming sunlight, equilibrium is achieved at the higher temperature. These so-called "greenhouse" gases can be identified in the laboratory. Carbon dioxide is one of them; methane is another, as is nitrous oxide, as are the chlorofluorocarbons (CFC's).

The principle has been in practice for decades. On a clear day in January, the earth and its adjacent air in Orange County California warm nicely, but the warmth radiates rapidly away during the clear nights, and frost may threaten the trees. Smudge pots, burning cheap oil on a windless night, produce substances, mainly carbon dioxide, that absorb the radiation and protect the trees with a blanket of warm air. Greenhouses, in contrast, mainly trap the air warmed by the earth's surface and keep it

*Department of Economics, University of Maryland, College Park, MD 20742.

from rising to be replaced by cooler air. The phenomenon should have been called the "smudgepot effect," but it is too late to do anything about it.

A first step in pursuing this phenomenon is to assess how much warming might go with an enhanced concentration of these gases. That cannot be done in the laboratory; there are too many feedbacks. A warmer atmosphere will contain more water vapor; water vapor itself is a greenhouse gas. Changes in temperature and humidity will change cloud cover; clouds can reflect or absorb incoming or outgoing light according to their composition and altitude. The average temperature is only one dimension; temperatures at different altitudes and different latitudes matter. But a starting point has been the change in average surface atmospheric temperature expected to accompany a specified increase in the concentration of greenhouse gasses; and arbitrarily, but reasonably, the base case is taken as a doubling of the concentration.

A moment on why a doubling is the benchmark. To compare estimates of warming, people must use the same hypothesized concentration of greenhouse gases in the atmosphere. (Alternatively, they could use the same hypothesized temperature increase and estimate the corresponding concentration.) Doubling, like a half-life in reverse, is a natural unit if it is within the range of practical interest, and it is. A doubling is expected sometime in the next century, so it is temporally relevant; and a doubling is estimated to make a substantial but not cataclysmic difference. If fixation on a doubling seems to imply an upper limit on any expected increase, the implication is unfortunate: enough fossil fuel exists to support several doublings.

In 1979, a committee of the National Academy of Sciences (NAS) (1979 p. 2) estimated the change in average temperature to accompany a doubling of carbon dioxide in the atmosphere: three degrees Celsius, with a range of 1.5 degrees to either side. (In the last 15 years other greenhouse gases have received attention; these other gases can be converted to their carbon dioxide equivalents and the original estimate applied to the mixture.) The NAS appointed another committee a few years later to reexamine that estimate, and the new committee saw no reason to change it (NAS, 1982 p. 51). An intergovernmental panel on climate change (IPCC), consisting of scientists from many nations, revisited the estimate in 1990 and concluded, from the several climate models they had examined, that "the models results do not justify altering the previously accepted range of 1.5 to 4.5 degrees C" (IPCC, 1990 p. xxv). Thus, the estimate appears to be robust over time, but the spread of uncertainty remains large: the upper limit is three times the lower limit. (No quantitative interpretation of these upper and lower "limits" has been made public. Both National Academy reports referred to them as "probable error.")

II

The uncertainties are even greater in translating a temperature change into climates. The media support a popular view that things will just get hotter; a news magazine cover was a sweating global face. But the laboratories that do the meteorology do not simply predict warming; they do not even predict that the most noticeable effects will necessarily be temperature changes. Among the great driving forces of weather and climate is the temperature differential between equatorial and polar regions; convection currents coupled with the rotation of the earth are engines of atmospheric circulation and, ultimately, ocean circulation. The models predict greater temperature change in the polar regions than near the equator. This change in gradient can drive changes in circulation. The results may be warmer in some places and colder in others, wetter in some places and drier in others, cloudier in some places and sunnier in others, stormier in some places and less stormy in others—generally a complex of changes that would bear no easy relation to an average change in global temperature.

The change in average temperature is useful as an *index* of climate change. It is thought, and the models demonstrate, that the greater the change in average temperature the greater the departure of current climates from what they are now. Thus, while it is wrong to think that what is going to happen can be readily characterized as "warming" it is not erroneous to take that average warming as a rough measure of the extent or severity of change to be expected. Unfortunately the widespread reference to "global warming" promotes the notion that things will simply get hotter. (Interestingly, virtually all public discussion is on hotter summers, not warmer winters; a hundred years ago popular discussion of a warming trend would likely have concentrated on the milder winters to be expected.)

If three degrees Celsius is taken as an index of climate change to come within the next century or so, how big is that compared with what has happened within the last century, or the last 10,000 years? From what I have just said, this cannot be answered in terms of whether anyone would notice the difference if every night and every morning, every winter and every summer, temperatures were exactly three degrees higher than they otherwise would have been. The question is: how would a three-degree change in a global average compare with what has been experienced in the past?

The answer is that for 10,000 years, since the disappearance of the last ice age, average temperature appears never to have varied over anything like three degrees. A band of one degree Celsius would cover the current estimates of what average temperatures have been since the dawn of recorded history. We will be moving into a climatic regime that has never been experienced in the current interglacial period.

"Mankind will undergo greater climate change in the next 100 years than has been experienced in the last 10,000." Properly qualified, the statement is true; what it neglects is that peoples have been migrating over great distances for at least several thousand years. Goths and Vandals, Huns, West Europeans who populated North and South America, Southerners who went North during the Great Depression, and Northeasterners who moved southwest after World War II all experienced changes in climate greater than any being forecast by the models. Almost everybody who attends this lecture in New Orleans will have undergone a greater change in the past few days than is expected to occur in any fixed locality during the coming century.

The changes that the models produce are *gradual* both in time and in space. The models do not produce discontinuities. Climates will "migrate" slowly. The climate of Kansas may become like Oklahoma's, Nebraska's like that of Kansas, South Dakota's like Nebraska's, but none of these is expected to become like the climates of Oregon, Louisiana, or Massachusetts.

A caution: the models probably cannot project discontinuities—just gradual change—because nothing goes into the models that will produce *catastrophes*. There may be phenomena that could produce drastic change, but they are not known with enough confidence to introduce them into the models. So the reassuring gradualness may be an artifact of the methodology. I will return to this point later.

This greenhouse problem, if problem it proves to be, is truly one of the "global common." A ton of carbon emitted anywhere on earth has the same effect as a ton emitted anywhere else. And carbon dioxide has a long residence time in the atmosphere: a century or more. There may be ways to remove it, but it doesn't disappear. The greenhouse influence on any national territory depends solely on the global concentration, not in any way on what part of the total is due to a nation's own emissions.

As I shall detail later, the costs of reducing carbon emissions will be large compared with any other emissions that have caused concern. The costs of phasing out CFC's will be in the billions of dollars per year for some years, and complete elimination is expected to be feasible. The cost of reducing sulfuric acid may be in the tens of billions of dollars. Proposals to hold emissions of carbon dioxide constant (with a linear increase of concentration in perpetuity) or to reduce emissions by 50 percent below what they would otherwise be, beginning perhaps in 2010, are expected to cost in the hundreds of billions in perpetuity.

There are a few numbers worth carrying in mind. There are 700 billion tons of carbon in the atmosphere. (Quotations are sometimes in tons of carbon dioxide, rather than carbon; the figure is then $3\frac{2}{3}$ times as large, about 2,600 billion.) Annual emissions are 6 billion tons. Close to half disappears somewhere, and a little over half remains in the atmosphere; so the concentration is increasing by one-half percent per year. It has increased 25 percent in the last hundred years. (Concentration is reported more often than tonnage; it is currently about 350 parts per million.) And there are upwards of ten trillion tons of carbon fuels out there to be burned; if it were all burned and half stayed in the atmosphere, the concentration could double at least three times.

If the carbon in the atmosphere has already increased by a quarter, has the average temperature gone up as predicted? And were the recent hot American summers that stirred popular interest harbingers of greenhouse summers to come?

To the first question, the answer is that average global temperature—summer and winter, both hemispheres, night and day—has apparently risen by half a degree in the last hundred years, but whether "as predicted" depends on what qualifications one reads into the predictions. The pattern differed between the Northern and Southern Hemispheres. The global average rose during the first 40 years of this century, was level for the next 40 years, and rose during the past decade. This pattern demonstrates that, whether or not we are witnessing the greenhouse effect, there are other decades-long influences that can obscure any smooth greenhouse trend. (The carbon concentration is not at issue; it is well measured and shows steady rise on a decade scale.) There are known phenomena that could account for the irregular temperature increase of the past century, and whether we are witnessing the "signal" probably depends on whether one wants high confidence to reject a null hypothesis or is about to bet money on whether, another 25 years from now, looking back, all doubt will have been removed. I don't know what bets are being placed by "greenhouse scientists," but they are cautious in public on the question.

To the second question—do the hot American summers of the past few years announce the arrival of a greenhouse, confirming predictions?—the answer is in two parts: maybe it's the greenhouse; but it's not what the greenhouse models predict. The global average in the four hot years of the past seven was only 0.2 degrees above the level of the preceding 40 years; and sudden hot American summers are not what the models predict.

III

In anticipating the impact on human welfare or natural systems, two kinds of uncertainty are unlikely to be dispelled soon. One is simply the question of what the changes will be in each region or locality. Current models are severely limited in their agreement with each other, in their handling of such topographical variables as mountain ranges, and in the fineness of the grids they superimpose on the globe. There is no great confidence that the models will be greatly improved within the next decade or two. A chaos-like process may defeat efforts to improve local predictions; and uncertainties in gross phenomena, such as the behavior of ocean currents under changed climatic conditions, may not be much better understood soon.

Even if we had confident estimates of climate change for different regions of the world, there would still be uncertainties about the kind of world it is going to be 50, 75, or 100 years from now. Imagine it were 1900 and the climate changes associated with a three-degree average temperature increase were projected to 1992. On what kind of world would we superimpose either a vaguely described potential change in climate or even a specific description of changes in the weather in all the seasons of the year, even for our own country. There would have been no way to assess the impact of changing climates on air travel, electronic communication, the construction of skyscrapers, or the value of California real estate. Most of us worked outdoors; life expectancy was 47 years (it is now 75); barely a fifth of us lived in cities of 50,000 or more. Anticipating the automobile, we might have been concerned with whether wetter and drier seasons would bring more or less mud, not anticipating that the nation's roads would become thoroughly paved. The assessment of effects on health would be without antibiotics or inoculation. And in contrast to most contemporary concern with the popular image of hotter summers to come, I think we would have been more concerned about warmer winters, later frost in autumn, and earlier thaw in the spring.

If the world, both North America and the other continents, is going to change as much in the next 90 years as it has changed in the 90 just past, we are going to be hard put to imagine the effects of climate changes.

Another thought experiment: suppose the kind of climate change expected between now and, say, 2080 had already taken place since 1900. Ask somebody 50, 60, or 80 years old what is different compared with when he or she was a child. Would the climate change be noticed? Even ask a 70-year-old farm couple living on the same farm where they were born: would the change in climate be among the most dramatic changes in either their farming or their lifestyle? I expect changing from horses to tractors and from kerosene to electricity, the arrival of the telephone and the automobile and the paving of roads, the development of pesticides and artificial fertilizer, the discovery of soy beans and the development of hybrid corn, and even improvements in outdoor clothing, veterinary medicine, and agricultural practices generally would swamp the climate change. And if instead of living and working conditions we inquire about changes in wildlife and natural ecosystems, changes in regional climates would have been competing, in their impact on nature, with population growth and economic development.

A conclusion we might reach is that a climate change would have appeared to make a vastly greater difference to the way people lived and earned their living in 1900 than to the way people live and earn their living today. Today very little of our gross domestic product is produced outdoors, susceptible to climate. Agriculture and forestry account for less than 3 percent of GDP, and little else is much affected. Some activities —tourism and holidays, professional sports, and school teaching—are seasonal, but many of the seasonalities are conventions that reflect the influence of climate in earlier times. (Children were needed in the fields in summer and could start school when the harvest was in; hockey and basketball used to be winter sports because one depended on ice and the other could fit in a building.)[1]

Manufacturing rarely depends on climate, and where temperature and humidity used to make a difference, air conditioning has intervened. When Toyota chooses among Ohio, Alabama, and Southern California for locating an automobile assembly, geographical considerations are important, but not because of climate. Minerals are extracted where they happen to occur, and oil fields and coal mines inhabit all kinds of climates and are little affected. The U.S. Postal Service's vow that neither snow nor rain nor heat nor gloom of night will "stay these couriers from the swift completion of their appointed rounds" sounds quaint in the era of e-mail and fax.

Finance is little affected by climate; similarly for health care, or education, or broadcasting. Transportation can be affected, but improvements in all-weather landing and take-off in the last 30 years are greater than any differences that climate makes. If the average effect is a warming, iced waterways and snow removal may decline in importance. Construction is affected, mainly by cold, and if the average effect is in the direction of warming, construction may benefit slightly.

It is really agriculture that is affected. But even if agricultural productivity declined by a third over the next half century, the per

[1] An imaginative discussion is in Jesse H. Ausubel (1991).

capita GNP we might have achieved by 2050 we would achieve only in 2051. Considering that in most of the developed countries—the United States, Japan, France, the United Kingdom, the Netherlands, and Israel—the agricultural problem has been protecting farmers, that agricultural productivity in most parts of the world continues to improve, and that many crops and cultivated plants will benefit directly from enhanced photosynthesis due to increased carbon dioxide, one cannot be certain that the net impact on agricultural productivity will be negative or, if negative, will be noticed in the developed world.

I conclude that in the United States, and probably Japan, Western Europe, and other developed countries, the impact on economic output will be negligible and unlikely to be noticed.[2] And there is no reason to believe that in these countries there could be a noticeable impact on health. Any influence of climate on health in this country would be more in the regional distribution of the population than in changes in local and regional climates.

Comfort is worth considering. Fortunately, the climate models predict a greater warming in winter than in summer. Most people in the United States, Japan, and Western Europe go south for vacation, both summer and winter; and when people move upon retiring in the United States they typically move toward warmer climates. In future years, elderly people may suffer more heat stroke in summer in St. Louis, but we can hope for fewer broken bones from ice in Boston. (Inhaling air richer in carbon dioxide has no effect on health.)

IV

This complacent assessment cannot be extended to the much larger population of the underdeveloped world. The livelihoods earned in agriculture and other climate-sensitive outdoor activities, 3 percent in the United States, comprise 30 percent and more of all livelihoods in most of the developing world. Reliable forecasts of likely climate changes in the different areas so dependent on agriculture are simply not available, so no assessment, region by region, of the effect on productivity can be provided. There is no strong presumption that the climates prevailing in different regions 50 or 100 years from now will be less conducive to food production. But there is also no assurance that climate changes will not be harmful, and even if on balance the impact is neutral, there may be large areas with large populations that suffer severely. Those people are vulnerable in a way that Americans, Western Europeans, and Japanese are not.

Nor can the impact on health be dismissed or readily subsumed among generally improving health conditions, as for the developed world. Numerous parasitic and other vector-borne diseases affecting hundreds of millions of people are sensitive to climate. Again, there is no strong presumption that malaria mosquitos, to take an example, will on balance benefit from climate changes, but the risk is there.

It is with the less-developed countries that we have to be most careful about superimposing the climates of the future on the economies and societies of today. As it was in our own country during this century, the trend in developing countries is to be less dependent on agriculture and less vulnerable to climate in transportation and other activities and health. If per capita income growth in the next 40 years compares with the 40 years just past, vulnerability to climate change should diminish, and the resources available for adaptation should be greater. I say this not to minimize concern about climate change, but to anticipate the question of whether developing countries should make sacrifices in their development to minimize the emission of gases that may change climate to their disadvantage. Their best defense against climate change may be their own continued development.

This is a point worth emphasizing. Some environmentalists argue that developing

[2] A comprehensive discussion of both impacts and costs of abatement is provided by William D. Nordhaus (1991a). A carefully argued opposing view is that of William R. Cline (1992).

countries should sacrifice some of their hopes for economic development in the interest of slowing the climate change that may prove disastrous. But the advice contains a contradiction Any disaster to developing countries from climate change will be a disaster to their economic development. What is desired is to optimize development by investing in greenhouse-gas abatement only when that appears, subject to all the uncertainties, to contribute more to their development in the future than the alternative direct investment in development. It is not economic growth versus environment; it is growth with the environment taken into account.

A related point: population growth is important for the climate change, in two respects. One is that carbon emissions in developing countries are positively driven by population; population growth does not merely dilute carbon emissions per capita, but for a number of reasons more people means more carbon. If China succeeds in holding population growth to near zero for the next couple of generations, it may do as much for the earth's atmosphere as would a heroic Chinese anticarbon program coupled with 2-percent annual population growth.

The other population effect is simply that the most likely adverse impact of climate change on human productivity and welfare would be on food production. In the poorest parts of the world, the adequacy of food depends on the number of mouths and stomachs. In a hundred years, adverse changes in climate for food production would be far more tragic if the countries we now associate with the developing world had populations totaling 12 billion than if they totaled 9 billion. For the developing world, the increasing concentration of people is probably more serious than the increasing concentration of carbon dioxide.

At this point, I appear to have reached the conclusion that the *developed* world has no self-interest in expensively curtailing carbon consumption and that the *developing* world cannot afford to incur economic penalties to slow the greenhouse effect. There is a mismatch between those who may be vulnerable to climate change and those who can afford to do anything about it.

V

Why should the rich developed countries care enough about climate change to do anything about it? The answer must depend partly on how expensive it is going to be to do anything about it. Abatement programs have been examined in a number of econometric models that suggest we might want to treat as pertinent the sacrifice of perhaps 2 percent of world GNP in perpetuity.

A strong argument for trying seriously to slow climate change is that the developing countries are vulnerable and we care. Developed countries are currently providing $50 billion per year of assistance to the developing world; we would be talking about expending or forgoing perhaps 4–8 times that much to slow emissions and slow climate change. Whether people in the developed democracies could be mobilized to contribute so much to benefit, half a century from now, the people in the countries we now call developing I do not know, but I believe that if the developed countries were prepared to invest, say, $200 billion per year in greenhouse-gas abatement, explicitly for the benefit of developing countries 50 years or more from now, the developing countries would clamor to receive the resources immediately in support of their continued development. There would undoubtedly be abatement opportunities so cheap that they could compete with direct aid to developing countries, but it would be hard to make the case that the countries we now perceive as vulnerable would be better off 50 or 75 years from now if 10 or 20 trillions of dollars had been invested in carbon abatement rather than in their economic development.

A second argument for an expensive program of carbon abatement is that, while our production of material goods and services may not suffer from climate change, our natural environment may be severely damaged. Natural ecosystems will be destroyed; plant and animal species will become extinct. Places of natural beauty will be degraded. Valuable chemistries of plant and animal life will be lost before we learn their genetic secrets. And the earth itself deserves our respect. For many people, something close to religious values are at stake.

This issue is doubly difficult to assess. It is difficult to know how to value what is at risk, and it is difficult to know just what is at risk. Even if climate changes at each point in time could be predicted accurately, the impacts on natural ecosystems could not yet be determined. And the benefits of slowing climate change by some particular amount would be even more uncertain. We know that carbon fuels are not going to be discontinued; the issue is the marginal gains, from carbon abatement and a slowing of climate change, in the survival of species and ecosystems and the preservation of enjoyable environments. This is an issue that simply has not been addressed.

The third argument for spending heavily to slow climate change is that the conclusions I reported earlier may be quite wrong. I said that the climate models predict that climates will change slowly and not much; the models do not produce discontinuities, surprises, *catastrophes*. What is known about weather and climate constitutes an equilibrium system.

The possibility has to be considered that if global temperature increases, not by the median estimate of three degrees Celsius for a doubling of carbon in the atmosphere, but by four or five degrees and continues to rise beyond the doubling because carbon fuels are still in use worldwide, some atmospheric or oceanic circulatory systems may switch to alternative equilibria, producing regional changes that are both sudden and extreme.

Have any such possibilities been thought of? One that was thought of but diminished upon further investigation was the possibility that the west Antarctic ice sheet might glaciate into the ocean and raise the sea level by 20 feet. As recently as 15 years ago, the best scientific judgment was that this could happen within 75 years as a result of global warming. This prospect naturally attracted attention, and further investigation with the help of newly available satellite sensing of glacial movement led to reassuring estimates that if that catastrophic rise in sea level were to happen it would take at least a few hundred years and be gradual, not sudden. But there isn't any scientific principle according to which all alarming possibilities prove to be benign upon further investigation.

A currently discussed likely source of discontinuous change is in the way oceans behave. Amsterdam is north of Newfoundland, yet is warmer, courtesy of the Gulf Stream. There is some indication that in earlier interglacial periods ocean currents may have pursued different courses. If a current like the Gulf Stream, or the Japanese Current for the United States, switched into an alternative pattern, the climatic consequences might be both sudden and severe. (Paradoxically, global warming might freeze Western Europe.)

Insurance against catastrophes is thus an argument for doing something expensive about greenhouse emissions. But to pay a couple percent of GNP as insurance premium, one would hope to know more about the risk to be averted. I believe research to improve climate predictions should be concentrated on the extreme possibilities, not on modest improvements to median projections.

I said that current estimates suggest that it might cost a couple percent of GNP to postpone the doubling of carbon in the atmosphere by several decades. Is 2 percent a big number or a small one?

That depends on your perspective and on what the comparison is. In recent years 100 billion dollars per year in budgets or taxes has been a politically unmanageable magnitude in the United States. On the other hand, subtracting 2 percent from GNP in perpetuity lowers the GNP curve by not much more than the thickness of a line drawn with a number-two pencil, or to formulate it as I did earlier, it postpones the GNP of 2050 until 2051. I say this not to belittle the loss of 10 trillion dollars from the American GNP over the next 60 years, but only to point out that the insurance premium, if we choose to pay it, will not send us to the poorhouse. The proper question is whether, if we were prepared to spend 2 percent of our GNP in the interest of protecting against damage due to climate change, we might find better use for the money.

I have mentioned one use: directly investing to improve the economies of the poorer countries. Another would be direct investment in preserving species, ecosystems, or wilderness areas. There is concern that many ecosystems could not migrate as rapidly as climate may change in the coming century; there has been little investigation of what might be done to facilitate the migration of ecosystems if the alternative is to invest 5 or 10 trillions of dollars in the reduction of carbon emissions.

VI

What can be done to reduce or offset carbon emissions? Reducing energy use and the carbon content of energy have received, I believe properly, most of the attention, especially the attention of economists. There are other possibilities to mention.

Trees store carbon. In growing, they take it out of the atmosphere. When they rot or burn it goes back into the atmosphere. A new forest will absorb carbon until it reaches maturity (i.e., maximum carbon density) in 75 or 100 years. If it then merely replenishes itself, with new growth replacing the oxidized dead trees, it holds its carbon but does not absorb more. If trees are harvested, the lumber that becomes house frames or furniture may last a hundred years or more; removing mature trees and storing them anaerobically is possible but expensive. The most recent report of the National Academy of Sciences considered that reforestation in the United States might sequester 2–3 percent of current global carbon dioxide emissions.[3] The prospects for that kind of reforestation in the rest of the world are not nearly so promising, and we should conclude that reforestation can contribute, but not greatly.

Stopping or slowing *deforestation* is important for reasons other than carbon emissions but is quantitatively more important than reforestation. Reforestation is unlikely to take up as much as 100 billion tons of carbon; deforestation, in areas where deforestation is likely, could contribute several hundred billion tons of carbon, partly because forest subsoils contain carbon typically greater than the amount in the trees themselves, and this carbon is subject to oxidation when the trees are removed.

Carbon can be "scrubbed" from stack gases, probably not with any known technology that would make such removal economically competitive with reducing emissions. (Part of the expense is disposing of sludge; where gaseous carbon might be pumped into the ocean or into underground cavities, economical disposal may prove feasible.) Parallel to reforestation is the idea of enhancing oceanic photosynthesis, by "fertilizing" the oceans, possibly with iron, if enough of the carbon residues from the enhanced growth will sink rather than remain near the surface. Experiments would probably be reversible and modest in scale; their political acceptability may be tested in the near future.

Finally—although nothing is final in a subject as new as the one we are talking about—there are numerous possibilities for putting substances or objects in orbit or in the stratosphere to reflect something like 1 percent of incoming sunlight to offset a large part of the radiation imbalance caused by greenhouse gases. Some of these are as apparently innocuous as stimulating cloud formation, and some are as dramatic as huge mylar balloons in low earth orbit. Until very recently these possibilities were nearly unmentionable, but they have recently been dignified by inclusion, along with caveats about "large unknowns concerning possible environmental side effects," in the 1991 report of the National Academy of Sciences. I shall not pursue them here, except for two observations. First, if in decades to come the greenhouse impact begins to confirm the more alarmist expectations, and if the economic sacrifices required to reduce

[3]Their estimate is 10 percent of U.S. emissions at "low to moderate cost" on economically or environmentally marginal crop and pasture lands and nonfederal forest lands in the United States (National Academy of Sciences, 1991 p. 57). A review of the issues in both afforestation and deforestation by Andrew Plantinga is in Joel Darmstadter (1991); an optimistic estimate of the afforestation option is that of Robert J. Moulton and Kenneth R. Richards (1990).

emissions prove unmanageable for economic or political reasons, some of these "geoengineering" options will invite attention. Second, if they do, and especially if they prove to be within the budgetary capabilities of individual nations, international greenhouse diplomacy will be transformed.

VII

What remains nearly certain is that the main responses to the greenhouse threat will be adapting to climate as climate changes and reducing carbon emissions. (CFC's are potent greenhouse gases and, if unchecked, might rival carbon dioxide in decades to come; but international actions are making good progress and are among the cheapest ways of reducing greenhouse emissions.)

Like estimates of warming, estimates of the costs of reducing emissions require some common but arbitrary objective to be comparable. A doubling of carbon became the conventional benchmark for warming estimates; no such benchmark for reduced carbon emissions has been adopted for estimating costs. (In principle, the estimates could adopt that doubling: the issue could be formulated as the cost of retarding the doubling time by a decade, two decades, or half a century.) Most estimates take as their target a reduction of emissions either to a specified fraction of what they would be in the absence of controls, or to some fixed ratio to the emissions of 1990 or the projected emissions of 2000 or 2010. The estimates examine minimum-cost trajectories, implicitly or explicitly assuming something like a uniform tax on the carbon content of fuel as the policy instrument. They typically make some assumption about a "fallback" energy technology, at least for electricity, available at some price in some decade of the next century. They have to project estimates of non-price-induced improvements in the use or avoidance of energy by industries and households. And if they deal with global emissions, they have to make some assumption about the distribution of abatement efforts among nations, especially among the developing countries, which, including China, account for about a quarter of emissions now and would be expected to account for half by the middle of the next century.

Any estimate of the cost of abatement needs therefore to specify at least half a dozen target assumptions. Furthermore, the estimates are produced by people and institutions that do not simultaneously estimate the costs associated with climate change, either damages or costs of adapting; the estimates do not optimize the combined costs of abatement and climate change. A "not unreasonable" target for reduction might be delaying a doubling by, say, four decades. One decade might be too trivial, a century too ambitious, and four decades an objective in which most audiences would be interested. But nobody who makes such an estimate wishes to be interpreted as proposing that when all the uncertainties about climate changes and their impacts have been resolved, if they ever are resolved, the optimum reduction in emissions will be found to retard doubling by 40 years, or any other specified period of time.

All I can do to summarize a multitude of estimates is to specify an order of magnitude that many economists and the Congressional Budget Office would not consider outrageous. That is the figure I mentioned earlier, possibly 2 percent of GNP for the developed countries and a similar, but even much more uncertain, percentage of GNP for the developing world. The uncertainty for the developing world is partly due to the estimates being mainly derived from the American economy.[4]

Two characteristics of these estimates need to be emphasized. One is that they tend to assume optimal technological adjustment, as in response to a carbon tax. To the extent that carbon emissions are controlled by direct regulatory measures, there may be the usual expected inefficiencies, and I leave the reader to make his own adjustment.

[4]Several critiques and surveys of different abatement-cost estimates are available (see Congressional Budget Office, 1990; Joel Darmstadter, 1991; William D. Nordhaus, 1991b; Energy Modeling Forum, 1992).

The second is that, since the early years of the energy crisis in the 1970's, there have been enthusiastic portrayals of currently available technologies, ranging from light bulbs to electric motors, double-glazed windows and improved internal-combustion engines, that for some reason have not been successfully marketed. The interest continues, and the recent National Academy of Sciences study gave sympathetic attention, but no analysis, to a number of proposals for residential, commercial, industrial, and transportation energy management and for improved electricity production and fuel supply and concluded that, including reductions in CFC's, "The United States could reduce or offset its greenhouse gas emissions by between 10 and 40 percent of 1990 levels at low cost or at some net saving, if the proper policies are implemented" (1991 p. 73).

All of these ideas are completely orthogonal to the econometric estimates. The Academy panel that produced the report was unable to offer an explanation for why these low-cost or negative-cost technologies have not caught on. Its quantitative assessment, including an allowance for elimination of CFC's, ranged from as little as 10 percent to as much as 40 percent of current U.S. emissions; CFC's aside, their range of possibility is from zero to about 30 percent. Whatever the correct figure, this is probably a once-and-for-all backlog of accumulated technologies, which once exploited may be permanent but not progressive. But the strong suggestion is that there is a lot to be accomplished in the next two or three decades.

VIII

With these qualifications, let us look at that 2 percent of GNP as a permanent reduction over the coming century. I consider it altogether improbable that the developing world, at least for the next several decades, will incur any significant sacrifice in the interest of reduced carbon (nor would I advise developing countries to do so). Anything done to reduce emissions in China, India, or Nigeria will be at the expense of the richer countries.

Financing energy conservation, energy efficiency, and switching from high-carbon to lower-carbon or noncarbon fuels in Asia and Africa would not only be a major economic enterprise but a complex effort in international diplomacy and politics. If successful, it would increase the costs to the developed world by at least another percent or two on top of the 2 percent I mentioned. It is furthermore not easy to hide the transfer of resources on the order of a couple of hundred billion dollars, dollars "budgeted" somehow or other, compared with hiding some of the costs due to regulation, such as automobile fuel-efficiency standards in the United States. The kind of thing we are talking about is inducing the Chinese, through our somehow offsetting their cost, to forgo a massive electrification based on coal and the cheapest coal-combustion technology. Without engaging in blackmail, the Chinese can assert that it is not in their interest to do that at their own expense, even if they are the keystone of a "social contract" and no other nation will do anything unless the Chinese fully participate.

I shall sketch what I can imagine as a major attack on the greenhouse problem. And I should be explicit about what I cannot imagine. For reasons that I would be delighted to elaborate but for which I cannot take space here, a universal uniform carbon tax is not a solution that I can imagine. My reason is simple. A carbon tax sufficient to make a big dent in the greenhouse problem would have to be roughly equivalent at least to a dollar per gallon on motor fuel, and for the United States alone such a tax on coal, petroleum, and natural gas would currently yield close to half a trillion dollars per year in revenue. No greenhouse taxing agency is going to collect a trillion dollars per year in revenue; and no treaty requiring the United States to levy internal carbon taxation at that level, keeping the proceeds, would be ratified by the Senate. Reduce the tax by an order of magnitude and it becomes imaginable, but then it becomes trivial as greenhouse policy.[5]

[5]A careful treatment of the universal carbon tax is provided by James M. Poterba (1991).

Tradable permits have been proposed as an alternative to the tax. There are two main possibilities: (i) estimating "reasonable" emissions country by country and establishing commensurate quotas or (ii) distributing tradable rights in accordance with some "equitable" criterion, such as equal emissions per capita (a possibility that has actually been discussed). Depending on how restrictive the aggregate of such tradable emission rights might be, the latter is tantamount to distributing trillions of dollars in discounted value and making, for a country like Nigeria, the outcome of its population census the country's major economic policy. If, instead, quotas are negotiated to correspond to every country's currently "reasonable" emissions level, they will surely be renegotiated every 5 or 10 years, and selling an emissions right will be perceived as evidence that a quota was initially too generous. It is unlikely that governments will engage in trades that acknowledge excessive initial quotas.

I do not foresee negotiated national quotas subject to serious enforcement, especially enforcement through financial penalties. I think any international regime for carbon abatement can seriously include only the developed countries, and I exclude from this category the countries that we used to call the Eastern Bloc. I can easily imagine institutional arrangements that are universalist, some kind of "framework agreement" to which every country subscribes, with specific commitments to be negotiated later. But I expect serious commitments to be undertaken only by the countries that can afford to, and I am undecided whether an institutional pretense of a universalist system has advantages or, instead, the developed world should proceed independently and unencumbered with the need for a universalist facade.

The model that I find most helpful in conceptualizing a greenhouse regime among the richer countries is the negotiations among the countries of Western Europe for distributing Marshall Plan dollars among themselves and the negotiations, beginning in 1951, on "burden sharing" in NATO. There was never a formula for distributing Marshall Plan dollars; there was never an explicit criterion, such as equalizing living standards, equalizing growth rates, maximizing aggregate output or growth, or establishing a floor under levels of living. Baseline dollar-balance-of-payments deficits were a point of departure, but the negotiations took into account investment needs, traditional consumption levels, war-induced capital needs, opportunities for import substitution and export promotion, and opportunities to substitute intra-European trade for trade with hard-currency countries.

The United States insisted that the recipients argue out and agree on shares. In the end, they did not quite make it, the United States having to make the final allocation. But all the submission of data and open argument led, if not to consensus, to a reasonable appreciation of each nation's needs. The negotiations were professional; they were assisted by a proficient secretariat. The resources involved for most recipient countries were immensely important. Good relations were observed throughout; and proficiency in debate, acceptance of criteria, and negotiating etiquette steadily improved.

That is the only model I find plausible, and I believe distribution of Marshall Plan and defense-support funds to Europe is the only model of multilateral negotiation involving resources commensurate with the cost of greenhouse abatement. (In the first year, Marshall Plan funds were about 1.5 percent of U.S. GNP and—adjusting for overvalued currencies—probably 5 percent of OEEC GNP).

What that model suggests is that the main participating countries in a greenhouse-abatement regime would submit for each other's scrutiny and cross-examination plans for reducing carbon emissions. The plans would be accompanied by estimates of emissions or emissions reduction from some projected level, but any commitments undertaken would be to the policies, not the emissions. And not all of the plans would necessarily be commitments.

The United States, for instance, could present a plan for the introduction of a new generation of nuclear power reactors beginning sometime in the next century, but it is difficult to see how the federal government can commit itself to what reactors public utilities will be purchasing 20 years from now. The United States can have a plan to mandate fuel-efficiency standards for automobiles, but it takes 10 years for the standards to work their way into the automobile fleet, and there is no accounting procedure that will estimate the effect on motor-fuel consumption of any level of average fuel efficiency a decade from now.

The current popular expectation is that participation in any greenhouse regime will take the form of commitments to specified percentage reductions of emissions below those of some specified year, like 1990 or 2000. I cannot help believing that adoption of such a commitment is an indication of insincerity. A serious proposal would specify policies, like taxes, regulations, and subsidies and would specify programs (like research and development), accompanied by very uncertain estimates of their likely effect on emissions. In an international public forum, governments could be held somewhat accountable for the policies they had or had not put into effect, but probably not for the emissions levels achieved.

Such a modest beginning will require finding a way to sublimate the current international enthusiasm for a new universalist greenhouse regime into institutional arrangements that are helpful but noncommittal when the U.N. Conference on Environment and Development convenes next June. This will require an understanding among the developed countries that it is initially up to them to find a way to mobilize their populations in support of national greenhouse policies.

IX

A major commitment to financing emissions abatement in the developing world is surely too far away to need specific plans now. A developing-world carbon-abatement effort would, in principle, be altogether different from foreign aid as we have known it since World War II. In principle it would all be directed, from whatever sources and through whatever channels, to protecting that same global common. There would be, for the first time, a single criterion: economizing carbon. In the abstract, aid recipients in the war on greenhouse gases would not compete; they would not make India–Pakistan comparisons, or Arab–Israel, or Poland–Czechoslovakia. All would in principle benefit equally from maximum carbon conservation, wherever it could be achieved. Trees may grow more rapidly, in carbon content, in Madras or Szechuan or Borneo or Alaska or South Carolina, but if someone were willing to finance the growth of a tree to absorb carbon dioxide, the citizens of those states should not have the slightest care where the tree were to be planted; they all benefit solely from the carbon fixed in the tree and benefit more, the faster the tree grows, no matter where it grows.

It wouldn't work that way, of course. Somebody gets the shade, or leases land for the tree; and if it's not a tree but a nuclear power plant to supplant coal, there are local impacts that make huge differences, and negotiations over sharing the cost differential between the coal and the nuclear plants. But it is worth noticing that if there were a "pure" carbon-abatement or carbon-absorbing technology, one that accomplished nothing else, there should be no dispute about locating it wherever it would be most effective. That is new in foreign aid and foreign investment.

If the developed countries ever manage to act together toward the developing countries, their bargaining position is probably enhanced by the fact that cleaner fuels and more efficient fuel technologies bring a number of benefits other than reduced carbon, and recipients of greenhouse aid will be actively interested parties, not merely neutral agents attending to the global atmosphere. At the same time, large nations like India and China will be aware of the extortionate power that resides in ambitious coal-development projects.

On a greatly reduced scale, there may be something constructive to do more immediately. There is a huge difference between transferring "technology" and transferring capital goods that embody technology or, going further, financing entire investments (local construction, etc.) in which the technology is embedded. The difference in cost is at least an order of magnitude. While the developed countries are feeling their way into some common attack on their own carbon emissions, a tangible expression of their interest and an effective first step would be to establish a permanent means of funding technical aid and technology transfer for developing countries, as well as research, development, and demonstration in carbon-saving technologies suitable to those countries. Eventually the rural Chinese household may cook more efficiently with nuclear-powered electricity, but for another generation or two what is important is less carbon-wasteful ways of cooking and heating.

Maybe there is a role here for the carbon tax. Western Europe, North America, and Japan will be burning 3 or 4 billion tons of carbon per year for the next decade. Taxing themselves, that is, contributing in proportion to the carbon they consume, at one, two, or three dollars per ton, they could contribute to a fund that might begin at $3 billion per year and grow to $10 billion. The carbon tax is a little arbitrary here, and a U.S. administration may be wary about a precedent that carries over when the tax rises an order of magnitude, but compared with alternative criteria for sharing costs it might not even be a bad precedent.

REFERENCES

Ausubel, Jesse H., "Does Climate Still Matter?" *Nature*, 25 April 1991, *350*, 649–52.

Cline, William R., *The Greenhouse Effect: Global Economic Consequences*, Washington, DC: Institute for International Economics, 1992 (forthcoming).

Darmstadter, Joel, "The Economic Cost of CO_2 Mitigation: A Review of Estimates for Selected World Regions," Discussion Paper ENR91-06, Resources for the Future, Washington, DC, 1991.

Moulton, Robert J. and Richards, Kenneth R., "Costs of Sequestering Carbon Through Tree Planting and Forest Management in the United States," General Technical Report WO-58, Forest Service, U.S. Department of Agriculture, Washington, DC, December 1990.

Nordhaus, William D., (1991a) "Economic Approaches to Greenhouse Warming," in Rudiger Dornbusch and James M. Poterba, eds., *Global Warming: Economic Policy Responses*, Cambridge, MA: MIT Press, 1991, pp. 33–66.

______, (1991b) "The Cost of Slowing Climate Change: A Survey," *Energy Journal*, 1991, *12* (1), 37–66.

Poterba, James M., "Tax Policy to Combat Global Warming: On Designing a Carbon Tax," in Rudiger Dornbusch and James M. Poterba, eds., *Global Warming: Economic Policy Responses*, Cambridge, MA: MIT Press, 1991, pp. 71–97.

Congressional Budget Office, *Carbon Charges as a Response to Global Warming: The Effects of Taxing Fossil Fuels*, Washington, DC: U.S. Government Printing Office, 1990.

Energy Modeling Forum, "Global Climate Change: Energy Sector Impacts of Greenhouse Gas Emission Control Strategies," Energy Modeling Forum Report No. 12, Stanford University, 1992 (forthcoming).

Intergovernmental Panel on Climate Change, *Climate Change: The IPCC Scientific Assessment*, J. T. Houghton, G. J. Jenkins, and J. J. Ephraums, eds., Cambridge: Cambridge University Press, 1990.

National Academy of Sciences, *Carbon Dioxide and Climate: A Scientific Assessment*, Washington, DC: National Academy of Sciences, 1979.

______, *Carbon Dioxide and Climate: A Second Assessment*, Washington, DC: National Academy Press, 1982.

______, *Policy Implications of Greenhouse Warming*, Washington, DC: National Academy Press, 1991.

[3]

Journal of Economic Perspectives—Volume 7, Number 4—Fall 1993—Pages 65–86

Global Environmental Risks

Graciela Chichilnisky and Geoffrey Heal

Managing climate risk is not a new activity. In medieval England, a peasant farmer's land was broken into many widely-dispersed parcels. Economic historians interpret this as a way of hedging climate risk (see references in Bromley, 1992). Land in different locations would be affected differently by droughts, floods and frosts. By spreading land holdings over different locations, as well as by organizing agricultural cooperatives and buying insurance, farmers have managed climate risk for many centuries.

Today's concerns about global climate change break new ground in two ways. One is the global scope of the potential changes considered, such as changes in the atmosphere of the planet. The second is that these changes appear to be driven by human activity, which has now reached levels at which it can affect the global climate. Climate has always been unpredictable, but the inclusion of these two new elements has extended this uncertainty both qualitatively and quantitatively. Classical formulations of uncertainty in economics no longer suffice. There is uncertainty about basic scientific relationships, such as the link between gaseous emissions and global mean temperature. There is also uncertainty about the connection between global mean temperature and climate. Clearly it is climate, a variable encompassing wind patterns, humidity and rain patterns, and not just temperature, that matters from a socio-economic perspective. The floods of 1993 in the United States and Bangladesh have reminded us of the profound vulnerability of human settlement to climate. Climatologists link these to El Niño, the ocean current off the coast of Chile, confirming the global linkages within the earth's climate system. Future emissions of greenhouse gases and future climate are also highly uncertain. In addition, these emissions can be driven by economic activity and by policy measures: hence the risks faced are endogenous.

■ *Graciela Chichilnisky is Professor of Economics, Columbia University, and Geoffrey Heal is Professor and Senior Vice Dean, Columbia Business School, both in New York City, New York.*

There are two standard ways in which societies can respond to the risks associated with such uncertainty. One way is *mitigation*. The other is *insurance*. We can think of them as broadly equivalent to prevention and cure respectively in the medical field.

Mitigation means taking measures to reduce the possible damage. One way of doing this is to take steps that minimize the damage if the harmful event occurs. Building levees, canals and flood drainage systems to reduce the impact of flood waters is an example. An alternative approach to mitigation is to reduce the incidence of harmful events. Of course, if steps are taken to reduce the risk of climate change, then the risks become endogenous, determined by our policy measures. This contrasts with most models of resource-allocation under uncertainty, in which probabilities are about acts of nature and are therefore exogenous.[1] The probabilities of states in an Arrow–Debreu framework may be subjective and an agent's subjective probabilities may be altered by learning. However the actual frequency of incidence of harmful events cannot be altered by agents: there is no scope for mitigation in the second sense of improving odds. The same is true in the classical models of insurance (see for example Malinvaud, 1972, 1973), where the incidence of harmful events is again taken to be exogenous. Mitigation acquires a new meaning when risks are endogenous.

Insurance by contrast does nothing to reduce the chances of damage due to climate change. It only arranges for those who are adversely affected to receive compensation after the event, as in the case of federal disaster relief for U.S. flood victims. Insurance is a major economic activity, involving both the insurance industry and large parts of the securities industry. Can the existing and very extensive private sector organizations provide those at risk from climate change with adequate insurance cover? If not, why not? What changes in market institutions might be appropriate in this case?

This paper is about these and related questions. In attempting to answer them, we deal with many different aspects of the theory of risk-bearing. Four key issues recur in our analysis.

The first issue concerns the difficulty in assessing risks. Most climate-related risks are difficult to quantify. Indeed, in a classical statistical sense the probabilities describing them are unknowable. We may never be able to observe enough experiments to approximate the probability of global climate change in the relative frequency sense: such events are inherently unique. It is possible to evaluate the frequency of occurrence of a health risk from morbidity or mortality data, as the outcomes of repeated experiments are available. However we cannot evaluate the risks from CO_2 emissions in this way.[2]

[1] Kurz (1974) set out a framework for endogenous uncertainty: recently this has been an active field, e.g., Kurz (1990, forthcoming 1993), Chichilnisky and Wu (1991), Chichilnisky, Dutta and Heal (1992), Chichilnisky (1992), Chichilnisky, Hahn and Heal (1992).

[2] In this respect there may be a difference between the various aspects of climate risk. There is historical data on the relation between atmospheric CO_2 and climate from tree ring and ice core studies. With ozone depletion the phenomenon is so new that such data is not available.

The second issue is endogenity of risks. The risks that we face are affected by our actions.[5]

The third issue is correlation of risks. Climate changes will affect large numbers of people in the same way. A rise in sea level, for example, will affect low-level coastal communities in most countries. Insurance in the traditional sense of risk-pooling works best for large numbers of small statistically independent risks. We thus have to ask what types of markets work best when risks are connected and collective.

Irreversibility is the final issue. In this area, many major economic decisions and their consequences are likely to be irreversible. Climate changes, the melting of ice caps, desertification, species extinction, are all processes not reversible, or at least not on relevant time scales.

In summary, we are dealing with risks that are *poorly understood, endogenous, collective and irreversible.* In policy terms, the nature and extent of uncertainty about global climate change implies that society's position will be dominated by two questions: What cost is it worth incurring to reduce the poorly-understood risk of climate change, or to improve our understanding of the risk? How may existing social institutions, such as insurance contracts and securities markets, be used to provide the most efficient allocation of the risks associated with global climate change?

Risk-Allocation in a General Equilibrium Framework

Economists have two standard models of risk-allocation in a market economy. The more general is that of Arrow and Debreu, in which agents trade "contingent commodities." The alternative is the model of insurance via risk-pooling in large populations. Neither case addresses the issue of mitigation via a reduction in the incidence of harmful events.

In the Arrow–Debreu framework there is a set of exogenous "states of nature" whose values are random and represent the sources of uncertainty. Classically one thinks of events such as earthquakes and meteor strikes. Agents in the economy are allowed to trade commodities contingent on the values of these exogenous variables. These are called "state-contingent commodities." As examples, think of "water in a drought" and "water in a flood." With a complete set of markets for state-contingent commodities, the first theorem of welfare economics holds for economies under uncertainty: a Pareto efficient allocation of resources can be attained by a competitive economy with uncertainty about exogenous variables.

[5]Endogeneity of risks leads to moral hazard when risks to agents depend on their actions, which cannot be observed by the insurers and will be influenced by the insurance offered. This leads to arguments for coinsurance (see for example Stiglitz, 1983). In the present context such moral hazard problems are not central to the analysis: asymmetric information is not a characteristic of climate risks. Endogenous uncertainty is more general than moral hazard.

Arrow (1953) showed that efficiency can in fact be attained by using a mixture of securities markets and markets for non-continent commodities, so that a complete set of contingent commodity markets is not required. This observation provides a natural and important role for securities markets in the allocation of risk-bearing. The securities used are contracts that pay one unit if and only if a particular state occurs. While the contingent contract approach is in principle all-inclusive and covers most conceivable cases of uncertainty, in practical terms there are cases where it can be impossible to implement. It can be very demanding in terms of the number of markets required (Chichilnisky and Heal, 1992). For example, if agents face individual risk (i.e., risks whose probabilities vary from individual to individual), then in a population of 100 similar agents each of whom faces two possible states, the number of markets required would be 2^{100}. The number of markets required is so large as to make the contingent contract approach unrealistic.

The use of insurance markets for pooling risks is a less general but more practical alternative. This requires that populations be large and that the risks be small, similar and statistically independent. The law of large numbers then operates and the frequency of occurrence of an insured event in a large sample of agents approximates its incidence in the population as a whole. There is thus a role for insurance companies to act as intermediaries and pool large numbers of similar but statistically independent risks. In so doing they are able via aggregation and the use of the law of large numbers to neutralize the risks faced by many similar agents. The main references on this are Arrow and Lind (1970) and Malinvaud (1972, 1973).

The insurance approach is at a disadvantage when risks are correlated. When large numbers are likely to be affected at once, risk-pooling will not work. However it does have the advantage relative to the contingent market approach of economizing dramatically on the number of markets needed. In the above example, only two mutual insurance contracts and 2809 securities would be needed instead of 2^{100} contingent contracts (Chichilnisky and Heal, 1992; Cass, Chichilnisky and Wu, 1991).

When risks are allocated by trading state-contingent securities, or by risk-pooling and insurance, *it is very important that agents know, or believe that they know, the relative frequencies of the states of nature, at least approximately*. This is obvious when trading insurance contracts. The actuarial calculations needed to set insurance premia can only be performed if the parties believe that the relative frequencies of the insured events are approximately known.

In the Arrow–Debreu approach, it suffices to think of agents maximizing expected utility to appreciate the need for them to know, or at least behave as if they know, the relative frequencies of exogenous states. These frequencies are the weights placed on their utilities from state-dependent consumption. The point is simple: if agents cannot assign relative frequencies then their preferences are not well-defined and they cannot act to maximize expected utility.

In the context of climate change this may be too demanding. Agents do not know the frequencies of different states, and recognize that they do not know them. They recognize that there are several different opinions about what these are, but feel unable to choose definitively between these alternatives. If they were expected utility maximizers, they would be uncertain about their own preferences. In such a case, it is natural to think of the frequency distribution over climate changes as a state of the world, a risk, in the Savage sense: we do not know what value it will assume, and whatever value this is, it affects economic activity. As shown below, ignorance then assumes the role of a collective risk, and can be treated by the use of state-contingent markets. One sometimes thinks of uncertainty about probabilities being resolved by learning. This is an avenue which is not open when scientific knowledge is incomplete and experiments are not possible. In this case an alternative approach is the opening of new markets (Chichilnisky and Heal, 1992, and the discussion below).

In sum: the Arrow–Debreu approach to risk allocation via state-contingent markets is in principle universally applicable. However, it is cumbersome and perhaps unrealistic in the case of risks with individual components. Insurance markets are more manageable, but leave uncovered collective or correlated risks such as the risk induced by ignorance of the true frequency distribution of harmful events. So it would be natural to allow agents to trade securities contingent on such collective risks, and cover the individual components of risks by mutual insurance contracts. This is precisely the approach that we develop below. Although new to the economics literature, it is by no means new in practice; we argue below that some of the oldest risk-bearing institutions recorded, agricultural cooperatives, have exactly this structure.

Ignorance as Collective Risk

Consider an economy in which agents face risks whose relative frequencies they know that they cannot evaluate. Such risks could derive from the impact of global climate change on income levels via floods, storms or droughts, or from the effects on health of ozone depletion, acid rain, or air pollution. What market structure would suffice to assure efficient allocations in this situation? There are widely differing opinions about these impacts, on which there is inadequate information.

Chichilnisky and Heal (1992) formalize this type of situation in a general equilibrium model. Each agent faces the risk of being in one of several states (e.g. healthy or sick, productive or unproductive). No one knows what will be the true frequency distribution of affected agents. A probability is assigned to each possible frequency. A typical probability distribution of this type might state for example that there is a 10 percent chance that 90 percent of the population will be harmed by global warming, a 25 percent chance that 50 percent of the population will be harmed, and so on. The probability distribution over alternative frequency distributions may be different from individual to individual.

In this framework, we have two levels of uncertainty. The first level of uncertainty is collective: what is the distribution of agents who are harmed in the economy? Will 90 percent be harmed, or only 30 percent? This is a question about the aggregate incidence of the phenomenon over the population as a whole. The second level of uncertainty is individual: it is uncertainty about whether a given agent is harmed or not by climate change. It devolves about questions such as: given that 90 percent of the population will be harmed, will a particular agent be harmed or not?

In our examples of the impact of the depletion of the ozone layer on cancer or the impact of climate change on agricultural productivity, the two levels of uncertainty are: first, uncertainty about the true relationship between ozone depletion and the incidence of individual disease in the population as a whole, or about the true relationship between climate change and agricultural productivity; and secondly, uncertainty about whether any given person or community will be affected.

Our ignorance of scientific processes (like the relation between ozone depletion and skin cancer or between CO_2 emission and climate change) causes the collective risk, by which we mean the uncertainty about the relative frequency of harmed agents in the population. Uncertainty about this frequency is central to the problem. When this is resolved we will still not know who is damaged and who is not, but we will at least know the frequency describing this. Once frequencies are known, actuarial calculations can be conducted and the problem is insurable.

We propose an institutional structure which uses two types of financial instruments which are tailored to these two aspects of the problem. These can lead to efficient allocation in the face of such risks. We follow a framework established by Cass, Chichilnisky and Wu (1991) and Chichilnisky and Heal (1992).

One instrument is a *mutual insurance contract* to deal with the risks faced by agents or communities contingent on each possible distribution of harmful effects worldwide. A mutual insurance contract is an agreement between parties subject to similar risks that those who are harmed will be compensated by the others. Examples are agricultural cooperatives of the type recorded in Europe at least since the fifteenth century, and the nineteenth century U.K. workers' associations and friendly societies. These involved agreements between a group of workers that if one were sick and unable to work, that worker would be compensated by the others who were not harmed. In the present context, one could think of groups of communities subject to the possible impact of climate change, with those unharmed compensating the others. Making the terms of such a mutual insurance contract contingent on the distribution of harmful effects worldwide means that there is a different compensation agreement between the parties for each possible aggregate distribution of harmful effects. To know what compensation is due in any particular case, the parties have first to assess the distribution of harmful effects globally, and on the basis of this decide which mutual insurance contract to apply.

Having dealt with individual risks by mutual insurance, we still face collective risks. We need *Arrow securities* to deal with these collective risks induced by uncertainty about the overall distribution of adverse effects. Arrow securities are usually defined as securities that pay one dollar if and only if a particular state of the world occurs. Here they pay one dollar if and only if there is a particular frequency of affected parties in the population. As already noted, the incidence of impacts on the population as a whole is being treated as a "state of the world" in the Arrow–Debreu sense. We treat each possible distribution of adverse affects as a distinct collective state, and use securities markets to enable parties to transfer wealth between these states. One Arrow security is needed for each possible distribution of adverse effects worldwide, because to attain Pareto efficiency each separate state must be covered by a security.

The following example will help to make this framework concrete. Consider a world of two countries 1 and 2, in which the climate may be in one of two states α or β. There are two possible probability distributions over these two climate states. These distributions are called A and B, with distribution A giving a probability of 0.1 to climate state α and a probability of 0.9 to climate state β. Distribution B gives the reverse probabilities, i.e., it gives probability 0.9 to climate state α and probability 0.1 to climate state β. The endowments of the two countries depend on the climate state, and are as follows: $\omega_1(\alpha)$ is country 1's endowment vector if the climate is in state α, and $\omega_2(\alpha)$ is the corresponding endowment for country 2. Similarly, endowments in climate state β are given by $\omega_1(\beta)$ and $\omega_2(\beta)$ respectively. Endowments satisfy $\omega_1(\alpha) > \omega_2(\alpha)$ and $\omega_1(\beta) < \omega_2(\beta)$, so that country 1 is relatively better off in state α and country 2 in state β.

To reach an efficient allocation of risks we need two Arrow securities. One, S_A, pays \$1 million if and only if the probability distribution over states of the climate is A. The other, S_B, pays \$1 million if and only if the probability distribution over states of the climate is B. In practice of course, probability distributions are not observable, and we cannot condition contracts on unobservable events. So conditioning on a probability distribution means conditioning on frequency distributions consistent with that probability distribution in a sampling sense.

Countries can spread the risk arising from not knowing which is the true distribution over states of the climate by trading these two securities. In addition they make mutual insurance contracts conditional on states of the climate. Such a contract could take the following form. If the distribution over climate states is A (distribution A gives probability 0.1 to climate state α and probability 0.9 to climate state β) then country 1 makes a transfer $\Delta^{\alpha}_{1,2}$ to country 2 if the state of the climate is α, and country 2 makes a transfer $\Delta^{\beta}_{2,1}$ to country 1 if the climate state is β. These transfers satisfy $0.1\Delta^{\alpha}_{1,2} + 0.9\Delta^{\beta}_{2,1} = 0$ so that the expected transfer is zero and the mutual insurance contract is actuarially fair. There would be a similar contract to cover the case when the distribution over climate states is B.

To summarize the argument: Our ignorance of the frequency of the impacts of climate change constitutes a collective risk. This collective risk can be allocated through markets for securities that pay off contingent on that frequency. For the individual risks that remain, it is more practical to use mutual insurance contracts: this is done by having a different individual insurance contract for each possible frequency of impacts.

There are two features of the results which are of general interest. One is the development of a framework for achieving efficient allocations in the face of uncertain individual risks. Given rapid changes in technology with potentially far-reaching environmental impacts and health effects, the problem of providing insurance against such risks is particularly important. It is a matter of very active concern in the insurance industry. The second interesting feature is the way a combination of securities markets and insurance markets can be used to provide a relatively simple institutional structure for dealing with unknowable and correlated risks.

An Institutional Framework

Our analysis suggests that although the risks associated with global climate change are very difficult to evaluate, there is nevertheless a market framework within which insurance can be provided. It involves first identifying the set of possible descriptions of the collective risks. Natural descriptions of risk are frequencies of occurrence of climate-related events such as floods, tropical storms or certain temperature patterns.

Next this framework involves introducing securities whose payoffs depend on which description of the risk is correct. This amounts to allowing agents to bet on which model of the risk is correct. Betting on which of several alternative descriptions of the way the world works is correct, is in effect what one does when choosing one research strategy over another. Corporations, individuals and governments all do this regularly. For example, a market for the securities of high-technology firms pursuing different research strategies towards the same goal is a financial market in which these bets are made. An additional example is provided by the Chicago Board of Trade, which is in the process of introducing securities called "catastrophe futures" whose payoffs depend inter alia on the incidence of tropical storms in the United States, as measured by the Insurance Service Organization's index. This is a particular example of the instruments initially proposed in Chichilnisky and Heal (1992).

Finally, our approach involves establishing compensation agreements between harmed and unharmed regions that depend on which description of the risk turns out to be correct. Mutual insurance contracts or mutual compensation agreements are already part of our institutional framework. In fact, they date back to the nineteenth century and beyond, and were the foundations of many current insurance companies and trade unions. Consider for example agricultural cooperatives, probably the oldest risk-allocation institutions in the world. In fact, one of the largest banks in Italy, the Monte dei Paschi di Siena, was founded to play this role in 1473. They have insured against weather risks since then,[4] and have provided mutual insurance contracts for their members, in that they have arranged transfers from the less to the more fortunate in any given season, the size of the transfer depending on the overall level of prosperity. They also hedge against the overall frequency of poor crop yields in their community by building up reserves to carry over from good to bad years. So they have actually fulfilled both of the insurance functions outlined above—making transfers between agents contingent on the overall incidence of negative events, and allowing a mechanism for transferring wealth between states in the sense of high or low overall incidences of negative events in the population.

Trading Risks

An interesting aspect of the markets just described is that they can provide a natural mechanism for reconciling differences in assessments of the likelihood of important climate changes between countries, and for testing the conviction behind publicly-stated positions.

Suppose for example the United States believes it most likely that there will be little climate change, and the European Community believes otherwise. Then through the market for securities whose payoffs depend on which description of climate change is correct, the United States will naturally sell insurance to the E.C. The United States would wish to be a seller of securities which pay if climate change is serious, because of its belief that this event will not occur, and a buyer of securities that pay if it is not, because of its belief that this will be the outcome. The E.C. would be on the opposite sides of these markets.

International markets for the risks of climate change would also provide an objective test of the seriousness with which countries adhere to their publicly-professed positions on the risk of climate change. It is possible that a country might publicly profess to a lack of concern about the risks of climate change, in spite of actually being concerned about these risks, in order to free ride on CO_2 abatement policies introduced by others. These issues are discussed in detail in Heal (1993) and the references cited there. The existence of markets for the risks of climate change would place such a country in a dilemma. The country's true beliefs would incline it to sell securities paying off in the event of climate change not being serious, and buy those paying off if it is serious. Consistency with its public positions would require that it be on exactly the opposite sides of these markets. There would therefore be a cash cost to convincing and consistent misrepresentation of true beliefs. These cash costs could offset some of the incentive to free ride on other countries' efforts to reduce greenhouse emissions.

Note that trading risks is different from the trading of emission permits, as discussed in the articles by Poterba and Weyant in this volume. The recognition of uncertainty suggests the need to consider state-contingent emission permits,

[4] They also supported part of this research and provided access to their records.

where the state is defined in terms of the frequency of climate-change related events. Permits to emit CO_2 would for example be contingent on the incidence of indicators of climate change. Such contingent emission permits could play the role of securities whose payoffs depend on the collective risk.

In the context of emission permits, it is worth noting the public good nature of the climate. Climate is a public good. However, it does not fit fully the conventional paradigm because emission abatement, which is the production of the public good "unchanged climate," is conducted independently in the various countries of the world. It is not produced in a central production facility, as assumed in the usual treatments of public goods: it is a privately produced public good. A consequence is that economic efficiency will only imply equalization of the marginal costs of emission abatement across countries if lump-sum transfers between countries are made to equalize the marginal utility of income in all countries, a somewhat unlikely contingency. Equalization of the marginal costs of emission abatement across countries is often taken as justification of the superiority of tradeable permits as a method for controlling emissions. This point is developed in Chichilnisky (1993) and Chichilnisky and Heal (1993). More generally, a key issue is that an efficient allocation of a public good such as unchanged climate must be supported by a Lindahl equilibrium and not a competitive equilibrium. In general competitive markets for tradeable emission permits may not decentralize Pareto efficient allocations of abatement.

Optimal Allocation with Endogenous Risks

What is it worth spending to reduce the probability of harmful climate change? Only if we can answer this question can we judge properly proposals for carbon taxes, alternative energy strategies, and CO_2-reduction protocols. Careful judgment is crucial, as all of these involve very considerable costs, as indicated by Weyant's paper in this symposium, Cline (1992) and others. Here we shall summarize one approach to this problem, based on Heal (1984, 1990). This is a model that examines the extent to which the consumption of fossil fuels should be curtailed because it increases the probability of a change in climate. The model in Heal (1984) is based on three assumptions. First, the atmosphere may be in one of two states, one favorable to economic activity and one unfavorable (there is a possibility of a future climate catastrophe). Secondly, the atmosphere moves stochastically from the favorable state to the unfavorable, and once there remains there forever, so that atmospheric change is irreversible. Finally, the probability of a transition from the favorable to the unfavorable state is endogenous and increases with the level of cumulative emissions from the use of fossil fuels. The probability that this change will definitely occur at some date is less than one, so there is a positive probability of the change never occuring.

Fossil fuels, capital equipment and the atmosphere are used to produce output, which may be consumed or reinvested to augment the capital stock. Production generates emissions, which affect the probability of a change in the state of the atmosphere. The atmosphere is a resource that enters into the economy's production function, which may be in a favorable or an unfavorable state. Initially the atmosphere is in the favorable state but may change stochastically to the unfavorable state, and once in this state will remain there forever. The source of emissions which affects the probability of climate change is the use of an exhaustible resource in production. The remaining input to production is the capital stock. An obvious example of this structure is the emission of CO_2 by the use of fossil fuels.

Consumption yields utility and the objective is to maximize the expected present discounted utility of consumption. There is a constraint on the total amount of the resource that can be used, as this is exhaustible: there is also a national income identity. The overall problem solved involves maximizing expected utility subject to the resource and national income constraints, where the expectation is over the process governing climate change.

Heal (1984) characterizes optimal paths of consumption, capital accumulation and use of fossil fuel for this problem. He compares these with those that are optimal in the absence of an atmospheric impact, and also studies the impact of changes in parameters such as the discount rate and degree of risk aversion and uses this to isolate the key parameters in determining the optimal rate of use of fossil fuels. The introduction of atmospheric impact makes a fundamental difference. The time profile of resource use which emerges is flatter than that which emerges from an optimal depletion problem with no atmospheric impact. Initial levels of resource use are lower, and they fall more slowly, than in the no-atmospheric-impact case. The difference depends on the degree of risk aversion and on the parameters of the probability distribution relating cumulative emission to climate change.

The behavior of the shadow price of the resource is also of interest. In the pure depletion case this price rises at the rate of discount: in Heal (1984) it may fall and even become negative. We can interpret the difference between the shadow price of the resource in the no-atmospheric-impact case and the current case as an optimal carbon tax. This tax depends on the country's degree of risk aversion and on the parameters of the probability distribution describing the risk of climate change as a function of carbon emissions, as well as on the damage resulting from climate change. The model thus leads to a distinctive approach to characterizing an optimal carbon tax and its evolution over time. Because it is an intertemporal model, it provides this characterization in terms of the internalization of probabilistic intergenerational externalities. Hartwick (1992) gives an analysis of carbon taxes using this framework.

The likelihood of climate change as a function of economic activity is a key relationship in evaluating the choices posed in this model. This is a functional relationship rather than a parameter. Global change R&D leads us to a better understanding of this relationship. It is worth stressing that proper economic analysis requires not just the likelihood of climate change as a result of one particular emission scenario, which is what most scientific analyses are providing, but rather a systematic evaluation of how the nature and likelihood of climate change varies with the pattern of economic activity. The study and characterization of this likelihood function is an important topic for interdisciplinary research.

It is not surprising that what it is worth paying to reduce the risk of climate change depends inter alia on a society's degree of risk aversion and discount rate. However, this has an interesting and important implication. Even if there were complete agreement about all of the scientific aspects of the global change problem, there could still be disagreement about policy responses. Because of the international externalities associated with climate, so that all countries "consume" the same climate, CO_2 abatement policies only make sense if coordinated internationally (Poterba's paper in this symposium; Barrett, 1990; Carraro and Siniscalco, 1991; Heal, 1993b).

The fact that different countries need not agree on policy choices even if they agree on the scientific evaluation of the problem could clearly make such agreement difficult to obtain. Different countries' positions with respect to measures to restrict greenhouse gas emissions depend on their discount rates and degrees of risk aversion. The United States, for example, has been against global abatement agreements, while Germany has been in favor. This fits with the conventional wisdom that the financial and industrial community in the United States has both a higher discount rate and a lower degree of risk aversion (greater willingness to take risks) than that in Germany. The differences in policy positions could then be attributed to differences in preferences rather than, or in addition to, different interpretations of the current scientific evidence.

Different perceptions of the risk involved do not however preclude efficient solutions. Economics is about differences in preferences leading to trade. In this case differences in attitudes towards risk could be grounds for the introduction of markets in which different risk positions are traded, with efficiency gains, as discussed earlier.

Option Values and Irreversibility

In valuing environmental resources such as current climate conditions, biodiversity, or complex ecological systems, the irreversibility of decisions and events can be central. A key aspect of these resources is that once altered they cannot easily be restored to their current conditions, at least on a relevant timescale. The decision not to preserve a rich reservoir of biodiversity such as the 60 million-year-old Korup forest in Nigeria is irreversible. The alteration or destruction of a unique asset of this type has an awesome finality, and analysts have sought to capture this in a framework for cost-benefit analysis. This has led to the concept of "option value:" preserving a unique asset in its present state allows us the possibility of changing our minds later. Altering it irreversibly does not. Preserving it has thus to be credited with an "option value" because it keeps open the option of reconsidering the decision.

A concept related to option value is that of non-use value or existence value. We may value environmental goods for which we have no immediate economic use. The existence of certain species is in this category: the Californian condor, the spotted owl, and various snails and fish come to mind. There is no sense in which we can currently use these species: possibly one could argue that the condor and the owl have consumption value for those willing to make the effort needed to see them, but few people come into this category. One doubts that this is a significant issue with the snail.

The two concepts, option value and non-use value, seem to overlap. Many goods which exemplify one also exemplify the other. At the same time, there are no doubt differences. Non-use values stem in some degree from ethical considerations, from a recognition that a species has a right to exist even if humanity places no direct value on it. But one suspects that behind many non-use valuations there lurks an option value: many non-use valuations stem from an unstated belief that a use value may emerge.

In this section we review two distinct formulations of this issue, one in which the returns to a preservation project are uncertain at present but will be revealed in the future, and one in which the preferences of future generations for environmental facilities are uncertain. The first framework is the one in which the issue of option values has traditionally been studied. We provide an outline of the argument in this case and illustrate the fact that one needs three conditions for an option value to exist. These are irreversibility, the acquisition of information with the passage of time, and an asymmetry of the underlying probability distribution. Similar results apply to the case of uncertainty about the preferences of future generations.

Waiting for Information

The option value of preserving an environmental or ecological asset has been explored in the context of uncertainty about the future benefits associated with its existence. A review of the literature is in Fisher and Krutilla (1985).[5] The central issue is that benefits will accrue in the future from the preservation of a resource, but these are currently unknown. If the resource is preserved into the future, then in the future the decision about whether to preserve it can be reconsidered in the light of better information then available about the benefits from its existence. If it is not preserved, then there is no chance of reconsideration when we have better information. In this case conventional decision rules will underestimate the value of preserving the asset. The follow-

[5]Amongst the studies of this issue are Weisbrod (1964), Krutilla (1967), Cichetti and Freeman (1971), Schmalensee (1972), Arrow and Fisher (1974), Henry (1974a, b) and Bohm (1975).

Figure 1
Uncertain Benefits from Preservation in Period 2

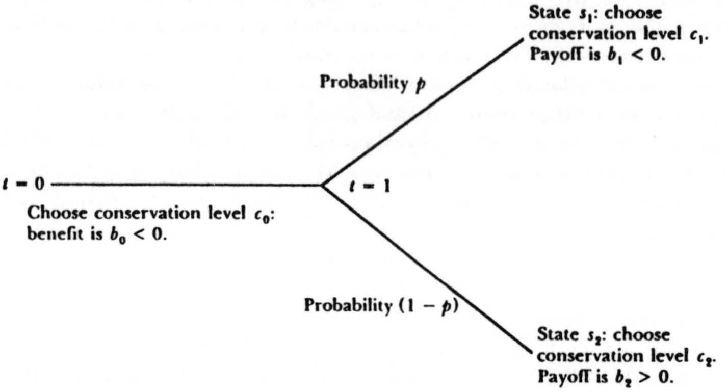

ing example (from Dasgupta and Heal, 1979) illustrates the key point in a simple framework.

We shall show that with irreversible decisions there is an option value to conservation in the initial period if and only if there is a positive expected payoff from conservation in that period given that we follow an optimal policy. We contrast this with the reversible case, in which we never conserve in the first period and there is no option value.

Consider two dates, the present $t = 0$ and the future $t = 1$. We have one unit of an environmental asset. The benefit from preserving this now at time $t = 0$ is b_0. At time $t = 1$ in the future there are two possible states of nature s_1 and s_2. The state of nature is revealed at time $t = 1$. If the state is s_1, the benefit of preserving the asset is b_1: if s_2 is the state, the benefit is b_2. The probabilities of s_1 and s_2 are p and $(1 - p)$ respectively. Decisions about preservation are made both currently at time $t = 0$ and in the future at $t = 1$. At $t = 0$ a decision is made on how much of the asset to preserve until $t = 1$: at that date we may either conserve everything conserved at $t = 0$, or conserve less. Given that destruction is irreversible, we cannot of course at $t = 1$ conserve more than was conserved at $t = 0$. Our options at $t = 1$ are therefore constrained by the decision made at $t = 0$. This situation is summarized in Figure 1.

We shall compare the case already described where the decision made at time $t = 0$ is irreversible with a hypothetical alternative case in which this decision can in fact be reversed. In this case the decision made at time $t = 0$ no longer constrains the options available at time $t = 1$. We look at this alternative case first, as it is simpler and provides a benchmark. Let c_0 be the amount of the resource conserved now at time $t = 0$, and c_1 and c_2 be the amounts conserved in the future at time $t = 1$ in states 1 and 2 respectively. The expected benefit from development (assuming a zero discount rate) is $b_0c_0 + pb_1c_1 + (1 - p)b_2c_2$. We have to choose conservation levels c_0, c_1 and c_2 to maximize (the expected benefit). We focus on the interesting case in which there is currently no benefit to preservation,[6] $b_0 < 0$, nor is there any benefit in state 1 in the future, $b_1 < 0$. However, there is the future possibility of state 2 in which there are positive benefits from preservation, $b_2 > 0$. If decisions are reversible, we preserve nothing at time $t = 0$, i.e., we set $c_0 = 0$. Then at time $t = 1$, we conserve nothing in state 1 and everything in state 2, i.e., we set $c_1 = 0$ and $c_2 = 1$. In the reversible case we *can* set $c_2 = 1$ because by assumption decisions made at $t = 0$ *are* reversible.

Now consider the real case in which the decision at time $t = 0$ cannot be reversed later. In this case the choice made at $t = 0$ does constrain the choices open at $t = 1$. We have to satisfy the constraint that what is conserved at time $t = 1$ cannot exceed that which was conserved initially, that is, $0 \leq c_1, c_2 \leq c_0 \leq 1$. In particular, if everything is destroyed in the first period, then we have no options in the second. What policies now maximize (the expected benefit)? Is there a value to carrying the option to conserve into the second period? Clearly if in the second period the state of the world is one in which there are positive benefits to conservation, then in that period we will conserve everything left to us by our earlier decision, that is we will always set $c_2 = c_0$. If however the state in the future is unfavorable to conservation, then we will conserve nothing and set $c_1 = 0$.[7]

It is optimal to conserve in the first period if and only if there is a positive expected payoff to future conservation, given that we choose optimally later, i.e., we set $c_1 = 0$ and $c_2 = c_0$. Contrast this with the decision in the reversible case, in which we never conserve and always chose $c_0 = 0$. These two decisions are different if the expected payoff to conservation in the first period is positive.[8] In this case there is an option value to conservation as a means of carrying the resource into the second period and taking advantage of future information.

Option Values

Note that the existence of an "option value" does not depend on risk aversion, as we assumed throughout the previous subsection that the maxi-

[6]If $b_0 > 0$, there are benefits to conservation in the first period, so that $c_0 = 1$, i.e., we conserve in the first period. We concentrate on the interesting case of $b_0 < 0$, when the only incentive to conserve in period 1 is the possibility of a positive return in period 2.

[7]Noting that $C_1 = 0$ and $C_2 = C_0$ in the irreversible case, the expression for expected benefit at the top of this page reduces to $(b_0 + (1 - p)b_2)c_0$. As the expression is linear in C_0, the initial conservation level is positive if and only if the derivative with respect to C_0 is positive, i.e. $(b_0 + (1 - p)b_2) > 0$. The inequality has a simple interpretation: the left-hand side is the expected payoff from conservation in the first period. It is the certain payoff in the first period plus the expected payoff from conservation in the second, given that if the state unfavorable to conservation occurs there will be no conservation in the second period. It is the expected payoff to conservation in period one given that an optimal policy is followed subsequently.

[8]An important simplifying assumption in this example is the linearity of payoffs in the level of preservation. Fisher and Krutilla (1985) discuss the role of linearity.

mand is the expected value of benefits.[9] The key issues here are: first, the irreversibility of the decision; second, the fact that delaying a decision can let one take advantage of better information, and third, that on average there will be benefits from conservation in the first period, provided that we choose optimally later (as shown in note 7).

There are important practical implications of the analysis that we have just completed. Climate change is likely to be irreversible if it occurs. So in a cost-benefit analysis of preventing climate change (i.e., preserving the atmospheric environment), it may be appropriate to credit preservation (preventing climate change) with an option value. This could be the case if the passage of time is likely to bring significant new information about the likelihood of climate change or about its consequences.

The most thorough study of the costs and benefits of reducing climate change is Cline (1992). It seems worth noting that although this study refers many times to the scientific uncertainties associated with predicting climate change, it at no point attributes an option value to preventing climate change. This means that Cline's study may systematically underestimate the benefit-cost ratio of preservation of the atmosphere in its status quo. There is also an analysis in Manne and Richels (1992) of the value of waiting for scientific information about the greenhouse effect. They consider two possibilities: acting strongly now to reduce the emission of greenhouse gases, or taking very limited action now and waiting until there is further scientific evidence. Taking major steps towards emission abatement now amounts to conserving the atmospheric environment in its present state, and should again be credited with an option value. Manne and Richels fail to do this, and so again underestimate the value of buying insurance against the greenhouse effect by acting strongly now. As the value of an option generally increases with increasing uncertainty about the future, and as uncertainty looms large in any projections regarding global warming, the extent of the underestimate could be important. It could for example be decisive in the endorsement of a global carbon tax.

Uncertainty about Future Generations

There are several ways of generalizing or refining the concept of option value. A key consideration seems to be the possibility that future generations will value environmental resources more than we do. If this is simply a statement that these resources will be scarcer, and so more valuable on the margin, then this effect is captured in the usual approach to cost-benefit analysis (Heal, 1993a).

It may, however, be a statement that future generations could have different preferences from us, and might value environmental assets differently.

[9]Pindyck (1991) considers a similar example in the case of irreversible investment decisions, and shows that the option value of delaying an investment decision to take advantage of information that will become available in the future, can be computed using the formula used in finance for valuing an option to buy a stock. See also Dixit (1992).

Because they might value them more, we should, it is argued, attribute a value to leaving them the option of high consumption levels. Solow (1992) argues that an important element in the definition of sustainability is recognizing the possibility that the preferences of future generations about environmental assets may be very different from ours (see also Chichilnisky, 1993b; Beltratti, Chichilnisky and Heal, forthcoming 1993). This seems close to the concept of option value set out above, and indeed it is, though there are some differences that are revealing. We next study this problem, drawing heavily on results in Beltratti, Chichilnisky and Heal (1992), and using a highly simplified version of that model.[10]

We shall illustrate the conclusion that uncertainty about future preferences alone is not sufficient to produce an "option value" case for increasing the resource left to the next generation. In addition to pure uncertainty, there must be asymmetry in the distribution of possible changes in preferences. Neutral uncertainty (increases and decreases in the intensity of preferences are equally likely) does not generate a case for leaving more to the future in case they value the resource more highly than us. Uncertainty makes a case for conservation only when the expected return to postponement of consumption is positive.

Consider a two period world where there is a fixed total stock of a natural resource to be consumed in the two periods. The initial stock is known to be s_0. The amounts consumed in the first and second periods are c_1 and c_2 respectively: these must obviously satisfy $c_1 + c_2 = s_0$. As the stock is irreplaceable, anything that is consumed in the first period is not available in the second, so that consumption here is an irreversible depletion of the stock. The utility from period one consumption is $u(c_1)$, which is an increasing strictly concave function. The utility from second period consumption is unknown: it may be either $(1 + \alpha)u(c_2)$ with probability p or $(1 - \beta)u(c_2)$ with probability $(1 - p)$. Here $0 < \alpha, \beta < 1$. So there is a probability p that the utility derived from future consumption will be "scaled up" by a factor α, and a probability $(1 - p)$ that it will be "scaled down" by a factor β.

Consider first as a benchmark the case in which there will be no change in preferences, so that we just have to pick c_1 and c_2 to maximize $u(c_1) + u(c_2)$. Figure 2 shows this situation: the length of the horizontal axis is s_0, the initial stock of the resource. Consumption in the first period c_1 is measured to the right from the left-hand origin and consumption in the second period to the left from the right-hand origin. Marginal utility in each period is plotted, and the optimal levels of c_1 and c_2 are those at which the marginal utility curves cross. In the absence of discounting, and with utility functions the same in each period, these curves will of course be symmetric, as shown in Figure 2.

[10]The full model is radically different from most other models in which option values have been studied. It is an infinite-horizon stochastic dynamic optimization model in which the maximand is the expected present value of utility and future preferences evolve stochastically.

Figure 2
Uncertainty about Future Preferences

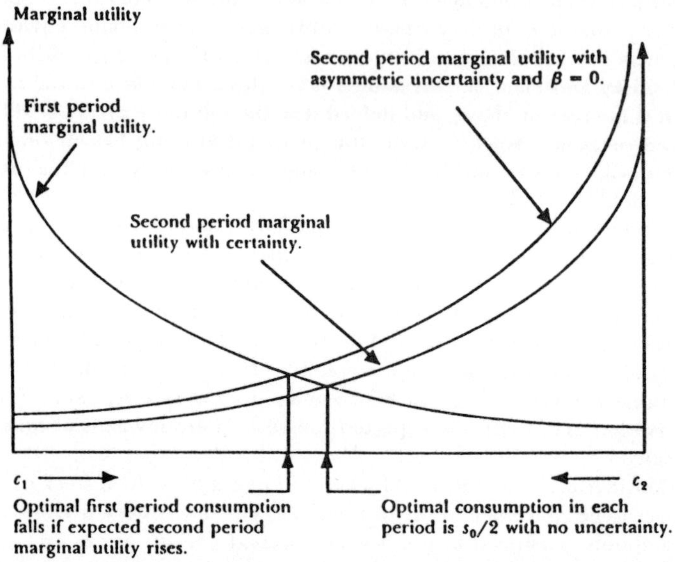

Now suppose that there is uncertainty about preferences in the second period—think of this as uncertainty about the preferences of a future generation. Also simplify matters by assuming that $\beta = 0$, so that the only possible change in preferences is a "scaling up" of the utility of consumption. This corresponds to the case that we mentioned at the start of this section, namely the possibility of an increase in the appreciation that people have for the resource.[11] The expected marginal utility curve for second period consumption is now above the certain second period marginal utility curve, as shown in Figure 2.

The optimal first period consumption level is now lower than before, as a result of the possibility of a shift in future preferences towards the natural resource. Generally one can show that the amount of this reduction depends on the probability distribution governing the change in preferences, the discount

[11]Now we have to choose c_1 and c_2 to maximize the expectation of utility, which as $\beta = 0$ is $u(c_1) + p(1 + \alpha)u(c_2) + (1 - p)u(c_2)$. The solution to this requires that the marginal utility of consumption in the first period equals the expected marginal utility in the second, i.e.,

$$\frac{\partial u}{\partial c_1} = (p\alpha + 1)\frac{\partial u}{\partial c_2}$$

Clearly $(p\alpha + 1) > 1$, so that the period two marginal utility curve is moved up by uncertainty, and period one consumption falls, as in Figure 2.

rate and the degree of risk aversion (Beltratti, Chichilnisky and Heal, 1992). Does this reduction in period one consumption reflect an "option value" in the sense of the previous section?

To understand this, we have to consider the more general case that we posed initially.[12] Suppose first that there is uncertainty about future preferences but on average we expect no net change, in the sense that increases and decreases balance out in probability terms. Then the second period marginal utility curve is unaltered by the uncertainty about future preferences. In this case uncertainty about future preferences will not lead to a reduction in present consumption. Indeed it will lead to no changes in any consumption levels, even if agents are strictly risk averse in the sense that their utility functions are strictly concave.[13]

If on the other hand there is an expectation of an increase in the utility of consumption in the second period, then the period two expected marginal utility curve will lie above that under certainty and consequently the optimal period one consumption level will be lower than under certainty. Conversely, if there is an expectation of a decrease in the utility of consumption in the second period, then there will be a decrease in the period one consumption relative to its level under certainty.

In conclusion, uncertainty about future preferences alone is not sufficient to produce an "option value" case for increasing the resource left to the next generation. In addition to pure uncertainty, there must be asymmetry in the distribution of possible changes in preferences. Neutral uncertainty with increases and decreases equally likely does not generate a case for leaving more to the future in case their preferences for the resource are stronger than ours. Uncertainty makes a case for conservation only when the expected return to postponement of consumption is positive.

One final observation and an indication of possible future research: the Beltratti, Chichilnisky and Heal model of option values summarized here is one in which utility is derived only from the flow of consumption of the environmental resource. In practice the stock may also enter as an argument of the utility function. For example, we value the current climate as an asset: we value the current stock of species or of rainforests. In this case there are likely to be two qualitatively different types of optimal consumption path, depending on

[12]Equalization of expected marginal utilities in the two periods requires in this case that

$$\frac{\partial u}{\partial c_1} = \{p(1+\alpha) + (1-p)(1-\beta)\}\frac{\partial u}{\partial c_2}.$$

Now the period two expected marginal utility curve may in general lie above or below the first period curve: it will be exactly the same as the first period curve, i.e., the curve in the absence of uncertainty, if and only if the coefficient of $\partial u/\partial c_2$ is one, i.e., $\{p(1+\alpha) + (1-p)(1-\beta)\} = 1$. This condition means that the expected shift in period two utility is zero.

[13]Technically, this is because uncertainty here is about the utility functions, and the maximand is linear in these. Society is not risk-averse about utility levels, even though it is about consumption levels.

the size of the initial stock of the resource. If this is large, the optimal path will involve the maintenance of positive stocks of the resource indefinitely: if it is small, then the entire stock will eventually be consumed. The size of the critical initial stock at which this qualitative change in solution occurs, will depend on preferences. In this case, it is possible that a change in uncertainty about future preferences will tip the economy from one optimal consumption regime to another. Such a phenomenon would make a dramatic difference to the computation of the option value.

Conclusions and Open Questions

While some of the foundations are in place for an understanding of the economics of global environmental risks, there are certain aspects that require more attention. From a policy perspective, the endogeneity of the risk faced is important. The purpose of many recommended policies is precisely to change the risks that we face. An example is the global carbon tax proposed by the OECD and reviewed in Chichilnisky (1993a). Only recently has there been systematic study of the welfare economics of markets with endogenous risks, and many questions remain open.

The policy implications of many of the issues that we have reviewed need further attention. How important, for example, are the option values associated with global changes? It would be interesting to see some of the large models reviewed by Weyant in this symposium address this question. We need to study further the institutional implications of trading environmental risks on financial markets such as contingent claims markets and mutual insurance markets. The public good aspect of the global climate has not been analyzed adequately. The structure of emissions permit markets needed for efficiency is still not understood. Our analyses of option values are still based on very limited assumptions, in spite of a long and distinguished literature in this area. Finally, economists clearly need more exchange of information with physical scientists: it is fair to say that the two areas fail to make any significant contact in a field that needs the skills of both.

The prospect of climate change induced by human activity faces societies with demanding issues in risk management and risk assessment: at the same time, it faces economics with challenges and opportunities. The challenge is to develop intellectual tools, communicate them to society at large and prove that they can add value to the analysis of a complex and possibly fundamental problem.

■ *Both Chichilnisky and Heal acknowledge financial support from the NSF (grants 92-16028 and 91-0460 respectively), the Fondazione Mattei and the Monte dei Paschi di Siena. Both authors appreciate valuable comments from Ken Arrow, Andrea Beltratti, Mordecai Kurz, Richard Sandor, and Bob Solow. We are particularly grateful to the editors of this issue, Dick Schmalensee, Carl Shapiro and Timothy Taylor, for valuable comments and suggestions which go way beyond the call of duty.*

References

Arrow, Kenneth J., "The Role of Securities in an Optimal Allocation of Risk-bearing." *Econometrie, Proceedings of the Colloque sur les Fondements et Applications de la Theorie du Risque en Econometrie, Centre National de la Recherche Scientifique*, Paris, 1953. English translation in *Review of Economic Studies*, April 1964, *31*, 91–96.

Arrow, Kenneth J., and Anthony C. Fisher, "Environmental Preservation, Uncertainty and Irreversibility," *Quarterly Journal of Economics*, May 1974, *88*:2, 312–19.

Arrow, Kenneth J., and Robert C. Lind, "Uncertainty and the Evaluation of Public Investments Decisions."*American Economic Review*, June 1970, *60*:3, 364–78.

Barrett, Scott, "The Paradox of International Environmental Agreements." Paris: Environment Directorate, OECD, 1990.

Beltratti, Andrea, Graciela Chichilnisky, and Geoffrey M. Heal, "Irreversibility, Uncertain Future Preferences and Option Values," Stanford Institute for Theoretical Economics, Technical Report No. 53, Stanford University, 1992.

Beltratti, Andrea, Graciela Chichilnisky and Geoffrey M. Heal, "Sustainable Growth and the Green Golden Rule," working paper, Columbia Business School, 1993. In Goldin, Ian and L. Alan Winters, *Economic Approaches to Sustainable Development*, OECD, Paris, forthcoming 1993.

Bohm, Peter, "Option Demand and Consumer's Surplus: Comment," *American Economic Review*, September 1975, *65*:4, 733–36.

Bromley, Daniel W., *Making the Commons Work*. San Francisco: ICS Press, 1992.

Carraro, Carlo, and Domenico Siniscalco, "Strategies for International Protection of the Environment," working paper, Fondazione Eni Enrico Mattei, Milano, 1991.

Cass, David, Graciela Chichilnisky, and Wu Ho-Mou, "Individual Risks and Mutual Insurance." CARESS Working Paper #91-27, Department of Economics, University of Pennsylvania, 1991.

Chichilnisky, Graciela, and Wu Ho-Mou, "Financial Innovation and Default in Incomplete Markets," working paper, Columbia University, Department of Economics, 1991.

Chichilnisky, Graciela, "Existence and Optimality of Equilibrium with Endogenous Uncertainty," working paper, Columbia University, Department of Economics, 1992.

Chichilnisky, Graciela, "The Abatement of Carbon Emissions in Industrial and Developing Countries." Paper presented at the International Conference on the Economics of Climate Change, OECD/IEA, Paris, June 1993a.

Chichilnisky, Graciela, "What is Sustainable Development?" Stanford Institute for Theoretical Economics Technical Report, 1993b.

Chichilnisky, Graciela, Jayasri Dutta, and Geoffrey M. Heal, "Price Uncertainty and Derivative Securities in a General Equilibrium Model," First Boston Working Paper, Columbia Business School, 1992.

Chichilnisky, Graciela, Frank H. Hahn, and Geoffrey M. Heal, "Price Uncertainty and Incomplete Markets," First Boston Working Paper, Columbia Business School, 1992. In Chichilnisky, Graciela, *The Formulation of Economic Theory: Essays in Honor of Kenneth Arrow*. Cambridge, U.K.: Cambridge University Press, forthcoming.

Chichilnisky, Graciela, and Geoffrey M. Heal, "Financial markets for unknown risks." First Boston Working Paper Series, Columbia Business School, 1992.

Chichilnisky, Graciela, and Geoffrey M. Heal, "Who Should Abate Carbon Emissions? An International Viewpoint." First Boston Working Paper, Columbia Business School, 1993.

Cichetti, Charles J., and A. Myric Freeman, III, "Option Demand and Consumer Surplus: Further Comment," *Quarterly Journal of Economics*, August 1971, *85*:3, 528–39.

Cline, William R, *The Economics of Global Warming*. Washington, D.C.: Institute for International Economics, 1992.

Dasgupta, Partha S., and Geoffrey M. Heal, *Economic Theory and Exhaustible Resources*. Cambridge: Cambridge Economic Handbooks, Cambridge University Press, 1979.

Dixit, Avinash K., "Investment and hysteresis," *Journal of Economic Perspectives*, Winter 1992, *6*:1, 107–32.

Fisher, Anthony C., and John V. Krutilla, "Economics of Nature Preservation." In Kneese, Alan V., and James L. Sweeney, ed., *Handbook of Natural Resource and Energy Economics*. Amsterdam, New York and Oxford: North Holland, 1985, 165–89.

Hartwick, John M., "Decline in Biodiversity and Risk-adjusted NNP," working paper, Queens University, Department of Economics, 1992.

Heal, Geoffrey M., "Interaction Between

Economy and Climate: A Framework for Policy Design under Uncertainty." In Smith, V. Kerry, and Ann Dryden White, eds., *Advances in Applied Microeconomics*. Greenwich: J.A.I. Press, 1984, 151–58.

Heal, Geoffrey M., "Economy and Climate: A Preliminary Framework for Microeconomic Analysis." In Just, Richard E., and Nancy Bockstael, eds., *Commodity and Resource Policies in Agricultural Systems*, Berlin, Heidelberg and New York: Springer, 1990, 196–212.

Heal, Geoffrey M., "Optimal Resource Use" In Kneese, Alan V., and James L. Sweeney, ed., *Handbook of Natural Resource and Energy Economics*. Amsterdam, New York and Oxford: North Holland, 1993a, 855–80.

Heal, Geoffrey M., "International Negotiations on Emission Control," *Economic Development and Structural Change*, 1993b, *1*:1, 1–19.

Henry, Claude, "Option Values in the Economics of Irreplaceable Assets," *Review of Economic Studies, Symposium on the Economics of Exhaustible Resources*, 1974a, 89–104.

Henry, Claude, "Investment Decisions Under Uncertainty: 'The Irreversibility Effect'," *American Economic Review*, December 1974b, *64*:6, 1005–12.

Krutilla, John V., "Conservation Reconsidered," *American Economic Review*, September 1967, *57*, 777–86.

Kurz, Mordecai, "The Kesten-Stigum Model and the Treatment of Uncertainty in Equilibrium Theory." In Balch, M. S., D. L. McFadden, and S. Y. Wu, *Essays on Economic Behavior Under Uncertainty*. Amsterdam: North Holland, 1974, 389–99.

Kurz, Mordecai, "On the Structure and Diversity of Rational Beliefs," Technical Report No. 39, Stanford Institute for Theoretical Economics, Stanford University, 1990.

Kurz, Mordecai, "General Equilibrium with Endogenous Uncertainty," working paper, Department of Economics, Stanford University, forthcoming 1993. In Chichilnisky, Graciela, ed., *The Formulation of Economic Theory: Essays in Honor of Kenneth Arrow*. Cambridge, U.K.: Cambridge University Press, forthcoming.

Malinvaud, Edmond, "The Allocation of Individual Risk in Large Markets," *Journal of Economic Theory*, April 1972, *4*:2, 312–28.

Malinvaud, Edmond, "Markets for an Exchange Economy with Individual Risk," *Econometrica*, May 1973, *41*:3, 383–410.

Manne, Alan S., and Richard G. Richels, *Buying Greenhouse Insurance*. Cambridge: MIT Press, 1992.

Pindyck, Robert S, "Irreversibility, Uncertainty and Investment," *Journal of Economic Literature*, September 1991, *XXIX*: *3*, 1110–48, 1991.

Schmalensee, Richard, "Option Demand and Consumer's Surplus: Valuing Price Changes under Uncertainty," *American Economic Review*, December 1972, *62*:5, 813–24.

Solow, Robert M., *Sustainability: an Economist's Perspective*, the Eighteenth J. Seward Johnson Lecture, Woods Hole Oceanographic Institution, Woods Hole, Massachusetts, 1992.

Stiglitz, Joseph E., "Risk Incentives and Insurance: The Pure Theory of Moral Hazard," *The Geneva Papers*, January 1983, *8*:26. 4–32.

Weisbrod, Burton A., "Collective-Consumption Services of Individual-Consumption Goods," *Quarterly Journal of Economics*, August 1964, *78*, 471–77.

Part II
Damages, Costs and the Search for an Optimal Response

A
Damage Estimates

[4]

The Economic Benefits of Limiting Global Warming

For purposes of public policy, it is desirable to determine the range of economic damage that may be expected from global warming.[1] If these damages are far smaller than the costs of reducing emissions from fossil fuels, deforestation, and other sources, then the case for preventive action is weak on grounds of traditional economic policy formation and must stand solely on ecological concern. If instead the economic damages equal or exceed the costs of limiting global warming, preventive action is warranted on economic grounds alone, reinforcing the case for action on the basis of concern about ecological risk.

For a time in the late 1980s, it appeared that policy action on the greenhouse problem would be based simply on the scientific view that global warming would take place. The international scientific review by the Intergovernmental Panel on Climate Change (IPCC, discussed in chapter 1) seemed to be the vehicle for resolving the issue. However, by 1990–91 it had become apparent that the scientific determinations were just the first step in the policy process. Policymakers were beginning to emphasize the need to weigh the costs of preventive action against the prospective damages that might be expected from global warming. For US officials at least, this calculus still found the case for action unconvincing, in part because of remaining scientific uncertainty (chapter 8).

The sole economic analysis that had attempted to place a valuation on the prospective damages from global warming had identified only a small economic impact and correspondingly counseled extremely modest action (Nordhaus 1990a, 1991a; see chapter 7, annex 7A, of this study). The striking implication of that study was that, even if the scientific predictions were entirely correct, the greenhouse effect would simply be too moderate in economic terms to warrant much preventive effort. Although that conclusion might have been alarming to those

1. An earlier version of this chapter appears in Cline (1992b).

in the scientific community and general public who felt that the expected ecological effects would be unacceptable, the implication was that policy action beyond a bare minimum would have to be justified on noneconomic grounds. Other estimates of the damages of global warming tended to be only qualitative or to be confined to individual effects rather than a comprehensive evaluation (IPCC 1990c; Environmental Protection Agency 1989a).

At the same time, as reviewed in chapter 4, several quantitative studies emerged on the cost side of the issue.[2] Typically based on energy-economic models, these calculations began to provide what appeared to be a relatively robust range of estimates of the price that economies would have to pay to limit emissions of greenhouse gases. This price was placed relatively high in the range of 2 percent to 3 percent of GDP on an ongoing basis (chapter 4). As the international community negotiated toward a convention on climate change in preparation for the United Nations Conference on Environment and Development in June 1992 (chapter 8), there was considerable risk that the seeming concreteness of cost estimates for preventive action, set against the only fragmentary measurements of the benefits of limiting global warming, would bias the policy process toward inaction.

This chapter seeks to provide an evaluation of the benefits of limiting global warming that is more comprehensive and satisfactory than those previously available. The estimates here first attempt to identify more thoroughly than before the economic damages to be expected from warming associated with a doubling of carbon-dioxide-equivalent, the conventional benchmark. However, what more distinctly sets the analysis apart from previous calculations is the incorporation of the dimension of rising economic damage from the much higher warming over the very long term of 250 to 300 years (chapter 2). It should be emphasized that the estimates must be seen as broad approximations rather than precise measurements, especially for the more intangible ecological effects (e.g., species loss). Nonetheless, the findings here seem sufficiently clear to suggest that previous examinations have tended to understate the scope of economic damage that may be expected from global warming. This conclusion applies especially to damages over the very long term.

Analytical Framework

Ideally, the measurement of economic benefits to be obtained from the limiting of global warming should include the following elements.

2. In particular, Manne and Richels (1990a), Edmonds and Barns (1990), Jorgenson and Wilcoxen (1990b), Congressional Budget Office (1990), and Nordhaus (1990a). For surveys, see Hoeller et al. (1990), Edmonds and Barns (1991), and Darmstadter and Plantinga (1991).

First, estimates should be prepared for each major actor in the international bargaining that will be required for action, as global warming will affect each country differently and in varying intensity. The countries and groups relevant for this purpose include at least the United States, the European Community, the Commonwealth of Independent States (former Soviet Union), Japan, and Canada, because of their large roles in global economic activity and emissions.[3] In addition, other key actors include China, because of its large coal reserves; Brazil and Indonesia, because of their large emissions from deforestation; Eastern Europe, where industrial emissions have been unusually high; Bangladesh, Egypt, and certain island nations, where severe damage from sea-level rise may be expected; and the Sahel and other developing regions already facing vulnerability of water supply.[4]

The high-resolution general circulation models (GCMS) have sufficient area detail to provide these geographical estimates.[5] The IPCC (1990b, 163–65) does warn that "confidence is low" for the regional (as opposed to global) calculations of the GCMS, especially with respect to precipitation and soil moisture. Nonetheless, as noted in chapter 2, the high-resolution GCMS have been relatively successful in simulating current-day regional climates and would seem to be the appropriate point of departure for analysis for individual countries and regions.

The estimates of this chapter attempt to distinguish effects among certain countries, especially in agriculture. However, much of the analysis here is oriented toward the United States, in part because of the use of underlying analysis by the US Environmental Protection Agency (EPA) but also because of the sheer scope of the research task involved. Future research would usefully extend the calculations more fully to other geographical areas, especially the key actors just listed. In lieu of geographically more comprehensive estimates, the benefit-cost analysis of chapter 7 below draws implicitly on the results found for the United States as an initial pattern for global effects.

Second, greenhouse damages should be identified by category of effect. Chapter 1 summarized the principal aspects of expected damage from global warming. Anticipating the empirical estimates of this chapter, table 3.4 in the overview

3. Some regional detail on at least the United States, the European Community, and Russia would also seem desirable, as their geographical expanses include areas likely to face significantly different impacts.

4. See table 8.1 for country-specific estimates of carbon dioxide emissions from the burning of fossil fuels and deforestation.

5. Thus, the UK Meteorological Office model, with resolution of 2.5° horizontal and 3.75° vertical, divides the earth's surface into 6,912 grids. At the equator, each grid has a north-south distance of 417 km and an east-west distance of 278 km; at higher latitudes the east-west distance is smaller (Cline 1990c).

section at the end of this chapter distinguishes 16 broad categories of economic effects of global warming.

Third, the damages from global warming need to be specified for different points in time. In practice, the approach of this chapter is to identify two temporal benchmarks: equilibrium warming from the doubling of carbon-dioxide-equivalent, and that from very-long-term warming. As noted in chapter 2, the corresponding dates for actual realized warming are set at 2050 and 2275, respectively. As developed in chapter 7 below, actual warming at intermediate periods may be estimated by linear interpolation. The corresponding damage associated with warming in intermediate years applies a geometric (i.e., exponential) relationship between damage and warming, providing what amounts to geometric interpolation between the two benchmarks (chapter 7, equations 7.5 and 7.6).

Fourth, the analysis should take account of adaptation.[6] Efforts to adapt reduce the damages of global warming and thus the benefits of limiting warming. However, adaptation also reduces the costs attributable to prevention measures, because these costs must be calculated on a net basis and therefore should appropriately deduct adaptation costs if estimated prevention benefits are correspondingly being reduced because of adaptation.

With regard to technological change more generally, it is true that unanticipated improvements in technology could reduce the extent of damages from global warming (e.g., through the development of miracle seed varieties resistant to drought). However, as argued in chapter 8, there is no reason to expect unforeseen technological breakthroughs to reduce greenhouse damages and the corresponding benefits of abatement policy action by more (or less) than they reduce the costs of action. There is therefore no presumption that the inability to fully incorporate future technical change biases the benefit-cost ratio for policy action either upward or downward.

Fifth, the benefits from preventive action must take account of the extent of the remaining warming not avoided by the policy measures in question. It is necessary to estimate a projection of both the warming damages under the baseline of no policy intervention and those still present under the policy scenario in question. The "benefit" of the policy is then the difference between the two projections over time, not the full amount of the baseline damage projection.

Sixth, analysis of benefits over time requires discounting to permit comparison of effects at different time periods (discussed in chapter 6).

Seventh, it is necessary to scale the estimated benefits. The variable used here (and in chapter 7) for scaling is gross domestic product (GDP). Absolute magnitudes of damage are first estimated against today's economic scale, and then

6. Thus, Schelling (1983, 453) calls attention to adaptation and technical change by warning that analysis of the greenhouse effect should not make the mistake of "superimposing a climate change that would occur gradually in the distant future on life as we know it today."

the resulting percentage of GDP is applied to future GDP levels. In reality some damages are likely to rise less than proportionately with GDP, and others more than proportionately. Agricultural effects would tend to rise less than GDP because of Engel's law (demand for food is income-inelastic). Such effects as electricity requirements for air conditioning might be expected to rise proportionately, or more than proportionately, with GDP, as they are likely to have unitary or above-unitary elasticity of demand with respect to per capita income. Still other effects, such as valuation on human life and the value of species loss, could be expected to rise more than proportionately with GDP. Human life valuation will reflect lifetime earnings capacity, which rises with per capita GDP; valuation of species and other ecological intangibles may be expected to be income-elastic, because economies have greater scope for allocating resources to these concerns as income rises further above subsistence levels. Overall, scaling by GDP would seem to be an appropriate approximation.[7]

Eighth, welfare weights might appropriately be applied to make policy comparisons across countries and regions. Thus, losses to a poor country such as Bangladesh might receive greater weight in global calculations than would be indicated solely on the basis of that country's measured GDP, because of the greater "marginal disutility" of losses at living standards close to subsistence levels.

Ninth, it is appropriate to include some form of weighting to take account of policymaker risk aversion. If policymakers are cautious, there should be relatively more weight attributed to the high-damage variants than to the low-damage variants flanking the central-case estimates.

Of these various dimensions, the analysis of this chapter concentrates on economic categories of damage, time horizon, and (to a lesser extent) geographical location and adaptation. The discussion of chapter 2 considers the component of damage that can be avoided through policy action. Chapter 6 examines time discounting; and chapter 7 synthesizes the elements of the analysis into an overall benefit-cost evaluation that incorporates risk aversion.[8]

7. Some stress that GDP itself is a misleading measure because it does not take account of environmental damage, the value of leisure time, and other important aspects of economic welfare. Although undoubtedly true to some extent, this critique is irrelevant to the question of gauging the impact of greenhouse damage against the GDP yardstick. Both the benefits of avoiding damage through abatement and the costs of abatement are expressed as percentages of GDP; therefore, any modification of GDP for factors such as those just mentioned would enter into both the numerator and the denominator of the benefit-cost ratio and thus wash out. In any event, such corrections seem unlikely to reduce measured GDP by more than, say, 10 percent to 15 percent; in this case, greenhouse damages would be 2.2 percent of "true" GDP rather than 2.0 percent (for example), with no major change in the qualitative interpretation. In sum, distaste for GDP is essentially a red herring if used as a critique against scaling greenhouse damages and abatement costs by this conventional measure.

8. Note that Cline (1992b) develops (but does not estimate) the following summary

Inherent Damage Limits?

Before turning to the specific estimates of this chapter, it is important to scrutinize an argument that stems solely from the "economic categories" dimension. Nordhaus (1991a) and Schelling (1991) have argued that, at least for industrial countries, the potential damage from global warming is greatly limited by the small share of climate-sensitive sectors in GDP. For example, in countries where agriculture accounts for less than 3 percent of GDP (according to the argument), even sharp reductions in agricultural output would represent relatively small percentages of GDP. There is some merit in this point, as one might expect a developing economy with a far higher share of agriculture to be more vulnerable than an industrial economy. However, the argument tends to understate potential damages generally and for industrial economies in particular.

Of the 16 categories identified in table 3.4 and analyzed below, only four are related to outdoor activity and thus directly affected by climate (agriculture, forestry, leisure, and construction). Eleven of the categories are not even tied to a specific production sector. Some of the effects are more likely to show a higher impact in an industrial economy than in a developing economy. Human life is one such effect, if standard approaches based on lifetime earnings (for example) are the criterion. Coastal defenses are likely to be another: the prospective investment in defending New York or Amsterdam is likely to exceed that of defending Recife, Brazil (abstracting from the different physical situations), because of higher structural buildup and property valuation in the countries that are more highly developed. The same may be said of the "pollution spillover" category.

Even in agriculture, the "inherent limitation" argument ignores the diamond-water paradox. Agriculture is only 3 percent of US GDP because agricultural production is abundant and prices are low. With serious cutbacks in agricultural

---

measure of benefits from greenhouse policy:

$$PVGB = \sum_i \sum_j \sum_t \lambda_j \, [w^l \, b' \, (i,j)^l_{p,t} + w^c \, b'(i,j)^c_{p,t} + w^u \, b' \, (i,j)^u_{p,t}]/(1 + \rho)^t$$

where PVGB is the discounted present value of benefits, b' is the entry in a matrix of benefits net of adaptation effects, i is the economic category of damage, j is the country or region, t is the year, ρ is the discount rate, w is the risk weight, l is the low-damage case, c is the central damage case, u is the upper-bound damage case, p is the policy scenario in question, and λ is the equity weight. For its part, $b'(i,j)$ is the ijth cell in the matrix:

$$B_p'{}_{,t} = D_{b,t} - D_{p,t} + C_{a,t} - (-\Delta D_{a,t})$$

where matrix D refers to damages, subscript b denotes the baseline, matrix C_a refers to the costs of implementing adaptation measures, and matrix ΔD_a is a matrix of changes in damages resulting from adaptation.

production, and given price-inelastic demand, ex post prices would be much higher. Proper evaluation of the loss of consumer surplus would mean a much larger damage potential than might be suspected from the small share of agriculture in present GDP.[9] In sum, it is misleading to cite the low shares of climate-sensitive sectors in the existing structure of production as prima facie evidence that the benefits of global warming abatement are likely to be small.

Agriculture

The specific estimates of the impact of the greenhouse effect may appropriately begin with the agricultural sector.

General Considerations

Global warming is likely to damage agriculture in many areas but aid it in others. The principal damage will arise from heat stress and decreased soil moisture. Warmer temperatures cause the growing cycle of the plant to accelerate, allowing less time for plant development before maturity. Thus, even under irrigation, corn yields are estimated to decline by as much as 50 percent in the Great Lakes region with temperature increases from a doubling of carbon dioxide as estimated with the Geophysical Fluid Dynamics Laboratory (GFDL) general circulation model (Ritchie et al. 1989, 1-1). Moreover, the continental interiors are expected to become drier with global warming. The principal offsets would be longer growing seasons in cold, northern climates and the fertilization effect of greater atmospheric carbon dioxide. Kane et al. (1989, 3) provide the following overview:

> The warming and increased precipitation in the northern latitudes [near 70°N] would improve the agricultural prospects in Canada, northern Europe, and the USSR. . . . The increases in temperature and drying of interiors of continents in the mid-latitudes [near 40°N and S] is expected to lead to reductions in agricultural productivity in the United States and western Europe. Projected coastal inundation of rice growing regions in parts of Southeast Asia, such as Bangladesh, combined with the projected movement of the Asian monsoon away from the Indian subcontinent could lead to reduced agricultural production.

On the basis of EPA studies and studies by the International Institute for Applied Systems Analysis and the United Nations Environment Programme, the three

9. The point may be seen by considering the extreme: if world agricultural production fell to zero, so would world population. The economic loss would equal the entirety of GDP, not just the ex ante share attributable to agriculture.

authors estimate that global warming from carbon-dioxide-equivalent doubling would reduce average yields by 20 percent in the United States and the European Community and by 18 percent in Canada, but increase yields by 15 percent in northern Europe and Australia, 10 percent in the former Soviet Union, and 4 percent in Japan.

Despite such estimates, it is an emerging stylized fact that the damage to agriculture from global warming will be moderate:

> because most farmers are not dumb, but rather are accustomed to continually adapting . . . [and] the projected impacts of climate are small in comparison with impacts associated with government subsidies, international trade, and other economic factors or . . . technological progress. (Ausubel 1990, 14)

In his review of the benefits of abatement, Nordhaus (1991a) places greenhouse damage (from a doubling of carbon-dioxide-equivalent) to US agriculture within a range of −$9.7 billion to +$10.6 billion annually (1982 prices), and thus a mean value of nearly zero, based on EPA estimates (discussed below).

For its part, the IPCC (1990c, 12) concludes:

> Negative impacts could be felt at the regional level as a result of changes in weather, diseases, pests and weeds associated with climate change. . . . There may be severe effects in some regions, particularly in regions of high present-day vulnerability that are least able to adjust technologically to such effects. There is a possibility that potential productivity of high and mid-latitudes may increase because of a prolonged growing season, but it is not likely to open up large new areas for production, and will be largely confined to the Northern Hemisphere. On balance, the evidence is that in the face of estimated changes of climate, food production at the global level can be maintained at essentially the same level as would have occurred without climate change; but the cost of achieving this is unclear. Nonetheless, climate changes may intensify difficulties in coping with rapid population growth.

Thus, the IPCC essentially adopts the benign interpretation, but with caution and recognition of likely damage to vulnerable regions.

There are several reasons to suspect that the benign view is too optimistic. Most important, it fails to consider effects over the very long term. As analyzed below, with eventual global warming on the order of 10°C, agricultural damage would be likely to rise nonlinearly. However, even for the standard benchmark of carbon-dioxide-equivalent doubling, there are grounds for skepticism. First, recent work on drought has concluded that the incidence of severe droughts would be greatly multiplied (Rind et al. 1990). Second, the moderate impact projections usually are premised on an important boost from carbon dioxide fertilization. Yet it is unclear that this phenomenon will prove as effective in the field—in farms around the world—as in the laboratory. Moreover, the amount of carbon dioxide buildup is considerably less than twofold for the doubling of carbon-dioxide-*equivalent* of all trace gases.

Consider drought. Rind et al. use results of the Goddard Institute for Space Studies (GISS) general circulation model to calculate that the incidence of severe droughts (measured by the Palmer index), which currently occur only 5 percent of the time, would rise to 50 percent frequency by the 2050s. Their finding is based on the difference between precipitation and potential evapotranspiration (E_p). They find that:

> E_p increases most where the temperature is highest, at low- to mid-latitudes, while precipitation increases most where the air is coolest and easiest to saturate by the additional moisture, at higher latitudes. . . . Higher temperatures increase the atmospheric water-holding capacity . . . of the order of 33% . . . and thus the ability of the atmosphere to draw moisture from the surface. (Rind et al. 1990, 9983).

The GISS GCM calculates E_p as a function of the difference between the specific humidity at ground level and that at a height of 30 meters (as well as wind velocity and turbulence). E_p rises strongly with ground temperature. Actual evaporation equals potential multiplied by an efficiency factor ($ET = \beta E_p$). The authors argue that the GCMs tend to understate the efficiency factor (at estimates of $\beta \approx 0.2$, versus the observed US range of 0.6 to 0.9), in part because of the absence of a vegetative canopy in the models. The overall thrust of the analysis by Rind et al. is that the GCMs understate the decline in soil moisture in projections of warmer climates. The implication is that agricultural estimates based on the GCM projections may understate the damage from global warming.

Similarly, Gleick (1987) uses GCM climate projections and a water-balance model for the Sacramento basin in California to calculate that summer soil moisture would fall by 8 percent to 44 percent under carbon-dioxide-equivalent doubling.

So far, however, the crop models have not explicitly incorporated the influence of drought in evaluating the impact of global warming on yields. In particular, the models typically begin each year afresh, with no model "memory" of water depletion in a preceding year. Yet, as shown by the experience of California in the late 1980s, it is the successive accumulation of back-to-back drought years that eventually has the most devastating impact. The economic losses from drought can be large. In the 1988 US drought alone, losses (primarily agricultural) amounted to $40 billion (Wilhite 1990).

The second area that may warrant skepticism is the heavy reliance on carbon dioxide fertilization in the benign assessments. Photosynthesis is the transformation of carbon dioxide and water into plant matter (carbohydrates) and oxygen.[10] The so-called C_3 crops (wheat, rice, soybeans, fine grains, legumes, root crops, most trees) have relatively inefficient photosynthesis (because of loss from induced respiration) and benefit substantially from increased exposure to carbon dioxide. The C_4 crops (maize, millet, sorghum, sugarcane) are more efficient and obtain relatively smaller photosynthesis gains from more carbon dioxide. In addition, both types experience stomatal (pore) closure from increased carbon dioxide, with the result that less water is lost to transpiration and the plants become more efficient in water use. The overall result is that a doubling of carbon dioxide (from 330 to 660 ppm) raises yields in greenhouse experiments by an average 34 percent for C_3 and 14 percent for C_4 crops (Schneider and Rosenberg 1989, 27–28).

However, these effects have not been documented under normal open-field farm conditions. The EPA (1989a, 110) suggests:

> Few studies have examined the interactive effects of CO_2, water, nutrients, light, temperature, pollutants, and sensitivity to daylight on photosynthesis and transpiration . . . [or] growth and development of the whole plant. Therefore, considerable uncertainty exists concerning the extent to which the beneficial effects of increasing CO_2 will be seen in crops growing in the field under normal farming conditions with climate change.

More recently, experiments by Bazzaz and Fajer (1992) have cast doubt on the carbon-fertilization benefits of the greenhouse effect:

> [A]ssumptions about the benefits of a world replete with CO_2 may be overstated. Studies have shown that an isolated case of a plant's positive response to increased CO_2 levels does not necessarily translate into increased growth for entire plant communities. . . . [P]hotosynthetic rates are not always greater in CO_2–rich environments. Often plants growing under such conditions initially show increased photosynthesis, but over time this rate falls and approaches that of plants growing under today's carbon dioxide levels. . . . When nutrient, water or light levels are low, many plants show only a slight CO_2 fertilization effect.
>
> Competitive interference and limited nutrients will influence not only natural ecosystems such as meadows and forests but also man-made ecosystems such as farms. *For this reason, we do not expect that agricultural yields will necessarily improve in a CO_2 rich future* [emphasis added]. Again, at first glance, harvests do seem to benefit from such an atmosphere: Bruce A. Kimball of the U.S. Department of Agriculture reviewed more than 700 agronomic studies and found that, on average, grain production increased 34 percent in high CO_2 conditions. But on closer examination, it becomes clear that such yields were dependent on the presence of fertilizers and water—resources often well supplied only on farms in the developed world. . . . Large quantities of fertilizers, biocides and water must be added to crops for them to be productive in a CO_2–rich environment. But of course today water

10. The photosynthesis reaction is $6CO_2 + 6H_2O \rightarrow C_6H_{12}O_6 + 6O_2$. The reverse direction is respiration.

is a limited and expensive resource. We fear that the costs of these inputs, in dollars and in environmental quality, may be too great. (Bazzaz and Fajer 1992, 68–71)

The authors' warning concerning the need for ample water to enable plants to respond to carbon fertilization is especially relevant in view of the likelihood of increased water scarcity as a consequence of global warming, as discussed below.

Other experts have also emphasized caution on whether open-field carbon fertilization (under changed temperature and rainfall conditions) would be as substantial as the laboratory experiments suggest (Parry 1990, 40–41; NAS 1991a, 37). Nonetheless, the majority view has stressed the large body of evidence supporting at least the laboratory results (Kimball and Rosenberg 1990), and some experts note that selected open-field experiments are beginning to confirm those findings.[11] Overall, it would seem risky to count on agriculture in general experiencing the same degree of benefits from carbon fertilization as has been observed in the laboratory experiments, especially in developing countries where the complementary water and fertilizers may be lacking.

A different type of limitation of assumed carbon fertilization is that it is expected to increase plant mass relative to nutritional content. The nitrogen content of plants would decrease as the carbon content increases, "implying reduced protein levels and reduced nutritional levels for livestock and humans" (Parry 1990, 41). The "effective yields" in nutritional terms could thus rise by less than weight yields. Bazzaz and Fajer (1992, 72) further point out that this same effect would be likely to induce greater crop consumption by insects, as grasshoppers and caterpillars would seek to compensate for lower nutritional quality through greater leaf consumption.

In addition, CO_2 fertilization tapers off relatively quickly with rising atmospheric concentration. Parry (1990, 38) estimates that for maize there is no further enhancement at concentrations above 550 ppm. For wheat, the rate of photosynthesis rises from about 50 mg CO_2 $dm^{-2}h^{-1}$ at 350 ppm (approximately current concentration) to 62.5 mg at 550 ppm; the rate rises only to 68 mg at 750 ppm and rises no further than a 70-mg ceiling for 850 ppm and higher.[12] As discussed below, the exhaustion of direct carbon dioxide enrichment in the face of above-linear increases in crop damage from heat stress and soil moisture decline mean that agricultural losses under very-long-term warming would be expected to far exceed those for the usual carbon-doubling benchmark.

11. US Department of Agriculture experiments in Arizona found that cotton grown in open-air fields with enhanced carbon dioxide levels showed large increases in yields (Norman J. Rosenberg, Resources for the Future, personal communication, June 1991).

12. The photosynthesis rates refer to milligrams of carbon dioxide per square decimeter per hour.

Another difficulty with interpretation of the typical agricultural results is the tendency to mistakenly link double carbon dioxide content (and its attendant enrichment) with the *warming* associated with carbon-dioxide-*equivalent* doubling. Because of other trace gases, there would be considerably less than double carbon dioxide at double total trace gas equivalent.[13] Specifically, the IPCC projects in its business-as-usual case that by the year 2025 the radiative forcing from all trace gases would be slightly higher than the equivalent of a doubling of carbon dioxide above preindustrial levels. At that time, atmospheric concentration of carbon dioxide would be enough to generate 2.88 watts per meter squared (wm^{-2}) radiative forcing. Using the IPCC–Wigley formula ($R = 6.3 \ln C/C_0$, where R is radiative forcing, C is carbon concentration, and $C_0 = 280$ ppm, the preindustrial level), we may calculate that the IPCC expects only 442 ppm carbon dioxide concentration at the equivalent-doubling benchmark.[14] Most of the agricultural studies use 330 ppm as the base level for atmospheric concentration and 660 ppm for the "doubled amount" in calculating yield impacts (EPA 1989a, 100). So *actual atmospheric carbon concentration is only expected by the IPCC to rise by about one-third the amount typically assumed in the agricultural estimates that incorporate "carbon dioxide doubling" enrichment.*

In sum, there are several reasons to be concerned that the present estimates of the agricultural effects of global warming are too optimistic and that, even for the modest benchmark of a doubling of carbon-dioxide-equivalent, agricultural damages would be larger, and gains smaller, than in the current mainstream calculations.

Damage Estimates: The United States

The EPA presented the following summary estimates of agricultural damage in its 1989 report to Congress. Combined producer and consumer economic effects of global warming at benchmark carbon-dioxide-equivalent doubling would range from −\$5.9 billion annually (GISS model) to −\$33.6 billion (GFDL model)

13. Nor is the force of this argument much affected by recent findings that chlorofluorocarbons may have little net warming effect because of their influence in stripping ozone, another greenhouse gas (chapter 1). As noted in chapter 2, by the year 2100 the exclusion of CFCs makes only minimal difference to the IPCC's projected radiative forcing, because a large substitution to HCFCs was already assumed. As for the date 2025, the time for benchmark carbon-equivalent doubling, the IPCC projection had included only 0.11 wm^{-2} radiative forcing from the CFCs out of 1.71 wm^{-2} for trace gases other than carbon dioxide and a total of 4.59 wm^{-2} including carbon dioxide (IPCC 1990b, table 2.7).

14. This calculation remains almost unchanged if account is taken of a faster phaseout of CFCs. If the CFC-HCFC radiative forcing in the IPCC report is taken from the "accelerated policies" rather than "business as usual" scenario, total radiative forcing by 2025 drops only from 4.59 wm^{-2} to 4.50 wm^{-2} (IPCC 1990b, table 2.7, 61).

at 1982 prices, before taking account of carbon fertilization. After including the effect of carbon fertilization at an increase from 330 ppm to 660 ppm, the estimates were +\$10.6 billion for GISS and −\$9.7 billion for GFDL (EPA 1989a, 104).[15] These are the estimates Nordhaus (1991a, 932) uses.

The EPA text, but not its summary table, cautions that the estimates incorporating carbon fertilization may be too optimistic because they assume *double carbon dioxide (from 330 ppm to 660 ppm)* even though the weather scenarios are for equilibrium $2 \times CO_2$ *equivalent* warming. It acknowledges that the concentration of carbon dioxide will be only about 450 ppm in 2030 when the doubling benchmark arrives, and it warns that, although the "carbon fertilization" variants apply 660 ppm, actual concentration would still be only 555 ppm in 2060, when the equilibrium warming effects of the 2030 radiative forcing are felt (after three decades for ocean thermal delay; EPA 1989a, 100).

Actually, the potential bias is even greater, because the concept under investigation is the *equilibrium* effect of the $2 \times CO_2$ equivalent benchmark. Whereas the 2060 climate in the EPA calculations (GISS Transient Scenario A) is relatively close in terms of temperature and precipitation to the IPCC's equilibrium climate for doubling carbon-dioxide-equivalent (as might be expected from the fact that the two diverge in time by about the amount required for ocean thermal lag), the transient EPA benchmark has considerably higher carbon concentration than the IPCC's corresponding value for equilibrium under "doubling" of carbon-dioxide-equivalent. It would be inappropriate to count even 555 ppm carbon as available for fertilization at the equivalent-doubling benchmark, because if carbon dioxide trends continue along the path that would take concentration to 555 ppm by 2060, then the warming commitment would have been increased substantially beyond the equivalent-doubling mark. Essentially, to combine even 555 ppm carbon concentration with the GCM warming projections is to mix a transient concept (concentration) with an equilibrium concept (equivalent doubling).

A proper interpretation of the EPA estimates thus requires a careful weighting of the two alternative sets: with carbon fertilization, and without. For measurement of the equilibrium effects of carbon-dioxide-equivalent doubling, the proper amount of carbon fertilization to accept corresponds to the IPCC's 440 ppm, not 660 or even 555 ppm. On this basis, it is appropriate to give a weight

15. The corresponding breakdown between consumers and producers was as follows. Without carbon fertilization: consumers −\$7.3 billion to −\$37.5 billion, producers \$1.5 billion to \$3.9 billion. With carbon fertilization: consumers −\$10.3 billion to +\$9.4 billion, producers \$0.6 billion to \$1.3 billion. In each case the GFDL model shows the less favorable results. The two models show comparable average temperature increase (GISS 4.5°C, GFDL 5°C) but GFDL shows serious reductions in precipitation, whereas GISS shows moderate increases in most of the five broad regions considered (EPA 1989b, volume 1, vi–vii).

of two-thirds to the estimates without carbon fertilization and one-third to the estimates with fertilization. That is, the actual increase in carbon concentration, from 330 ppm to 440 ppm, is only one-third of the way to the 660 ppm used in the carbon-enriched calculation. Even this calculation gives the benefit of the doubt to the carbon fertilization effect because it makes no allowance for the possibility (some would say likelihood) that open-field results will be less favorable than those in the laboratory, nor for possible lesser nutritional content per plant mass.[16]

It should be recognized that this adjustment by linear interpolation may give different results from those that would be obtained by rerunning the crop models with a $2 \times CO_2$ climate (temperature, rainfall, soil moisture) and a 440-ppm carbon dioxide concentration. The interactions of climate and carbon fertilization can be nonlinear. Nonetheless, linear interpolation is the best available shorthand method for adjusting the crop estimates so that they represent an equilibrium rather than a transient concept. The resulting central estimate for the United States is that benchmark warming from carbon-dioxide-equivalent doubling would impose total (consumer and producer) damages of $13 billion annually at 1982 prices,[17] or $17.5 billion annually at 1990 prices.[18]

An alternative central estimate based on drought incidence might be formulated as follows. Suppose the Rind et al. estimate of increase in incidence of severe drought from 5 percent to 50 percent is accepted. Suppose the $40 billion figure for the 1988 drought is applied as the cost of severe drought. On this basis, the expected annual damage from benchmark warming would be (50 percent − 5 percent) × $40 billion = $18 billion annually. This alternative approach gives an almost identical estimate to that of the appropriately adjusted

---

16. Note also that the carbon-equivalent "doubling" represented by the approach here is above the preindustrial level (280 ppm) used by the IPCC rather than the 1958 level (330 ppm) used by the EPA.

17. This amount equals $(2/3)[(-5.9 - 33.6)/2] + (1/3)[(10.6 - 9.7)/2]$.

18. Note that there is no further adjustment for the fact that the underlying GCMs have higher climate sensitivity parameters than the IPCC's "best guess" ($\Lambda = 2.5°C$). The IPCC (1990b, 64, table 5.1) estimates that central North America (35–50°N, 80–105°W) would experience 2.8°C actual mean annual warming by 2030 with $\Lambda = 2.5°C$ and global mean realized (transient) warming of 1.8°C, for a regional warming coefficient of 1.56 (= 2.8/1.8). With this coefficient applied to $\Lambda = 2.5°C$, equilibrium warming for the United States would be 3.9°C for carbon-dioxide-equivalent doubling. In contrast, the simulation runs of the EPA models applied (GISS, GFDL) have an average US mean equilibrium warming (for 48 states) of 4.7°C for the same benchmark (EPA 1989b, vi). The EPA estimates might thus be seen as overstating damage by some 20 percent (or somewhat more considering nonlinearity). Nonetheless, influences omitted in the calculations seem likely to cause offsetting understatement of damages. These include lesser carbon fertilization in the open field than in the laboratory, and omission in the crop models of the effects of weeds and pests, heat damage to pollination (e.g., for maize), and the impact of sequential drought years.

EPA figure, although the concept is very different because the damage under the drought-probability approach is an expected average reflecting good and bad years, whereas the EPA figure is for a median year.

In other important recent work, Resources for the Future has attempted to do an in-depth calculation for the Missouri-Iowa-Nebraska-Kansas (MINK) area (Rosenberg and Crosson 1991). The study is unique in that it seeks not only to incorporate carbon dioxide fertilization but also to take account of farmer adaptation (earlier planting, use of longer season varieties, changes in tillage to conserve water). The study uses the actual climate conditions of the 1930s as an analog for a 2030s climate.

The MINK results find that if climate change alone is considered, warming by 2030 cuts agricultural production in the area by 17.1 percent. If carbon fertilization (specified at 100 ppm, from 350 ppm today to 450 ppm) is incorporated, the production loss is only 8.4 percent. With on-farm adaptations but without carbon fertilization, the production loss is 12.1 percent. With on-farm adaptations and carbon fertilization, production declines by only 3.3 percent (Rosenberg and Crosson 1991, 11–12).[19] Equilibrium long-term losses in the region's forests are 20 percent to 50 percent even with carbon fertilization, and stream flows decline by 28 percent in the Missouri and Upper Mississippi basins and 7 percent in the Arkansas basin, suggesting that forestry and water supply may be areas of relatively greater damage.

The results of the MINK study, which is among the most careful of all the agricultural studies, might seem at first glance to imply that, with carbon fertilization and farmer adaptation taken into account, agricultural damages in the United States as a result of global warming would be limited. A second look suggests otherwise. The 1930s climate was only 0.9°C warmer than that of the 1951–80 base period (Rosenberg et al. 1991, table 3.2). In contrast, the warming associated with benchmark carbon-dioxide-equivalent doubling for the MINK area would be 5°C for the average of three high-resolution GCMS (IPCC 1990a, figure 5.4, summer and winter average, as shown in figures 2.5 and 2.7, chapter 2, of the present study). This warming would be 3.4°C if scaled down by the ratio of the IPCC's best-guess climate sensitivity parameter, $\Lambda = 2.5°C$, to the average Λ for these three models ($\Lambda = 3.7°C$; IPPC 1990b, 87, table 3.2). Benchmark equilibrium warming is thus more than three times that assumed by the MINK authors. Moreover, the MINK estimates may be favorably biased by using for carbon dioxide enrichment the concentration (450 ppm) that would be associated with benchmark trace-gas-equivalent doubling, even though the warming considered in the study is less than half that standard. Considering these factors, and even without allowance for nonlinearity in the damages, one

19. The estimates do not appear to incorporate valuation changes on consumer surplus.

would expect the damage associated with the normal warming benchmark to be at least three times the MINK estimate. On this basis, *even with carbon fertilization and farm adaptation included, the MINK estimates imply agricultural losses of about 10 percent from equilibrium warming of carbon-dioxide-equivalent doubling.*

The adjusted MINK agricultural losses are even higher if the study's alternative estimate taking account of indirect effects on livestock production is considered. The estimates just reviewed refer to a narrow "export" concept for grains use. However, the study also provides estimates in which grains are instead considered as inputs into regional livestock. In this variant, induced losses in animal production and meat packing raise total losses to 5.7 percent of crop and animal output value, even when the beneficial effects of carbon fertilization and adaptation are taken into account (Rosenberg and Crosson 1990, 16). Applying the threefold expansion just suggested to translate the MINK results to an equilibrium climate with $2 \times CO_2$–equivalent, the resulting losses amount to about 17 percent of agricultural production. Moreover, because the livestock industry could not count on easily importing grain inputs from outside the region in such a world (because of comparable losses elsewhere), the regionally integrated estimates may be more relevant than the "export" estimates.

US agricultural production amounted to an estimated $93 billion in 1990 (*Economic Report of the President*, February 1991). Although there would tend to be gains from warming in Minnesota, Wisconsin, and the Dakotas, the agricultural losses in the South and West would tend to be considerably larger proportionately than those in the MINK states (EPA 1989a, 101; EPA 1989b, volume 1, section 5). For the United States as a whole, losses would thus conservatively be on the order of $10 billion to $17 billion annually, reformulating the MINK results for compatibility with the usual warming benchmark and applying the range of 10 percent to 17 percent loss just identified. This estimate ranges from about two-thirds to about the same as that obtained here after adjusting the EPA central estimate, even though the MINK study incorporates farm adaptation.

Explicit treatment of loss of consumer surplus could raise all the estimates further. In a simple model of agricultural supply and demand, annex 3A below suggests that as an approximation the total consumer and producer surplus lost from lower output is at least as large as the yield decrease multiplied by the initial value base. That result obtains for the horizontal supply curve. With the geometrically rising supply curve implied by the Ricardian concept of declining fertility of marginal land, combined producer and consumer surplus losses would tend to be higher, and perhaps much higher (annex 3A).

To recapitulate, for the United States the agricultural damages from benchmark CO_2–equivalent global warming would amount to some $18 billion annually (both the adjusted EPA basis and the Rind et al. drought approach), with a floor of perhaps $10 billion (MINK-adjusted, low "export" variant).

International Yield Estimates

For other countries, estimates tend to be in the form of calculated changes in yields rather than value-based estimates. However, by combining yield impacts with the estimated value base, it is possible to obtain a minimum estimate of producer and consumer losses (under the proposition just stated). For Canada, Parry (1990) estimates that spring wheat yields will fall by 15 percent to 37 percent, although winter wheat should improve. Drought frequency would multiply 13-fold in Saskatchewan. Maize and soybean yields would fall in Ontario. The IPCC places spring wheat yield reductions at 18 percent (IPCC 1990c, 2–15), based on estimates by G. D. V. Williams and by B. Smit.

It is expected that in southern Europe there will be a large decline in summer soil moisture and a major reduction in agricultural potential (Parry 1990). In the former Soviet Union, production could improve in Siberia, but soils there are relatively unproductive (Flavin 1989). Studies in the former Soviet Union suggest that decreased soil fertility, increased salinity, and increased soil erosion would more than offset prospective warming benefits, and that in the Leningrad and Perm areas yields of winter rye and spring wheat would decline because of premature ripening of crops (S. E. Pitovranov, as cited in Parry and Carter 1989).

For northern Japan, yields of rice, corn, and soybeans have been predicted to rise by some analysts (M. Yoshino et al., as cited in Parry 1990, 121). However, as noted below, preliminary estimates by an international team assembled by the EPA find negative yield effects.

The Scandinavian countries are seen as standing to gain the most from global warming effects on agriculture (IPCC 1990c, 2–16). For China, the IPCC cites mixed results, with a stronger summer monsoon threatening greater flooding in southern China; possible increases in rice, maize, and wheat yields nationally; but maize decreases in the eastern and central regions (IPCC 1990c, 2–19).

Most analysts see Central America as vulnerable to decreases in soil moisture (Parry 1990; *New York Times*, 17 October 1990). In the Sahel, an increase in evapotranspiration is expected to exceed the increase in precipitation and therefore to reduce soil moisture (Parry 1990). Drier soils are expected to reduce yields in Northwest and West Africa, the Horn of Africa, and southern Africa (*New York Times*, 17 October 1990). The Middle East is similarly considered vulnerable.

Cline (1992b, annex table A-1) provides a summary of prospective agricultural effects as identified by Working Group II of the IPCC. More specific estimates are available from the underlying studies consulted by the IPCC and from more recent studies. Table 3.1 presents an overview of the available yield estimates. Most are drawn from the summary provided by Kane et al. (1989). For Australia, China, and Japan, the estimates are identified merely as "negative," based on

Table 3.1 Estimates of agricultural yield changes from global warming, various countries

Country	Crop	Yield change (percentages)
Australia[a]	Various	Negative
Belgium	Wheat, spelt	7
Canada (Saskatchewan)	Spring wheat	−18
China[a]	Maize	Negative
Denmark	Wheat, spelt	10
European Community	Various	−20
Finland		
Helsinki	Spring wheat	10
	Barley	9
	Oats	18
Oulu	Spring wheat	20
	Barley	14
	Oats	13
France	Wheat, spelt	−11
Germany	Wheat, spelt	−5
Iceland	Hay	64
	Pasture	48
Italy	Wheat, spelt	−1
Japan[a]	Rice	Negative
Netherlands	Wheat, spelt	1
Soviet Union		
Leningrad	Rye	−13
Cherdyn	Spring wheat	−3
Saratov	Spring wheat	13

a. See text for further details.

Source: Sally Kane, John Reilly, and Rhonda Bucklin. 1989. "Implications of the Greenhouse Effect for World Agricultural Commodity Markets." Washington: US Department of Agriculture (June).

preliminary results of an international team conducting research under EPA auspices (EPA 1990). If confirmed, these initial findings will represent important reversals for China and Japan in comparison with the previous IPCC surveys.

Table 3.2 applies the estimates of table 3.1 to the value of the agricultural base in each country in question to obtain an impression of agricultural effects globally.[20] The final two columns apply the central percentage yield change to the initial agricultural GDP base with a multiple ranging from $\alpha\beta = 0.75$ to $\alpha\beta = 1.25$, where α is the ratio of the output reduction after incorporation of low-cost adaptational changes to the initial yield reduction; and β is the ratio of total change in consumer and producer surplus to the reduction in output

20. For the United States, the table entries are based on the agricultural loss estimated above, expressed as a percentage of base output. For Australia, China, and Japan, the magnitudes are simply postulated, pending final results of the EPA international study. Note that the negative impact in these preliminary EPA results is based on my interpretation that only 440 ppm carbon fertilization should be incorporated for purposes of analyzing equilibrium carbon-dioxide-equivalent warming.

Table 3.2 Rough estimate of the impact of carbon-dioxide-equivalent doubling on agriculture, selected countries

	GDP, 1988 (billions of dollars)			Implied value change	
Country	Total	Agriculture	Yield change (percentages)	$\alpha\beta = 0.75$	$\alpha\beta = 1.25$
Australia	205	8.2	−10.0[a]	−0.6	−1.0
Belgium	145	3.0	7.0	0.2	0.3
Canada	440	17.6	−18.0	−2.4	−4.0
China	360	120.8	−10.0[a]	−9.1	−15.2
Denmark	95	4.1	10.0	0.3	0.5
Finland	90	5.2	15.0	0.6	1.0
France	900	30.8	−11.0	−2.5	−4.2
Germany	1,130	16.5	−5.0	−0.6	−1.0
Iceland	5	2.0[b]	50.0	0.8	1.3
Italy[b]	765	31.1	−10.0	−2.3	−3.8
Japan	2,850	65.4	−5.0[a]	−2.5	−4.2
Netherlands	215	8.5	1.0	0.1	0.2
Soviet Union	1,000[c]	200.0[c]	0.0	0.0	0.0
United States[d]	4,885	89.8	−20.0	−13.1	−21.9
Total	12,815	603.0	−6.9	−31.1	−52.0

α = ratio of net output effect at a given product price after adaptation to the initial effect; β = ratio of total producer and consumer surplus change to net output effect evaluated at the base price.

Note: the estimates do not take into account importer versus exporter status.

a. Postulated; see text for details.

b. Based on Parry (1990, 87).

c. Rough estimate.

d. Based on text estimate of $17.5 billion for 1990.

Sources: Table 3.1, this volume; World Bank, *World Development Report 1990*; International Monetary Fund, *International Financial Statistics, Yearbook 1990*. World Bank and International Monetary Fund data reprinted with permission.

volume at base-period prices. As suggested above, β is likely to be substantially greater than unity. It should be noted that the estimates of table 3.2 do not take account of trade patterns and essentially treat all consumer and producer surplus effects as proportionate to the production base, whereas the presence of international trade means that consumer surplus losses would tend to be relatively greater for food importers, and producer surplus effects relatively greater for food exporters. Nonetheless, the estimates provide a first approximation of international effects.

The estimates of table 3.2 suggest that, worldwide, the central expectation would be for net agricultural losses from benchmark greenhouse warming. This result is less optimistic than the IPCC suggestion that world agricultural output (or costs) could change by ±10 percent, depending on climate, soil moisture, and carbon dioxide enrichment effects (IPCC 1990c, 2–24). One reason appears to be the reliance of the IPCC on favorable effects for China in the rough global estimate.

The central estimate of table 3.2, in contrast, is that the adverse effect of $2 \times CO_2$–equivalent warming amounts to a loss of some 7 percent of world agricultural production (or about $40 billion annually for the countries considered, midpoint of the two variants). The loss is largest in both absolute and relative terms in the United States. The large proportionate gains in northern countries operate on a sufficiently small base that they do not compensate for losses in the United States, China, Canada, Italy, and Japan.

Agriculture Under Very-Long-Term Warming

The central estimate for US agricultural losses from benchmark $2 \times CO_2$ warming is set at approximately $18 billion annually (1990 US economic scale). For very-long-term warming, losses could be expected to be much greater. Most agricultural experts confronted with a 10°C warming scenario indicate that agricultural losses would be devastating.[21]

The crop models used in many of the estimates of table 3.1 could ideally be tailored to examine the impact of global warming of 10°C for very-long-term warming. If they were, it seems likely that the results would show extremely severe reductions in crop yields. Crops typically cannot grow at temperatures above 35°C to 45°C; wheat, barley, oats, and rye, for example, cannot grow above 35°C, while corn, rice, and sorghum cannot grow above 45°C (Office of Management and Budget and US Department of Agriculture 1989). Yet these ranges could be reached in summertime in a wide range of US agricultural areas under very-long-term warming (chapter 2). As noted above, the carbon dioxide fertilization effect is exhausted at about 800 ppm (and earlier for maize). Evapotranspiration tends to rise more than linearly with temperature and thus outstrips precipitation increases, so that soil moisture under very-long-term warming would be likely to be dramatically reduced.

The broad expectation would be that agricultural losses would be more than linear with respect to temperature rise. Suppose the relationship were only mildly nonlinear and less than quadratic, for example, that $D = k(\Delta t)^{1.2}$, where D is damage, k a constant, and Δt is the change in temperature. Then if Δt = 10°C instead of 2.5°C, the damage would be $(10/2.5)^{1.2}$ = 5.3 times as large. The initial benchmark world agricultural loss of 7 percent would rise to 37 percent. Under these conditions, the consumer surplus losses could rise dramatically as agricultural output became scarce in the face of inelastic demand.

21. For example, Paul Waggoner, of the Connecticut Agricultural Experiment Station, commented to this effect at a Resources for the Future conference on greenhouse nonlinearities held in Washington, 23–24 March 1992. However, he was skeptical that 10°C warming could occur.

In short, with a central estimate for agricultural losses on the order of \$18 billion annually for the United States and \$40 billion worldwide for the normal benchmark of $2 \times CO_2$-equivalent warming (2.5°C), the corresponding estimates for very-long-term warming would be in the range of \$95 billion and \$212 billion, respectively, even before allowing for greater proportionate consumer welfare losses (and without scaling to future economic size).[22]

Forests

Global warming is expected to cause a poleward migration of forests and a change in their composition. Sedjo and Solomon (1989) use the Holdridge Life Zone classification system together with GCM projections of changed climate by grid area to calculate global forest changes from $2 \times CO_2$ warming. They find that boreal forests would decline by 40 percent and temperate forests by 1.3 percent, whereas tropical forests would rise by 12 percent (measured by biomass; area figures are comparable). The net change would amount to a decline of 3.7 percent globally in biomass and 5.8 percent in area.

However, these estimates are equilibrium measures after enough time has passed for complete migrational adjustment to the new climate. The potential northern range of forest species in the United States is expected to shift northward by 600 to 700 km over the next century from global warming, and the southern boundary could move north by 1,000 km. However, the actual migration pace could be as low as 100 km over the same period because of slow migration capacity (EPA 1989a, 71). The implied result would be a "temporary" decline in forested area for some three centuries or more (because of tree death on the southern boundary in excess of additional growth on the northern boundary) and thus much larger forest loss than after eventual equilibrium is reached.

Studies cited by EPA (1989a, 83–84) indicate that over the next 100 years US forests could lose 23 percent to 54 percent of standing biomass in the Great Lakes region and 40 percent in western forests. On this basis, it might be esti-

22. The estimates of agricultural impact in this study place no weight on the view that global warming will have little impact on agriculture because it will come primarily at night. Although new data indicate that warming in this century has been greater at nighttime than during the daytime, the likely explanation is that increased urban pollution has provided an offset (Beardsley 1992). Reflection of solar radiation from sulfate aerosols and additional clouds from pollution would tend to reduce the daytime warming otherwise expected. However, it is implausible in the extreme that, over the long term, nations would increase their urban pollution apace with increased carbon emissions; on the contrary, most nations have begun major programs to reduce pollution. Moreover, the sulfate aerosol effect is regional and may be expected to have diminishing returns with increasing sulfur dioxide concentration.

mated that global warming from the doubling of carbon-dioxide-equivalent could cause a loss of 40 percent of US forests. Some portion of this loss could be avoided by increased planting to locate the replacement species in the areas of in-migration sooner than they would naturally arrive. But the EPA (1989a, 86), expressing doubts about this response, points out that "seedlings that would appear to be favored on some northern sites several decades in the future may not survive there now because of constraints imposed by temperatures, day length, or soil conditions."

In 1989, the US logging industry produced $13 billion in gross output (*U.S. Industrial Outlook* 1990, 6–2). The wage bill for 92,000 employees amounted to $1.7 billion. If, as in most US industries, the ratio of wages to capital services was at least two to one, total factor inputs cost approximately $2.6 billion. The value of the wood extracted was thus approximately $10 billion. In view of the prospective 40 percent forest loss, these estimates would imply greenhouse warming damages on the order of $4 billion annually for the United States just for the commercial portion of forests. It is unlikely that these losses would be offset by higher cutting rates, as the current rates of extraction already reflect optimum considerations.[23]

Neither the EPA nor other authors appear to have placed quantitative estimates on forest loss.[24] The EPA does suggest that, despite doubts about successful anticipatory planting for migrating species, US reforestation efforts would need to double or triple above their current rate of 0.4 percent of forest area annually at a cost of $400 million annually (EPA 1989a, 89). On this basis, some $600 million in additional reforestation costs could be expected. If the increased reforestation limited the potential forest loss by one-third, the net annual loss would amount to (2/3 × $4 billion) + $600 million, or a total of $3.3 billion.[25]

The estimates here, based on the EPA results, do not incorporate possible beneficial effects on forests from carbon fertilization.[26] The EPA authors argue that higher carbon dioxide concentrations "may not significantly affect forest

23. Similarly, one does not expect a major reduction in a livestock herd to call forth higher slaughter rates; the calculus of the slaughter (extraction) rate remains, as before, a function of the interest rate and the ratio of present to future price. Over the time horizon here, future prices would not fall very fast, so the extraction rate would be unlikely to rise.

24. Thus, Nordhaus (1990b) merely refers to such losses as "small."

25. The resulting extent of forest product loss (2/3 × 40 percent = 26 percent) would be approximately the same as identified by Binkley (1987), who estimates that global warming would cause income from timber sales to decline by 20 percent in the East and 26 percent in the West by 2030 (cited in Regens et al. 1989, 308).

26. As noted, most trees are C_3 plants and thus should benefit from carbon fertilization.

productivity" because of the limiting factors of water, nutrients, and light under the canopy (EPA 1989a, 81). Bazzaz and Fajer (1992, 71) indicate that competition among trees for such "scarce resources" may account for the findings in experiments in two deciduous temperate forests and one Mexican rain forest, which showed that "tree seedling communities were not more productive in a CO_2–rich atmosphere when different species were grown together."

Similarly, although the estimates here do make some allowance for limitation of losses through forest management, it may be asked whether a higher fraction of the losses could be avoided than the one-third assumed here. Intensively managed plantation systems could increase water and nutrients to take greater advantage of potential carbon fertilization. However, in the United States such managed plantations account for only a small part of forested area.[27]

The estimate of forest loss developed here could be understated from two other standpoints. First, as noted, the estimate makes no allowance for public valuation of forested area for its own sake (esthetic or for camping and hiking use), but focuses instead on a narrow valuation of lumber loss. Second, the underlying EPA estimates do not incorporate losses from "possible increases in fires, pests, disease outbreaks, wind damage, and air pollution" (EPA 1989a, 71).

In the very long term, two factors would work in opposite directions on the value of forest losses. Eventual completion of migration would potentially reverse the major loss associated with the difference between potential and actual migration rates. However, in the absence of policy changes, warming could be expected to increase without interruption over the next 250 to 300 years, so there would be a continual lag of forest migration behind that of potential boundaries. Thus, little could be expected from achieving a new equilibrium (because over the time frame none would be attained). At the same time, much higher temperatures and drier conditions would be likely to cause additional damage to the forests. A plausible conclusion is that under business as usual there would be no reduction in the annual losses even over the very long term, and that instead there would be further increases. In the absence of more formal analysis, something on the order of twice the loss associated with $2 \times CO_2$-equivalent might be appropriate for the very-long-term loss.

Internationally, the shift toward tropical, and away from boreal and temperate, forests suggests that most developing countries would experience relatively less forest loss than would the mid- to high-latitude countries. In this important

27. Smith (1991) indicates the following breakdown of US forests (in millions of hectares): unmanaged natural, 120; managed natural, 190; suburban, 80; urban, 40; intensively managed plantation, 20.

instance, the pattern of relative gains from global warming by such northern areas as Scandinavia would thus be likely to be reversed.[28]

Research in progress for the EPA finds that globally approximately half of all land area would show a shift in the vegetation life zone as a consequence of benchmark carbon-dioxide-equivalent doubling. Tundra and desert area would decline by approximately 40 percent and 20 percent, respectively, whereas dry forest would increase by about one-third and wet forest would show little change (positive for some GCMS, negative in others; Thomas M. Smith et al., in EPA 1990). Although the expansion of forest at the expense of desert would be felicitous, the estimates are equilibrium changes. Thus, major forest loss over the period of perhaps two to three centuries would still mean damages over the relevant horizon.

Species Loss

Analyses of the extent of species loss from global warming are sparse, and so far quantification in the corresponding economic valuation remains nonexistent. Yet some argue that overall biosphere viability depends on biodiversity because of lessened ability to respond to shock with a narrower biological range.

The EPA (1989a) has noted in general terms the risk of increased species extinction because of changes in habitat and predator-prey relationships as well as physiological changes. The EPA indicates major reductions in population, but not species loss, for shellfish, fish, and waterfowl as sea-level rise causes saltwater intrusion into wetlands. The study cites estimates that human activity is already reducing the some 10 million species by a rate at least 1,000 times as fast as would occur from natural forces (EPA 1989a, 152). It cites the poleward migration of forests as a major reason to expect stress on species, especially in view of natural and man-made barriers to animal migration.[29] Coral reefs could be expected to suffer coral death from warmer waters, with extinction of some coral species (as suggested by trends in recent years, including the effect of El Niño warming; Glynn and de Weerdt 1991).

Economists have identified the economic value of species as being of three types: use, option, and existence. Direct use is the most obvious and would include, for example, the use of certain species in the production of medicines.

---

28. Moreover, in some northern countries, such as Finland, the value of forestry production considerably exceeds that of agriculture, so that forestry losses could dominate agricultural gains.

29. One interesting response might be government purchase of land to establish migratory corridors (Peters, as cited in Goklany 1989).

Option value refers to the economic value of preserving a species to retain the option that it may be of economic use in the future. Existence value is a benefit inherent in the existence of the species, like the value society places on the existence of a particular great work of art. Much of the concern with biological diversity seems to derive from existence value (Elliott Sober, as cited in Batie and Shugart 1989).

The yew tree provides a good example of the option value of a species. In 1991, medical researchers found that the drug taxol, derived from the bark of the relatively rare Pacific yew tree, showed impressive effects in causing remission of ovarian and breast cancer.[30] Society could be well advised to pay a fee to avoid the extinction of species (i.e., purchase an "option" to retain them) because some, like the yew tree, might eventually be found to have medicinal or agricultural value. Indeed, plant and animal species already used in traditional and modern medicine number over 5,000 in China, 2,000 in the Amazon Basin, and 2,500 in the former Soviet Union; in the United States, one-fourth of all prescriptions contain ingredients from plants, and over 3,000 antibiotics are derived from microorganisms (Reid et al. 1992, 1–3). Similarly, natural plant varieties are important for enriching agricultural varieties.

Weitzman (1991) has proposed an elegant approach to conceptualizing the value of biodiversity. His method depicts the evolution of a genealogical tree of different species from a common ancestor. He defines biodiversity as, essentially, the total length of all branches on this family tree. If a species must be sacrificed, the greatest possible biodiversity will be maintained by choosing to give up that species that is the shortest final branch of the tree and is thus most closely related to another species on the tree. The caveat is that if the remaining closely related species is unique and thus on a long branch separate from the other species derived from the same ancestor, and is thus extremely distant from them (and hence valuable for biodiversity), the probability that this remaining close relative may itself be lost must be taken into account. This line of reasoning leads to the implication that it is better to concentrate scarce environmental resources on assuring survival of such species rather than dispersing these resources in an attempt to save a larger number of species that are only slightly differentiated by type.

Weitzman's analysis is potentially rich, but it would seem to require extension in some dimensions. First, some weight might be given to the "salience" of species. Public willingness to pay for a species' preservation is likely to be influenced by its perceived importance. For example, the bald eagle would likely be considered more worth saving than, say, a variety of earthworm. The scientific assessment could differ (e.g., because of the role of the earthworm in soil fer-

30. Sedjo (1992); *Journal of Commerce*, 20 December 1991; "Trees That Yield a Drug for Cancer Are Wasted," *New York Times*, 19 January 1992, 1.

tility). Second, some weight might appropriately be given to species close to humans (primates) in the expectation that their option value is high because of their genetic similarity to humans and therefore prospective utility in medicine. Third, there may be "keystone" species whose extinction would threaten the collapse of entire related species structures.

Society is already incurring certain costs that may provide an initial idea of the value of species preservation. As a recent major example, the US government declared in 1991 that 11.6 million acres of timberland in the Northwest would be put off limits to logging to preserve the habitat of the spotted owl (*Wall Street Journal*, 29 April 1991). This area corresponds to 1.6 percent of total forest land in the United States (738 million acres; EPA 1989a, 72). As developed above, a working estimate for the commercial value of forest resources is $10 billion annually. Therefore the land set aside for the spotted owl represents an opportunity cost of $160 million annually.[31]

Many species are already endangered or threatened in the United States (including the bald eagle, the grizzly bear, the Florida panther, the whooping crane, and numerous other fauna and flora; Council on Environmental Quality 1991, 138). An extremely conservative estimate might be that the incremental endangerment of these and all other species from benchmark $2 \times CO_2$ warming amounts to, say, 25 times the present endangerment of the spotted owl alone. On this basis, the public might be prepared to pay at least 25 times as much toward species preservation under the species-threatening conditions of a doubling of carbon-dioxide-equivalent as for the present case of the spotted owl. If so, the annual value of the prospective species damage from benchmark warming would amount to 25 × $160 million = $4 billion. However, the appropriate value could just as easily be an order of magnitude larger, or $40 billion annually. The central estimates here use the more conservative figure, but accompany it by some explicit additional but unmeasured value (Δ_s in table 3.4). For very-long-term warming with temperature increases four times as great, a simple linear extrapolation would once again be a conservative estimate, because species loss would probably rise nonlinearly.

Before more satisfactory estimates of the benefits of avoiding species loss can be made, it will be necessary to obtain a better idea of the prospective extent of loss, ideally with numerous concrete illustrations. With such estimates in hand, it might be possible to use the "contingent evaluation" survey technique for evaluating a public good (see, e.g., Brookshire et al. 1982).

31. The illustration is complicated by the fact that, for many, the motive in saving the spotted owl is primarily to preserve its habitat—forests that include trees centuries or even thousands of years old. US environmental law protects species but not ancient forests, yet many in the public set an "existence" value on such forests. The problem is essentially one of externality: such value (which may or may not be large) is not taken into account when the ancient tree is sold for lumber in the private market.

Sea-Level Rise

Global warming should raise the level of the seas, as a result of thermal expansion of water and melting of land-based glaciers and ice sheets.[32] The IPCC has estimated that, under "business as usual," actual global warming would amount to 4.2°C by the year 2100 (and warming "commitment" even more). The group's central estimate for corresponding sea-level rise is 66 cm (IPCC 1990b, figure 9.6).

US Estimates

James Titus of the EPA and his coauthors note that the IPCC estimate was lower than those of most previous studies and was based on the assumption that the Antarctic would accumulate more ice rather than be a source of melting over the next century—a view not universally accepted by glaciologists (Titus et al. 1992). The EPA thus continues to use a sea-level increase of one meter by the year 2100 as its central estimate.

It should be noted that a point estimate for 2100 is not an equilibrium estimate, because the melting continues over lengthy periods. Titus et al. note that, in the last interglacial period 100,000 years ago, when temperatures were 1°C warmer, the sea level was approximately six meters higher than it is today. The vast disparity between six meters for 1°C of warming and one meter for 4°C suggests that the ultimate sea-level increase would be much higher than the point estimate at 2100. Unfortunately, neither the IPCC nor the EPA appears to have specified the very-long-term sea-level rise associated with a specific equilibrium temperature change. Ideally, one would identify the time path of sea level over, say, the next 300 years as the consequence of no warming, benchmark $2 \times CO_2$ warming (= 2.5°C), and very-long-term warming (e.g., 10°C).

In its 1989 report to Congress, the EPA estimated that, for the United States, a one-meter rise in sea level by the year 2100 would require $73 billion to $111 billion in cumulative capital costs to protect developed areas through the building of bulkheads and levees, the pumping of sand, and the raising of barrier islands (EPA 1989a, 123). It estimated that in addition there would be a loss of dry land of 4,000 to 9,000 square miles.

The agency has recently updated its estimates to incorporate the economic value of the land lost. Morgenstern (1991) indicates that the revised estimates place the economic losses associated with a one-meter sea-level rise at $10.6 billion annually.[33]

32. The melting of sea ice would have no influence, as the volume of its submerged portion is equal to that of its water equivalence.

33. He notes that this figure is twice the amount used by Nordhaus (1990b), which was based on the earlier EPA estimates.

Titus et al. (1992, 8–11) summarize the damages as follows.

> A rise in sea level would inundate wetlands and lowlands, accelerate coastal erosion, exacerbate coastal flooding, threaten coastal structures, raise water tables, and increase the salinity of rivers, bays, and aquifers. . . . Coastal marshes and swamps . . . collect sediment and produce peat upon which they can build [so that] most wetlands have been able to keep pace with the past rate of sea-level rise [but would be unable to do so] if sea level rose too rapidly. . . . Moreover, [where] people have built bulkheads just above the marsh . . . the wetlands would be squeezed between the estuary and the bulkhead. . . . Such a loss would reduce available habitat for birds and juvenile fish. . . . [For barrier islands t]ypically . . . the bay side is less than a meter above high water. Thus, even a one meter rise in sea level would threaten much of this valuable land with inundation. . . . a one meter rise in sea level would generally cause beaches to erode 50/100 meters from the Northeast to Maryland; 200 meters along the Carolinas; 100/1000 meters along the Florida coast; and 200/400 meters along the California coast. . . . [Yet] most US recreational beaches are less than 30 meters . . . wide at high tide. . . . [A] one meter rise in sea level would . . . enable a 15-year storm to flood many areas that today are only flooded by a 100-year storm. . . . [A] rise in sea level would enable saltwater to penetrate farther inland and upstream in rivers, bays, wetlands, and aquifers, which would be harmful to some aquatic plants and animals, and would threaten human uses of water.

The authors estimate that if the shoreline retreats naturally, a one-meter sea-level rise would inundate 7,700 square miles of dry land (an area the size of Massachusetts). Seventy percent of the losses would be in the Southeast (Florida, Louisiana, North Carolina). The eastern shores of the Chesapeake and Delaware bays would also lose sizable areas.

The updated EPA analysis by Titus et al. reaches the following costs of a one-meter sea-level rise for the United States if developed areas are protected. Dry land amounting to 6,650 square miles would be lost, and 49 percent of today's wetlands would be lost. The value of land lost would be $17 billion to $128 billion for wetlands, $21 billion to $71 billion for undeveloped dry land, and $14 billion to $48 billion for land for dikes. Coastal defenses would cost $27 billion to $146 billion for sand, $62 billion to $170 billion to elevate structures, and $11 billion to $33 billion to construct dikes. The total costs would reach $270 billion to $475 billion. The amount would be $128 billion to $232 billion for a 50-cm rise and $576 billion to $880 million for a rise of two meters (Titus et al. 1992, table 9).

The central estimate from the updated EPA estimates is thus a capital cost of $370 billion for a one-meter sea-level rise. Morgenstern's translation of this cost into an annual cost of $10.6 billion implies a discount rate of 2.9 percent. There are important conceptual questions in obtaining the annualized damage value. Essentially the costs of sea-level rise divide into two types: capital costs of protective constructions, and the recurrent annual cost of forgone land services.

The capital costs would be spread over a 100-year period. The land "rent" values would also phase in over the same period. But because the rental opportunity cost is likely to be set at a higher market interest rate than that appropriate for evaluation of the capital costs, the two components should probably be treated individually.

Of the $370 billion total, $224 billion is capital costs for sand, structure elevation, and dike construction. An appropriate way to evaluate this portion is to consider the present value of a 100-year stream of a $2.24 billion capital outlay annually. Discounting at 1.5 percent (the social rate of time preference suggested in chapter 6), the average annual capital outlay for construction amounts to $1.2 billion. Essentially, the entire capital outlay is not made at the outset, so the annual economic cost is considerably smaller than implied by application of even a 3 percent interest rate to the principal.

In contrast, the economic costs associated with land loss can be higher. With a base of 13,000 square miles of wetlands, the 49 percent loss amounts to 6,440 square miles. Titus et al. note that wetland preservation programs typically cost $30,000 per acre. Using a more conservative $10,000 per acre value for wetland, and a real land rental rate of 10 percent, each acre of wetland lost costs $1,000 annually. The loss of 6,440 square miles thus generates $4.1 billion in annual losses.[34] For dry land, the median price in US coastal states is approximately $2,000 per acre (US Department of Agriculture 1990). Allowing twice the median for greater value near the coast, a conservative valuation would be $4,000 per acre (still lower than the minimum $6,000 in the Titus et al. study). With a land rental opportunity cost of 10 percent real, this land typically contributes $400 annually to economic activity. The annual cost of the 6,650 square miles of dry land loss (midpoint) thus amounts to $1.7 billion.

Overall, this alternative approach yields US damage from sea-level rise at $7 billion annually. This estimate is of the same order of magnitude as Morgenstern's $10.6 billion, but it takes explicit account of the long phase-in of the capital expenditures. An important implication is that the composition of the total costs resulting from sea-level rise would be much more heavily weighted toward the loss of annual economic services of the land than toward capital costs of coastal defense construction than is implied in the capital values approach of Titus et al. and Morgenstern.[35]

Over the very long term, damage from sea-level rise could be far greater. As noted in chapter 1, over 300 to 500 years the West Antarctic ice sheet could disintegrate, raising sea level six meters—an event not considered likely with

---

34. 6,440 sq mi × 640 acres per square mile × $1,000 per acre.

35. In the Titus approach, wetland and dryland loss accounts for 40 percent of damage costs, and coastal defense amounts to 60 percent. In the approach here, the composition is 75 percent and 25 percent, respectively.

just benchmark warming of 2.5°C but perhaps considerably more likely with very-long-term warming of 10°C. Even without such a catastrophe, the damage would be likely to rise more than linearly with warming, suggesting much larger losses with very-long-term warming. One reason is that higher warming would be expected to push the Antarctic from a zone where increased snow buildup from more moist, warmer air offsets increased melting (temperature below − 12°C; IPCC 1990b, 271) to a range where the Antarctic could become a large source of water. In addition, not only would the extent of sea-level rise be larger, but the economic damage from a given increment of sea-level rise would increase, imposing further nonlinearity. Thus, Yohe (1992, table II.1) estimates that a doubling of sea-level rise causes approximately a tripling of economic damage. In view of nonlinearity, the estimate in table 3.4 sets US losses from sea-level rise associated with very-long-term warming at $35 billion, a conservative five times the amount for benchmark carbon dioxide doubling.

Other Countries

As noted in chapter 2, Rosenberg et al. (1989a) estimate that a one-meter rise in sea level would cause a loss of approximately 3 percent of the world's land area. The IPCC (1990c) provides a useful review of the impact of sea-level rise by major country affected. The study cites estimates that a one-meter rise would inundate 12 percent to 15 percent of Egypt's arable land and 17 percent of Bangladesh. It notes that in flat deltaic areas, such a rise would cause shores to retreat several kilometers, displacing populations. The most vulnerable deltas include the Nile in Egypt; the Ganges in Bangladesh; the Yangtze and Hwang Ho in China; the Mekong in Vietnam; the Irrawaddy in Burma; the Indus in Pakistan; the Niger in Nigeria; the Parana, Magdalena, Orinoco, and Amazon in South America; the Mississippi in the United States; and the Po in Italy (IPCC 1990c, 6–3, 6–4).

A study prepared by Delft Hydraulics Laboratory for the IPCC estimated the costs of coastal protection shown in table 3.3. These costs do not include the value of lost land (although they may overstate dike construction and other defense costs because they assume that any area with a population density of 10 per km² or more would be defended). It is evident from the table that the most severe costs of shoreline defense would be in numerous island nations, with the highest proportionate costs in the Maldives. Thus it is not surprising that many of these nations have formed the Alliance of Small Island States to press for international agreements limiting emissions (*New York Times*, 5 February 1991). Also according to the Delft study, other countries with coastal defense costs of at least 0.5 percent of GDP annually would include Argentina,

Table 3.3 Costs associated with a one-meter rise in sea level, selected countries and territories

Country or territory	Annual cost (percentages of GNP)[a]	Length low coast (km)	Beach length (km)
Maldives	34.33	1	25
Kiribati	18.79	0	0
Tuvalu	14.14	0	0
Tokelau	11.11	0	0
Anguilla	10.31	30	5
Guinea-Bissau	8.15	1,240	0
Turks and Caicos	8.10	80	5
Marshalls	7.24	0	0
Cocos (Keeling) Island	5.82	1	0
Seychelles	5.51	1	25
Falklands	4.75	1	0
French Guiana	2.96	540	0
Belize	2.93	500	0
Papua New Guinea	2.78	6,400	0
The Bahamas	2.67	400	200
Liberia	2.66	2,200	0
The Gambia	2.64	400	0
Mozambique	2.48	10,015	25
St. Kitts and Nevis	2.33	40	10
Nieu	2.18	6	0
Guyana	2.12	1,040	0
Suriname	1.94	2,800	0
Sierra Leone	1.86	1,835	0
Aruba	1.85	1	15
Pitcairn Island	1.71	0	0
Fiji	1.53	11	25
São Tome and Principe	1.46	3	5
Nauru	1.25	3	0
British Virgin Islands	1.24	1	10
Tonga	1.14	4	0
Cayman Islands	1.04	1	25
Cook Islands	1.03	0	0
Equatorial Guinea	1.02	7	0
Antigua and Barbuda	1.01	50	10
Sri Lanka	0.89	9,770	30
Togo	0.87	300	20
St. Lucia	0.82	25	10
Burma	0.77	7,470	0
Benin	0.74	485	10
Micronesia, Fed. St.	0.73	0	0
New Zealand	0.70	14,900	100
Palau	0.69	7	0
Grenada	0.67	6	5
Netherlands Antilles	0.66	25	30
Senegal	0.65	1,345	0
Ghana	0.64	2,400	20
Somalia	0.62	700	0
Western Samoa	0.59	11	0
Madagascar	0.56	1,190	0
St. Vincent and Grenadines	0.55	7	5

a. This estimate does not represent the total cost of sea-level rise, only the cost of erecting shore protection structures.

Source: Intergovernmental Panel on Climate Change. 1990. *Potential Impacts of Climate Change: Report Prepared for IPCC by Working Group III.* World Meteorological Organization and United Nations Environment Programme (June).

Bangladesh, Mozambique, and Vietnam (*Change*, The Hague, Netherlands, April 1990).

Work in progress in the EPA's international studies program indicates that, in China, a one-meter rise in sea level could displace over 72 million people in the absence of measures to hold back the sea (Han, Hou, and Wu, in EPA 1990). The losses would be concentrated in the Lower Liao River Delta, the North China Coastal Plain, the East China Coastal Plain, and the Pearl River Deltaic Plain. Shanghai and other major and ancient cities would be completely submerged, and over 125,000 km^2 of agricultural land would be lost. Another study in the same series estimates that, for Egypt, a one-meter rise would eliminate one-fourth of the agricultural land on the Nile delta and displace 8 million people (El-Raey, in EPA 1990).

Other nations at high risk include Thailand and Indonesia (*New York Times*, 5 February 1991). Among the developed countries, the United States would appear to be the most affected by sea-level rise (and it has the highest absolute costs in the Delft study, in view of its long and highly developed coastline; *Change*, The Hague, Netherlands, April 1990). In Europe, beaches along the Mediterranean (Greece, Yugoslavia, Italy, Turkey) would be adversely affected, along with tourism. Venice, which has already experienced a fourfold rise in floods over the last century, would be subjected to increased flooding. The low-lying portion of Poland's coast along the Baltic would be seriously threatened, along with a population of nearly half a million people. In Australia, the Great Barrier Reef would be threatened, as would hundreds of deltas, bays, estuaries, and islands (IPCC 1990c, 6–14, 6–15).

Elsewhere in Europe, the costs of sea-level rise would appear to be relatively moderate. Rosenberg et al. (1989a, 6) point out:

> Along the coast from France to Denmark sea walls and dikes already stand at least 16 meters above mean sea level . . . to reduce the probability of exceedance to less than 1 in 10,000 years. A 1 m SLR [one-meter sea-level rise] will not be particularly difficult to contend with where such structures already exist.

Similarly, the costs for the former Soviet Union would appear relatively limited.

Space Cooling and Heating

The EPA (1989a, chapter 10) has estimated the increased demand for electricity for space cooling (air conditioning), net of reduced demand for heating to be expected as a consequence of global warming. The estimates assume the GISS scenarios, indicating 1.2°C warming in the United States by 2010 and 3.7°C by 2055. The EPA estimated that the resulting increase in electricity demand above baseline would amount to $4.5 billion in "annual costs for capital, fuel, and operation and maintenance" by 2010 and $53 billion by 2055 (midpoint estimates, 1986 dollars; EPA 1989a, 192). In addition, there would be cumulative capital cost increases of $36 billion by 2010 and $224 billion by 2055 (midpoints). The EPA study identified the largest increases in demand for electricity to be in the South and midcontinent (20 percent to 30 percent above 2055 baseline) and predicted intermediate increases in California and the mid-Atlantic states (10 percent to 20 percent), more moderate increases in the mountain and north central states (0 percent to 10 percent), and net reductions in the demand for electricity in four northwestern states and the three northeasternmost states (0 percent to − 10 percent; EPA 1989a, 193).[36]

As indicated in the methodological section above, it is useful to state damages against an initial base-year GDP size and then deal separately with scaling for future years. Against a 1990-scale economy (rather than the future baseline), expected warming would increase operating expenses for electricity by $3.6 billion by the year 2010 and $18.6 billion by 2055; the corresponding increases in cumulative capital costs would be $29 billion and $89 billion (at 1990 prices; Cline 1990a).

Considering ocean thermal lag, the EPA-GISS estimates for 2055 would approximately correspond to equilibrium estimates for $2 \times CO_2$ equivalent. However, the GISS model yields somewhat higher warming than the IPCC's $\Lambda = 2.5°$ (with GISS, $\Lambda = 4.2°$; IPCC 1990b, table 3-2a). Scaling back to the IPCC's standard, adjusted estimates (at the 1990 economic scale and prices) for the year 2055 would stand at $11.2 billion for annual operating and capital costs (i.e., 60 percent of the amount before adjustment), with cumulative capital costs placed at $53 billion.[37] The annual damage of $11.2 billion would constitute one of the largest impacts of benchmark $2 \times CO_2$ warming, larger than the damage from sea-level rise and about two-thirds the size of agricultural losses (table 3.4).

The EPA estimates, even after scaling back to 1990 economic size, indicate significant nonlinearity. The operating costs mushroom from $4.5 billion for 1.2°C warming to $18.6 billion for 3.7°C warming. If we apply the form $D = k(\Delta t)^{\gamma}$ and estimate from the two "observations" just mentioned, then

36. A further review of the EPA estimates is presented in Cline (1990a).

37. This central estimate is several times the Nordhaus (1990b) estimate of $1.65 billion annual cost at 1981 prices, even though the Nordhaus study is based on the draft version of the same EPA study used here. Considering that the lowest figures in the EPA table are $3 billion to $6 billion "annual costs" for 2010 at 1986 prices, there is no evident way the Nordhaus estimate could be so low, unless perhaps the annual costs were misinterpreted as capital costs. If this occurred, and the $36 billion midpoint capital costs for 2010 are added and a 6 percent discount rate is applied (the Nordhaus rate), and prices are converted to 1981 levels, the result would be $2.0 billion—annually close to the Nordhaus range. However, this approach, besides confusing capital and annual costs, would have applied a 2010 benchmark rather than the more appropriate 2055 level.

$\gamma = 1.26$. On this basis, in the case of 10°C very-long-term warming with Δt four times as large as for $2 \times CO_2$ equivalent warming, additional US annual space-cooling costs (operating expenses) would amount to \$11.2 billion $\times 4^{1.26}$ = \$64 billion (at the 1990 economic scale and prices).

The EPA electricity estimates are already net of savings on space heating. However, these savings do not include those for oil and natural gas heating. Nordhaus (1990b) has estimated savings on heating costs at \$1.16 billion annually (1981 prices), on the assumption that heating costs would decline by 1 percent. However, he apparently includes electricity savings in the estimate, whereas the EPA estimates of the demand for electricity are already net of savings on heating. Estimates by the Energy Information Administration (EIA) of the US Department of Energy (1990a, 30) place nonelectric energy consumption for heating (residential and commercial) at 7.3 quadrillion Btus for 1990. At an average price of \$2.60 per million Btus for natural gas, residual oil, and distillate oil in 1982 dollars (US Department of Energy, EIA 1990a, 233), or \$3.50 at 1990 prices, each quad costs \$3.5 billion. Nonelectric heating thus amounted to some \$25.6 billion in 1990. If it is assumed that benchmark warming would reduce heating costs by an optimistic 5 percent, the annual savings would amount to \$1.3 billion.[38]

Internationally, the IPCC summarizes energy demand effects as follows. For the United States (based on the EPA study), the capacity for electricity would need to rise 14 percent to 23 percent above baseline by 2055. For Japan, 3°C warming would cause a 5 percent to 10 percent rise in electricity demand above baseline. In contrast, for Germany a 1°C warming is calculated to reduce energy consumption (apparently including nonelectric consumption) by 13 percent for older single-family homes and up to 45 percent for new homes, while increasing

38. Howard Gruenspecht (then of the Council of Economic Advisers, personal communication, 29 April 1991) has questioned the result here that increased space-cooling costs far exceed reduced heating costs from global warming. He argues that, for major US cities, some warming would move the population-weighted average closer to, rather than farther from, the ideal indoor temperature. However, this observation is not inconsistent with the results here, if two considerations are taken into account. First, the extent of temperature increase even for a doubling of carbon-dioxide-equivalent seems likely to go beyond the slight amount of warming that might move temperatures closer to an ideal comfort level. This consideration is of course dramatically reinforced if the analysis is extended to very-long-term warming. Second, air conditioning is by electricity, which has large efficiency losses, whereas heating is primarily by gas and oil. In this regard, James Titus (EPA, personal communication, 14 May 1991) has made the following observations. Efficiency losses are large in the conversion of oil and gas into electric power through generation and transmission—on the order of 60 percent to 70 percent. For a given change in the number of Btus, the cost is thus far higher for an increase in electric air conditioning than is the saving from a decrease in gas or oil heating. Moreover, the greater the temperature differential, the less efficient the air conditioner. In addition, air conditioning in automobiles would increase with warming, whereas automobile heating requires no additional fuel because the engine must be cooled in any event.

energy consumption for air conditioning by 12 percent to 38 percent. The result is a net decrease in energy demand for space heating and cooling by 12 percent in 2010. Similarly, for the former Soviet Union a 1°C warming is expected to generate larger savings in heating costs than increases in air conditioning costs. However, whereas this milder warming might move these countries toward, rather than away from, optimal comfort temperatures, warming from benchmark carbon dioxide doubling could go well beyond that point and turn net savings into net losses.

The international group noted that a smaller percentage of electricity consumption is devoted to residential and commercial use in developing countries than in industrial countries (28 percent versus 55 percent for Japan, for example). It thus conjectured that incremental electricity needs from global warming would be less than 10 percent in developing countries (IPCC 1990c, 5–19, 5–20).

Human Amenity

Personal comfort could conceivably improve from very mild global warming; indeed, in the United States internal migration has been toward the warmer states in recent decades. However, it would seem likely that personal amenity from outdoor comfort would begin to deteriorate as warming moved closer to the 2.5°C central estimate for benchmark carbon-dioxide-equivalent doubling, and would undergo severe deterioration from the very-long-term warming suggested above. As noted in chapter 2, with very-long-term warming on the order of a 10°C global mean increase, of the 66 largest US cities the number with average maximum daily temperatures above 90°F in July would rise from 18 to 62, and the number with average daily maximums above 100°F in July would rise from 2 to 42.

Mearns et al. (1984) use statistical distributions of current temperatures to explore the impact of global warming on extreme temperature events for the United States. In their base case, they increase mean temperature by 3°F (1.7°C) and hold the variance and autocorrelation of daily temperatures constant. Under these assumptions, they calculate that the frequency of heat waves (defined as five consecutive days with maximum temperature at least 95°F) would multiply threefold (estimated for Des Moines, Iowa). Again, there is reason to believe that people would be willing to pay something to avoid a threefold increase in heat waves.

Global warming could reduce the disamenity of severe winters in colder areas, providing some offset to the increased disamenity of heat waves and generally hotter summers. One way to approach estimation would be the survey technique (contingent evaluation), with questions formulated to elicit a meaningful evaluation of what the respondent would be prepared to pay to avoid the change (or, for more clement winters, what he or she would be prepared to pay to enjoy the change). In the absence of such estimates, valuation of human disamenity from global warming remains unestimated for the central calculations here, but with the presumption that this damage valuation could be extremely large in the case of very-long-term warming. For purposes of establishing an order of magnitude, if people were willing to pay just 0.25 percent of personal income to avoid the sharp increase in heat waves and other effects of a benchmark $2 \times CO_2$ warming in outdoor comfort, the damage of such disamenity would stand at some $10 billion annually for the United States.

Disamenity damages in other countries would depend on location and income level. For high-latitude countries, the effect would probably be favorable (i.e., negative disamenity damage). For mid- and low-latitude countries, disamenity effects might dominate, and almost certainly would do so under very-long-term warming. Valuation would tend to be higher for higher income countries, even relative to GDP, under the assumption that amenity is an income-elastic service.

Life and Morbidity

In the United States, an estimated 1.1 million of the 1.7 million deaths annually are from diseases that are potentially sensitive to weather (cerebrovascular, pulmonary, diabetes, heart, pneumonia and influenza; EPA 1989a, 221). Heat waves as well as very cold weather increase the incidence of stroke and heart attacks. Air pollution increases the occurrence of respiratory diseases (emphysema and asthma), and longer, warmer summers are expected to increase the severity of air pollution (as discussed below).

As cited in EPA (1989a, 224), Kalkstein has carried out a statistical analysis relating deaths to weather variables (including temperature and humidity). He has then applied changed weather conditions under carbon-dioxide-equivalent doubling to calculate the change in number of deaths for 15 major US cities. The estimates assume one case with no acclimatization and another case with full acclimatization, the latter estimated on the basis of corresponding mortality statistics for a control city with climate conditions comparable to those in the warmer future for the city in question. The estimates with acclimatization yield some anomalies (i.e., higher deaths with than without acclimatization to the new climate, or lower deaths after acclimatization than at present). However, the majority of estimates are in the directions expected. Moreover, increased summer deaths substantially exceed decreased winter deaths. For the unweighted sample, summer deaths rise from 1,156 at present to 7,402 under warming without acclimatization and 2,198 with acclimatization. Winter deaths fall from 243 at present to 52 without acclimatization and 159 with acclimatization (the latter result is spurious, because in winter the population should be able to do at least as well with acclimatization as without).

The Kalkstein estimates may be reinterpreted by enforcing as the postwarming summer deaths the *minimum* of the following: present level, postwarming without acclimatization, and postwarming with acclimatization. Similarly, postwarming winter deaths may be set as the *maximum* of present, future with acclimatization, and future without acclimatization. When this screening is performed, and when the resulting net death rates per 1,000 population of the individual cities are weighted proportionately to city population, the overall result is that benchmark warming would increase mortality by 0.0397 person per 1,000 population. Applied to the present US population base, this result indicates that global warming under carbon-dioxide- equivalent doubling would cause some 9,800 additional deaths annually.

Life valuation methods for public policy often take as a point of departure the lifetime earnings of an individual, under the assumption that earnings are what the society is willing to pay the individual (and thus the person's "marginal product"). Average nonagricultural wages in the United States amounted to $17,994 in 1990 (*Economic Report of the President*, February 1991, 336). If a working life span is set at 45 years and a 1.5 percent discount rate is applied, the value of average lifetime wages is $595,000. If this rate is applied to the estimated increase in deaths, annual mortality costs from global warming would amount to $5.8 billion for the United States.

The value of damage could be much greater on the basis of some of the higher life-valuation estimates in the literature (Cropper and Oates 1990). On the basis of the statistical relationship of wages to risk of death by occupation and industry, estimates of the "value of a statistical life" have been placed in the range of $2 million and even up to $6 million. Contingent valuation studies that ask workers what wage differentials would be required for more dangerous work generate results in the same range ($2 million to $3 million). On the other hand, studies that measure actual behavior in hazard avoidance (e.g., the purchase of smoke detectors) place the value of a statistical life at a much lower range of $500,000 to $600,000. Moreover, for environmental risks that are longer term and have a lower probability of death (e.g., exposure to hazardous waste), the value of a statistical life is even lower ($200,000 to $400,000), in part because the risk is some 20 years or more in the future.

One issue in valuing a life in the range of $2 million to $3 million is that the implied value of the entire US population would be more than 100 times the size of GDP. Since each worker's productive life is only some 45 years, this implied aggregate amount would be at least twice what American society could afford to pay for its own survival. This seeming anomaly can be resolved, however, if it is considered that for events that would only cause the death of a small fraction of the population, people might well be prepared to pay a higher per-death insurance cost than implied by the budget constraint of per capita life span GDP, given risk aversion and the opportunity for risk pooling.

The incidence of increased deaths would tend to be disproportionately among the elderly. It might be argued that the lifetime earnings approach would consequently apply a substantial reduction to this estimate. However, in other areas of health policy the US public has shown no inclination to value the lives of the elderly at any less than (for example) those in the 20- to 30-year age group. Accordingly, no reduction is applied here for the age distribution of the incremental deaths.

A related issue is that increased deaths from warming might primarily be instances in which the date of death is "merely" advanced by a few years. There might be some grounds for reducing the estimate from this standpoint. However, from another perspective the estimate is already low, as it uses the average wage rather than total income per capita and implicitly excludes the deceased's human and financial capital income from lifetime valuation. Use of GDP per capita would increase the loss estimate by about one-fourth.

On balance, the rate used here is probably an understatement of the value of a life saved, compared with most of the estimates in the literature. However, there is a more fundamental question about the estimates of global warming effects. Many in the medical profession are reportedly skeptical about the statistical estimates relating death to temperature and tend to doubt that there would be any major mortality consequences from warming; instead, they tend to be more concerned about morbidity effects (Rob Coppock, National Academy of Sciences, personal communication, 26 March 1991). It is unclear, for example, whether the statistical tests have adequately controlled for the tendency of older people to live in warmer cities in retirement. In lieu of more concrete analysis of the medical basis for mortality effects, the estimates here may be seen as compensating for a possible overstatement of mortality effects by applying a conservative per-life valuation.

Morbidity could increase in the area of "vector borne diseases," where the vectors are such carriers as ticks and mosquitoes and the diseases include Rocky Mountain spotted fever, Lyme disease, malaria, and dengue fever. Simulations for the United States on a possible increase of malaria under warming are inconclusive (EPA 1989a, 230). However, there would be a general migration of the vectors (carriers) from low to middle latitudes.

Placing a value on health effects is difficult, and no estimate of health losses is included in the central damage calculations reported in table 3.4. However, US health expenditures are so large (12 percent of GDP, or some $700 billion) that even a small increment could imply substantial costs. For example, an increase in medical expenses by just ½ percent would impose annual costs of over $3 billion.

Migration

Those who approach the greenhouse issue primarily from the standpoint of adaptation rather than abatement often stress migration as a means of response (see, e.g., Schelling 1983, 455–56). This raises two types of costs: utility sacrificed by the migrant, and cost imposed on the target host country.

In the abstract, it seems curious indeed that migration cost would be construed implicitly as minimal or even negative by some in the adaptation school. To state the issue starkly, peoples have often fought wars to avoid being forced to leave their homelands against their will. There is little reason to suppose that, when the cause is climate change rather than an invading army, they will instead be indifferent between staying and moving. There is also a question of the ultimate feasibility of the migration response. As the amount and quality of land "opened up" by warming becomes smaller relative to the land closed off, the migration option becomes progressively less viable. Only so many people will fit onto Iceland.

Long before people move en masse toward more favorable regions, however, they will confront political obstacles to cross-border migration. This raises the second type of migration cost: that imposed on, or at least perceived by, the target country. For migration from developing to industrial countries, neoclassical economic analysis from a global standpoint might attribute a gain rather than a cost for each additional immigrant. Labor would be reallocated away from countries where the marginal product of labor is low to countries where it is high.[39] However, in terms of political economy, revealed behavior indicates that the perception of imposed costs from increased immigration dominates; otherwise the borders of industrial countries would be completely open. Moreover, as the rise in immigration would stem solely from an adverse "push" on source nations, it is also impossible to argue that citizens of those nations would benefit from increased migration associated with global warming.

For the United States, it is not difficult to envision increased pressure of legal and illegal immigration from global warming. Increased hurricane damage and sea-level rise could intensify the push of immigration from the Caribbean. Increased drought could do the same for immigration from Mexico.

Present-day stocks and flows of immigrants provide a rough basis for a sense of the costs that might be imposed by warming-induced migration. Approximately 1.8 million illegal immigrants, 70 percent of them from Mexico, applied for legalization under the 1986 Immigration Reform and Control Act. In the period 1975–80, an estimated 130,000 undocumented aliens from the Western Hemisphere entered the United States each year (Goering 1990). The annual

39. Such a gain would abstract from the distributional effects between capital and labor.

flow of legal immigrants has been approximately 640,000 (*Statistical Abstract of the United States*, 1991, 9). The base of legal and illegal immigrants is thus approximately 800,000 annually.

Suppose that benchmark warming ($2 \times CO_2$) increases illegal immigration by 25 percent and legal immigration by 10 percent. Suppose that, on a base of 800,000 immigrants, the share of illegals is 20 percent. Then warming would increase annual immigration by about 100,000. Infrastructure spending (education, roads, police, sanitation) in the United States is primarily at the state and local level. State and local government spending in 1989 amounted to $762 billion (*Economic Report of the President*, February 1991, 383), or approximately $3,000 per capita. Suppose that an immigrant only begins to pay taxes that cover his incremental social infrastructure costs after a period of 18 months (for illegals the period could be longer). Then each additional immigrant imposes a cost of $4,500 on the United States.[40] On this basis, an extra 100,000 immigrants annually would represent a cost of $450 million as an adverse effect of benchmark warming for the United States.

Outside the United States, there are indications that concern about increased immigration induced by global warming is especially high in Europe, where immigration problems already have intensified in the past decade as the consequence of political change in Eastern Europe and North Africa. As semiarid African states are among those expected to be seriously affected by global warming, such concern would appear well founded.

Hurricane Damage

Emanuel (1987, 485) argues that tropical cyclones, or hurricanes, are "particularly sensitive to sea surface temperature as the latent heat content of air at fixed relative humidity is a strongly exponential function of temperature, approximately doubling for each 10°C increment of temperature above 0°C." He calculates that for a rise of 3°C in sea surface temperature, there is a 30 percent to 40 percent rise in the maximum drop of barometric pressure. Wind velocity rises as the square root of the drop in barometric pressure. Wind pressure against obstructions rises with the square of wind speed. Applying a 2.3°C-to-4.8°C range of GCM estimates for increased August temperatures in the tropics under

40. This calculation makes no allowance for the possibility that subsequently the taxes paid by the immigrant would exceed his share of infrastructure costs. Precisely because higher-skilled immigrants are likely to have high incomes and above average taxes, immigration quotas tend to favor them over low-skilled applicants. For illegal immigrants, however, skills (and earning potential) tend to be lower, and evasion of formal taxation tends to be higher. For both reasons, the taxes they pay might never come to equal their share in infrastructure costs.

$2 \times CO_2$, Emanuel estimates that the destructive potential of hurricanes would rise by 40 percent to 50 percent.

Similarly, Hansen et al. (1989, 70–71) find in $2 \times CO_2$ simulations with the GISS model that moist static energy (the sum of sensible heat, latent heat, and geopotential energy) rises by 10 kJ/kg at the earth's surface and 8 kJ/kg at the upper troposphere (approximately 40,000 feet). Because of the increase in this gradient, and thus greater vertical penetration of moist convection (as well as increased evaporation and higher temperatures at low levels), they conclude that global warming would bring "increased intensity" of "both ordinary thunderstorms and mesoscale tropical storms."

Wendland (1977) has provided empirical support for the relationship of hurricanes to ocean surface temperatures. From monthly data for 1971–81, the frequency of hurricanes is closely related to the size of ocean area with a temperature over 26.8°C, and the relationship is exponential.

For its part, the IPCC (1990b, section 5.3.3, 162) surveys these and other studies and concludes: "There is some evidence from model simulations and empirical considerations that the frequency per year, intensity and area of occurrence of tropical disturbances may increase, though it is not yet compelling." In part, this tentativeness is attributable to the fact that, except for the high-resolution models, the GCMs have grid areas too large to capture "smaller scale disturbances such as hurricanes."[41]

The Emanuel study provides a basis for the quantification of prospective damages. Over the last 40 years, US hurricanes have caused an annual average of $1.5 billion in damages at 1989 prices (Cline 1990b). For the United States the most costly storm on record was Hurricane Hugo in 1989 ($7 billion). On the basis of the 50 percent increase suggested by Emanuel, the increased damage to be expected from benchmark warming would be $750 million annually. Hurricanes also cause loss of life, typically a total of 50 to 100 lives annually.

For other countries, especially island states, hurricane damage from global warming could be more severe. In 1970, tides from a cyclone killed hundreds of thousands in what is now Bangladesh. Hurricane Fifi in 1974 killed some 5,000 in Central America. Hurricane Gilbert caused $8 billion in damage in Jamaica alone in 1988 (IPCC 1990c, 5–10).

41. The IPCC (1990b, section 5.3.3, 161) also noted, however, that climate models tend to show a "reduction in day-to-day and interannual variability in the mid-latitude storm tracks in winter" (i.e., *not* hurricanes), because of "the reduction in the equator-to-pole temperature gradient." Nonetheless, if this judgment is meant to imply lesser storm intensity, it would not appear to take account of the other gradient—the vertical one just noted (Hansen et al. 1989). Thus, the IPCC reports one GCM result indicating that the temperature at an altitude of 22 km would fall by 4°C and the temperature at an altitude of 0 to 10 km rise by 4°C at latitude 40°N under $2 \times CO^2$ (IPCC 1990b, figure 5.2). The increase in vertical temperature gradient would seem likely to increase storm intensity.

The exponential relationship of latent heat content of air (at fixed relative humidity) to temperature, and the threshold effect whereby ocean surfaces above 26.8°C become subject to hurricane conditions, suggest that increased hurricane damage from global warming would be more than linear in relation to temperature rise. Thus, damages under very-long-term warming could be far greater proportionately than the corresponding ratio of temperature increase to that under benchmark $2 \times CO_2$ warming.

The Construction Sector

It is often assumed that global warming would benefit the construction sector because "lengthening of the construction season" would be "likely to increase productivity" (Nordhaus 1990b). However, the increased incidence of heat waves (discussed above) would seem likely to eliminate some summer construction days. More important, although construction is adversely affected by frost, it is also inhibited by rainfall (Jones 1964 and Russo 1966, as cited in Solomou 1990). For example, gravel digging has to stop under conditions of heavy rain.

Solomou (1990) has examined data on UK construction and rainfall in the period 1856–1913 and found a close negative correlation between the two. The GCMS typically predict an increase in global mean precipitation by about 8 percent as the consequence of benchmark $2 \times CO_2$ warming, and the estimates range as high as 13 percent to 15 percent (IPCC 1990b, table 3-2, 87). As noted above, there would be increased precipitation in winter in midlatitudes and year-round in the high latitudes and tropics. The IPCC similarly notes an adverse effect of rainfall on industry more generally and cites Parry and Read to the effect that "rainfall is responsible for more delays than any other climatic variable" for UK industry (IPCC 1990c, 5–34).

More work is required before a judgment can be offered on whether the adverse effects of increased rainfall on construction (and other industries) would be greater or smaller than the beneficial effects of warming on construction (and other activities) during winter months.

Leisure Activities

Studies for Canada cited by the IPCC indicate that the ski industry would lose 40 percent to 70 percent of skiable days as a consequence of warming from carbon dioxide doubling, even after taking account of the availability of snow-making machinery (IPCC 1990c, 5–35). Losses would presumably be larger in relative terms in US ski areas, where the temperature base is already higher. Other outdoor leisure activities would be much less affected. For example, golf might benefit in cold areas but be adversely affected in warm regions because of high temperatures and along heavily populated coastal areas where rainfall would tend to increase. Increased camping is sometimes cited as a benefit of warming, but it is unclear whether those who hold this view have taken into account the impact of forest loss and increases in precipitation.

Tourism could also be affected in areas where coral reefs are attractions. Coral bleaching and coral death were observed in the Pacific after the 1982–83 El Niño and in the Caribbean after unusually high water temperatures in the summer of 1987 (EPA 1989a, 157; *In Depth: Consumer Reviews for Sports Divers*, Austin, TX, September 1990). Glynn and de Weerdt (1991) indicate that the 1982–83 event may have brought extinction to two coral species (out of 12) along the Pacific coast of Panama. Coral bleaching results from the expulsion of symbiotic algae under environmental stress. As the algae are the primary source of food for the coral, the result can be coral death. Stress on coral reefs is expected to occur under global warming, in part because of higher temperatures but also because vertical accretion of reef flats may be unable to keep up with sea-level rise.

Placing a dollar value on impacts in leisure activities is difficult. However, as a point of departure, the US ski industry accounted for an estimated $5.6 billion in economic activity in 1988, when there were 53 million skier visits to ski areas (Waters 1990). If we suppose that global warming of 2.5°C would reduce ski activity by 60 percent in the United States, and it is assumed that half of the gross loss is offset by the difference between the value of released labor and other variable factors of production on the one hand and the loss of consumer surplus on the other, then the net loss would be on the order of $1.7 billion annually. This estimate makes no allowance for tourism losses associated with coral reef death. As noted, there is no clear case for attributing offsetting net gains in other leisure activities.

Water Supply

Warming is expected to put stress on water supply. A smaller share of precipitation in the form of snow, combined with earlier snowmelt, would mean higher runoff in the winter. Runoff in the spring and summer would correspondingly be reduced. In those areas or periods where precipitation declines, the combination of higher evaporation (from warmer temperatures) and lower precipitation would reduce soil moisture and water levels and flows. At the same time, the demand for water would tend to rise with warming, because of increased needs for irrigation and for cooling in electric power production (EPA 1989a, 165–70).

Average precipitation in the United States is 4,200 billion gallons per day (bgd). Of this amount, 2,765 bgd evaporates, 338 is withdrawn for use, and 1,435 bgd goes to surface and groundwater flows. Of the withdrawals, 140 bgd is for irrigation, 131 for thermoelectric power, 36 for domestic use, and 31 for industry and mining. Each of these sectors returns a portion of the flows, so that the shares of net consumption are different: 76 bgd for irrigation, 4 for electric power, 7 for domestic use, and 5 for industry and mining (EPA 1989a, 166).

Using a water balance model for the Sacramento basin, Gleick (1987) estimates that a 4°C increase in temperature would decrease summer runoff by 55 percent even if there were a 10 percent rise in precipitation. If there were a 10 percent decline in precipitation in the region, the reduction in summer runoff would amount to 65 percent. Winter runoff would increase in both cases (but would decrease if precipitation fell 20 percent).

The EPA estimates that warming from a doubling of carbon-dioxide-equivalent would reduce annual water deliveries in California's central valley basin by 200,000 to 400,000 acre-feet, or by 7 percent to 16 percent, even as demand for water rises by some 1.4 million acre-feet by as early as 2010 (EPA 1989a, 251). The price of water in California in a normal year is approximately $250 per acre-foot (*New York Times*, 6 March 1991). In this one instance, then, the costs of increased water scarcity would amount to $75 million annually (300,000 acre-feet × $250).

Nationwide withdrawals of water amount to 0.378 billion acre-feet annually.[42] Assuming $250 per acre-foot for domestic and industrial use (59 percent) and $100 for irrigation (41 percent), the annual value of water withdrawals is on the order of $70 billion annually. If water availability were to decline just 10 percent with climate change (based on the Gleick and especially the EPA–California estimates), the annual cost would amount to $7 billion.

Inefficient water allocation could make the costs higher. After five years of drought, in early 1991 California authorities cut off irrigation water to a wide range of agricultural producers. The first desalinization plant for residential water supply in the United States is now planned for Catalina Island in Southern California. Other communities are considering turning to salinization plants. Costs are expected to amount to $1,000 per acre-foot (*New York Times*, 6 March 1991). If such sources represent the marginal cost of water, the estimate just suggested could substantially understate the value of water lost to climate change.

---

42. 338 bgd × 365 days = 123,370 bgd = 123.4×10^{12} gallons. One acre-foot = 326,000 gallons. Annual withdrawals = $(123.4/326) \times 10^{12-3}$ gallons = 0.378×10^{9} gallons.

One way to ease prospective shortages would be to reform pricing practices so that agricultural uses pay a price closer to that for residential and commercial use. In the past, politics has prevented this outcome. Moreover, as discussed above in the analysis of agriculture, the demand for irrigation is highly likely to rise from global warming, especially if the projections of sharply higher drought incidence are accurate.

The IPCC reports on various methods for estimating the impact of global warming on water supply. The group stresses that river basin runoff is very sensitive to small variations in climatic conditions, largely because runoff is a residual between precipitation and soil absorption or evaporation, and small changes in any of the underlying variables can cause a much larger proportionate impact on this residual. For the United States, water basin simulation models show that in a warmer, drier climate (+2°C, −10 percent precipitation), water supply in 18 major water regions (covering the bulk of national supply) would decline by approximately one-third (calculated from IPCC 1990c, figure 4.2). Under the same conditions, annual river runoff in regions with relatively low precipitation would decline by 70 percent (IPCC 1990c, 4–7). These estimates suggest that the 10 percent decline applied above is conservative.

For Canada, several studies based on GCM estimates for global warming indicate that, in the Great Lakes region, runoff to the lakes could decline by 8 percent to 11 percent (for warming of 3.1°C to 4.8°C and a precipitation change of −3 percent to +8 percent). Farther to the north, in the James Bay region the studies show increased runoff because of much higher precipitation (+15 percent to +17.5 percent); and for the Canadian prairie (Saskatchewan River), estimates diverge (IPCC 1990c, 4–6).

For Europe, several studies based on water balance models and GCM projections suggest that there would be reduced precipitation and runoff in the south (Spain, Portugal, Greece), possible decreases in runoff in the central region, and significant increases in the north (the United Kingdom, the Netherlands, and Belgium; IPCC 1990c, 4–11, 4–12).

For the former Soviet Union, studies in part based on paleoclimatic analogues have suggested that global warming would tend to increase water resources, as a 2°C global warming would increase annual runoff by 10 percent to 20 percent on all the large rivers. An important exception is that in the south of the forest zones of the European part of the former Soviet Union and in western Siberia annual runoff could fall by 80 percent or more (IPCC 1990c, 4–15, 4–16).

The IPCC reports that, for Japan, normal GCM projections are especially inadequate because they do not provide information on typhoons. Japanese experts anticipate that with global warming there would tend to be longer periods of drought, interrupted by intense precipitation. Water shortages could become more prevalent.

Detailed studies in New Zealand conclude that with global warming there would be a large increase in annual runoff through most of the country but a large decrease along the eastern shores of the two islands. The studies indicate a sharp increase in flooding (IPCC 1990c, 4–19; figure 4-5).

The broad picture that emerges is that, as Gleick (1987) suggests, stress on water supply may be one of the most important consequences of global warming. The principal exception would be in the high latitudes and tropics, where increased precipitation would be concentrated (chapter 1; Rind et al. 1990). The prospective decrease in supply in many areas (especially the United States) would confront an increase in demand that could pose severe price increases. Rosenberg et al. (1989a) recall that as early as the 1960s there were discussions in the United States of large-scale water transfers from Alaska and northern Canada to the Southwest and high plains of the United States, and in the Soviet Union there was even consideration of reversing the northerly flow of some major rivers (until the large economic and environmental costs became more apparent). Water shortages could revive some of these proposals (at least for the United States), with the certain effect of stirring political conflicts over water transfers between regions and among sectors (especially away from agriculture).

If global benchmark warming of 2.5°C would cause problems for water supply, very-long-term warming in the range of 10°C would seem likely to cause much more dramatic difficulties. It is useful to recall the point made by Rind et al.: with a warmer atmosphere, more water is in the atmosphere in the form of water vapor and less is left in the surface land, lakes, and rivers (i.e., not all of the water transferred to the atmosphere comes from the oceans). As Rind et al. note, "from the Claussius-Clapeyron equation, an incremental increase in temperature produces a greater increase in atmospheric moisture-holding capacity when the temperature is warmer" (1990, 9983). One would thus expect evapotranspiration to rise more than linearly with temperature, causing a nonlinear rise in water availability problems.

Nonlinearity is confirmed in the simulations of Gleick (1987). He finds that summer runoff decreases by 12 percent when temperature rises by 2°C and precipitation rises by 10 percent. Summer runoff decreases by 49 percent when temperature rises by 4°C and precipitation rises by 20 percent (Gleick 1987, 146). Thus, a doubling of both the temperature and the precipitation increases leads to a quadrupling of the percentage cutback in summer runoff. Yet there is reason to believe that the result would be even worse, because, as noted above, summertime precipitation would be unlikely to rise in midlatitudes.

In sum, a damage estimate on the order of $7 billion annually might be appropriate and perhaps conservative for the United States for water supply under global warming with the doubling of carbon-dioxide-equivalent. Water supply thus ranks as another of the preeminent categories of economic damage from global warming, in the same range as sea-level rise (table 3.4). For warming four times as large over the very long term, damage would be likely to be much more than four times as high. In the formulation $D = k(\Delta t)^{\gamma}$, one might expect $\gamma = 1.5$ (for example). In that case, the very-long-term damage would amount to \$7 billion $\times\ 4^{1.5}$ = \$56 billion (against the 1990 economic scale).

Urban Infrastructure

The EPA has examined the impact of global warming on urban infrastructure costs. For coastal cities, sea-level rise or more frequent droughts would increase the salinity of coastal aquifers and tidal surface waters, requiring a response where these are the sources of water. New Orleans illustrates the point. In the drought of 1988, when the Mississippi River was far below normal levels, it was necessary to build a temporary nine-meter silt wall to halt the upriver advance of saltwater that threatened the city's water supply. In the Philadelphia-Wilmington-Trenton area, a sea-level rise of just 0.3 meters could require a 12 percent increase in reservoir capacity to prevent saltwater from advancing past water intakes on the Delaware River (EPA 1989a, 239). In addition, more frequent and intense storms would likely overload existing storm sewer systems. Flooding and the release of untreated waste into watercourses from storm and wastewater systems could require new sewer systems.

A study for New York City found that, with benchmark warming, increased water use for cooling large buildings and for lawn watering could increase annual water demand by 5 percent, while increased evaporation of water in reservoirs could cut supply by 10 percent to 24 percent. Saltwater infiltration from rising sea level could place some intakes below the salt line during the summers with mild droughts, reducing supply further. The EPA summary of various studies for New York estimated that, for water supply adjustments alone, the city would need to invest an additional \$3 billion as a consequence of $2 \times CO_2$ warming (EPA 1989a, 243).

In Miami, apart from costs for levees to deal with rising sea level (considered separately above), capital outlays for canal control, drainage, and raising streets (otherwise subject to collapse from water table infiltration of their base, for about one-third of the streets) would amount to some \$580 million. In contrast, for Cleveland the savings on snow and ice control and municipal heating would offset higher air-conditioning costs, and additional dredging and water supply costs would be negligible (EPA 1989a, 241–42).

Urban infrastructure investments, primarily those related to adjustments in water supply, sewer, and drainage systems, might amount to something on the order of \$10 billion, based on the New York and Miami estimates. Divided over

75 years[43] and discounted, a $10 billion capital cost amounts to approximately $100 million annually.

Pollution

The benefits of action to reduce global warming would include two types of gains that stem from the interrelationship of global warming to more traditional problems of pollution. First, there would be a direct "damage avoidance" similar conceptually to the avoidance of damage to agriculture (for example). This effect would arise because a warmer climate could aggravate the problem of air pollution, as discussed below. Second, there would be a spillover benefit from the action undertaken to reduce global warming to the extent that the action involved a cutback in the burning of fossil fuels that presently contribute to air pollution. For example, if the use of coal is reduced as a consequence of policy action on global warming, there is a spillover benefit in the form of reduced pollution associated with coal burning (for example, emissions of sulfur dioxide and resulting acid rain). The discussion that follows will focus on the first of these two concepts, but the second is also important.

A warmer climate would aggravate urban pollution. One piece of evidence is that in the hot summer of 1988 the extended stagnation periods and high temperatures caused 76 cities in the United States to exceed the national ambient air quality standard (NAAQS) for low-level (tropospheric) ozone pollution by 25 percent or more (EPA 1989a, 200).

Air pollution involves total suspended particulates (TSP), sulfur dioxide, carbon monoxide, nitrous oxide, ozone, and lead. Of these, the most severe problem in terms of the number of persons living in areas with air quality indexes that exceed (violate) the NAAQS is that of ozone (75 million in 1986); TSP and carbon monoxide also remain leading problems (41 million each). Ozone has proven the most difficult to reduce. Its concentrations declined only 13 percent from 1979 to 1986, compared with reductions of 23 percent for TSP and 37 percent for sulfur dioxide (EPA 1989a, 201–2).

Numerous studies confirm that ozone concentrations rise with temperature (see, e.g., IPCC 1990c, 5–46, 5–47). Models estimated by Morris indicate that a 4°C rise in temperature (approximately the $2 \times CO_2$ benchmark for the United States) would increase the number of person-hours of exposure to ozone in excess of the NAAQS ceiling of 120 ppb from 661,000 to 2.1 million in central California and from 29.8 million to 47.5 million in the Midwest and Southeast

43. Some of the effects here are for carbon-dioxide-equivalent doubling, and thus within a time horizon of 50 years for effective warming. Others refer to one-meter sea-level rise, whose occurrence is on a 100-year time scale.

(EPA 1989a, 214). The EPA has summarized the various estimates with the conclusion that a 4°C rise in temperature could cause an increase in peak ozone concentrations by 10 percent. The result would be to double the number of cities exceeding the standards from 68 to 136, causing most midsize and some small cities in the Midwest, South, and East to be added to the list of those presently in violation.

The EPA has applied its past models relating ozone concentrations to emissions of volatile organic compounds (VOC) to estimate that, to offset benchmark $2 \times CO_2$ warming, it would be necessary to reduce VOC in the United States by 700,000 tons from a base of 6 million tons expected for the year 2000. At an estimated cost of $5,000 per ton, the agency calculates that the resulting costs would amount to $3.5 billion annually (EPA 1989a, 215).

Warming could affect other parts of the pollution problem as well. At higher temperatures, there is a greater concentration of hydrogen peroxide, the agent in the conversion of atmospheric sulfur dioxide into sulfuric acid (acid rain). Higher demand for electricity for air conditioning (discussed above) would mean greater pollution from power plants. A 10 percent rise in electric power demand would mean a 30 percent rise in sulfur dioxide emissions (IPCC 1990c, 5–47). If total cloud cover declined with global warming (chapter 1), there would be a greater incidence of sunlight available to cause greater ozone production. The rise in water vapor associated with warming would accelerate the reaction rates of VOCs and increase production rates of ozone, hydrogen peroxide, and sulfates (EPA 1989a, 207).

One way to pursue a quantitative grasp of the issue is to consider the amount being spent to abate pollution already. The EPA has estimated that, in 1990, the US economy spent $115 billion in pollution abatement, the vast bulk of it in the private sector (Roberts 1991a). Of this total, 29 percent was for air pollution abatement. Therefore something like $33 billion annually was already being spent in the United States to combat air pollution. The Clean Air Act of 1990 was expected to increase annual expenses by at least $25 billion. For example, lower emissions required for automobiles were likely to cost some $500 per vehicle (about $5 billion annually), and reduction of sulfur dioxide emissions in power plants was expected to cost an additional $2 billion to $4 billion annually (*New York Times*, 23 October 1990). The relevant base for environmental spending on air pollution is thus in the range of some $60 billion annually.

The potential impact of global warming policy on these costs may then be visualized by asking by what percentage these costs would be changed. On the side of direct damage avoidance, if a warmer climate raised total abatement costs by *Y* percent to hold air quality standards constant, then the "benefit" of avoiding the warming amounts to $60 billion × *Y* percent. On the side of spillover benefits of greenhouse action, if lower fossil fuel burning (for example) cut the size of the air pollution problem by Z percent, there would be a \$60 billion × Z percent benefit from the action.[44] Thus, just a 10 percent reduction in urban pollution as a side effect of a shift away from fossil fuels would contribute a benefit of \$6 billion annually. Although not included in the central estimates here, this illustrative estimate suggests that favorable side effects in reducing pollution from a program of emissions abatement could be in the same general range as, say, benefits from limiting sea-level rise or water losses (as reviewed above). Indeed, aggressive cutbacks in fossil fuel emissions might reduce existing air pollution by much more than 10 percent, suggesting a beneficial side effect on the order of tens of billions of dollars annually rather than \$6 billion.

The IPCC "impacts" report (IPCC 1990c) has only two pages on air pollution, out of a volume of over 300 pages, suggesting that relatively little is known about this area. It reports very little on the impacts outside the United States. However, in view of the severe pollution problems of Eastern Europe, the former Soviet Union, and some cities in developing countries (e.g., Mexico City and São Paulo), there could be relatively wide geographical participation in any benefits in air pollution avoidance (and reduction) associated with policy action to abate global warming.

For much higher warming over the very long term, the effects in the area of air pollution are likely to be nonlinear (i.e., they would rise more than proportionately with temperature). For example, although the underlying relationship of ozone concentration to temperature is approximately linear (EPA 1989a, 205), the problem involves thresholds, so that increased concentrations can push cities over the acceptable ceilings and require action where none was necessary before (i.e., an infinite percentage increase in expense from a zero base).

Overview

Table 3.4 presents a summary of the damage estimates developed in this study for the case of the United States. For the very-long-term estimates, the term γ

44. There is an interaction term that must be handled carefully. The policy action would reduce the base for the pollution problem, thereby reducing the benefit by Y percent × \$60 billion. In the extreme, if the greenhouse policy action eliminated all air pollution problems, the base of \$60 billion would shrink to zero. The second, spillover benefit would equal Z percent × \$60 billion = 100 percent × \$60 billion = \$60 billion, but it would be double-counting at that point to attribute an additional X × \$60 billion for the further damage avoided.

Table 3.4 Estimates of annual damage from global warming to the US economy at 1990 scale (billions of 1990 dollars)

	$2xCO_2$ (2.5°C)	Very-long-term warming (10°C)
Agriculture	17.5	95.0
Forest loss	3.3	7.0
Species loss	$4.0 + \Delta_s$	$16.0 + \Delta'_s$
Sea-level rise		35.0
Construction of dikes, levees	1.2	
Wetlands loss	4.1	
Drylands loss	1.7	
Electricity requirements	11.2	64.1
Nonelectric heating	−1.3	−4.0
Human amenity	X_a	Y_a
Human life	5.8	33.0[a]
Human morbidity	X_m	Y_m
Migration	0.5	2.8[a]
Hurricanes	0.8	6.4[b]
Construction	$\pm X_c$	$\pm Y_c$
Leisure activities	1.7	4.0
Water supply	7.0	56.0
Urban infrastructure	0.1	0.6[a]
Air pollution		
Tropospheric ozone	3.5	19.8[a]
Other	X_o	Y_o
Total	61.6 $+X_a+X_m+X_o+\Delta_s\pm X_c$	335.7 $+Y_a+Y_m+Y_o+\Delta'_s\pm Y_c$

a. $\gamma = 1.25$.

b. $\gamma = 1.5$.

refers to the exponential pace of the expansion of damage as temperature rises ($D = k[\Delta t]^{\gamma}$), where specific text estimates are not made.[45]

The estimates here indicate that, even for benchmark warming of $2 \times CO_2$-equivalent, damages for the United States could be relatively large. The $61 billion estimate is approximately 1.1 percent of GDP.[46] Moreover, it excludes several of the unestimated effects, and the potentially important species loss may be underestimated. The table's estimates for damage from very-long-term warming are more striking and, on average, are approximately 5½ times the estimates for $2 \times CO_2$-equivalent.

The very-long-term damages require appropriate scaling. Most of the effects would rise proportionately with GDP, and some would rise more than proportionately (e.g., species loss, if existence value to society is an income-elastic service). The long-term damages are thus on the order of at least 6 percent of GDP.

The central damage estimates of this study, based on estimates for the United States, are that warming associated with benchmark doubling of carbon-dioxide-equivalent (2.5°C) would impose losses of at least 1 percent of GDP and that very-long-term warming of 10°C would cause damage amounting to at least 6 percent of GDP. As indicted in table 3.4, numerous damages are excluded in these central figures. Incorporation of some of the "softest" estimates could substantially raise calculated losses.

Thus, as noted above, the annual value of species loss for benchmark doubling might well be $40 billion rather than $4 billion; annual human amenity costs might amount to $10 billion, health damages $3 billion, and the side benefits of pollution reduction from emissions abatement at least $6 billion. If these estimates are included, the total damages from permitting unabated global warming reach $117 billion annually for benchmark doubling, or 2 percent of GDP.

The benefit-cost analysis below accordingly uses a range of 1 percent to 2 percent of GDP for damages from benchmark warming (the doubling of carbon-dioxide-equivalent, with warming at 2.5°C). As it turns out, this is the same range proposed by Nordhaus (1991a). Although his direct calculations reach only ¼ percent of GDP, he suggests the range of 1 percent to 2 percent of GDP to take account of effects he does not measure.

Even the range of 1 percent to 2 percent of GDP for carbon-dioxide-equivalent doubling, and correspondingly 6 percent to 12 percent of GDP for very-long-

45. Note, however, that the estimate for leisure activity rises less than linearly in the long-term estimate, because the ski industry is already cut by more than half in the $2 \times CO_2$ scenario.

46. This estimate is approximately seven times the direct estimate of Nordhaus (1990b), which is $6.6 billion at 1981 prices. However, it is consistent with his "medium damage" estimate, placed at 1 percent of GDP to allow for unmeasured effects in his direct estimate.

term warming, does not reflect the damage associated with upper-bound warming (climate sensitivity parameter $\Lambda = 4.5°C$). With a damage function exponent of $\gamma = 1.3$, damage for upper-bound warming would reach levels 2.15 times as high ($= [4.5/2.5]^{1.3}$). Thus, with upper-bound warming, damage would be expected to reach 2.1 percent to 4.3 percent of GDP for carbon-dioxide-equivalent doubling, and 13 percent to 26 percent of GDP for very-long-term warming. On this basis, even for $2 \times CO_2$, the high-damage case considered here has approximately twice the highest damage considered by Nordhaus (1991a).

Conversion of the damages in table 3.4 into avoidance benefits requires a sense of the portion of damage that could be avoided with an aggressive anti-greenhouse policy. As analyzed in chapter 2, over the very long term it is likely that as much as 2.5°C out of a total 10°C potential warming (central estimate) cannot be avoided even with aggressive action. The 2.5°C minimum is equivalent to $2 \times CO_2$-benchmark warming. On this basis, the amount of long-term damage that can be avoided through energetic policy measures should equal the difference between the first and second columns of table 3.4 (stated, in both cases, against a 1990-scale US economy). This difference, which amounts to 5 percent of GDP, is the basis for the (somewhat more conservative) "avoidance proportion" applied in the benefit-cost analysis of chapter 7.

This analysis has the sobering implication that much of the damage usually considered is already unavoidable. It also emphatically underscores the importance of the very-long-term analysis. If all that were in store were the damages typically associated with benchmark $2 \times CO_2$-warming, there would be some basis for largely abandoning attempts at abatement on the grounds that a large portion of the damage is unavoidable (although the slower speed of the warming with abatement might still warrant action on ecological grounds). Attention would then more appropriately focus on adaptation. But if the much larger stakes in the very long term are valid, there are correspondingly much stronger grounds for abatement policy action.

Annex 3A

The Welfare Effects of Agricultural Yield Reductions Resulting from Global Warming

Global warming is expected to reduce agricultural yields in many regions, even after allowance is made for the fertilization effect of higher atmospheric carbon dioxide and for partially offsetting changes in agricultural practices. To translate yield reductions into estimates of economic damage, however, it is generally insufficient merely to apply the percentage yield changes to the prospective baseline value of agricultural output. This note sets forth a simple methodology for calculating the welfare effects, taking account of consumer and producer surplus on the basis of underlying supply and demand curves.

The Case of Infinitely Elastic Supply

It is useful to begin with a special case: land is completely homogeneous and the supply curve horizontal, with constant unit cost. Let λ be the ratio of post-warming yields to original yields ($\lambda < 1$). Then, as shown in figure 3A.1, the horizontal supply curve will shift upward from the original price P_0 to $P_1 = P_0(1/\lambda)$. That is, each hectare will produce only λ much as before at the same capital and labor costs, so the price per bushel will have to be $1/\lambda$ times as high for the sales to cover the same costs as before.

In this special case, the welfare effect of lower yield approximates what might intuitively be expected: for a given percentage cut in yields, the welfare change is approximately equal to this percentage multiplied by the original value of production. Consider figure 3A.1. The welfare change equals the change in consumer surplus plus the change in producer surplus. With a horizontal supply curve, there is no producer surplus, either before or after the yield decrease. For its part, consumer surplus declines by the area $a + b$. If the demand curve is steep in the relevant region, this area will approximate $(P_1 - P_0)Q_0 = [(1/\lambda) - 1]P_0Q_0$. This amount equals the percentage decline in yield multiplied by the original value of output.

A Generalized Ricardian Supply Curve

In practice, land is not wholly homogeneous, and the supply curve is more likely to be upward sloping than horizontal. The following analysis will demonstrate that, under production conditions of a classical Ricardian agricultural

Figure 3A.1 Welfare effects of lower agricultural yields under constant costs

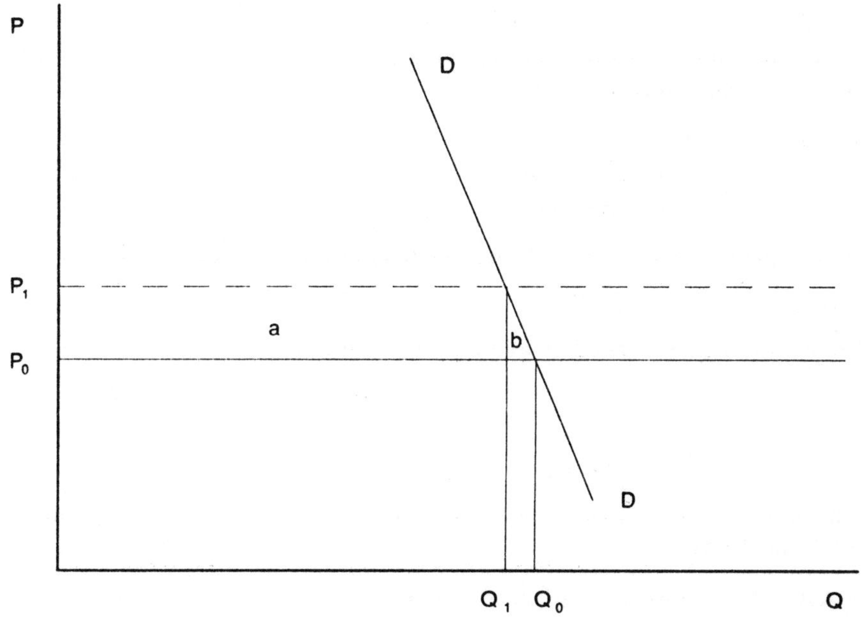

model, and with a plausible demand function, the welfare losses from yield reduction are likely to be larger than the percentage yield reduction times the initial value of production.

Consider a Ricardian agricultural supply function, based on the declining productivity of land incorporated at the margin. Let the amount of labor and capital required per hectare be fixed (for the moment) at L and K, respectively. Then variable costs per hectare are:

(3A.1) $c = wL + rK$

where w is the wage rate and r the rental cost of capital. Let n denote the nth hectare (or million-hectare block) brought into production. Let q denote yield per hectare (e.g., bushels) and P price (e.g., dollars per bushel). Then profit earned on the nth hectare is:

(3A.2) $\pi_n = Pq_n - c.$

Define P_n as the price required to call forth production on n hectares. This price will be such that profit just falls to zero on the marginal hectare, so that:

(3A.3) $\pi_n = 0 = P_n q_n - c; \; P_n = c/q_n.$

Let yield on the marginal hectare be a declining function of the number of hectares brought into production. Then:

(3A.4) $q_n = q_0 - \alpha n \,;\, n < q_0/\alpha.$

Total production equals the number of hectares times the average yield over all qualities of land in production, or:

(3A.5) $$\begin{aligned} Q_n &= 0.5\, n\, [q_0 + q_n] \\ &= 0.5\, n\, [2q_0 - \alpha n] \\ &= nq_0 - 0.5\, \alpha n^2. \end{aligned}$$

From equations 3A.3 and 3A.4, the supply price may be rewritten as:

(3A.6) $P_n = c/[q_0 - \alpha n].$

To find the supply of hectares as a function of the price, we may write:

(3A.7) $$\begin{aligned} q_n &= c/P_n; \\ q_0 - \alpha n &= c/P_n; \\ n &= [1/\alpha][q_0 - (c/P_n)]. \end{aligned}$$

Agricultural output supply as a function of price, in turn, is (from equations 3A.7 and 3A.5):

(3A.8) $$\begin{aligned} Q &= [1/\alpha][q_0 - (c/P_n)]q_0 - 0.5\alpha\{[1/\alpha][q_0 - (c/P_n)]\}^2 \\ &= a - bP_n^{-2} \end{aligned}$$

where a is $0.5q_0^2/\alpha$ and b is $0.5c^2/\alpha$.

The Impact of Changes in Yields

Now consider the effect of multiplying all yields by λ ($\lambda < 1$). Denoting the new situation by primes, we have:

(3A.9) $q_n' = \lambda q_n$

(3A.10) $P_n' = c/[\lambda q_n] = [1/\lambda]P_n.$

That is, the price required to call forth n hectares into production will now be $1/\lambda$ times as high as before.

Total production becomes (from equation 3A.8):

(3A.11) $$\begin{aligned} Q &= 0.5[\lambda q_0]^2/\alpha - [0.5c^2/\alpha][(1/\lambda)P_n]^{-2} \\ &= \lambda^2 a - \lambda^2 b P_n^{-2} \\ &= \lambda^2 Q \end{aligned}$$

The result of the yield multiplication by λ is thus to shift the supply curve to the left at any given price by the multiple λ^2.

Figure 3A.2 Welfare effects of lower agricultural yields with diminishing returns

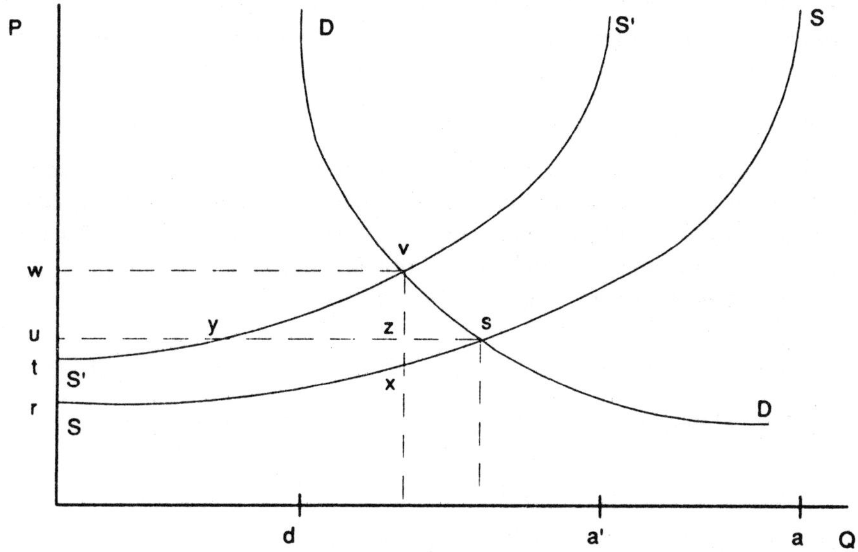

Figure 3A.2 shows the original and shifted supply curves. As indicated by equations 3A.8 and 3A.11, they turn vertical at *a* and *a'* (which is equal to $\lambda^2 a$), respectively, as the final term involving the inverse of the square of the price vanishes when the price rises to infinity. Essentially, the supply curve turns vertical when the full amount of productive land is exhausted, that is, where $n = q_0/\alpha$ (and $q_n = 0$).

It is immediately evident from the comparison of the original supply curve, *SS*, and the new curve after yield decrease, *S'S'*, that the price may potentially rise by more than $[(1/\lambda) - 1]$, the increment when the supply curve is horizontal. Indeed, if there were an attempt to hold production at fractionally below *a'*, the new price would have to approximate infinity and thus would be infinitely times as high as the original price rather than just $1/\lambda$ times as high.

In fact, the bare minimum that the price may rise is by the multiple $1/\lambda$. Consider the intercept on the price axis (*r* in the initial supply curve and *t* in the shifted curve). This is the farmers' reservation price for bringing the first hectare into production. With yield λ times as high as before, the initial reservation price will be $1/\lambda$ times as high ($t = r/\lambda$). So the revised supply curve lies along a path on which the price turns infinitely higher than before at quantity *a'* and is at least $1/\lambda$ times as high as before when quantity is nearly zero. At all intermediate points, the price along the new curve will be more than $1/\lambda$ times the price on the original curve for the same output.

Equilibrium and Welfare Loss

The actual market equilibrium will depend on the demand curve. For ease of solution, it is helpful to specify demand also as a function of the inverse of the square of price:

(3A.12) $D = d + e P^{-2}$.

In this curve, an infinite price makes the second term vanish, and demand, d, remains at the absolute minimum (for example, the amount of food needed to avoid starvation). At lower prices, the second term takes effect, and if the price goes to zero the second term becomes infinitely large.

Equilibrium under the original conditions occurs at:

(3A.13) $D = Q = d + eP^{-2} = a - bP^{-2};\ P^* = \sqrt{[e + b]/[a - d]}$.

Correspondingly, the equilibrium after a decrease in yield occurs at:

(3A.14) $P_1 = \sqrt{[e + \lambda^2 b]/[\lambda^2 a - d]}$

and the ratio of the new equilibrium price to the old price is:

(3A.15) $$P'/P^* = \sqrt{\frac{ea + \lambda^2 ba - ed - \lambda^2 bd}{\lambda^2 ea + \lambda^2 ba - ed - bd}}.$$

It is evident from equation 3A.15 that the price has risen ($P'/P^* > 1$), because the first term in the numerator is larger than the first in the denominator, the second and third terms are identical, and the fourth term in the numerator is negative and smaller than that in the denominator, so that the numerator is strictly larger than the denominator. How much larger depends on the supply and demand parameters (a, b, d, and e) as well as the yield multiple (λ).

The welfare effects of the yield decrease as shown in figure 3A.2 are as follows. Consumer surplus declines by the area *usvw*. Producer surplus changes from *rsu* to *tvw*. Net welfare change is thus:

(3A.16) $\Delta W = tvw - rsu - usvw$.

As suggested in the diagram, the loss of original producer surplus (*rsu*) may approximate the gain in new producer surplus (*tvw*), leaving the bulk of the consumer surplus loss (*usvw*) as net welfare loss (deadweight loss). As suggested above, this loss is likely to exceed the percentage of yield reduction multiplied by original output value.

Table 8.1 The top 50 carbon-emitting countries

	Carbon dioxide emissions (millions of tons of carbon-equivalent)			Economic indicators		Emissions performance				
	Fossil fuels	Land use change	Total emissions	1989 population (thousands)	1989 GNP per capita	Share of world emissions (percentages) A	Share of world population (percentages) B	Share of world GNP (percentages) C	A/B	A/C
United States	1,222	5	1,227	248,800	20,690	17.52	4.78	25.74	3.67	0.68
Soviet Union	1,034	0	1,034	282,830	7,545[a,d]	14.77	5.43	10.67	2.72	1.38
European Community	786	0	786	342,489	13,294	11.22	6.58	22.76	1.71	0.49
China	596	0	596	1,113,900	1,153[d]	8.52	21.40	6.42	0.40	1.33
Brazil	54	390	444	147,300	4,409[d]	6.34	2.83	3.25	2.24	1.95
India	155	118	273	832,500	980	3.89	15.99	4.08	0.24	0.95
Japan	248	0	248	123,100	15,710	3.54	2.36	9.67	1.50	0.37
Indonesia	35	185	220	178,200	1,468[d]	3.14	3.42	1.31	0.92	2.40
Myanmar (Burma)	1	126	128	40,000	984[b,d]	1.82	0.77	0.20	2.37	9.26
Poland	129	0	129	37,900	4,980	1.84	0.73	0.94	2.53	1.95
Colombia	14	101	115	32,300	2,655[d]	1.64	0.62	0.43	2.64	3.83
Canada	110	0	110	26,200	19,230	1.57	0.50	2.52	3.13	0.62
Mexico	82	27	109	84,600	3,763[d]	1.55	1.63	1.59	0.95	0.97
Thailand	16	79	95	55,400	4,190	1.35	1.06	1.16	1.27	1.16
Cote d'Ivoire	1	84	85	11,700	1,700	1.22	0.22	0.10	5.43	12.28
South Africa	78	0	78	35,000	4,327[d]	1.11	0.67	0.76	1.66	1.47
Laos	0	71	72	4,100	735[d]	1.02	0.08	0.02	12.97	67.79
Philippines	10	57	67	60,000	2,280	0.96	1.15	0.68	0.83	1.40
Nigeria	16	49	64	113,800	1,290	0.92	2.19	0.73	0.42	1.25
Czechoslovakia	66	0	66	15,600	5,425[d]	0.94	0.30	0.42	3.13	2.21
Australia	65	0	65	16,800	14,290	0.92	0.32	1.20	2.86	0.77
Romania	58	0	58	23,200	6,083[a,d]	0.83	0.45	0.71	1.87	1.18
Vietnam	5	49	54	64,800	n.a.	0.77	1.24	0.00	0.62	n.a.
Korea (South)	48	0	48	42,400	6,720	0.68	0.81	1.42	0.84	0.48
Peru	6	38	44	21,200	2,362	0.63	0.41	0.25	1.55	2.53
Malaysia	11	32	43	17,400	3,951[d]	0.62	0.33	0.34	1.85	1.80
Venezuela	26	15	41	19,200	4,303[d]	0.59	0.37	0.41	1.59	1.42
Korea (North)	40	0	40	20,600	n.a.	0.57	0.40	0.00	1.44	n.a.
Iran	40	0	40	53,300	4,430	0.57	1.02	1.18	0.55	0.48
Ecuador	4	33	37	10,300	2,378[d]	0.53	0.20	0.12	2.67	4.31

Continued overleaf

Turkey	37	0	37	55,000	4,610	0.53	1.06	1.27	0.50	0.42
Yugoslavia	34	0	34	23,700	2,920	0.49	0.46	0.35	1.08	1.42
Bulgaria	34	0	34	9,000	5,320	0.48	0.17	0.24	2.78	2.01
Zaire	1	29	30	34,500	943[d]	0.43	0.66	0.16	0.66	2.67
Cameroon	2	29	30	11,600	2,070	0.43	0.22	0.12	1.94	3.59
Argentina	30	0	30	31,900	3,951[d]	0.43	0.61	0.63	0.70	0.68
Sudan	1	23	24	24,500	1,428[d]	0.34	0.47	0.17	0.72	1.93
Hungary	21	0	21	10,621	6,200	0.30	0.20	0.33	1.46	0.91
Egypt	21	0	21	51,000	3,160	0.29	0.98	0.81	0.30	0.36
Madagascar	0	19	20	11,300	700	0.28	0.22	0.04	1.29	7.07
Algeria	19	0	19	24,400	4,038[d]	0.28	0.47	0.49	0.59	0.56
Saudi Arabia	18	0	18	14,400	7,907[d]	0.25	0.28	0.57	0.90	0.44
Nicaragua	1	14	15	3,504	2,018[c,d]	0.21	0.07	0.04	3.16	6.01
Sweden	15	0	15	8,500	15,670	0.22	0.16	0.67	1.35	0.33
Pakistan	15	1	15	109,900	1,700	0.22	2.11	0.93	0.10	0.23
Finland	15	0	15	5,000	15,230	0.21	0.10	0.38	2.18	0.55
Malawi	0	13	14	8,200	660	0.19	0.16	0.03	1.23	7.18
United Arab Emirates	14	0	14	1,500	16,859[d]	0.20	0.03	0.13	6.96	1.59
Costa Rica	1	13	13	2,700	3,466[d]	0.19	0.05	0.05	3.68	4.07
Iraq	13	0	13	17,600	5,232[b,d]	0.19	0.34	0.46	0.56	0.41
Subtotal	5,244	1,600	6,844	4,533,744		97.77	87.09		1.12	
Others	456	0	156	672,356		2.23	12.91		0.17	
World	5,700	1,600	7,000	5,206,100		100.00	100.00		1.00	

ICP = (United Nations) International Comparison program; n.a. = not available.

a. GNP estimate is that of PlanEcon; population is from World Bank, *World Development Report*, 1990, 1991.

b. 1988 GDP from International Monetary Fund, *International Financial Statistics*; population from World Bank, *World Development Report* 1990, 1991.

c. 1987 GNP from World Bank, *World Debt Tables*, 1991.

d. ICP estimates of per capita GDP from the equation $\ln(\text{ICP}) = 3.085925 + (0.676735 \times \ln[\text{GDP per capita}])$.

Sources: data on carbon dioxide emissions from World Resources Institute, *World Resources 1990–91*, table 24-1; population and GNP from World Bank, *World Development Report 1990, 1991*, tables 1 and 30, except where noted. World Resources Institute and World Bank data reprinted with permission.

References

Ausubel, Jesse H. 1990. "Conventional Wisdom About Impacts of Global Change." New York: Rockefeller University (mimeographed, 1 April).

Batie, Sandra S., and Herman H. Shugart. 1989. "The Biological Consequences of Climate Changes: An Ecological and Economic Assessment." In Norman J. Rosenberg, William E. Easterling, Pierre Crosson, and Joel Darmstadter, eds., *Greenhouse Warming: Abatement and Adaptation*, 121–32. Washington: Resources for the Future.

Bazzaz, Fakhri A., and Eric D. Fajer. 1992. "Plant Life in a CO_2–Rich World." *Scientific American* 266, no. 1 (January): 68–74.

Brookshire, David S., Mark A. Thayer, William D. Schulze, and Ralph C. D'Arge. 1982. "Valuing Public Goods: A Comparison of Survey and Hedonic Approaches." *American Economic Review* 72, no. 1 (March): 165–77.

Cline, William R. 1990a. "Economic Stakes of Global Warming in the Very-Long Term." Washington: Institute for International Economics (mimeographed, November).

Cline, William R. 1990b. "Global Warming and the Costs of Hurricane Damage." Washington: Institute for International Economics (mimographed, July).

Cline, William R. 1990c. "Greenhouse Gas Emissions and Global Warming: Parameters and Timetables." Washington: Institute for International Economics (mimeographed, December).

Cline, William R. 1992b. *Global Warming: The Benefits of Emissions Abatement*. Paris: Organization for Economic Cooperation and Development (May).

Congressional Budget Office. 1990. *Carbon Charges as a Response to Global Warming: The Effects of Taxing Fossil Fuels*. Washington: Congressional Budget Office (August).

Cropper, Maureen L., and Wallace E. Oates. 1990. "Environmental Economics: A Survey." *Discussion Papers* QE90–12. Washington: Resources for the Future (January).

Darmstadter, Joel, and Andrew Plantinga. 1991. "The Economic Cost of CO_2 Mitigation: A Review of Estimates for Selected World Regions." *Discussion Papers* ENR91–06. Washington: Resources for the Future (January).

Edmonds, Jae, and David W. Barns. 1990. "Factors Affecting the Long-Term Cost of Global Fossil Fuel CO_2 Emissions Reductions." Washington: Pacific Northwest Laboratory (mimeographed, December).

Emanuel, Kerry A. 1987. "The Dependence of Hurricane Intensity on Climate." *Nature* 326, no. 2 (April): 483–85.

Environmental Protection Agency. 1989a. *The Potential Effects of Global Climate Change on the United States* (Joel B. Smith and Dennis Tirpak, eds.). Washington: Environmental Protection Agency.

Environmental Protection Agency. 1989b. *The Potential Effects of Global Climate Change on the United States*, Appendix C: Agriculture, volumes 1 and 2. Washington: Environmental Protection Agency.

Environmental Protection Agency. 1990. *Progress Reports on International Studies of Climate Change Impacts (Draft)*. Washington: Environmental Protection Agency (6 November).

Flavin, Christopher. 1989. "Slowing Global Warming: A Worldwide Strategy." *Worldwatch Papers* 91. Washington: Worldwatch (October).

Gleick, Peter H. 1987. "Regional Hydrologic Consequences of Increases in Atmospheric CO_2 and Other Trace Gases." *Climatic Change* 10: 137–61.

Glynn, P. W., and W. H. de Weerdt. 1991. "Elimination of the Reef-Building: Hydrocorals Following the 1982–83 El Niño Warming Event." *Science* 253 (5 July): 69–71.

Goering, John M. 1990. "The Causes of Undocumented Migration to the United States: A Research Note." *Working Papers* 52. Washington: Commission for the Study of International Migration and Cooperative Economic Development (July).

Goklany, Indur M. 1989. "Climate Change Effects on Fish, Wildlife and Other DOI Programs." In John C. Topping, Jr., ed., *Coping with Climate Change*, 273–81. Washington: The Climate Institute.

Hansen, J., et al. 1989. "Regional Greenhouse Climate Effects." In John C. Topping, Jr., ed., *Coping with Climate Change*, 68–81. Washington: The Climate Institute.

Hoeller, Peter, Andrew Dean, and Jon Nicolaisen. 1990. "A Survey of Studies of the Costs of Reducing Greenhouse Gas Emissions." *OECD Working Papers* 89. Paris: Organization for Economic Cooperation and Development.

Intergovernmental Panel on Climate Change. 1990a. *IPCC First Assessment Report, Volume I: Overview and Summaries.* New York: World Meteorological Organization and United Nations Environmental Programme (August).

Intergovernmental Panel on Climate Change. 1990b. *Scientific Assessment of Climate Change: Report Prepared for IPCC by Working Group I.* New York: World Meteorological Organization and United Nations Environment Programme (June).

Intergovernmental Panel on Climate Change. 1990c. *Potential Impacts of Climate Change: Report Prepared for IPCC by Working Group II.* New York: World Meteorological Organization and United Nations Environment Programme (June).

Jorgenson, Dale W., and Peter J. Wilcoxen. 1990b. "The Cost of Controlling US Carbon Dioxide Emissions." Cambridge, MA: Harvard University (mimeographed, September).

Kane, Sally, John Reilly, and Rhonda Bucklin. 1989. "Implications of the Greenhouse Effect for World Agricultural Commodity Markets." Washington: US Department of Agriculture (June).

Kimball, Bruce, and Norman J. Rosenberg, eds. 1990. *The Impact of CO_2, Trace Gases, and Climate Change. Publication* no. 53. Madison, WI: American Society of Agronomy.

Manne, Alan S., and Richard G. Richels. 1990a. "CO_2 Emission Limits: An Economic Cost Analysis for the USA." *The Energy Journal* 11, no. 2.

Mearns, Linda O., Richard W. Katz, and Stephen H. Scheider. 1984. "Extreme High-Temperature Events: Changes in their Probabilities with Changes in Mean Temperature." *Journal of Climate and Applied Meterology* 23 (December): 1601–13.

Morgenstern, Richard D. 1991. "Towards a Comprehensive Approach to Global Climate Change Mitigation." *American Economic Review* 81, no. 2 (May): 140–45.

Nordhaus, William D. 1990a. "An Intertemporal General-Equilibrium Model of Economic Growth and Climate Change." In David O. Wood and Yoichi Kaya, eds., *Proceedings of the Workshop on Economic/Energy/Environmental Modeling for Climate Policy Analysis, October 22–23, 1990*, 415–33. Cambridge, MA: MIT Center for Energy Policy Research (January).

Nordhaus, William D. 1990b. "To Slow or Not to Slow: The Economics of the Greenhouse Effect." New Haven: Yale University (mimeographed, 5 February).

Nordhaus, William. 1991a. "To Slow or Not to Slow: The Economics of the Greenhouse Effect." *The Economic Journal* 101, no. 6: 920–37.

Parry, Martin. 1990. *Climate Change and World Agriculture*. London: Earthscan.

Parry, M. L., and T. R. Carter. 1989. "An Assessment of the Effects of Climatic Change on Agriculture." *Climatic Change* 15: 95–116.

Regens, James L., Frederick W. Cubbage, and Donald C. Hodges. 1989. "Climate Change and U.S. Forest Markets." In John C. Topping, Jr., ed., *Coping with Climate Change*, 303–09. Washington: The Climate Institute.

Reid, Walter, Charles Barber, and Kenton Miller, 1992. *Global Biodiversity Strategy: A Policy-Maker's Guide*. Washington: World Resources Institute.

Rind, D., R. Goldberg, J. Hansen, C. Rosenzweig, and R. Ruedy. 1990. "Potential Evapotranspiration and the Likelihood of Future Drought." *Journal of Geophysical Research* 95, no. D7: 9983–10004.

Ritchie, J. T., B. D. Baer, and T. Y. Chou. 1989. "Effect of Global Climate Change on Agriculture: Great Lakes Region." In Environmental Protection Agency, *The Potential Effects of Global Climate Change on the United States*, chapter 1. Washington: Environmental Protection Agency.

Roberts, Leslie. 1991a. "Costs of a Clean Environment." *Science* 251 (8 March): 1182.

Rosenberg, Norman J., and Pierre R. Crosson. 1991. *Processes for Identifying Regional Influences of and Responses to Increasing Atmospheric CO_2 and Climate Change: The MINK Project—An Overview* (document no. DOE/RL/01830T-H5). Washington: US Department of Energy (August).

Rosenberg, Norman J., Pierre Crosson, William E. Easterling, Kenneth Frederick, and Roger Sedjo. 1989a. "Policy Options for Adaptation to Climate Change." *Discussion Papers* ENR 89–05 (March). Washington: Resources for the Future.

Rosenberg, Norman J., et al. 1991. *Processes for Identifying Regional Influences of and Responses to Increasing Atmospheric CO_2 and Climate Change—The* MINK *Report I Background and Baselines* (document no. DOE/RL/01830T-H6). Washington: US Department of Energy (August).

Schelling, Thomas C. 1983. "Climatic Change: Implications for Welfare and Policy." In National Research Council, *Changing Climate*, 449–86. Washington: National Academy Press.

Schelling, Thomas C. 1991. "Economic Responses to Global Warming: Prospects for Cooperative Approaches." In Rudiger Dornbusch and James M. Poterba, eds., *Global Warming: Economic Policy Responses*, 197–221. Cambridge, MA: MIT Press.

Schneider, Stephen H., and Norman J. Rosenberg. 1989. "The Greenhouse Effect: Its Causes, Possible Impacts, and Associated Uncertainties." In Norman J. Rosenberg, William E. Easterling, Pierre Crosson, and Joel Darmstadter, eds., *Greenhouse Warming: Abatement and Adaptation*, 7–34. Washington: Resources for the Future.

Sedjo, Roger A. 1992. "Preserving Biodiversity as a Resource." *Resources* 106 (Winter): 26–29. Washington: Resources for the Future.

Sedjo, Roger A., and Allen M. Solomon. 1989. "Climate and Forests." In Norman J. Rosenberg et al., eds., *Greenhouse Warming: Abatement and Adaptation*, 105–19. Washington: Resources for the Future.

Smith, William H. 1991. "Air Pollution and Forest Damage." *Chemical and Engineering News* (11 November).

Solomou, Solomos. 1990. "The Impact of Climatic Variations on Construction Cycles and Economic Activity, 1856–1913." Cambridge, England: Cambridge University (mimeographed, August).

Titus, James G., et al. 1992. "Greenhouse Effect and Sea Level Rise: Potential Loss of Land and the Cost of Holding Back the Sea." *Coastal Management* (forthcoming).

Waters, Somerset. 1990. *The Travel Industry World Yearbook: The Big Picture*. Madrid: World Tourism Organization.

Weitzman, Martin L. 1991. "On Diversity." *Discussion Papers* 1553. Cambridge, MA: Harvard University Institute of Economic Research (May).

Wendland, Wayne M. 1977. "Tropical Storm Frequencies Related to Sea Surface Temperatures." *Journal of Applied Meteorology* 16, no. 5 (May): 477–81.

Wilhite, Donald A. 1990. "The Enigma of Drought: Management and Policy Issues for the 1990s." *International Journal of Environmental Studies* 36: 41–54.

Yohe, Gary W. 1992. "Sorting Out Facts and Uncertainties in Economic Response to the Physical Effects of Global Climate Change." Middletown, CT: Wesleyan University (mimeographed, February).

[5]

The Impact of Global Warming on Agriculture: A Ricardian Analysis

By ROBERT MENDELSOHN, WILLIAM D. NORDHAUS, AND DAIGEE SHAW*

We measure the economic impact of climate on land prices. Using cross-sectional data on climate, farmland prices, and other economic and geophysical data for almost 3,000 counties in the United States, we find that higher temperatures in all seasons except autumn reduce average farm values, while more precipitation outside of autumn increases farm values. Applying the model to a global-warming scenario shows a significantly lower estimated impact of global warming on U.S. agriculture than the traditional production-function approach and, in one case, suggests that, even without CO_2 fertilization, global warming may have economic benefits for agriculture. (*JEL* Q10, Q25)

Over the last decade, scientists have extensively studied the greenhouse effect, which holds that the accumulation of carbon dioxide (CO_2) and other greenhouse gases (GHG's) is expected to produce global warming and other significant climatic changes over the next century. Numerous studies indicate major impacts on agriculture, especially if there is significant mid-continental drying and warming in the U.S. heartland.[1] Virtually every estimate of economic impacts relies on a technique we denote the production-function approach.

This study compares the traditional production-function approach to estimating the impacts of climate change with a new "Ricardian" approach that examines the impact of climate and other variables on land values and farm revenues. The traditional approach to estimating the impact of climate change relies upon empirical or experimental production functions to predict environmental damage (hence its label in this study as the production-function approach).[2] This approach takes an underlying production function and estimates impacts by varying one or a few input variables, such as temperature, precipitation, and carbon dioxide levels. The estimates might rely on extremely carefully calibrated crop-yield models (such as CERES or SOYGRO) to determine the impact upon yields; the results often predict severe yield reductions as a result of global warming.

While providing a useful baseline for estimating the impact of climate change on farming, these studies have an inherent bias and will tend to overestimate the damage. This bias is sometimes called the "dumb-farmer scenario" to suggest that it omits a

*Mendelsohn: School of Forestry and Environmental Studies and the Department of Economics, Yale University, 360 Prospect Street, New Haven, CT 06511; Nordhaus: Department of Economics and Cowles Foundation, Yale University, P.O. Box 1972, Yale Station, New Haven, CT 06511; Shaw: Institute of Economics, Academia Sinica, Taipei, Republic of China. We thank the National Science Foundation and Economic Research Service, USDA, for funding this project, although all views are the authors' alone. We also thank John Miranowski, Daniel Hellerstein, John Reilly, Katherine Segerson, Paul Waggoner, and referees for their assistance and advice, and we thank Susan Helms for producing the color maps in this paper. Correspondence regarding the paper should be addressed to Mendelsohn at the above address.

[1]See particularly the reports of the Intergovernmental Panel on Climate Change (1990) and the National Academy of Sciences Panel on Greenhouse Warming (1992).

[2]Important studies include John Callaway et al. (1982), W. Decker et al. 1986, Richard Adams et al. (1988, 1990), Adams (1989), D. Rind et al. (1990), and Cynthia Rosenzweig and Martin L. Parry (1994). For useful surveys, see National Research Council (1983), Joel Smith and Dennis Tirpak (1989), National Academy of Sciences (1992), and William Cline (1992).

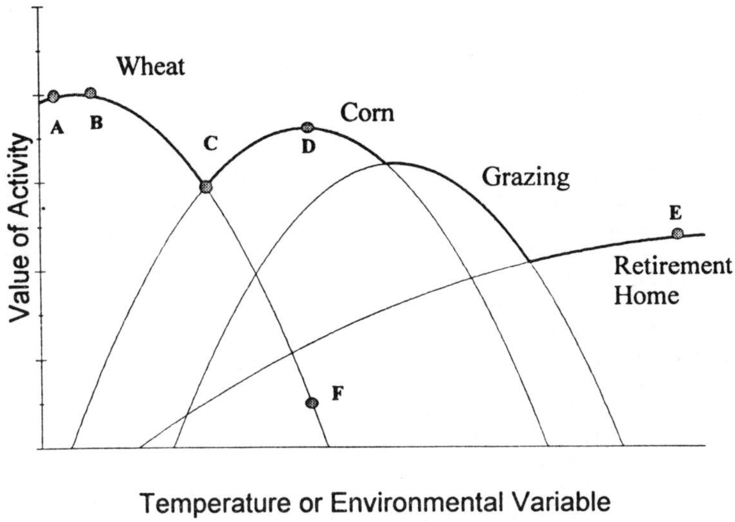

FIGURE 1. BIAS IN PRODUCTION-FUNCTION STUDIES

variety of the adaptations that farmers customarily make in response to changing economic and environmental conditions. Most studies assume little adaptation and simply calculate the impact of changing temperature on farm yields. Others allow limited changes in fertilizer application, irrigation, or cultivars (see William Easterling et al., 1991). None permits a full adjustment to changing environmental conditions by the farmer. For example, the literature does not consider the introduction of completely new crops (such as tropical crops in the south); technological change; changes in land use from farming to livestock, grassland, or forestry; or conversion to cities, retirement homes, campsites, or the 1,001 other productive uses of land in a modern postindustrial society.

By not permitting a complete range of adjustments, previous studies have overestimated damages from environmental changes. Figure 1 shows the hypothetical values of output in four different sectors as a function of a single environmental variable, temperature, in order to illustrate the general nature of the bias. In each case, we assume that the production-function approach yields an accurate assessment of the economic value of the activity as a function of temperature. The four functions provide a simplified example of how the value of wheat, corn, grazing, and retirement homes might look as a function of the temperature. For example, the curve to the far left is a hypothetical "wheat production function," showing how the value of wheat varies with temperature, rising from cold temperatures such as point A, then peaking at point B, finally falling as temperatures rise too high. A production-function approach would estimate the value of wheat production at different temperatures along this curve.

The bias in the production-function approach arises because it fails to allow for economic substitution as conditions change. For example, when the temperature rises above point C, adaptive and profit-maximizing farmers will switch from wheat to corn. As temperature rises, the production-function approach might calculate that the yield has fallen to F in wheat, but wheat is in reality no longer produced; the realized value is actually much higher, at point D where corn is now produced. At a slightly higher temperature, the land is no longer optimally used for corn but switches to grazing, and production-function estimates that do not allow for this conversion will again overestimate the losses from climate change. Finally, at point E, even the best agricultural model will predict that the land is unsuitable for farming or grazing and that the damage is severe. A more complete approach might find that the land has been converted to retirement villages, to which old folks flock so they can putter around in the warm winters and dry climates.

All this is of course illustrative. However, it makes the crucial point that the production-function approach will overestimate the damages from climate change because it does not, and indeed cannot, take into account the infinite variety of substitutions, adaptations, and old and new activities that may displace no-longer-advantageous activities as climate changes.

In this study, we develop a new technique that in principle can correct for the bias in the production-function technique by using economic data on the value of land. We call this the *Ricardian approach*, in which, instead of studying yields of specific crops, we examine how climate in different places affects the net rent or value of farmland. By directly measuring farm prices or revenues, we account for the direct impacts of climate on yields of different crops as well as the indirect substitution of different inputs, introduction of different activities, and other potential adaptations to different climates. If markets are functioning properly, the Ricardian approach will allow us to measure the economic value of different activities and therefore to verify whether the economic impacts implied by the production-function approach are reproduced in the field.

The results of the Ricardian approach can be seen in Figure 1. We assume that the "value" measured along the vertical axis is the net yield per acre of land; more precisely, it is the value of output less the value of all inputs (excluding land rents). Under competitive markets, the land rent will be equal to the net yield of the highest and best use of the land. This rent will in fact be equal to the heavy solid line in Figure 1. We label the solid line in Figure 1 the "best-use value function."

In general, we do not observe market land rents because land rent is generally a small component of the total profits. However, with farms, land rents tend to be a large fraction of total costs and can be estimated with reasonable precision. Farm value is the present value of future rents, so if the interest rate, rate of capital gains, and capital per acre are equal for all parcels, then farm value will be proportional to the land rent. Therefore, by observing the relationship of farm values to climatic and other variables, we can infer the shape of the solid, best-use value function in Figure 1.[3]

This study measures the effect of climatic variables on agriculture. We examine both climatic data and a variety of fundamental geographical, geophysical, agricultural, economic, and demographic factors to determine the intrinsic value of climate on farmland. The units of observation are U.S. counties in the lower 48 states. We examine the effect of climatic variables as well as nonclimatic variables on both land values and farm revenue, and the analysis includes a number of urban variables in order to measure the potential effect of development upon agricultural land values. The analysis suggests that climate has a systematic impact on agricultural rents through temperature and precipitation. These effects tend to be highly nonlinear and vary dramatically by season. The paper concludes with a discussion of the impacts of global warming on American farms.

I. Measuring the Effect of Climate on Agriculture

Using the Ricardian technique, we estimate the value of climate in U.S. agriculture. Agriculture is the most appealing application of the Ricardian technique both because of the significant impact of climate

[3]The analytical basis for the present empirical study is presented in Mendelsohn et al. (1993).

on agricultural productivity and because of the extensive county-level data on farm inputs and outputs.

Sources and Methods

The basic hypothesis is that climate shifts the production function for crops. Farmers at particular sites take environmental variables like climate as given and adjust their inputs and outputs accordingly. Moreover, we assume perfect competition in both product and input markets. Most important, we assume that the economy has completely adapted to the given climate so that land prices have attained the long-run equilibrium that is associated with each county's climate.

For the most part, the data are actual county averages, from the *1982 U.S. Census of Agriculture*, so that there are no major issues involved in obtaining information on these variables.[4] The *County and City Data Book* (U.S. Bureau of the Census, 1988) and the computer tapes of those data are the sources for much of the agricultural data used here, including values of farm products sold per acre, farm land and building values,[5] and information on market inputs for farms in every county in the United States. In addition, in many specifications, we include social, demographic, and economic data on each of the counties; these as well are drawn from the *County and City Data Book*.

Data about soils were extracted from the National Resource Inventory (NRI) with the kind assistance of Daniel Hellerstein and Noel Gollehon of the U.S. Department of Agriculture. The NRI is an extensive survey of land characteristics in the United States. For almost 800,000 sites, NRI has collected soil samples, or land characteristics, each providing a measure of salinity, permeability, moisture capacity, clay content, sand content, flood probability, soil erosion (K-factor), rain erosion (R-factor), slope length, wind erosion, whether or not the land is a wetland, and numerous other variables that are not used in this analysis. Each sample also contains an expansion factor, which is an estimate of the amount of land the sample represents in that county. Using these expansion factors, we aggregate these data to yield an overall county estimate for each soil variable.

Climatic data pose more difficult issues. They are available by meteorological station rather than by county, so it was necessary to estimate county-average climates. To begin with, climate data were obtained from the National Climatic Data Center, which gathers data from 5,511 meteorological stations throughout the United States. These stations form a dense set of observations for most regions of the United States, with the exception of some of the desert Southwest. The data include information on precipitation and temperature for each month from 1951 through 1980. Since the purpose of this study is to predict the impacts of climate changes on agriculture, we focus on the long-run impacts of precipitation and temperature on agriculture, not year-to-year variations in weather. We consequently examine the "normal" climatological variables —the 30-year average of each climatic variable for every station. In this analysis, we collect data on normal daily mean temperature and normal monthly precipitation for January, April, July, and October. We focus on these four months in order to capture seasonal effects of each variable. For example, cold January temperatures may be important as a control on insect pests, warm-but-not-hot summers may be good for crop growth, and warm October temperatures may assist in crop harvesting.

In order to link the agricultural data which are organized by county and the climate data which are organized by station, we conduct a spatial statistical analysis that examines the determinants of the climate of each county. Although the specific climatic variables we analyze in this study have been measured frequently, there are some counties with no weather stations and others with several. Some of the weather stations

[4]Appendix A contains complete descriptions and definitions of the variables used in this study.

[5]The definition and source of the farm value variable is critical to this study, and its derivation is described in Appendix B.

TABLE 1—INTERPOLATION FOR COUNTY CLIMATE MEASURES (FRESNO, CA)

	Temperature			Precipitation		
Independent variable	April	July	October	April	July	October
Constant	131,535	231,764	124,970	−58,846	−184,063*	16,551
Longitude	−32.8*	−59.6*	−29.2	26.7	45.2*	1.96
Latitude	−13.2	−18.2	−16.8	−19.6	21.7*	−16.33
Latitude squared	1.9×10^{-4}	2.8×10^{-4}	4.1×10^{-4}	1.6×10^{-3}	-3.1×10^{-4}	$1.6\times10^{-3*}$
Longitude squared	$2.0\times10^{-3*}$	$3.8\times10^{-3*}$	1.7×10^{-3}	-2.3×10^{-3}	$-2.7\times10^{-3*}$	-3.9×10^{-4}
Longitude × latitude	1.8×10^{-3}	2.8×10^{-3}	2.1×10^{-3}	1.5×10^{-3}	$-2.9\times10^{-3*}$	1.1×10^{-3}
Altitude	−0.56*	−1.44*	−1.00*	0.525	1.28*	1.48*
Altitude squared	$-1.6\times10^{-6*}$	$-3.0\times10^{-6*}$	$-2.3\times10^{-6*}$	$-3.7\times10^{-6*}$	$-6.5\times10^{-7*}$	$-2.4\times10^{-6*}$
Latitude × altitude	4.3×10^{-5}	8.8×10^{-5}	$7.7\times10^{-5*}$	-4.8×10^{-5}	$-1.1\times10^{-4*}$	$-1.1\times10^{-4*}$
Longitude × altitude	6.2×10^{-5}	$1.8\times10^{-4*}$	$1.1\times10^{-4*}$	-4.6×10^{-5}	$-1.5\times10^{-4*}$	$-1.7\times10^{-4*}$
Distance	−40.4*	−74.5*	−35.2	−5.47	59.4*	−26.6
Distance squared	2.6×10^{-3}	4.2×10^{-3}	2.2×10^{-3}	2.9×10^{-3}	$-4.9\times10^{-3*}$	$4.8\times10^{-3*}$
Distance × longitude	$5.2\times10^{-3*}$	$9.6\times10^{-3*}$	4.2×10^{-3}	-1.3×10^{-3}	$-6.7\times10^{-3*}$	2.6×10^{-3}
Distance × latitude	2.0×10^{-3}	3.7×10^{-3}	2.3×10^{-3}	4.3×10^{-3}	$-4.9\times10^{-3*}$	2.7×10^{-3}
Distance × altitude	6.7×10^{-5}	1.3×10^{-4}	$9.7\times10^{-5*}$	-1.9×10^{-4}	$-7.0\times10^{-5*}$	$-2.3\times10^{-4*}$
Adjusted R^2:	0.999	0.998	0.999	0.796	0.777	0.706
Standard error:	0.13	0.24	0.13	0.54	0.13	0.30
Number of observations:	331	331	331	525	525	525

Notes: Temperature is measured in Fahrenheit, and precipitation is in inches per month.
*Statistically significant at the 5-percent level.

are not in representative locations, such as the station on the top of Mt. Washington. Furthermore, some counties are large enough or contain sufficient topographical complexity that there is variation of climate within the county. We therefore proceeded by constructing an average climate for each county.

First, we assume that all the weather stations within 500 miles of the geographic center of the county provide some useful climate information. The 500-mile circle invariably draws in many stations, so that our measure does not depend too heavily on any one station.

Second, we estimate a climate surface in the vicinity of the county by running a weighted regression across all weather stations within 500 miles. The weight is the inverse of the square root of a station's distance from the county center because we recognize that closer stations contain more information about the climate of the center. We estimate a separate regression for each county since the set of stations within 500 miles and the weights (distances) are unique for each county. The dependent variables are the monthly normal temperatures and precipitation amounts for January, April, July, and October for the 30-year period. The independent variables include latitude, longitude, altitude, and distance from closest shoreline. The regression fits a second-order polynomial over these four basic variables, including interactive terms, so that there are 14 final variables in the regression, plus a constant term. Eight regressions (4 seasons × 2 measures) for each of 3,000 counties leads to over 24,000 estimated regressions.

Third, we calculate the predicted value of each climatic variable for the geographic center of the county. The predicted values of normal precipitation and temperature from the climate regressions are the independent variables for climate in the property-value regressions. This complicated interpolation procedure is intended to provide geographically accurate estimates of the climatic variables for each county.

The estimates of the climate parameters for individual counties are too numerous to present, but we show two selected counties in Tables 1 and 2. These show the indepen-

TABLE 2—INTERPOLATION FOR COUNTY CLIMATE MEASURES (DES MOINES, IOWA)

	Temperature			Precipitation		
Independent variable	April	July	October	April	July	October
Constant	6,425	5,006	8,967	−32,243	77,324*	41,650
Longitude	−0.919	−1.12	−2.55	7.72	−15.8*	−9.61
Latitude	−2.48	−0.829	−1.55	10.0	−32.9*	−16.32
Latitude squared	2.5×10^{-4}	2.0×10^{-5}	3.2×10^{-5}	-9.7×10^{-4}	3.2×10^{-3}*	1.6×10^{-3}
Longitude squared	3.7×10^{-5}	8.1×10^{-5}	2.0×10^{-4}	-4.9×10^{-4}	6.8×10^{-4}	5.9×10^{-4}
Longitude × latitude	2.0×10^{-4}	1.0×10^{-4}	2.4×10^{-4}	-9.9×10^{-4}	3.8×10^{-3}*	1.8×10^{-3}
Altitude	−0.13	0.046	0.34*	0.353	3.02*	2.09*
Altitude squared	-1.2×10^{-6}	-1.3×10^{-6}*	1.6×10^{-6}*	1.1×10^{-5}*	-1.5×10^{-6}	2.1×10^{-5}*
Latitude × altitude	2.1×10^{-5}	-1.6×10^{-5}	-6.9×10^{-5}*	-1.2×10^{-4}	-5.7×10^{-4}*	-2.8×10^{-4}*
Longitude × altitude	1.1×10^{-5}	-9.7×10^{-6}	-4.9×10^{-5}*	-3.1×10^{-5}	-3.6×10^{-4}*	-3.2×10^{-4}*
Distance	1.14	−1.17	−0.564	−0.150	26.8	18.6
Distance squared	1.8×10^{-4}	-3.1×10^{-4}	-1.9×10^{-4}	5.8×10^{-4}	-1.2×10^{-3}	1.4×10^{-3}
Distance × longitude	-4.4×10^{-5}	1.9×10^{-4}	-1.2×10^{-4}	-4.1×10^{-4}	-2.7×10^{-3}	-1.9×10^{-3}
Distance × latitude	-3.6×10^{-4}	2.2×10^{-4}	9.0×10^{-5}	4.2×10^{-4}	-5.4×10^{-3}*	-3.8×10^{-3}
Distance × altitude	-2.2×10^{-5}	3.2×10^{-5}	9.9×10^{-5}*	-1.7×10^{-4}	6.9×10^{-4}*	3.6×10^{-4}*
Adjusted R^2:	0.999	0.999	0.999	0.989	0.987	0.976
Standard error:	0.04	0.04	0.04	0.14	0.17	0.15
Number of observations:	928	928	928	1,477	1,477	1,477

Notes: Temperature is measured in Fahrenheit, and precipitation is in inches per month.
*Statistically significant at the 5-percent level.

dent variables as well as the coefficients and summary regression statistics for Fresno, California, and Des Moines, Iowa. Note that more coefficients are significant in the Fresno regressions than in the Des Moines regressions. There is more variation across the sample in Fresno because of the effects of the coast and nearby mountain ranges. Although there are more significant coefficients in the California regression, the Iowa regression has a better overall fit and smaller standard errors. In general, the fit east of 100 degrees longitude (the east slope of the Rocky Mountains) was tighter than in the West.

In order to gain some sense of the reliability of this geographic approximation method, we predicted the climate for each of the weather stations. Dropping the weather station itself, we predicted the climatic variables for the station from all stations within 500 miles in the manner explained above. Comparing these results with the actual measurements from each station reveals that the approximation method predicts between 87 percent and 97 percent of the variation in precipitation in the continental United States and between 97 percent and 99 percent of the variation in temperature. It should be noted that, even in a statistically stationary environment, the observations of "climate" themselves contain error because they contain only 30 observations. Depending upon the relative importance of idiosyncratic error in climate versus misspecification error in our equation, the predictions might actually be superior to the recorded observations themselves. In any case, the predictions serve as sophisticated interpolations of the climate between stations.

II. Empirical Analysis

The Ricardian approach estimates the importance of climate and other variables on farmland values. As noted above, land values are the expected present value of future rents. There is little reason for the riskless interest rate to vary across counties in the United States, but the risk and capital-gains components of land value might vary considerably. For example, California agricultural land near growing cities might well have a larger capital-gains component than would rural land in an economically stagnant coal-mining region of Appalachia. Moreover, there are major potential errors in measurement of land values since values are estimated by farmers, and such estimates are often unreliable. However, there is no reason to believe that the errors of measurement are correlated with independent data such as temperature or precipitation. The major effect of measurement errors will be imprecision of the econometric estimates rather than bias in the estimation of the coefficients or in the estimate of the economic value of climate on agriculture.

We regress land values on climate, soil, and socioeconomic variables to estimate the best-value function across different counties. There are 2,933 cross-sectional observations. The means have been removed from the independent variables in this regression. The quadratic climate variables are consequently easier to interpret. The linear term reflects the marginal value of climate evaluated at the U.S. mean, while the quadratic term shows how that marginal effect will change as one moves away from the mean.

We present several regressions in Table 3. In order to give a sense of the importance of the nonfarm variables in the model, we begin with a model that contains only climate variables. The first set of regressions in Table 3 is a quadratic model that includes the eight measures of climate (four months of precipitation and temperature). For each variable, linear and quadratic terms are included to reflect the nonlinearities that are apparent from field studies.

In the remainder of regressions, we include urban, soil, and other environmental variables to control for extraneous factors influencing land values and farm revenues. This raises the question of how the counties should be weighted. A first set of regressions uses the *cropland weights*, in which observations are weighted by the percentage of each county in cropland. Counties with a large fraction of cropland should provide a better reading on price determination because other influences, such as cities or forests, are minimized; these results are particularly useful for the grain belt. A second set of regressions uses *crop-revenue weights*; that is, observations are weighted by the aggregate value of crop revenue in each county. This second weighting scheme emphasizes those counties that are most important to total agricultural production, even though some of the counties might have their land values affected by large neighboring cities; it also places greater weight on counties where more valuable crops are grown. On the whole, the cropland measure tends to emphasize the corn, wheat, and soybean belt and therefore reflects the influence of climate on the grains. The crop-revenue weights, by contrast, give more influence to the truck farms and citrus belt of the coast lands, and the crop-revenue regressions thus reflect a broader definition of agriculture.

The results of this analysis are shown in columns (ii)–(v) of Table 3. The squared terms for most of the climate variables are significant, implying that the observed relationships are nonlinear. However, some of the squared terms are positive, especially for precipitation, implying that there is a minimally productive level of precipitation and that either more or less precipitation will increase land values. The negative quadratic coefficient implies that there is an optimal level of a climatic variable from which the value function decreases in both directions.

The overall impact of climate as measured by the marginal impacts is largely the same across the different models, although the quantitative estimates vary. All models suggest that higher winter and summer temperatures are harmful for crops; that higher fall temperatures and higher winter and spring rainfall are beneficial for crops; but that higher summer or fall rainfall is harmful. The two weighting schemes differ, however, in terms of their assessment of the relative importance of winter versus summer temperature. The cropland model finds higher winter temperatures less harmful, valuing a 1°F increase by between \$89 and \$103 per acre, whereas the crop-revenue model finds this effect more harmful, with estimated impacts between \$138 and \$160

TABLE 3—REGRESSION MODELS EXPLAINING FARM VALUES

	Cropland weights			Crop-revenue weights	
	1982	1982	1978	1982	1978
Independent variables	(i)	(ii)	(iii)	(iv)	(v)
Constant	1,490	1,329	1,173	1,451	1,307
	(71.20)	(60.18)	(57.95)	(46.36)	(52.82)
January temperature	−57.0	−88.6	−103	−160	−138
	(6.22)	(9.94)	(12.55)	(12.97)	(13.83)
January temperature squared	−0.33	−1.34	−2.11	−2.68	−3.00
	(1.43)	(6.39)	(11.03)	(9.86)	(14.11)
April temperature	−137	−18.0	23.6	13.6	31.8
	(10.81)	(1.56)	(2.23)	(1.00)	(2.92)
April temperature squared	−7.32	−4.90	−4.31	−6.69	−6.63
	(9.42)	(7.43)	(7.11)	(9.44)	(11.59)
July temperature	−167	−155	−177	−87.7	−132
	(13.10)	(14.50)	(18.07)	(6.80)	(12.55)
July temperature squared	−3.81	−2.95	−3.87	−0.30	−1.27
	(5.08)	(4.68)	(6.69)	(0.53)	(2.82)
October temperature	351.9	192	175	217	198
	(19.37)	(11.08)	(11.01)	(8.89)	(9.94)
October temperature squared	6.91	6.62	7.65	12.4	12.4
	(6.38)	(7.09)	(8.93)	(12.50)	(15.92)
January rain	75.1	85.0	56.5	280	172
	(3.28)	(3.88)	(2.81)	(9.59)	(7.31)
January rain squared	−5.66	2.73	2.20	−10.8	−4.09
	(1.86)	(0.95)	(0.82)	(3.64)	(1.72)
April rain	110	104	128	82.8	113
	(4.03)	(4.44)	(5.91)	(2.34)	(4.05)
April rain squared	−10.8	−16.5	−10.8	−62.1	−30.6
	(1.17)	(1.96)	(1.41)	(5.52)	(3.35)
July rain	−25.6	−34.5	−11.3	−116	−5.28
	(1.87)	(2.63)	(0.94)	(6.06)	(0.34)
July rain squared	19.5	52.0	37.8	57.0	34.8
	(3.42)	(9.43)	(7.54)	(8.20)	(6.08)
October rain	−2.30	−50.3	−91.6	−124	−135
	(0.09)	(2.25)	(4.45)	(3.80)	(5.15)
October rain squared	−39.9	2.28	0.25	171	106
	(2.65)	(0.17)	(0.02)	(14.17)	(11.25)
Income per capita		71.0	65.3	48.5	47.1
		(15.25)	(15.30)	(6.36)	(7.39)
Density		1.30	1.05	1.53	1.17
		(18.51)	(16.03)	(18.14)	(17.66)
Density squared		-1.72×10^{-4}	-9.33×10^{-5}	-2.04×10^{-4}	-9.38×10^{-5}
		(5.31)	(3.22)	(7.47)	(4.57)
Latitude		−90.5	−94.4	−105	−85.8
		(6.12)	(6.95)	(5.43)	(5.33)
Altitude		−0.167	−0.161	−0.163	−0.149
		(6.09)	(6.41)	(4.72)	(5.20)
Salinity		−684	−416	−582	−153
		(3.34)	(2.20)	(2.59)	(0.81)
Flood-prone		−163	−309	−663	−740
		(3.34)	(6.98)	(8.59)	(11.99)
Wetland		−58.2	−57.5	762	230
		(0.47)	(0.51)	(4.41)	(1.72)
Soil erosion		−1,258	−1,513	−2,690	−2,944
		(6.20)	(8.14)	(8.21)	(11.23)
Slope length		17.3	13.7	54.0	30.9
		(2.91)	(2.49)	(6.24)	(4.54)
Sand		−139	−35.9	−288	−213
		(2.72)	(0.77)	(4.16)	(3.95)
Clay		86.2	67.3	−7.90	−18.0
		(4.08)	(3.47)	(0.22)	(0.63)
Moisture capacity		0.377	0.510	0.206	0.450
		(9.69)	(14.21)	(3.82)	(10.07)
Permeability		−0.002	−0.005	−0.013	−0.017
		(1.06)	(2.53)	(5.58)	(8.61)
Adjusted R^2:	0.671	0.782	0.784	0.836	0.835
Number of observations:	2,938	2,938	2,941	2,941	2,941

Notes: The dependent variable is the value of land and buildings per acre. All regressions are weighted. Values in parenthesis are *t* statistics.

per acre. However, a 1°F increase in summer temperature decreases farm values by only \$88–\$132 according to the crop-revenue model but by between \$155 and \$177 in the cropland model. Except for spring rains, the crop-revenue model suggests that rain has a much larger effect on land value than the cropland model. For example, the crop-revenue model suggests that winter rain increases farm values between \$172 and \$280 per monthly inch, whereas the cropland model suggests an effect between \$57 and \$85 per monthly inch.

The predicted overall effects from the existing climate across the United States are shown in Figures 2 and 3. These maps show the *Ricardian values of climate* by county in 1982, that is, the partial effect of climate on property values. To construct each map, we begin with the difference between the estimated climate for each county and the national average climate. We then multiply these differences by the estimated coefficients in Table 3 and sum them across the climate variables. Figures 2 and 3 show the estimated contribution of climate to the farmland value in each county. The results match folk wisdom about farm values (for example, the infamous 100th meridian of American history can be seen sharply in Figure 2). The most valuable climates are along the west coast, the corn belt near Chicago, and the northeast. The least valuable areas are the southwest and southeast regions. Both figures show almost identical geographic patterns, indicating that the results are stable; similar results were also found using 1978 data.

The control variables in Table 3 provide a rich set of results in and of themselves. Economic and soil variables play a role in determining the value of farms. Farm values are higher in denser, growing, and wealthier counties because of higher local demand for food and the potential for conversion of land to nonfarm uses. Farm values respond as expected to other environmental factors such as solar flux (latitude) and altitude. Salinity, likelihood of flooding, presence of wetlands, and soil erosion all act negatively as expected. Slope length was slightly beneficial to land values. Irrigation is left out of the regressions shown in Table 3 because irrigation is clearly an endogenous reaction to climate. However, when included, irrigation is a strongly positive variable, increasing land values substantially; which is not surprising, given the crucial importance of irrigation in many areas of the arid West.[6]

One hypothesis suggested in the theory section is that the impacts of environmental effects would be exaggerated by a gross-revenue model. We explore this hypothesis in Table 4 by regressing the same climate and contol variables on the gross revenue earned from crops. The marginal effects in Table 4 for the farm-revenue model suggest similar seasonal patterns as the farm-value equation with the exception of spring. Warmer Aprils reduce farm revenues, whereas they increase farm values. Wetter springs, good for farm values, reduce farm revenues according to the cropland model but increase farm revenues according to the crop-revenue model.

The magnitude of damages predicted by the gross-revenue model, however, are generally larger than the effects predicted by the Ricardian model. To compare the two approaches, we need to translate the annual rents into land value using the discount rate defined in Section II. Based on asset returns and farm earnings, a real discount rate of 5 percent per annum appears most suitable.[7] At this discount rate, the marginal coefficients in Table 4 should be multiplied

[6] Including irrigation does not significantly change the results of the paper.

[7] According to Roger Ibbotson and Gary Brinson (1987), farmland prices over the period 1947–1984 had a compound annual return (income and capital gains) of 9.6 percent while the GNP deflator rose at an average of 4.4 percent annually. This produces an average real yield of 4.99 percent per annum. By comparison, all real-estate investments had an average real yield of 4.4 percent per annum over this period. Another comparison is the rate of profit on farms, defined as the net income of farms divided by total value of farms and farmland. For the three census years of 1974, 1978, and 1982, the average rate of profit on farms was 5.02 percent per annum.

TABLE 4—REGRESSION MODELS EXPLAINING FARM REVENUES

	Cropland weights		Crop-revenue weights	
Independent variables	1982 (i)	1978 (ii)	1982 (iii)	1978 (iv)
Constant	180 (31.37)	143 (28.09)	213 (16.61)	186 (16.27)
January temperature	−11.6 (5.00)	−6.65 (3.21)	16.1 (3.19)	16.4 (3.55)
January temperature squared	−0.048 (0.88)	0.006 (0.13)	0.867 (7.80)	0.659 (6.71)
April temperature	−23.5 (7.89)	−20.3 (7.63)	−47.7 (8.62)	−39.3 (7.83)
April temperature squared	−1.31 (7.67)	−1.12 (7.43)	−2.74 (9.43)	−2.26 (8.55)
July temperature	−27.2 (9.85)	−21.5 (8.66)	−10.0 (1.90)	−7.20 (1.49)
July temperature squared	0.053 (0.32)	−0.166 (1.14)	1.27 (5.52)	0.341 (1.65)
October temperature	51.3 (11.43)	41.4 (10.43)	−2.12 (0.21)	2.92 (0.32)
October temperature squared	0.637 (2.62)	0.598 (2.85)	−0.025 (0.06)	0.569 (1.58)
January rain	30.1 (5.29)	21.4 (4.26)	−28.9 (2.42)	−11.5 (1.06)
January rain squared	−4.10 (5.49)	−2.93 (4.49)	−4.08 (3.36)	−3.33 (3.04)
April rain	−22.5 (3.67)	−23.2 (4.29)	47.5 (3.28)	16.0 (1.24)
April rain squared	−2.46 (1.12)	4.65 (2.39)	−5.73 (1.24)	2.65 (0.63)
July rain	−3.29 (0.97)	2.12 (0.70)	−64.5 (8.25)	−33.3 (4.61)
July rain squared	10.8 (6.93)	6.74 (5.23)	22.8 (8.03)	13.2 (5.02)
October rain	−40.2 (6.93)	−16.1 (3.17)	−44.4 (3.32)	−16.3 (1.35)
October rain squared	27.2 (7.73)	17.4 (5.62)	33.8 (6.84)	9.32 (2.15)
Income per capita	0.568 (0.47)	0.803 (0.73)	3.37 (1.08)	8.24 (2.81)
Density	0.172 (9.46)	0.133 (8.47)	0.457 (13.28)	0.280 (9.14)
Density squared	2.86×10^{-6} (0.34)	2.92×10^{-6} (0.43)	-4.47×10^{-5} (3.99)	-1.92×10^{-5} (2.03)
Latitude	−24.3 (6.28)	−15.4 (4.44)	−72.6 (9.15)	−41.6 (5.59)
Altitude	−0.049 (6.91)	−0.033 (5.03)	−0.096 (6.78)	−0.059 (4.47)
Salinity	−156 (2.97)	−149 (3.23)	−502 (5.44)	−427 (4.90)
Flood-prone	29.8 (2.36)	25.4 (2.27)	−40.7 (1.29)	−1.45 (0.05)
Wetland	70.9 (2.21)	64.8 (2.32)	234 (3.31)	115 (1.86)
Soil erosion	−169 (3.18)	−74.5 (1.60)	−413 (3.08)	−360 (2.98)
Slope length	−1.18 (0.73)	−1.21 (0.85)	−15.3 (4.33)	−13.5 (4.31)
Sand	28.7 (2.18)	32.3 (2.84)	70.3 (2.49)	46.7 (1.88)
Clay	11.1 (1.99)	12.3 (2.49)	−48.1 (3.32)	−31.8 (2.43)
Moisture capacity	0.062 (6.10)	0.050 (5.49)	0.101 (4.57)	0.058 (2.79)
Permeability	0.001 (2.22)	0.001 (2.15)	−0.001 (6.94)	−0.005 (5.30)
Adjusted R^2:	0.525	0.509	0.800	0.762
Number of observations:	2,834	2,443	2,834	2,443

Notes: The dependent variable is the gross value of crop revenue per acre per year. All regressions are weighted. Values in parenthesis are t statistics.

TABLE 5—PREDICTED IMPACT OF GLOBAL WARMING ON FARMLAND VALUES AND FARM RENTS

		Change in farmland values (billions of dollars, 1982 prices)		Change in farmland rents (percentage of 1982 farm marketings)	
Year	Weight	Impact	Truncated impact	Impact	Truncated impact
1982	Cropland	−$125.2	−$118.8	−4.4	−4.2
1978	Cropland	−$162.8	−$141.4	−5.7	−4.9
1982	Crop revenue	$34.5	$34.8	1.2	1.2
1978	Crop revenue	−$14.0	$21.0	−0.5	0.7

Notes: The global-warming scenario is a uniform 5°F increase with a uniform 8-percent precipitation increase. The "impact" column shows the estimated loss; the "truncated impact" columns show the impact when the loss in farmland value in each county is limited to the original value of the land. The last two columns are annualized impacts, as explained in the text, as a percentage of 1982 farm marketings.

by 20 to make them comparable with the present-value estimates in Table 3. Making this adjustment, a 1°F increase in summer temperature decreases the present value of farms by between $140 and $540 according to the gross-revenue model but only between $88 and $177 according to the Ricardian models.

One concern with the Ricardian approach to climate effects is that the results may not be robust over time and that the weather and economic factors in a given year may have distorted the results. We consequently estimated the model again using data from 1978. These values have been converted to 1982 dollars using the GNP deflator obtained from the *1991 Economic Report of the President.* The 1978 results are surprisingly similar to the findings using the 1982 data. The control variables have similar impacts in both years. Evaluating the marginal effects of climate in 1978 at the national mean and comparing the results with 1982 shows that the climatic variables are also similar in 1978 and 1982 with few exceptions. The pattern of climate effects on agriculture is stable over time, but apparently some factors can alter the magnitude of the effects from year to year.

III. Implications for Greenhouse Warming

The Ricardian analysis in the previous section shows that climate has complicated effects on agriculture, highly nonlinear and varying by season. An important application of this analysis is to project the impact of global warming on American agriculture. For this projection, we take a conventional CO_2 doubling scenario, which is associated with a 5°F increase in global mean surface temperature (see Intergovernmental Panel on Climate Change, 1990; National Academy of Sciences Panel on Greenhouse Warming, 1992). According to most projections, such an increase will occur sometime in the second half of the next century if current trends continue. According to the survey by the Intergovernmental Panel on Climate Change, a 5°F temperature increase will be accompanied by an 8-percent average increase in precipitation. These changes are applied uniformly by season and region to the United States in the calculations that follow. In principle, they show the impact of climate change including all adaptations, although they omit the impact of CO_2 fertilization and price effects.

Table 5 shows the results of this experiment for the two years and sets of weights. The "impact" columns show the estimated impact of global warming on farmland values; the "truncated impact" columns truncate these losses if they drive land values below zero. This truncated impact is the preferred economic measure. The estimates diverge dramatically depending upon whether cropland or crop revenues are used for weighting. Under the cropland weights, the loss in land value from warming ranges from \$119 billion to \$141 billion; assuming that the annual crop loss is 5 percent of this value,[8] the annual loss ranges from \$6 billion to \$8 billion (in 1982 prices at 1978 or 1982 levels of output). Relating this value to gross farm income in 1982 of \$164 billion, the annual damage is in the neighborhood of 4–5 percent. The cropland model emphasizes the unattractiveness of a warmer climate for an agriculture that emphasizes grains, which have relatively low value per acre and thrive in the relatively cool climate of the northern United States.

Strikingly different results emerge if we use the crop-revenue approach. For these, the net impact of warming (again without CO_2 fertilization) is slightly positive, suggesting an increase of \$20–\$35 billion in farmland values. Annualizing these capital values, this suggests a gain of between \$1 billion and \$2 billion per year. As a fraction of 1982 revenue, this amounts to about a 1-percent gain. The differing results arise because the crop-revenue approach weights relatively more heavily the irrigated lands of the West and South that thrive in a Mediterranean and subtropical climate, a climate that will become relatively more abundant with a warming. Including this broader set of crops and adaptations paints a more optimistic picture because the gains from the sunbelt crops tend to offset the losses in the marginal grain regions.

The striking difference between the crop-revenue and cropland approaches is a useful reminder of how we can be misled by our mental images. The specter of global warming calls up the vista of corn blistering on the stalk or desiccated wheat fields. Yet the major grains so vulnerable to drought—wheat and corn—represented only \$22.5 billion of the \$143 billion of farm marketings in 1982. Our results suggest that the vulnerability of American agriculture to climate change may be exaggerated if the analysis is limited to the major grains. A broader vision should also include the warm-weather crops such as cotton, fruits, vegetables, rice, hay, and grapes in addition to other sectors such as livestock and poultry. Whereas past production-function studies focus ominously on the vulnerable cool-weather grains, the comprehensive crop-revenue Ricardian model reminds us that the irrigated warm-weather crops may be a silver lining behind the climate-change cloud.

Figures 4 and 5 provide geographic detail for these global-warming scenarios. According to the cropland model shown in Figure 4, warming will be particularly harmful for the entire southern part of the United States and will only be beneficial to the northern fringe of the country. The crop-revenue model of Figure 5 suggests, by contrast, that global warming will be beneficial to California and the citrus belt of the Southeast as well as the corn and wheat belts of the Midwest. Global warming will be harmful, in this model, only to the relatively unimportant mountainous regions of Appalachia and the Rocky Mountains.

It will be useful to compare these estimates with results from other studies. In its analysis, Smith and Tirpak (1989) surveyed a number of different climate and agriculture models to estimate the impact of CO_2 doubling. Omitting CO_2 fertilization, the EPA concluded that the impact would lie in the range of \$6 billion to \$34 billion per year (in 1982 prices). Cline (1992) used two different approaches, the EPA estimate and a modification of Rind et al. (1990), both of which project losses of \$20 billion per year without CO_2 fertilization. It is instructive to note that these studies all rely on the production-function approach and apply it to grains; these estimates therefore are closest to our cropland model, and as was predicted in the theoretical section above, they show a higher estimate of damage for that universe than the Ricardian approach—approximately triple the estimates in Table 5. By excluding the nongrain, warm-weather crops, these studies further bias upward the estimates of damage, as is shown by the

[8]See the discussion of this issue in the last section and in footnote 6.

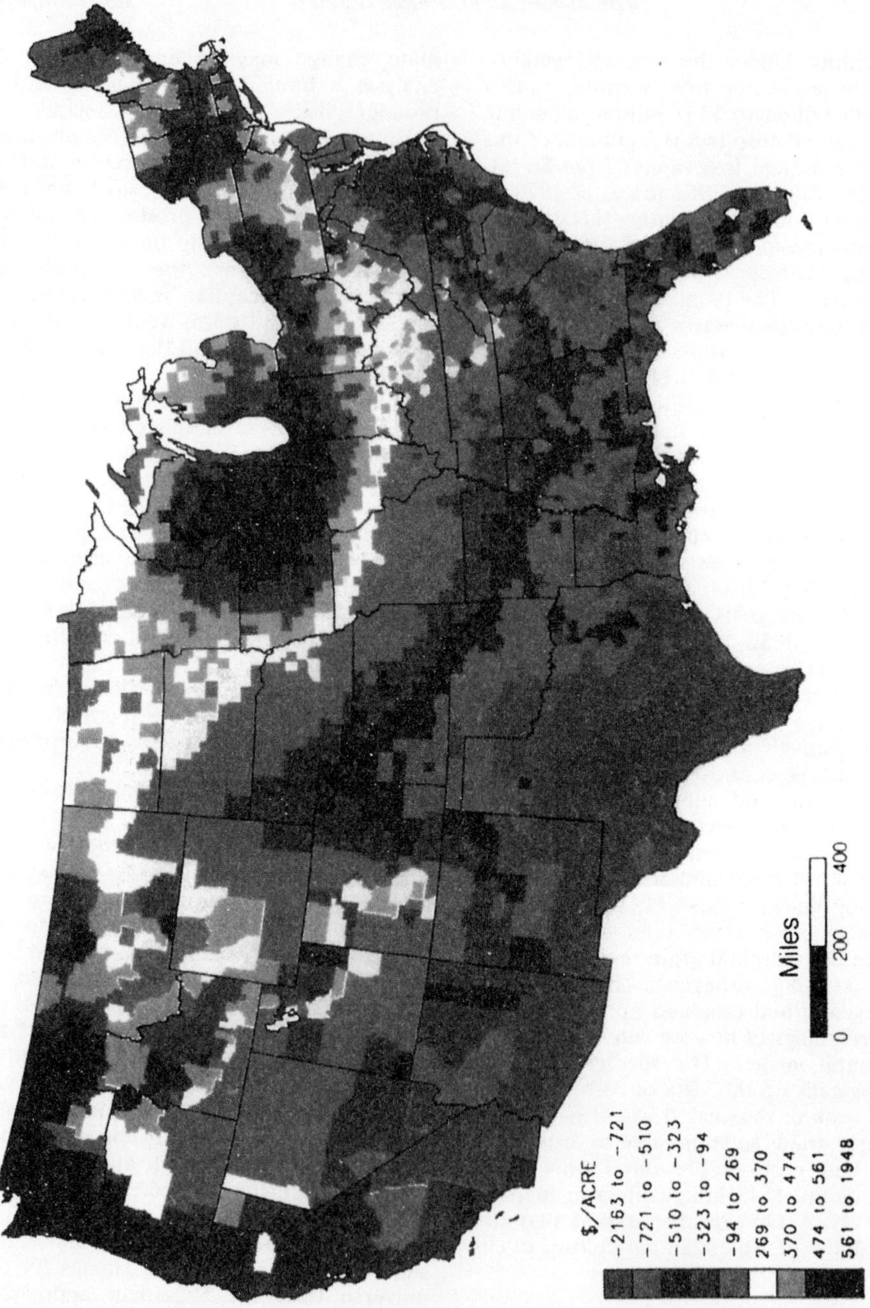

FIGURE 2. INFLUENCE OF CURRENT CLIMATE ON FARM VALUES: CROPLAND WEIGHTS

Note: Farm value is measured as the difference in dollars per acre from the sample average, 1982 prices.

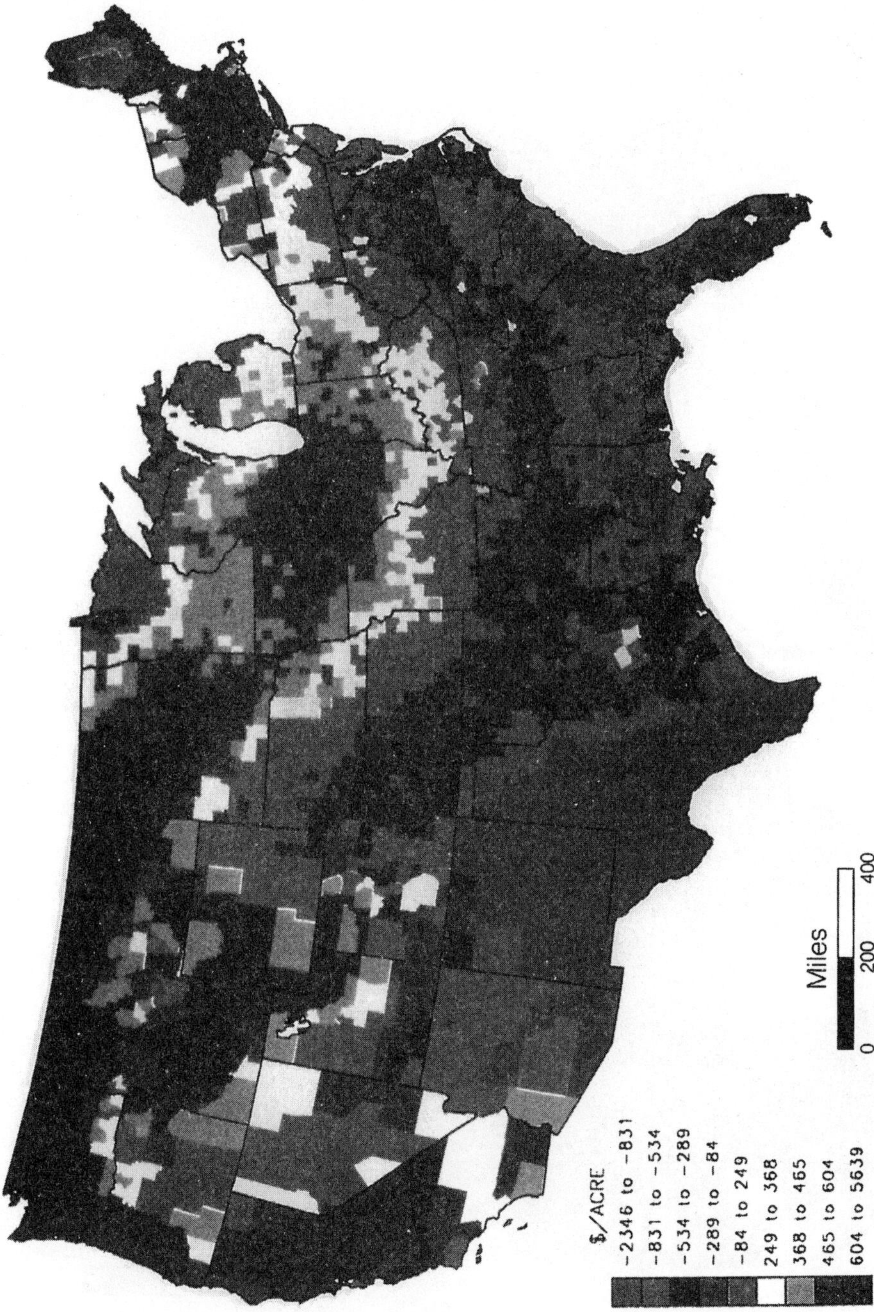

FIGURE 3. INFLUENCE OF CURRENT CLIMATE ON FARM VALUES: CROP-REVENUE WEIGHTS
Note: Farm value is measured as the difference in dollars per acre from the sample average, 1982 prices.

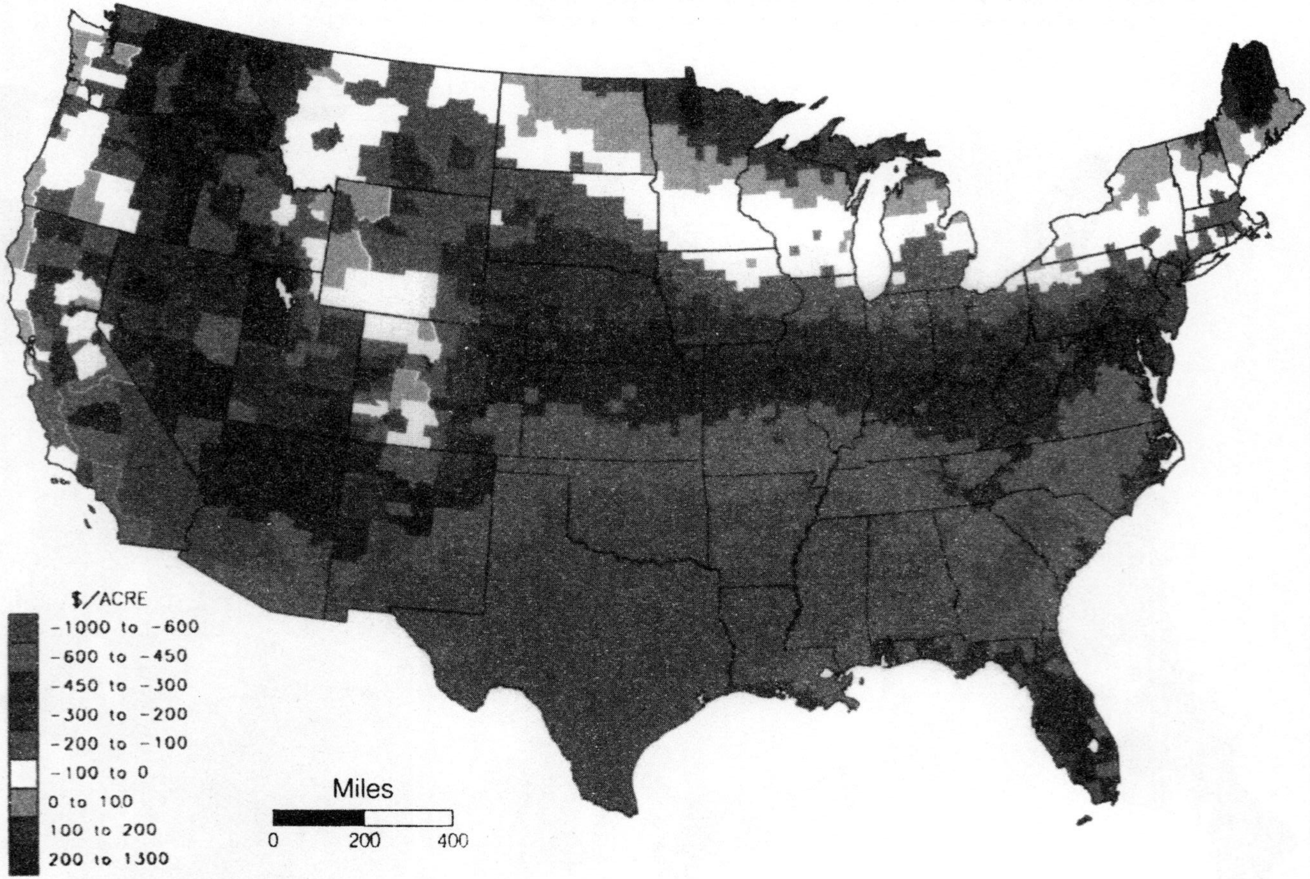

FIGURE 4. CHANGE IN FARM VALUE FROM GLOBAL WARMING: CROPLAND WEIGHTS

Note: The map shows the change in terms of dollars per acre for a 5° F uniform warming and an 8-percent increase in precipitation. 1982 prices.

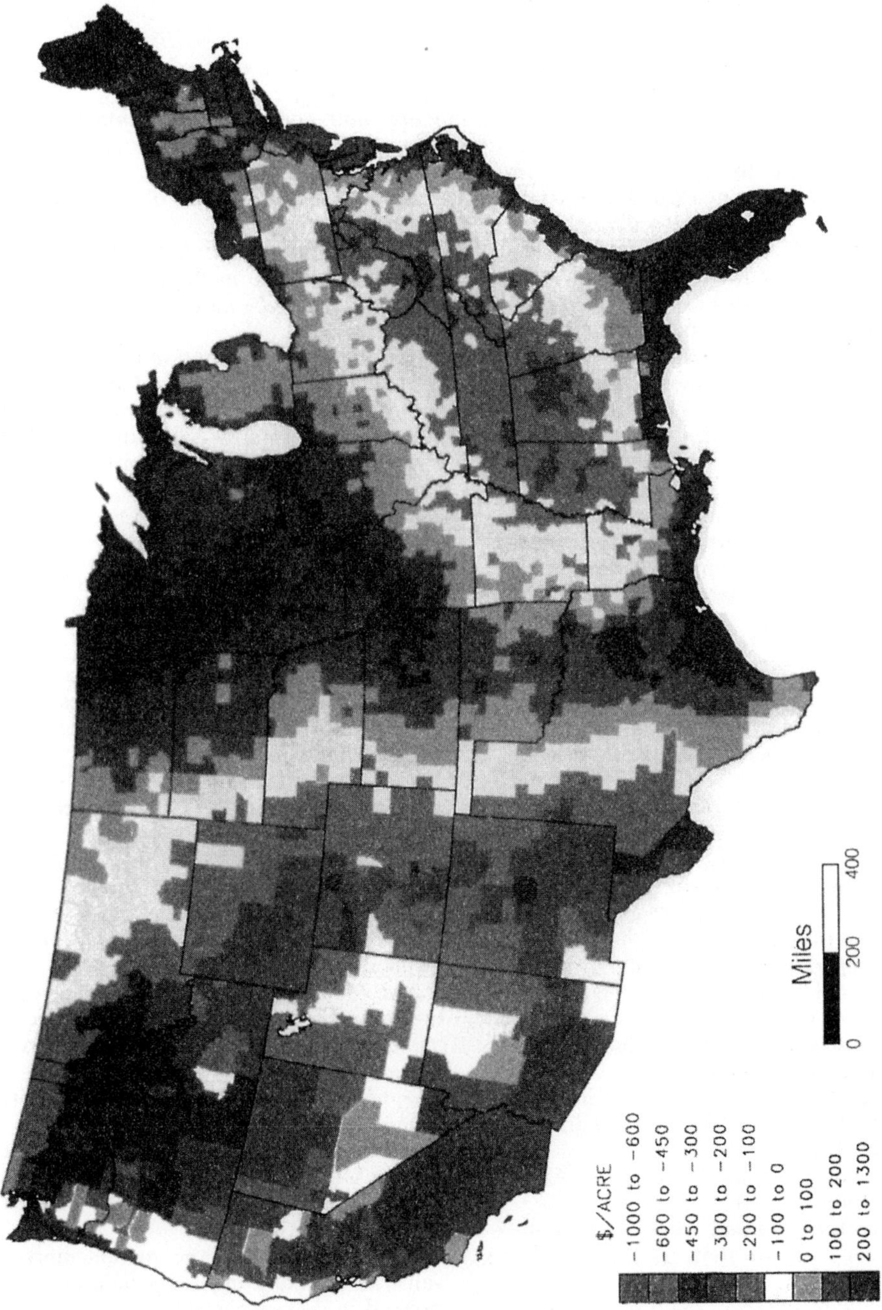

FIGURE 5. CHANGE IN FARM VALUE FROM GLOBAL WARMING: CROP-REVENUE WEIGHTS

Notes: The map shows the change in terms of dollars per acre for a 5°F uniform warming and an 8-percent increase in precipitation, 1982 prices.

comparison between the cropland and the crop-revenue models.

The results in Table 5 are based on a highly stylized global-warming scenario and are therefore quite tentative. In research underway, we are drawing estimated global-warming results from large-scale general circulation models; these should allow differentiation among broad regions of the United States. In addition, the effects of CO_2 fertilization should be included, for some studies indicate that this may produce a significant increase in yields. Other omitted variables are the effect of extremes and ranges in climatic variables as well as the effect of changes in irrigation. Notwithstanding these omissions, the present paper does provide a benchmark for projecting the impact of global warming on American agriculture. Using the narrow definition of crops, the negative impact is estimated to lie between 4 percent and 6 percent of the value of farm ouput. Using a more inclusive definition that weights warm-weather crops and irrigated agriculture more heavily, our projections suggest that global warming may be slightly beneficial to American agriculture.

Appendix A: Definitions of Major Variables Used in this Study

Constant: a term equal to 1

January temperature: normal daily mean temperature (°F) from 1951 to 1980 in the month of January

January temperature squared: value of January temperature squared

April temperature: normal daily mean temperature (°F) from 1951 to 1980 in the month of April

April temperature squared: value of April temperature squared

July temperature: normal daily mean temperature (°F) from 1951 to 1980 in the month of July

July temperature squared: value of July temperature squared

October temperature: normal daily mean temperature (°F) from 1951 to 1980 in the month of October

October temperature squared: value of October temperature squared

January rain: normal precipitation (inches) from 1951 to 1980 in the month of January

January rain squared: value of January rain squared

April rain: normal precipitation (inches) from 1951 to 1980 in the month of April

April rain squared: value of April rain squared

July rain: normal precipitation (inches) from 1951 to 1980 in the month of July

July rain squared: value of July rain squared

October rain: normal precipitation (inches) from 1951 to 1980 in the month of October

October rain squared: value of October rain squared

Income per capita: annual personal income per person in the county, 1984

Density: resident population per square mile, 1980

Density squared: value of density squared

Latitude: latitude measured in degrees from southernmost point in United States

Altitude: height from sea level (feet)

Migration: net of incoming people minus outgoing people from 1980 to 1986 for the county

Salinity: percentage of land that needs special treatment because of salt/alkaline minerals in the soils

Flood prone: percentage of land that is prone to flooding

Irrigated: percentage of land where irrigation provides at least 50% of water needs

Wetland: percentage of land considered wetland

Soil erosion: K-factor soil (erodibility factor) in hundredths of inches

Slope length: length of slope (feet) (not steepness)

Wind erosion: measure of wind erosion (hundredths of inches)

Farm value: estimate of the current market value of farmland including buildings expressed in dollars per acre, 1982

Farm revenue: gross revenue from crops sold in 1982 in dollars per acre

Sdist: linear distance from the nearest shoreline

Long: longitude measured in degrees from the easternmost point of the United States
Permeability: soil permeability (inches per hour)
Moisture capacity: available water capacity (inches/pound)

Appendix B: Data on Farms and Value of Land and Buildings[9]

The data on farms and on farmland values are central to this study. This appendix describes the definition and sources of the data. The current definition of a farm, first used for the *1974 Census of Agriculture* final reports, is any place from which $1,000 or more of agricultural products were sold or normally would have been sold during the census year. Land in farms is an operating-unit concept and includes land owned and operated as well as land rented from others. The acreage designated as "land in farms" consists primarily of agricultural land used for crops, pasture, or grazing. It also includes woodland and wasteland not actually under cultivation or used for pasture or grazing, provided it was part of the farm operator's total operation.

The land is defined to lie in the operator's principal county, that is, the county where the largest value of agricultural products was raised or produced. Irrigated land includes land watered by any artificial or controlled means, such as sprinklers, furrows or ditches, and spreader dikes. Cropland includes land from which crops were harvested or hay was cut, land in orchards, citrus groves, vineyards, nurseries, and greenhouses, land used only for pasture or grazing that could have been used for crops without additional improvement, and all land planted in crops that were grazed before the crops reached maturity. Also included were all cropland used for rotation pasture and lands in government diversion programs that were pastured.

[9]This description is drawn from the *City and County Data Book*, and the underlying data are from *1982 Census of Agriculture* (U.S. Bureau of the Census, 1984).

Respondents were asked to report their estimate of the current market value of land and buildings owned, rented, or leased from others, and of land rented or leased to others. Market value refers to the respondent's estimate of what the land and buildings would sell for under current market conditions. If the value of land and buildings was not reported, it was estimated during processing by using the average value of land and buildings from a similar farm in the same geographic area.

The value of products sold by farms represents the gross market value before taxes and production expenses of all agricultural products sold or removed from the place regardless of who received the payment. In addition, it includes the loan value received in 1982 for placing commodities in the Commodity Credit Corporation loan program.

REFERENCES

Adams, Richard. "Global Climate Change and Agriculture: An Economic Perspective." *American Journal of Agricultural Economics*, December 1989, *71*(5), pp. 1272–79.

Adams, Richard; McCarl, Bruce; Dudek, Daniel and Glyer, David. "Implications of Global Climate Change for Western Agriculture." *Western Journal of Agricultural Economics*, December 1988, *13*(2), pp. 348–56.

Adams, Richard; Rosenzweig, Cynthia; Pearl, Robert; Ritchie, Joe; McCarl, Bruce; Glyer, David; Curry, Bruce; Jones, James; Boote, Kenneth and Allen, Hartwell. "Global Climate Change and U.S. Agriculture." *Nature*, 17 May 1990, *345*(6272), pp. 219–24.

Callaway, John; Cronin, F.; Currie, J. and Tawil, J. "An Analysis of Methods and Models for Assessing the Direct and Indirect Economic Impacts of CO_2-Induced Environmental Changes in the Agricultural Sector of the U.S. Economy." Pacific Northwest Laboratory (Richland, WA) Working Paper No. PNL-4384, 1982.

Cline, William. *The economics of global warming*. Washington, DC: Institute of International Economics, 1992.

Decker, W.; Jones, V. and Achutuni, R. *The impact of climate change from increased atmospheric carbon dioxide on American agriculture*, DOE/NBB-0077. Washington, DC: U.S. Department of Energy, 1986.

Easterling, William, III; McKenney, Mary; Rosenberg, Norman and Lemon, Kathleen. *Processes for identifying regional influences of and responses to increasing atmospheric CO_2 and climate change—The MINK project—Report IIB*, DOE/RL/01830T-H8. Washington, DC: U.S. Department of Energy, 1991.

Ibbotson, Roger and Brinson, Gary. *Investment markets*. New York: McGraw-Hill, 1987.

Intergovernmental Panel on Climate Change. *Climate change: The IPCC scientific assessment* (J. T. Houghton, G. J. Jenkins, and J. J. Ephraums, eds.). New York: Cambridge University Press, 1990.

Mendelsohn, Robert; Nordhaus, William and Shaw, Dai Gee. "The Impact of Climate on Agriculture: A Ricardian Approach," in Yoichi Kaya, Nebojsa Nakicenovic, William Nordhaus, and Ferenc Toth, eds., *Costs, impacts, and benefits of CO_2 mitigation*. Laxemburg, Austria: International Institute of Applied Systems Analysis, 1993, pp. 173–208.

National Academy of Sciences Panel on Greenhouse Warming. *Policy implications of greenhouse warming: Mitigation, adaptation, and the science base*. Washington, DC: National Academy Press, 1992.

National Research Council. *Changing climate*. Washington, DC: National Academy Press, 1983.

Rind, D.; Goldberg, R.; Hansen, J.; Rosenzweig, Cynthia and Ruedy, R. "Potential Evapotranspiration and the Likelihood of Future Drought." *Journal of Geophysical Research*, June 1990, *95*(D7), pp. 9983–10004.

Rozenzweig, Cynthia and Parry, Martin L. "Potential Impact of Climate Change on World Food Supply." *Nature*, 13 January 1994, *367*(6459), pp. 133–38.

Smith, Joel and Tirpak, Dennis, eds. *The potential effects of global climate change on the United States: Report to Congress*, EPA-230-05-89-050. Washington, DC: Environmental Protection Agency, 1989.

U.S. Bureau of the Census. *1982 Census of Agriculture*. Washington, DC: U.S. Government Printing Office, 1984.

______. *County and city data book*. Washington, DC: U.S. Government Printing Office, 1988.

[6]

Potential impact of climate change on world food supply

Cynthia Rosenzweig* **& Martin L. Parry**†

* Columbia University and Goddard Institute for Space Studies, 2880 Broadway, New York, New York 10025, USA
† Environmental Change Unit, Oxford University, 1a Mansfield Road, Oxford OX1 3TB, UK

A global assessment of the potential impact of climate change on world food supply suggests that doubling of the atmospheric carbon dioxide concentration will lead to only a small decrease in global crop production. But developing countries are likely to bear the brunt of the problem, and simulations of the effect of adaptive measures by farmers imply that these will do little to reduce the disparity between developed and developing countries.

RECENT research assessing the potential effects of climate change on agriculture has focused on regional and national evaluations[1-3]. These efforts have, for the most part, treated each region or nation in isolation, without relation to changes in agricultural production elsewhere. Recent work[4,5] emphasizes the important role of international trade in the adjustment of the world food system to climate change-induced changes in crop yields. Crop growth models have been used in an extensive international collaboration[6] to determine the effect of various climate change scenarios on crop yields for individual countries and geographical regions (Fig. 1). In this article we combine the data from these individual studies to obtain a global picture of the simulated change in crop yield associated with the different climate change scenarios. We then use a world food trade model to simulate the economic consequences of these potential changes in crop yields; we estimate changes in world food prices, and in the number of people at risk of hunger (defined as a measure of food energy availability—which depends on income and food price levels—relative to nutritional requirements) in developing countries.

The major finding of the study is that there appears to be a large disparity in agricultural vulnerability to climate change between developed and developing countries. This occurs even though simulated global agricultural production of major grain crop declines are only small to moderate under the climate change conditions tested. The analysis included the combined effects of climate change and increasing CO_2 on crop yields and water use. Although projected temperature change in low latitudes (where many developing countries are located) tends to be lower than the global average in the general circulation model (GCM) scenarios tested, modelled yield changes are primarily negative there, in contrast to predominantly positive yield changes in middle and high latitudes where many developed countries are located. This result has significant implications for potential future distributional aspects of the world food system.

Studies such as this explore the sensitivity of important human systems (in this case world food supply), as currently understood, to projected levels of global climate change. Such studies are initial demonstrations of the comprehensive, interdisciplinary research needed to improve understanding of the interactive biophysical and socio-economic effects that may result from global environmental change.

Climate change scenarios

Scenarios were developed from climate conditions predicted by three GCMs for doubled atmospheric CO_2 levels (Table 1). The temperature changes of these GCM scenarios (4.0–5.2 °C) are near the upper end of the range (1.5–4.5 °C) projected for doubled CO_2 warming by the IPCC[7,8]. Mean monthly changes in temperature, precipitation and solar radiation from the appropriate GCM gridbox were applied to observed daily climate records (1951–80, or as many years of daily climate records as available) to create climate change scenarios for each study site.

Because atmospheric concentrations of other greenhouse gases besides CO_2 (for example, methane (CH_4), nitrous oxide (N_2O) and the chlorofluorocarbons (CFCs)) are also increasing, an 'effective CO_2 doubling' has been defined as the combined radiative forcing of all greenhouse gases having the same forcing as doubled CO_2, usually taken to be ~600 p.p.m. CO_2 level is important when estimating potential impact on crops because crop growth and water use have been shown to benefit from increased levels of CO_2 (refs 9, 10). For the crop model simulations, climate changes from the doubled CO_2 GCM simulations were used with an associated level of 555 p.p.m. CO_2; these conditions are assumed to occur in AD2060. The 555 p.p.m. level is based on the GISS GCM trace gas scenario A (ref. 11), in which the simulated climate had warmed to the effective doubled CO_2 level of ~4 °C by AD2060.

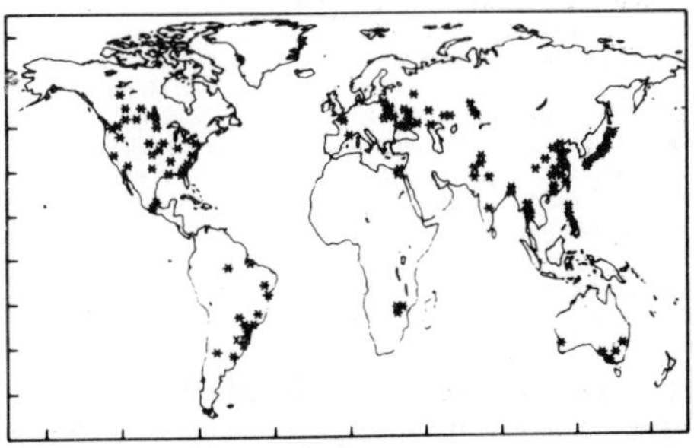

FIG. 1 Crop model sites[6]. Countries and participants were as follows: Argentina, O. E. Sala and J. M. Paruelo; Australia, B. D. Baer, W. S. Meyer and D. Erskine; Bangladesh, Z. Karim, M. Ahmed, S. G. Hussain and Kh. B. Rashid; Brazil, O. J. F. de Siqueira, J. R. B. Farias and L. M. A. Sans. Canada, M. Brklacich, R. Stewart, V. Kirkwood and R. Muma; China, Z. Jin, D. Ge, H. Chen, J. Fang and X. Zheng; Egypt, H. M. Eid; France, R. Delécolle, D. Ripoche, F. Ruget and G. Gosse; India, D. G. Rao; Japan, H. Seino; Mexico, D. Liverman, M. Dilley, K. O'Brien and L. Menchaca; Pakistan, A. Qureshi and A. Iglesias; Philippines, C. R. Escaño and L. Buendia; Thailand, M. L. C. Tongyai; Russia, G. Menzhulin, L. Koval and A. Badenko; USA, C. Rosenzweig, B. Curry, T.-Y. Chou, J. Ritchie, J. Jones and R. Peart; Uruguay, W. E. Baethgen; Zimbabwe, P. Muchena.

ARTICLES

TABLE 1 GCM doubled-CO_2 climate change scenarios

GCM	Year*	Resolution (lat. × long.)	CO_2 (p.p.m.)	Change in average global temperature (°C)	Change in average global precipitation (%)
GISS†	1982	7.83 × 10	630	4.2	11
GFDL‡	1988	4.4 × 7.5	600	4.0	8
UKMO§	1986	5.0 × 7.5	640	5.2	15

* When calculated.
† Goddard Institute for Space Studies[20].
‡ Geophysical Fluid Dynamics Laboratory[21].
§ United Kingdom Meteorological Office[22].

Crop yield change methods

Agricultural scientists in 18 countries estimated potential changes in national grain crop yields using compatible crop models and the GCM scenarios at 112 sites (Fig. 1). The crop model linkages were developed by the US Agency for International Development[12]. Simulations were carried out in regions representing 70 75% of the current world production of wheat, maize and soybean; rice production was less well represented (48% of current world production).

Site-specific estimates of yield changes were aggregated to national levels based on current regional production. The regional yield estimates represent the current mix of rainfed and irrigated production, and today's crop varieties, nitrogen management and agricultural soils. The national crop yield changes were then extrapolated to provide estimates of yield changes for the other countries and crops included in the world food trade model. National yield changes of other crops and commodity groups and regions not simulated were estimated based on three criteria: (1) similarities to modelled crops and growing conditions; (2) results from ~50 previously published and unpublished regional climate change impact studies; and (3) projected temperature and precipitation changes from the GCM scenarios. The primary sources of uncertainty in the estimates lies in the sparseness of the crop modelling sites and the lack of explicitly modelled yield changes for subsistence crops such as millet and cassava, which may respond differently to both climate change and increases in CO_2.

Estimates were made of yield changes with and without the direct physiological effects of CO_2 on crop growth, that is increased rates of net photosynthesis and reduced stomatal openings as reported from experimental results[13]. The photosynthesis ratios (555 330 p.p.m. CO_2) for soybean, wheat and rice, and maize were 1.21, 1.17, and 1.06, respectively. Changes in stomatal resistance were based on experimental results[14] (49.7 34.4 s m^{-1} for C3 crops, 87.4 55.8 s m^{-1} for C4 crops). This method of simulating the physiological CO_2 effects on crops may provide a positive bias to yield estimates, as plants grown in experimental settings are often subject to fewer environmental and competitive stresses than are likely to be encountered in farmers' fields.

GCM scenarios and direct CO_2 effects

Figure 2 shows estimated potential changes in average national grain crop yields for the GISS GFDL and UKMO doubled-CO_2 climate change scenarios (Table 1) with and without physiological (direct) CO_2 effects on plant growth. The maps are created from nationally averaged yield changes for wheat, rice, coarse grains and protein feed estimated for each country or group of countries. When climate change is considered without direct CO_2 effects on crop growth and water use, averaged national crop yields declined everywhere, although reductions were less at middle and high latitudes. In the simulations with direct CO_2 effects, yields were positive at middle and high latitudes, and negative at low latitudes for the GISS and GFDL scenarios which produced yield changes ranging from +30 to −30%. The UKMO scenario caused average national crop yields to decline almost everywhere (up to −50% in Pakistan).

Several factors contributed to the latitudinal differences in simulated yields. At some sites near the high-latitude boundaries of current agricultural production, increased temperatures benefited crops otherwise limited by cold temperatures and short growing seasons. In many middle and high latitude areas, where current temperature regimes tend to be cooler, increased temperatures exerted a negative influence on yields through shortening of crop development stages, but did not significantly increase heat or water stress levels. In these regions beneficial CO_2 effects dominated. The climate-change-induced warming at low latitudes brought not only accelerated growing periods for crops, but also greater heat and water stress, resulting in greater yield decreases than at higher latitudes, despite beneficial CO_2 direct effects.

Farm-level adaptations

In each participating country, the agricultural scientists used the crop models to test possible responses to the worst climate-change scenario (this was usually, but not always, the UKMO scenario; adaptation simulations were done with all three GCMs at some sites). These adaptations included changes in planting date, variety and crop, and applications of irrigation and fertilizer. Irrigation simulations assumed automatic irrigation to field capacity when plant available water dropped to 50% and 100% irrigation efficiency. These optimistic assumptions imply that water supply for irrigation would be fully available at all locations under climate change conditions. All adaptation possibilit-

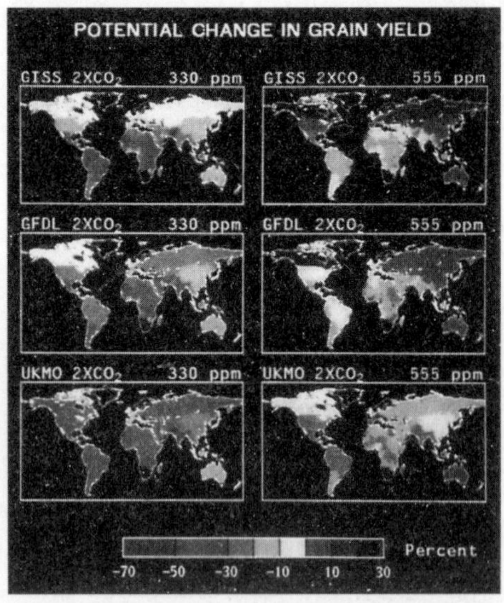

FIG. 2 Estimated change in average national grain yield (wheat, rice, coarse grains, and protein feed) for the GISS, GFDL, and UKMO climate change scenarios. The left-hand column shows the direct physiological effect on grain yield of current (330 p.p.m.) CO_2 concentration. The right-hand column shows the direct physiological effect of 555 p.p.m. Results shown are averages for countries and groups of countries in the Basic Linked System (BLS) world food trade model; regional variations within countries are not reflected.

ies were not simulated at every site and country: choices of adaptations to be tested were made by the participating scientists, based on their knowledge of current agricultural systems.

For the economic analysis, crop model adaptation results were grouped into two levels of adaptation. Level 1 implies little change to existing agricultural systems, reflecting responses to a changing climate that should be easily available to individual farmers. Level 1 adaptations included shifts in planting date (±1 month) that do not imply major changes in crop calendar, additional application of irrigation water to crops already under irrigation, and changes in crop variety to currently available varieties more adapted to the altered climate.

Adaptation level 2 implies more substantial change to agricultural systems, possibly requiring resources beyond the farmers' means, investment in regional and national agricultural infrastructure, and policy changes, although these types of changes were beyond the scope of the analysis. Level 2 adaptations included large shifts in planting date (>1 month), increased fertilizer application (included here because of implied costs for farmers in developing countries), installation of irrigation systems, and development of new varieties (tested by manipulation of genetic coefficients in crop models). Level 2 represents a fairly optimistic assessment of world agriculture's response to the changed climate conditions tested.

To extend the adaptation site results to national yield change estimates for all the countries in the world food trade model and to the other GCM scenarios, a simplifying assumption was made based on the crop modelling results to halve the negative impact if adaptations partially compensated for the negative effects of climate change, and to set yield changes to zero if compensation was full. If yield changes were positive, adaptation to produce even greater yield increases was not included, with the assumption that farmers would lack incentive to adapt further. This unrealistic assumption tends to underestimate potential disparities between countries able to improve productivity under the climate change scenarios and countries that cannot fully adapt. The adaptation estimates were developed only for the scenarios that included direct physiological CO_2 effects as these were judged to be most realistic.

improves realism, but many critical uncertainties remain. Although crop models allow testing of some potential improvements in agricultural production, they do not include yield-enhancing technological developments induced by negative climate change impacts. Level of adoption and efficacy of adaptive practices are also uncertain. There may be social or economic reasons why farmers are reluctant to implement adaptation measures, for example, increased fertilizer application and improved seed stocks may be capital-intensive and/or not suited to indigenous agricultural strategies. Furthermore, such measures may not necessarily result in sustainable production increases (for example, irrigation may eventually lead to soil salinization).

Yield estimates for the two adaptation levels are shown in Fig. 3. Level 1 adaptation compensated incompletely for the climate change scenarios, particularly in the developing countries. For the GISS and GFDL scenarios, level 2 adaptation compensated almost fully for negative climate change impacts. With the high level of global warming projected by the UKMO climate change scenario, neither level 1 nor level 2 adaptation fully overcame the negative climate change effects on crop yields in most countries, even when direct CO_2 effects are taken into account.

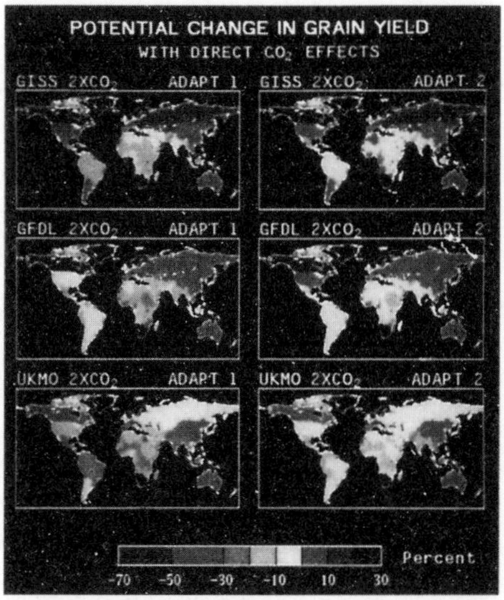

FIG. 3 Estimated changes in average national grain yield (wheat, rice, coarse grains, and protein feed with direct 555 p.p.m. CO_2 effects) under two levels of adaptation for the GISS, GFDL and UKMO doubled-CO_2 climate change scenarios[8]. Adaptation level 1 signifies minor changes to existing agricultural systems; adaptation level 2 signifies major changes. Results shown are averages for countries and groups of countries in the BLS world food trade model; regional variations within countries are not reflected.

World food trade model

The basic linked system (BLS) consists of a set of linked national agricultural sector models[15]. It is comprised of 16 national (including the EU) models with a common structure, 4 models with country-specific structure and 14 regional group models. The political changes as well as changes in national boundaries of the very recent past are not in the BLS, although the model formulation has been adjusted, away from centrally planned economies to more market-oriented behaviour. The 20 models in the first two groups cover ~80% of world agricultural production: the remaining 20% is covered by 14 regional models for countries which have broadly similar attributes (for example African oil-exporting countries, Latin American high-income exporting countries, Asian low-income countries). The BLS is a general equilibrium model system, with representation of all economic sectors, empirically estimated parameters and no unaccounted supply sources or demand sinks. Countries are linked through trade, world market prices and financial flows.

The BLS does not incorporate any climate relationships *per se*. Effects of changes in climate were introduced to the model as changes in the average national or regional yield per commodity as described above. Internal economic adjustments occur as increased agricultural investment, reallocation of agricultural resources according to economic returns (including crop switching), and reclamation of additional arable land as a response to higher commodity prices. Improvements in agricultural technology are represented by annual yield trends for developed and developing countries based on historical trends. The BLS contains yield-fertilizer production functions to capture the effects that changes in fertilizer prices and subsequent changes in fertilizer applications may have on yields at the national level. The economic adjustments to climate change simulated by the BLS are assumed not to alter the basic structure of yield-fertilizer functions, even though some of these relationships (for example, yield responses to nitrogen fertilization) may be altered in a changed climatic regime and under elevated CO_2 conditions.

Summary indicators of the world food system's sensitivity to the climate change scenarios include world cereal production, world cereal prices and population in developing countries (excluding China) at risk of hunger. The 'risk of hunger' indicator in the BLS was developed from estimates of the number of undernourished people in developing countries made by the FAO[16]: a cross-country regression was estimated, explaining the number of people at risk of hunger by a measure of food energy availability relative to nutritional requirements[15]. Food availability, in turn, depends on income and price levels. Average and marginal budget shares of consumption categories besides food (for example, housing and clothing) are included in the analysis. We include the risk of hunger indicator to show the possible trends in future food security, realizing that more comprehensive measures incorporating other socioeconomic variables may be devised.

Here we limit our analysis to results relating to the major cereal food crops, even though the BLS explicitly represents efficiency of feed use and trends in production and consumption of alternative livestock commodities. Thus, the reference case simulates the shift to higher-efficiency feed conversion (poultry over beef) occurring in some developed countries. Beyond the indirect effects of changes in feed costs, livestock production is a significant component of the global food system that is potentially sensitive to climatic change because of changes in rangeland and animal productivity.

World food trade model scenarios

The reference scenario (a future without climate change). The reference scenario projects the agricultural system to the year 2060 with no climate change and no major changes in the political or economic context of world food trade (Table 2). Population growth rates were exogenously specified from the medium projections of the United Nations to 2025, and from World Bank projections thereafter, resulting in ~10.3 billion people by 2060 (refs 17, 18).

Economic growth rates in the BLS are endogenously determined in most of the national models, yielding a moderate projection of world economic growth for the reference scenario. A 50% trade liberalization in agriculture (for example, removal of import restrictions) is introduced gradually by 2020. The analysis of trade liberalization is restricted to removal of distortions between trade prices and domestic prices at the level of agricultural raw materials.

Technology is projected to increase yields over time, but at a slowing rate based on historical trends. The rate of exogenous

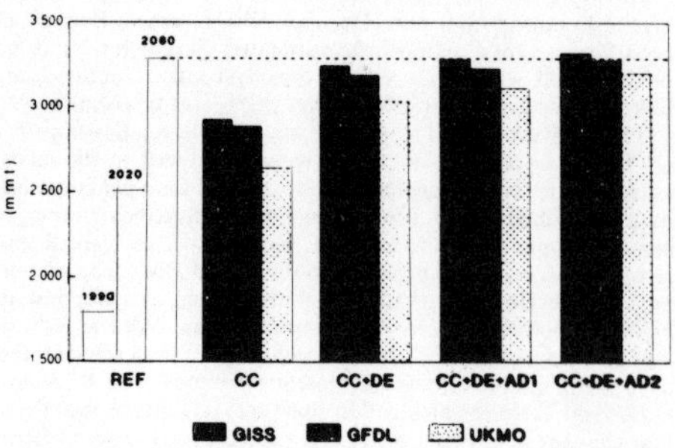

FIG. 4 World cereal production projected by the BLS for the reference, GISS, GFDL and UKMO doubled-CO_2 climate change scenarios, with (CC - DE) and without (CC) direct CO_2 effects on crop yields, and with adaptation levels 1 and 2 (AD1 and AD2). Adaptation level 1 implies minor changes to existing agricultural systems; adaptation level 2 implies major changes. (m.m.t., million metric tons).

TABLE 2 World growth rates in the BLS reference scenario

Growth rate %	1980–2000	2000–2020	2020–2040	2040–2060
Population	1.7	1.3	0.8	0.5
GDP	2.9	2.0	1.5	1.1
Cereal yield	1.2	0.7	0.5	0.4
Agricultural production	1.8	1.3	1.0	0.7

All growth rates refer to world average annual per cent growth during the indicated period.

technical progress starts from historical values (1.3% in the 1980s) and for cereal crops approaches 0.5% per annum by 2060. Availability of arable land for expansion of crop production is based on FAO[19] data.

Climate change scenarios. The food trade simulations for the three GCM scenarios were started in 1990. The yield changes estimated from the crop model simulations were applied linearly up to 2060 to the yields simulated by the BLS, which include the effect of technological improvement. Although the testing of climate change impacts without farm-level adaptation is unrealistic, it is done for the purpose of establishing a baseline with which to compare the effects of farmer response.

Scenarios including the effects of farm-level adaptations. The next step in the analysis involved adjusting the climate-induced yield changes assuming adaptation levels 1 and 2 described above. We can safely assume that at least some farm-level adaptations will be adopted, especially techniques similar to those tested in adaptation level 1. Policy, cost and water resource availability were assumed not to be barriers to adaptation.

World food trade results

The reference scenario. Assuming that population growth, economic growth, technological progress and trade liberalization proceed as specified above without climate change, world cereal production (wheat, rice, maize, millet, sorghum and minor grains) is estimated to grow to 3,286 million metric tons (m.m.t.) in year 2060 (compared with 1,795 m.m.t. in 1990). Cereal production in developing countries grows to exceed production in developed countries by 2020. Despite slowing gains in yield increases, food production (measured as net calories produced) is projected to exceed population growth throughout the simulation period of the reference scenario.

Cereal prices are estimated at an index of 121 (1970 value, 100) for the year 2060, reversing the falling trend of real cereal prices over the past 100 years. The standard reference scenario has two phases of price development. During 1980 to 2020, while trade barriers and protection are still in place but are being reduced, there are increases in relative prices. This occurs in the short-to-medium term because a removal of subsidies leads to lower farm-gate prices and therefore disincentives to production, while consumers benefit from somewhat lower retail prices, an incentive to demand. In the longer term, price decreases follow due to efficiency gains and technical progress. The number of people at risk of hunger is estimated at ~640 million or ~6% of total population in 2060 (compared with 530 million in 1990, ~10% of total current population).

Climate change scenarios. Without direct CO_2 effects on crop yields, world cereal production is reduced by 11 to 20%, and their inclusion brings yield decreases to between 1 and 8% (Fig. 4). The world production changes mask a disparity in response to climate change between developed and developing countries (Fig. 5). The largest negative changes occur in developing regions, though the extent of decreased production varies greatly by country depending on the projected climate. By contrast, in developed countries, production is estimated to increase under all but the UKMO scenario.

Price increases resulting from climate-induced decreases in yield are estimated to range between ~24–145% (Fig. 5). These increases in price affect the number of people at risk of hunger. Their estimated number increases ~1% for each 2–2.5% increase in prices (depending on climate change scenario). People at risk of hunger increase by 10% to almost 60% in the scenarios tested, resulting in an estimated increase of between 60 million and 350 million people in this condition (above the reference scenario projection of 640 million) by 2060.

Scenarios including the effects of farm-level adaptations. Globally, both minor and major levels of adaptation help restore world production levels (when CO_2 effects are included), compared to the climate change scenarios with no adaptation (Fig. 4). Averaged global cereal production decreases by up to ~160 m.m.t. (0 to −5%) from the reference scenario projection of 3,286 m.m.t. with minor level 1 adaptations. With adaptations implying major changes, global cereal production responses range from a slight increase of 30 m.m.t. to a slight decrease of ~80 m.m.t. (+1% to −2.5%).

Level 1 adaptation largely offsets the negative climate change yield effects in developed countries, improving their comparative advantage in world markets (Fig. 5). In these regions, cereal production increases by 4 to 14% over the reference scenario. However, developing countries are estimated to benefit little from this level of adaptation (−9 to −12% change in cereal production). More extensive adaptation virtually eliminates global negative cereal yield impacts derived under the GISS and GFDL climate scenarios, and reduces impacts under the UKMO scenario to one third.

Under adaptation level 1, price increases range from 10 to 100% (Fig. 5). Under adaptation level 2, cereal price responses range from a decline of ~5% to an increase of 35%. As a consequence of climate change and adaptation level 1, the number of people at risk of hunger increases by ~40 million to 300 million (6–50%) from the reference scenario of 641 million (Fig. 5). With a more significant amount of adaptation by farmers, the number of people at risk of hunger is altered by between −12 million for the GISS scenario and 120 million for the UKMO scenario (−2% and +20%). These results indicate that, except for the GISS scenario under adaptation level 2, the simulated farm-level adaptations did not mitigate entirely the negative effects of climate change on the number of people at risk of hunger, even when economic adjustment, that is, the production and price responses of the world food system, are taken into account.

Discussion

Several major points emerge from this study. Climate change scenarios near the high end of the IPCC range of doubled-CO_2 warming exerted (in most cases) a slight-to-moderate negative effect on simulated world cereal production, even when the beneficial direct effects of CO_2, farm-level adaptations and future technological yield improvements were taken into account. The only scenario that increased global cereal production was one involving major, and possible costly, changes in current agricul-

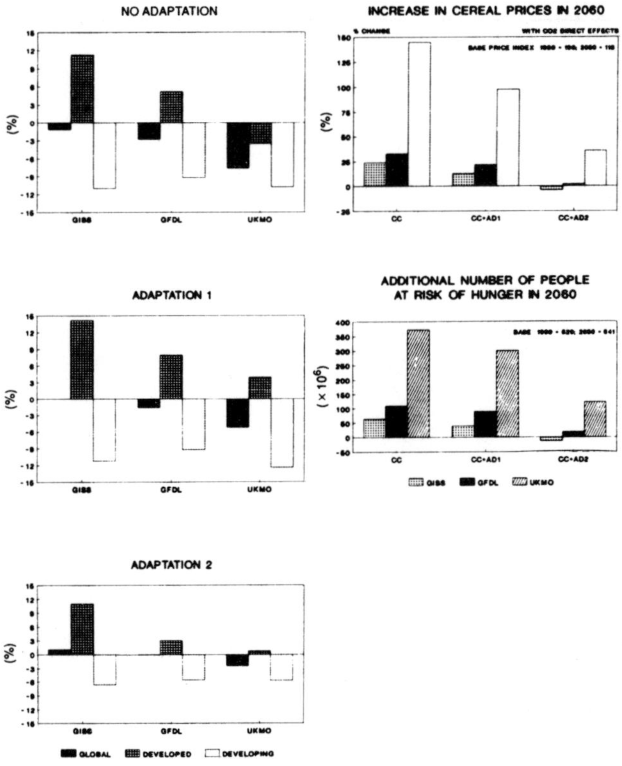

FIG. 5 Change in global, developed country, and developing country cereal production (with direct CO_2 effects), cereal prices, and people at risk of hunger in 2060 for climate change scenarios (CC), and with adaptation levels 1 and 2 (AD1 and AD2). Reference scenario for 2060 assumes no climate change, and projects global cereal production to be 3,286 m.m.t., developed country production to be 1,449 m.m.t., and developing country production to be 1,836 m.m.t.

ARTICLES

tural systems, for example, installation of irrigation. Availability of irrigation water, however, was not included explicitly in the study; supplies may be limited under climate change conditions not only for expanding irrigation but for maintaining the current extent of irrigation in some areas. Although the overall results of the study are relatively benign, they depend strongly on the full realization in the field of beneficial direct physiological CO_2 effects on crop growth and water use as currently measured in experimental settings.

Climate change was found to increase the disparities in cereal production between developed and developing countries. Whereas production in the developed world benefited from climate change, production in developing nations declined. Adaptation at the farm-level did little to reduce the disparities, with the developing world suffering the losses. Cereal prices, and thus the population at risk of hunger, increased despite adaptation. Even a high level of farm-level adaptation in the agricultural sector did not entirely prevent such negative effects. Thus, while some countries in the temperate zones may reap some benefit from climate change, many countries in the tropical and subtropical zones appear more vulnerable to the potential impacts of global warming. □

Received 9 July; accepted 2 December 1993.

1. Parry, M. L., Carter, T. R. & Konijn, N. T. (eds) *The Impact of Climatic Variations on Agriculture* Vols 1 & 2 (Kluwer, Dordrecht, 1988).
2. Adams, R. M. *et al. Nature* **348,** 219–224 (1990).
3. Intergovernmental Panel on Climate Change *Climate Change: The IPCC Impacts Assessment* (eds Tegart, W. J. McG., Sheldon, G. W. & Griffiths, D.C.) (Aust. Govt Publishing Serv., Canberra, 1990).
4. Kane, S., Reilly, J. & Tobey, J. Report AER-No. 647 (US Dept of Agric. Econ. Res. Serv., Washington DC, 1991).
5. Reilly, J. & Hohmann, N. *Am. econ. Ass. Pap. Proc.* **83,** 306–312 (1993).
6. Rosenzweig, C. & Iglesias, A. (eds) *Implications of Climate Change for International Agriculture: Crop Modeling Study* (US Environmental Protection Agency, Washington DC, in the press).
7. *Climate Change: The IPCC Scientific Assessment* (eds Houghton, J. T., Jenkins, G. J. & Ephraums, J. J.) (Cambridge Univ. Press, 1990).
8. IPCC *Climate Change 1992* (eds Houghton, J. T., Callander, B. A. & Varney, S. K.) (Cambridge Univ. Press, 1992).
9. Acock, B. & Allen, L. H. Jr in *Direct Effects of Increasing Carbon Dioxide on Vegetation* (eds Strain, B. R. & Cure, J. D.) 33–97 (US Dept. of Energy DOE/ER-0238, Washington DC, 1985).
10. Cure, J. D. & Acock, B. *Agric. For. Met.* **38,** 127–145 (1986).
11. Hansen, J., *J. geophys. Res.* **93,** 9341–9364 (1988).
12. International Benchmark Sites Network for Agrotechnology Transfer (IBSNAT) *Decision Support System for Agrotechnology Transfer Version 2.1* (DSSAT V2.1) (Dept Agron. & Soil Sci., Coll. Trop. Agric. & Hum. Resources, Univ. Hawaii, Honolulu, 1989).
13. Peart, R. M., Jones, J. W., Curry, R. B., Boote, K. & Allen, L. H. Jr. in *The Potential Effects of Global Climate Change on the United States*. Appendix C Vol. 1 (eds Smith, J. B. & Tirpak, D. A.) (US Envir. Protection Ag., Washington DC, 1989).
14. Rogers, H. H., Bingham, G. E., Cure, J. D., Smith, J. M. & Surano, K. A. *J. envir. Qual.* **12,** 569–574 (1983).
15. Fischer, G., Frohberg, K., Keyzer, M. A. & Parikh, K. S. *Linked National Models: A Tool for International Food Policy Analysis* (Kluwer, Dordrecht, 1988).
16. Food and Agriculture Organization *Fourth World Food Survey*. (UN, Rome, 1984); *Fifth World Food Survey* (UN, Rome, 1987).
17. United Nations *World Population Prospects 1988* (UN, New York, 1989).
18. International Bank for Reconstruction and Development/World Bank *World Population Projections* (Johns Hopkins Univ. Press, Baltimore, 1990).
19. *AGROSTAT/PC* (Food and Agricultural Organization, United Nations, Rome, 1991).
20. Hansen, J. *et al. Mon. Weath. Rev.* **111(4),** 609–662 (1983).
21. Manabe, S. & Wetherald, R. T. *J. atmos. Sci.* **44,** 1211–1235 (1987).
22. Wilson, C. A. & Mitchell, J. F. B. *J. geophys. Res.* **92(13),** 315–343 (1987).

ACKNOWLEDGEMENTS. We acknowledge all our collaborators, and especially thank the IBSNAT crop modelling team for worldwide scientific and technical support, G. Fischer and K. Frohberg for their work with the BLS, and A. Iglesias for development of yield estimation techniques. We also thank C. Shashkin and R. Goldberg for project support. This work was funded by the US Environmental Protection Agency and the US Agency for International Development.

B
Cost of Control Estimates

[7]

Cutting CO_2 Emissions:

The Effects of Alternative Policy Approaches*

*John Whalley*** and Randall Wigle****

This paper starts from the premise that attempts to curtail global emissions of greenhouse gases are likely to be made in the next few decades. We discuss some of the possible international effects which could result from attempts to achieve such a cutback, and illustrate a methodology which we hope to extend, in subsequent work, to further evaluating the consequences of responding to the problem of global warming. We identify possible magnitudes of effects of cutting global CO_2 emissions, and illustrate ways in which inter-country terms-of-trade effects and changes in trade patterns may occur.

INTRODUCTION

Rather than debate the wisdom of curtailing greenhouse gas emissions, this paper starts from the premise that attempts to curtail emissions of greenhouse gases are likely to be made in the next few decades.[1] Given this, we discuss some of the possible international effects which could result from attempts to achieve such a cutback, and illustrate a methodology which we hope to extend, in subsequent work, to further evaluating the consequences of responding to the problem of global warming. In the paper, we concentrate

The Energy Journal, Volume 12, Number 1.

*The authors would like to thank, without implicating, T. Hertel, M. Rothschild, G. Harrison and C. Waddell, as well as two anonymous referees, for helpful comments. Earlier drafts of this paper were presented to the Annaul Meetings of the Canadian Economics Association, Quebec City, June 5, 1989, and to a conference on Applied General Equilibrium Modelling, San Diego, September 8-9, 1989. A further version of this paper is to be subsequently published in the conference proceedings. Wigle acknowledges support from the Social Sciences and Humanities Research Council of Canada.

**Department of Economics, University of Western Ontario
***Department of Economics, Wilfrid Laurier University

1. See the final statement from the Environment Canada Conference in Toronto in June 1988, entitled *The Changing Atmosphere: Implications for Global Security*, which we discuss below.

on identifying possible magnitudes of effects of cutting CO_2 emissions, and illustrate the ways in which terms-of-trade effects and changes in trade patterns may occur.

One point which emerges is that a given reduction target can be achieved via a number of alternative policy instruments. Not surprisingly, the burden of the resulting adjustments by country is sensitive to the allocations of cutbacks among the countries. Somewhat less obvious is that the choice between production and consumption taxes affects the allocation of the burden of adjustment.

ALTERNATIVE APPROACHES TO CUTTING EMISSIONS

A number of issues will arise with the implementation of any future cut of global CO_2 emissions. The first is the depth of cut to be adopted. Once the target proportionate cut is chosen, a given cut could be achieved in one of several ways, but two 'polar cases' may be illustrative. The most direct way to achieve a 20% reduction would be to have all countries agree to reduce their own emissions by 20%. An alternative would be to have countries agree to reduce their emissions to an equal per capita amount, with the level set so as to reduce total emissions by 20%.

These proposals are, however, deceptively different from one another. The equi-proportionate reduction approach effectively confers squatters' rights on the heaviest-emitting countries, while forcing reductions by the least-offending countries. By contrast, the equal per capita standard would require significant cutbacks by high-emissions countries, with no cutbacks at all (even in projected growth to 2005) by the lowest-income countries. The first scheme is the more regressive in impact, since the low-emissions countries also tend to be the lowest-income countries.

Targets for cutbacks could be set in terms of global or national consumption, although proposed cutbacks could be achieved in a number of different ways. Taxes could be applied to either fossil fuel use, or production. Quotas could be placed on either the use, or production, of fossil fuels. Further, the revenues from any taxes or quota sales could be collected either by national governments, or by an international body such as the proposed World Atmospheric Fund.

If taxes are used, or if quota rights are sold, the disposition of these revenues also becomes a significant issue. If consumption taxes are used and national governments receive the revenues, high-emission countries would receive the bulk of the revenues. In contrast, if quotas were sold and revenues collected by an international agency, the issue of the distribution of the revenues would become a significant one. (Below, we estimate that revenues could amount to several hundred billions of dollars annually.) Once again the most obvious scheme would be to have national governments collect and redistribute the revenues as they wish, while a less obvious and dramatically different alternative would be either to allocate revenues from quotas on an equal per capita basis between all countries around the world, or to allocate them more than proportionately to lower-income countries.

Restrictions on the use of "greenhouse" energy would cause the producers' price of fossil fuels to fall, with the users' price rising. Energy-intensive products would increase in price relative to those which are not. Associated with these would be changes in the terms of trade against fossil-fuel exporters, higher prices for consumers of energy-intensive products, and lower returns for producers of energy-intensive products. One would also expect the price of energy from non-greenhouse sources to rise.

Since the size of each of these effects is hard to estimate in isolation, we use a simple numerical general equilibrium model of global trade and production to provide an indication of the possible incidence, terms of trade, and transfer effects associated with a number of emission reduction proposals, all loosely modelled after the Toronto proposal outlined above.

A NUMERICAL GENERAL EQUILIBRIUM MODEL FOR ASSESSING THE EFFECTS OF CO_2 EMISSION REDUCTIONS

We have used a single-period computable general equilibrium model to evaluate the possible effects of alternative CO_2 emission cuts. In it, the world is divided into three regions, as indicated in Table 1. The Developed Region includes all countries with per capita income, in 1986, greater than $2,000 US, and with less than 25% of their exports made up of fuels. The Developing Region includes all countries with per capita incomes below $2,000 US, and with fuel exports accounting for less than 25% of total exports. The Oil Exporting Region includes all countries where fuel exports account for more than 25% of exports.

The nested functional structures used in the model to represent production and demand in each region are set out in Table 2. Each region is endowed with three non-traded primary factors: (i) primary factors, exclusive of energy resources, (ii) greenhouse energy resources (deposits of oil, gas and coal), and (iii) other energy resources (hydroelectric capacity and nuclear). Both energy resources are converted into the relevant energy products through a refining/extraction process, which uses other resources (primary factors). There are three internationally traded commodities in the equilibrium system: greenhouse energy products, energy-intensive manufactures, and other goods (all other GNP). Energy-intensive manufactures, other goods, and the composite energy product (greenhouse and non-greenhouse energy) are the commodities which enter final demands.

For each of five produced goods in each region (listed in Table 2), production is represented by a constant elasticity of substitution (CES) function. Greenhouse and non-greenhouse energy products use the respective energy resources and primary factors. Non-greenhouse energy products are non-traded, since hydroelectric, solar and nuclear power are not traded in

Table 1. Regions in the Equilibrium Model Used to Evaluate Options for Reductions in CO_2 Emissions

Developed World:

All countries with per capita income (1986) above $2,000 US, with less than 25% of exports in fossil fuels.

Developing World:

All countries with per capita income (1986) below $2,000 US, with less than 25% of exports in fossil fuels.

Oil Exporters:

All countries with fuel more than 25% of their exports.

significant quantities between the regions as defined here[2]. A domestic energy composite is produced by a third (energy conversion) industry, using inputs of the two energy products. The two final-goods (energy-intensive manufactures, and other goods) use primary factors and the composite domestic energy product as inputs. Perfect competition is assumed throughout for all sectors and in all regions.

Equilibrium, in this model, involves full global market clearing in all three traded goods (greenhouse energy products, energy-intensive manufactures, and other goods). For the two non-traded goods (non-greenhouse energy products, and composite energy) there is domestic market clearing within each economy. Since prices in this system are treated as completely flexible, they will adjust to the levels required to clear the relevant international and domestic markets.

The introduction of carbon taxes into this system will raise prices of greenhouse energy products, and hence the price of composite energy to users, and reduce prices to producers. This will have the effect of cutting both consumption and production of greenhouse energy products, with prices adjusting until the global market in greenhouse energy products clears. If prices to consumers increase fully by the amount of the tax, the tax will have been fully passed on to users of carbon. If prices to consumers change little, then the tax will have been passed backwards to producers of greenhouse energy products. Substitution will also occur between greenhouse and non-greenhouse energy products, and between the energy composite product, and other inputs. The strength of all of these effects depends on the elasticity values used at the respective nodes in the nesting of substitution possibilities set out in Table 2.

2. While there is trade in radionuclides, this is ignored.

Table 2. Structure of Equilibrium Model Used to Evaluate Options for Reductions in CO_2 Emissions

Factors and Goods:

Endowments

"Greenhouse" Energy Resources (GR), Other Energy Resources (ER), Primary Factors (PF)

Produced Goods

Greenhouse Energy Products (GP), Other Energy Products (EP), Composite Energy (E), Energy-Intensive Goods (EI), Non-Energy Intensive Goods (NEI)

Structure of Production:

Stage 1 Production of energy products

```
PF    GR        PF    ER
 \    /          \    /
   GP              EP
```

Stage 2 Production of composite energy

```
GP    EP
 \    /
   E
```

Stage 3 Production of energy-intensive and non-energy intensive products

```
E     PF        E     PF
 \    /          \    /
   EI              NEI
```

Arguments in Final Demands

EI, NEI, E

To evaluate effects of alternative carbon tax schemes on the three regions, we use the model in what we term counterfactual mode[3]. This involves constructing a base (or reference) case solution to which the equilibrium model is calibrated[4] (i.e., parameters are chosen such that the model will reproduce the base case data as a full equilibrium solution prior to any policy change being introduced). The base case data thus fully describes a no-policy change model solution. Different carbon tax schemes can then be introduced, with a new equilibrium (counterfactual equilibrium) solution computed for each. Using a pair-wise comparison (base case versus

3. This is a technical term used in the applied general equilibrium modelling literature; see Shoven and Whalley (1984, and forthcoming) for more details.

4. See Mansur and Whalley, 1984.

counterfactual), international incidence effects are evaluated following each such calculation.

An important characteristic of the model is the ease, specified through substitution elasticities, with which outputs can be shifted between the two energy products, relative to the ease of substitution between the two energy products in consumption. The former is assumed to be relatively low, while the latter is relatively high. This configuration reflects energy resources being non-tradeable and in fixed supply. As a result, the elasticity of input substitution in the energy products sectors is set to 0.2. The "energy" sector in each region uses inputs of the two energy products, and substitution between them is relatively easy (the elasticity of substitution is set to 4). This reflects the fact that inputs of coal or hydroelectricity in the proper ratio are close substitutes for the purpose of heating a house, or running an industrial plant. Final demand is for the energy-intensive and non-intensive goods plus the (region-specific) energy aggregate. All of the traded goods produced in each of the three regions are perfect substitutes (i.e., we do not adopt the Armington structure for traded goods).

This structure allows us to readily identify the ease of shifting outputs between the greenhouse and other energy products sector as well as the ease of substitution in consumption between the final demand products.

Data

The model is calibrated to a 1986 global benchmark equilibrium data set built using available sources (see Table 3). Data on GNP, population and exchange rates are obtained from the World Tables. Energy consumption production and trade data are obtained from a secondary source (Filion, 1988). Finally, trade data assembled for another purpose from the UN trade tapes are used to complete the trade data for the non-energy products.

One of the most striking differences between the developed and developing countries is that the developing countries have dramatically higher per capita emissions than developed countries. As we split the countries, the GNP of developed countries is about three times that of the developing countries, while emissions are (roughly) equal. Developing countries account for a higher share of emissions than GNP because they tend to use less energy-efficient technologies, and a higher share of coal in total energy products. The production of one BTU of energy with coal generates significantly more CO_2 than with either oil or natural gas.

Developed countries are net exporters of both energy-intensive goods and non energy-intensive goods, and are significant importers of fossil fuels. The developing countries import both fossil fuels (greenhouse energy products) and energy-intensive goods, and export non-energy-intensive goods. Finally, the "oil" region is a significant net exporter of fossil fuels and has correspondingly large imports of the two other traded goods. The goods

Table 3. Global Benchmark data for 1986 Used in the General Equilibrium Model

Production and Consumption of Energy by Region:

Bloc	Greenhouse Energy (Billions of US$) Production	Consumption	Share of Other Energy To Total Energy Use
Developed	189.8	339.8	15.4%
Developing	247.4	287.0	7.1%
Oil Exporters	232.6	48.4	11.2%

Trade Pattern by Region (New Exports in Billions of US$):

	Developed	Low-Income	Oil
G Energy	-150.0	-39.6	189.7
E Intensive	38.0	-26.8	-11.2
Other (non-intensive)	+112.0	66.5	-178.5

Population and GNP by Region:

	Product	Share of World's Population	Emissions
Developed	74.8%	15.6%	50.7%
Developing	22.2%	80.6%	42.8%
Oil Exporters	3.0%	3.7%	6.4%

Sources: GNP, production, population and non-energy trade data were based on The World Bank's World Tables. Data for the USSR and People's Republic of China were generated from estimates in Holzman (1976) and Rosefielde (1973). Energy use and trade flows are based on Filion, ed., *The Canadian World Almanac*, 1989. See pages 419 and 461. Trade data for energy-intensive goods are based on UN trade tapes.

which we call energy-intensive are mostly what might be called primary manufactures (for example, metals, basic products, glass, ceramics, fabricated goods). The net trade between these large blocs is a relatively small fraction of world trade, since much of recorded trade takes place within our regions.

As emphasized earlier, the developed bloc accounts for much higher shares of the world's emissions and production, than population. Trade was forced to balance without capital flows. The energy-intensive goods sector includes what is often referred to as primary manufacturing. Notably, it includes primary metal manufacturing, as well as the glass, pottery, and plastics industries. These are identified as having the highest energy input requirements in Ross (1989), and the input ratios used in that paper were used to infer energy input requirements for the respective industries.

SOME COUNTERFACTUAL POLICY EXPERIMENTS USING THE MODEL

As mentioned earlier, two proposals made at the Toronto conference on the changing atmosphere called for a 20% cut and a 50% cut in CO_2 emissions, respectively, by 2005. In what follows we model the effects of a 50% cut in emissions, abstracting from the impact this would have on atmospheric CO_2 concentrations. The limitations inherent in using a static general equilibrium model for this purpose are discussed in the conclusion.

The main experiments we consider are the following:

1. Production taxes collected by national governments to achieve a 50% reduction in their respective production of greenhouse energy, and hence CO_2 emissions.

2. Consumption taxes collected by national governments to achieve a 50% reduction in their own consumption of greenhouse energy, and hence CO_2 emissions.

3. A global consumption (or production) tax collected by an international body such as the "UN," with revenues distributed to countries according to population, and achieving the same 50% reduction.

4. A reduction of global emissions by 50% achieved by establishing a ceiling on per capita emissions.

Implementation[5]

As a first approximation, we assume that the rate of CO_2 emissions is proportional to the use of greenhouse energy products. In particular, if reductions are achieved by scaling back proportionally on the use of all fossil fuels, this assumption is reasonable. Switching between coal and natural gas, for a given level of energy consumption, would reduce CO_2 emissions.

The evaluation of national production and consumption taxes is achieved by imposing (respectively, production or consumption) quotas in each country to correspond to the required cut. Since industries are all modelled as perfectly competitive, the quota values determined in this way correspond to the taxes required to bring about the specified reduction in emissions. For the case of "global" taxes, revenue is distributed on an equal per capita basis. Production and consumption taxes are formally equivalent in this case because all consumers pay the same price and all producers receive the same price.

5. The model was implemented using MPS/GE. See Rutherford, 1988.

Given that the revenue distribution is exactly the same for the two taxes, the size of tax required to reduce global consumption (and thus production) by 50% is equivalent.

National Taxes

The effects shown by the model, from imposing national production taxes on greenhouse energy products such as oil, gas and coal, to achieve a target of 50% cut in emissions, are shown in Table 4. The proportionate effects on consumption are similar in all the regions, but welfare effects are dissimilar. Also, while the reductions in greenhouse energy trade are easy to understand, there is a dramatic increase in trade of energy-intensive for non-intensive products. The other energy sectors in each economy expand by over 58%.

These welfare effects by region may at first seem surprising, since the oil region experiences a welfare gain, even though they are the largest producers (and are thus subject to the most tax). The welfare effects reflect two considerations. First, the national production taxes raise revenue which goes to countries proportional to their own production. This means that revenues accrue disproportionately to the net exporters, so that the national production taxes operate much like an export tax. This explains why the loss of the developed region is so large, while the oil region experiences a small welfare (and terms of trade) gain. We further note that the revenue collected by such a tax is significant, amounting to almost $200 billion in the developed region alone. Implied tax rates from these restrictions are about 80%.

The effects of achieving 50% reductions in CO_2 emissions through national consumption taxes are shown in Table 5. Results for national consumption taxes are different from those for national production taxes in a number of ways. The most striking are the welfare effects. The revenues generated by these consumption taxes accrue to countries proportional to their consumption. This means that the bulk of the revenue is received by the developed region (in effect acting somewhat like an import tariff from the point of view of that region). As a result, the oil region experiences a large welfare loss in this case, and collects little of the revenue. Most of the burden of the emission reduction effort is borne by the oil-exporting region.

In both experiments, dramatic changes in the pattern of trade occur. This occurs partly because of the simple production structure used, but also because trade in the energy intensive and non-energy intensive goods are such a small share of total production and consumption (see Table 3).

A Global Tax

Table 6 shows the effects of a global tax on either consumption or production of greenhouse energy products. The tax is computed by imposing

Table 4. Reduction of CO_2 Emissions by 50% Through National Production Taxes

Energy Quantities and Prices (% Changes):[a]

Region	Greenhouse Energy Production	Consumption	Other Energy Production	Price
Developed	-50.0	-72.6	7.0	58.9
Developing	-50.0	-28.0	9.5	101.2
Oil Exporters	-50.0	-17.7	9.8	108.0
Price of Greenhouse Energy			-56.3	

Trade Pattern (% Changes in Trade Volume):[b]

	Developed	Developing	Oil
Greenhouse energy	+ + +	109.15	-57.31
Energy-Intensive	---	+ + +	+ + +
Non-Energy-Intensive	1419.87	---	99.36

Welfare and Revenue Effects (Billions of US$):

	Welfare[c] (%)	Revenue Generated	Tax Rate
Developed	-179.70 (-1.68)	170.54	80.42
Developing	-151.23 (-4.76)	223.26	80.78
Oil Exporting	46.99 (10.80)	209.92	80.78
World Total	-283.94 (-1.98)		

a. Price effects are % changes in producers' prices.

b. % changes in trade volumes (net exports) expressed as a percentage of original net exports. + + + denotes that the region has changed from an importer to an exporter, while --- indicates the converse.

c. Welfare effects are in billions of US$ (% of changes in brackets).

a global quota on production and calculating the value of one unit's worth of quota. The quota is sold by a UN agency which, in turn, redistributes the revenues proportionally to each region's population.

The implementation of this scheme leads to a reduction in the producer's price of greenhouse energy products and, at the same time, a terms of trade loss to the oil-exporting region. Further, the majority of the revenues associated with the plan accrue to the developing region, which is,

Table 5. Reduction of CO_2 Emissions by 50% Through National Consumption Taxes

Energy Quantitites and Prices (% Changes):[a]

Region	Greenhouse Energy Production	Consumption	Other Energy Production	Price
Developed*	-55.7	-50.0	7.9	73.4
Developing	-48.0	-50.0	9.7	106.9
Oil Exporters	-47.4	-50.0	10.0	113.1
Price of Greenhouse Energy				-56.9

Trade Pattern (% Changes in Trade Volumes):[b]

	Developed	Developing	Oil
Greenhouse Energy	-42.76	-62.34	-46.86
Energy-Intensive	---	265.68	+ + +
Non-Energy-Intensive	-45.69	67.07	-3.70

Welfare and Revenue Effects (Billions of US$):

	Welfare[c] (%)	Revenue Generated	Tax Rate
Developed	-67.39 (-0.63)	283.99	79.51
Developing	-120.60 (-3.80)	299.55	82.89
Oil Exporting	-108.13 (-24.85)	46.44	83.41
World Total	-269.12 (-2.07)		

a. Price effects are % changes in producers' prices.

b. % changes in trade volumes (net exports) expressed as a percentage of original net exports. + + + denotes that the region has changed from an importer to an exporter, while — indicates the converse.

b. Welfare effects are in billions of US$ (% of changes in brackets).

in effect, rewarded for its low emissions by a significant transfer. The oil region's terms of trade losses exceed their share of the revenues, resulting in a welfare loss. In the case of the developing region, the terms of trade work against them, but a significant transfer in their favour leads to a sizable gain. This contrasts with similar sized welfare losses under either of the alternative "national" tax schemes. This reinforces the significant implications of the redistributive scheme suggested here. Note that total tax revenue amounts to $601 billion, of which $484 billion would be received by the developing region.

Table 6. Reduction of CO_2 Emissions by 50% Through Global Production Taxes

Energy Quantities and Prices (% Changes):[a]

Region	Greenhouse Energy Production	Consumption	Other Energy Production	Price
Developed	-54.3	-72.8	6.9	58.2
Developing	-48.1	-24.5	9.6	102.8
Oil Exporters	-48.5	-39.9	9.1	91.8
Price of Greenhouse Energy				-56.8

Trade Pattern (% Changes in Trade Volumes):[b]

	Developed	Developing	Oil
Greenhouse Energy	-96.11	122.3	-50.45
Energy-Intensive	---	+ + +	+ + +
Non-Energy-Intensive	479.24	---	29.05

Welfare and Revenue Effects (Billions of US$):

	Welfare[c] (%)	Revenue Generated	Tax Rate
Developed	-252.92 (-2.36)	93.85	80.59
Developing	+93.98 (+2.96)	484.49	
Oil Exporting	-122.84 (-28.23)	22.61	
World Total	-281.78 (-1.97)		

a. Price effects are % changes in producers' prices.

b. % changes in trade volumes (net exports) expressed as a percentage of original net exports. + + + denotes that the region has changed from an importer to an exporter, while --- indicates the converse.

c. Welfare is measured in billions of US$ (% changes in brackets).

Emissions Ceilings

We finally consider reductions in emissions through per capita ceilings on levels of emissions. As we mentioned earlier, this differs from previous cases where all regions experience similar (or the same) reductions in consumption or production of greenhouse energy-intensive products. In this case, we assume that a per capita standard is established in such a way as to reduce total emissions by 50%. As a practical matter, this would mean that the developed region reduces consumption by more than 50%, with smaller cuts in both the developing and oil region. This follows, since the per capita emissions in the developed region are approximately six times those of the rest of the world combined.

We consider two cases: first, emissions of the developing and oil region are restricted to 75% of their original value (a 25% cut), with the developed region required to cut by 74% to achieve the required overall reduction. This case is shown in Table 7. We also evaluate a case where the required reductions correspond exactly to both a 50% reduction in emissions, and equal per capita emissions. This case is presented in Table 8. This involves an 84% cut for the developed region, a 6% cut by the developing region and a 69% cut by the oil-exporting region.

The pattern of gainers and losers under both ceilings cases is the same as that under the global tax case. This highlights the importance of revenues, since under the latter proposal revenues generated by the taxes used to fulfil the "national" ceilings will be distributed exactly the same way as the revenue from a global tax. Correspondingly, the welfare effects in these two cases (Tables 6 and 8) are similar. The developing region has a small welfare gain in the case of the first ceiling because, although it loses from having to cut greenhouse-energy consumption by 25%, it has a revenue gain. In the second "ceiling" experiment it has a much smaller cut in consumption and almost the same revenue gain.

A further aspect of the policy approaches used here deserves further comment. In all the experiments involving taxes or limits on consumption of greenhouse energy, we have taken consumption to include both intermediate and final consumption of energy. One reason for choosing this approach is that such a tax would be much easier to collect than one which distinguished between these uses. In the case where limits are stricter in the developed than in the developing region, the pattern of production of energy-intensive products changes dramatically. The developed region becomes a significant net importer of these goods in both "ceilings" experiments. An unanticipated effect of treating intermediate use the same as final demand is thus to move the location of intermediate consumption of fossil fuels (as well as the intended effect of reducing total consumption).

SUMMARY AND CONCLUSIONS

While the model used in this paper has some obvious weaknesses (some of which we discuss below), a few general observations about the implications of policies aimed at responding to the global warming problem do suggest themselves from the model results.

First, the rule for distributing tax or quota revenues between regions, and particularly between the developing and developed regions, is important. In the two extreme cases (national collection and distribution of taxes versus

Table 7. Reduction of CO_2 Emissions by 50% Through Per Capita Emissions Ceilings

Energy Quantities and Prices (% Changes):[a]

Region	Greenhouse Energy Production	Consumption	Other Energy Production	Price
Developed	-51.6	-74.0	7.4	65.6
Developing	-46.5	-25.0	9.4	99.4
Oil Exporters	-52.0	-25.0	7.7	69.7
Price of Greenhouse Energy				-56.5

Trade Pattern (% Changes in Trade Volumes):[b]

	Developed	Developing	Oil
Greenhouse Energy	+ + +	108.96	-58.14
Energy-Intensive	---	+ + +	+ + +
Non-Energy-Intensive	1420.68	---	40.50

Welfare and Revenue Effects (Billions of US$):

	Welfare[c] (%)	Revenue Generated	Tax Rate
Developed	-188.28 (-1.76)	170.14	81.56
Developing	0.41 (0.01)	377.95	80.14
Oil Exporting	-91.73 (-21.08)	45.75	76.55
World Total	-279.60 (-1.95)		

a. Price effects are % changes in producers' prices.

b. % changes in trade volumes (net exports) expressed as a percentage of original net exports. + + + denotes that the region has changed from an importer to an exporter, while --- indicates the converse.

c. Welfare is measured in billions of US$ (% changes in brackets).

collection and distribution by an international agency), the developing region receives transfers from the revenues equal to $223 billion or $484 billion, respectively. This leads to a welfare loss of about 5% of GNP in the case where consumption tax revenues are collected and distributed by national governments, and a welfare gain as high as 3% in the case of revenue distribution proportional to population.

Table 8. Reduction of CO_2 Emissions by 50% Through Alternative Per Capita Ceilings

	Welfare[a] (%)	Revenue Generated	Tax Rate
Developed	-247.28 (-2.31)	190.12	88.97
Developing	54.61 (1.72)	357.25	75.34
Oil Exporting	-122.81 (-28.22)	36.68	86.42

a. Welfare is measured in billions of US$ (% changes in brackets).

Second, the choice between national production and consumption taxes is of some significance -- and to the oil region in particular -- as production taxes act to some extent like export taxes, benefiting the oil bloc through positive terms of trade effects. National consumption taxes act to some degree like an import tariff, hurting the oil region and benefiting the developing region.

A third observation is that the oil-exporters bear a significant burden from responding to CO_2 reductions under all plans, except under the production tax scheme. This should not be surprising as in these calculations the target for reduction is greenhouse energy products (fossil fuels), of which this group is a dominant exporter. In the case of the national production taxes they generate significant revenue, leading to a small welfare gain, but in all other cases the welfare loss they suffer exceeds 20% of GNP.

Finally, while the alternative of per capita limits on emissions may have appeal from an equity point of view, such a policy could cause a severe distortion of trade patterns, with the production of energy-intensive products shifting from the developed to the developing countries. The trade restructuring we show is partly due to the simple production structure used, but the incentives generated by differential cutbacks are quite clear. The asymmetric restrictions on the use of fossil fuels in the developed world generate significant incentives for this type of restructuring. In terms of the sum of the region welfare changes criteria, the most equitable of the "per capita ceilings" approaches (Table 8) fares the worst in terms of efficiency, perhaps partly due to the implied distortion.

REFERENCES

Filion, John, ed., (1988). *The Canadian World Almanac, 1989.* Toronto: Global Press.

Mansur, A. and J. Whalley (1984). "Numerical Specification of Applied General Equilibrium Models: Estimation, Calibration and Data." In J.B. Shoven and H. Scarf, eds., *Applied General Equilibrium Analysis*. Cambridge, MA: Cambridge University Press.

Ross, Marc (1989). "Improving the Efficiency of Electricity Use in Manufacturing." *Science* 244(4902): 311-317.
Rutherford Thomas F. (1988). "General Equilibrium Modelling with MPS/GE." London, Ontario: Department of Economics, University of Western Ontario.
Shoven, J. and J. Whalley (1984). "Applied General Equilibrium Models of Taxation and International Trade: An Introduction and Survey." *Journal of Economic Literature* (September): 1007-1051.
Shoven, J. and J. Whalley (forthcoming). *Applying General Equilibrium*. Cambridge, MA: Cambridge University Press.
The Climate Institute (1988). "Toronto Conference Calls for Sharp Cuts in Carbon Dioxide Emissions." *Climate Alert* 1(3).
The World Bank (1987). *World Tables 1960-1986*. Washington: International Bank for Reconstruction and Development.

[8]

Resource and Energy Economics 15 (1993) 7–25. North-Holland

Reducing US carbon emissions: An econometric general equilibrium assessment*

Dale W. Jorgenson

Harvard University, Cambridge, MA 02138, USA

Peter J. Wilcoxen

University of Texas at Austin, Austin, TX 78712, USA

Final version received November 1992

This paper describes research we conducted as part of Energy Modeling Forum 12, a recent study of the costs of limiting carbon dioxide emissions organized by the Energy Modeling Forum at Stanford University. It discusses how our approach differed from that of other participants in EMF-12 and presents several important findings. In particular, we show that in the United States the effects of a carbon tax will be very similar to the effects of a tax placed solely on coal. Outside the coal sector, the principal effect of carbon tax will be to raise the cost of electricity and to shift base load electric generating capacity toward fuels other than coal. At the aggregate level, higher energy prices will cause gross national product to fall unless the revenue from tax is used to reduce high marginal tax rates elsewhere in the economy.

1. Introduction

This paper describes research we conducted as part of Energy Modeling Forum 12, a recent study of the costs of limiting carbon dioxide emissions organized by the Energy Modeling Forum at Stanford University.[1] EMF-12 involved more than a dozen models built by research groups from the United States, Europe and Japan. However, our approach had three unusual

Correspondence to: Professor D.W. Jorgenson, Dept. of Economics, Harvard University, Littauer Center, Room 122, 1875 Cambridge St., Cambridge, MA 02138, USA

*Our work has benefit from many insightful discussions with Alan Manne, Richard Richels, David Montgomery, Thomas Rutherford, Robert Shackleton, James Sweeney, John Weyant and the other participants of EMF 12. We are also grateful to an anonymous referee for several helpful comments. This project has been supported by the Environmental Protection Agency under contracts 68-W8-0113 and 68-W1-0009 and the National Science Foundation under grants SES-90-11463 and SES-91-10231. The authors retain sole responsibility for the views expressed in this paper.

[1]A complete description of all findings of EMF-12 is contained in Gaskins and Weyant, *Reducing Carbon Emissions from the Energy Sector: Cost and Policy Options.*

features which set it apart from the other studies. These features enabled us to explore several questions not addressed by most of the other participants.

First, our results are based on a highly disaggregated intertemporal general equilibrium model of the United States. Using a disaggregated model allowed us to examine the effects of carbon taxes on narrow segments of the economy, such as particular industries or types of household. Second, all parameters in our model were obtained by econometric estimation using a data set spanning 39 years. Thus, the response of industries and consumers to changes in prices will be consistent with the historical record. Third, we model productivity growth at the industry level and allow it to be an endogenous function of relative prices. Allowing productivity growth to differ across industries permits the model to reflect a conspicuous feature of historical data.

Together, these features enabled us to reach several important conclusions. First, in the United States the effects of a carbon tax will be very similar to the effects of a tax placed solely on coal. Of all fossil fuels, coal is the least expensive per unit of energy and produces the most carbon dioxide when burned. Thus, a tax levied on carbon emissions will raise the cost of coal-based energy far more in percentage terms than the price of energy derived from oil or natural gas. In response to this price change, the demand for coal will fall substantially. The demands for oil and natural gas will also decline, but by much smaller percentages.

Almost all coal consumed in the US is used to generate electric power. As the price of coal rises, electric utilities will convert some generating capacity to other fuels. However, substitution possibilities are fairly limited, particularly in the short run, so the tax will raise the price of electricity significantly. Consumers and firms will substitute other inputs for electricity, leading to a fall in electricity demand.[2] Higher energy prices will lead to slower productivity growth, reduced capital formation, and a reallocation of labor to lower-wage industries, all of which will cause gross national product to be lower than it would have been in the absence of the tax.

The tax rate needed to achieve a fixed absolute emissions target, such as maintaining emissions at 1990 levels, will depend on how fast emissions grow in the absence of the tax. Baseline emissions growth, in turn, will depend on the rate of productivity growth, the rate of capital accumulation, the rate of growth of the labor force, any energy-saving biases in technical change, and the path of world oil prices. More rapid economic growth will generally lead to higher baseline emissions and will thus require higher tax rates if emissions are to be held at a fixed absolute level. Moreover, deeper absolute cuts in emissions will require sharply increasing tax rates.

A carbon tax large enough to have much effect on emissions will raise tens

[2]In other words consumers and firms will conserve energy as it becomes more expensive.

to hundreds of billions of dollars annually. How this revenue is used will affect the overall economic burden of the tax. By using the revenue to lower highly distortionary taxes elsewhere in the economy, the tax could actually increase output and economic welfare. The remainder of the paper presents our model and discusses these findings in more detail.

2. An overview of the model

Our results are based on a set of simulations we conducted using a detailed model of the United States economy designed specifically for examining the effects of energy and environmental policies. One feature of our approach which distinguishes it from many others is that we use a general equilibrium model. General equilibrium models are constructed by dividing the economy into a collection of interdependent sectors which interact through product and factor markets. The behavior of each sector is represented by an appropriate submodel. Prices and wages adjust until demands and supplies are equated in every market and the economy reaches equilibrium. Our model is composed of thirty-five producing sectors, a consumer sector, an investment sector, a government sector and a foreign sector. In this section we present an overview of the model by describing the submodels used to represent each of these sectors. We also discuss our base case simulation.[3]

2.1. Production

Production is disaggregated into the thirty-five industrial sectors listed in table 1. Most of these industries match two-digit sectors in the Standard Industrial Classification (SIC). Each industry produces a primary product and may produce one or more secondary products. This level of industrial detail makes it possible to measure the effect of changes in tax policy on relatively narrow segments of the economy. Since most anthropogenic carbon dioxide emissions are generated by fossil fuel combustion, a disaggregated model is essential for capturing differences in the response of each sector to a carbon dioxide control policy.

We derive the behavior of each of the industries from a hierarchical tier-structured transcendental logarithmic cost function. At the highest level, the cost of each industry's output is assumed to be a function of the prices of energy, materials, capital services and labor. At the second level, we take the price of energy to be a function of prices of coal, crude petroleum, refined petroleum, electricity, and natural gas, and the price of materials to be a

[3]For more detail on the specification of the model or the base case simulation, see Jorgenson and Wilcoxen (1990b).

Table 1
Industry definitions.

Num	Description
1	Agriculture, forestry and fisheries
2	Metal mining
3	Coal mining
4	Crude petroleum and natural gas extraction
5	Nonmetallic mineral mining
6	Construction
7	Food and kindred products
8	Tobacco manufactures
9	Textile mill products
10	Apparel and other textile products
11	Lumber and wood products
12	Furniture and fixtures
13	Paper and allied products
14	Printing and publishing
15	Chemicals and allied products
16	Petroleum refining
17	Rubber and plastic products
18	Leather and leather products
19	Stone, clay and glass products
20	Primary metals
21	Fabricated metal products
22	Machinery, except electrical
23	Electrical machinery
24	Motor vehicles
25	Other transportation equipment
26	Instruments
27	Miscellaneous manufacturing
28	Transportation and warehousing
29	Communication
30	Electric utilities
31	Gas utilities
32	Trade
33	Finance, insurance and real estate
34	Other services
35	Government enterprises

function of the prices of all other intermediate goods. Given this structure we derive factor demands for capital services, labor and intermediate inputs from each of the thirty-five industries.

We estimated the parameters of each industry submodel econometrically, using a set of consistent inter-industry transactions tables constructed for the purpose. The tables describe the US economy for the period 1947 through 1985.[4] Estimating the production parameters over a long time series ensures

[4]Data on inter-industry transactions are based on input-output tables for the US constructed by the Bureau of Economic Analysis (1984). Income data are from the US national income and product accounts, also developed by the Bureau of Economic Analysis (1986). The data on capital and labor services are described by Jorgenson (1990). Additional details are given by Wilcoxen (1988, Appendix C), and Ho (1989).

that each industry's response to changes in prices is consistent with historical evidence.[5]

An unusual feature of our model is that productivity growth is determined endogenously.[6] Other models used to study global warming, for example, Manne and Richels (1992), take productivity growth to be exogenous. In our model the rate of productivity growth in each industry is determined endogenously as a function of input prices. In addition, each industry's productivity growth can be biased toward some inputs and away from others. Biased productivity growth is a common feature of historical data but is often ignored when modeling production. By allowing for biased productivity growth, our model is able to capture the evolution of industry input patterns much more accurately.

Although we allow for biased productivity growth, we do not impose that technical change be energy-saving in every industry. Most other participants in EMF-12 introduce energy-saving technical change through an exogenous parameter giving the rate of 'Autonomous Energy Efficiency Improvements.' The extent to which technical change is energy-saving is a source of considerable controversy. Manne and Richels (1992) suggest that engineering analysis shows technical change to energy-saving, while Hogan and Jorgenson (1991) present econometric evidence that aggregate technical change may be slightly energy-using. However, these positions are not necessarily inconsistent. After the sharp increases in energy prices of the 1970s, the US economy became markedly less energy-intensive as producers and consumers substituted away from energy. To the engineers reported by Manne and Richels, this would have appeared as a shift toward energy-saving technology. Hogan and Jorgenson's study reports biases in technical change *after* accounting for movements induced by substitution. Our model is based on the same data set as Hogan and Jorgenson's work, and so has the technical change properties they describe.

In sum, the salient features of our model of producer behavior are as follows. First, production is disaggregated into thirty-five industries. Second, all parameters of the model are estimated econometrically from an extensive historical data base developed specifically for this purpose. This allows the model to incorporate extensive historical evidence on the price responsiveness of input patterns, including changes in the mix of fossil fuels. Third, the model determines rates of productivity growth endogenously and allows for biased productivity change in each industry.

[5]A detailed discussion of our econometric methodology is presented by Jorgenson (1984, 1986).

[6]Our approach to endogenous productivity growth was originated by Jorgenson and Fraumeni (1981).

2.2. Consumption

We represent consumer behavior by assuming that households follow a three-stage optimization process. At the first stage, each household allocates full wealth (the sum of financial and human wealth, plus the imputed value of leisure time) across different periods.[7] We formalize this decision by introducing a representative agent who maximizes an additive intertemporal utility function subject to an intertemporal budget constraint. The portion of full wealth allocated to a particular period is called full consumption. At the second stage, households allocate full consumption to goods and leisure in order to maximize an indirect utility function. This allows us to derive demands for leisure and goods as functions of prices and full consumption. The demand for leisure implicitly determines labor supply, while the difference between current income and consumption of goods implicitly determines savings.

The third stage of the household optimization problem is the allocation of total expenditure among capital services, labor services, and the thirty-five commodities. At this stage, we relax the representative consumer assumption in favor of the approach of Jorgenson et al. (1982) to derive separate systems of demand functions for households of different demographic characteristics. We distinguish among 1344 household types according to demographic characteristics such as the number of household members and the geographic region in which the household is located. The spending patterns of each household type are derived from a hierarchial tier-structured indirect utility function. This allows us to derive household demands for individual commodities.

As with production, the parameters of the behavioral equations for all three stages of our consumer model are estimated econometrically.[8] Our household model incorporates extensive time series data on the price responsiveness of demand patterns by consumers and also makes use of detailed cross-section data on the effects of demographic characteristics on consumer behavior. In addition, an important feature of our approach is that we do not impose that household demands be homothetic. As total expenditure increases, spending patterns may change even in the absence of price changes. This captures an important feature of cross-sectional expenditure data which is often ignored.

2.3. Investment and capital formation

We assume there is a single capital stock in the economy which is in fixed

[7]Full wealth is defined to be the sum of financial wealth, the present value of future labor earnings, and the imputed value of leisure time.

[8]Details on the econometric methodology are given by Jorgenson (1984). Additional details are provided by Wilcoxen (1988) and Ho (1989).

total supply in the short run. However, we also assume that capital is perfectly malleable and can be reallocated among industries and between industries and final demand categories at zero cost. Thus, the price of a unit of capital services will be equal in every industry and there will be a single economy-wide rate of return on capital.

In the long run the supply of capital is determined by investment. Our investment model is based on the assumption that investors have rational expectations and that arbitrage occurs until the present value of future capital services is equated to the purchase price of new investment goods. This equilibrium is achieved by adjustments in prices and the term structure of interest rates. New capital goods are produced from individual commodities according to a model identical to those for the industrial sectors so the price of new capital will depend on commodity prices. We estimated the behavioral parameters for new capital goods production using final demand data for investment over the period 1947–1985. Thus, the model incorporates substitution among inputs in the composition of the capital.

2.4. Government and foreign trade

The two remaining parts of the model are the government and foreign sectors. To specify government behavior, we begin by computing total government spending on goods and services. We apply exogenous tax rates[9] to taxable transactions in the economy and then add the capital income of government enterprises and non-tax receipts to obtain total government revenue.[10] Next we assume the government budget deficit can be specified exogenously and add the deficit to total revenue to obtain total government spending. To arrive at government purchases of goods and services, we subtract interest paid to holders of government bonds together with transfer payments to domestic and foreign recipients. We then allocate spending among commodity groups according to fixed shares constructed from historical data.

In the modeling the foreign sector we begin by assuming that imports are imperfect substitutes for similar domestic commodities.[11] The mix of goods purchased by households and firms reflects substitution between domestic and imported products. We estimate the price responsiveness of this mixture econometrically from historical data. In effect, each commodity is assigned a separate elasticity of substitution between domestic and imported goods.

[9]These tax rates are based on data extending through 1985 and hence predate the 1986 tax reform. We are in the process of updating our description of the tax code.

[10]The capital income of government enterprises is endogenous while non-tax receipts are exogenous.

[11]This is the Armington (1969) approach. See Wilcoxen (1988) and Ho (1989) for further details on our implementation of this approach.

Table 2
Domestic production and heat content of fossil fuels.

Fuel	Unit	Domestic production	MBTU per unit	Total QBTU
Coal	ton	916.9×10^6	21.94	20.1
Oil	bbl	3033.2×10^6	5.80	17.6
Gas	kcf	17.8×10^9	1.03	16.8

Since the prices of imports are given exogenously, intermediate and final demands implicitly determine the quantity of imports of each commodity.

Exports are determined by a set of isoelastic export demand equations, one for each commodity, that depend on foreign income and the foreign prices of US exports.[12] Foreign prices are computed from domestic prices by adjusting for subsidies and the exchange rate. The demand elasticities in these equations are estimated from historical data. Without an elaborate model of international trade it is impossible to determine both the current account balance and the exchange rate endogenously so we take the current account to be exogenous and the exchange rate to be endogenous.

2.5. Carbon dioxide emissions

For EMF-12, the most important remaining feature of the model is the way in which carbon dioxide emissions are calculated We begin by assuming that carbon dioxide is produced in fixed proportion to fossil fuel use. For each fuel, table 2 gives total domestic production, heat content per unit and total heat produced.[13] Heat production is measured in quadrillion BTU.

We calculated the carbon content of each fuel by multiplying the numbers in table 2 by figures from the Environmental Protection Agency on the amount of carbon emitted per million BTU produced from each fuel.[14] Total carbon emissions were then calculated using figures on total fuel production. Table 3 shows data for each fuel in 1987. For comparability with other studies, we measure CO_2 emissions in tons of contained carbon.[15]

All prices in our model are normalized to unity in 1982, so quantities do not correspond directly to physical units. Moreover, the model has a single sector for oil and gas extraction. To convert the oil and gas data in table 3 into a form appropriate for the model we added carbon production for crude petroleum and natural gas and divided by the industry's output for 1987 to obtain the industry's carbon coefficient. Similarly, the coefficient for coal was obtained by dividing total carbon production from coal by the model's 1987

[12] We take foreign income to be exogenous.
[13] The data is taken from Department of Energy (1990).
[14] Environmental Protection Agency (1990) internal memorandum.
[15] To convert to tons of carbon dioxide, multiply by 3.67.

Table 3

Carbon emissions data for 1987

Item	Unit	Fuel		
		Coal	Oil	Gas
Unit of measure	ton	bbl	kcf	
Heat content	10^6 BTU/unit	21.94	5.80	1.03
Emissions rate	kg/10^6 BTU	26.9	21.4	14.5
	kg/unit	590.2	124.1	14.9
Domestic output	10^9 units	0.92	0.30	17.8
Carbon emissions	10^6 tons	595.3	414.1	268.6

value for coal mining output. These coefficients were used to estimate carbon emissions in each simulation.

2.6. The base case

To assess the effect of a carbon tax we must first determine the future path of the US economy in the absence of the tax. To construct such a scenario, which we will call a 'base case', we adopted a set of default assumptions about the time path of each exogenous variable in the absence of changes in government policy. Since our model is based on agents with perfect foresight values must be specified far into the future. Through 1990–2050, we forecast values of the exogenous variables on the basis of their behavior in the sample period. After 2050 we assume the variables remain constant at their 2050 values to allow the model to converge to a steady state by the year 2100.[16]

Our projections for 1990–2050 were made as follows. First, all tax rates are set to their values in 1985, the last year in our sample period. Next, we assume that foreign prices of imports in foreign currency remain constant in real terms at 1985 levels. We then project a gradual decline in the government deficit through the year 2025, after which the nominal value of the government debt is maintained at a constant ratio to the value of the national product. Finally, we project the current account deficit by allowing it to fall gradually to zero by the year 2000. After that we project a current account surplus sufficient to produce a stock of net claims on foreigners by the year 2050 equal to the same proportion of national wealth as in 1982.

Some of the most important exogenous variables are those associated with growth of the US population and corresponding changes in the economy's time endowment. We project population by age, sex and educational attainment through the year 2050, using demographic assumptions consistent with Social Security Administration projections.[17] After 2050 we hold

[16]A more detailed discussion of the base case is given by Jorgenson and Wilcoxen (1992).

[17]Our breakdown of the US population by age, sex, and educational attainment is based on the system of demographic accounts compiled by Jorgenson and Fraumeni (1989).

population constant, which is roughly consistent with Social Security projections. In addition, we project the educational composition of the population by holding the level of educational attainment constant beginning with the cohort reaching age 35 in the year 1985. We transform our population projection into a projection of the time endowment by assuming that the pattern of relative wages across different types of labor remains as it was in 1985. Since capital formation is endogenous in our model, our projections of the time endowment effectively determine the size of the economy in the more distant future.

3. A summary of results

One policy often proposed for slowing global warming is to stabilize greenhouse gas emissions at 1990 levels. This corresponds to EMF-12 scenario 4, so we begin our discussion there. Subsequent sections will examine the effects of more stringent emissions targets, the consequences of using carbon tax revenue in different ways, and the sensitivity of results to assumptions embodied in the base case scenario.

3.1. Stabilizing emissions

To implement scenario 4 we introduced a tax on the carbon content of primary fossil fuels. The tax was applied to both domestic and imported sources: domestic fuels were at the mine mouth or well head while imported fuels were taxed at the point of import. The tax rate varied from year to year but was always chosen to be exactly enough to hold US carbon dioxide emissions at their 1990 value of 1576 million tons. We returned the revenue raised by the tax to households as a lump-sum rebate. The carbon tax we obtained is shown as a function of time in fig. 1. Base case emissions increase over time so the tax grows gradually over the next few decades. By 2020 our forecast of the US population crests and growth begins to slow, reducing the rate of carbon tax growth.[18] The tax produces significant reductions in carbon emissions, as shown in fig. 2. By 2020 emissions are sixteen percent lower than they would have been without the tax. The tax also produces significant revenue: $31 billion annually by 2020.[19]

The direct effect of the tax is to increase purchasers' prices of coal and crude oil. By 2020, for example, the tax reaches $22.71 per ton of carbon. This amounts to a tax of $14.75 per ton of coal, $3.10 per barrel of oil or $0.37 per thousand cubic feet of gas. Fig. 3 shows the effect of the tax on

[18]As noted above, our population projection is based on forecasts made by the Social Security Administration in which population growth approaches zero early in the next century.

[19]All dollar amounts are in 1990 prices.

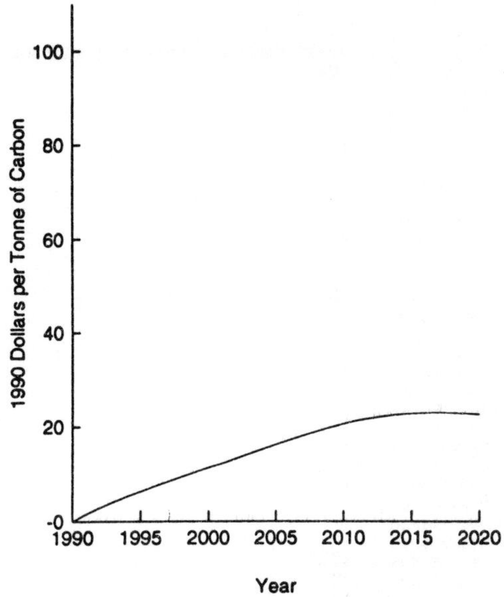

Fig. 1. Carbon tax under scenario 4.

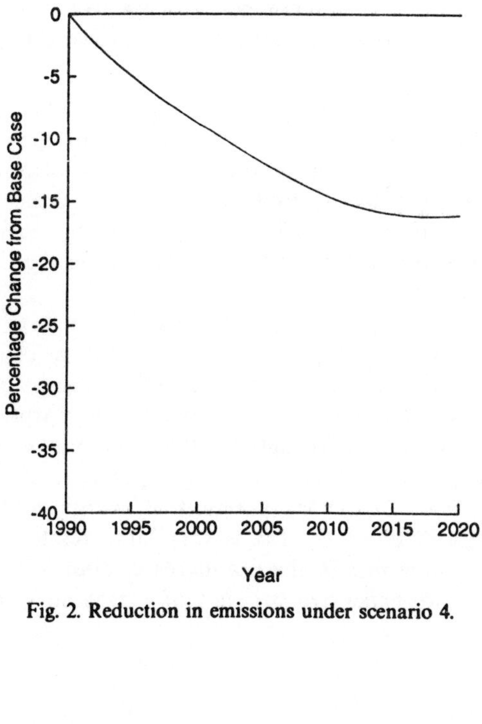

Fig. 2. Reduction in emissions under scenario 4.

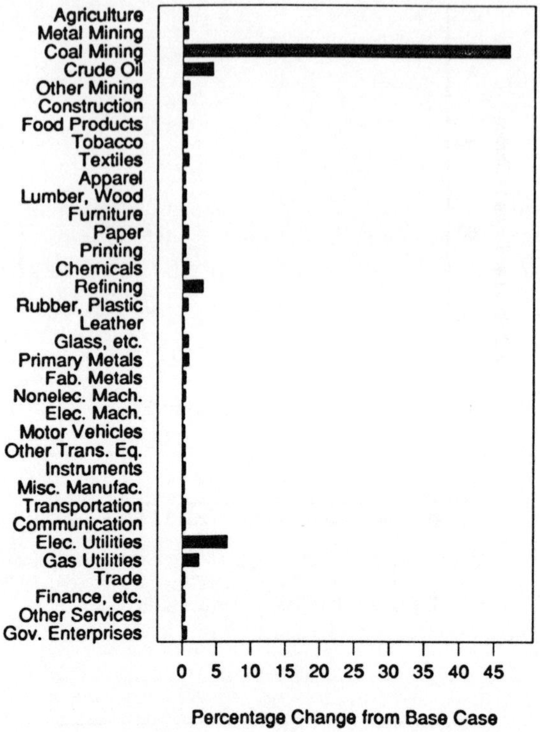

Fig. 3. Changes in prices at 2020 under scenario 4.

purchasers' prices of each industry's output in 2020. The price of coal rises by forty-seven percent, the price of electricity rises by almost seven percent (coal accounts for about thirteen percent of the cost of electricity), and the price of crude oil rises by around four percent. Other prices showing significant effects are those for refined petroleum and natural gas utilities. These rise, directly or indirectly, because of the tax on the carbon content of oil and natural gas.

Changes in relative prices affect demands for each good and lead to changes in industry outputs. Fig. 4 presents percentage changes in industry outputs in 2020 relative to the base case. Most sectors show only small changes in output. Coal mining is an exception: its output falls by almost thirty percent. Coal is affected strongly for three reasons. First, coal emits the more carbon dioxide than oil or natural gas per unit of energy produced. Thus, the absolute level of the tax per unit of energy content is higher on coal than other fuels. Second, the tax is very large relative to the base case price of coal: at the mine mouth the tax increases coal prices by around fifty percent. Oil is far more expensive per unit of energy so in percentage terms

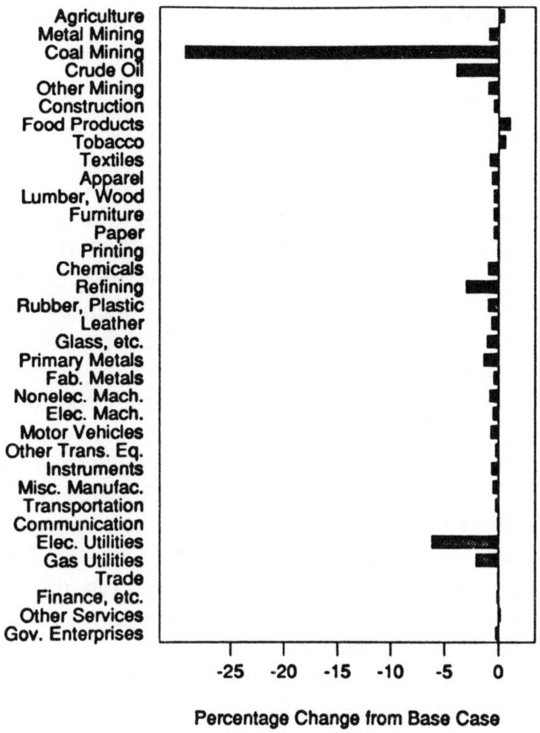

Fig. 4. Change in output at 2020 under scenario 4.

its price is less affected by the tax. In fact, the price of crude oil rises only about ten percent. Third, the demand for coal is relatively elastic. Most coal is purchased by electric utilities which can substitute other fuels for coal when the price rises. Moreover, the demand for electricity itself is relatively elastic so when the price of electricity rises, demand for electricity (and hence demand for coal) falls substantially.

For the rest of the economy, the main result of the tax is to increase the prices of electricity, refined petroleum and natural gas, each by a few percent. This would have two effects. First, higher energy prices would mean that capital goods (which are produced using energy) would become more expensive. To the extent that domestic saving does not rise enough to compensate, higher prices for capital goods mean a slower rate of capital accumulation and lower GNP in the future. Second, higher energy prices discourage technical change in industries in which technical change is energy-using. Together, these two effects cause the capital stock to drop by 0.7% and GNP to fall by 0.5% (both relative to the base case) by 2020. Average annual GNP growth over the period 1990–2020 is 0.02 percentage points lower than in the base case. About half of this is due to slower productivity growth and half due to reduced capital formation.

3.2. Larger reductions

EMF-12 scenarios 2 and 3 specify more ambitious goals for carbon dioxide control. In scenario 2, emissions are required to decrease gradually until 2010, when they must be twenty percent below 1990 levels. In scenario 3, emissions must fall by twenty percent in 2010 and then by fifty percent by 2050. Fig. 5 shows the paths of carbon taxes needed over time in scenarios 2, 3 and 4 while fig. 6 shows the paths of carbon emissions in each case. Table 4 compares key long-run results for these two simulations with results for scenario 4.

Comparing scenarios 2 and 3 with 4 shows that increasing the stringency of the policy rapidly increases its cost. Moving from scenario 4 to 2 doubles the effect of the policy: emissions fall by 32.90% instead of 16.12%. However, this comes at the cost of tripling the carbon tax and the loss of output. Moving from scenario 4 to scenario 3 raises costs even more sharply: the carbon tax and loss of GNP go up by more than a factor of four while the carbon reduction rises by only a factor of 2.4.

Comparing industry results across each of the scenarios shows that the more stringent policies have progressively larger effects on sectors other than coal mining. Doubling the emissions reduction by moving from scenario 4 to scenario 2 does not cause the reduction in coal output to double. Coal users, particularly electric utilities, find it increasingly difficulty to substitute away from coal. Thus, larger reductions in carbon emissions require progressively larger reductions in oil use. From scenario 4 to scenario 2, for example, the drop in oil use more than triples.

3.3. Use of carbon tax revenues

Even a modest carbon tax would raise substantial revenue. In the simulations above we assumed the proceeds were returned to households by a lump-sum rebate. This is probably not the most likely use of the revenue, nor is likely to be the best use given that the federal budget deficit is large and that other government revenue is raised by distortionary taxes. Using the revenue to reduce a distortionary tax would lower the net cost of the carbon tax by removing inefficiency elsewhere in the economy.

To determine how large this efficiency improvement might be we constructed three simulations based on EMF-12 scenario 7. In each simulation

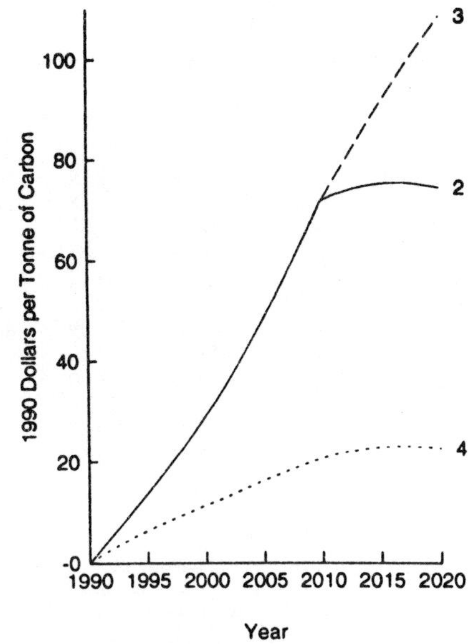

Fig. 5. Carbon taxes under scenarios 2–4.

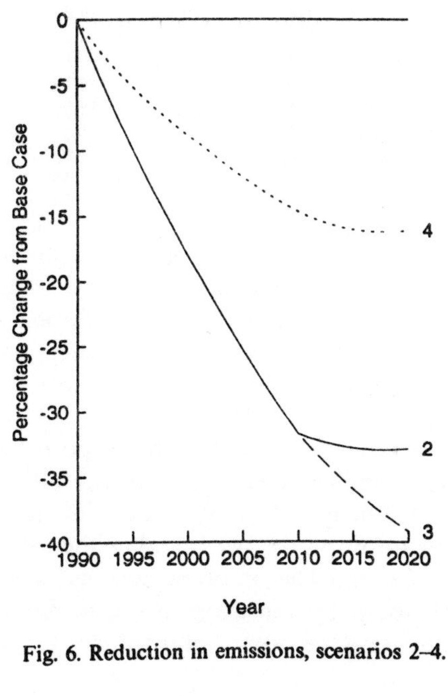

Fig. 6. Reduction in emissions, scenarios 2–4.

Table 4

Selected results for scenarios 2–4 in 2020.

Variable	Unit	Scenario		
		2	3	4
Carbon emissions	%Δ	−32.90	−39.19	−16.12
Carbon tax	$/t	74.49	108.78	22.71
Price of capital	%Δ	1.10	1.40	0.40
Capital stock	%Δ	−2.35	−2.95	−0.83
Tax revenue	$B	82.52	109.16	31.41
Real GNP	%Δ	−1.71	−2.36	−0.55
Coal price	%Δ	149.86	216.09	46.99
Coal output	%Δ	−55.03	−62.47	−29.28
Electricity price	%Δ	19.46	26.90	6.60
Electricity output	%Δ	−16.43	−21.74	−6.17
Oil price	%Δ	15.06	22.37	4.45
Oil output	%Δ	−12.35	−17.63	−3.90

Table 5

Selected results revenue experiments in 2020.

Variable	Unit	Revenue policy		
		Lump	Labor	Capital
Carbon emissions	%Δ	−32.24	−32.09	−31.65
Carbon tax	$/t	64.83	64.83	64.83
Price of capital	%Δ	0.97	−1.86	0.23
Capital stock	%Δ	−2.13	−1.36	1.89
Tax revenue	$B	79.65	79.82	80.35
Real GNP	%Δ	−1.70	−0.69	1.10
Coal price	%Δ	143.49	140.57	142.06
Coal output	%Δ	−54.14	−54.19	−53.45
Electricity price	%Δ	18.57	15.97	16.99
Electricity output	%Δ	−15.93	−15.37	−14.66
Oil price	%Δ	14.20	12.28	14.55
Oil output	%Δ	−11.92	−11.54	−11.39

we impose a carbon tax of $15 per ton in 1990 with the rate rising by 5% annually in subsequent years. The simulations differ in how the revenue was used. In the first simulation the revenue was returned to households by a lump-sum rebate; in the second it was used to lower taxes on labor; and in the third it was used to lower taxes on capital. Table 5 reports results from the three simulations.

The GNP results in table 5 show that the disposition of revenue from a carbon tax has a very significant effect on its overall impact. In the lump-sum case, output in 2020 drops by 1.70% relative to the base case. When the revenue is returned by lowering the tax on labor the loss of GNP is less than half as much: only 0.69%. The improvement is due to an increase in employment brought about by the drop in the wedge between before- and after-tax wages. If the revenue were returned as a reduction in taxes on capital, GNP would actually increase above its base case level by 1.10%. In this case, the gain is due to accelerated capital formation generated by an increase in the after-tax rate of return on investment. These results suggest that a carbon tax would provide an opportunity for significant tax reform.

3.4. Sensitivity to base case assumptions

Most EMF-12 scenarios required carbon dioxide emissions to be held indefinitely at a specified absolute level, such as that prevailing in 1990. In general, uncontrolled emissions will rise at the overall rate of economic growth less adjustments for increasing fossil fuel prices and energy-saving technical change. Achieving a fixed target in the face of continuously rising baseline emissions will require a growing carbon tax. At each point in time, the magnitude of the tax will depend on the size of the gap between uncontrolled emissions and the target. The more rapid the growth of uncontrolled emissions, the larger will be the gap and hence the larger will be the tax needed to keep emissions constant.

The growth of carbon emissions is determined by productivity growth, the rate of capital accumulation, the rate of labor supply growth, any energy-saving biases in technical change, and the path of world oil prices. For most models participating in EMF-12, many of these factors were exogenous. In fact, the EMF-12 study design includes a specified, exogenous GNP growth rate for each region. In our model, however, productivity growth, capital accumulation and biases in the rate of technical change are all endogenous. Moreover, productivity growth and biased technical change occurs at the level of individual industries rather than at the level of the economy as a whole. Thus, GNP growth is fundamentally endogenous in our model.

This difference in the treatment growth makes direct comparisons of our results with those of the other EMF-12 participants difficult. Our base case estimate of US growth is substantially below the rate specified in the EMF-12 study design. As a result, our base case path of carbon emissions rises more slowly than it would if growth were faster. Thus, the carbon taxes needed to hold emissions at a fixed absolute level will tend to be lower in our model. To allow our results to be compared more easily with those of the other participants, and to explore the importance of GNP growth in general, EMF-12 scenario 12 was designed. Scenario 12 was identical to scenario 2 (a twenty percent reduction of emissions below 1990 levels), except that the growth rate of US GNP was set to the value predicted by our base case simulation. Thus, the results of other participants could be directly compared with ours.

The principal finding of scenario 12 was that the carbon taxes and GNP loss predicted by other participants dropped sharply. Comparing our results for scenario 2 to the scenario 12 results of others, our carbon tax moves from being particularly low to being near the middle of the group. Our prediction for loss of GNP due to the tax moves from the middle of the group to among the highest. This is precisely what might be expected given that our initial growth rate was relatively low. Our model incorporates stronger intertemporal links between current carbon taxes, capital accumulation and productivity growth than most of the other models. Thus, our model should generate higher GNP losses than the others. Scenario 12 shows that after adjusting for base case emissions growth, this is in fact the case. More generally scenario 12 emphasizes that overall economic growth will have a profound effect on the cost of controlling carbon dioxide emissions.

4. Conclusion

In summary, our approach to EMF-12 has three features that distinguish it from others: our model is disaggregated to the level of individual industries, our behavioral parameters are obtained by econometric estimation, and productivity growth and biased technical change are endogenous in our model. These features allow us to investigate several aspects of carbon taxes that other EMF-12 participants were unable to examine. In particular, we find that a carbon tax will have sharply differing effects in different industries. The tax will fall most heavily on coal mining, leading to a substantial drop in coal use by electric utilities and other industries. Outside the coal industry, the main effect of the tax will be to increase the price of electricity. This will reduce productivity growth and slow the rate of capital formation, leading to slower growth of US output.

The tax needed to attain any particular emissions target will depend heavily on how fast carbon dioxide emissions would have grown in the absence of the tax. Higher rates of GNP growth will lead to higher baseline emissions and will thus require higher carbon taxes. In addition, to achieve larger reductions in carbon emissions will require sharply increasing carbon taxes and will produce markedly higher losses of output.

Finally, a carbon tax large enough to have much effect on carbon emissions will raise considerable revenue. How this revenue is used will affect the overall economic burden of the tax. If the revenue were used to reduce distortionary taxes elsewhere in the economy, the impact of the tax on GNP would be reduced. In fact, it is possible that GNP would actually increase if the revenue were used to reduce taxes on capital.

References

Armington, P.S., 1969, The geographic pattern of trade and the effects of price changes, IMF Staff Papers 16, 176–199.

Bureau of Economic Analysis, 1984, The input–output structure of the U.S. economy, 1977, Survey of Current Business 64, 42–79.

Bureau of Economic Analysis, 1986, The income and product accounts of the United States, 1929–1982: Statistical tables, U.S. Department of Commerce, Washington, DC.

Energy Information Administration, 1990, Annual energy review 1990, U.S. Department of Energy, Washington, DC.

Environmental Protection Agency, 1990, internal memorandum on carbon emissions coefficients.

Gaskins, D. and J.P. Weyant, forthcoming, Reducing carbon emissions from the energy sector: Cost and policy options (MIT Press, Cambridge, MA).

Ho, M.S., 1989, The effects of external linkages on U.S. economic growth: A dynamic general equilibrium analysis, Ph.D. Dissertation, Harvard University.

Hogan, W.W. and D.W. Jorgenson, 1991, Productivity trends and the costs of reducing carbon dioxide emissions, The Energy Journal 12, 67–85.

Jorgenson, D.W., 1984, Econometric methods for applied general equilibrium analysis, in: H.E. Scarf and J.B. Shoven, eds., Applied general equilibrium analysis (Cambridge University Press, Cambridge, MA) 139–203.

Jorgenson, D.W., 1986, Econometric methods for modeling producer behavior, in: Z. Griliches and M.D. Intriligator, eds., Handbook of econometrics, Vol. 3 (North-Holland, Amsterdam), 1841–1915.

Jorgenson, D.W., 1989, Capital as a factor of production, in: D.W. Jorgenson and R. Landau, eds., Technology and capital formation (MIT Press, Cambridge, MA) 1–35.

Jorgenson, D.W., 1990, Productivity and economic growth, in: E.R. Berndt and J. Triplett, eds., Fifty years of economic measurement (University of Chicago Press, Chicago, IL) 19–118.

Jorgenson, D.W. and B.M. Fraumeni, 1981, Relative prices and technical change, in: E.R. Berndt and B. Field, eds., Modeling and measuring natural resource substitution (MIT Press, Cambridge, MA) 17–47.

Jorgenson, D.W. and B.M. Fraumeni, 1989, The accumulation of human and nonhuman capital, 1948–1984, in: R.E. Lipsey and H.S. Tice, eds., The measurement of saving, investment, and wealth (University of Chicago Press, Chicago, IL) 227–282.

Jorgenson, D.W. and P.J. Wilcoxen, 1990a, environmental regulation and U.S. economic growth, Rand Journal of Economics 21, 314–340.

Jorgenson, D.W. and P.J. Wilcoxen, 1990b, Intertemporal general equilibrium modeling of U.S. environmental regulation, Journal of Policy Modeling 12, 715–744.

Jorgenson, D.W. and P.J. Wilcoxen, 1992, Reducing US carbon dioxide emissions: The cost of different goals, in: John R. Moroney, ed., Advances in the economics of energy and natural resources, Vol. 7 (JAI Press, Greenwich, CT) 125–158.

Jorgenson, D.W. and K.-Y. Yun, 1986, The efficiency of capital allocation, Scandinavian Journal of Economics 88, 85–107.

Jorgenson, D.W. and K.-Y. Yun, 1991, The excess burden of U.S. taxation, Journal of Accounting, Auditing, and Finance 6, 487–509.

Jorgenson, D.W., L.J. Lau and T.M. Stoker, 1982, The transcendental logarithmic model of aggregate consumer behavior, in: R.L. Basmann and G. Rhodes, eds., Advances in econometrics, Vol. 1 (JAI Press, Greenwich, CT) 97–238.

Manne, A.S. and R.G. Richels, 1992, Buying greenhouse insurance: The economic costs of CO_2 emissions limits (MIT Press, Cambridge, MA).

Wilcoxen, P.J., 1988, The effects of environmental regulation and energy prices on U.S. economic performance, Ph.D. Dissertation, Harvard University.

[9]

The Cost of Slowing Climate Change: a Survey

*William D. Nordhaus**

Policies to deal effectively and efficiently with the threat of greenhouse warming must balance the costs of slowing climate change against the potential damages. This survey discusses one half of this question, the costs of slowing climate change by reducing greenhouse gas (GHG) emissions. The analysis provides estimates of the cost of reducing chlorofluorocarbon and CO_2 emissions, and inquires into the costs of using forestry options to remove CO_2 from the atmosphere. A promising new approach, the use of geoengineering, is discussed qualitatively.

BACKGROUND CONSIDERATIONS

At the outset, it must be noted that analysis of the costs of slowing climate change is difficult because of the complexity of the problem and because technology and the economy are evolving. Therefore, the "correct" estimates will change over time. This paper represents a snapshot of estimates as of late 1990.

Measurement of CO_2-Equivalent Emissions and the Contribution of Different Greenhouse Gases

The concern about the greenhouse effect occurs today because human activities are currently raising atmospheric concentrations of greenhouse gases and threatening a significant and undesirable climate change. The major anthropogenic greenhouse gases (GHGs) are carbon dioxide, methane, nitrous oxides, and chlorofluorocarbons (CFCs). Scientific monitoring has firmly established the buildup of the major GHGs. Table 1 shows the important GHGs, recent and projected concentrations, and the past and estimated future growth rates of major GHGs.

One of the major difficulties that arises in analyzing the cost of GHG reduction is the multitude of greenhouse gases in addition to CO_2 that are contributing to greenhouse warming. In the analysis that follows, it will be

The Energy Journal, Volume 12, Number 1.

*Department of Economics, Yale University, New Haven, Connecticut

Table 1. Estimated Concentrations of Important Greenhouse Gases

Atmospheric Concentration of Greenhouse Gas					
	Level (parts per billion)			Growth (percent per year)	
Greenhouse Gas	1850	1986	2100	1850-1986	1986-2100
Carbon Dioxide (000)	290	348	630	0.16	0.52
Methane	880	1675	3100	0.56	0.54
Nitrous Oxides	285	340	380	0.15	0.10
CFCs (11 and 12)	0	0.618	0.700	—	0.11

Sources: Wuebbels and Edmunds (1988) and EPA (1989). Data for CFCs assume that international efforts succeed in stabilizing atmospheric concentrations at 1990 levels.

important to consider equivalent units of GHGs. For this purpose we transform each of the GHGs into its CO_2 equivalent. The CO_2 equivalent of a GHG is the quantity of CO_2 that would lead to the equivalent amount of total warming.

Often, studies of greenhouse warming use an "instantaneous" measure of warming. That is, warming is defined as the increase in atmospheric concentration of a gas times the instantaneous or current radiative impact of that gas. This calculation is defective because it ignores the fact that substances have different lifetimes and are transformed over time into other substances. A more satisfactory index of the importance of different gases measures the *total warming potential* of a GHG in terms of the warming over the indefinite future.

To calculate the total warming potential, we need to add together the warming in each future period. A complete dynamic analysis would also incorporate discounting to take into account that the cost of warming is different depending upon the time at which the warming occurs. Table 2 shows a comparison of the instantaneous and total warming potential of major GHGs and their source of emission in the mid-1980s.[1] The major result of this calculation is to show the dominance of CO_2 in long-term warming from GHG emissions over the next century. According to the conversion factors used by Lashof and Ahuja, the increase in CO_2 concentrations is currently contributing about 80 percent of the total potential global warming of the major GHGs. This figure is likely to rise over the next few decades as the CFCs are phased out.

Table 3 shows the estimates of CO_2-equivalent emissions of each of the major GHGs in 1985. The difference between the estimates is that the

1. A non-technical discussion is provided in Nordhaus, 1990a. The estimates used here rely on the more complete analysis of Lashof and Ahuja, 1990.

Table 2. Estimated Contribution of Different Greenhouse Gases to Global Warming for Concentration Changes, mid-1980s

Greenhouse Gas	Relative Contribution: Instantaneous	Total	Source of Emission
CO_2	53.2%	80.3%	Largely from combustion of fossil fuels.
Methane	17.3%	2.2%	Unknown. From a wide variety of biological and agricultural activities.
CFC11 and 12	21.4%	8.8%	Wholly industrial, from both aerosols and non-aerosols.
Nitrous Oxides	8.1%	8.7%	From fertilizers and energy use.

Source: Emission from EPA (1989), vol. I. Estimates of the ratio of total warming potential to instantaneous warming potential from Lashof and Ahufa (1990), with calculations and data explained in Nordhaus (1990a).

column on emissions shows the CO_2 equivalent of the total production in 1985. The second column reflects the fact that some production does not end up in the atmosphere so that the increase in atmospheric concentrations is less than production. The difference between these two estimates is small, however, as both show that CO_2 is approximately 80 percent of the total global CO_2-equivalent emissions of around 8 billion tons. In the analysis that follows, we will use these totals to estimate the cost of reducing GHG emissions and to compare the cost-effectiveness of different strategies.

Table 3. Global CO_2-equivalent Emissions, 1985 (Millions of metric tons, carbon content of CO_2 per year, total warming potential)

	Emission Weights	Concentration Weights
Carbon Dioxide	6500	6500
Methane	612	181
Nitrogen Oxides	549	703
Chlorofluorocarbons	261	714
Total, CO_2 Equivalent	7922	8098

Source: Emission and concentration weights are from EPA (1989). Estimates of global warming potential are described in Table 2 and text.

One important note on measurement units. In this study, we measure CO_2 in terms of its *carbon content*. The ratio of CO_2 weight to carbon weight is (12+16+16)/12=3.67.

Costs of Reduction: Theory and Methods

Before surveying different cost estimates, it will be useful to lay out the general points. Greenhouse warming is an example of an economic externality, which occurs because the emissions of GHGs are not reflected in market prices. In the case at hand, we can analyze the costs of slowing greenhouse warming in terms of a fundamental function: the *GHG-reduction cost function*, which describes the costs that must be paid to reduce GHG emissions. The cost function would include, for example, the cost of changing from fossil to non-fossil fuels, the substitution of different substances for CFCs, or the costs of removing CO_2 by growing trees.

In what follows, I concentrate upon *efficient* strategies to reduce the costs of climate change. An efficient strategy is one in which all steps to reduce GHG emissions are taken in a way that minimizes costs. Figure 1

Figure 1. Cost Function for GHG Reduction (Total and Marginal Costs of Reduction)

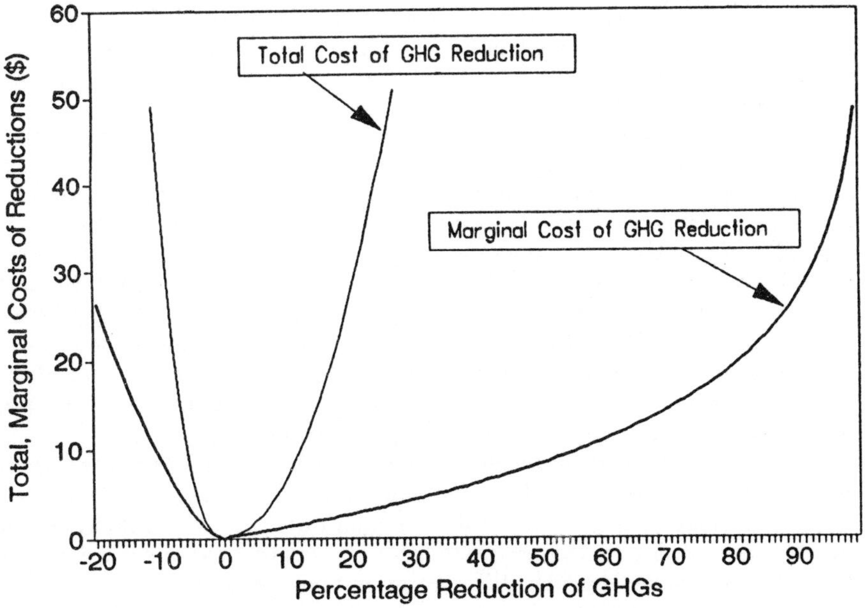

shows schematically, a U-shaped total cost curve as the GHG-reduction cost function. This represents the minimum cost that must be incurred at different levels of GHG concentrations. The horizontal axis measures the level of GHG reduction relative to the uncontrolled quantity. This variable has a value of 0 when GHGs are uncontrolled (i. e. in an unregulated environment like that of today). As GHG emissions are reduced, resources must be devoted to using alternative fuels, to insulating walls, and so forth. Because restricting GHG emissions is costly, the cost-reduction curve is upward sloping as GHG reductions increase.

From economic theory, we know a fair amount about the shape of the reduction cost function in a competitive economy with no other externalities and where controls are efficiently designed. First, we know that it has a minimum of zero at the uncontrolled point. Second, we know that the function is convex. Third, society can always do worse than the reduction cost function by inefficiently designing regulations.

The total cost function can be used to produce the crucial *marginal cost curve for reducing GHGs*, also sketched in Figure 1. The marginal cost curve indicates the incremental cost of reducing CO_2 or other GHGs by one unit. From economic theory, we know that, in an efficient program, the marginal cost of reduction of GHGs is an increasing (strictly, a non-decreasing) function to the right of the laissez-faire point at 0. This signifies that it becomes increasingly costly to reduce GHGs as the amount of reduction is larger.

A surprising result, however, is the following: *The first units of reduction of CO_2 and other greenhouse gases are virtually free. This is the result of the zero market price on the emission of trace gases*. This result is seen by the fact that the marginal cost curve touches the horizontal axis at zero reduction in Figure 1.

One final analytical point should be noted: Studies of the cost of GHG reduction must distinguish between the average, marginal, and total cost of reducing GHGs. This distinction is important because the marginal cost is likely to be far in excess of the average cost. For example, assume that the total cost of reducing CO_2 emissions follows the equation: $C(R) = \beta R^{\alpha}$, where $C(R)$ is the total cost function, β is a constant, and α is an exponential term. For this equation, marginal cost is $MC = C'(R) = \alpha\beta R^{\alpha-1}$, average cost is $AC = C(R)/R = \beta R^{\alpha}/R = \beta R^{\alpha-1}$, and therefore the ratio of marginal to average cost is α. Estimates below indicate that α lies between 1.5 and 3, so marginal cost is likely to be 1.5 to 3 times larger than average cost. This point is important in our reconciliation of different estimates.

Another issue, to which we return below, concerns the assumptions used in this diagram that all markets are competitive and that there are no other external effects. Neither of these conditions is likely to hold in practice, and the deviation of actual markets from our competitive ideal is important to remember as we consider different greenhouse policies.

The purpose of this paper is to estimate, crudely at least, the cost curves in Figure 1. It must be emphasized that this depiction is highly oversimplified; a complete description of costs needs to incorporate the temporal dynamics of the economy, the life cycles of CO_2 and other gases, and the changing technology of the energy and other sectors. No puny task.

ESTIMATES OF COSTS OF GHG REDUCTION FOR DIFFERENT GASES

Estimation of Mitigation Costs for CO_2

As Tables 1 through 3 show, the most important source of reduction of GHGs is for CO_2. In contrast to the estimates for other GHGs, however, estimation of the costs of CO_2 reduction is complicated by the variety of sources, uses, forms of substitution, and potential for new low-CO_2 or no-CO_2 energy technologies. In this section we review some of the different studies of the costs of CO_2 reduction and estimate a cost-reduction schedule.

The estimates below can be grouped into three general categories: estimates for specific technologies, econometric or elasticity studies, and mathematical-programming or optimizing models. Some models are combinations of these, but we will group approaches in terms of their general structure.

Note the "costs" in these estimates in principle represent the *gross* resource cost or dead-weight loss to the economy of imposing CO_2 constraints. They are gross in the sense that they do not include any correction, plus or minus, for improvements in economic welfare due to slower climate change, nor do they include adjustments for any other externality, such as that due to CFC accumulation and ozone depletion. In addition, the costs are distinct from the budget impact of any carbon or energy taxes that might be levied.

In what follows, we will show the marginal cost of reducing CO_2 emissions as a function of the percentage reduction from a baseline path. It can be shown that this relationship is the same as the relationship between a uniform carbon tax and the CO_2 reduction.

Specific technologies

One approach has been to estimate the cost of CO_2 reduction by finding low-CO_2 technologies that can be substituted directly for high-CO_2 technologies. Examples include CO_2 scrubbing and substitution of methane for oil and coal.

Kram and Okken estimate that the costs of CO_2 removal from stack gases is in the order of 30 to 65 Dutch florins per ton CO_2 ($55 to $120 per ton C). These estimates are also consistent with the assumptions used in the larger model of Manne and Richels (1990). A major difficulty with these estimates is that they assume that the CO_2 will be disposed in gas or oil fields at low cost. This allows for only limited storage, after which the costs are likely to rise significantly.

An alternative technological substitution that has been investigated by Edmunds et al. (1989) is direct substitution of methane for oil and gas in electrical generation. This option turns out to be extremely expensive -- with their estimates being from $300 to $700 per ton of C removed. The high cost of this option arises from the fact that the economic penalty from use of methane is quite large in many large power plants. Other studies of the feasibility of substituting methane for coal sometimes overlook the limited resources of methane, a factor which is included in many energy-sector models.

In the estimates that follow, we will not include these options separately. The first option is indirectly included in the Manne-Richels study, while the second is allowed for in any model with fuel substitution. A third direct option, reforestation, is discussed in the last part of this section.

Econometric or elasticity studies

A second approach, which relies largely on behavioral estimates of supply and demand functions, is one we call "econometric." In these studies, estimates are made of the structure of demand and supply on the basis of observed market data on prices and quantities. The models are often energy models which have been extended to include CO_2 emissions. To obtain the impact of policies upon CO_2 emissions, a variety of policies are examined, such as imposing carbon taxes on consumption or production. There are a number of different models that take this approach, and we summarize briefly each of the studies that is used in the overall estimate (Table 4 and Figure 2 on pages 50 and 51, respectively, summarize the quantitative results of a carbon tax.):

[#8] *Nordhaus-Yohe*[2]. In a study of future emissions of CO_2 for the National Academy study, *Changing Climate*, Nordhaus and Yohe (1983) developed a simple econometric model of the global economy that was used to estimate the uncertainties of future CO_2 paths. That study also provided a methodology for estimating the impact of carbon taxes on CO_2 emissions. The assumptions underlying this model were that there are two kinds of fuels (carbon and non-carbon); two inputs into the aggregate production function (energy and non-energy); and several different kinds of technological change. Production is modeled as a series of nested CES production functions, and a simple resource availability constraint was included.

There are a number of reservations about the use of the Nordhaus-Yohe model for this purpose. The most important is that the model is highly oversimplified on the supply side, with only two technologies. It shares with other econometric approaches a difficulty in that it omits information about

2. The number in bracket before the name of the model refers to the identification number in Table 4 and Figure 2.

specific energy technologies. In this respect, it is likely to be most useful for small taxes and least useful for large taxes.

[#8] *Revised Nordhaus-Yohe*. Because the original Nordhaus-Yohe model included only a small number of tax cases, for this survey I simplified the original version of the model and calculated the impact of different carbon taxes in the simplified model. The changes from the original model were to assume that the elasticities of substitutions between all factors are unity and that the output elasticities of different inputs are constant. With these simplifying assumptions, I then estimated the impact of carbon taxes for a range of $0 to $300 per ton of carbon, also shown in Table 4 and Figure 2. The reservations about the revised Nordhaus-Yohe model are the same as those for the original discussed above.

[#2] *Edmunds-Reilly*. Edmunds and Reilly have developed a series of increasingly detailed models of the world energy system. These models are generally based on behavioral equations for energy demand and a combination of supply equations and specific technological assumptions for energy production. The advantage of this approach is that it contains a wealth of detail about individual regions and sectors. In addition, a version of the model is available for use on personal computers.

The estimates shown in Table 4 and Figure 2 are from Edmunds and Reilly (1983), and these estimates are on the low side of the range. In addition, Edmunds has provided me estimates of the impact of carbon taxes on CO_2 emissions from the most recent version of the model. More recent estimates give higher estimates of the cost of reducing CO_2, although the exact reason for the difference has not been identified.

[#1] *Jorgenson-Wilcoxen*. A recent study by Jorgenson and Wilcoxen has applied the multisectoral model used by Jorgenson and collaborators in a number of different contexts. This model has the advantage of having an extremely detailed interindustry structure and having been carefully refined over several years and generations of models. In this study, which applied to the United States, Jorgenson and Wilcoxen estimate the long-run impact of different carbon taxes on emissions.

The Jorgenson-Wilcoxen estimates are at the low end of the range of costs of CO_2 reduction. These estimates share the shortcoming of the econometric approaches in using a behavioral approach to energy supply, one which does not necessarily respect the long-run availability of different energy resources. In addition, this model differs from all the others in this study in assuming that technological change is endogenous and affected by factor prices, an assumption that clouds the interpretation of costs.

[#3] *Whalley-Wigle*. A final econometric study reviewed here is the multi-region computable general equilibrium (CGE) model constructed by Whalley and Wigle (1990). This model is similar in philosophy to other CGE models used for estimating the impacts of taxation and tariffs. It is a one-period model for the period 1990-2030, and combines six different regions. The major advantage of this approach as compared with the other econometric models is that it explicitly allows for changing trade as a means of adjusting to carbon taxation. This model has the disadvantage, as do most other econometric studies, of a supply structure that does not reflect resource availabilities and actual technologies, a shortcoming that is probably particularly serious for large rates of carbon taxation.

The case examined here is the efficient version in which all regions impose the same rate of carbon taxation. This model provides extremely high estimates of the cost of CO_2 reduction, with a 50 percent reduction requiring rates of carbon taxation in the order of $430 per ton (1989 prices). It is likely that this high cost of CO_2 reduction results from the low supply elasticities assumed for energy supply; in the long run, to which these estimates apply, energy-supply elasticities are probably well in excess of the unitary supply elasticity assumed by Whalley and Wigle.

Mathematical programming or optimization approaches

A third approach to estimating the impact of CO_2 reductions on the economy is to draw upon programming or optimization models of the energy sector. In this approach, the energy sector of the economy is represented in terms of a combination of demand functions and technological activities. Using a mathematical-programming or other algorithm, the models then solve for the "optimal" trajectory of prices, outputs, fuel mix, and technologies. It can be shown that the "optimal" trajectory would correspond to the outcome of perfectly competitive markets. In addition, the prices (or shadow prices) represent not only the competitive prices but also the scarcity prices, that is the impact of reducing the activity or factor upon the objective function of the maximization.

To compare these scenarios with the tax runs, we examine model runs that impose a CO_2 constraint upon the energy system. The dual variable or shadow price associated with the CO_2 constraint is exactly equivalent to the tax rate that would, in perfectly competitive markets, induce the CO_2 constraint in the model. We can therefore identify the shadow price on CO_2 emissions with the tax rate in the earlier runs. We have examined four different models for the results of CO_2 constraints, as follows:

[#9] *Nordhaus energy models*. The first study to use this technique for calculating the cost of CO_2 reduction was presented in Nordhaus (1977) and (1979). These estimates are based upon a large, multiregion, dynamic mathematical programming model of the global economy. They incorporate four end-use sectors and a large number of different fuels and potential production techniques and calculate trajectories over several decades. Unlike many of the econometric studies, these models keep track of the availability of energy resources. They also assume explicitly that nuclear power will constitute a backstop technology for electricity production. To estimate the marginal cost of CO_2 reduction, alternative constraints were placed upon the maximum concentrations of CO_2, and the efficient pricing and production of energy subject to those constraints was then calculated.

The results of these calculations are shown in Table 4 and Figure 2. The estimates are for different years and different strategies and do not represent a smooth curve because of the impact of resource exhaustion and timing upon the costs of CO_2 reduction. These models are in principle well designed to estimate the long-run cost of reducing CO_2 for they include both demand responses and alternative technologies. The shortcoming of this approach is that it probably tends to underestimate the costs of CO_2 control because it assumes that both policy and economic activity are efficiently organized. To the extent, however, that carbon taxes are able to stimulate new technologies, the estimates may be high.

[#9] *Argonne model.* The earlier versions of the Nordhaus model were reestimated and changed slightly by a team of scientists including the original author, Thomas Daly, Noriyuki Goto, and Richard Kosobud. The basic structure of the model was unchanged, although the data base was thoroughly updated, including more realistic estimates of the cost of nuclear power. In addition, the structure of capital accumulation was included to assume a "putty-clay" structure of technology.[3] The results of the revised model are surprisingly close to the earlier estimates, and are included in Table 4.

[#7] *Manne-Richels.* A recent study of the cost of CO_2 reduction for the United States employs an extension of earlier work of Manne to include CO_2 constraints and (unlike most other models) explicitly incorporates carbon removal technologies. This study has a relatively thorough specification of technologies for electricity generation but highly aggregated assumptions for non-electric technologies. It includes the possibility of carbon removal from stack gases, but does not allow for the difficulties of disposal and therefore probably underestimates the costs of CO_2 removal.

Table 4 and Figure 2 show the estimates of the CO_2 shadow price in the Manne-Richels study. The convergence of the estimate toward $250 per ton carbon is mainly a product of the assumption about the cost of carbon-free technologies.

[#5] *EC study.* A team of experts has prepared estimates of the cost-effective options for reducing CO_2 emissions from the energy sector. The estimates are based on a model dubbed "EFOM-ENV," which is a standardized model used for five European countries with country-specific data. The model has a number of subsystems, and a least-cost trajectory for attaining different CO_2 constraints is estimated.

An important development in this model is to introduce many alternative energy-use technologies and to compare the competitive approach with one in which decisions are not optimized. The study contains two sets of estimates of the costs: one which corresponds to the competitive assumptions used in other models ("Ref"), and a second which allows for inefficiencies in energy-using sectors ("Mure"). The second set contains a segment of the cost

3. See Daly et al., 1984 for a full discussion.

function with "negative costs," that is, with reductions of CO_2 emissions of 10 to 30 percent of a country's emissions by 2010 with no net cost.

In the estimates that follow, I examine only the Mure case for the U. K. In addition, I assume that the Mure outcome will be attained (thereby reducing CO_2 emissions by 10 to 30 percent, depending upon the country). I then examine the costs of additional restraints beyond the minimum-cost Mure trajectory. This correction is by no means obvious and raises troublesome questions that are discussed in a later section. The estimates of the cost function are at the high end of the range of estimates; the high estimate may, however, arise from difficulties in interpretation.

[#4] *Bodlund et al.* Although designed for another purpose, a study of the Swedish energy system provides estimates of the cost of CO_2 reduction in the electricity and electricity-using sectors (Bodlund et al., 1989). This study estimates the implications of both cost-minimization (called "economic dispatch") and a technology that minimizes CO_2 emissions (called "environmental dispatch"). The technique is a scenario analysis of different ways to meet environmental and economic objectives.

Estimates of the cost of CO_2 reduction are shown in Table 4. These studies tend to have low *levels* of emissions of CO_2 relative to other studies because of aggressive inclusion of a number of novel technologies. However, the cost of reducing CO_2 emissions beyond the low level in the baseline scenario is not out of line with other studies.[4]

[#6] *Netherlands model.* A recent estimate of the costs of CO_2 reduction has been made by a team of scientists in the Netherlands. These estimates use a national economic optimizing LP model of the Netherlands (IEA-MARKAL). This model includes capital and energy inputs along with a large number of different energy technologies.

The results of the CO_2 reductions in the Netherlands parallel those for Sweden and the European Community in their reliance upon new technologies, which gives a relatively flat or even declining baseline path for CO_2 emissions. This study has an added feature of including carbon removal techniques.[5]

4. Robert Williams has pointed out that this treatment may be misinterpreted because it normalizes the estimates by excluding what is called the "negative cost region." This issue is considered below.

5. Some of the studies present the average rather than the marginal costs of reducing CO_2 emissions. For these studies, I have converted the average to the marginal cost by the following technique. Examining those studies that provided both average and marginal costs, I estimated the marginal cost function and calculated the ratio of marginal to average cost at every level of CO_2 reduction; that ratio varied with the level of reduction. I then took the estimated ratio of marginal to average cost at each level of reduction and multiplied that figure times the average cost to obtain the given estimate of marginal cost. To illustrate, for the Netherlands study, the average reduction of the two scenarios was 28 percent. At that level of reduction, the other studies had a ratio of marginal to average cost of 1.86. I therefore multiplied the average cost estimate from the Netherlands study by this factor to obtain marginal cost.

Free Lunches and Cheap Lunches

A problem that arises in analysis of the costs of reducing GHG emissions concerns the underlying assumptions that, aside from the climatic externality, the economy is efficiently organized. This assumption rests on the three subsidiary premises: (1) that there are no externalities other than those that involve greenhouse warming, (2) that all markets are perfectly competitive, and (3) that all economic agents have adequate information and optimize their economic decisions.

Of these assumptions, few would argue that there are no externalities, so to the extent that other externalities are included, the "net social costs" of GHG reductions might differ from the estimates given here. Among important externalities are ozone depletion, pollution from coal burning, the impacts of tropical deforestation, and national security external effects from oil imports. To the extent that tropical deforestation causes ecological problems and species extinction as well as CO_2 emissions, slowing tropical deforestation will kill three birds with one stone. The analysis presented here would be usefully extended to include other external effects.

The second and third assumptions are also unlikely to hold in reality. There is extensive evidence of market and informational imperfections. In some cases, institutional obstacles may prevent economical energy-conserving capital from being adopted. For example, subsidies to energy prices in many countries lead to much higher levels of energy consumption than would be the case were prices to reflect realistic market prices or world prices. In addition, the public and private agenda is so full that cost-effective investments are sometimes simply pushed aside in favour of more pressing concerns. Finally, many households do not possess either the information or the funds to buy energy-using capital goods that are most economical from a social point of view. In all these cases, it seems technically feasible *in principle* to increase energy efficiency and reduce CO_2 emissions at little or even negative social cost. In the colloquialism of economics, this analysis suggests not only that there are free lunches, but that in some restaurants the you can get paid to eat!

To go from principle to practice, however, requires an act of faith that is not warranted by economic evidence and raises a number of important questions. First, it is not clear that energy use has more negative than positive externalities; transportation, which is an energy-intensive activity, is often pointed to as one of the social overhead investments that are preconditions for economic development. Secondly, it might be that when all the costs of decisionmaking and investments are included, the costs of energy conservation may be higher than is calculated. Third, although the discount rates implicit in many energy conservation investment appear high, this may reflect liquidity constraints.

Most important are questions about the policies that would be required to implement the negative-cost options. They generally involve regulatory strategies that attempt to outsmart market failures. Yet regulatory approaches are hardly new and have a spotty record in the United States. Advocates of regulation often overlook the fact that socialist economies, which have high degrees of command and control, also show very high levels of energy use, great pollution, and low levels of economic efficiency. Much work needs to be done to demonstrate that the negative-cost approaches have some relevance for economic policy.

Overall Results

Figure 2 shows the different estimates of the marginal cost of emissions reduction collected in a single figure. For the present study, we have used a "consensus" estimate by fitting an equation to the different estimates. The estimate assumes that the cost function (relating the marginal cost of CO_2 reduction or the carbon tax to the percentage reduction) takes a semi-logarithmic form:

$$\ln[1-R] = \alpha + \beta \, MC$$

where R = reduction of CO_2 emissions from an uncontrolled path, MC = marginal cost of removal or tax rate on CO_2 (in 1989 dollars per ton of carbon), while α and β are parameters. Fitting this curve to the data yields the following equation:

$$\ln[1-R] = -.0223 - .0054 \, MC$$

The line fitted by this equation is shown in Figure 2 by the solid line, which is also marked "Regression." This equation appears reasonable in terms of its consistency with the estimates presented above. On the other hand, it should be interpreted as purely a curve-fitting exercise and does not have any particular probabilistic interpretation.

In addition, we show in Figure 3 the total cost of reducing CO_2 emissions. For this estimate, we normalize to the 1989 level of global economic activity and emissions and assume that the cost function shown in Figure 2 applies to this normalized economy.

Short-Run Costs of Reducing Emissions

It is important to note that the estimates of the cost of reduction of CO_2 emissions in this section are by and large ones that calculate the *long-run* costs of reduction. That is, they measure the costs when all factors of production are allowed to adjust fully.

Estimates of the short-run costs of reducing CO_2 emissions are much more fragmentary. On theoretical grounds, we know by the envelope theorem that the short-run costs are higher than the long-run costs. To take a simple

Table 4. Estimated Percentage Reduction in CO_2 at Given Carbon Tax or Marginal Cost (Dollar per Ton Carbon)

Tax or Cost [$/tC]	Percentage Reduction in CO_2					
	Econometric Models		Optimization Models			
	Small	Large	Nordhaus	Manne-Richels	Other	Regression
0	0	0	0	0	0	0.0
10	3 [8]	10 [1]	5 [9]			6.6
20	7 [8]	18 [2]	9 [9]		10 [5]	10.9
30	10 [8]	40 [1]	17 [9]			15.0
40	13 [8]	36 [1]	20 [9]		28 [6]	18.9
50	15 [8]					22.6
60						26.1
70						29.5
80						32.8
90		40 [2]	43 [9]		20 [5]	35.9
100	27 [8]					38.8
110						41.6
120						44.3
130		50 [2]				46.8
140					30 [5]	49.3
150	36 [8]		60 [9]			51.6
160					63 [4]	53.8
170						56.0
180						58.0
190			92 [9]			59.9
200	43 [8]					61.7
210						63.5
220						65.2
230						66.8
240			78 [9]			68.3
250					74 [4]	69.8
260			90 [9]	82 [7]		71.1
270			94 [9]			72.5
280						73.7
290						74.9
300						76.1
310						77.2
320						78.2
330						79.2
340						80.2
350				76 [7]		81.1
360						82.0
370						82.8
380						83.6
390						84.3
400						85.1
410						85.7
420						86.4
430		50 [3]	98 [9]			87.0

Figures in brackets refer to model number. For interpretation see notes to Table 4 and Figure 2 in the Appendix.

Figure 2. Marginal Cost of CO_2 Reduction (Cost per ton CO_2, 1989 prices)

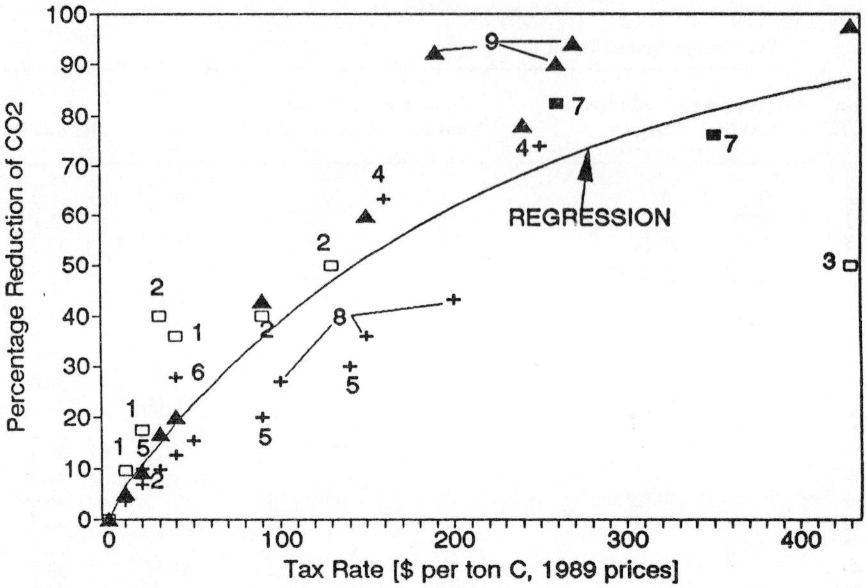

Figure 3. Global Cost of CO_2 Reduction (Cost at Different Reduction Rates)

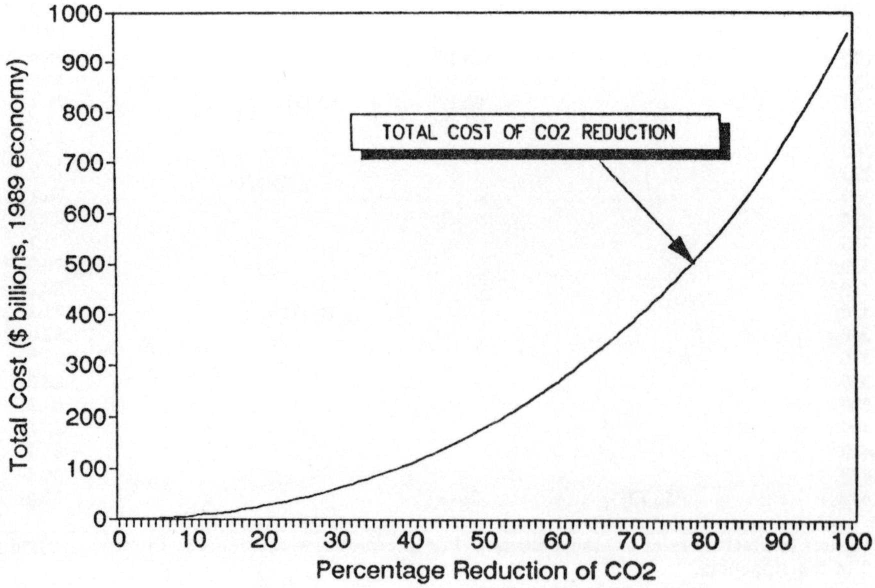

example, assume that all the response is on the demand side. If the short-run price elasticity is -0.1 and the long run elasticity of demand is -1.0, it can be shown that the short-run costs are *ten times* the long-run costs. This example also shows that the short-run costs will depend on the length of the run. Note, however, that the short run gradually turns into the long run, so the high short-run costs of a surprise increase in prices soon becomes the lower long-run costs.

Three studies have results in which the short-run costs can be calculated. In the Manne-Richels study, the price of the carbon constraint rises to around $600 per ton C in the short run because of difficulties of the system to adapt to stringent constraints before a full adaptation of the capital stock and the technology. In a paper by Blitzer and others (1990) for Egypt, carbon constraints are also found to be very costly in the short run, which may arise because of fixity of capital in the short run. A third example comes in the Congressional Budget Office (CBO) study of the impact of carbon taxes in the DRI quarterly model of the U. S. economy. This simulation presents a completely different approach from the others in this survey, for the CBO simulation allows deviations of output from its potential. The short-run impact of carbon taxes are not far from the estimates in Figure 2, but the output losses are several times larger than those implicit in the estimates in Figure 2 because the carbon tax leads to lower utilization of potential. The CBO simulation is a reminder that carbon taxes have price-raising effects and that these must be considered in weighing the short-run consequences of CO_2 restraints.

Estimation of Mitigation Costs for CFCs

In estimating mitigation costs for CFCs, the major difficulty arises from the abundance of chemical elements. Table 5 collects the different characteristics of the CFCs and uses them to estimate the importance of CFCs as a whole for global warming. Column (8) of Table 5 shows the cumulative total warming for the major CFCs in production in the mid-1980s. Almost 99 percent of total warming is estimated to come from four substances (CFC-12, CFC-11, CFC-113, and HCFC-22).

In the estimates that follow, we assume that there are two sources of mitigation of global warming from CFCs. First, there are assumed to be substitutes with less powerful warming effects. These substitutes are assumed to be HFC-134a as a substitute for CFC-12 and HCFC-123 for CFC-11. We estimate that the cost of substitute chemicals have about one-twelfth of the total warming potential of the average of CFC-11 and CFC-12.[6]

6. A recent study indicates that HFC-134a has around 8 percent of the global warming potential of CFC-12, while HCFC-123 has around 0.5 percent of the global warming potential of CFC-12. See AFEAS, 1989.

Table 5. Basic Data on CFCs

Chemical Name (a)	Formula (b)	Prod'n (c)	Concent. (d)	Growth in Concent. (e)	Forcing (f)	Lifetime (g)	CO_2 Equiv. (h)
CFC12	CF2Cl2	480	392.0	0.04	0.08	111.0	532
CFC11	CFCl3	350	226.0	0.04	0.07	75.0	182
CFC113	C2Cl3F3	166	50.0	0.07	0.03	90.0	36
HCFC22	CHClF2	240	100.0	0.07	0.03	20.0	16
Others		2113	n/a	n/a	n/a	n/a	11
Total		3349					777

Column notes:

a Common name for chemical.
b Chemical formula.
c Production (1986 actual or estimate) in millions of kgs.
d Atmospheric concentrations (1986 actual or estimate), parts per trillion by volume.
e Growth in concentrations, during 1980s or estimated 1980-2030, per annum.
f Radiative forcing (no feedback, $w/m^2/ppb$).
g Lifetime (calculated as time for atmospheric value to decline to (1/e) or .367 times original value).
h Total warming potential from CFC measured in million tons of carbon. This is calculated as the growth rate in e times the radiative forcing in f divided by [1/T], where T is the lifetime in g. It is then converted into CO2 equivalents according to Table 3.

Second, we assume that the demand for CFCs or their substitutes has the usual economic property of being sensitive to price. Hence, as regulations on CFCs tighten, firms and consumers will recycle the more expensive CFCs and substitute other substances or goods and services for CFCs that contribute substantially to warming. A key assumption behind this proposition is that the regulation or tax is proportional to the total warming of a substance (rather than a flat tax per unit weight or bans on certain kinds of chemicals). As a result of higher taxation, then, the total emissions into the atmosphere will decrease according to the demand elasticity. Although there are no well established elasticities for this kind of chemical, estimates of general price elasticities of industrial materials, as well as estimates of the response of recycling to prices, suggests price elasticities in the range of -0.1 to -0.4.

The technological data are described in Tables 5. The pertinent price data for standard chemicals was provided by duPont and refers to deliveries in truckload quantities. For new substances (HFC-134a and HCFC-123), we obtained order-of-magnitude estimates of price from Dr. Robert Owens of the Freon Division of duPont. These estimates are speculative since the current price for small quantities are around $90 per pound for HFC-134a and $6 per pound for HCFC-123.

Table 6 shows the response of the index of total greenhouse warming to taxation (or the regulatory equivalent) on CFCs. A substantial decrease in

Table 6. Impact of Carbon-equivalent Taxes on Global Demand for CFCs

Tax rate***	Total Greenhouse Warming* (Millions of tons of CO2 equivalent) For different price elasticities**:		
(per ton CO_2 equivalent, C weight)	-0.10	-0.20	-0.50
0	777	777	777
1	621	590	507
2	490	454	362
3	372	340	264
4	261	240	192
5	155	149	134
10	151	143	121
20	147	134	104
30	143	128	94
50	139	120	80
100	131	108	63
200	124	96	47
300	119	89	39
500	114	81	31

* Total greenhouse warming is measured as the total potential warming over the indefinite future given the lifetime and radiative properties of the four major CFCs and their potential future substitutes.

** The price elasticity refers to the percentage change in demand for CFCs per percent change in the price, including the tax.

*** The tax rate is calculated per ton of CO_2 equivalent warming, where the index of warming is similar to that described in the first footnote in this table. A $1 per ton tax on CO_2 (carbon weight) is equivalent to a $1.30 per kg tax on CFC-11 and a $3.70 per kg tax on CFC-12.

global warming potential can come from the substitution induced by relatively modest taxes (up to $5 per ton of CO_2 equivalent). After that point, the reduction is purely demand-side and much more modest, although the exact degree of substitution depends upon the demand elasticity.

Forestry Options

A number of studies have proposed using trees as a method of removing carbon from the atmosphere. Among the major proposals are slowing the deforestation of tropical forests; reforesting open land, thereby increasing the amount of carbon locked into the biosphere; a "tree bounty," which subsidizes the sequestration of wood in durable products; and a "tree pickling" program, wherein trees are stored indefinitely.

No detailed study of the economics of tropical forests has been undertaken here. However, deforestation may be adding ½ to 3 billion tons of carbon per year to CO_2 emissions (this amounting to between 7 and 40 percent of non-forest GHG emissions). Much deforestation is thought to be uneconomic in tropical regions even without invoking greenhouse considerations. If so, the cessation of uneconomical deforestation can significantly and inexpensively slow greenhouse warming.

In this section, we analyze three different kinds of options: (i) providing subsidies to harvesting of wood, (ii) storage of carbon through a lumber purchase program, and (ii) afforestation of open land. It should be emphasized at the outset that the analysis is even more difficult in this area than in many of the other areas of prevention. The difficulty arises from the poor data on forest cover, the heterogeneity of species, and the difficulty of collecting accurate data on afforestation costs. Nonetheless, the possibility of using forests to sequester carbon is sufficiently attractive to warrant a quick look.

A Simple Model of Wood Harvesting

The first option analyzed is the possibility of subsidizing *productive* uses of wood that effectively sequester carbon for long periods of time. This is an activity that harvests large quantities of carbon in wood and uses them in productive economic capital like houses, railroad ties, and so on. The idea is that this is a renewable carbon-removal industry, wherein trees are cut, the wood is sequestered, and the new growth removes carbon from the atmosphere.

This option is in principle attractive given the large volume of carbon in the biosphere -- in the order of 800 billion tons of carbon in forests worldwide. If a substantial fraction of the annual growth could be economically harvested, a reasonable amount of the annual additions to GHGs could be offset.[7]

There are to date no forestry models of the kind that have been developed for the energy sector. Calculations of the economics of reforestation are based on engineering estimates and generally do not reflect the existing economics of forest products. The approach here is to construct a simple aggregate global forest model as a first step toward creating more accurate estimates of the cost of carbon sequestration. The model contains four basic equations: demand for wood, supply of annual wood harvested, an ecological equation representing the growth of forests, and an accounting identity for the inventory of forests.

The first equation represents the demand for wood products. In this model, we assume that all demand and supply can be represented in terms of raw harvests (measured in metric tons of wood). The demand is assumed to

7. Most of the data on wood and the global harvest as well as on rates of carbon fixation are from EPA, 1989.

be constant over time at 800 million tons per annum and has a price elasticity of demand of minus one:

$$Q(t) = A(t)\ p(t)^{-\epsilon} \tag{1}$$

where Q(t) = annual harvest for wood (millions of metric tons); A(t) = intercept of demand function (taken constant over time); p(t) = price of wood (in $ per ton) = $100 per metric ton in the base case; and ϵ = price elasticity of demand for wood = -1.

The supply equation follows traditional studies in assuming that, for a given inventory of forests, the cost of production is an increasing marginal cost of the harvest. We further assume that, at constant costs, the harvest increases proportionally with the size of the forest. The equation representing harvests is then:

$$Q(t) = B(t)\ S(t)\ \{p(t)[1+s(t)]\}^{\eta} \tag{2}$$

where B(t) = the fraction of the stock that is harvested annually, = .001 in original equilibrium; S(t) = the total stock of forests (in billions of metric tons) = 800 in the initial condition; s(t) = subsidy to wood production; and η = price elasticity of supply of harvesting = 1.

The next two equations represent the natural net growth and accumulation of forest stock. Following the modelling used by Binkley, we assume that the growth of forests follows a quadratic function. This gives:

$$g[S(t)] = .8 + \alpha_1\ S(t-1) - \alpha_2\ S(t-1)^2 \tag{3}$$

where g[S(t)] = the accumulation or net growth in forest stock in year t (in billions of metric tons); 0.8 = the current harvest; α_1 = growth parameter = 0.05; α_2 = growth parameter fit so that equilibrium stock is 800 billion metric tons at harvest of 0.8 billion tons per year.

Finally, the net stock of wood accumulates according to the inventory equation:

$$S(t) = S(t-1) + g[S(t)] - Q(t) \tag{4}$$

The model is fit by calibrating its annual harvests, prices, and stocks to estimated data.

Cost Estimates for Forestry Options

Costs for a "Tree Bounty"

The next stage is to estimate the impact of a policy to increase the harvesting of trees and sequestering the carbon in houses or in other wood products. The particular policy examined is to provide a "tree bounty" wherein those who harvest trees and sequester the carbon are provided a percentage subsidy, akin to the oil-depletion allowance, related to the market price. Subsidies from 5 percent to 200 percent of the price were investigated. A rough estimate of the size of these subsidies is obtained by noting that the total value of harvest is in the order of 800 million metric tons of harvest at an average price of $50 per cubic meter of wood and $200 per metric ton of carbon. The total subsidies (before output responses) are therefore in the order of $2 billion to $80 billion.

To estimate the economic costs of the tree bounty, we employ standard dead-weight loss analysis. We use the standard approximation that the dead-weight loss is equal to one-half the subsidy times the change in quantity produced. This calculation assumes that there is no deadweight loss in providing the subsidy (i. e., that the funds are raised with no efficiency losses). The estimates assume that 75 percent of the incremental harvest is indefinitely sequestered. In these estimates, we examine only the steady state, that is the period when the stock of trees has reached its new ecological equilibrium (which requires several decades).

Table 7 shows the results for this experiment. Two major conclusions can be drawn from this estimate. First, for low levels of the "tree bounty"--between 5 and 10 percent of tree prices--the cost per ton of carbon removed is relatively low. Not surprisingly, as the size of the subsidy grows, the distortions and efficiency losses grow apace.

The other and more important conclusion is that the quantitative importance of forest harvesting at these modest efficiency costs is quite small. To take the case of the subsidy of 25 percent of the value of the harvest (which would be around $10 billion worldwide), this would accomplish an annual sequestration of only 75 million metric tons of carbon annually. This is approximately 1 percent of the total carbon equivalent of GHGs annually emitted into the atmosphere today. Even moving to the highly costly 100 percent bounty would yield a sequestration rate of only 2½ percent of annual GHG emissions.

It should be noted, as well, that these calculations are likely to overestimate the attractiveness of the forest bounty option for a number of reasons. First, the assumption about the sequestration rate is relatively high; estimates today are that about 25 percent of past forest production remains in the stock of wood products. Second, we have omitted the energy costs of harvesting the timber, which is likely to offset some of the carbon sequestration. Third, the assumption of a non-distortionary source of the revenues for the tree bounty is unrealistic and underestimates costs. Fourth, an accelerated harvesting of forest lands may have undesirable environmental and ecological impacts.

Costs for a "Tree Pickling" Option

A second option for sequestering large quantities of wood, which might involve less government funds, is to purchase lumber and simply store

Table 7. Costs and Carbon Sequestration from a Tree Bounty

Subsidy rate (percent)	Annual Sequestration (10^9 tC/yr)	Cost ($ per t Carbon)
0	.000	0
5	.016	19
10	.031	38
25	.075	97
50	.142	195
75	.204	293
100	.261	391

Note: Estimates assume a 8 percent cost of capital and that 75 percent of additional harvest is sequestered.

it in a manner that sequesters the carbon for long periods of time. This use differs from the first in that it is not necessary to subsidize the entire industry; rather the government only purchases the amount of wood that it desires to store. The disadvantage of this approach relative to the "tree bounty" is that the lumber serves no economic purpose other than to store carbon. That is, the useful economic purposes to which wood can be put are not exploited.

To estimate the costs and benefits of the lumber purchase program, we again used the simple forest-products model described above. The government is assumed to purchase a fraction of the crop of harvested products, and the purchases (in the form of logs or in whatever form was most advantageous) were then assumed to be transported and stored costlessly in a remote location.

The economic costs of the program are equal to the government revenues plus whatever gains or losses in consumer and producer surplus occur. In this case, however, to a first approximation, the costs are the direct costs of the government purchases. The estimates of the yield and costs of such a program indicate that a log set-aside program is enormously inefficient as a method of sequestering carbon and not worth pursuing further. The reason is straightforward: with lumber selling on the market at a price of around $200 per ton carbon, it simply does not pay to buy and store logs in an economically unproductive manner.

Afforestation of Open Areas

A final approach is the afforestation or reforestation of open lands. The potential attractiveness of this approach is that it expands the natural accumulation of carbon in forest products, which requires relatively little in the way of scarce inputs other than land. The data for this option are largely drawn from the EPA on policy options to slow greenhouse warming (1989).

The key assumptions, some of which are also shown quantitatively in Table 8, are the following:

(a) In keeping with the general methodology of this survey, I assume existing technology and ignore the possibility of a "miracle tree" that would store carbon more efficiently. In other words, the estimates assume the average rates of carbon fixation that are observed in current forest areas. The average annual rate of removal is estimated to be 1.6 tons of carbon per hectare per year in tropical areas and .82 tons of carbon per hectare per year in temperate areas. The mature stands are assumed to fix 50 tons of carbon per hectare in tropical areas and 30 tons of carbon in temperate areas.

(b) The growth curve follows a logistics equation, $S(t) = a/[b+\exp\{c+dt\}]$, where S(t) is the stand in tons, a depends upon location, $b=1.0$, $d=-.0055$, and $c=5$. In this equation, the trees are assumed to mature in 40 years. The importance of this equation comes from the delay of carbon fixation behind investment.

(c) The amount of land available for sequestration is drawn from EPA (1989) and is the most problematical data. We assume that 510 million hectares of forestable land are available at a price ranging from $20 per hectare to $200 per hectare; this amounts to a land area approximately one-third the size of the United States.

(d) Costs of forestation include between $400 and $450 of planting cost and between $10 and $20 per year in management costs.

The results of this calculation are shown in columns (9) and (11) of Table 8. The total costs of carbon sequestration range between $42 and $114 per ton of carbon. The total amount of carbon that can be sequestered in this manner is estimated to be 21.3 billion tons of carbon globally, or about three years emissions. This is a one-shot reduction in the amount of carbon in the atmosphere rather than a renewable option. Assuming that this program is spread out over 75 years, this program yields about .28 billion tons of carbon removed per year. Therefore, carbon sequestration through afforestation appears to be a moderately economic option (relative to extreme levels of control on carbon fuels). The major drawback of this approach is that it has but limited potential for offsetting the growth of greenhouse gases over the next century.

In summary, a small amount of removal of carbon in wood products, through afforestation and a subsidy to wood-using products, appears feasible and has costs that are in the mid-range of different strategies to slow greenhouse warming. On the other hand, the quantitative significance of these options appears to be quite limited.

Geoengineering Options

There is an old German saying which translates freely into, "Use brains, not brawn." The preventive strategies of planting billions of trees or reducing fossil-fuel use through trillions of dollars of investments are the

Table 8. Estimates of the Cost-Effectiveness of Forestation Option in Sequestration of Carbon

(1) Location of Land	(2) Removal rate (tc/h/y)	(3) Area (million ha)	(4) Plant cost ($/ha)	(5) Land cost ($/ha)	(6) Manage cost ($/ha/yr)	(7) Stand, mature forest (tc/ha)	(8) Growth period (yr)	(9) Total cost of removal ($/tc)	(10) Mature stand (Billion tons C)	(11) Average removal (Billion tc/yr)
Tropical land, low cost	1.60	100.00	400.00	20.00	10.00	50.00	40.00	41.59	5.000	0.067
Tropical land, med. cost	1.60	100.00	400.00	50.00	10.00	50.00	40.00	43.88	5.000	0.067
Tropical land, high cost	1.60	100.00	400.00	100.00	10.00	50.00	40.00	47.70	5.000	0.067
U.S. unused land	0.82	10.00	450.00	20.00	20.00	30.00	40.00	91.58	0.300	0.004
Temperate, marginal	0.82	100.00	450.00	100.00	20.00	30.00	40.00	101.75	3.000	0.040
U.S. marginal land	0.82	100.00	450.00	200.00	20.00	30.00	40.00	114.47	3.000	0.040

Notes:

Column 2: Different regions are provided in EPA (1989); U.S. and temperate use the average growth U.S. commercial forests 1977; see EPA (1989), p. VII-233. Tropical uses figure for unmanaged forest; average figures given in EPA (1989), p. VII-223.
Column 3: Rough estimates of the amount of land available from EPA (1989), p. VII-238, and by author.
Column 4: Estimates from EPA (1989), pp. VII-245-246.
Column 5: Some estimates from EPA (1989); others are estimates of land prices by author.
Column 6: Estimates from author.
Column 7: Estimates from EPA (1989) and elsewhere.
Column 8: Estimates from EPA (1989) and elsewhere.
Column 9: Assumes a 6 percent real cost of capital; that carbon buildup is according to logistics equation; and that no offsite sequestration is undertaken.
Column 10: Total annual rate of sequestration in reforestation activities.
Column 11: Assumes that the reforestation in column 10 is spread over period of 75 years.

brawn philosophy. A promising new approach to the threat of greenhouse warming is to use our brains to find a way to offset greenhouse warming through climatic engineering; this is the global equivalent of turning on an air conditioner. Potential approaches are changing the albedo (reflectivity) of the earth or increasing the rate of removal of greenhouse gases.

Careful analysis of these proposals is just beginning, but a number have already been identified that appear much more cost-effective than plugging the oil wells and shutting down the coal mines. One approach would be to create a sunscreen by sending tiny particulates into the stratosphere to cool the earth. These particles could be shot up with 16" naval rifles, lifted by hydrogen balloons, or deposited by tuning the engines of aircraft to burn somewhat richer than normal. One estimate finds that 100,000 kilograms of carbon can be offset by 1 kilogram of particles. That's real leverage!

A further intriguing possibility raised by John Martin and his associates is placing minute quantities of iron in the North Pacific and Antarctic ocean regions. Studies suggest that this would increase the limiting nutrients, foster much greater photosynthesis, and would increase the rate of carbon precipitation to the ocean floor. Preliminary estimates indicate that this option might well annually remove from the atmosphere a quantity equal to current CO_2 emissions.[8]

Preliminary estimates suggest that the cost of the geoengineering options is in the order of $0.10 to $10 per ton carbon equivalent, which is far less than many perennial favorites like reforestation or fuel switching or energy conservation. These technologies may be the panacea for global warming. At the same time, they pose unknown risks by perturbing a complex system that is already being perturbed. They deserve careful study.

ESTIMATES OF THE OVERALL COST OF REDUCING GHG EMISSIONS

Finally, we collect together the three sets of estimates of the cost of GHG reduction (excepting geoengineering). Figure 4 displays the estimates of the marginal cost curves for each option and calculates a total marginal cost curve for reducing GHGs. The curve marked "All GHGs" in Figure 4 is calculated as the most efficient combination of the three options; that is, it represents the marginal cost curve of all options.

Figure 5 shows the same result in terms of the total global cost of GHG reduction (in billions of 1989 dollars at the 1989 level of world economic activity) for different levels of reduction of GHG emissions. Table 9 provides a table of estimated marginal, average, and total costs of reducing GHGs. These estimates suggest that a modest reduction of greenhouse gas emissions can be obtained at low cost. It is possible to reduce GHG emissions by about 10 percent for a cost of less than $10 per ton CO_2 equivalent -- most of this

8. The technological options discussed here were derived from communications with Robert Frosch of General Motors, to whom I am most grateful for clarifying several issues.

Figure 4. Marginal Cost of GHG Reduction

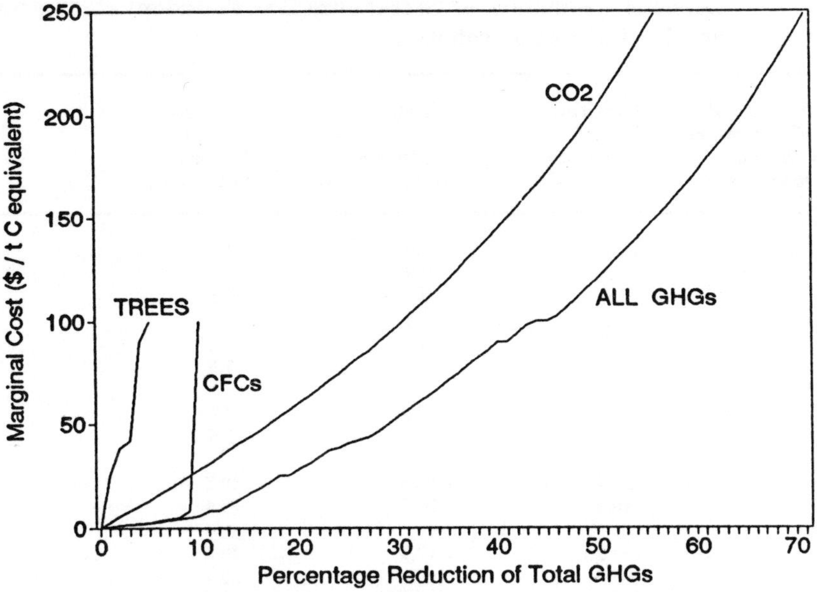

Figure 5. Total Cost of GHG Reduction

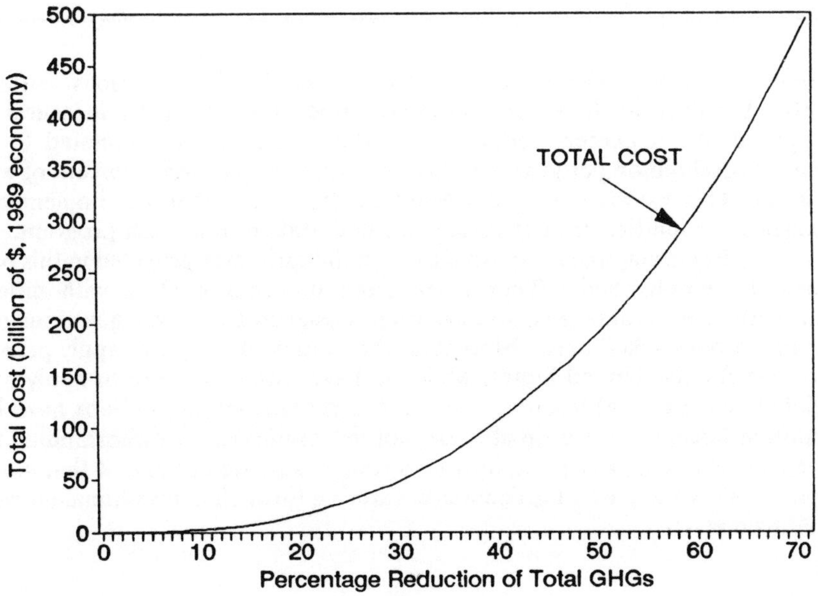

Table 9. Estimated Marginal, Total, and Average Global Costs for Different Levels of Reductions of Greenhouse Gas Emissions (1989 prices and level of global economy)

Reductions of GHG Emissions (as percent of baseline)	Marginal Cost of Reduction ($ per t C)	Total Cost of Reduction (billions of $)	Average Cost of Reduction ($ per t C)
0	0.0	0.0	n/a
1	0.5	0.0	0.5
2	1.0	0.1	0.8
3	1.5	0.2	1.0
4	2.0	0.4	1.3
5	2.6	0.6	1.5
10	5.3	2.2	2.8
15	16.3	6.8	5.6
20	27.9	16.3	10.2
25	40.2	30.7	15.4
30	53.3	49.5	20.6
40	89.9	107.9	33.7
50	119.9	190.8	47.7
60	171.7	308.7	64.3
70	240.4	474.8	84.8
80	342.4	708.8	110.8

coming because of the great cost-effectiveness of CFC reductions. After 20 percent reduction, however, the curve rises as more costly measures are required. A 50 percent reduction in GHG emissions is estimated to cost almost $200 billion per year in today's economy, or around 1 percent of world output. This estimate is understated to the extent that the implementing policies are inefficient or that they are undertaken in a crash program.

In closing, the tentative nature of the estimates provided in this paper must be emphasized. They come from different models, with differing strengths and weaknesses, and are often designed to answer questions other than the ones asked here. Moreover, the empirical estimates apply primarily to data for the United States, while we have extended these to apply to the global economy. Moreover, some of the geoengineering options have been omitted because cost estimates are not yet available. Notwithstanding their weaknesses, these estimates provide a rough estimate of the cost that must be paid to slow the pace of greenhouse warming through conventional emissions reductions.

APPENDIX

Legend for Figure 2 and Table 4

The symbols in Figure 2 represent estimates of the costs of CO_2 reduction from the following models or studies. In addition, the estimates are shown in Table 4 and labelled according to the following list:

#1: Jorgenson and Wilcoxen (1990). Interpretation of results is straightforward.

#2: Edmunds and Reilly (1983). The estimate is derived by taking the carbon equivalent of the energy taxes imposed in simulations.

#3. Whalley and Wigle (1990). Interpretation of results is straightforward.

#4. Bodlund et al. (1989). Estimates compare the total cost and the carbon emissions for two scenarios. There appears to be no ambiguity in the results.

#5. European Community (1990). Estimates use the results for the U. K., which are difficult to convert into the metric used here. Conversion assumes that the discount rate used is 6 percent, that costs are centered on 2005, that the exchange rate is $1.9 per £, and that the data are tons of CO_2 rather than C.

#6. Kram et al. (1989). Similar difficulties arise in converting the results in this study as in 5.

#7. Manne and Richels (1990). Because the model is dynamic, there is ambiguity about the cost. We have taken average estimates for costs after the short-run adjustments costs have disappeared. Otherwise, interpretation of results is straightforward.

#8. All the rectangles are from Nordhaus and Yohe (1983) or the version revised for this survey. Interpretation is straightforward.

#9. All of the starred points are from one of the vintages of the Nordhaus model or its descendant, the Argonne model. The same ambiguity about cost arises here as in Manne-Richels because of the dynamic nature of the simulation, but in fact the difference between different runs or time periods is small. Interpretation of results is straightforward.

REG = regression fit to the different estimates. The regression was estimated using ordinary least squares fitted to the estimates from different models. The equation is described in the text.

REFERENCES

AFEAS (1989). *Scientific Assessment of Stratospheric Ozone: 1989.* World Meteorological Organization, Global Ozone Research and Monitoring Project, vol II.

Blitzer, Charles R., Richard S. Eckaus, Supriya Lahari, and Alexander Meeraus (1990). "A General Equilibrium Analysis of the Effects of Carbon Emission Restrictions on Economic Growth in a Developing Country." Paper presented at a Workshop on Economic/Energy/Environmental Modelling for Climate Policy Analysis, Washington, D. C. (October).

Bodlund, Birgit, Evan Miller, Tomas Karlsson, and Thomas B. Johansson (1989). "The Challenge of Choice: Technology Options for the Swedish Electricity Sector." In Thomas B. Johansson, Birgit Bodlund and Robert H. Williams, eds., *Electricity: Efficient End-Use and New Generation Technologies, and Their Planning Implications.* Lund, Sweden: Lund University Press.

Congressional Budget Office of the United States (1990). *Carbon Charges as Response to Global Warming: the Effects of Taxing Fossil Fuels.* Washington (August).

Daly, T. A., N. Goto, R. F. Kosobud and W. D. Nordhaus (1984). "CO_2 Forecasting and Control: A Mathematical Programming Approach." In John Weyant and Dorothy Sheffield, eds., *The Energy Industries in Transition, 1985-2000.* Published by International Association of Energy Economists, printed by Westfield Press.

Edmunds, J. A. and J. M. Reilly (1983). "Global Energy and CO_2 to the Year 2050." *The Energy Journal* 4: 21-47.

EPA (1989). U. S. Environmental Protection Agency, *Policy Options for Stabilizing Global Climate.* Draft Report to Congress (February).

European Community (1990). Report for the Commission of the European Community: "Cost-Effectiveness Analysis of CO_2 Reduction Options." Paper presented at a Workshop on Economic/Energy/Environmental Modelling for Climate Policy Analysis, Washington, D. C. (October).

Jorgenson, Dale W. and Peter J. Wilcoxen (1990). "The Cost of Controlling U. S. Carbon Dioxide Emissions." Paper presented at a Workshop on Economic/Energy/Environmental Modelling for Climate Policy Analysis, Washington, D. C. (October).

Kram, T. and P. A. Okken (1989). "Two 'Low CO_2' Energy Scenarios for the Netherlands." Paper prepared for IEA/OECD Expert Seminar on Energy Technologies for Reducing Emissions of Greenhouse Gases, Paris (April).

Lashof, Daniel A. and Dilip R. Ahuja (1990). "Relative Global Warming Potentials of Greenhouse Gas Emissions." Forthcoming in *Nature.*

Manne, Alan and Robert Richels (1990). "CO_2 Emissions Limitations: An Economic Cost Analysis for the United States." *The Energy Journal* (April).

National Research Council (1983). *Changing Climate.* Washington, D.C.: National Academy Press.

Nordhaus, William D. (1979). *The Efficient Use of Energy Resources.* New Haven, CT: Yale University Press.

______ (1977). "Strategies for the Control of Carbon Dioxide." *Cowles Foundation Discussion Paper* 443 (January).

______ (1977a). "Economic Growth and Climate: The Case of Carbon Dioxide." *The American Economic Review* (May).

______ (1990), "To Slow or Not to Slow: The Economics of the Greenhouse Effect." Paper presented to the annual meetings of the American Association for the Advancement of Science, New Orleans (February).

______ (1990a). "Contribution of Different Greenhouse Gases to Global Warming: A New Technique for Measuring Impact." Processed February 10.

Nordhaus, William D. and Gary Yohe (1983). "Future Carbon Dioxide Emissions from Fossil Fuels." In National Research Council-National Academy of Sciences, *Changing Climate.* National Academy Press.

Whalley, John and Randall Wigle (1990). "The International Incidence of Carbon Taxes." University of Western Ontario (October).

Wuebbles, Donald J. and Jae Edmonds (1988). *A Primer on Greenhouse Gases.* Prepared for the Department of Energy, DOE/NBB-0083 (March).

[10]

Journal of Economic Perspectives—Volume 7, Number 4—Fall 1993—Pages 27–46

Costs of Reducing Global Carbon Emissions

John P. Weyant

Projecting the costs of reducing carbon emissions is extremely important, and exceedingly difficult. Such projections are an integral component of cost-benefit analyses of alternative policies; for example, for the global climate policy analysis described by William Nordhaus in this symposium. Moreover, the nature of the climate change problem dictates that costs be incurred long before benefits can be realized or accurately measured. Consequently, the political spotlight predictably focuses on the cost estimates whenever specific policy options are contemplated.

Projecting the costs of reducing carbon emissions is difficult because many assumptions must be made about how the world will evolve over a very long period of time with and without a control program. Typically, some of these assumptions, like population growth, are taken from outside sources. Other factors, like the response of energy demand to changes in economic output or energy prices, are the result of modeling the behavior of economic actors in response to exogenous stimuli. Considerable uncertainty exists about both the exogenous factors and the best way to model the behavioral responses.

The global nature of the climate change problem adds yet another complication to the cost analysis. Scientists have concluded that it is the global concentration of carbon in the atmosphere that influences climate. Moreover, since carbon in the atmosphere gets mixed more or less uniformly, emissions anywhere change concentrations everywhere, but with a long lead time. Current global carbon emissions are estimated to be only about 4 percent of the stock of CO_2 in the atmosphere attributable to human activity (and about

■ *John P. Weyant is Professor of Engineering-Economic Systems and Director of the Energy Modeling Forum, Stanford University, Stanford, California.*

1 percent of total atmospheric carbon). Thus, the cost of reducing global carbon concentrations depends on the collective emissions of all the countries of the world over long periods of time. This makes an analysis of the costs of controlling emissions in any one country or region at any point in time an important guide to appropriate policies for that particular country, but an incomplete one.

For the purposes of surveying a number of assumptions and models, the recently completed Energy Modeling Forum 12 study entitled "Controlling Global Carbon Emissions; Costs and Policy Options," which included a comparison of results from 14 models for 13 standardized scenarios, is helpful in this endeavor (Gaskins and Weyant, 1993a, b). A number of more specialized study groups were also organized during EMF 12 to study dimensions of emissions control policy not considered in the original study design. In total, results from 16 models were included in EMF 12, and over two dozen control scenarios were considered. In addition, several other major studies completed over the last three years are relied upon in developing our conclusions here, in particular the excellent surveys by Nordhaus (1991), Congressional Budget Office (1990), Beoro, Clarke, and Winters (1991), Cline (1992), and Dean et al. (1993).

An important limitation of the existing studies, including those represented in EMF 12, is that they focus almost exclusively on reducing carbon dioxide emissions, and include very little detail on the developing countries. Although carbon dioxide is the most important greenhouse gas, reductions in emissions of other greenhouse gases like methane may ultimately prove more cost effective (Blitzer et al., 1992; Schmalensee, 1993).

Many models are now available for studying emissions reductions in the United States and other OECD countries, and a growing number of global analyses offer some regional detail. In the global studies, however, developing countries are typically treated in a simplistic fashion, or aggregated together because of lack of appropriate data and institutional information for the countries involved, or computational limits on model size. A number of developing countries (including China, India, Brazil, Egypt, Nigeria, Indonesia, and others) have substantial emissions at present and/or emissions levels that are likely to grow rapidly during the first half of the next century. This has led to the development of models that can provide a more realistic picture of the potential for emission reductions in these individual countries (Blitzer et al., 1992, 1993; Pachauri and Khanna, 1992), and valuable insights for the global models.

The next section of this paper discusses the key dimensions of any projection of the cost of reducing carbon emissions. The paper then discusses the projections that have been made, including long-, medium- and short-range time horizons. Finally, the conclusion summarizes what we know and don't know about the costs of controlling carbon emissions and recommends an agenda for future research.

Key Determinants of Cost of Control Projections

Hundreds of projections of the cost of reducing carbon emissions have been published over the past decade, a pace that seems likely to continue. Each projection has four key elements: (1) the baseline input assumptions to the analysis; (2) the specification of the control scenario being considered; (3) the structure of the model employed to make the projection; and (4) the cost measure(s) reported. Knowing a few basics about how these elements are handled in a particular study can go a long way towards helping one understand how its results relate to other available projections.

Baseline Input Assumptions

The assumptions included in the baseline scenario are important because the incremental cost of carbon emission controls are measured relative to them. Any analysis of the cost of reducing carbon emissions must rely on input assumptions in three areas: population and economic activity; energy resource availability and prices; and technology availabilities and costs.

Most of the groups projecting the costs of reducing carbon emissions have relied on population and economic assumptions from other groups. The worldwide population growth projections are commonly from demographic models of the World Bank or the United Nations. For economic assumptions, such studies generally rely on economic growth projections made by others, or on external assumptions about labor force participation, and productivity growth. For either population or economic growth, there is less uncertainty about projections for the developed countries than for the developing countries. Birdsall (1992) offers a discussion of the dependence of the results of global warming analyses on population projections.

A second key set of input assumptions to the models concerns the price and/or availability of energy resources. The prices of fossil fuels like oil, natural gas, and coal tend to be important because they are what generally need to be substituted away from when carbon emissions are restricted. Since gas combustion produces about 60 percent of the carbon produced by coal combustion and 80 percent of the carbon produced by oil combustion per unit of energy consumed, optimistic assumptions about natural gas availability can make carbon emissions targets easier to achieve in the short run. Another key dimension of the energy supply picture is the assumption made about the response of the oil exporters to the imposition of a carbon tax or other policies that reduce the demand for oil imports.

A final set of key assumptions about inputs are those made about the costs and efficiencies of current and future technologies, both for energy supply and energy use. Most analysts use a combination of statistical analysis of historical data on the demand for individual fuels, and a process analysis of individual technologies in use or under development to represent trends in energy technologies. At some point these two approaches tend to look quite similar though, as the end-use process analysis usually runs out of new technology concepts after a couple of decades, and it is then assumed that the efficiency of the most efficient technologies for which there is an actual proposed design will continue to improve as time goes on. Particularly important, but difficult, here is projecting technological progress. Attempts to estimate empirically the dependence of future trends in productivity on factor prices, including energy, are rare. Jorgenson and Wilcoxen (1991) present one prominent attempt to do so.

In the long run, the assumptions made about the cost of substitutes for conventional oil and gas determine the cost of controlling carbon emissions. It is generally assumed that the long-run carbon-based fuel alternatives (for example, coal generated electricity and coal-based liquid synthetic fuels) will be less expensive than non-carbon alternatives (like solar-generated electricity, and liquid fuels made from biomass). If this were not the case, there would be no projected long-run costs of controlling carbon emissions.

Existing emissions controls on fossil fuel combustion, like those on sulfur dioxide included in the U.S. Clean Air Act and its amendments, are represented in the models via their projected impact of fuel choice trends, or by using higher costs for technologies that emit sulfur dioxide in the process analysis approach. Representing these costs can be an important element of the baseline scenario because coal use generally leads to high sulfur *and* carbon emissions. Thus, restricting sulfur emissions can lead to some reductions in coal use and, hence, carbon emissions.

Control Scenario

A control scenario translates a carbon emission control program into model inputs. Several dimensions of the control scenario considered are important. First, the control program must specify a goal, which can be an annual carbon emissions target, an atmospheric CO_2 concentration limit, a rate of carbon taxation, or some combination of these elements. The goal must also include a rate at which the objective is to be achieved—because of the time required to replace energy-producing and -consuming capital equipment, it may be easier to reach a more stringent annual emissions target over a longer period of time than a less stringent one over a shorter period of time.

The instrument to be used to achieve the objective must also be specified. For example, an emissions target may be met via a carbon tax, an emissions permit trading system, energy efficiency standards, or a variety of other policies. For control scenarios involving carbon taxation, some assumption needs to be made about the use of the revenues generated, and about the tax treatment of imports of energy and energy intensive products.

A final crucial element of the control program is the allocation of the allowable emissions among countries or regions. Since developing countries are poorer to begin with and likely to grow faster than the developed countries, they will probably not agree to proportional cuts in emissions by all countries.

The paper by James Poterba in this symposium offers an excellent discussion of "North-South" negotiations.[1]

Model Structure

The models considered here can be classified into six distinct classes, depending on the way the relevant markets are represented. The first three categories are *energy-economy models*: the distinguishing feature of such models is that they treat the energy sector in a more aggregate fashion than the energy models, and instead contain much more detail on markets and/or agents in the rest of the economy or economies. The other three categories are *energy models*, which focus on the various markets for energy—fossil fuels, renewables, electric power, and refined fuels—and on specific energy conversion and/or end-use technologies. They treat the rest of the economy in a more aggregated fashion. (See Beaver, 1992, for a more detailed structural comparison of the EMF 12 models.)

The salient characteristics of each class of model are summarized in Table 1, along with examples of models that are included in each class, the time horizon considered, the amount of foresight assumed, and whether the behavioral parameters included in the model are empirically or judgmentally determined. None of the models considered here incorporates uncertainty about future events in representing decision-making by individual participants in the economy.

One distinction of importance in the three energy-economy approaches lies in the level of disaggregation of the non-energy sectors. Both disaggregated and aggregated economic equilibrium models allow investment levels to be chosen endogenously. However, the disaggregated equilibrium models consider markets by industry category, and allow GDP to be determined, in part, by these interindustry interactions. The aggregate equilibrium models, on the other hand, usually include markets for broad sectors like capital, labor, and energy, and determine GDP with an aggregate production function. The main advantage of the disaggregated approach is that it allows a closer look at non-energy markets; the main advantage of the aggregate approach is that it allows more detail in looking at energy markets.

The disaggregated energy-economy models can be further subdivided into long-run growth and short-run macroeconomic adjustment variants. The macroeconomic adjustment models include multiple sectors and explicitly represent the processes by which unemployment and inflation are created and influence the level of economic output.

In the three classes of energy models—energy-sector equilibrium, optimization, and energy demand simulation models—the major distinction is the manner of modeling prices in the energy sector. An energy-sector equilibrium model examines simultaneous equilibria in all energy sector markets: fossil fuel, renewable energy, electricity, and so on. Contrary to the first three model

[1]See also Chichilnisky (1993).

Table 1
Model Types and Distinguishing Characteristics

Market Representation	*Model Author(s)*	*Scope*	*Time Horizon*	*Fore-sight*	*Param*
Disaggregated Economic Equilibrium	CRTM (Carbon Rights Trade Model) Rutherford (1992)	World	2100	M	J
	DGEM (Dynamic General Equilibrium Model) Jorgenson/Wilcoxen (1991)	U.S.	2050	P	E
	GOULDER Goulder (1993)	U.S.	2030	P	J
	GREEN (General Equilibrium Environmental) Burniaux/Martin/Nicoletti/Oliveira-Martins (1991)	World	2050	M	J
Disaggregated MacroEmetric Adjustment	DRI McGraw-Hill Brinner, Shelby, Yanchar, Christofaro (1992)	U.S.	2010	M	E
	LINK Kaufman/Li/Pauly/Thompson (1991)	World	2010	M	E
Aggregate Economic Equilibrium	CETA (Carbon Emissions Trajectory Assessment Model) Peck/Teisberg (1992a)	World	2100	P	J
	GLOBAL 2100 Manne/Richels (1992a)	World	2100	P	J
Energy-sector Equilibrium	GEMINI Decision Focus, Inc. (1990)	U.S.	2030	P	J
	FOSSIL2 AES Corporation (1990)	U.S.	2030	M	J
	GLOBAL MACRO ICF, Inc. (1988)	World	2100	M	J
	ERM (Edmonds-Reilly Model) Edmonds/Reilly (1985); Edmonds and Barns (1991)	World	2095	M	J
	MWC (Model of Warming Commitment) Mintzer/Schaper (1992)	World	2095	M	J
	EDS (IEA Energy Demand System) International Energy Agency (1991)	World	2005	M	E
Energy-sector Optimization	MARKAL (MARket ALlocation Model) Morris/Solomon/Hill/Goldstein (1990)	U.S.	2025	P	J
Energy Demand Simulation	T-GAS (Trace Gas Accounting System) Piccot, et al. (1990)	World	2010	M	E

Key: M is myopic; P is perfect; J is judgmental; E is econometrically estimated.

types, this type of model does not include any markets for non-energy goods. These models generally offer a regional analysis of the nation or globe, along with great detail on energy technology.

The objective of an energy-sector optimization model is minimization of total U.S. energy sector costs, given the demands for energy services from the rest of the economy. Although this optimization approach contains within it an implicit equilibrium, it does not actually use explicit market prices. This approach usually allows energy use to be reduced only through the introduction of more efficient technologies, and employs a great deal of detail about such technology. However, it does not examine economic effects outside the energy sector—even questions like what effect higher energy prices would have on the quantity of energy demanded.

The energy demand simulation approach uses projections of energy prices and economic activity, together with econometrically estimated parameters representing the behavioral response of industry and consumers to these factors, to estimate how much energy will be consumed. This approach has no explicit representation of supply, and thus does not model an explicit equilibrium in energy markets.

Two other features of the models are important in understanding and assessing differences in model projections: the assumption made about the foresight employed by economic agents in making decisions, and the source of key parameters employed in representing economic behavior. In some models it is assumed that economic agents base current consumption, production, and investment decisions on current prices only (myopic expectations), while others assume that they base these decisions on perfect information about the price projections produced by the model.

The substitution parameters used in the models are either judgmental (although generally calibrated to the available empirical literature in some way) or econometrically estimated with historical data. Reflecting the typical practice of including only a limited number of aggregate parameters whose values can easily be changed in sensitivity analyses in them, the models with judgmentally determined parameters are sometimes also referred to as "parametric" models.

Cost Measure

The cost of carbon emission reductions are reported using at least four alternative cost measures: (1) the *carbon tax* required, which measures the marginal cost of the last ton of emissions reduced; (2) the *total direct cost*, which measures the marginal cost of the reductions integrated over all emissions reductions; (3) the loss in *gross domestic product* (GDP), which measures the reduction in the total goods and services produced by the economy, and thus may include the impact of the carbon emission reductions on capital accumulation and technological progress, and (4) *compensated income variation*, which measures how much consumers would have to be compensated to make them as well off in terms of their own utility after the imposition of the emissions reduction program as in the baseline scenario. Not all models can produce results for all cost measures, which further complicates comparisons. For example, energy sector models cannot produce compensated income variation results (since they do not examine the broader economy in that way), and the total direct cost is generally not reported by general equilibrium modelers. These measures tend to move together, but departures from this general rule can be significant. For example, general equilibrium models may project lower carbon taxes required to achieve an emissions target, but greater GDP losses than energy sector models if they include a strong negative influence of higher energy prices on economic growth. This lowers baseline emissions, but increases the share of emissions reductions that comes from lower economic growth rates rather than reductions in energy intensity or fuel switching.

Along with the problem of measuring the cost of emissions reductions along different dimensions, cost measures are often discussed according to their time frame. The next section of this paper discusses "long-run" costs, defined as those costs which remain after a transition to primary reliance on non-fossil fuel sources of energy can be completed. The following section will address "short-to-intermediate-run" costs, which occur until that transition is completed. Both sorts of costs are likely to be important to policy-makers. Although the total annual costs of carbon emission reductions are generally projected to be higher in the long run, the marginal costs of control can be higher in the short run. In addition, the relative importance of short- and long-run costs depends on the rate at which future costs are discounted, which is an ongoing debate as mentioned by Richard Schmalensee in his introduction to this symposium. Throughout the discussion here, the goal is to describe results that appear to be robust across all sets of input assumptions and models, and when the results differ, to figure out why.

Long-Run Cost Projections

The transition period to a low-carbon world is likely to take 40 to 60 years, even if it started today. This results from the long life of energy-producing and -consuming capital stock, and the long lead time required to develop, refine, and phase in new carbon-free technologies. Every carbon emission control study must come to grips with this fact.

Common Conclusions

Three major general conclusions can be drawn about the cost of controlling carbon emissions in the existing long-run economic studies, defined here as those extending beyond 2040: controlling global carbon emissions over this time horizon will require a significant amount of international cooperation; if such cooperation can be achieved, a control program that is phased in gradually need not slow world economic growth appreciably, but will be expensive in absolute terms; and a system of international emissions trading can improve the efficiency of a global control program and significantly reduce its cost. Each of these points is explored in turn.

A significant amount of international cooperation will be required to control carbon emissions in the long run. Manne and Richels (1992a) observe that it will probably be impossible to limit global carbon emissions to 1990 levels by the middle of the next century if China does not participate in the control program—even if *all* other countries reduce their emissions to zero! Although China is responsible for only about 11 percent of global carbon emissions today (650 million tons out of world total of 6 billion tons) that percentage is expected to grow as the Chinese economy develops and uses its substantial coal reserves as part of that development process. If carbon emissions in China grow at 4 percent per year, which is not a particularly high baseline projection, in 56 years the emissions in China alone will equal total world emissions today. Efforts to limit global carbon emissions may start with the developed countries, but they cannot end there.

The costs of stabilizing global carbon emissions appear likely to be in the range of about 4 percent of GDP per year by the year 2100. For example, consider four key control scenarios from the EMF 12 study. Each includes participation by all major carbon emitters. In the "Stabilize Global Emissions" scenario, the countries of the OECD and the former Soviet Union are assumed to reduce emissions by 20 percent relative to 1990 levels by 2010 and maintain them at that level through 2100, while the rest of the world is restricted to increasing their emissions by no more than 50 percent relative to 1990 levels through 2100. This allocation of emissions rights leads to global emissions in 2010 and thereafter that are no greater than those in 1990. In the "Reduce Global Emissions" scenario the OECD and former USSR reduce emissions by 50 percent relative to 1990 by 2050, and the rest of the world is again limited to no more than a 50 percent increase relative to 1990 levels; this leads ultimately to about a 15 percent reduction in global carbon emissions. In the "Slower Growth in Global Emissions" scenario, the developed countries limit emissions to 1990 levels by 2000 and the developing countries are again limited to no more than 50 percent more than today's emissions levels, resulting in about an 18 percent increase in global emissions relative to 1990. Finally, the "Phased-In Tax" scenario includes a carbon tax imposed by all countries that increases from $15 per metric ton of carbon in 1990 at 5 percent per year. This is approximately the tax trajectory that would result from limiting *cumulative* world emissions over the next 50 years to 50 times 1990 emissions.

By the year 2040, all four scenarios would cause the world GDP on average to drop by about 2 percent of GDP per year, plus or minus half a percent. By 2100, they would reduce global GDP by 4 percent per year, plus or minus 1 percent. However, all of these scenarios represent a significant departure from the historical trend in carbon emissions. The baseline projection, without any of these policy changes, is that carbon dioxide emissions will have doubled by 2040, and increased by a factor of nearly six-fold by 2100. The "Phased-In Tax" scenario would reduce emissions rather sharply by the year 2100, while the other three scenarios would keep carbon emissions within about 20 percent of their present level over the next century.

The cost of a long-run global carbon emissions control program designed to limit emissions to their 1990 level (about 6 billion metric tons), is projected to be about 2.5 percent of world GDP by the year 2043, plus or minus 1 percent. If world GDP were projected to grow at 2.3 percent per year over the next 50 years in the absence of a control program, a control program phased in over that period would only lower that growth rate to 2.25 percent per year. Thus, if climate change turns out to be a very serious problem, it can be controlled without eliminating or even noticeably reducing long-run economic growth. On the other hand, 2.5 percent of the almost $90 trillion dollar world economy in 2043 (in 1990 dollars) would amount to about $2.25 trillion dollars per year (2.5 percent of current world GDP is over half a trillion dollars). Given other pressing world problems, this sum represents a substantial long-run commitment of societal resources.

These cost estimates have so far generally assumed that the emissions reductions are achieved efficiently in each region as, for example, through the use of a tax on the carbon content of fossil fuels. Given institutional constraints in both the developed and developing countries, this assumption may ultimately turn out to be overly optimistic, but the results obtained for the carbon tax cases provide useful benchmarks on what might be achievable.

A growing number of analysts have studied the potential for international trading of emissions rights to reduce the cost of a global carbon emissions control program. In such a program, one country can pay for emissions reductions in another country, if that is less expensive than the available domestic options. In EMF 12, an emissions trading program was considered relative to the global emissions stabilization scenario, where the initial allocation of emissions rights is 20 percent lower than current emissions for the developed countries and 50 percent more than current emissions levels for the developing countries. Results from four models show that this strategy can reduce the costs of control by one-third or more, by equalizing the marginal cost of control across all nations.[2] Allocation of emissions rights by population or GDP per capita would lead to even greater benefits from international emissions trading (Yamaji and Okada, 1992; Edmonds et al., 1993). At some point, the technology of energy production and use may become homogeneous enough to eliminate these advantages, but until that point an international emissions trading program will be worth serious negotiating efforts; in fact, it may be the only way to solve the participation problem noted above.

[2]For example, the four models included in EMF 12 that attempted the International Emissions Trading Scenario—ERM, OECD-GREEN, Global 2100, and CRTM—project a decrease of 30 percent to 40 percent in the initial cost of Stabilize Global Emissions scenario with trading with respect to the no trading case.

Significant Differences in Assumptions and Models

Although the long-run cost of controlling carbon emissions seems fairly similar in many published studies, there are some differences as well. First, the estimated cost of reducing emissions to achieve any CO_2 emission or concentration target depends on the baseline emissions projection of carbon emissions, which in turn depends on the projected rates of economic growth, income elasticities for energy demand, and rates of non-price induced shifts in energy efficiency. Second, the degree to which "backstop" technology assumptions—advanced technologies with high cost and unlimited supply—are employed has a large impact on the cost of control estimates. Third, in lieu of backstops, some analysts use resource supply curves with large flat segments at higher prices, and differences in these assumptions can be significant. Again, each of these three points will be examined in turn.

Most long-run projections of the costs of controlling carbon emissions foresee a major role for coal in the future baseline energy mix, and since coal is abundant on a global basis, this assumption limits future energy price increases. As a result, the major determinants of long-run energy demand turn out to be population growth, economic growth, the income elasticity of demand for energy, and the non-price-induced rate of change in energy demand per unit of economic output. Very few modelers make their own projections of long-term worldwide economic growth, worldwide income elasticities, or long-term energy efficiency trends. Thus, most researchers rely on the few estimates that are available, most of which are for just the United States or the other members of the Organization for Economic Cooperation and Development (OECD).

Even when the assumptions about GDP growth rates are standardized, a wide range of reference case emissions projections are produced by the models. By the year 2100, projections of world CO_2 emissions from eight different models range all the way from a tripling to a seven-fold increase over 1990 levels.[3] The reason for this variation is simple enough: relatively small differences in model parameters lead to large differences when their effects are compounded over more than a century. Much of the difference in projections from the models for 2100 can be explained by differences in the assumed rate of decrease in energy use per unit of economic output independent of energy price changes. The Global 2100 model uses a value of .5 percent per year for this parameter, while the Edmonds-Reilly model employs 1.0 percent per year assumption. This single difference explains virtually all of the nearly factor of two difference in the baseline emissions projections from these two models in 2090 (Dean et al., 1993; Gaskins and Weyant, 1993b). The more disaggregated assumptions made in the Global Macro model imply about a 1.25 percent rate of decrease in energy use (per unit of output). Estimates of this aggregate parameter based on historical data range from a rate of decrease of about .5 percent per year to an *increase* at about that rate (Jorgenson and Wilcoxen,

[3]The eight models are ERM, Glob-Macro, Global 2100, OECD, IEA, CETA, MWC, and CRTM.

1991; Manne and Richels, 1992a). Researchers who have attempted to extrapolate the types and efficiencies of energy-using equipment into the future have argued that the potential exists for a rate of decrease in energy use per unit of economic output from 1 percent per year to over 2 percent per year without any increase in energy prices.

Some analysts use the backstop technology concept (introduced by Nordhaus, 1973) to represent the idea that at a high enough price for energy, one or more technologies will combine to provide a virtually unlimited supply of energy. The inclusion of such technologies puts a cap on the ultimate cost of carbon emissions reductions. In fact, Manne and Richels (1990), who include both a carbon-based backstop (like synthetic oil from coal) and a carbon-free energy backstop (like synthetic oil from biomass) in their analysis, have observed that the long-run carbon tax in such a set-up can be calculated on the back of an envelope, as the difference between the costs of the carbon-free and carbon-based backstops divided by the carbon emissions rate of the carbon-based backstop.[4]

Analyses that do not include backstop technologies per se generally include long-run supply curves for carbon free resources that become very flat at high cost levels. The resources included and extent of high-cost carbon-free resource availability vary from one analysis to another, but each serves to limit the long-run cost of carbon emission reductions.

Short-to-Intermediate Run Costs

In the short-to-intermediate run, the economy must face the costs of transition away from primary reliance on carbon-based energy. Again, some commonalities among the results of these transition studies are reviewed, and then some differences.

Common Results

Several common themes emerge from a review of the cost of control projections for the short-to-intermediate term. First, if the emissions target requires moving faster than the natural rate of capital stock turnover and technology development, significant additional adjustment costs are likely to be incurred. Second, recycling the revenues obtained from carbon taxation in a way that reduces less efficient ways of raising government revenues can significantly reduce the net cost of the control program. Third, if some major carbon emitters fail to participate in the control program, the downward pressure on oil prices and the supply of carbon-intensive goods in the participating coun-

[4]For example, in EMF 12 a carbon-based non-electric backstop costing \$50/BBL, a non-carbon non-electric backstop at \$100/BBL, and a carbon emission for the carbon-based backstop of .24 metric tons carbon/barrel oil were specified. Thus, the long-run carbon tax produced by the Global 2100 model is (\$100–\$50)/.24 = \$208/metric ton carbon.

tries will cause increases in carbon emissions by non-participants relative to their baseline levels, somewhat offsetting any gains.

Studies of the cost of controlling carbon emissions show that incremental reductions in allowable emissions cost more as the absolute level of allowed emissions in any particular year is reduced. This finding is especially true in the short- and intermediate-run, up to about 2040, before old fossil-fuel based energy producing and consuming equipment can be fully retired and new carbon-free technologies can be fully introduced. For example, in EMF 12 the cost of stabilizing U.S. carbon emissions (reducing them an average of 30 percent relative to the no control baseline) ranged from .2–.75 percent of GDP in 2010, while the cost of reducing emissions by 20 percent in that year (or an additional 15 percent relative to baseline emissions on average) range from .9–1.7 percent of GDP. In other words, a 20 percent additional reduction in emissions relative to 1990 levels (and, on average, a 15 percent reduction relative to 2010 baseline emissions) could more than double the cost.

Existing studies also demonstrate that trying to reach fixed emissions targets faster—say by 2010 rather than 2040—imposes significant additional adjustment costs. Short- to intermediate-run carbon tax projections from the models that significantly exceed their long-run levels provide the most dramatic evidence of situations where these additional adjustment costs are likely to be incurred.[5] This result is observed in results from models that take a long-term perspective; it is still larger in the medium-term models that try to track year by year energy-using and energy-producing capital stock adjustments over the next 20–40 years; and amplified still further in the short-run econometrically estimated demand models that focus primarily on projecting energy demand over the next 5–15 years. These adjustment costs result from a combination of empirically and judgmentally estimated adjustment rate parameters, as well as actual data on existing technologies and capital equipment in place. Despite the heterogeneity in modeling approach adopted, each leads to a projection that substantial adjustment costs are likely to be incurred in this scenario. Extensive sensitivity analyses by Petersen et al. (1992) show that even a ten-year delay in the date by which the global stabilization target must be achieved can reduce the peak in the carbon tax projections dramatically. In addition, Manne and Richels (1992b) show that the peak is virtually eliminated in Global 2100 model by limiting total emissions over a 50-year time period to fifty times the 1990 levels, and letting the model decide when the emissions should occur to maximize the discounted utility of consumers. This optimization results in more emissions than required to stabilize annual emissions in the short run, and less emissions than required to stabilize annual emissions in the long run.

[5]In EMF 12, the average carbon tax required to reduce U.S. carbon emissions in 2010 to 20 percent below their 1990 level is about $350 per metric ton, while the same goal can be achieved in 2050 with an average carbon tax of slightly over $200 per metric ton.

If a carbon tax is used to reduce emissions, the way in which revenues from that tax are used has an important impact on the projected GDP loss. Models that do not explicitly represent the tax system often assume tacitly that the revenues simply accrue to the government, with no particular incentive affects, like the revenue from a lump sum tax. However, carbon tax revenues could be used to reduce existing taxes that discourage economic activity, particularly capital formation. This is not a prescription to overhaul fiscal policy to minimize the impact of a carbon tax on the economy, but a more general observation regarding the relationship between taxation and economic growth. It emerges in the carbon emission control context because it helps explain differences in model results and identifies the way in which carbon tax revenues are recycled as an important consideration in evaluating them as a potential policy instrument. As part of the EMF 12 study, simulations with four models of the U.S. economy (DRI McGraw-Hill, LINK, DGEM, and Goulder) indicate that from 35 percent to as much as 100 percent of the U.S. GDP losses for the phased-in carbon tax could ultimately be offset by recycling revenues through cuts in existing taxes (Shackleton et al., 1992). On the other hand, if the revenues are used to fund low return government projects, the cost of achieving the carbon emission reductions could be substantially *greater* than when they are recycled in lump sum fashion (Nordhaus, 1993).

If the OECD or any other group of countries unilaterally implements a carbon reduction program, resulting changes in international energy prices will cause carbon emissions in other countries to increase relative to baseline levels (Rutherford, 1992; Nicoletti and Oliviera-Martins, 1992; Horton, Rollow, and Ulph, 1992; Rutherford et al., 1992). Increased carbon emissions by non-participating regions occur both as a result of increased energy intensity of economic activity domestically (that is, through the use of more energy intensive production processes and demands for more energy intensive products) and through the migration of energy-intensive production into unconstrained regions. Carbon restrictions place countries who control at a competitive disadvantage in energy-intensive industries. Thus, the cost to countries who control emissions increases with the level of cutback, but the impact on global emissions may drop off sharply if large groups of countries fail to cooperate. In comparing model results, we will generally assume universal participation because that is what most existing studies have assumed. Edmonds et al. (1993) and Nicoletti and Oliviera-Martins (1992) also consider some of the implications of partial participation on control costs.

Significant Differences

The most significant differences in short-to-intermediate term cost projections result from differences in the rate of achieving emission targets, introduction rates for new technologies, and retirement rates for existing facilities; economic growth projections; the treatment of pre-existing taxes on energy and on other inputs to the economy; and assumptions about natural gas availability and price.

If a fixed annual emissions rate target is specified, cumulative costs can be reduced with some increase in short-term emissions either by allowing more time for reaching the target, or by phasing in more gradually the instrument(s) used to achieve the target (like a carbon tax). The cost reduction can be particularly significant if the target date and rate of implementation are set to allow new carbon-free technologies to be phased in smoothly. The timing issue has been explored through extensive sensitivity analysis by Petersen et al. (1992) and Manne and Richels (1992b). Similar issues involving the rates of technology introduction and retirement have also been explored through sensitivity analysis by Petersen et al. (1992).

Different projections of economic growth can lead to different estimates of carbon control costs. The EMF 12 study design includes a 2.2 percent growth rate in GDP for the U.S. over the next 30 years. However, one of the models included in the study (DGEM) produced an independent GDP projection of 1.4 percent per year over that time frame (for a description of the methodology employed see Jorgenson and Wilcoxen, 1991, 1992). The lower growth rates result in lower carbon taxes being required to meet any particular emissions target. Interestingly, though, the computed GDP losses from such carbon taxes are not significantly less than in the other models, because higher carbon taxes lead to higher energy prices, which significantly diminish capital accumulation and productivity growth.[6]

Models that include preexisting energy taxes generally require a larger increase in energy prices and, therefore, a higher carbon tax to reach a fixed emissions target (for example, Kaufman, 1992). In countries with higher existing energy taxes, a larger carbon tax must be utilized to achieve a given percentage increase in retail energy price levels. Moreover, most models exhibit increasing marginal costs of energy demand reductions with respect to decreases in the level of energy demand. In addition, Goulder (1993) has observed that general equilibrium models can include preexisting taxes on all factors of production which results in carbon taxes being more costly to the economy. Carbon taxes lead to less energy use which makes other factors less productive, but if these factors are already taxed, the marginal productivity loss will be greater than if there were no preexisting taxes on the other factors. In effect, a carbon tax operates as an implicit factor tax; higher preexisting taxes elsewhere increase the propensity of the carbon tax to generate further distortions in factor markets. This effect also serves to blunt the effectiveness of

[6]The lower GDP growth rate was adopted for a "Low GDP Growth" sensitivity scenario. This scenario does lead to a significant reduction in the cost of control because it directly reduces the reference level of emissions projected by each model. In addition, when all the models are run with the low GDP growth rate assumptions, they produce carbon taxes that are more closely consistent with those projected by the lower growth models.

revenue recycling as a means of reducing the net costs of imposing a carbon tax to reduce carbon emissions.

Finally, the cost of the transition to the non-carbon-based energy technologies can be significantly affected by the availability of natural gas resources. Since natural gas has a lower carbon emissions rate than oil or gas, more fossil energy can be consumed within any emissions constraint if the use of natural gas can be increased. The "High Natural Gas Resources" scenario included in EMF 12 postulates a quadrupling of natural gas resources in each region, an optimistic assumption that thus gives an upper bound on the importance of this factor. In any event, this assumption does lead to a 30 to 40 percent reduction in the discounted cost of stabilizing emissions over the next 20 years.

Conclusions and Directions for Future Research

Regardless of one's view on the magnitude of the global climate change problem and preferred carbon emissions control objective, two things appear to be paramount in any attempt to minimize the cost of any control program—international cooperation and time for existing energy-related facilities to be retired and for new technologies to be phased in. A carefully designed revenue recycling program and set of trade policies can also reduce the costs of the control program significantly. The long-run cost of stabilizing global carbon emissions at 1990 levels appears to be about 4 percent of world GDP per year (plus or minus 1 percent). Although this cost is very high in absolute terms, it implies only a minor reduction in the rate of economic growth if the control program is phased in gradually. On the other hand, if policy were to focus on some convenient but arbitrary earlier goal—like stabilizing or decreasing emissions rates by 2000 or 2010 relative to 1990 levels—the increase in costs could be substantial.

Given the timing and coordination issues discussed here (involving coordination with fiscal and trade policies, as well as with control programs in other nations), a look at the cost side of the equation suggests the advantages of a more gradual control policy, unless the damages resulting from the small amount of additional emissions over the next half century can be argued to be worth the relatively large incremental cost of eliminating them. Reducing the growth rate of emissions might not only be beneficial itself, but also put us in a better position to control emissions more aggressively in the future, if that should become necessary.

Additional research in several areas could significantly improve our ability to understand the costs of reducing carbon emissions, and thus to evaluate alternative policy options.

1. Information about the potential for technologies that improve energy efficiency is incomplete and often inconsistent, and there is little conclusive analysis regarding their likely rate of adoption. However, this factor is a major

determinant of baseline emissions levels, a major source of differences in emissions projections between the models, and a major determinant of the cost of achieving any emissions target. Particularly important here are assessments of market imperfections or distortions that impede the introduction of more efficient technologies. For example, energy pricing in China, the developing countries, and the former Soviet Union has been far below world market levels.

2. The existing long-run population and economic projections display some considerable differences. In particular, relatively little is known about likely economic and population growth in the future in such important and diverse countries as China, India, Brazil, Nigeria, and the independent states of the former Soviet Union.

3. The adjustment costs that result from any control strategy depend on the availability and cost of energy sources that do not emit carbon. Many of these are "renewable" resources like solar or wind power. Such technologies are not currently cost-effective, but they could easily become competitive in the future—say, if carbon taxes reach $100 per metric ton. The models included in this study all represent these adjustment costs in one way or another, but often the approaches lead to markedly different results. This suggests the value of additional work on data and models of new technology availability dates and introduction rates, as well as of technology transfer to the developing countries.

4. Some possible offsets to carbon emissions are not directly part of energy production or use; for example, tree planting or slowing deforestation, or reducing other greenhouse gas emissions like methane from natural gas system leaks, coal bed seams, or ruminants. These approaches might be as effective as carbon emissions reductions in slowing climate change.[7]

5. The value of additional work on the linkages between environmental policies and international trade in energy and non-energy goods is high. See McKibbin and Wilcoxen (1992) and Manne and Rutherford (1992) for promising new work in this direction.

■ *I am grateful for research support to the U.S. Department of Energy, the U.S. Environmental Protection Agency, and the National Science Foundation and for valuable comments to Darius Gaskins, Jim McMahon, Don Rosenthal, Richard Schmalensee, Bob Shackleton, Carl Shapiro, Jim Sweeney, and Timothy Taylor.*

[7] Herzog et al. (1993) argue that a large scale *long-run* program of carbon sequestration may not ultimately prove feasible. A small interim program could, however, provide much needed time for the world's economies to make the adjustments necessary to transition to non-carbon based energy fuels. See, for example, Callaway et al. (1993) and Richards et al. (1993).

References

AES Corporation, "An Overview of the Fossil2 Model." Prepared for the U.S. Department of Energy, Office of Policy and Evaluation, Arlington, Virginia, July 1990.

Beaver, R., "Structural Comparison of EMF 12 Models," Working Paper 12.3, Energy Modeling Forum, Stanford, California, October 1992.

Birdsall, N., "Another Look at Population and Global Warming." Paper presented to the United Nations Expert Group on Population and Environment, United Nations, New York, January 1992.

Blitzer, C., R. Eckaus, S. Lahiri, and A. Meeraus, "Comparative Economic Effects of Greenhouse Emission Restrictions in Egypt and India," working paper, MIT, May 1992.

Blitzer, C., R. Eckaus, S. Lahiri, and A. Meeraus, "Growth and Welfare Losses From Carbon Emissions Restrictions: A General Equilibrium Analysis for Egypt," *The Energy Journal*, 1993, *14*:1, 57–82.

Boero, G., R. Clarke, and L. A. Winters, "Controlling Greenhouse Gases: A Survey of Global Macroeconomic Studies," working paper, Department of the Environment, Environmental Economics Research Series, University of Nottingham, U.K., August 1991.

Brinner, R., M. G. Shelby, J. M. Yanchar and A. Cristofaro, "Optimizing Tax Strategies to Reduce Greenhouse Gases Without Curtailing Growth." Washington, D.C.: U.S. Environmental Protection Agency, 1992.

Burniaux, J. M., J. D. Martin, G. Nicoletti, and J. Oliveira-Martins, "Green—A Multi-Region Dynamic General Equilibrium Model for Quantifying the Costs of Curbing CO_2 Emissions: A Technical Manual," Technical Report No. 104, OECD Department of Economics and Statistics, Resource Allocation Division, Paris, June 1991.

Callaway, J. M., B. McCarl, R. Adams, and C. Chang, "From Crop Subsidies to Carbon Subsidies: The Welfare Consequences of Replacing Farm Programs With Carbon Sequestration Programs in The U.S." Paper presented at the 68th Annual Western Economics Association Conference, Lake Tahoe, June 23, 1993.

Chichilnisky, G., "Global Environment and North-South Trade," Stanford Institute for Theoretical Economics, Technical Report No. 31, Stanford, California, 1993.

Cline, W. R., "The Economics of Global Warming," Institute for International Economics, Washington, D.C., June 1992.

Congressional Budget Office, "Carbon Charges as a Response to Global Warming: The Effects of Taxing Fossil Fuels," U.S. Congress. Washington, D.C.: U.S. Government Printing Office, 1990.

Dean, A., J. D. Martin, J. M. Burniaux and J. Oliveira-Martins, "Costs of Cutting Carbon Emissions: Evidence From Global Models," Technical Report No. 132, OECD Department of Economics and Statistics, Resource Allocation Division, Paris, 1993.

Decision Focus, Inc, "Gemini: an Energy-Environmental GEMS Model," Report Number C2023001.01, Los Altos, California, November 1990.

Edmonds, J., and D. W. Barns, "Factors Affecting The Long-Term Cost of Global Fossil Fuel CO_2 Emissions Reductions," Pacific Northwest Laboratory, Washington, D.C., April 1991.

Edmonds, J., D. W. Barns, M. Ton, "Carbon Coalitions: The Cost and Effectiveness of Energy Agreements to Alter Trajectories of Atmospheric Carbon Dioxide Emissions." Report prepared for the Office of Technology Assessment, Pacific Northwest Laboratories, Washington, D.C., February 1993.

Edmonds, J., and J. Reilly, "Global Energy: Assessing the Future." New York: Oxford University Press, 1985.

Gaskins, D. W., and J. P. Weyant, "Model Comparisons of the Costs of Reducing CO_2 Emissions," *American Economic Review Papers and Proceedings*, May 1993a, *83*:2, 318–30.

Gaskins, D. W., and J. P. Weyant, eds., "Reducing Global Carbon Emissions: Costs and Policy Options." A Report on the Work of Energy Modeling Forum Study Number 12, Stanford, California, forthcoming 1993b.

Goulder, L. H., "Effects of Carbon Taxes in an Economy With Prior Tax Distortions," working paper, Stanford University, January 1993.

Herzog, H., E. Drake, and J. Tester, "Research Needs Assessment for the Capture, Utilization, and Disposal of Carbon Dioxide from Fossil Fuel-Fired Power Plants." Report to the U.S. Department of Energy, Grant No. DE-FG02-92ER30194.A000, from MIT Energy Laboratory, Cambridge, March 1993.

Horton, G. R., J. M. C. Rollo, and A. Ulph, "The Implications for Trade of Greenhouse Gas Emission Control Policies." Report prepared for the U.K. Department of Trade and Industry, Environment Division, by National Economic Research Associates, London, March 1992.

ICF Incorporated, "Global Macro-Energy Model: Summary," Fairfax, Virginia, October 1988.

International Energy Agency, "Carbon Taxes and CO_2 Emissions Targets: Results from the IEA Model," Paris, August 1991.

Jorgenson, D. W., and P. J. Wilcoxen, "Reducing U.S. Carbon Emissions: The Cost of Different Goals." In Moroney, John R., ed., *Energy, Growth and the Environment.* Greenwich: JAI Press, 1991. 125–38.

Jorgenson, D. W., and P. J. Wilcoxen, "Reducing U.S. Carbon Emissions: An Econometric General Equilibrium Assessment," Working Paper 12.14, Energy Modeling Forum, Stanford, California, October 1992.

Kaufmann, R. K., "Is the Forecast for Carbon Taxes Forecast by TGAS an Outlier; Why the Forecasts by Other Models are too Low," Working Paper 12.16, Energy Modeling Forum, Stanford, California, October 1992.

Kaufmann, R., H. Y. Li, P. Pauly, and L. Thompson, "Global Macroeconomic Effects of Carbon Taxes: A Feasability Study," discussion paper, University of Toronto, July 1991.

Manne, A. S., and R. G. Richels, "CO_2 Emission Limits: An Economic Cost Analysis for the USA," *The Energy Journal,* April 1990, *11*: 2, 51–74.

Manne, A. S., and R. G. Richels, "Buying Greenhouse Insurance: The Economic Costs of CO_2 Emission Limits." Cambridge: MIT Press, 1992a.

Manne, A. S., and R. G. Richels, "The Impact of Timing on the Cost of Achieving Carbon Emission Targets," EMF Working Paper 12.23, Energy Modeling Forum, Stanford, California, October 1992b.

Manne, A. S., and T. F. Rutherford, "International Trade in Oil, Gas and Carbon Emission Rights—an Intertemporal Equilibrium Model. Paper presented at the International Workshop on "Costs, Impacts and Possible Benefits of CO_2 Mitigation," International Institute for Applied Systems Analysis, Vienna, September 1992.

McKibbin, W. J., and P. J. Wilcoxen, "G-Cubed: A Dynamic Multi-Sector General Equilibrium Model of the Global Economy," Discussion Paper No. 98 in International Economics, The Brookings Institution, Washington, D.C., November 1992.

Mintzer, I., and V. Schaper, "The Model for Warming Commitment," working paper, Stockholm Environment Institute, January 1992.

Morris, S. C., et al., "A Least Cost Energy Analysis of U.S. CO_2 Reduction Options." Paper presented at MIT Conference, Energy and the Environment in the 21st Century, Cambridge, March 26–28, 1990.

Nicoletti, G., and J. Oliveira-Martins, "Global Effects of the European Carbon Tax," Technical Report No. 125, OECD Department of Economics and Statistics, Resource Allocation Division, Paris, 1992.

Nordhaus, W. D., "The Allocation of Energy Resources," *Brookings Papers on Economic Activity,* 1973, *3,* 529–76.

Nordhaus, W. D., "A Survey of Estimates of the Cost of Controlling Greenhouse Gases," *The Energy Journal,* April 1991, *12*:1, 37–66.

Nordhaus, W. D., "Optimal Greenhouse-Gas Reductions and Tax Policy in the 'Dice' Model," *American Economic Review Papers and Proceedings,* May 1993, *83*:2, 313–17.

Pachauri, R. K., and N. Khanna, "Review of Costs to Developing Countries." Paper presented at the Workshop on "Costs, Impacts, and Possible Benefits of CO_2 Mitigation" at the International Institute for Applied Systems Analysis in Vienna, September 1992.

Peck, S. C., and T. J. Teisberg, "CETA: A Model for Carbon Emissions Trajectory Assessment," *The Energy Journal,* January 1992, *13*: 1, 55–77.

Peck, S. C., and T. J. Teisberg, "Cost-Benefit Analysis and Climate Change," Working Paper 12.7, Energy Modeling Forum, Stanford, California, October 1992.

Petersen, E., S. Belanger, D. Cohan, A. Diener, J. M. Drozd, and A. Gjerde, "The Transition to Reduced Levels of Carbon Emissions," Working Paper 12.6, Energy Modeling Forum, Stanford, California, October 1992.

Piccot, S., et al., "Development of Trace Gas Accounting System (T-Gas) for 14 Countries," draft report to the U. S. Environmental Protection Agency, June 1990.

Richards, K. R., D. H. Rosenthal, J. A. Edmonds, and M. Wise, "The Carbon Emissions Game: Playing the Net." Paper presented at the 68th Annual Western Economics Association Conference, Lake Tahoe, June 23, 1993.

Rutherford, T., "The Welfare Effects of Fossil Carbon Restrictions: Results From a Recursively Dynamic Trade Model," working paper, OECD, Paris, January 1992.

Rutherford, T., J. Oliviera-Martins, P. Pauly, and R. Kaufman, "International Trade Implications of Carbon Taxes," Working Paper 12.11, Energy Modeling Forum, Stanford, California, October 1992.

Schmalensee, R., "Comparing Greenhouse Gases for Policy Purposes," *The Energy Journal*, 1993, *14*:3, 245–55.

Shackleton, R., et al., "The Efficiency Value of Carbon Tax Revenues," Working Paper 12.8, Energy Modeling Forum, Stanford, California, October 1992.

Yamaji, K., and K. Okada, "A Simulation Study on Tradeable CO_2 Emission Permits." Paper presented at the International Workshop on Costs, Impacts and Possible Benefits of CO_2 Mitigation, International Institute for Applied Systems Analysis, Vienna, September 1992.

C
Deriving Optimal Solutions

[11]

Resource and Energy Economics 15 (1993) 27–50. North-Holland

Rolling the 'DICE': An optimal transition path for controlling greenhouse gases*

William D. Nordhaus

Yale University, New Haven, CT 06520, USA

Final version received November 1992

Economic analyses of efficient policies to slow climate change require combining economic and scientific approaches. The present study presents a dynamic integrated climate-economy ('DICE') model. This model can be used to investigate alternative approaches to slowing climate change. Evaluation of five policies suggest that a modest carbon tax would be an efficient approach to slow global warming, while rigid emissions-stabilization approaches would impose significant net economic costs.

1. Introduction

The threat of climate change has become a major economic and political issue, symbolic of growing concerns that humans are making irreversible and potentially calamitous interventions in life-support systems. Climatologists and other scientists have warned that the accumulation of carbon dioxide (CO_2) and other greenhouse gases (GHGs) is likely to lead to global warming and other significant climatic changes over the next century. Many scientific bodies, along with a growing chorus of environmentalists and governments, are calling for severe curbs on the emissions of greenhouse gases, as for example the reports of the Intergovernmental Panel on Climate Change [IPPC (1990)] and the Second World Climate Conference (October 1990). Governments have recently approved a 'framework treaty' on climate change to monitor trends and national efforts, and this treaty formed the centerpiece of the 'Earth Summit' held in Rio in June 1992.

To date, the calls to arms and treaty negotiations have progressed more or

Correspondence to: Professor William D. Nordhaus, Cowles Foundation, Yale University, Box 1972, Yale Station, New Haven, CT 06520, USA.

*This research was supported by the National Science Foundation. The author is grateful for helpful suggestions from Jesse Ausubel, Richard Cooper, William Hogan, Allan Manne, Thomas Schelling, Stephen Schneider, Paul Waggoner, and John Weyant, and for patient research assistance from Zili Yang.

0928-7655/93/$06.00

less independently of economic studies of the costs and benefits of measures to slow greenhouse warming. Over the last few years, however, a growing (but not unanimous) body of evidence has pointed to the likelihood that greenhouse warming will have only modest economic impacts in industrial countries, while programs to cut GHG emissions will impose substantial costs. Like two ships passing in the night, the economic studies and the treaty negotiations seems to be proceeding independently under their own steam.

Notwithstanding the difficulties of marrying the economic analysis with the policy process, the need to address the potential issues raised by future climate change is daunting for those who take policy analysis seriously. It raises formidable issues of data, modeling, uncertainty, international coordination, and institutional design. In addition, the economic stakes are enormous, involving investments on the order of hundreds of billions of dollars a year to slow or prevent climate change.

In earlier studies, I developed a simple cost–benefit framework for determining the optimal 'steady-state' control of CO_2 and other greenhouse gases.[1] This earlier study came to a middle-of-the-road conclusion that the threat of greenhouse warming was sufficient to justify modest steps to slow the pace of climate change, but I found that the calls for draconian cuts in GHG emissions by 50% or more were not warranted by the current scientific and economic evidence on costs and impacts.

The earlier studies had a number of shortcomings, but one of the most significant from an analytical point of view was the inadequate treatment of the dynamics of the economy and the climate. The steady-state approach is unsatisfactory primarily because of the extraordinarily long time lags involved in the reaction of the climate and economy to greenhouse gas emissions. Current scientific estimates indicate that the major GHGs have an atmospheric residence time over 100 years; moreover, because of the great thermal inertia of the oceans, the climate appears to have a lag of several decades behind the changes in GHG concentrations; and there are long lags in introduction of new technologies in human economies to changing economic conditions. It would appear, therefore, that the dynamics are of the essence and that an examination of the steady state may provide misleading conclusions for the steps that we should take at the dawn of the age of greenhouse warming.

The plan of the present study is to develop a dynamic, global model of both the impacts of and policies to slow global warming. It is an integrated model that incorporates both the dynamics of emissions and impacts and the economic costs of policies to curb emissions. We call it the *DICE model* as

[1]The latest version appears in abbreviated form in Nordhaus (1991a) and in greater detail in Nordhaus (1991b).

an acronym for a 'Dynamic Integrated Climate-Economy model'.[2] This new model is an advance over earlier studies in that it allows for different policies in the transition path from those in the ultimate steady state. It does this through the extension of the standard tools of modern optimal economic growth theory and adding to this analysis both a climate sector and a closed-loop interaction between the climate and the economy. The model is sufficiently small as to be transparent (or at least translucent), to allow a range of sensitivity analyses, and to be available for a number of further extensions.

The purpose of this paper is to lay out in details the structure of the model and the nature of the assumptions. The first section lays out the algebra of the model in simplified form. The following sections derive the parameters of the model in detail. The final sections then show some empirical runs of the model and provide estimates of the optimal policy along with some alternative approaches.

2. Methodology

Existing empirical studies of the interaction between climate change and economic growth have generally been of a partial-equilibrium or static nature. Much economic work has to date analyzed the costs of different GHG restrictions. Estimating the economic and other impacts of greenhouse warming has proved extremely difficult. I have attempted to summarize the results of studies for the United States [see Nordhaus (1991c)], but these remain incomplete in a number of respects.

The present study constructs a dynamic optimization model for estimating the optimal path of reductions of GHG gases. The basic approach is to use a Ramsey model of optimal economic growth with certain adjustments and to estimate the optimal path for both capital accumulation and GHG-emission reductions. The resulting trajectory can be interpreted as either (i) the most efficient path for slowing climate change given initial endowments or (ii) as the competitive equilibrium among market economies where the externalities are internalized using the appropriate social shadow prices for GHGs. We first describe the approach verbally and then present the model in equation form.

In intuitive language, the approach is the following. Begin with the market sector of the economy. The global economy is assumed to produce a composite commodity. This means that countries can differ in their quantita-

[2]The complete model is presented in a forthcoming book [Nordhaus (1993a)]. In addition, the theoretical underpinnings of the model are developed in a companion paper, Nordhaus (1993b), which developed an optimal growth model in which to analyze the issue of the optimal response to the threat of climate change under conditions of certainty.

tive attributes, but there cannot be large differences in the composition or relative proportions of different commodities. While this is a restrictive assumption, preliminary work with a more complete multi-country model suggests that aggregation does not affect the major conclusions.

Our composite economy is endowed with an initial stock of capital, labor, and technology, and all industries behave competitively. Each country maximizes an intertemporal objective function, identical in each region, which is the sum of discounted utilities of per capita consumption times population. Output is produced by a Cobb–Douglas production function in capital, labor, and technology. Population growth and technological change are exogenous, while capital accumulation is determined by optimizing the flow of consumption over time.

The next part of the model introduces a number of relationships that attempt to capture the major forces affecting climate change. This part includes an emissions equation, a concentrations equation, a climate equation, a damage relationship, and a cost function for reducing emissions. Emissions represent all GHG emissions, although they are most easily interpreted as CO_2. Uncontrolled emissions are a slowly declining fraction of gross output; this assumption is consistent with a complex set of assumptions about the underlying production functions. GHG emissions can be controlled by increasing the prices of factors or outputs that are GHG-intensive.[3]

Atmospheric concentrations are increased with emissions, with concentrations reduced with an atmospheric residence time of 120 years. Climate change is represented by realized global mean surface temperature, which uses an equilibrium relationship drawn from the consensus of climate modelers and a lag given by a recent coupled ocean-atmospheric models. The economic impacts of climate change are assumed to be increasing in the realized temperature increase.

We note that this model has one major shortcoming as a representation of economic and political reality. It assumes that the public goods nature of climate change is somehow overcome. That is, it assumes that, through some mechanism, countries internalize in their *national* decision making the *global* costs of their emissions decisions. This seems unlikely, but the current solution has the virtue of calculating the equilibrium that would emerge were each country to behave in such a farsighted and altruistic fashion.

[3]One concern that has arisen about this set of assumptions is whether it is consistent with fundamental laws of physics. Is it possible, in other words, to have an indefinite decline in emissions-output ratios? In principle, the answer is clearly yes because output is measured in 'utils' (or, more precisely, in ratios of marginal utilities), not in physical terms. Reductions arise if the composition of output moves toward goods and services that have lower intensities of constrained physical units (away from copper wire and toward moving a few electrons in fibers). Moreover, the waste of energy in current economic activity is prodigious.

3. Model

3.1. Basic outline[4]

In estimating the efficient path of capital accumulation and emissions reduction, we use the following model and assumptions. The different regions of the world are aggregated together and we analyze the optimal policy for the average individual. Clearly, this assumption misses much of the current dilemma and debate between developed and poor countries, and it also averages out the losers and the winners from climate change. The defense of this assumption is that this study is concerned with the efficient *intertemporal* policies, not with the issues concerning the distribution of income across countries or people.

The model operates in steps of 10 years centered on 1965, 1975, 1985, ..., 2095, ... The model is calibrated by fitting the solution for the first three decades to the actual data for 1965, 1975 and 1985 and is then optimizing for capital accumulation and GHG emissions in the future. This approach assumes that it is desirable to maximize a social welfare function that is the discounted sum of the utilities of per capita consumption. The major decision is about the level of consumption today, where abstaining from consumption today increases consumption for future generations. In technical language, we desire to maximize the objective function:

$$\max_{\{c(t)\}} \sum_t U[c(t), P(t)](1+\rho)^{-t}, \tag{1}$$

which is the discounted sum of the utilities of consumption, $U[c(t), P(t)]$, summed over the relevant time horizon. Here U is the flow of utility or social well-being, $c(t)$ is the flow of consumption per capita at time t, $P(t)$ is the level of population at time t, and ρ is the pure rate of social time preference.[5]

The maximization is subject to a number of constraints. The first set

[4]This section presents the bare bones of a longer study which documents the data, assumptions, and literature more fully. The longer study is currently available in Nordhaus (1992) and will be forthcoming as a monograph from MIT Press in Nordhaus (1993a).

[5]No issue is more controversial than the role of discounting in long-term environmental problems. Much confusion arises because of the failure to distinguish between time discounting and goods discounting, a distinction that is well handled in the Ramsey model. Time discounting refers to the trade-off between the utility or well-being of different generations and is represented by the pure rate of time preference, ρ, in the objective function. However, most social decisions involve goods and services rather than well-being. Society must decide whether to make an investment aimed at reducing today's consumption in order to increase consumption in the future; this approach is one of goods discounting, and the intertemporal price is reflected in the real interest rate, r. In the framework used here, with no population growth, with a growth rate of real income at rate g, and in steady state, goods discounting is related to time discounting by the equilibrium formula $r = \rho + \alpha g$.

represents economic constraints, while the second is the novel set of climate-emissions constraints.

3.2. Economic constraints[6]

The first set of constraints is those relating to the growth of output. Economists will recognize these as a standard model of optimal economic growth known as the 'Ramsey model' [Ramsey (1928); Solow (1988)]. The first equation is the definition of utility, which is equal to the size of population [$P(t)$] times the utility of per capita consumption $u[c(t)]$. We take a power function to represent the form of the utility function:

$$U[c(t), P(t)] = P(t)\{[c(t)]^{1-\alpha} - 1\}/(1-\alpha). \tag{2}$$

In this equation, the parameter α is a measure of the social valuation of different levels of consumption, which we call the 'rate of inequality aversion'. When $\alpha = 0$, the utility function is linear and there is no social aversion to inequality; as α gets large, the social attitude becomes increasingly egalitarian. In the experiments reported here, we take $\alpha = 1$, which is the logarithmic or Bernoullian utility function.

Output [$Q(t)$] is given by a standard Cobb–Douglas production function in capital in technology [$A(t)$], capital [$K(t)$], and labor, which is proportional to population. The term $\Omega(t)$ relates to climatic impacts and will be described in eq. (12):

$$Q(t) = \Omega(t)A(t)K(t)^{\gamma}P(t)^{1-\gamma}, \tag{3}$$

where γ is the elasticity of output with respect to capital. We assume constant returns to scale in capital and labor.

The next equation shows the disposition of output between consumption [$C(t)$] and gross investment [$I(t)$]:

$$C(t) = Q(t) - I(t). \tag{4}$$

This simply notes that output can be devoted either to investment or to consumption.

The next equation is the definition of per capita consumption:

$$c(t) = C(t)/P(t). \tag{5}$$

Finally, we have the capital balance equation for the capital stock:

[6]This section presents a summary of a more extensive analysis of the DICE model. The full documentation is given in Nordhaus (1993a).

$$K(t)=(1-\delta_K)K(t-1)+I(t), \quad (6)$$

where δ_K is the rate of depreciation of the capital stock.

3.3. Climate-emissions-damage equations

The next set of constraints will be unfamiliar to most economists and consists of a simple representation of the relationship between economic activity, emissions, concentrations, and climate change. As with the economic relationships, these equations are highly aggregated.

3.3.1. Emissions

The first equation links greenhouse-gas emissions to economic activity. In the analysis that follows, we translate each of the GHGs into its CO_2 equivalent. To aggregate the different GHGs, we use a measure of the total warming potential, which is the contribution of a GHG to global warming summed over the indefinite future. Approximately 80% of the total warming potential is due to CO_2, and we therefore put most of our effort into analyzing that gas.

In modeling GHG emissions, I assume that the ratio of uncontrolled GHG emissions to gross output is a slowly moving parameter represented by $\sigma(t)$. In what follows, we assume that the exogenous decline in σ is 1.25% per annum. GHG emissions can be reduced through a wide range of policies. We represent the rate of emissions reduction by an 'emissions control factor', $\mu(t)$. This is the fractional reduction of emissions relative to the uncontrolled level. One of the key questions investigated here is the optimal trajectory of emissions control. The emissions equation is given as:

$$E(t)=[1-\mu(t)]\sigma(t)Q(t). \quad (7)$$

In this equation, $E(t)$ is GHG emissions, $\sigma(t)$ is determined from historical data, and it is assumed that GHG emissions were uncontrolled through 1990. The variable $\mu(t)$ is determined by the optimization.

3.3.2. Concentrations

The next relationship in the economy–climate nexus represents the accumulation of GHGs in the atmosphere. For the non-CO_2 GHGs, the issues are relatively straightforward issues of estimating the atmospheric lifetimes or chemical transformations. We concentrate here on CO_2 because that is likely to be the most important gas for greenhouse warming. I assume that CO_2 accumulation and transportation can be represented as a system of boxes in which each of the boxes is well mixed. The background study [Nordhaus (1992a, 1993)] shows that this can be represented by the following equation:

$$M(t)=\beta E(t)+(1-\delta_M)M(t-1), \tag{8}$$

where $M(t)$ is the change in concentrations from pre-industrial times, β is the marginal atmospheric retention ratio, and δ_M is the rate of transfer from the quickly mixing reservoirs to the deep ocean. This equation is the GHG analog of the capital accumulation equation. Atmospheric concentrations in a period are determined by last period's concentrations $[M(t-1)]$ times $(1-\delta_M)$, where δ_M is the rate of removal of GHGs. We have estimated this relationship on historical data (1860–1985) and derived the following equation:

$$M(t)-0.9917M(t-1)=\underset{(0.015)}{0.64}E(t) \qquad R^2=0.803, \text{SEE}=0.519 \tag{8'}$$

This is the equation we use in the model.

3.3.3. Climate change

The next step concerns the relationship between the accumulation of greenhouse gases and climate change. Climate modelers have developed a wide variety of approaches for estimating the impact of rising GHGs on climatic variables. On the whole, existing models are, unfortunately, much too complex to be included in economic models. Another difficulty with current general circulation models (GCMs) is that they have generally been used to estimate the equilibrium impact of a change in CO_2 concentrations upon the level of temperature and other variables. For economic analyses, it is essential to understand the dynamics or transient properties of the response of climate to GHG concentrations.

The basic approach is to develop a small model that captures the summary relationship between GHG concentrations and the dynamics of climate change. In what follows, we represent the climate system by a multi-layer system; more precisely, there are three layers – the atmosphere, the mixed layer of the oceans, and the deep oceans – each of which is assumed to be well mixed. The accumulation of GHGs warms the atmospheric layer, which then warms the mixed ocean, which in turn diffuses into the deep oceans. The lags in the system are primarily due to the thermal inertia of the three layers. We can write the model as follows:

$$T_1(t) = T_1(t-1) + (1/R_1)\{F(t) - \lambda T_1(t-1) - (R_2/\tau_2)[T_1(t-1) - T_2(t-1)]\},$$
$$T_2(t) = T_2(t-1) + (1/R_2)\{(R_2/\tau_2)[T_1(t-1) - T_2(t-1)]\}, \qquad (9)$$

where $T_i(t)$ = temperature of layer i in period t (relative to the pre-industrial period); $i=1$ for the atmosphere and upper oceans (rapidly mixed layer) and $i=2$ for the deep oceans; $F(t)$ = radiative forcing in the atmosphere (relative to the pre-industrial period); R_i = the thermal capacity of the different layers; τ_2 = the transfer rate from the upper layer to the lower layer; and λ = feedback parameter.

The next step is to find the appropriate numerical representation of the simplified climate model in (9). We estimate the parameters in (9) by calibrating the smaller model to transient runs from larger GCMs and by comparing the predictions with historical data. Unfortunately, the models disagee by a wide margin, and the historical data are even further at variance from the climate models. In the study here, we use the results from a study by Schlesinger and Jiang (1990) for calibration purposes. This study has a temperature–CO_2 sensitivity of 3°C for CO_2 doubling, which is close to that of the scientific consensus [see National Academy of Sciences (1992)].

3.3.4. Impacts

The next link in the chain is the impact of climate change on human and natural systems. Estimating the damages from greenhouse warming has proven extremely difficult. An early discussion is contained in Schelling (1983), EPA summarized a number of studies (1989), and I put those studies into the context of the national-income accounts in Nordhaus (1991b). The overall assessment of the cost of greenhouse warming in the U.S. was that the net economic damage from a 3°C warming is likely to be around 0.25% of national income for the United States in terms of those variables could be quantified. This figure is clearly incomplete, for it neglects a number of areas that are either inadequately studied or inherently unquantifiable. As a rough adjustment, I increased this number to around 1% of total U.S. output to allow for these unmeasured and unquantifiable factors. Making adjustments for output composition in different countries, I further raised the estimated impact to 1.33% of global output for all countries. In addition, there is evidence that the impact increases nonlinearly as the temperature increases, and we assume that the relationship is quadratic. Therefore, the final relationship between global temperature increase and income loss is:

$$d(t) = 0.0133[T(t)/3]^2 Q(t), \qquad (10)$$

where $d(t)$ is the loss of global output from greenhouse warming. Although there is much controversy about the parameters of (10), using alternative estimates of impacts does not change the basic results markedly.[7]

3.3.5. Cost of emissions reduction

The last major link in the chain is the costs of reduction of greenhouse gases. This is the one area that has been extensively studied and, while not without controversy, the general shape of the cost function has been sketched on a number of occasions. There are numerous estimates, particularly for CO_2, of the cost of reducing GHGs; see the extensive surveys in EPA (1990), Nordhaus (1991c), Dean and Hoeller (1992), Amano (1992), EC (1992a, b) and the results of EMF-12.[8] Using current annual emissions of 8 billion tons of CO_2 equivalent, my survey suggested that a modest reduction of GHG emissions can be obtained at low cost. After 10% reduction, however, the curve rises as more costly measures are required. A 50% reduction in GHG emissions is estimated to cost almost $200 billion per year in today's global economy, or around 1% of world output. This estimate is understated to the extent that policies are inefficient or are implemented in a crash program. The final form of the equation used in the model is:

$$TC(t)/GNP(t) = b_1 \mu(t)^{b2} = 0.0686\mu(t)^{2.887} \quad (11)$$

where μ is the fractional reduction in GHG emissions and TC/GNP is the total cost of the reduction as a fraction of world output.[9]

Combining the cost and damage relationships, we have the Ω relationship in the production function as follows:

$$\Omega(t) = [1 - b_1 \mu(t)^{b2}]/[1 + d(t)]$$

[7]A thorough review of impacts by Cline (1992) finds an estimated impacts of 1.1% of GNP for a 2.5°C warming as opposed to the estimate of 1% for 3°C warming by the present author. A more recent unpublished study by Fankhauser (1992) estimates total impacts of a doubling of CO_2 would lead to a 1.3% cost to the US, a 1.4% cost to the OECD, and a 1.5% cost to the world. Because estimating the impacts of climate change has proven extremely difficult, the present author is in the process of undertaking a survey of experts on the economic impacts of climate change on human and non-human systems. At the mid-point of the survey, the 'trimmed mean' of the experts' estimate of the impact of a 3°C warming is approximately 20% higher than the estimate used in the DICE model while the experts' estimate of the impact of a 6°C warming is about 30% lower than that in the DICE model. One major concern of most respondents is that the impact is thought to be considerably higher for low-income countries than in high-income countries.

[8]The most systematic study is the model comparison study of the Stanford Energy Modelling Forum 12 under the general direction of John Weyant. The results of this study have been presented informally and conform for the most part to the survey in Nordhaus (1991c).

[9]Alternative views of the costs of mitigation are contained in National Academy of Sciences (1992) and Cline (1992). The major difference among studies concerns the possibility of zero-cost or low-cost mitigation in areas where informational deficiencies or market failures hinder reaping all cost-beneficial investments in energy conservation. National Academy of Sciences (1992) estimates these to range between 10 and 40% of U.S. GHG emissions, whereas this study is at the lower end of that range.

$$=[1-0.0686\mu(t)^{2.887}]/[1+0.00144T(t)^2]. \tag{12}$$

4. Data, calibration, and sensitivity analysis[10]

4.1. Fitting the model

This section describes briefly the sources of the data used for the model and the calibration of the model to the data. Data on the major variables were collected for three years, 1965, 1975 and 1985, while future periods are estimated by the calculations described above. Data on population, GNP, consumption, and investment are obtained from existing data sources of the World Bank, UNESCO, the OECD and the U.S. and other national governments.

The parameters of the Cobb–Douglas production function are obtained by assuming that the output-elasticity of capital is 0.25 and then by estimating the level and rate of Hicks-neutral technological change directly as a residual. The utility function is assumed to be logarithmic, and the rate of social time preference is taken to be three percent per year. This preference function leads to predictions of the rate of return on capital and the gross savings rate that are close to observed levels.

Assumptions about future growth trends are as follows: the rate of growth of population is assumed to decrease slowly, stabilizing at 10.5 billion people in the 22nd century. The rate of growth of total factor productivity is calculated to be 1.3% per annum in the 1960–1989 period. This rate is assumed to decline slowly over the coming decades.

The model has been run using the 486 version of the GAMS algorithm on various 386 compatible machines. The canonical runs presented below use a 40-period calculation with terminal valuations (or transversality conditions) on carbon, capital, and atmospheric temperature; these terminal valuations were obtained from a 60-period run and are sufficient to stabilize the solution for the first 20 periods. The canonical 40-period run can be solved in about two minutes on an Intel 486/33 processor.

Optimization models of the kind analyzed here have proven extremely resistant to conventional econometric estimation. In the place of a formal statistical procedure, we have simply chosen parameters so that the values taken by the model in the first three periods are tolerably close to actual data. The current solution matches global GNP, emissions, GHG concentrations, and even estimates of global temperature change reasonably well for the historical periods.

[10]The details of the data are described in Nordhaus (1993a).

4.2. Sensitivity analyses

The DICE model contains many parameters and assumptions that will affect the projections and policy conclusions. The present study presents the central or 'best-guess' case and does not at this stage include sensitivity analyses. In work underway, however, an extensive and systematic attempt to evaluate the major uncertainties has been undertaken, and that analysis will be presented in Nordhaus (1993a). A brief description of the approach and tentative results of the sensitivity analysis will be provided here.

The sensitivity analysis first screens each of the parameters and major model components in the DICE model to determine which are important for economic, environmental, and policy variables. Among the two dozen parameters, nine which have the most impact upon key outcomes are then selected for detailed analysis – these nine including parameters such as population and productivity growth, the parameters of the climate, emissions, and concentrations modules, and the rate of time preference. A Monte Carlo run then provides estimates of the uncertainty due to each parameter as well as the underlying uncertainty about the major variables (such as temperature, GHG concentrations, and GHG control rates). The major results of this stage to data are that the results are modestly sensitive to alternative values of major variables. Among the important variables that produce large uncertainties are population and productivity growth, the trend in the GHG emissions–output ratio, and the pure rate of time preference.[11]

The final stage of the analysis is to determine the impact upon the optimal policy of uncertainty. Should we pay an 'insurance premium' to reflect the impact of uncertainty, non-linearity, and risk aversion upon our optimal policy? While the research on the issue is still underway, it appears that the uncertainty about the size and impacts of future climate change would add a significant risk premium to, perhaps even doubling, the 'best guess' policies analyzed here.

5. Policy experiments

We now describe the different scenarios or policy experiments to which the model is applied.

[11]Many readers of this and associated studies have expressed concern that the results are inherent in assuming too high a 'discount rate'. While there is some merit in this point, the whole story is more complicated. In the first point, the distinction between time discounting and goods discounting is often obscured. While people may raise ethical objections to time preference in the form of a high pure rate of time preference, there is no analogous objection to a high rate of goods discounting in the form of a high real interest rate where that reflects rapid growth of living standards. Second, when the pure rate of time preference is lowered to 0.1% per year while adjusting the rate of inequality aversion to maintain the same real rate of interest, there is only a modest change in the optimal policy (either the carbon tax or the GHG control rate). The change in the optimal policy occurs because an increase in the net savings rate drives down the real interest rate.

5.1. No controls ('baseline')

The first run is one in which there are no policies taken to slow or reverse greenhouse warming. Individuals would adapt to the changing climate, but governments would take no steps to curb greenhouse-gas emissions or to internalize the greenhouse externality. This policy is one which has been followed for the most part by nations through 1989.

5.2. Optimal policy

The second case undertakes to construct economically efficient or 'optimal' policies to slow climate change. This run maximizes the present value of economic welfare; more precisely, this case maximizes the discounted value of utility in (1) subject to the constraints and relationships in (2) to (12). This policy can be thought of as one in which the nations of the world gather to set the efficient policy for internalizing the greenhouse externality. It is assumed that the policy is efficiently implemented, say through uniform carbon taxation, in the decade beginning 1990.

5.3. Ten-year delay of optimal policy

This policy is one which delays implementing the optimal policy for ten years. This policy examines the issue of the costs and benefits of delaying implementing policies until our knowledge about the greenhouse effect, along with its costs and benefits, is more secure. This approach has been advocated by the U.S. government during the Bush administration. In this scenario, we assume that sufficient information is in hand so that the optimal policy is implemented beginning in the decade starting in 2000.

5.4. Twenty percent emissions reductions from 1990 levels

Many environmentalists and some governments are proposing a substantial cut in CO_2 or GHG emissions. One target that has been prominently mentioned is a 20% cut in emissions. This is interpreted here as a 20% cut of the combination of CFC and CO_2 emissions from 1990 levels, where these are converted to a CO_2-equivalent basis. In quantitative terms, this represents an emissions limitation of 6.8 billion tons per year of CO_2 equivalent. This policy has no particular analytical, scientific, or economic merit, but it has the virtue of simplicity; it implies a growing percentage reduction in the future given a growing uncontrolled emissions path.

Table 1

Impact of alternative policies on discounted consumption.

Run no.	Description	Discounted value of consumption, 1990 [trillions of 1989 US $]	Difference from no controls [billions of $]	Percent difference
1	No controls	731.694	0	0.000
2	Optimal policy	731.965	271	0.037
3	Ten-year delay	731.937	243	0.033
4	Stabilize emission	720.786	(10,908)	−1.491
5	Geoengineering	737.296	5,602	0.766

5.5. Geoengineering

A final policy would be to determine the benefit of a technology which would provide costless mitigation of climate change. This could occur, for example, if some of the geoengineering options proved technically feasible and environmentally benign. Two interesting proposals include shooting smart mirrors into space with 16-inch naval rifles or seeding the oceans with iron to accelerate carbon sequestration.[12] An alternative interpretation would be that the greenhouse effect has no harmful economic effects. This scenario is useful as a baseline to determine the overall economic impact of greenhouse warming and of policies to combat warming.

6. Results and conclusions

We now summarize the overall results for the five scenarios described above. A longer description of the model and a presentation of the numerical results is contained in Nordhaus (1993a).

6.1. Overall results

Table 1 shows the overall evaluation of the different policies. The first column shows the discounted value of consumption for the five paths. This is calculated as the present value of consumption after 1990 discounted at the market rate of return on capital (discounted back to 1990 in 1989 prices).

The optimal policy in row 2 has greater value than the three other policies in rows 1, 3 and 4.[13] The optimal policy has a net benefit of $271 billion relative to a policy in which no controls are undertaken. This number is

[12]The issues of geoengineering are discussed in National Academy of Sciences (1992).

[13]The values in column 3 for run *i* are equal to the present value of consumption for run 1 plus the algebraic difference in the attained value of the objective function from run *i* to run 1.

absolutely large, although it is only 0.037% of the discounted value of the consumption. The cost of delaying the optimal policy by ten years is estimated to be $28 billion.

The environmentalists policy of reducing emissions 20% below 1990 levels is extremely costly. We estimate that this policy will cost $10.9 trillion in present value terms. This constitutes 1.5% of the total discounted value of consumption.

The last row shows the overall economic impact of climate change. The net damage from global warming is estimated to be $5.9 trillion relative to the optimal policy and $5.6 trillion relative to a policy of no controls. These represent 0.81 and 0.77% of the discounted value of consumption.

In general, these numbers are mind-numbing in absolute size – largely because we are considering global output over the indefinite future. On the other hand, with the exception of the policy of stabilizing emissions, the numbers are modest relative to the total size of the global economy.

6.2 Emissions and concentrations

We next show some of the details of the model runs. Figs. 1 and 2 show the emissions control rates in different scenarios. These show the extent to which GHG emissions are reduced below their uncontrolled levels. In the optimal path, the rate of emissions reduction is approximately 10% of GHG emissions in the near future, rising to 15% late in the next century. (Recall that this is primarily CO_2 emissions.) The environmental path of a 20% cut in emissions from the 1990 level shows steeply rising control rates, with the rate of control reaching 70% by the end of the next century.

Fig. 3 shows projected CO_2-equivalent atmospheric concentrations in billions of tons CO_2 equivalent (again, this includes both CO_2 and CFCs). The impact of the optimal control strategy is noticeable, reducing concentrations by a little more than 100 billion tons at the end of the next century. Note that even with emissions stabilized at 80% of 1990 levels, the atmospheric concentrations of CO_2-equivalent concentrations continue to rise. The ten-year delay in implementing greenhouse gas restraints show virtually no difference from the optimal path and is not included in the graph.

6.3. Global temperature

Fig. 4 shows the resulting projected increase in realized mean global surface temperatures (relative to temperatures in the 19th century). The uncontrolled path shows an initial increase of around 0.6°C today, rising to 3.1°C by 2100.

The optimal path shows a modest decline in the growth rate of global

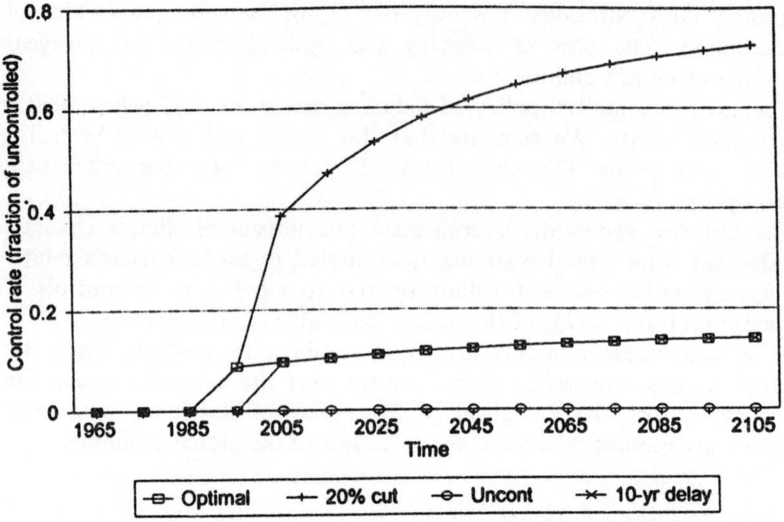

Fig. 1. Greenhouse-gas control rate (reduction in GHG emissions).

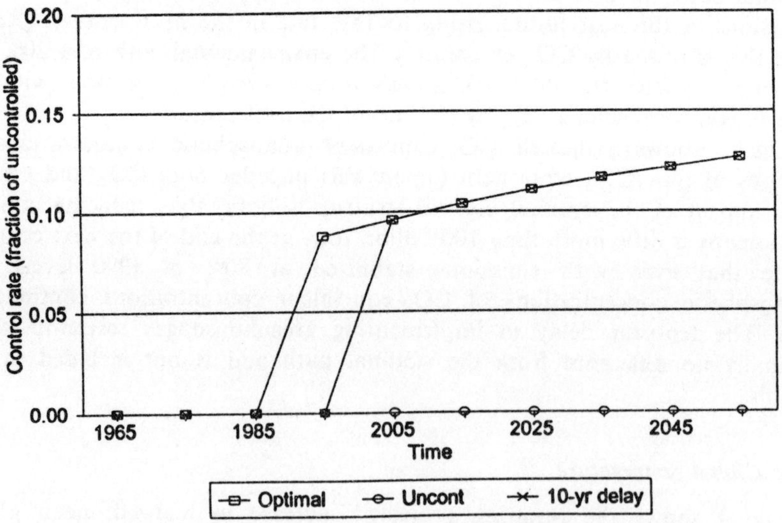

Fig. 2. Greenhouse-gas control rate (reduction in GHG emissions).

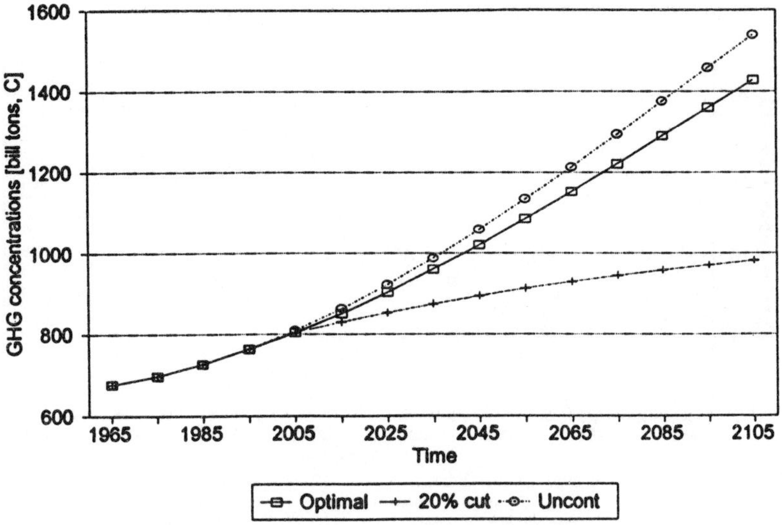

Fig. 3. Atmospheric greenhouse gas concentrations (billion tons CO_2 equiv., C weight).

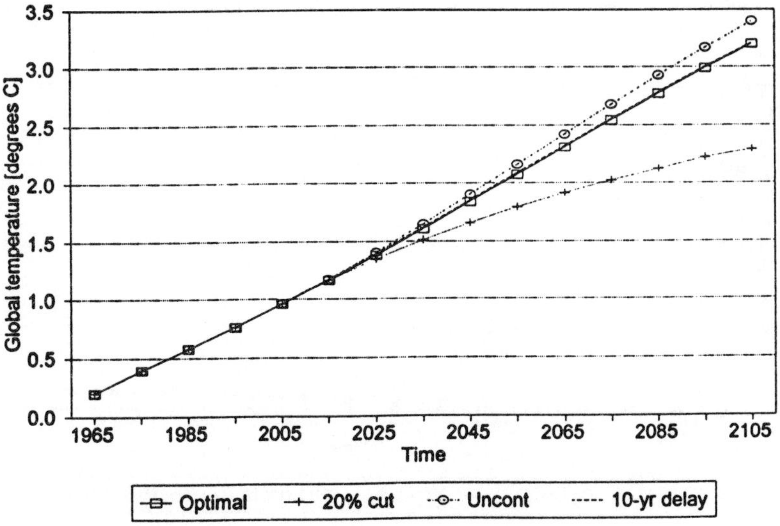

Fig. 4. Global mean temperature (°C, difference from 1860).

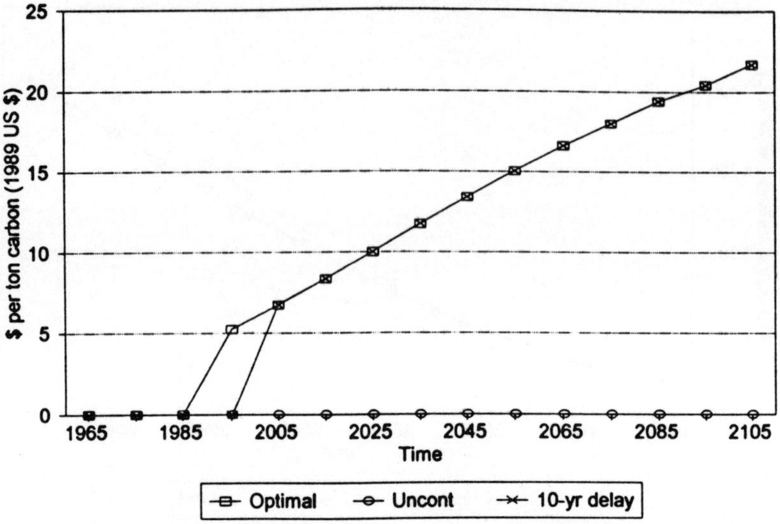

Fig. 5. Carbon tax, different scenarios (tax in $ ton C equiv.).

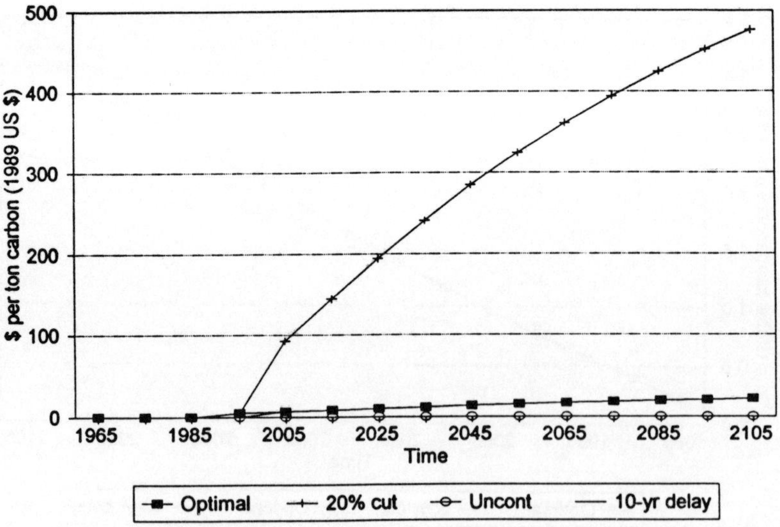

Fig. 6. Carbon tax, different scenarios (tax in $ ton C equiv.)

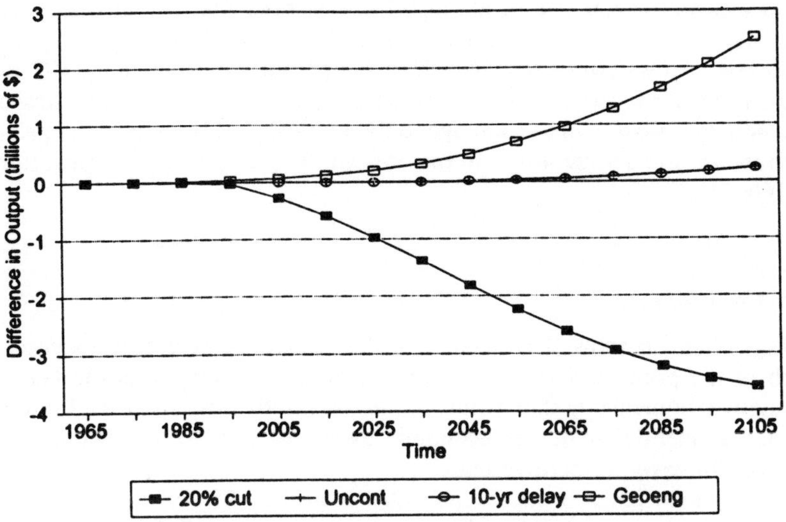

Fig. 7. Differences in global output (from baseline, trillions 1989 US $).

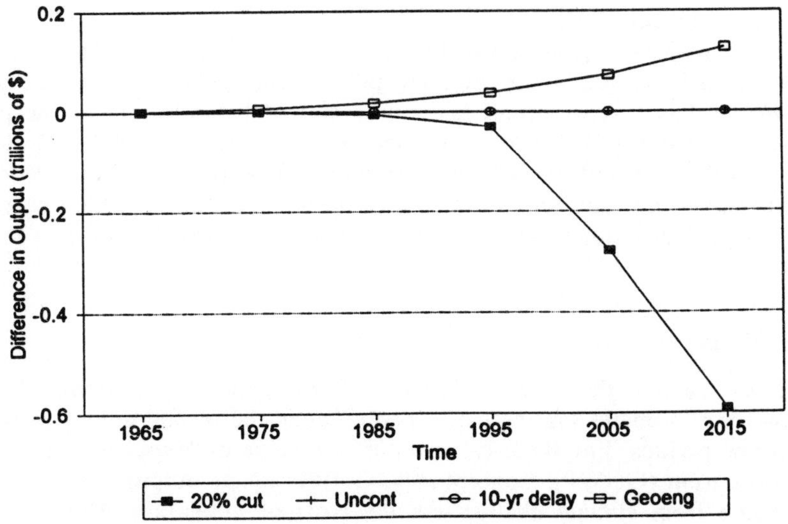

Fig. 8. Differences in global output (from baseline, trillions 1989 US $).

temperatures, with a rise of about 0.2°C less than the uncontrolled path by the end of the next century. The policy that cuts emissions to 80% of the 1990 level shows continued growth in temperatures, rising to 2.25°C by the end of the next century. This surprising result shows that even draconian policies will slow climate change only modestly. The reason is primarily because of the momentum in the system from existing concentrations of GHGs.

6.4. Carbon taxes

Figs. 5 and 6 show the carbon tax that would be necessary to implement each of the policies. The carbon tax should be thought of as the tax (or its regulatory equivalent) that would be necessary to raise fossil fuel and other prices sufficiently to induce economic agents to substitute other goods and services for carbon-intensive ones.

The optimal path shows a carbon tax of around $5 per ton carbon (or the equivalent in other GHGs) for the first control period, 1990–1999. For reference, a $10 per ton carbon tax will raise coal prices by $7 per ton, about 25% at current U.S. coal prices. The carbon tax increases gradually over time to around $20 per ton carbon by the end of the next century. The rising tax primarily reflects the rising level of global output rather than increasingly stringent control efforts.

The ten-year delay has a zero tax in the fourth period, but then is virtually indistinguishable from the optimal policy. The policy of no mitigation obviously has a zero carbon tax. Fig. 6 shows the trajectory of the policy that cut emissions 20% from 1990 levels. This tax reaches about $100 per ton early in the 21st century and climbs to almost $500 per ton by the end of the next century. Clearly, very substantial fiscal or regulatory steps are necessary to bring about a trajectory with constant CO_2 emissions.

6.5. Output

Figs. 7 and 8 show the impact of different policies on output. The first shows the estimates for the entire period while the second zooms in on the first few periods. For these calculations, the value of output is 'green' gross world output (GGWP). Conceptually, GGWP equals output less the flow of damages from climate change less the costs of mitigation. The surprising result of these figures is that the difference between a policy of no controls and the optimal policy is relatively small through the next century. The flow impact, relative to the optimum, is somewhat less than one percent of real output at the maximum. Of course, the actual damage (equal to the

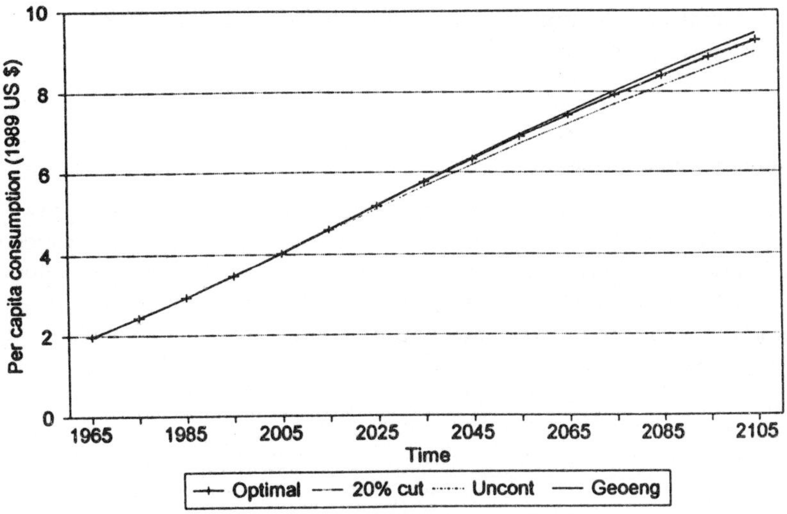

Fig. 9. Per capita consumption (1989 US $).

difference between the 'no controls' and the 'geoengineering') is much larger than the cost of no controls relative to the optimum. But the latter appears small because a fair amount of economic cost would occur even in the optimal trajectory.

While the difference between the no-controls and the optimal policies is small, there are big stakes in both the geoengineering option and in the environmental option. The impact of a geoengineering solution would be quite substantial – because it would cut the costs of both climate damage and of mitigation.

There is also potential for a major waste of resources if the greenhouse policies go too far. Fig. 7 shows the impact on green world output of going too far in the control of greenhouse gases – leading to *net* losses in output of over $3 trillion annually by the end of the next century.

Finally, fig. 9 shows the trajectory of real consumption per capita in the four cases. The striking feature of this figure is that, even though there are differences among the cases studied here, the overall economic growth projected over the coming years swamps the projected impacts of climate change or of the policies to offset climate change. In these scenarios, future generations may be worse off as a result of climate change, but they are still likely to be much better off than current generations. In looking at this graph, I was reminded of Tom Schelling's remark a few years ago that the difference between a climate-change an a no-climate-change scenario would be thinner than the line drawn by a number 2 pencil used to draw the curves. Thanks to the improved resolution of computerized graphics, we can now barely spot the difference!

7. Summary and conclusion

The present study has investigated the implications of economic growth on the environment as well as the economic impact of different environmental control strategies upon the global economy. This study takes the approach that an efficient strategy for coping with greenhouse warming must weigh the costs and benefits of different policies in an intertemporal framework. Using this approach, the major results and reservations are the following.

This study has examined five different approaches to GHG control: no control, an economic optimization, geoengineering, stabilization of emissions, and a ten-year delay in undertaking climate-change policies. Among these five, the rank order from a pure economic point of view at the present time is geoengineering, economic optimum, ten-year delay, no controls, and stabilizing emissions. The advantage of geoengineering over other policies is enormous, although this result assumes the existence of an environmentally benign geoengineering option. The policies of no controls, the economic optimum, ten-year delay, and emissions stabilization have differential impacts that are less than one percent of discounted consumption.

It is instructive to compare these results with those from other economic studies. The studies of Manne and Richels (1990, 1992), Peck and Teisberg (1991), and Kolstad (1992) find conclusions that are roughly similar to those reported here. All these studies contain explicit or implicit relationships between emissions control rates and carbon taxes; the relationships are broadly similar to those found in figs. 1, 2 and 5 of this paper, although papers with more detailed energy sectors have more complex dynamics than those seen here. The studies by Jorgenson and Wilcoxen (1991 expecially) show a lower set of carbon taxes needed to reduce GHG emissions than those shown here in part because of the induced innovation in the Jorgenson–Wilcoxen model.

Three other studies – those of Cline (1992), Peck and Teisberg (1991), Kolstad (1992) as well as earlier studies by the present author (1979, 1991a, b) – also determine the optimal emissions control rates and carbon taxes. With the exception of Cline (1992), all the earlier studies show optimal policies in the general range of those determined here. A study by Hammitt et al. (1992) traces out alternative control strategies to attain certain temperature constraints; while not determining an optimal path, this study concludes that a 'moderate reduction strategy' is less costly than an 'aggressive' approach if either the temperature–concentrations sensitivity ($1/\lambda$) is low or if the allowable temperature change is above 3°C. The study of Cline (1992), by contrast, has much higher control rates. The more stringent controls in the Cline study are due to a number of features – primarily, however, because the Cline result is not grounded in explicit intertemporal optimization and assumes a rate of time preference that is lower than would be consistent with observed real interest rates.[14]

It must be emphasized that the present analysis has a number of important qualifications. The most important shortcoming is that the damage function, particularly the response of developing countries and natural ecosystems to climate change, is poorly understood at present; moreover, the potential for catastrophic climatic change, for which precise mechanisms and probabilities have not been determined, cannot currently be ruled out. Furthermore, the calculations omit other potential market failures, such as ozone depletion, air pollution, and R&D, which might reinforce the logic behind greenhouse gas reduction or carbon taxes. Issues of sensitivity analysis with respect to either parameters or components of the model have not been addressed in this study, although an examination of these issues is underway, as discussed above. And finally, this study abstracts from issues of uncertainty, in which risk aversion and the possibility of learning may modify the stringency and timing of control strategies. Notwithstanding these qualifications, the optimal-growth approach may help clarify the questions and help identify the scientific, economic and policy issues that must underpin any rational decision.

[14]See the discussion in footnote 5 above.

References

Amano, A., 1992, Economic costs of reducing CO_2 emissions: A study of modeling experience in Japan, Paper presented to the International Workshop on Costs, Impacts, and Possible Benefits of CO_2 Mitigation, IIASA, Laxenburg, Austria, September.

Cline, W., 1992, The economics of global warming (Institute of International Economics, Washington, DC).

Dean, A. and P. Hoeller, 1992, Costs of reducing CO_2 emissions: Evidence from six global models, OECD/GD (92) 140 (Organisation for Economic Co-operation and Development, Paris) processed.

EPA, 1989, U.S. Environmental Protection Agency, The potential effects of global climate change on the United States: Report to Congress, EPA-230-05-89-050, Dec.

EPA, 1990, D.A. Lashof and D.A. Tirpak, eds., Policy options for stabilizing global climate (Hemisphere, New York).

EC, 1992a, The economics of limiting CO_2 emissions, Commission of the European Communities, Special Edition no. 1 (ECSC-EEC-EAEC, Brussels).

EC, 1992b, The climate challenge: Economic aspects of the community's strategy for limiting CO_2 emissions, Commission of the European Communities, no. 51 (ECSC-EEC-EAEC, Brussels) May.

Fankhauser, S., 1992, The economic costs of global warming: Some monetary estimates, Paper presented to the International Workshop on Costs Impacts, and Possible Benefits of CO_2 Mitigation, IIASA, Laxenburg, Austria, Sept.

Hammitt, J.K., R.J. Lempert and M.E. Schlesinger, 1992, A sequential-decision strategy for abating climate change, Nature 357, 28 May, 315–318.

IPCC, 1990, Intergovernmental panel on climate change, climate change: The IPCC scientific assessment, J.T. Houghton, G.J. Jenkins and J.J. Ephraums, eds. (Cambridge University Press, New York).

Jorgenson, D.W. and P.J. Wilcoxen, 1991, Reducing U.S. carbon dioxide emissions: The cost of different goals, in: J.R. Moroney, ed., Energy, growth and the environment (JAI Press, Greenwich, CT) 125–128.

Kolstad, C.D., 1992, Looking vs. leaping: The timing of CO_2 control in the face of uncertainty and learning, Paper presented to the International Workshop on Costs, Impacts, and Possible Benefits of CO_2 Mitigation, IIASA, Laxenburg, Austria, Sept.

Manne, A.S. and R.G. Richels, 1990, CO_2 emission limits: An economic cost analysis for the U.S.A., The Energy Journal 11, no. 2, 51–74.

Manne, A.S. and R.G. Richels, 1992, Buying greenhouse insurance: The economic costs of CO_2 emission limits (MIT Press, Cambridge, MA).

National Academy of Sciences, 1992, Committee on Science, Engineering and Public Policy, policy implications of greenhouse warming: Mititagion, adaptation, and the science base (National Academy Press, Washington, DC).

National Research Council, 1983, Changing climate (National Academy Press, Washington, DC).

Nordhaus, W.D., 1979, The efficient use of energy resources (Yale University Press, New Haven, CT).

Nordhaus, W.D., 1991a, A sketch of the economics of the greenhouse effect, The American Economic Review 81, no. 2, 146–150.

Nordhaus, W.D., 1991b, To slow or not to slow: The economics of the greenhouse effect, The Economic Journal 101, July, 920–937.

Nordhaus, W.D., 1991c, A survey of the costs of reduction of greenhouse gases, The Energy Journal 12, no. 1, 37–65.

Nordhaus, W.D., 1992, Explaining the 'DICE': Background paper on dynamic integrated model of climate change and the economy, Mimeo. (Yale University, New Haven, CT).

Nordhaus, W.D., 1993a, Managing the global commons: The economics of global warming (MIT Press, Cambridge, MA) forthcoming.

Nordhaus, W.D., 1993b, How much should we invest to slow climate change?, in: H. Giersch, ed., Policies on economic growth (Springer-Verlag, Berlin) forthcoming.

Ramsey, F.P., 1928, A mathematical theory of saving, The Economic Journal, Dec., 543–559.

Peck, S.C. and T.J. Teisberg, 1991, CETA: A model for carbon emissions trajectory assessment, The Energy Journal 13, no. 1, 55–77.

Schelling, T.C., 1983, Climatic change: Implications for welfare and policy, in: National Research Council, 1983, 449–482.

Schlesinger, M.E. and X. Jiang, 1990, Simple model representation of atmosphere–ocean GCMs and estimation of the time scale of CO_2-induced climate change, Journal of Climate 3, Dec., 12.

Solow, R.M., 1988, Growth theory: An exposition (Oxford University Press, New York).

[12]

Benefit-Cost Synthesis

General Approach

This chapter draws upon the findings of the previous analysis of the benefits of global warming abatement (chapter 3) and the corresponding costs (chapters 4 and 5) to arrive at an overall benefit-cost evaluation. In this process, the preceding discussion of the discount rate (chapter 6) informs the choice of this crucial parameter.

The fundamental policy examined is one of a worldwide program of aggressive action to limit global warming. Its centerpiece is the cutback of carbon emissions to 4 billion tons annually, and a subsequent permanent freeze at this level. As developed below, afforestation and reduced deforestation play a central role in this cutback for the initial decade or two, but thereafter increasingly deep reductions in fossil fuel emissions become necessary as well. Those reductions initially draw upon the backlog of low-cost improvements perceived by the technology community, but increasingly require the more costly curtailments analyzed by the cadre of carbon-energy-economic modelers.

The overall pattern that emerges is one in which there is a phase of initially low-cost carbon reductions, followed by a period when these costs rise to a peak in the vicinity of 3½ percent of world GDP, a level that then tapers off to some 2½ percent of GDP as the passage of time permits a wider range of technological alternatives. On the benefits side, the pattern is one of initially minimal gains (because of the long delays in warming), followed by a relatively steady, geometric rise. Depending on the scenario chosen, after some decades the benefits of damage avoidance exceed the cost of action. Thereafter this wedge of net gain continues to expand. A horizon through the year 2275 is chosen for the analysis because of the prospect of partial warming reversal thereafter through deep-ocean mixing of carbon dioxide from the atmosphere (chapters 1 and 2). The overall benefit-cost ratio of aggressive policy action is obtained by discounting the stream of benefits and the stream of costs and taking the ratio of the former to the latter. A ratio of unity or above indicates the desirability of action, from a social economic benefit-cost standpoint.

Table 7.1 provides a summary view of the benefit-cost analysis, and figure 7.1 illustrates its results. Figure 7.1 depicts the costs of aggressive action (a 4-GtC annual emissions ceiling) as a percentage of world GDP. It also shows two alternative "benefits" curves: the base and high cases (a principal "low" case is also evaluated but not shown in the figure). In both cases, the costs of action exceed the benefits in the initial decades. The possibility of a favorable evaluation of action is considerably greater in the high-benefits case. For all cases, the comparison of benefits and costs depends crucially on the discount rate.

Because there are alternative assumptions in each of several dimensions of the analysis, the number of possible outcomes is large (over 200). The analysis identifies the preferred or central evaluation. However, the results reported for the array of alternative estimates provide a menu of outcomes for the reader's inspection, so that he or she may choose alternative entries according to taste on such key issues as the discount rate, the likelihood of greater or lesser warming, and the chances of more or less severe economic damage associated with a given amount of warming.

The Benefit-Cost Model

The model for calculating benefits and costs derives from the analysis of the preceding chapters. The following discussion first sets forth the structure of the calculations and then turns to a description of the values applied for the estimated results. The reader not concerned with the algebra of the model may wish to proceed directly to the subsequent section on assumptions for values used in the calculations.

The analysis first calculates real per capita income (y) and population (P) for developed (D) and less developed (L) countries for the year in question (t), based on assumed growth rates (g for per capita income, n for population) from the base year (year 0).[1]

(7.1) $y_t^D = y_0^D(1 + g_D)^t$

(7.1′) $y_t^L = y_0^L(1 + g_L)^t$

(7.2) $P_t^D = P_0^D(1 + n_D)^t$

(7.2′) $P_t^L = P_0^L(1 + n_L)^t$

Gross world product (Y) is then the sum of developed- and developing-area per capita incomes times their respective populations:

1. As discussed below, these growth rates taper off over future time periods.

Table 7.1 Central features of the benefit-cost analysis

Aggressive abatement	A global ceiling of 4 gigatons of annual carbon emissions permanently. Achieved initially by reduced deforestation, increased afforestation, and a move to best practices in energy use. Subsequently achieved through cutbacks in carbon dioxide emissions throughout the economy.
Very-long term warming	In the absence of action, global warming reaches a central estimate of 10°C by 2275; the upper bound is 18°C by 2275.
Damage function	Damage d_o is 1 percent of GDP at carbon-dioxide-equivalent doubling for 2.5°C warming, central case, or 2 percent, high-damage case. Damage rises exponentially with warming (at exponent 1.3, central, or 2.0, high).
Benefits	Equal 80 percent of damage, the fraction that is assumed avoidable through action
Costs	Forestry measures are low cost; general economy-wide emission cutbacks are free for the first 22 percent, from a move to best practices. For larger cutbacks, costs rise according to a formula based on carbon-energy-economic models
Discounting	All investment effects are converted to consumption equivalents using a capital shadow price. All consumption-equivalent magnitudes are then discounted at a social rate of time preference (SRTP), with a base case value of 1.5 percent annually.
Economic scale	Damage, benefit, and cost estimates are estimated as percentages of GDP and then applied to future world GDP. Prospective GDP is based on world population that reaches 10.5 billion by the middle of the 21st century, and per capita income growth averaging 1 percent annually over the next 2½ centuries
Sensitivity	Tests are included for alternative values of the SRTP climate sensitivity parameter, damage function exponent, and damage at 2.5°C warming.

Figure 7.1 Costs and benefits of aggressive abatement of greenhouse warming, 1990–2270

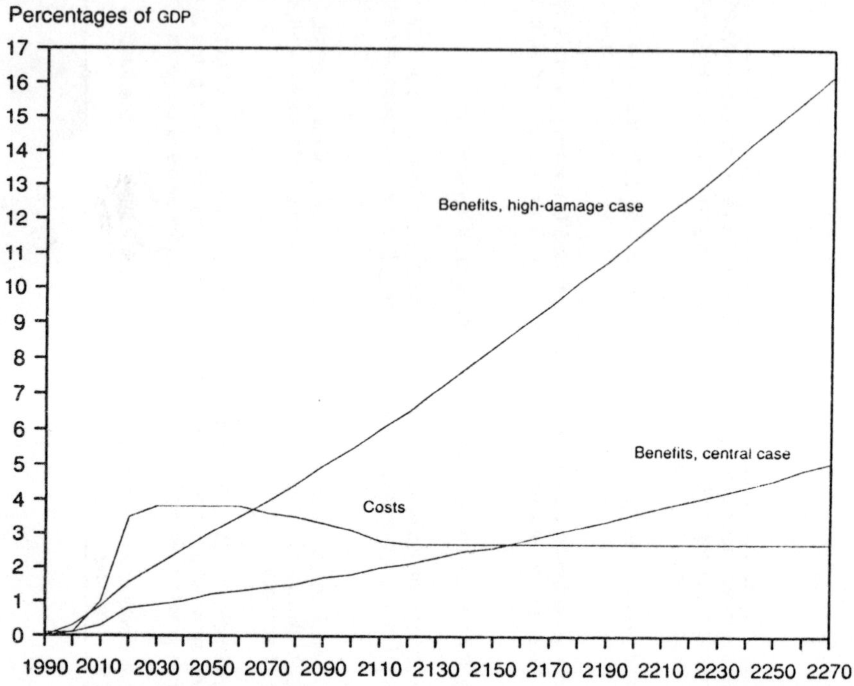

(7.3) $Y_t = y_t^D P_t^D + y_t^L P_t^L$

An exogenous projection of global carbon emissions (E) is then compared for each year to the target ceiling emission (K) set in the abatement program. The difference is the amount by which emissions must be reduced (R):

(7.4) $R_t = E_t - K$

The amount of global warming at a future year t, expressed in degrees Celsius, is W_t. The path of warming is linear (chapter 2). It passes through the amount S, the climate sensitivity parameter for carbon-dioxide-equivalent doubling, in the year 2050 (for 2025 concentration doubling plus an ocean thermal lag of 25 years); and through the amount W^* for very-long-term warming, set in the year 2275. For years before 2050 ($t = 60$), warming is set by linear interpolation between zero at the outset (with $t = 1$ at 1991) and S. Accordingly:

(7.5*a*) $W_t = S + [(W^* - S)/225](t - 60),\ t \geq 60;$

(7.5*b*) $W_t = S\ (t/60),\ t < 60$

The amount of economic damage from warming is assumed to be geometrically related to the amount of warming. Benchmark economic damage is estimated for carbon-dioxide-equivalent doubling, set at d_0 percent of world GDP. As the benchmark is calibrated for a climate sensitivity of 2.5°C, the resulting damage function is:

$$(7.6) \quad d_t = d_0 \left[\frac{W_t}{2.5} \right]^{\gamma}$$

where γ is the exponent in the geometrical relationship of damage to warming. The function expresses damage (d_t) as a percentage of world GDP. The linkage to GDP is discussed further below.

The benefit of the abatement policy by year t equals that portion of the damages that can be avoided as a result of the policy. The avoidance proportion is set at μ, so that:

$$(7.7) \quad b_t = \mu d_t$$

The costs of abatement arise from the need to set aside land for afforestation (FA), to curtail frontier agricultural land use and thereby carbon release from deforestation (FD), and to reduce fossil fuel emissions and therefore output (Q) in the rest of the economy. The amount of carbon that can be reduced through afforestation is set at a constant (R_0^{FA}) over a 30-year growing period, and then falls to zero as the new forest reaches its steady state. Thus, afforestation reduces emissions by:

$$(7.8) \quad R_t^{FA} = R_0^{FA}, \ 1 \leq t \leq 30; \ = 0, t > 30$$

The reduction of deforestation to a lower rate and the commitment to hold it indefinitely at that rate provides a constant annual flow of emissions reductions:

$$(7.9) \quad R_t^{FD} = R_0^{FD}$$

The reductions in emissions from afforestation and reduced deforestation only partially meet the need to cut back emissions from the baseline. The remaining reduction that must be obtained from the rest of the economy is thus:

$$(7.10) \quad R_t^{Q} = R_t - R_t^{FA} - R_t^{FD}$$

The cost of afforestation is calculated in a manner that incorporates all current and future costs within the initial growing period on a constant annualized basis, at C_0^{FA}. Expressed relative to world GDP (and in percentage points), this cost is thus:

$$(7.11) \quad c_t^{FA} = 100 \left[\frac{C_0^{FA}}{Y_t} \right], \ 1 \leq t \leq 30; \ = 0, t > 30$$

For its part, the cost of forgone deforestation (C_0^{FD}) continues indefinitely. Expressed as a percentage of world GDP, it is:

$$(7.12) \quad c_t^{FD} = 100 \left[\frac{C_0^{FD}}{Y_t} \right]$$

The cost of carbon reduction through fossil fuel cutbacks depends on the extent of the proportionate reduction (chapter 4) as well as initial gains from improved technical efficiency (chapter 5). The percentage reduction in carbon from economywide fossil fuel cutbacks is:

$$(7.13) \quad z_t^Q = 100 \left[\frac{R_t^Q}{(E_t - E_0^F)} \right]$$

where E_0^F is the base-year emission contribution from deforestation so that the right-hand side denominator is the baseline value of all other (e.g., economywide fossil fuel) emissions.

As developed in chapter 5, the cost function for carbon reductions in the general economy depends on the percentage cutback, but with adjustments that provide for an initial "free" reduction (z_0) and take account of a downward swiveling over time of the curve relating cost to percentage cutback because of improving technological opportunities. Thus (and with parameter β less than zero), we have:

$$(7.14a) \quad c_t^Q = [z_t^Q - z_0] \, [\alpha + \beta \, (t - 35)], \; t \geq 35;$$

$$(7.14b) \quad c_t^Q = \alpha \, [z_t^Q - z_0], \; t < 35$$
$$s.t. \; c_t^Q \geq 0; \; c_t^Q \geq 2 \; for \; z_t^Q \geq 50$$

Note that in the second formulation (equation 7.14b) there is no time-dependent downward shift in the cost curve before the year 2025 ($t = 35$), as the various model estimates on which the cost curve is based are for that year or later. Note further that the final set of restrictions places a floor of 2 percent of GDP on the cost of carbon reductions in excess of 50 percent. Without this floor, the time-dependent cost curve would fall to much lower levels over more distant periods, but the underlying model estimates do not include such distant years, and there is no assurance that costs would continue to decline.

The calculations then introduce the gains from alleviation of the excess burden of the existing tax system (discussed below). The required carbon tax rate (τ, in dollars per ton) is estimated as:

$$(7.15) \quad \tau_t = [z_t^Q - z_0] \, h_0, \; 0 \leq \tau \leq H^*$$

where for each additional percentage-point cutback in carbon a specified incremental dollar amount of tax per ton, h_0, must be imposed, up to a ceiling long-range tax of H^* based on the backstop noncarbon energy source (as discussed below).

Because the emission of only K tons of carbon annually is permitted to continue, tax revenue (T) will equal the tax rate times this base:[2]

(7.16) $T_t = \tau_t K$

Where the existing structure of taxation is inefficient and each dollar of additional tax revenue causes a partially offsetting loss equal to the fraction ρ of a dollar (i.e., ρ is the marginal efficiency cost of taxation), the new source of revenue from carbon taxes will permit a benefit from reduction of the excess tax burden amounting to:

(7.17) $B_t^{\tau} = T_t \rho$

The carbon tax will of course have its own excess burden, but that cost is already incorporated on the cost side of the analysis (equations 7.11 to 7.14).

The overall cost of the abatement program equals world GDP multiplied by the sum of the three cost sources (converted from percentages of GDP to fractions): economywide (Q), afforestation (FA), and reduced deforestation (FD). This cost is expanded by the proportion ω to take into account likely costs associated with curtailing *all* greenhouse gases in a way commensurate with reducing carbon dioxide emissions. The cost is further increased by considering the portion of the cost—essentially production sacrificed—that would have gone into investment (π_k^c), and applying a shadow price of capital (p_k^*) to this component to convert it to consumption-equivalents (chapter 6). Thus, the absolute magnitude of global abatement cost in year t is:

(7.18) $C_t = .01\, Y_t\, [c_t^Q + c_t^{FA} + c_t^{FD}]\, [1 + \omega]\, [(1 - \pi_k^c) + p_k^* \pi_k^c]$

For its part, the shadow price of capital is estimated as set forth in chapter 6, annex 6A:

(7.19) $p_k^* = A_{r,N} \psi_{i,N} N$

where $A_{r,N}$ is the annual payment of an annuity with lifetime N years and interest rate r; $\psi_{i,N}$ is the present discounted value of a stream of unity annually over N years discounted at interest rate i, divided by N; and the two discount rates are the rate of return on private capital (r) and the social rate of time preference (i), respectively.

On the benefits side, the gains from avoidance of warming equal the percentage of GDP benefit (converted to a fraction) multiplied by world GDP. This

2. Note that because some emissions from deforestation will continue, the implication is that the tax will be imposed on this source as well.

magnitude is further expanded to take into consideration the fact that some of these gains accrue to production going into investment. A fraction ϕ_{bc} is consumption-specific (e.g., human amenity from avoided heat waves). For the remainder, the fraction $\pi_k{}^b$ represents production that goes into investment, and this component is expanded by the shadow price of capital (symmetrically with the treatment on the cost side). Finally, the absolute benefit from reduction of the excess tax burden is included. The overall benefits are thus:

$$(7.20) \quad B_t = .01\, Y_t b_t \{\phi_{bc} + (1 - \phi_{bc})\, [(1 - \pi_k{}^b) + p_k{}^* \pi_k{}^b]\} + B_t{}^T$$

The present discounted value of benefits from the base year 1991 through the terminal year 2275, discounting at the social rate of time preference, is thus:

$$(7.21) \quad V_b = \sum_{t=1}^{285} \frac{B_t}{(1 + i)^t}$$

Similarly, the present discounted value of costs of abatement over the same horizon amounts to:

$$(7.22) \quad V_c = \sum_{t=1}^{285} \frac{C_t}{(1 + i)^t}$$

The ratio of the present discounted value of benefits to that of costs is thus:

$$(7.23) \quad q = \frac{V_b}{V_c}$$

If q exceeds unity, the program for abatement of global warming is socially efficient in economic terms; if q is below unity, the program of aggressive abatement is socially inefficient because the benefits do not warrant the costs.

Variable Baselines and Parameter Values

Population and Income Growth

Table 7.2 reports the values applied for the parameters of the benefit-cost model and for the baseline projections of key variables. The world is divided into developed countries (including the former Soviet Union and Eastern Europe) and developing countries. Their populations in 1990 were an estimated 1,235 million and 4,075 million, respectively. The projected population growth rates in table 7.2 are derived from the assumptions used by the Intergovernmental Panel on Climate Change (IPCC) through the year 2100 (IPCC 1990d, 20).[3] World

3. The IPCC estimates in turn were based on Zacharia and Vu (1988).

population reaches a plateau by the mid-21st century at about 10½ billion, and after the year 2100 population growth is zero in both areas.

Per capita income in 1990 constant dollars is set at $14,100 for the developed region. This level is considerably lower than the World Bank's estimate of $19,000 for member countries of the Organization for Economic Cooperation and Development (OECD; World Bank, *World Development Report 1991*, 205). The figure here is calibrated to reflect the much lower per capita income of the former Soviet Union and Eastern Europe.[4]

The estimated per capita income of $3,400 for developing countries is based on a purchasing power parity concept. Kravis et al. (1982, 15) have estimated that at cross-sectional purchasing power exchange rates, which take account of lower-priced nontradeables (especially labor-intensive services) in developing countries, real per capita income in these countries is about one-sixth (1/5.6) the level in industrial countries (excluding the former Soviet Union and Eastern Europe). The resulting world GDP for 1990 is $31.3 trillion, of which $17.4 trillion is in the developed countries and $13.9 trillion in the developing countries.[5]

For the period 1990–2100, growth of per capita income by region is set equal to the lower alternative applied in the IPCC emissions scenario (IPCC 1990d, chapter 2). Beyond the year 2100, per capita income growth is assumed to fall gradually below 1 percent, to rates of ½ percent or lower in the 23rd century. Per capita income growth remains somewhat higher over a longer period in the developing countries than in the developed countries. Over the entire period, average per capita income growth is close to 1 percent in both regions. Although the rate may appear modest, the power of compound interest is such that it implies per capita incomes of $185,000 by the year 2275 for developed countries and $63,000 for developing countries. By that year, global GDP stands at a remarkable $837 trillion (at 1990 prices), or about 26 times today's global output.

The possibility of such high per capita income in the long term raises the question of why the present generation should make any sacrifice at all to avoid a greenhouse-imposed loss to future generations, so long as that loss does not exceed the difference between per capita income today and in the long term. The formal answer to this question is that the analysis already takes account of higher future income by discounting the future on the basis of expected rising per capita consumption (chapter 6). As developed below, the loss of long-term

4. World Bank, *World Development Report 1991*, places OECD population at 722.6 million and GDP at $13.8 trillion in 1989. Collins and Rodrik (1991, 6–8) estimate the combined population of the former Soviet Union and Eastern Europe at 422.8 million and 1988 GDP for the area at $2.4 trillion. The combined "developed country" estimates amount to a population of 1.15 billion and per capita income of $14,150 for 1988–89.

5. Without the correction for purchasing power parity in developing countries, world GDP is approximately $20 trillion.

per capita income from global warming could be on the order of 6 to 20 percent or more, even with no unexpected catastrophe. With global per capita annual income in 2275 at about $80,000, versus about $6,000 today (purchasing power parity), a 15 percent loss of 2275 income would be a reduction by some $12,000. If, say, only 2 percent of today's income need be paid to avoid this loss (as discussed below), then the trade-off is $12,000 near the end of the 23rd century in exchange for $120 today. This trade-off of 100 to 1 can be attractive, unless the discount rate is high.

Some might argue that there should be more emphasis on the greater relative value of an extra dollar to today's person with $6,000 than to his descendant with $80,000. The extreme version of this view would hold that no person should ever be asked to give up a single dollar to a descendant who will be at a higher per capita income than himself, regardless of how large the resulting increase in the descendant's income could be. In formal terms, that approach would represent a value of infinity for the elasticity of marginal utility of income (discussed in chapter 6). Any small increase in income would cause marginal utility to plunge to zero, so that even a highly leveraged bargain such as the one just discussed would be dismissed out of hand. In contrast, the analysis here makes some allowance for greater- than-unitary elasticity of marginal utility (the elasticity is set at 1.5) but leaves sufficient marginal utility even at high future incomes that the leveraged bargain just outlined may make sense.

It seems likely that, in the broader public discussion of the greenhouse effect, there is a much different wager at stake. Many of those who stress the dangers of the problem appear to have a much grimmer future in mind. In their future scenario, on a scale of two to three centuries the typical inhabitant of the earth may not be much richer than today even without global warming, and the scope of possible greenhouse damage is catastrophic rather than limited to, say, 10 percent to 20 percent of income. Indeed, in this type of approach, the possible damage is so great that the world's future inhabitants are worse off than those today. The grounds for discounting future income disappear or the discount rate even turns negative (with falling per capita consumption, marginal utility is rising rather than falling and an extra dollar is worth more in the future than today). The equity case for a transfer of income to the more impoverished becomes an argument for shifting income from the richer, present generation to the poorer, future one.

Without going to the extreme of fearing a secular decline in future per capita income, the choice between the forecast of a prosperous or a stagnant future is not as obvious as it might seem simply on the basis of past growth. Suppose that the average rate of growth of per capita income is ½ percent instead of 1 percent, and the growth of population is ½ percent higher than expected. Neither proposition is implausible, yet the consequences over two to three centuries are dramatic. By the late 23rd century, world GDP—and thus (to a first approximation) carbon emissions—would reach just as high as in the base case. However, world per capita income would reach only about $24,000 rather than the $80,000 under the baseline scenario.

The thrust of the growing attention to "sustainable development" (e.g., Bruntland et al. 1987) is precisely that lower per capita growth than in the past may be difficult to avoid. Multiplication of global real GDP by some 25-fold as in the baseline here would raise serious doubts of feasibility in the minds of many in the sustainable development school, which stresses limits of the global resource and environmental base. Moreover, this approach argues that past observed growth has been substantially overstated by the failure to take account of resource depletion and environmental deterioration (cf. Repetto 1989). On this basis, a projection of past growth trends in per capita income that is accurately measured might show future rates considerably below the baseline used here. If so, any equity objections to taking even small amounts from the present to avoid even large losses in a richer future would be further ameliorated.

Although the base case expectaton is that per capita income will keep rising, there is no guarantee that future generations will be richer than the present one. To take into account the possibility of stagnation of living standards over the long term, the analysis below includes one variant for a "Malthusian scenario" in which population growth is high and per capita income remains unchanged (zero growth in living standards). This scenario has somewhat lower emissions (especially at first) but also a lower social rate of time preference, set at zero because of zero growth in per capita income.

Nonetheless, the baseline projections of approximately 1 percent annual growth in per capita income do not seem high, at least on the basis of US experience over the past century and using data unadjusted for resource depletion and environmental deterioration. From 1870 to 1930, an index of US production showed annual growth of 3.7 percent (Schumpeter 1942, 64), or per capita growth of 1.7 percent. From 1929 to 1990, real US GDP grew by an annual average of 3 percent (calculated from *Economic Report of the President*, February 1991, 288), or by 1.8 percent annually per capita.[6] This sustained growth confirmed the optimistic half-century forecast of the famous economist Joseph Schumpeter (1942), but it proved insufficient to achieve his corresponding prediction that "this would do away with anything that according to present standards could be called poverty, even in the lowest strata of the population" (Schumpeter 1942, 66). The complication of uneven income distribution adds further to grounds for caution about the assumption of a (uniformly) richer future.

6. US population rose from 39.8 million in 1870 to 122.8 million in 1930 and approximately 251 million in 1990 (*Statistical Abstract of the United States*, 7; International Monetary Fund, *International Financial Statistics*, 1991).

One further word is warranted on the implications of the baseline incomes. They do suggest that international implementation of an abatement policy should include special measures to minimize the current losses imposed on the developing countries, whether through compensatory transfers, the sale of developing-country emission permits, or otherwise. The sharp contrast between present and future expected per capita income in the developing area implies a steeper trade-off for marginal utility over time than in the industrial countries.[7] This divergence adds to the case for differential burden sharing based solely on present income differences between the two regions.

Carbon Emissions and Abatement

As shown in table 7.2, global emissions rise from 6.7 billion tons of carbon (GtC) in 1990 to approximately 14 GtC in 2050, 22 GtC in 2100, 39 GtC in 2200, and 57 GtC by the end year, 2275. This path is the average of the projections in chapter 2 based (through 2100 and then extended) on three leading energy-economic models (Reilly et al. 1987, Nordhaus and Yohe 1983, Manne and Richels 1990c), plus 1 GtC annually from deforestation. For the first century, the projection is approximately the same as that indicated by the IPCC for its "2030 High-Emissions Scenario," which "depicts a world in which few or no steps are taken to reduce emissions in response to concerns about greenhouse warming" (IPCC 1990d, 10).[8]

It may be asked whether this emissions baseline is consistent with the economic growth baseline just discussed, in view of the fact that the growth assumptions here adopt the lower of two IPCC variants (rather than an average). The ratio of emissions to global GDP provides a gauge of this consistency, as a rising or even constant emissions-production ratio would suggest overstatement of future carbon emissions. Instead, the baseline of table 7.2 shows a steady decline in this ratio, from 0.21 billion tons of carbon per trillion (1990) dollars of world GDP in 1990 to 0.13 in 2050, 0.10 in 2100, 0.07 in 2200, and 0.07 in 2275. For the period as a whole, carbon emissions grow at 0.7 percent annually, whereas world GDP grows at 1.2 percent annually. This divergence provides room for both autonomous energy efficiency growth and response to rising real

7. Thus, the ratio of per capita income in 2275 to per capita income in 1990 is 12.8 for developed countries, but 17.3 for less developed countries. When we apply the generalized utility function of chapter 6 with an elasticity of marginal utility of -1.5, the trade-off ratio between a dollar in the year 2275 and a dollar today is about 60 percent higher for the developed countries (at 2.18 cents on the dollar) than for the less developed countries (at 1.39 cents on the dollar; calculated from equation 6.15).

8. That scenario places 2050 carbon emissions at 15.2 GtC and 2100 emissions at 22.4 GtC.

prices of fossil fuel resources. In short, the carbon baseline appears compatible with the economic growth assumptions. Moreover, as shown in chapter 2, it is consistent with the availability of fossil fuel reserves.

The abatement policy examined cuts global emissions to a ceiling of 4 GtC annually (*K*, table 7.2). This level is approximately the same as that identified by the IPCC for an aggressive abatement stance of "accelerated policies."[9]

Greenhouse Warming, Damage, and Abatement Benefits

The climate sensitivity parameter (*S*) for warming at carbon-dioxide-equivalent doubling is set at the three benchmarks identified by the IPCC: 1.5°C for the lower bound, 2.5°C for the "best guess," and 4.5°C for the upper bound (IPCC 1990b, chapter 5). The corresponding estimates for warming in the very long term (see chapter 2 above) are 6°C, 10°C, and 18°C, respectively. As noted, the calculations set the years 2050 for $2 \times CO_2$-equivalent warming and 2275 for very-long-term warming ($t = 285$, table 7.2). Warming in intermediate years is determined by linear interpolation.

The damage estimates are based on those for the United States developed in chapter 3. The use of US patterns as a proxy for global effects should not be seriously misleading and provides a basis for a first estimate that should ideally be followed up with regional estimates. As discussed in chapter 3, damages could be higher in developing countries because of a greater share of agriculture in GDP (but with some offset because warming should be less in the lower latitudes where developing countries tend to be located). Damages could be proportionately lower at the high latitudes because of possible favorable effects on agriculture, less relevance of air conditioning, and lesser water supply reduction in view of increased precipitation.

The central estimate for economic damage from global warming is set at 1 percent of world GDP at benchmark $2 \times CO_2$ warming, as developed in detail in chapter 3 above. Because of the likelihood that this estimate understates the value of some key effects (e.g., species loss) and simply omits others (human disamenity, side benefits of reduced pollution), the alternative estimate of 2 percent of GDP is examined as well. Nordhaus (1991a) also uses 1 percent of

9. In its "accelerated policies" scenario, the IPCC sets carbon emissions at 5.6 GtC in the year 2000, 5.1 GtC in 2025, 2.9 GtC in 2050, 3.0 GtC in 2075, and 2.7 GtC in 2100. For global warming there should be little difference between the plateau of 4 GtC used here and the 3 GtC applied in the later portion of the IPCC scenario. Only half the difference would accumulate in the atmosphere, so that over 200 years the additional atmospheric stock would amount to 100 GtC. By comparison, present atmospheric stock is approximately 750 GtC. On the other hand, reducing the ceiling to 3 GtC after 2050 would not raise costs much, as the percentage cutback from baseline would not differ radically (86 percent cutback by 2100, versus 82 percent using $K = 4$).

Table 7.2 Variables and parameter values

Variable or Parameter	Concept	Value
t	Time (years)	1991 = 1
y_0^D	Initial per capita income, developed countries	$14,100
y_0^L	Initial per capita income, less developed countries[a]	$3,400
P_0^D	Initial population, developed countries	1,235 million
P_0^L	Initial population, less developed countries	4,065 million
g_D	Per capita income growth, developed countries (percentages per year)	1.9, 1.2, 1.2, 1.2, 1.2, 1.0, 0.8, 0.5, 0.2[b]
g_L	Per capita income growth, less developed countries (percentages per year)	1.6, 1.3, 1.5, 1.5, 1.0, 1.0, 1.0, 0.5, 0.5[b]
n_D	Population growth, developed countries (percentages per year)	0.4, 0.4, 0, 0, 0, 0, 0, 0, 0[b]
n_L	Population growth, less developed countries (percentages per year)	1.6, 1.4, 0.8, 0.3, 0.2, 0, 0, 0, 0[b]
E_t	Global carbon emissions (GtC)	6.7, 7.1, 10.9, 13.8, 17.3, 21.6, 25.5, 29.7, 34.3, 39.2, 44.7, 50.6, 57.2[c]
K	Aggressive abatement carbon ceiling (GtC)	4
S	Climate sensitivity parameter	Base: 2.5°C; low: 1.5°C; high: 4.5°C
W^*	Very-long-term warming	Base: 10°C; low: 6°C; high: 18°C
d_0	Greenhouse damage at 2.5°C warming (percentages of GDP)	Base: 1.0; high: 2.0
γ	Damage function exponent	Base: 1.3; high: 2.0
μ	Damage avoidance potential	0.8
R_0^{FA}	Emissions reduction from afforestation	1.6 GtC per year, 1991–2020
R_0^{FD}	Emissions reduction from lower deforestation	0.7 GtC per year, 1991–2275
E_0^F	Base year deforestation emissions	1.0 GtC
C_0^{FA}	Annualized cost of afforestation	$17.2 billion, 1991–2020; zero thereafter

R_0^{FD}	Annual cost of reduced deforestation	$4.5 billion
α	Abatement cost parameter	0.0678 (chapter 5)
β	Abatement cost parameter	−0.00039 (chapter 5)
z^0	Percentage carbon reduction at zero cost	22 (chapter 5)
h_0	Carbon tax parameter	$5 per ton per percentage-point cut
H^*	Carbon tax, long-term backstop	$250 per ton
ρ	Marginal efficiency cost of tax burden	0.33
ω	Proportionate cost expansion for other greenhouse gases	0.2
P_k^*	Shadow price of capital	1.6; depends on social rate of time preference, rate of return on capital
π_k^c	Investment share of cost	0.2
π_k^b	Investment share, nonspecific benefits	0.2
ϕ_{bc}	Consumption-specific share of benefits	0.5
i	Social rate of time preference (percentage per year)	Base: 1.5; Other: 1, 2, 3, 5, 10
N	Capital life	15 years

GtC = gigatons of carbon.

a. At purchasing power parity.

b. In 1990–2000, 2001–2025, 2026–2050, 2051–2075, 2076–2100, 2101–2150, 2151–2200, 2201–2250, 2251–2275, respectively.

c. In 1990, 2000, 2025, 2050, 2075, 2100, 2125, 2150, 2175, 2200, 2225, 2250, 2275, respectively.

GDP as the "medium" estimate of greenhouse damage for carbon dioxide doubling, and 2 percent of GDP as the "high" estimate.[10]

When the high-damage case of 2 percent of GDP at the central climate sensitivity parameter ($S = \Lambda = 2.5°C$) is translated into damage for upper-bound warming ($S = 4.5°C$), economic damage reaches 4.3 percent of GDP for $2 \times CO_2$ (= 2 percent $\times [4.5/2.5]^{1.3}$). The central and high-damage estimates for very-long-term warming with $S = 2.5°C$ (that is, 10°C warming) are 6 percent and 12 percent of GDP, respectively. With upper-bound very-long-term warming (18°C), damage reaches 13 percent to 26 percent of GDP (that is, 6 percent to 12 percent multiplied by $[4.5/2.5]^{1.3}$) under the central damage exponent case ($\gamma = 1.3$).

The exponent in the geometric function relating damage to warming is set at 1.3 in the base case and 2.0 in the high-damage case. The parameter 1.3 is derived from the weighted average for various effects, discussed in chapter 3. The higher, quadratic damage function is easily plausible, however. It is widely recognized that there are strong reasons to expect damage to be nonlinear in warming, and the degree of nonlinearity suggested in chapter 3 may be an understatement. Thus, in one model variant Nordhaus (1990a) applies a quadratic damage function (i.e., $\gamma = 2$).

Application of the damage function to world GDP raises the question of whether the impact of a given amount of warming in the distant future would be proportionate, relative to GDP, to the impact imposed on today's world economy. As discussed in chapter 3, there is no strong reason for expecting the overall effect to rise less or more than proportionately with GDP. Some components would tend to rise by less than GDP (for a given warming)—for example, agricultural damage in view of the likely decline of the share of agriculture in GDP over time. Others could rise more than proportionately, including the valuation of human lives lost and of species loss.

Abatement benefits are determined by the amount of prospective damage that can be avoided through policy action. The damage avoidance fraction μ is set at 0.8. Thus, even with aggressive policy action, it is assumed that 20 percent of the prospective baseline damage is unavoidable. This fraction is based on the evaluation that, in the very long term, about the best that can be expected from aggressive abatement policy is to hold the ceiling global warming to the amount associated with carbon dioxide doubling, or 2.5°C in the central case. The amount of damage avoided, expressed as a percentage of GDP, thus becomes the difference between baseline damage in the very long term and that for a doubling of carbon-dioxide-equivalent. As discussed in chapter 3, the base-case analysis places very-long-term warming damage at 6 percent of GDP, so that this difference amounts

10. The differences between the Nordhaus damage estimates and those of this study are discussed below.

to 6 − 1 = 5 percent of GDP. Forceful-action can thus avoid five-sixths of the potential damage. The use of 80 percent here provides a conservative estimate of the fraction of damage avoided, especially for the case of the quadratic damage function. In that more extreme case (but still applying only the central IPCC value for climate sensitivity, *S*), very-long-term damage reaches 16 percent of GDP (1 percent × [10°C/2.5°C]^2). In that case, the damage avoided would be 16 − 1 = 15 percent, so the avoidance fraction would be as high as 94 percent.

Cost Parameters

The estimate for emissions reduction from afforestation, 1.6 GtC annually for 30 years, is developed in chapter 5. A worldwide program for planting 260 million hectares in new forest would be required for this purpose. The estimated cost, including land purchase, would amount to $17.2 billion (1990 dollars) annually during the first 30 years and zero thereafter.[11] Similarly, the estimate of 0.7 GtC for annual emissions reduction from curtailment of deforestation is developed in chapter 5, on the basis of potential effects in just three major developing countries (Brazil, Indonesia, and Côte d'Ivoire). This program would cost only $4.5 billion annually, but the cost would recur over the entire time horizon. Base-year deforestation emissions (and, in the baseline, constant deforestation emissions thereafter) are set at 1 GtC.[12]

Table 7.2 next reports the parameters in the cost function for reduction of emissions in industry and elsewhere in the economy (excluding forestry). As developed in chapter 5, the estimates assume that a reduction of up to 22 percent of emissions can be achieved at zero cost, on the basis of the body of engineering estimates. As noted, the term that shifts cost downward over time (reflecting improved technological opportunities) is restricted from lowering cost below 2 percent of GDP, where large cuts (over 50 percent) are involved even in the distant future.

11. An alternative formulation could use the cost of land rental rather than purchase, with lower costs initially but recurrent and rising costs over the whole, long-term horizon.

12. The IPCC (1990d, chapter 2) no-abatement baseline ("high scenario") uses 0.7 GtC in 1985, 1 GtC in 2000, 1.4 GtC in 2025–50, 0.7 GtC in 2075, and 0.4 GtC by 2100, thereby allowing for the impact of forest exhaustion. The IPCC's use of a lower figure for its emissions projections than the central value of 1.6 GtC identified in its scientific volume (IPCC 1990b, chapter 1) reflects a wide range of uncertainty regarding the current extent of deforestation.

Tax Parameters

The carbon tax parameters are based on the models surveyed in chapter 4. From those models, it may be estimated that, for the period from approximately 2025 to 2050, for each percentage point reduction in the economy's carbon emissions a carbon tax of $5 per ton (1990 dollars) is required. To provide a conservative estimate of tax revenue, it is assumed that the relevant "percentage cut" is the total less that which can be achieved at zero cost (or 22 percent, z_0). The maximum tax is $250 per ton, which reflects the cost differential for backstop noncarbon energy sources and thus is sufficient for high cutbacks over an indefinite range.

To estimate tax revenue, the tax rate is applied to the volume of emissions still permitted, or 4 GtC. On this basis, the maximum tax revenue is $1 trillion annually.[13] The marginal efficiency cost of existing taxes, ρ, is set at 0.33. Thus, for each dollar of existing revenue that can be replaced by carbon tax revenue, there is a savings of 33 cents for the economy because of reduction in the excess tax burden—the loss to inefficiencies from tax-induced distortions in resource allocation. This parameter is based on the estimate by Jorgensen and Yun (1990) that, in the United States (after the Tax Reform Act of 1986), the marginal efficiency cost of the tax system was 0.383.[14] With $\rho = 0.33$ and maximum carbon-tax revenue at $1 trillion, the maximum benefit from tax burden reduction reaches $330 billion in 2060, or about 0.25 percent of world GDP.

Other Greenhouse Gases

The factor for expansion of cost estimates to incorporate reduction of noncarbon greenhouse gases (ω) is set at 0.2. This range is consistent with the following considerations. First, under "business as usual," carbon dioxide is expected to account for about two-thirds of the increase in radiative forcing from all greenhouse gases over the next century (IPCC 1990b, table 2.7). Second, Nordhaus (1990b) has emphasized that the reduction of chlorofluorocarbons is the lowest-cost means of reducing global warming. Third, carbon dioxide has spectrum-based saturation effects that makes its radiative forcing logarithmic in concentration, whereas with lower saturation the radiative forcing of other trace gases is closer to linear (chapter 1). The implication of this last point is a high payoff

13. Note, however, that revenue does not reach this maximum until approximately 2060, when it amounts to only 0.8 percent of world GDP.

14. The marginal efficiency cost falls to 0.335 if tax rates are cut by 20 percent, and to 0.307 if they are cut by 30 percent. The estimates here suggest that carbon revenue would be only a modest percentage of GDP, so that the lower marginal efficiency costs at still steeper general tax reductions are not relevant.

to prevention of expansion of other trace gases relative to that for carbon dioxide. Fourth, as discussed in chapter 8, many of the measures that cut carbon emissions should also reduce emissions of other greenhouse gases.

These four considerations together imply that the ratio of abatement cost expansion for other trace gases to that for carbon dioxide is substantially less than implied by the relative shares in the buildup of radiative forcing (about 50 percent, i.e., ⅓ ÷ ⅔). On this basis, expansion by 20 percent would appear reasonable, or perhaps even excessive, to translate carbon abatement costs into costs for a corresponding reduction in total trace gases.

Shadow Pricing of Capital

Table 7.2 next reports the shadow price of capital for the central value of the social rate of time preference (*i*), estimated on the basis of equation 7.19 above (and using a capital lifetime of 15 years). At $i = 1\frac{1}{2}$ percent, this shadow price is approximately 1.6, so that one unit of displaced investment is worth 1.6 units of displaced consumption. The share of investment (π_k) is set at 20 percent on both the side of abatement cost (*c*) and the side of non-consumption-specific benefits (*b*). It is assumed, however, that half of the benefits of greenhouse damage avoidance are consumption-specific.

Adaptation

The National Academy of Sciences (NAS 1991a, 1991b) has stressed the importance of adaptation as an approach to the greenhouse problem. Adaptation reduces greenhouse damage but it also imposes a new cost in the "baseline." The benefit-cost calculus after incorporation of adaptation thus involves a reduction in the benefit from abatement (less damage to be avoided). However, it also involves a reduction in the net costs of the abatement policy, because the costs of the *alternative* strategy of adaptation may now be deducted in assessing the abatement option.

This formulation of mutually exclusive alternatives of abatement and adaptation is useful conceptually, but in practice public policy on global warming is likely to involve a mixture of both. What is really at issue, then, is a comparison between two broad strategies: one with heavy emphasis on abatement and lesser reliance on adaptation, and the other the reverse. The implicit assumption of this study is that, in view of the possible extent of very-long-term warming, the amelioration possible through adaptation will be limited, and the primary strategy if a meaningful response is to be mobilized, will have to be abatement.

The discussion of agriculture in chapter 3 suggests that adaptation can help but is unlikely to fundamentally alter the qualitative dimensions of potentially large greenhouse damage, especially in the very long term.[15] More generally, the range of changes in the benefit-cost calculation resulting from consideration of adaptation seems likely to be modest compared with the already wide range of uncertainty regarding the underlying warming effects and damage functions.

In the absence of explicit measures for the impact of adaptation, the approach taken here is simply to register that the benefit-cost ratio of a particular abatement scenario may be a modest overstatement for lack of direct inclusion of adjustments for the adaptation alternative.

Benefit-Cost Results

Table 7.3 reports the results of the benefit-cost calculations for 36 principal cases using combinations of alternative assumptions for key parameters. It should be noted that, whereas the discount rate selected affects both the cost and the benefit side, variation of the other three parameters—climate sensitivity (S), the damage function exponent (γ), and benchmark damage (d_0)—affects only the benefit side.[16]

The broad picture that emerges is that, under the combinations of lower rates of social time preference and greater greenhouse damage, it will be socially efficient to undertake aggressive abatement action, whereas with higher discount rates and lesser damage, action will not be warranted. From the standpoint of the analysis in chapter 6, the most meaningful results are those that apply a social rate of time preference of 1½ percent (with a corresponding shadow price of capital of 1.56). Of the 11 cases reported for this discount rate, six show a benefit-cost ratio (q) in excess of unity for aggressive abatement action. With the IPCC's best-guess climate sensitivity parameter (S = 2.5°C), three of the four cases have $q > 1$, and with the upper-bound warming (S = 4.5°C), all three cases show $q > 1$. In contrast, if the lower-bound climate sensitivity parameter is applied (S = 1.5°C), the benefit-cost ratio is strictly less than unity, although it approaches unity in the case with the highest economic damage assumptions (d_0 = 2 percent of GDP at 1.5°C warming and damage function exponent γ = 2.0).

If a single case were to be chosen as the central estimate, it would be case 1. Here the climate sensitivity parameter is at the IPCC central value, and the economic damage parameters are set at the central magnitudes developed in chapter 3. The ratio of the present discounted value of benefits to the present discounted value of costs for this case is 0.74. If it were certain that these were the stakes of global warming, the implication would be that abatement is too costly relative to the prospective damage and no action should be undertaken.

---

15. See, in particular, the interpretation in chapter 3 of the MINK results.

16. The climate sensitivity parameter S is that referred to as Λ in chapters 1 and 2.

As discussed below, however, the outlook is far from certain. A more appropriate approach to decision making is to take into account a weighted average of the alternative scenarios. The degree of policymaker risk aversion will also enter into this calculation. For the moment, however, it is useful to consider the further implication of the finding of $q = 0.74$ for the base case. This result indicates that, of every dollar spent on greenhouse abatement, there would be an outright loss of about 26 cents. More specifically, as the abatement costs are on the order of 2½ percent of GDP over time, the implication is that about 0.6 percent of GDP would be a pure loss. One way to think about this outcome is that this is the price of purchasing greenhouse insurance—an insurance policy that yields benefits covering only three-quarters of its costs but provides protection against the contingency of higher, unexpected losses.

The remaining cases in table 7.3 are reported for purposes of sensitivity analysis with higher social rates of time preference. As expected, there are fewer instances of benefit-cost ratios in excess of unity when this rate is set at 3 percent or 5 percent instead of the preferred 1½ percent. However, even with these discount rates, the benefits of action outweigh the costs in several cases where the warming and damage functions are high. This is the outcome for 3 of the 11 cases estimated with a discount rate of 3 percent (cases 16, 22, 23) and for one of the cases with the discount rate at 5 percent (case 34). The principal implication of this finding is that, even with higher discount rates than proposed in this study, abatement action may be warranted if emphasis is given to the possible outcomes with higher damage.

The potential scope of the damage is very large in several of the scenarios considered in table 7.3. Whereas in the base case (case 1) greenhouse damage in the very long term amounts to 6 percent of GDP, if the damage function exponent is raised from 1.3 to 2.0 this estimate increases to 16 percent of GDP (1 percent × [10°C/2.5°C]^2). For climate sensitivity $S = 2.5$°C, the maximum very-long-term damage reaches 32 percent of GDP (e.g., case 4); with $S = 4.5$°C, this maximum reaches 52 percent of GDP (e.g., case 11). The highest possible damage combination ($S = 4.5$°C, $\gamma = 2$, $d_0 = 2$) is ruled out, because it gives very-long-term damage in excess of 100 percent of GDP.[17]

Policy Choice Under Uncertainty with Risk Aversion

Cases 1, 5, and 9 in table 7.3 represent the central calculations of this study for the three alternative warming scenarios of the IPCC: best-guess, lower-bound,

17. For the seven other combinations involving $S = 2.5$°C or 4.5°C, $\gamma = 1.3$ or 2.0, and $d_0 = 1$ or 2, greenhouse damage at very-long-term warming amounts to 6 percent of GDP in one, 12 percent to 16 percent of GDP in three, 26 percent to 32 percent in two, and 52 percent in one.

Table 7.3 Results of benefit-cost analyses of abatement of greenhouse warming, principal cases[a]

	Key parameters				Benefit-cost
Case	i	S	γ	d_0	ratio (q)
1	1.5	2.5	1.3	1	0.74
2	1.5	2.5	1.3	2	1.42
3	1.5	2.5	2.0	1	1.33
4	1.5	2.5	2.0	2	2.60
5	1.5	1.5	1.3	1	0.41
6	1.5	1.5	1.3	2	0.76
7	1.5	1.5	2.0	1	0.51
8	1.5	1.5	2.0	2	0.97
9	1.5	4.5	1.3	1	1.52
10	1.5	4.5	1.3	2	2.99
11	1.5	4.5	2.0	1	4.18
12	1.5	4.5	2.0	2	n.a.
13	3	2.5	1.3	1	0.44
14	3	2.5	1.3	2	0.80
15	3	2.5	2.0	1	0.58
16	3	2.5	2.0	2	1.09
17	3	1.5	1.3	1	0.26
18	3	1.5	1.3	2	0.45
19	3	1.5	2.0	1	0.26
20	3	1.5	2.0	2	0.44

21	3	4.5	1.3	1	0.86
22	3	4.5	1.3	2	1.63
23	3	4.5	2.0	1	1.71
24	3	4.5	2.0	2	n.a.
25	5	2.5	1.3	1	0.33
26	5	2.5	1.3	2	0.57
27	5	2.5	2.0	1	0.33
28	5	2.5	2.0	2	0.56
29	5	1.5	1.3	1	0.22
30	5	1.5	1.3	2	0.34
31	5	1.5	2.0	1	0.18
32	5	1.5	2.0	2	0.26
33	5	4.5	1.3	1	0.61
34	5	4.5	1.3	2	1.12
35	5	4.5	2.0	1	0.85
36	5	4.5	2.0	2	n.a.

i = social rate of time preference, in percentages per year; S = climate sensitivity factor (degrees Celsius warming for a doubling of carbon-dioxide-equivalent); γ = damage function exponent; d_0 = damage from 2.5°C warming (in percentages of GDP); n.a. = not applicable.

a. All cases *include* the benefits from tax burden reduction and apply a shadow price of capital ranging from 1.56 (i = 1.5 percent) to 1.21 (i = 5 percent).

and upper-bound, respectively. Cases 10 and 11 represent upper-bound warming in combination with high-damage functions (d_0 = 2 and γ = 2, respectively). Suppose that the probability of the best-guess case is 50 percent, and the probability of either the upper- or the lower-bound is 25 percent. Suppose further that policymakers are risk averse, so that they attach more weight to avoiding unpleasant surprises than to enjoying pleasant ones. An informal way to arrive at the policy decision is to weight the (present discounted values of) benefits of the alternative scenarios by a combination of their underlying probabilities and a factor that takes account of risk aversion. Note that the costs of the alternative cases are all identical, because their abatement cutbacks of carbon emissions are the same.

Suppose that, because of risk aversion, the weight for the upper-bound warming case should be three times the weight for the lower-bound case. Suppose further that the weight for the central case is unity times its probability of occurrence. On this basis, the respective weights might be, for example, 0.5 for the central case, 0.125 for the lower-bound case, and 0.375 for the upper-bound case.

For the upper-bound damage case, an average of scenarios 9 and 10 may be used. In case 9, damage is 1 percent of GDP at benchmark CO_2 doubling and rises to 13 percent in the very long term; in case 10, damage is 2 percent of GDP at benchmark warming and reaches 26 percent of GDP in the very long term. These two cases suggest that the most meaningful estimate for high damage with very long term warming is a loss of 20 percent of GDP annually (the average of the two cases). With the weights just discussed applied to the present discounted value of benefits of cases 1 (central), 5 (lower-bound), and an average of cases 9 and 10 (upper-bound), respectively, and then divided by the present discounted value of costs, the resulting weighted benefit-cost ratio is 1.26. Thus, *using the central assumptions of this study, if there is weighting of outcomes to take account of risk aversion, the benefit-cost analysis finds that aggressive abatement action is warranted.*[18]

This result may be seen further by returning to figure 7.1. The figure shows the abatement-cost curve (as a percentage of GDP) and two alternative benefit curves: that for the central case and the average of high-damage cases 9 and 10 (the low case is omitted from the diagram). Once the magnitude of growing world GDP is taken into account, on the one hand, and values are discounted at the social rate of time preference, on the other, the effect is that the area under the cost curve in the figure is less than a weighted average of the areas

18. Note also that this approach is conservative in at least one regard. If the highest-damage case (11) is used to represent the high-risk outcome rather than the less severe high-damage alternatives (cases 9 and 10), the risk-weighting procedure gives a benefit-cost ratio of 1.98.

under the two alternative benefit curves, so that the weighted benefits exceed costs (present discounted value).[19]

Sensitivity Analysis

Table 7.4 reports the benefit-cost ratios of the abatement program under alternative assumptions. The table indicates the sensitivity of the outcome to the discount rate, the shadow price of capital, and inclusion or exclusion of tax burden effects. The first 40 cases (37 through 76) apply traditional discounting with no special conversion of capital effects into consumption- equivalents (i.e., $p_k^* = 1$).[20] Perhaps the biggest surprise of the tests is that, even with a discount rate of 10 percent, one variant of the high-damage outcome shows a benefit-cost ratio for abatement in excess of unity (case 74, with $q = 1.07$). The benefit-cost ratio is also above unity for the corresponding high-damage case using a discount rate of 5 percent (case 70, with $q = 1.15$). These results suggest that, even with the higher discount rates frequently used in medium-term public investment projects, a program of aggressive abatement is warranted under the high-damage assumptions.[21] At the other end of the spectrum, if a discount rate

19. Note that visual inspection of the central case shows the area under the benefits curve to be closer in size to the area under the cost curve than the benefit-cost ratio for this case, $q = 0.74$. (The ratio of the strictly nonoverlapping components of these two areas and thus the difference between benefits and costs as a percentage of GDP—may be seen by visual inspection to be smaller, but it is the ratio of total benefits to total costs that matters.) The contrast between the measured q and the visual comparison indicates that the net effect of the conflicting influences of time discounting and GDP growth is dominated by the former. That is, because the diagram reports results as percentages of GDP, a corresponding diagram of absolute amounts would show much larger areas toward the right of the figure (from growing absolute GDP); but once discounted by the time discount rate, these larger absolute areas would be more than proportionately shrunk in comparison with the areas of net cost in the earlier decades. In part the reason for this outcome is that GDP is growing more slowly in the latter part of the time horizon than in the earlier part.

20. In terms of the discussion of chapter 6, the range of 2 percent to 3 percent may be considered to represent the approaches of the Congressional Budget Office and the General Accounting Office, whereas the extreme of 10 percent may be seen as representing the approach of the Office of Management and Budget.

21. A nuance in this regard is that, with high discount rates, earlier damages matter more than eventual damage magnitudes. For example, case 74 has a higher benefit-cost ratio than case 75 even though very-long-term damage is higher in the latter (52 percent of GDP rather than 26 percent). The reason is that the damage occurs earlier ($d_0 = 2$ percent in case 74 and 1 percent in case 75). The higher discount rates also emphasize an anomaly of the earlier years. The higher the damage exponent (γ), the lower the damage prior to 2050. In these years, the ratio $W_t/2.5$ in equation 7.6 is less than unity, so a higher exponent γ causes lower damage. Thus, the benefit-cost ratio is lower in case 75 than in case 73. However, this difference is essentially spurious and would reverse if linear in-

Table 7.4 Sensitivity analysis of benefits and costs of abatement of greenhouse warming[a]

Case	Key parameters i	S	γ	d_0	Benefit-cost ratio (q)
37	1	2.5	1.3	1	0.96
38	1	2.5	1.3	2	1.86
39	1	2.5	2.0	1	1.87
40	1	2.5	2.0	2	3.70
41	2	2.5	1.3	1	0.64
42	2	2.5	1.3	2	1.20
43	2	2.5	2.0	1	1.03
44	2	2.5	2.0	2	2.00
45	3	2.5	1.3	1	0.46
46	3	2.5	1.3	2	0.84
47	3	2.5	2.0	1	0.61
48	3	2.5	2.0	2	1.13
49	5	2.5	1.3	1	0.34
50	5	2.5	1.3	2	0.59
51	5	2.5	2.0	1	0.34
52	5	2.5	2.0	2	0.58
53	10	2.5	1.3	1	0.33
54	10	2.5	1.3	2	0.56
55	10	2.5	2.0	1	0.24
56	10	2.5	2.0	2	0.37
57	1	4.5	1.3	1	2.00
58	1	4.5	1.3	2	3.94
59	1	4.5	2.0	1	5.96
60	1	4.5	2.0	2	n.a.
61	2	4.5	1.3	1	1.29
62	2	4.5	1.3	2	2.50
63	2	4.5	2.0	1	3.19
64	2	4.5	2.0	2	6.32
65	3	4.5	1.3	1	0.89
66	3	4.5	1.3	2	1.70
67	3	4.5	2.0	1	1.78
68	3	4.5	2.0	2	n.a.
69	5	4.5	1.3	1	0.62
70	5	4.5	1.3	2	1.15
71	5	4.5	2.0	1	0.87
72	5	4.5	2.0	2	n.a.
73	10	4.5	1.3	1	0.59
74	10	4.5	1.3	2	1.07
75	10	4.5	2.0	1	0.54
76	10	4.5	2.0	2	n.a.
77	1.5	2.5	1.3	1	0.66
78	1.5	2.5	1.3	2	1.28
79	1.5	2.5	2.0	1	1.19
80	1.5	2.5	2.0	2	2.34
81	1.5	4.5	1.3	1	1.37
82	1.5	4.5	1.3	2	2.69
83	1.5	4.5	2.0	1	3.76
84	1.5	4.5	2.0	2	n.a.
85	1.5	2.5	1.3	1	0.68
86	1.5	2.5	1.3	2	1.37
87	1.5	2.5	2.0	1	1.27
88	1.5	2.5	2.0	2	2.54
89	1.5	4.5	1.3	1	1.47
90	1.5	4.5	1.3	2	2.93
91	1.5	4.5	2.0	1	4.12
92	1.5	4.5	2.0	2	8.24

i = social rate of time preference, in percentages per year; S = climate sensitivity factor (degrees Celsius warming for a doubling of carbon-dioxide-equivalent); γ = damage function exponent; d_0 = damage from 2.5°C warming (in percentages of GDP); n.a. = not applicable.

a. All cases except cases 85 to 92 *include* the benefits from tax burden reduction. The shadow price of capital is 1 for cases 37 to 76, 3 for cases 77 to 84, and 1.56 for cases 85 to 92.

of only 1 percent is applied (e.g., on grounds that intergenerational comparison calls for a low discount rate), the benefit-cost ratio exceeds unity in five out of six of the cases considered, and reaches nearly 6:1 in the highest-damage variant.

Cases 77 to 84 in table 7.4 examine the influence of the shadow price of capital. These cases are identical to their counterparts in the initial section of table 7.3 (using the central social rate of time preference, i = 1.5 percent), except that they arbitrarily assign a higher shadow price of capital (p_k* = 3.0 rather than 1.56). The higher shadow price does not markedly alter the benefit-cost ratios. Thus, for the central case (case 1, table 7.3) the shift to the higher shadow price reduces the benefit-cost ratio only from 0.74 to 0.66 (case 77). In the high-damage range, the change is also modest (from 4.18, case 11, to 3.76, case 83).

Cases 85 to 92 report the influence of omitting the reduction in excess tax burden from the benefits side of the analysis, for example on grounds that governments might independently shift to optimal tax structures with no help from carbon tax revenue. Otherwise the assumptions are as in the central cases (with the social rate of time preference at 1.5 percent and the shadow price of capital according to equation 7.19 above). It is evident that the influence of the tax revenue effect is minor. Thus, in the central case its omission lowers the benefit-cost ratio only from 0.74 to 0.68 (case 1 versus case 85), and in the highest-damage case from 4.18 to 4.12 (case 11 versus case 91).

Overall, the tests in table 7.4 show that the benefit-cost ratio is highly sensitive to the social rate of time preference, but that the desirability of abatement holds up more robustly than expected under high discount rates if high-damage variants are considered. The tests also show that the shadow price of capital and the inclusion of excess tax burden effects have only minor influences on the results. In short, the sensitivity tests broadly confirm the findings of the central analysis.

Catastrophe Analysis

The main analysis of this study presumes steadily rising greenhouse damage from modest levels by the first half of the 21st century, with varying degrees of severity over the very long term. These scenarios might be described as "moderate progressive decay," and they are superimposed on a baseline of rising per capita income so that the end result is to leave future incomes higher than today but lower than attainable in the absence of global warming damage. Evaluation with relatively low (but theoretically justifiable) discount rates tends to find

terpolation were used for damage in the early years. The phenomenon becomes apparent only when high discount rates are used, because they tend to make all but the early years of little importance.

favorable benefit-cost ratios for abatement action; calculation with higher rates (5 percent to 10 percent) tends to find that the costs of action outweigh the benefits except in the high-damage cases.

It is useful to consider an alternative approach that applies a relatively conventional discount rate of, say, 5 percent and then asks the following question: how high would the probability of catastrophe have to be to make the expected (i.e., weighted) benefit-cost ratio greater than unity? As discussed in chapter 1, possible catastrophic effects could include the release of vast reservoirs of carbon-equivalent trapped in methane clathrates, the disintegration of the West Antarctic ice shelf under very-long-term warming, and fundamental shifts in ocean currents.

Let us postulate a catastrophe that is large and relatively sudden. Suppose that it is sufficient to reduce per capita incomes worldwide by 20 percent from present levels and freeze them at that plateau thereafter. Suppose that the catastrophe arrives in the year 2050.

Now return to the risk-weighted analysis considered above, but with reformulations to discount at a higher rate but also incorporate the chance of a catastrophe. First, we apply a discount rate of 5 percent. Second, we calculate the risk-weighted benefit-cost ratio as above. Third, we set a probability p for the catastrophe occurring and $(1 - p)$ for the (risk-weighted) principal scenarios. We may then examine how high the probability of the catastrophe would have to be to raise the overall benefit-cost ratio for aggressive abatement action to above unity.

Cases 29, 25, and 34 in table 7.3 represent the low, central, and high-damage cases with 5 percent discounting. Using the same technique as discussed above, the risk-weighted benefit-cost ratio for these three cases is 0.61. The catastrophe scenario would have to have a probability of 7.8 percent to raise this ratio to unity.[22] The implication of this exercise is that, if a discount rate of 5 percent is applied, it requires only a relatively modest probability of the catastrophe scenario to boost the (risk-weighted) benefit-cost ratio calculated from the principal scenarios to the threshold of unity required for social profitability of an aggressive abatement program.

This conclusion is sharply reinforced if it is kept in mind that there should be an interdependency between the discount rate and the probability of catastrophe. If future incomes are lower than today's because of catastrophe, the social rate of time preference should be lower and in fact, under the analysis of

22. The present value of the cost of abatement over 285 years, discounting at 5 percent, is \$18.3 trillion. The risk-weighted present value of benefits of the three noncatastrophe scenarios is \$11.3 trillion. The present value of benefits of abatement in the case of the catastrophe scenario is \$101 trillion. Thus, $(0.078 \times 101) + (0.922 \times 11.3) = 18.3$.

chapter 6, negative.[23] A second version of the catastrophe analysis may thus appropriately apply a weighted average in which the 5 percent discount rate holds for the three noncatastrophe scenarios, but the catastrophe evaluation applies a discount rate of zero. If this approach is adopted, then the probability of catastrophe needs to be only one-hundredth of one percent to raise the weighted-average benefit-cost ratio to unity.[24]

My own expectation is that, even in the very long term, global warming is unlikely to be so catastrophic that it would reduce world income sharply below today's levels. Thus, my preference is for a methodological approach that applies the central value of the social rate of time preference developed here (1.5 percent), in which case the risk-weighted analysis finds aggressive abatement action socially attractive even without considering the possibility of catastrophe. Others may prefer to impose a higher discount rate such as the 5 percent just examined.[25] The principal point is that, even under that approach, abatement remains attractive if just a small probability is attached to catastrophic consequences.

A Malthusian Scenario

Approximately two hundred years ago the English economist Thomas Malthus predicted that population growth would surpass that of production (the first being geometric and the second arithmetic, he argued). The result would be chronic poverty among the lower classes unless population growth was checked by "moral restraint" or war. In the event, technological change has permitted production to outstrip population. However, as the discussion of sustainable development in chapter 8 suggests, concern remains that the global economy could stagnate over the long term and fail to increase standards of living above those of today for much of the world's population. If that occurred, the con-

23. With $i = g\theta$, where i is the social rate of time preference, g is the rate of per capita consumption growth, and θ is the elasticity of marginal utility of consumption (absolute value), an initial value of $i = 5$ percent where g is only 1 percent (the baseline consumption growth of the analysis) implies that the agency or analyst believes the elasticity of marginal utility is a high value, 5. In the catastrophe case, $g = -0.3$ percent for 1991–2050 and zero thereafter. Consistency would then require discounting at -1.5 percent, at least over the first 60 years.

24. With $i = 0$, the present discounted value of benefits under catastrophe is $72.5 quadrillion. Following note 22, we have: $(72{,}532 \times 0.000097) + (11.3 \times 0.999903) = 18.3$.

25. Indeed, my own expectation is that whereas the probability of the catastrophe specified above is likely to be less than the threshold 7.8 percent indicated in the calculation, this probability could easily exceed the level of 0.01 percent required if the catastrophe analysis is conducted in the theoretically consistent way.

sequence would be a slow-motion economic catastrophe comparable in impact to the more abrupt natural catastrophe postulated and examined above. It is useful to consider the implications of a Malthusian scenario for the benefit-cost analysis of greenhouse policy.

The central features of a Malthusian world are rapid population growth, minimal technological change, and constant rather than rising per capita income. Population projections are thus the place to begin in formulating such an outcome. As noted above, in the basic projections here, world population stabilizes by the year 2100 at 10.5 billion. One recent study suggests, however, that under unfavorable assumptions world population could reach 28 billion by 2150 (Harrison 1992). An alternative, somewhat less extreme formulation, is to reconsider the population growth rates of the projections here and simply assume that in developing countries these rates do not fall to zero.

In table 7.1, population growth in developing countries is shown to decline from 1.6 percent annually at present to 1.4 percent, 0.8 percent, 0.3 percent, and 0.2 percent in the successive 25-year periods of the next century, and to zero thereafter. If instead it is assumed that the rate remains constant at 0.8 percent annually from the middle of the 21st century on, world population reaches 15.6 billion by 2100, 22.6 billion by 2150, 33 billion by 2200, and 48 billion by 2250. The 2150 level is actually smaller than the 28 billion suggested in the study just cited.

Because there is no growth in production per capita, the economy expands in size strictly in proportion to population. With no technical change, the ratio of carbon emissions to population (and thus to output) remains constant. Today the global average of carbon emissions amounts to approximately 1 ton per person per year. On this basis, the scenario calls for carbon emissions to rise to 15.6 GtC by 2100, considerably below the baseline amount of 20.6 GtC. However, eventually emissions nearly catch up with those in the baseline. Thus, by 2250 the Malthusian case shows emissions at 48.8 GtC, versus 50.9 GtC in the main projections. The absence of technical change and the much higher population growth eventually compensate for the absence of growth in output per capita in generating baseline carbon emissions.

With zero growth in per capita income, the social rate of time preference becomes zero. There is no longer any reason to give a higher weight to consumption today than to consumption in the future, because people in the future will no longer be richer than today. However, the shadow price of capital rises, because the discount rate at which capital earnings over the life of equipment is discounted falls to zero. With the social rate of time preference at zero and the rate of return on capital at 8 percent, the shadow price of capital rises to 1.75 (equation 6.1, chapter 6).

When the benefit-cost model is run applying these projections and parameters, it turns out that the discounted present value of benefits of an aggressive action program exceeds the discounted present value of the costs even for the central case (where $\Lambda = 2.5°C$, $d_0 = 1$ percent of GDP, and $\gamma = 1.3$)—that is, the analogue of case 1 in table 7.3 but with the social rate of time preference i = zero. The central Malthusian case shows the benefit-cost ratio at $q = 1.09{:}1$. If the higher benchmark damage alternative is used ($d_0 = 2$ percent), the benefit-cost ratio rises to 2.07. If in addition warming is set at the upper bound ($\Lambda = 4.5°C$), the benefit-cost ratio rises to 4.28:1.

The simulations for the Malthusian scenario thus reinforce the main findings above that aggressive action to limit global warming seems warranted. The effect of somewhat lesser and later warming, as the consequence of a slower rise in carbon emissions from slower per capita income growth, is more than offset by the fact that the proper rate to use for the social rate of time preference is zero under these circumstances. Although this scenario is not the principal case of this study, it has considerable importance because it represents what is perhaps a major theme of public concern: that future generations will be no better off than the present, and therefore that burdening them with greenhouse damage would be inequitable.

Comparison With the Nordhaus Results

Annex 7A to this chapter examines the optimization models developed by William D. Nordhaus for analysis of greenhouse policy. The central findings of Nordhaus's earlier, best-known greenhouse model were the following (1991a, 936):

> For the low damage function—which includes only identified costs and uses a middle discount rate [1 percent pure rate of time preference]—we estimate the marginal damage of greenhouse gases to be about $1.83 per ton of C [carbon] in CO_2 equivalent, which suggests very little CO_2 abatement. For the medium damage function, which assumes damage from greenhouse warming of 1% of GNP, the cost is reckoned at $7.33 per ton carbon; in this case, the efficient reduction is 11% of total GHG [greenhouse gas] emissions. In this case, CFCs are substantially reduced, and CO_2 emissions are reduced by about 2%. In the high damage case, with damages taken to be 2% of total output and with no discounting, GHG emissions are reduced by about one-third.

The main thrust of the Nordhaus analysis is that little action needs to be taken on greenhouse abatement unless one is pessimistic about damages. In contrast, the analysis of this study suggests that an aggressive course of abatement is warranted, at least with risk aversion. A major difference between the two studies is that Nordhaus focuses on the impact of a doubling of carbon-dioxide-equivalent rather than very-long-term warming.

A central problem with the Nordhaus analysis is that it assumes a future in which resources are in a "steady state"; there is no population growth and no rise in per capita income except from technical change. By implication, greenhouse gas emissions are also in a steady state. Yet as the projections of chapter 2 indicate, emissions could rise some tenfold over the very long term. By the year 2100, a 2 percent cut in carbon dioxide emissions (the central Nordhaus result above) would still leave them at more than three times their present level.

As discussed in annex 7A, Nordhaus has subsequently stated that the steady-state formulation was inadequate, and has constructed a dynamic model (the DICE model) that allows for rising population, per capita income, and baseline emissions. The revised analysis considers a 400-year horizon and allows for nonlinear damage from warming, both key improvements compatible with the approach of the present study. Nonetheless, the initial results of the DICE model continue to find relatively modest optimal abatement, with cutbacks from baseline on the order of 10 percent initially and 15 percent by 2100, and carbon taxes of only $5 to $20 per ton. As discussed in annex 7A, however, it seems likely that, with arguably appropriate parameter modifications and explicit attention to upper-bound warming and risk aversion, the DICE model would yield optimal abatement much closer to the aggressive policy examined here.[26]

In short, the new Nordhaus results may be less indelibly contrary to stiff abatement measures than were the earlier results. In at least one regard, even the previous estimates were similar to those here. As discussed in chapter 3, Nordhaus arrived at direct estimates of damage from global warming equal to ¼ percent of GDP for benchmark carbon-dioxide-equivalent doubling, and he suggested 1 percent of GDP as a central damage estimate and 2 percent for high damage. The estimates developed in the present study also amount to 1 percent to 2 percent of GDP for benchmark doubling.

However, there is an important distinction. In the analysis of this study, warming at the upper-bound climate sensitivity parameter (Λ = 4.5°C) is considered as well as warming at the central parameter (Λ = 2.5°C). With the upper-bound variant, the high-damage estimate at benchmark doubling in the present study amounts to 4.3 percent of GDP (= 2 percent × $[4.5/2.5]^{1.3}$). In this sense, then, the high-damage case in this study indicates about twice as much damage as the high case in the Nordhaus analysis, even without considering very-long-term warming.

When in addition the impact of very-long-term warming is taken into account, the differences between the initial Nordhaus results and those here become evident. At 10°C, the central estimate of damage is 6 percent of GDP, and the higher-damage cases amount to some 20 percent of GDP (the average for cases

26. One such modification would be removal of the 3 percentage points Nordhaus incorporates into the discount rate for pure (myopic) time preference.

9 and 10, table 7.3) or even higher (case 11). With a linear damage function, the early, steady-state Nordhaus results did not reach damages anywhere near this high.

Nordhaus's analysis does include a sophisticated treatment of other trace gases, and he usefully examines the rank ordering of cost-effectiveness in reducing greenhouse gas emissions through cutting CFCs, planting trees, and reducing fossil fuel emissions. As discussed in chapter 5, the analysis of the present study finds forestry measures relatively more attractive than does Nordhaus. His findings on the high cost-effectiveness of reducing CFCs would have made eminent sense had it not been for the subsequent scientific findings suggesting that their warming effect has been relatively neutral because of their ozone-stripping impact (chapter 1), such that focusing attention on CFCs does not go far toward reducing global warming.

In sum, for several reasons, but especially because of the inclusion of more dramatic effects associated with nonlinear damage and very-long-term warming, the policy conclusion in this study differs from that found in the Nordhaus steady-state analysis. The results here indicate that a program holding global carbon emissions to 4 GtC per year—which would amount to a 71 percent reduction from baseline by 2050, an 82 percent reduction by 2100, and a 90 percent reduction by 2200—is warranted under risk aversion. The broad finding of aggressive action contrasts with Nordhaus's conclusion that only very limited action is optimal. It remains unclear whether appropriate reformulation of Nordhaus's more recent dynamic model (the DICE model) would reach a conclusion closer to that of the present study.

Insurance and the Value of Information

The approach of this chapter (and the more recent Nordhaus analysis) is to treat the damages from global warming as a continuous (geometric) function and to consider whether these damages are likely to warrant aggressive abatement measures.[27] Manne and Richels (1990b) offer an alternative approach that attempts to circumvent a lack of estimates on the damages (benefits) side by simplifying the policy choice to one between inaction after scientific progress shows the greenhouse effect is not dangerous, and intense action if new research discovers that greenhouse damage would be catastrophic.

Annex 7B outlines the conceptual framework and principal results of the Manne-Richels insurance analysis. Essentially, it says that policymakers will be

27. Nordhaus seeks to identify the optimal level of intervention; the analysis here specifies a particular, aggressive action scenario and seeks to determine whether it has a favorable benefit-cost ratio.

better off to take only moderate action during an initial period of three decades, and thereafter—when more scientific information is available—to take massive steps if necessary, or otherwise to phase out abatement efforts altogether. The authors suggest that this strategy could save the United States a total of $90 billion (present discounted value), compared with an approach that attempts to pursue abatement from the start.

As noted in annex 7B, because the probability of near-catastrophic consequences may be higher than suggested by Manne and Richels and for other reasons—the same model might more appropriately place the value of "waiting and learning" at, say, one-third of the $90 billion total indicated by the authors. But even the full $90 billion is a small gain if considered against the possible damages.

The wait-and-learn strategy assumes that in the second, catch-up phase, strong measures can fully deflect the path of trace gas buildup back to one that converges at the permissible ceiling. However, it would seem more realistic to conclude that a "go slow" initial phase would inevitably leave the total eventual buildup and warming higher than under a policy of more aggressive action undertaken from the start. One reason is the irreversibility of the atmospheric accumulation. Another is the political implausibility of a sudden intensification of international effort as opposed to a steady but gradual tightening of measures.

The 30-year waiting period suggested by Manne and Richels would by itself be enough for a global "commitment" to the doubling of carbon-dioxide-equivalent (chapter 1). The implication is that the total warming commitment expected over a 300-year horizon under the program of aggressive action described in this chapter would already be fully exhausted after the first 30 years (or at least nearly so, depending on just how minimal the Manne-Richels measures are during this first period). As complete avoidance of further buildup after the first three decades would seem highly unlikely, the net result would seem capable of boosting global warming by perhaps at least 1°C, compared with the alternative of earlier forceful action.

If ultimate warming rises from 2.5°C to 3.5°C, the damage function of this chapter would indicate additional damage of about ½ percent of GDP.[28] Half of 1 percent of US GDP is approximately $30 billion even at today's economic scale. The implication is that the entire present discounted value of the gains from "waiting" in the Manne-Richels analysis ($90 billion according to the authors, $30 billion if scaled down as suggested here) could be offset by just one to three years' worth of damages from the likely net increase in warming.

There is nonetheless an important element of truth in the Manne-Richels insurance analysis. It correctly implies that, during at least some initial period,

28. With d_0 at 1 percent for S = 2.5°C and γ = 1.3 (table 7.2), damage reaches 1.5 percent of GDP instead of 1 percent ($1.0[3.5/2.5]^{1.3}$ = 1.5).

the phase-in of measures might appropriately be gradual and that considerable effort should be dedicated to research. An initial phase-in period of 10 years would seem considerably safer than 30 years, however.

These issues are explored further in the discussion of international policy in chapter 8 below. However, in terms of the analysis provided in the present chapter, it may be noted that the initial stage of the response relies primarily on low-cost initiatives (reduced deforestation, afforestation, and a move toward state-of-the-art technology). This initial modest level of costs helps resolve the tension between the need to begin a relatively aggressive program of action sooner rather than later, on the one hand, and the high return to further scientific verification, on the other.

Overview

In sum, benefit-cost analysis suggests that the benefits of an aggressive program of abatement warrant the costs of reducing greenhouse gas emissions if policymakers are risk averse, or if one is pessimistic and concentrates attention on high-damage cases (especially even a slim chance of an economic catastrophe).

Some will argue that the issue cannot be decided in this way, and in particular that it is the unquantifiable ecological effects that should dominate the analysis rather than the measurable impacts considered here. However, that argument reinforces the result here rather than reversing it, because the analysis concludes that aggressive abatement action is justifiable on economic grounds alone and thus would be all the more warranted if further ecological considerations were added.

These findings contradict the principal conclusion of the only other benefit-cost analysis of global warming to date. Nordhaus (1991a, 936) finds instead that the optimal amount of cutback in greenhouse gases amounts to only 11 percent; that most of this reduction should be in CFCs; that carbon emissions should be reduced by only 2 percent; and that the optimal carbon tax is only $7.33 per ton. The principal reasons for the divergence between the two results are that the estimates here consider much greater warming for the very long term and associated nonlinear damage, whereas the Nordhaus analysis focuses on carbon-dioxide-equivalent doubling; and that the central discount rate applied here is lower.

Annex 7A

The Nordhaus Optimal Emissions Models

Nordhaus (1991a, 1990b) has published results of an optimization model that has received widespread attention among economists and policymakers. This model, which for purposes of clarity we may refer to as NORD1, found that only modest reductions in carbon emissions from baseline were economically optimal. In subsequent unpublished work, Nordhaus (1992a; 1992b) has acknowledged that NORD1:

> had a number of shortcomings, but one of the most significant from an analytical point of view was the inadequate treatment of the dynamics of the economy and the climate. The earlier work examined a "resource steady state". . . . The previous approach is unsatisfactory primarily because of the extraordinarily long time lags involved . . . [such that] dynamics are of the essence. . . . (Nordhaus 1992a, 2–3)

Because of the great policy influence of NORD1, especially in the United States, it seems useful to review more specifically its results and problems. This annex first examines NORD1 and then considers the results of the more recent Nordhaus model, DICE (for Dynamic Integrated Climate-Economy model).

The NORD1 Model

Table 7A.1 reports the equations of the earlier Nordhaus model. T is cumulative greenhouse warming; subscript t indicates the year; M is the atmospheric stock of anthropogenic greenhouse gases (carbon-equivalent); and the overdot indicates a change in the variable in question. The first equation states that the change in temperature is a function of the difference between equilibrium warming for the given stock of M (as indicated by function g) and warming to date. The second equation indicates that the change in M is determined by the fraction β of annual emissions E retained in the atmosphere, less a decay rate δ for absorption into the deep ocean. The third equation linearizes the first.

Equation 7A.4 states that per capita consumption c depends on a steady-state level y^* for constant resources, as augmented by technological change at the rate h (e in the equation denotes the base of the natural logarithm). In addition, consumption varies positively with the steady-state level of emissions, E^*, and negatively with the amount of warming in the steady state, T^*. In equation 7A.5, the objective function to be maximized, V, is the discounted sum of utility of consumption, discounting at the pure time preference rate ρ.

The model then posits a one-time burst of emissions, ΔE, at the initial year. The resulting change in temperature in any given future year is shown in equa-

Table 7A.1 Nordhaus model equations

(1) $\dot{T}_t = \alpha\{g(M_t) - T_t\}$

(2) $\dot{M}_t = \beta E_t - \delta M_t$

(3) $\dot{T}_t = \alpha\{\mu M_t - T_t\}$

(4) $c_t = y^* e^{ht}\{g(E^*) - \phi(T^*)\}$

(5) $V = \int_0^\infty u(c_t)e^{-\rho t}\,dt$

(6) $\Delta T_t = \dfrac{\Delta E \mu \beta \alpha [e^{-\delta t} - e^{-\alpha t}]}{(\alpha - \delta)}$

(7) $\Delta c_t = y^* g'(E^*)\,\Delta E,\ t = 0$

$= -y^* e^{ht} \phi'(T^*)\,\Delta T_t,\ t > 0$

$\rightarrow y^* g'(E^*)\,\Delta E = \int_0^\infty y^* e^{ht} \phi'(T^*)\,\Delta T_t e^{-rt}\,dt$

(8) $g'(E^*) = \mu\beta\phi'(T^*)\dfrac{\alpha\left[\dfrac{1}{r+\delta-h} - \dfrac{1}{r+\alpha-h}\right]}{(\alpha - \delta)}$

(9) $g'(E^*) = \mu\beta\phi'(T^*)\,\Gamma$

Source: William D. Nordhaus. 1991. "To Slow or Not to Slow: The Economics of the Greenhouse Effect." *The Economic Journal* 101, no. 6: 920–37. Reprinted with permission.

tion 7A.6. The first line of equation 7A.7 reports the gain in consumption in the initial year from the extra emissions. The second line reports the loss in consumption in any future year as a result of the corresponding additional warming. The final line of equation 7A.7 sets the initial consumption gain equal to the present discounted value of future consumption losses to identify the optimal size of the one-shot burst of extra emissions. This optimal level occurs where the marginal consumption gain equals the present discounted value of the future marginal consumption losses, as shown in the left- and right-hand sides of equation 7A.8, respectively. Equation 7A.9 is a restatement that consolidates discount and decay terms into a single multiplier Γ.

As Nordhaus has now recognized, the fundamental problem with NORD1 is that it attempts to use a steady-state or comparative static framework to investigate a problem that is inherently dynamic. In the simplest of terms, the baseline for emissions identified in chapter 2 of this study finds a rise from about 6 GtC annually today to approximately 50 GtC by late in the 23rd century—hardly a static, steady-state world.

In policy terms, the static-versus-dynamic problem may be seen as follows. NORD1 concludes under central parameters that greenhouse gases should be cut back by 11 percent, primarily from CFC reduction, and carbon dioxide emissions should be reduced by only 2 percent (through a carbon tax of only \$7.33 per ton). The implication is that these cuts are from baseline. But in a real, dynamic world, the baseline is rising rather than steady-state. By the year 2100, for example, a 2 percent cut in carbon emissions would still leave them at about 19½ GtC annually, or more than three times today's level rather than 2 percent below it.

There are other problems with NORD1. It is evident in equation 7A.7 that the model has an asymmetry between its treatment of the gains from extra emissions and the costs from extra warming. The gains are treated as occurring instantaneously and are therefore not discounted. The damages, on the other hand, are spread over time and discounted. In reality, the forgoing of emissions would also occur over time, so that losses from a restraint program (or gains from extra emissions) should be phased over time and discounted. The asymmetry biases the results toward higher optimal emissions.

A crucial problem with NORD1 is that it sets a linear damage function, Φ. This function is based on 1 percent of world GDP loss from 3°C warming (benchmark carbon-dioxide-equivalent doubling), or \$200 billion. Each degree Celsius of warming imposes damage of \$(200/3) billion. This formulation misses the more-than-linear damage to be expected with warming, and thus the much greater damage for warming beyond 3°C.

A closely related problem is that NORD1 is specified for benchmark carbon-dioxide-equivalent doubling. It thereby misses the damages associated with much higher very-long-term warming, as emphasized in this study.

The DICE Model

Nordhaus's new, dynamic model is a major improvement over NORD1. The DICE model allows for growing population and per capita income over time. It maximizes the discounted sum of utility of per capita consumption multiplied by population, with a logarithmic utility function. Production in the economy is a Cobb-Douglas function of capital and labor (population), multiplied by a term for technological change. In addition, production at any given year is shrunk by a summary factor that takes account of the cost of emissions cutbacks as well as damage to the economy from global warming. This factor is:

$$\Omega_t = \frac{1 - .0686\,(\mu_t)^{2.887}}{[1 + .00144\,(T_t)^2]}$$

Here the term μ_t is the proportionate cutback in greenhouse gas emissions from baseline in the year in question, and T is the amount of global warming. Thus, both the abatement cost and the warming damage functions are nonlinear (the former nearly cubic, and the latter quadratic).

Importantly, the DICE model extends the time horizon to 400 years, a crucial change compatible with the much longer term horizon suggested in the present study. Nonetheless, in the results reported in the model version of Nordhaus (1992a, 19), the optimal policy is again found to be a relatively mild reduction in emissions from baseline (as in the static NORD1 model). Optimal emissions are only 10 percent below baseline in the near future, "rising to 15 percent late in the next century." These cutbacks are accomplished by a carbon tax of $5 per ton initially, rising to $20 per ton by 2100.[1]

It seems quite possible that reasonable alternative parameters and modest respecification of the DICE model could yield results much closer to those found here, for several reasons.

First, the results just cited do not take account of a probabilistic distribution of lower- and upper-bound warming (in contrast to the risk-averse treatment in the present study). DICE applies 3°C for benchmark carbon-dioxide-equivalent doubling and a damage of 1.33 percent of GDP for that level. This damage does not consider even the 2 percent high case applied in NORD1. Nor does it consider upper-bound warming (climate sensitivity Λ = 4.5°C). If the DICE damage function is applied to upper-bound warming, all damages are 2.25 times as high (= $[4.5/3.0]^2$). If Nordhaus's earlier high-damage case of 2 percent of GDP is considered in conjunction with upper-bound warming, all damages are 3.4 times as high (= $[2.0/1.33] \times 2.25$).

1. However, the study does not report cutbacks and carbon taxes for subsequent years, or more than two-thirds of the horizon.

Second, the specification of transient (i.e., actual time-dependent) warming as the consequence of specified atmospheric concentration of carbon-dioxide-equivalent may yield an understatement of warming. Thus, in the DICE baseline, global warming is only 3°C by the year 2100, whereas the IPCC (1990b, figure 6.11) places realized baseline warming by that year at 4.2°C (despite a lower climate sensitivity, $\Lambda = 2.5°C$). The DICE model's 50-year thermal lag may be overstated.[2]

Third, whereas the abatement cost function seems reasonable for cutbacks of 50 percent from baseline (about 1 percent of GDP), for the very high cutbacks that are eventually required for stiff emissions limits the function may overstate costs. Thus, at a 90 percent cutback, abatement costs amount to almost 7 percent of GDP. This level is far above the 2½ percent plateau identified in the present study. One reason is that the overall cost curve is linear in percentage cutback in the present study, and the degree of nonlinearity in the DICE formulation may be overstated. Another reason is that the DICE cost curve makes no allowance for declining cost over time with widening technological alternatives, in contrast to the treatment here.[3]

Fourth, the DICE model applies not only a logarithmic utility function, which is an implicit means of discounting future income by the growth rate of per capita income (as demonstrated in chapter 6 of this study), but in addition it applies a 3 percent discount rate for "pure" time preference (i.e., myopia, which means discounting even at unchanged per capita income). The net result is much higher discounting than in the methodology applied here. Curiously, whereas the earlier steady-state analysis contained sensitivity analysis with the pure rate of time preference equal to zero (Nordhaus 1991a, 927), the DICE results reported in Nordhaus (1992a) do not.

In sum, the reformulated Nordhaus analysis might be considerably more consistent with the results found here if the DICE model were further extended for risk aversion with attention to upper-bound warming; declining abatement costs with wider technological choice over time; shorter thermal lag; and elimination of the myopia component of discounting. Even as formulated, the model

2. Note further that DICE continues the ambiguity of NORD1 with respect to whether the atmospheric stock M is total greenhouse gas or merely the anthropogenic component. Its value is set at 677 GtC for 1965 (Nordhaus 1992b, 21), whereas anthropogenic carbon-equivalent in the atmosphere today is only about 150 GtC for carbon dioxide and, say, 200 GtC for total greenhouse gases. The atmospheric buildup equation (Nordhaus 1992b, 21) yields the curiosity that, if emissions cease immediately, total atmospheric stock falls to its preindustrial level within about 25 years. If emissions remain at zero, within a century the atmospheric stock of carbon dioxide is down to nearly half its natural level. As this instance suggests, the model's parameters would seem appropriately subject to revision.

3. Note further that the DICE model makes no allowance for "free" emissions reductions from a move to best practices, or for low-cost forestry measures.

contains some disturbing implications. As developed in chapter 2 of this study, baseline warming may be expected to be approximately linear in time under business as usual. Even with just 3°C warming by 2100, the DICE model thus implies warming of about 9°C by 2300, similar to the very-long-term warming of 10°C identified in the present study. The model's damage function would generate annual losses of 12 percent of GDP for this much warming (and an implied 27 percent of GDP for Λ = 4.5°C).

Overall, the DICE model would seem to hold promise for policy guidance, although parameter values in its initial application would seem to lead to an understatement of optimal abatement.

Annex 7B

The Manne-Richels Model of Greenhouse Insurance

Manne and Richels (1990b) have attempted an indirect measurement of the merits of policy action to avoid the greenhouse effect. Their approach incorporates in-depth estimates on the cost side of carbon restraints, based on their Global 2100 carbon-energy-economic model. However, on the benefits side, instead of a continuum of warming damages avoided, they apply a dichotomy between two outcomes: a catastrophe that must be avoided at all cost, on the one hand, or a mild annoyance that may safely be permitted to run its course, on the other.

Their initial example considers the merits of a "learn, then act" strategy. They examine a two-period horizon where fraction x of cutback is taken in the first and $1 - x$ in the second (if necessary). The initial probability of catastrophe is denoted p. By the end of the first period, science knows for sure whether the catastrophic threat is real. If it is, the rest of the action is undertaken; otherwise, no further action is required. With a cost function that rises with the square of action, expected cost is $C = 0.5x^2 + 0.5p(1 - x)^2$ and optimal $x = p/(1 + p)$.[1] Thus, if the chances of catastrophe are 50-50, the optimal hedge is to commit 33 percent (0.5/1.5) of the required reduction in the near term and 67 percent later, if necessary. With optimal $x = 0.33$, expected cost is $C = 0.167$. If the catastrophe were known for certain, it would be optimal simply to do half of the adjustment in the first period and half in the second ($x = 0.5$), in which case $C = 0.125$. The value of perfect information is thus 25 percent of total cost. The simple initial model thus provides two lessons. First, if there is any uncertainty at all, it pays to do less early and (contingently) more later. Second, there is value to information.

The authors then adopt a threefold world: greenhouse warming is harmless (case A); it is sufficiently serious to warrant a 20 percent cutback in carbon emissions from the 1990 level (case B); or it is so serious that the cutback should be 50 percent from that level (case C).

Manne and Richels allow 30 years to find out which state of the world is true. In 2020, policymakers either make up the shortfall of needed carbon dioxide cutbacks (case B or C is true) or eliminate restraints (case A is true). In case A, US emissions rise to nearly 3 billion tons of carbon annually by 2050; in case C, they are cut from their current 1.25 billion tons to 0.62 billion. In a

1. Cost is minimized at $dC/dx = 0$, or $x - p(1 - x) = 0$. Rearranging, $x + px = p$, $x(1 + p) = p$, and $x = p/(1 + p)$.

hedged "act first" approach, emissions rise only slightly through 2020, then diverge depending on whether case A, B, or C is true.

The authors calculate the value of improved information by considering how much cost could be saved through earlier knowledge of the true state (A, B, or C). If the probability of A is 60 percent, that of B 27 percent, and that of C 13 percent, in the extreme the savings from complete certainty by the first year in the forecast (2000) amount to $90 billion based on the Global 2100 model (discounting at 5 percent). This savings arises from being able to avoid the abrupt shifts from a hedged emissions path to the ideal one at the end of 30 years of research. Under either case B or case C, the long-term tax on carbon is $250 per ton, but there is a temporary bulge of taxes up to $500 per ton if case C is true and the strategy does not achieve cost-smoothing through improved information.

Manne and Richels then explore the value of information under alternative assumptions. One key influence is the probability of case A (that global warming is no problem). The value of perfect information peaks at $90 billion if this probability is 60 percent, but is only $20 billion if it is 5 percent and $30 billion if it is 95 percent. That is, if it is highly likely that something will have to be done about the greenhouse effect, there is less value in delaying for purposes of further confirmation; and if it is highly likely that the greenhouse effect is no problem, the initial hedged strategy without further information will not be far different from business as usual (so that a complete shift to business as usual would provide only limited gains). Hence the authors' third lesson: information is worth less if the probability is high that the greenhouse effect is either very severe or very mild, essentially because the proper action in each case is obvious and needs little further investment in information.

The value of information also depends on the likelihood that the new information is accurate (lesson four). If there are two chances in five of the research being incorrect, the value of information drops from a maximum of $90 billion to only $10 billion. Technology optimism also matters (lesson five). If future availability of low-cost nonrenewable energy (e.g., biomass) is twice the base-case assumption (10 quads annually rather than 5) and the autonomous energy efficiency increase twice as high (1 percent annually rather than 0.5 percent), the maximum value of perfect information falls from $90 billion to $45 billion. That is, if one expects it to be easy to dispense with fossil fuels, the cost of reducing carbon emissions will be low, so that savings from avoiding abrupt future cutbacks through improved knowledge will also be low. Conversely, if these two technology changes are pessimistically set at zero, perfect information value rises to nearly $150 billion (for p in the range of 60 percent).

The key policy conclusion of Manne and Richels (1990b, 25) is that "there can be a big payoff to reducing climate-related uncertainties—something of the order of $100 billion for the U.S. alone." However, this conclusion appears somewhat exaggerated even in terms of the model itself. Probabilities of only about 20 percent for case A and 40 percent each for cases B and C would seem to be a distribution more in keeping with the findings of the Intergovernmental Panel on Climate Change (1990a). This assessment sharply reduces the value of waiting for more information. Technology optimists would further cut these gains, as just discussed (note that the Global 2100 model does not include the initial range of zero-cost improvements suggested by the engineering estimates; see chapter 5 above). Furthermore, the model does not provide a persuasive bridge from the maximum value of perfect information available immediately to a plausible value related to how much we will know and by when.

More fundamentally, the structure of the exercise, although conceptually illuminating, seems inadequate to the task. A threefold classification is insufficient for evaluating the size of warming damage. The idea of waiting 30 years while additional research is carried out also seems questionable, especially if policymakers erroneously interpret the authors as saying there need be no action in the interim (instead, even their second-best strategy is the fixed hedge based on *a priori* probabilities, with considerably less emissions than under business as usual). In the absence of action, carbon-dioxide-equivalent doubling is likely to occur by 2025 (chapter 1).

Overall, the Manne-Richels insurance analysis is an important qualitative reminder that there is a value to obtaining further scientific information before embarking on draconian measures. However, its specific quantitative estimates may be misleading. In particular, the economic damages from delaying action too long could be much larger than the Manne-Richels waiting period of 30 years would imply, as suggested in the illustrative calculation in the main text of this chapter.

References

Bruntland, Gro Harlem, et al. 1987. *Our Common Future*. Oxford, England: Oxford University Press for World Commission on Environment and Development.

Harrison, Paul. 1992. *The Third Revolution*. London: I. B. Tauris.

Intergovernmental Panel on Climate Change. 1990a. *IPCC First Assessment Report, Volume I: Overview and Summaries*. New York: World Meteorological Organization and United Nations Environmental Programme (August).

Intergovernmental Panel on Climate Change. 1990b. *Scientific Assessment of Climate Change: Report Prepared for IPCC by Working Group I*. New York: World Meteorological Organization and United Nations Environment Programme (June).

Intergovernmental Panel on Climate Change. 1990d. *Formulation of Response Strategies: Report Prepared for IPCC by Working Group III*. New York: World Meteorological Organization and United Nations Environment Programme (June).

Jorgenson, Dale W., and Kun-Young Yun. 1990. "The Excess Burden of Taxation in the U.S." *Discussion Papers* 1528. Cambridge, MA: Harvard Institute of Economic Research (November).

Kravis, Irving B., Alan Heston, and Robert Summers. 1982. *World Product and Income: International Comparisons of Real Gross Product*. Baltimore: Johns Hopkins Press.

Manne, Alan S., and Richard G. Richels. 1990b. "Buying Greenhouse Insurance." Stanford, CA: Stanford University (mimeographed, November).

Manne, Alan S., and Richard G. Richels, 1990c. "Global CO_2 Emission Reductions: The Impacts of Rising Energy Costs." Stanford, CA: Stanford University (mimeographed; subsequently published in *The Energy Journal* 12, no. 1: 88–107, 1991).

National Academy of Sciences. 1991a. *Policy Implications of Greenhouse Warming*. Washington: National Academy Press.

National Academy of Sciences. 1991b. *Policy Implications of Greenhouse Warming: Report of the Adaptation Panel*. Washington: National Academy Press (prepublication manuscript, 23 August).

Nordhaus, William D. 1990a. "An Intertemporal General-Equilibrium Model of Economic Growth and Climate Change." In David O. Wood and Yoichi Kaya, eds., *Proceedings of the Workshop on Economic/Energy/Environmental Modeling for Climate Policy Analysis, October 22–23, 1990*, 415–33. Cambridge, MA: MIT Center for Energy Policy Research (January).

Nordhaus, William D. 1990b. "To Slow or Not to Slow: The Economics of the Greenhouse Effect." New Haven: Yale University (mimeographed, 5 February).

Nordhaus, William. 1991a. "To Slow or Not to Slow: The Economies of the Greenhouse Effect." *The Economic Journal* 101, no. 6: 920–37.

Nordhaus, William D. 1992a. "Rolling the 'DICE': An Optimal Transition Path for Controlling Greenhouse Gases." New Haven: Yale University (mimeographed, February).

Nordhaus, William D. 1992b. "The 'DICE' Model: Background and Structure of a Dynamic Integrated Climate Economy Model of the Economics of Global Warming." New Haven: Yale University (mimeographed, February).

Nordhaus, William D., and Gary W. Yohe. 1983. "Future Carbon Dioxide Emissions from Fossil Fuels." In National Research Council, *Changing Climate*, 87–153. Washington: National Academy Press.

Reilly, J. M., J. A. Edmonds, R. H. Gardner, and A. L. Brenkert. 1987. "Uncertainty Analysis of the IEA/ORAU CO_2 Emissions Model." *The Energy Journal* 8, no. 3: 1–29.

Repetto, Robert. 1989. *Wasting Assets: Natural Resources in National Income Accounts*. Washington: World Resources Institute.

Schumpeter, Joseph A. 1942/1976. *Capitalism, Socialism and Democracy*. New York: Harper.

Zachariah, K. C., and M. T. Vu. 1988. *World Population Projections, 1987–1988 Edition*. Baltimore: Johns Hopkins University Press for the World Bank.

[13]

Butterworth Heinemann

Energy Policy, Vol. 23, No. 1, pp. 17–34, 1995
Elsevier Science Ltd
Printed in Great Britain
0301–4215/95 $10.00 + 0.00

MERGE

A model for evaluating regional and global effects of GHG reduction policies

Alan Manne
Department of Operations Research, Stanford University, Stanford, CA 94305, USA

Robert Mendelsohn
School of Forestry and Environmental Studies, Yale University, New Haven, CT 06511, USA

Richard Richels
Electric Power Research Institute, 3412 Hillview Avenue, Palo Alto, CA 94303, USA

MERGE provides a framework for thinking about climate change management proposals. The model is designed to be sufficiently flexible to be used to explore alternative views on a wide range of contentious issues, eg costs, damages, valuation and discounting. We begin with a description of the model's individual components and show how they fit together. We then provide an initial application to illustrate how the framework can be used in the assessment of alternative policy options. Given the level of uncertainty which pervades the climate debate, it would be unrealistic to expect cost–benefit analysis to lead to consensus on a bottom line – at least any time soon. Rather, models such as MERGE should be viewed as research tools capable of providing insights into which aspects of the debate may be most important. In this way, they can help focus the discussion and identify the areas where additional research may have the highest pay-off.

Keywords: Climate change; Integrated assessment; Cost–benefit analysis

To date, much of the greenhouse debate has focused on the costs of emissions abatement. At one end of the spectrum, there are those who believe that substantial reductions can be achieved at negligible costs. All that is needed is to remove energy subsidies, and to dismantle the artificial barriers which limit investment in cost-effective conservation and renewables. If this view is correct, emissions abatement represents an inexpensive hedge against the possibility of unacceptable climate changes.

Toward the other end of the spectrum, there are those who believe that energy markets are already operating efficiently. This group is considerably less sanguine about the costs of greenhouse insurance. Its members find difficulties in believing that economically attractive alternatives currently exist, and are not automatically entering the marketplace. If this view is correct, emission abatement will be far from costless.

As with most policy issues, the answer is apt to lie somewhere between the polar points of view. There are certainly cost-effective opportunities for improving energy efficiency and reducing greenhouse gas emissions. These 'no regrets' options would be worthwhile in themselves. But the global abatement targets currently under discussion are likely to entail substantial reductions below a no regrets emissions path. Few believe that this can be accomplished at a negligible cost.

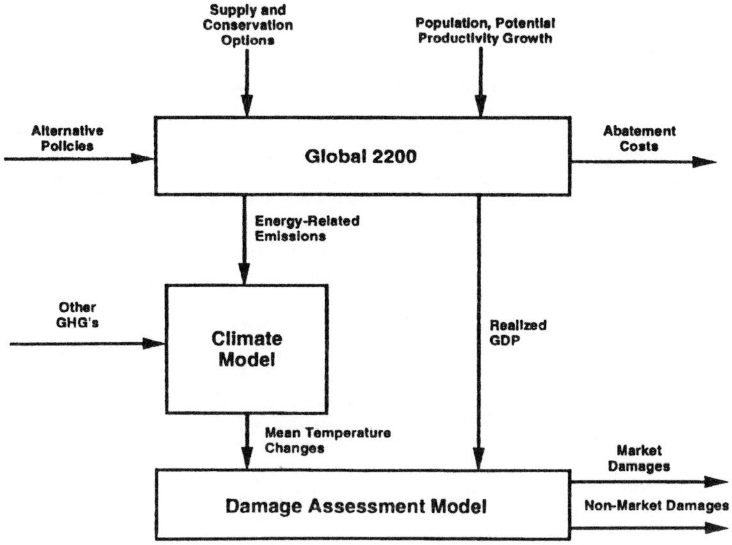

Figure 1 An overview of MERGE (model for evaluating regional and global effects of GHG reductions)

Once we recognize that there are limits to the possibilities for 'free lunch', we are forced to make difficult choices. Limited resources must be allocated among competing ends. It then becomes essential to analyze the trade-offs, and to ask what the proposed reductions will buy in terms of reducing the undesirable impacts of global climate change. There are many claimants for scarce resources – health, education and welfare expenditures. How much of today's limited budget is it worthwhile to spend on the purchase of greenhouse insurance?

Climate cost–benefit analysts face a daunting task. In recent years, considerable attention has been devoted to the economy-wide costs of emission abatement proposals. Although the range of uncertainty is narrowing, we have yet to arrive at a consensus. It is even more difficult to assess what such measures will buy in terms of reduced environmental damages.

Why is it so hard to assess the benefits? To begin with, there remain important gaps in our understanding of the science of global climate change. Increased concentrations of greenhouse gases are likely to lead to global warming. But by how much? Over what time frame? And what will be the impacts on different regions of the globe? Without a better understanding of these issues, it will be difficult to assess the effectiveness of various countermeasures.

There is also the difficult problem of valuation. Critics often complain that economists count what they can count, and not necessarily what counts. It is far easier to value damages to agriculture, energy, and coastal structures than to biodiversity, environmental quality and human health. Yet both categories are important and need to be incorporated into global climate cost–benefit analyses. Unfortunately, there is no widely accepted standard for translating global environmental damages into their dollar equivalents.

And there is the issue of discounting. The greenhouse debate is one of intergenerational trade-offs. Many proposals require substantial near-term costs. Benefits (in terms of reduced damages) do not begin to accrue for half a century. Unless one places a high weight on the welfare of future generations, it will be difficult to justify costly near-term actions. Among policy analysts, there is an enormous range of disagreement as to what constitutes an appropriate discount rate.

This paper will not attempt to resolve these issues: costs, benefits, valuation and discounting. Rather, our goal is to provide a sufficiently transparent integrating framework to explore the implications of alternative viewpoints. The framework is called MERGE (model for evaluating regional and global effects of GHG reduction policies). It consists of a series of linked modules representing the major processes of interest. These include: (1) the costs of reducing the emissions of radiatively important gases, (2) natural system disposition and reactions to the emissions of these gases, and (3) the reaction of human and natural systems to changes in the atmospheric/climate system.

We begin with a description of the model's individual components and show how they fit together. We then provide an initial application to illustrate how the framework can be used in the assessment of alternative policy options. Four alternative policies are compared with a business as usual scenario. In each case, we provide estimates of the costs of abatement and the resulting benefits.

This work builds upon the earlier contributions of a number of researchers. The pioneering effort in this area was done by Nordhaus (1991). Peck and Teisberg (1992) followed up the Nordhaus analysis by showing how the optimal global emissions path might change under alternative assumptions about economy-wide costs and the nature of the damage function. Falk and Mendelsohn (1993) developed a global model of carbon control which emphasizes the optimal set of carbon taxes over time. Cline (1992) has also conducted analyses at the global level, reporting cost–benefit ratios for a wide range of scenarios. Other researchers have developed regional models for exploring the pay-offs from various levels of international cooperation: see eg Eykmans *et al* (1992), Dowlatabati and Morgan (1993) and Hope *et al* (1993).

Global 2200

Figure 1 provides an overview of the principal components of MERGE and highlights the major linkages. There are three major submodels: (1) Global 2200; (2) the climate submodel; and (3) the damage assessment submodel. Each is described in turn.

Global 2200 is used to assess the economy-wide costs of alternative emission constraints at the regional and global level. It is an extension of the Global 2100 model of Manne and Richels (1992). Like its predecessor, Global 2200 divides the world into five major geopolitical regions: the USA, other OECD nations (Western Europe, Japan, Canada, Australia and New Zealand), FSU (the former USSR), China and the ROW (rest of world). Unlike Global 2100, Global 2200 is a fully integrated applied general equilibrium model. Each of the regions is viewed as an independent price taking agent, and is subject to an intertemporal budget constraint. At each point in time, supplies and demands are equilibrated through the prices of the internationally traded commodities: oil, gas, coal, carbon emission rights and a numeraire good. This numeraire represents a composite of all items produced outside the energy sector. All prices are expressed in terms of US dollars of constant 1990 purchasing power.

The model is benchmarked with energy and economic statistics for 1990. Global 2200 is based upon look ahead rather than recursive dynamics. This seems particularly important for the evolution of the prices of exhaustible resources such as oil, gas and coal – and for their eventual replacement by backstop technologies.

To facilitate computations, the model employs time intervals of unequal length. There are 10 year time steps from 1990 to 2050, and 25 year steps during the following century and a half. For purposes of current decision making, it is important to provide details on the near future but sufficient to take a more aggregate view of future events.

Intertemporal optimization

For each region, there is a single representative producer–consumer. Savings decisions are modeled by choosing each region's consumption sequence so as to maximize the sum of the discounted 'utility' of consumption. That is,

$$\operatorname*{Max} \sum_{t=1}^{T} U(c(t))(1+\rho)^{-t} \qquad (1)$$

where U is the single period level of utility or social well-being, $c(t)$ is the flow of consumption at time t, and ρ is the rate of time preference for utility.

For optimizing the pattern of investment and consumption over successive time periods, we take the special case where the utility function is the logarithm of consumption. That is

$$\operatorname*{Max} \sum_{t=1}^{T} \log c(t)(1+\rho)^{-t} \qquad (2)$$

This implies that marginal utility is always positive, but is a diminishing function of the aggregate level of consumption. There is a unitary elasticity of substitution between consumption in each time period.

It is important to distinguish between the rate of time preference for utility and the marginal productivity of capital. The former applies to the discounting of the utility of different generations, the latter to the discounting of goods and services. For the logarithmic form of the utility function, the following equation will hold along an optimal steady-state growth path:

$$r = g + \rho \qquad (3)$$

where r is the marginal productivity of capital and g is the annual growth rate. For a proof of this proposition, see Chakravarty (1969, p 65). Our approach has been to choose a value of ρ such that the marginal productivity of capital will remain close to its current level over the

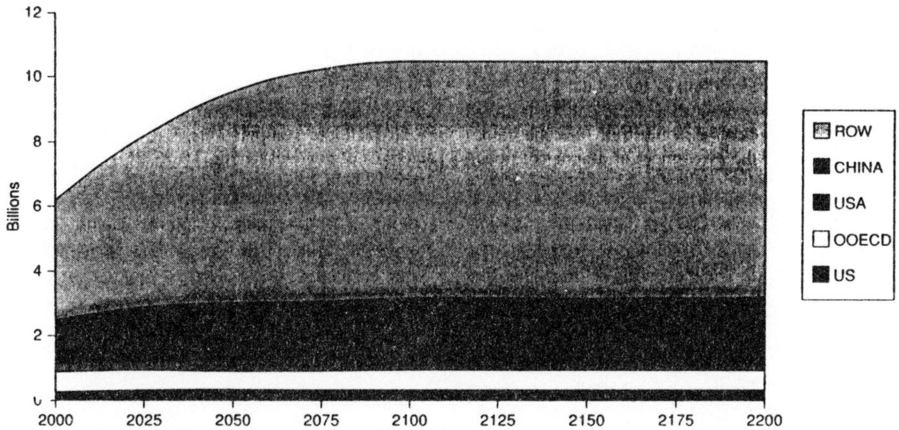

Figure 2 Population projections

entire planning horizon. For example, in order to simulate an economy in which the growth rate g = 2%, and r is 5%, we set ρ at 3%.

Note that a lower or a zero rate of utility time preference would not provide a good description of the collective outcome of individual choices. It would also imply an unrealistically rapid increase in the near-term rate of investment and capital formation (see eg the numerical results reported in Manne (1992)). In that report, see also the implications of choosing a more general isoelastic utility function rather than the logarithmic form.

The model is benchmarked, and the utility discount rates are chosen so that – when expressed in terms of the international numeraire good – the net real rates of return on capital are identical (5% per year) in all regions. At first glance, this simplification appears unrealistic. For practical purposes, however, this seemed preferable to incorporating the complexities required to define the rate of change of each region's real foreign exchange rate relative to the US dollar.

Production and consumption

In order to focus upon the long-run issues of energy–economy interactions, resource exhaustion and the introduction of new technologies, each region is described in highly aggregated terms. Outside the energy sector, all economic activity is represented in terms of dollars of real purchasing power. Within the energy sector, only two end products are distinguished: electric and non-electric energy.

Electric and non-electric energy are supplied by the energy sector to the rest of the economy. Like the material balance equations of an input-output model, aggregate economic output (Y) is allocated between interindustry payments for energy costs (EC) and 'final demand' for current consumption (C) and investment (I). Thus:

$$Y = C + I + EC \qquad (4)$$

For the economy-wide production function in each region, we assume that gross output (Y) depends upon four inputs: K, L, E, N – capital, labor, electric and non-electric energy. To minimize the number of parameters that require either calibration or econometric estimation, the long-run static production function is described by a nested non-linear form:

$$Y = [a(K^{\alpha}L^{1-\alpha})^{\gamma} + b(E^{\beta}N^{1-\beta})^{\gamma}]^{1/\gamma} \qquad (5)$$

where $\gamma = (\sigma - 1)/\sigma$ for $\sigma \neq (0,1,\infty)$.

Equation (5) is based on the following assumptions:

- there are constant returns to scale in terms of these four inputs;
- there is a unit elasticity of substitution between one pair of inputs – capital and labor – with α being the optimal value share of capital within the pair;
- there is a unit elasticity of substitution between the other pair of inputs – electric and non-electric energy – with β being the optimal value share of electricity within the pair;
- there is a constant elasticity of substitution between these two pairs of inputs – the constant being denoted by σ (also known as ESUB);
- the scaling factors a and b are determined so that the energy demands in the base year are consistent

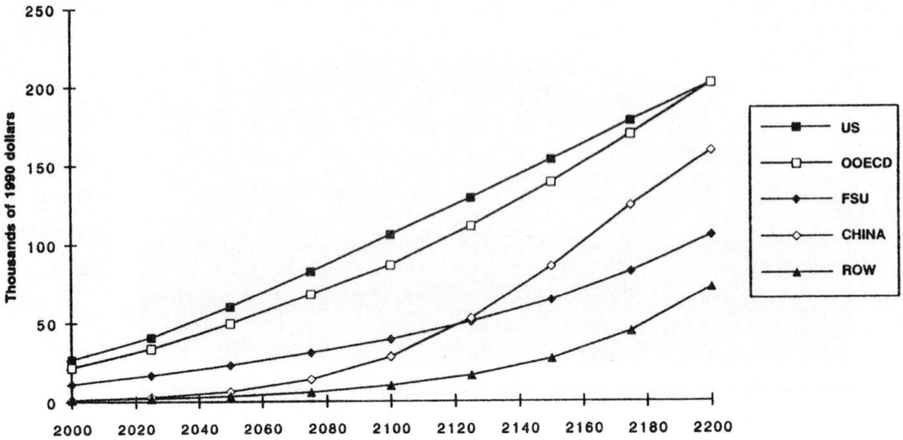

Figure 3 Per capita GDP

with the 'reference price' for non-electric energy; and

- there are autonomous energy-efficiency improvements (AEEI) that are summarized by the scaling factor *b*.

Key demand-side parameters

The rate of GDP growth is a key determinant of energy demands. The rate depends on both population and per capita productivity trends. Figure 2 shows our estimates of the projected population. For the 21st century, these are taken from Zachariah and Vu (1988). It is assumed that the population will stabilize in all regions during the 22nd century. At that time, 88% of the world's population will be located in the regions currently described as developing countries: China and the ROW.

Figure 3 presents our per capita GDP projections. These are termed 'potential' because they are based upon constant energy prices. Up to 2100, these rates represent the average of the higher and lower growth cases adopted by the IPCC's Working Group III (1990). From 2100 onward, it is assumed that per capita growth rates will decline over time, but that the gap between the industrialized countries and the ROW will continue to be substantial. If we follow the IPCC's projections through to 2100, it is difficult to believe that the ROW's growth rate will accelerate rapidly enough so that its per capita incomes will converge to those of the other regions between 2100 and 2200.

Because of energy–economy interactions, the potential GDP growth rates do not uniquely determine the realized rates. Energy costs represent just one of the claims on the economy's output. Tighter environmental standards and/or an increase in energy costs will reduce the net amount of output available for meeting current consumption and investment demands. The realized GDP will then fall short of the potential.

Energy consumption need not grow at the same rate as the GDP. Over the long run, they may be decoupled. In Global 2200, these conservation possibilities are summarized through two macroeconomic parameters: ESUB (the elasticity of price induced substitution) and AEEI (autonomous energy-efficiency improvements).

If there is sufficient time for the adaptation of capital stocks, most analysts would agree that there is a good deal of possible substitutability between the inputs of capital, labor and energy. The degree of substitutability will affect the economic losses from energy scarcities and from price increases. The ease or difficulty of these trade offs is summarized by the elasticity of price induced substitution (ESUB). The higher its value, the less expensive it is to decouple energy consumption from GDP growth during a period of rising energy prices.

When energy costs are a small fraction of total output, ESUB is approximately equal to the absolute value of the price elasticity of demand. In Global 2200, this parameter is measured at the point of secondary energy production: electricity at the busbar, crude oil and synthetic fuels at the refinery gate. On the basis of a 'backcasting' experiment for the USA, the reference case values for ESUB have been set at 0.40 both for the USA and for the other OECD nations. These countries have already demonstrated their ability to use the price mechanism for decoupling energy from GDP growth.

Table 1 Identification of electricity generation technologies

	Identification	Earliest possible introduction date[a]	Estimated US cost[b] (mills/kWh)
Existing technologies			
HYDRO	Hydroelectric, geothermal, and other renewables		
GAS-R	Remaining initial gas fired		
OIL-R	Remaining initial oil fired		
COAL-R	Remaining coal fired		
NUC-R	Remaining initial nuclear		
New technologies			
GAS-N	Advanced combined cycle, gas fired	1995	34.3[c]
COAL-N	New coal fired	1990	51.0[d]
ADV-HC	High cost carbon free	2010	75.0
ADV-LC	Low cost carbon free	2020	50.0

[a] Estimated year when the technology could provide 0.1 trillion kWh (approximately 20 GW of installed capacity at 60% capacity factor). For the other OECD region, we assume that the technology could provide 0.2 trillion kWh by this date.
[b] Based on 1990 dollars.
[c] Based on price of gas in 1990. Gas prices are projected to rise over time.
[d] Estimated costs are assumed to be 10 mills/kWh higher in the other OECD region due to higher fuel costs.

In other regions of the world, price induced conservation is more problematic. It depends on long-term structural changes in the political and economic system. We have therefore set ESUB at 0.30 for these other three regions.

Along with the reductions in energy demand induced by rising energy prices, there is also the impact of autonomous energy-efficiency improvements (AEEI). Non-price efficiency improvements may be brought about by deliberate changes in public policy eg speed limits for automobiles. Energy consumption may also decline as a result of shifts in the basic economic mix away from manufactured goods and toward more services. Thus, the AEEI summarizes all sources of reductions in the economy-wide energy intensity per unit of output.

For most regions and time periods, we have assumed that the AEEI will be 0.5% per year. That is, regardless of whether the GDP growth is 2.5% or 1.0% per year, total primary energy consumption will grow at a rate that is 0.5% lower than that of the GDP.

The numerical value of the AEEI is highly controversial. Clearly the cost of emissions abatement is sensitive to the value of this parameter. Suppose, for example, that one were quite optimistic and projected the AEEI at 1.5% per year. By the year 2100, the economy would then consume only 20% of the amount of energy required with an AEEI of zero. This is entirely apart from the conservation induced by rising energy prices in response to the depletion of conventional oil and gas resources.

In addition to the ESUB and AEEI, one other numerical parameter affects the decoupling between energy consumption and economic growth. The reference price of non-electric energy, PNREF, is employed in benchmarking the aggregate production function. Its numerical value is chosen so as to allow for differences in the extent to which domestic policies have insulated consumers from international energy price movements.

As a consequence, the economies of some regions may not have completed the long-term adjustment to the oil price shocks of the 1970s. If PNREF is less than the actual 1990 base year oil price of $3.70 per gigajoule ($22 per barrel), there will be some near-future price induced conservation in addition to that determined by the AEEI factor. The importance of this effect will depend upon what is assumed with respect to the persistence of tax and subsidy wedges between domestic and international prices.

In calculating these once and for all efficiency gains, the value of PNREF is taken to be $3.00 per gigajoule in the USA, $7.70 (including taxes of $4.00) in the other OECD region, $1.00 in the former USSR and $2.00 both in China and the ROW. For purposes of the simulations reported here, domestic taxes are maintained at the 1990 level within the other OECD region, but subsidies are eliminated at a rapid pace elsewhere.

Key supply-side parameters

Table 1 identifies the alternative sources of electricity supply. The first five technologies represent existing sources: hydroelectric and other renewables, gas, oil and coal fired units, and nuclear power plants. The second group of technologies includes the new electricity generation options that are likely to become available during the coming decades. They differ in terms of their projected costs, carbon emission rates, and dates of

Table 2 Non-electric energy supplies

Technology name	Description	Carbon emission coefficient (tons of carbon per GJ of crude oil equivalent)	Unit US cost per GJ of crude oil equivalent (1990 dollars)
OIL-MX	Oil imports or exports	0.0199	3.70 in 1990 rising over time
CLDU	Coal – direct uses	0.0241	2.00
OIL-LC	Oil – low cost	0.0199	2.50
GAS-LC	Natural gas – low cost	0.0137	2.75[a]
OIL-HC	Oil – high cost	0.0199	3.00
GAS-HC	Natural gas – high cost	0.0137	4.25[a]
RNEW	Renewables	0.0000	6.00
SYNF	Synthetic fuels	0.0400	8.33
NE-BAK	Nonelectric backstop	0.0000	16.67

[a] To allow for gas distribution costs, $1.25 per GJ is added to the wellhead price.

Source: The source of most of these carbon emission and cost coefficients is Energy Modeling Forum 12 (1993).

introduction. The introduction date for a new technology is defined as the earliest year in which the technology could provide 0.1 trillion kWh in each region (approximately 20 GW of installed capacity at a 60% capacity factor).

ADV-HC and ADV-LC, respectively, refer to advanced high- and low-cost carbon free electricity generation. Any of a number of technologies could be included in these categories: solar, wind, nuclear, biomass and others. Given the enormous disagreement as to which of these will ultimately win out in terms of economic attractiveness and public acceptability, we have chosen to refer to them generically.

Table 2 identifies the nine alternative sources of non-electric energy within Global 2200. In the absence of a carbon constraint or a limit on coal resources, the SYNF technology (coal or shale based synthetic liquid fuels) places an upper bound on the future cost of non-electric energy. Table 2 includes two broad categories of carbon free alternatives: RNEW (low-cost renewables such as ethanol from biomass) and NE-BAK (high-cost backstops such as hydrogen produced through photovoltaics and electrolysis). The key distinction is that RNEW is in limited supply, but NE-BAK is available in unlimited quantities at a constant but considerably higher cost.

Upper bounds are imposed upon the contribution of each of these supply technologies during each time period. There are also constraints that limit the rate of expansion from one period to the next. As a result of these constraints, there may be an erratic time path for the prices of alternative forms of energy. There may be a temporary phase during which prices overshoot their long-term backstop equilibrium levels.

Oil, gas and coal are viewed as exhaustible resources. Proved reserves are depleted by current production and augmented by new discoveries out of the remaining stock of undiscovered resources. At any one time, production is a fixed fraction of remaining reserves. New discoveries may not exceed a fixed fraction of the remaining undiscovered resources. Thus, Global 2200 combines some of the desirable economic attributes of Hotelling's model of depletable resources and the geological attributes of Hubbert's model. Under our standard assumptions, it turns out that conventional oil and gas resources are largely depleted by 2050, but that the world's coal supplies could provide for an additional century's worth of synthetic fuels.

Global 2200 is written in the GAMS language (see Brooke *et al*, 1988). In order to solve this computable general equilibrium model, we used the sequential joint maximization technique originated by Rutherford (1992). There are 2800 linear and non-linear constraints, and there are 3800 variables – many of them subject to individual upper bounds. On an IBM RS/6000-375 work station, the solution time is about half an hour per case.

Climate submodel

In MERGE, we focus on three of the most important anthropogenic greenhouse gases: carbon dioxide (CO_2), methane (CH_4) and nitrous oxide (N_2O). The concentrations of these gases in the earth's atmosphere have been increasing since the industrial revolution – primarily as a result of human activities. In this section, we describe the relationship between man-made emissions and atmospheric concentrations and the resulting impact on temperature.

The emissions of each gas are divided into two categories: energy and non-energy. Global 2200 projects energy related emissions of CO_2, CH_4 and N_2O by fuel type for each period up to 2200. Emissions from other sources are exogenous inputs into MERGE. Table 3 presents a summary of key assumptions.

Without the withdrawal of carbon from the atmosphere, observed atmospheric concentrations of CO_2 would have risen much more rapidly than indicated by

Table 3 Summary of assumptions regarding key greenhouse gases

Trace gas	CO_2	CH_4	N_2O
Atmospheric concentration			
Pre-industrial (ppmv)	280	0.8	0.288
1990 (ppmv)	353	1.72	0.310
Energy related emissions			
1990 (billion tons)	6.0	0.08	0.0001
Growth rate, post-1990		(determined endogenously)	
Non-energy related emissions			
1990 (billion tons)	0.2	0.454	0.0139
Growth rate, post-1990	0	0.8	0.2

Sources: IPCC (1990; 1992).

the quantity of carbon emissions. In recent history, approximately 50% of the industrial emissions of CO_2 appeared to have been absorbed. Most of this absorption is by the oceans but not all of the withdrawals can be explained in this way. We assume that prior to the industrial revolution, natural additions were offset exactly by natural removal. That is, the stock of carbon was in steady state.

Future atmospheric CO_2 concentrations are modeled using a reduced form carbon cycle model developed by Maier-Reimer and Hasselmann (1987). Carbon emissions are divided into five classes, each with different atmospheric lifetimes. Specifically, $G(t)$, the impulse response function to an instantaneous injection of CO_2 to the atmosphere, is expressed as a weighted sum of the exponentials

$$G(t) = \sum_{i=1}^{5} a_i \exp(-t/\tau_i) \tag{6}$$

where a_i are scaling factors; $\sum a_i = 1$; and τ_i are decay constants. The above linear response expression was fitted by least squares to the computed response of a full scale ocean carbon cycle model (Maier-Reimer and Hasselmann, 1982).

For CH_4 and N_2O, the atmospheric stock in year $t+1$ equals the fraction of the stock in year t remaining in the atmosphere plus new emissions. That is,

$$S_{G,t+1} = k_G * S_{G,t} + E_{G,t} \tag{7}$$

where $S_{G,t}$ is the stock of gas G in year t; k_G is the retention factor for gas G ($k_{CH_4} = 0.9$ and $k_{N_2O} = 0.9933$); and $E_{G,t}$ are the emissions of gas G in year t.

The next step is to assess the impact of future concentrations of greenhouse gases on the earth's radiative forcing balance, that is, the balance between the energy absorbed by the earth and that emitted by it in the form of infrared radiation.

According to the IPCC (1990), atmospheric concentrations of CO_2, CH_4 and N_2O have the following impacts on radiative forcing relative to their 1990 levels (respectively indicated by CO_{2_0}, CH_{4_0} and N_2O_0):

$$\Delta F_{CO_2} = 6.3 \ln (CO_2/CO_{2_0}) \tag{8}$$

$$\Delta F_{CH_4} = 0.036\,(CH_4^{0.5} - CH_{4_0}^{0.5}) - f(CH_4, N_2O_0) + f(CH_{4_0}, N_2O_0) \tag{9}$$

$$\Delta F_{N_2O} = 0.14\,(N_2O^{0.5} - N_2O_0^{0.5}) - f(CH_{4_0}, N_2O) + f(CH_{4_0}, N_2O_0) \tag{10}$$

where ΔF_G is the change in net flux (W/m^2) corresponding to a volumetric concentration change for gas G relative to the 1990 level and the CH_4–N_2O interaction term $f(CH_4, N_2O) = 0.47 \ln[1 + 2.01*10^{-5}*(CH_4*N_2O)^{0.75} + 5.31*10^{-15}*CH_4*(CH_4*N_2O)^{1.52}]$.

The aggregate effect is obtained by summing the radiative forcing effect of each gas. Aggregate radiative forcing is assumed to have the following impact on the change in global potential surface temperature, ΔPT:

$$\Delta PT = d * \Delta F \tag{11}$$

where the potential temperature is the long-run temperature which will occur if a specific level of forcing is maintained indefinitely; ΔPT is measured relative to 1990 temperature; d = 0.555 °C/W/m^2; and $\Delta F = \Delta F_{CO_2} + \Delta F_{CH_4} + \Delta F_{N_2O}$.

Since oceans take a long time to warm up, the actual temperature, AT, will lag behind the potential temperature. We model this lag process as follows:

$$\Delta AT_{t+1} - \Delta AT_t = c_1 * (\Delta PT_t - \Delta AT_t) \tag{12}$$

where $c_1 = 0.05$ represents a 20 year mean lag, and ΔAT_t measures the actual change in temperature in year t relative to 1990.

According to global circulation model analyses, temperatures will increase more rapidly as one moves toward the poles. We assume that the temperature

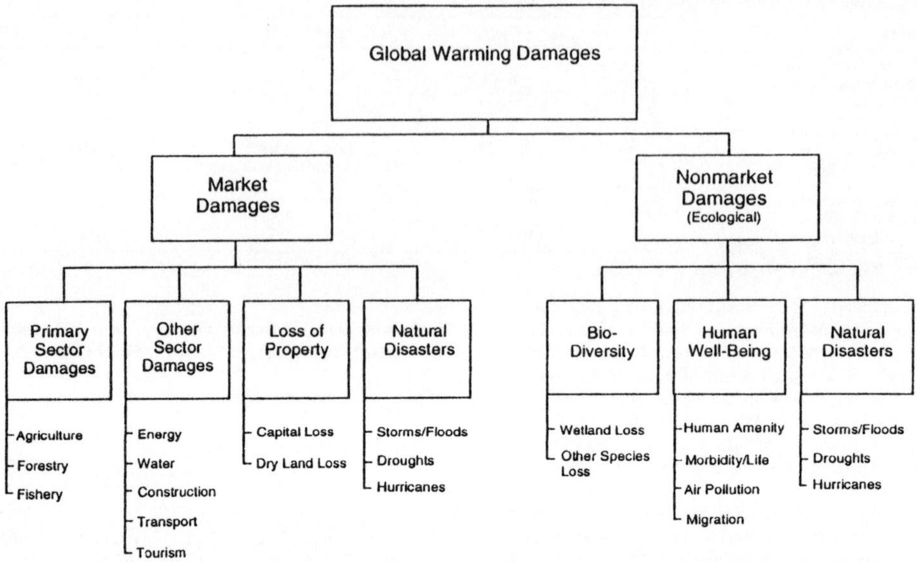

Figure 4 Overview on global warming impacts

change in temperate countries is the global mean and that the change in tropical countries is half the global mean.

Damage assessment submodel

In a critique of attempts to quantify the impacts of climate change, Grubb (1993) identifies the question that analysts need to address: 'What may be the impact of increasing greenhouse gas concentrations on the overall welfare of future generations globally, expressed in terms of present day monetary equivalence?' It is easy to pose this question in qualitative terms, but difficult to arrive at a consensus on the quantitative impacts.

Part of the problem is what to include in the evaluation. It is far easier to quantify effects on goods sold in markets such as food and energy than on services such as biodiversity, environmental quality and human health which have no market values. Both categories are important and need to be incorporated into global climate cost–benefit analyses.

Figure 4 (adapted from Fankhauser (1993)) presents a classification of climate change impact categories. Market effects reflect categories that are included in conventionally measured national income and can be valued using prices and observed demand and supply functions. Non-market effects have no observable prices, and so they must be valued using alternative revealed preference or attitudinal methods.

Consistent with Figure 4, we divide damages into two categories: market and non-market. $D_{t,n}$, market damages for period t in region n, are defined by the following relationship:

$$D_{t,n} = d_{1,n} * \Delta AT_{t,n}^{d_{2,n}} * GDP_{t,n} \quad (13)$$

Much of the discussion has focused on damages at a single point in time – when temperature increases by 2.5 °C above 1990 levels. It is also important to specify what happens before and after this date. There has been considerable speculation about the shape of the damage function (see, eg, Peck and Teisberg (1992)). For the present analysis, we adopt the assumption of Nordhaus (1992) that damages rise quadratically with temperature change. That is, in Equation (13), $d_{2,n} = 2$.

Next we turn to non-market damages. Few issues in the greenhouse debate are more challenging or have engendered more controversy. What value does society place on biodiversity? Environmental quality? Human health? Many cost–benefit calculations ignore the issue altogether and confine their analyses to economically measurable consequences only. As noted by Dorfman and Dorfman (1976), 'there is discretion in declining to attempt the nearly impossible, but there is also danger

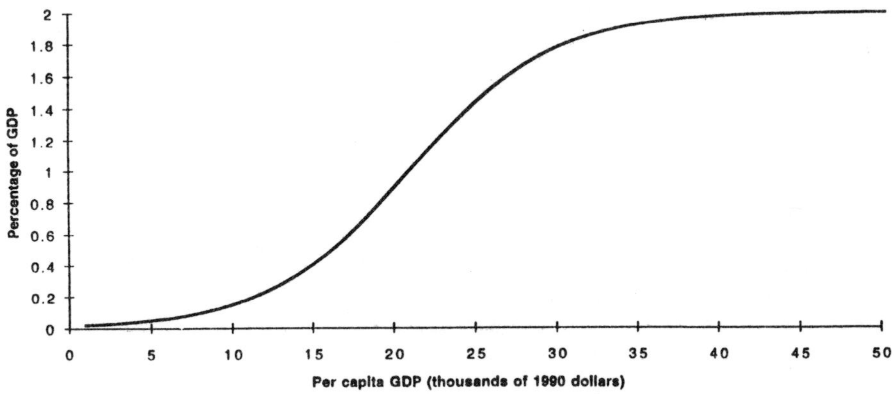

Figure 5 Willingness to pay to avoid 2.5 °C change in temperature: non-market (ecological) damages

since effects omitted from the benefit-cost calculation tend to be given insufficient weight in making decisions based upon the calculation'.

In this paper, we take an approach based on willingness to pay (WTP). The issue is framed in the context of how much consumers in each region would be willing to pay to avoid ecological damages. The fraction depends both on temperature change and GDP per capita:

$$WTP_{t,n} = d_{3,n} * \Delta AT_{t,n}^{d_{4,n}} / (1+100 * \exp(-0.23*GDP_{t,n}/POP_{t,n}) \quad (14)$$

The relationship between *WTP* for non-market goods and per capita income is assumed to be S-shaped, and is calibrated so that it does not exceed 100% of GDP. At low incomes, people are not willing to pay much to avoid non-market effects. However, as incomes climb, *WTP* increases rapidly and then plateaus at a constant proportion of income. For example, Figure 5 shows the WTP function for $d_{3,n} = 0.0032$ and $d_{4,n} = 2$. This means that when per capita incomes approach 40 (thousands of 1990 dollars), consumers are willing to pay 2% of GDP to avoid a 2.5 °C increase in temperature, and 8% of GDP to avoid a 5 °C increase.

According to Equation (14), each region values ecological damages independently of where such damage occurs. That is, consumers in a particular region place the same value on wetland losses whether they occur within or outside their own boundaries. The same is true for human health and wildlife. The S-shaped function implies, however, that lower income regions will place a lower value on such losses than higher income regions. As a result, non-market damages will be higher for developed countries.

Rather than advocate any particular set of numbers, we will explore the implications of alternative valuations. We seek to address the following question: how much must an informed global community be willing to pay to avoid ecological damages in order for a particular policy to make sense?

Emissions, concentrations and temperature change under five alternative policy scenarios

Numerous proposals have been put forward for controlling greenhouse gas emissions. These have ranged from a modest slowing in the rate of growth to drastic reductions below present levels. To illustrate how MERGE can be used for integrated cost–benefit analysis, we will apply it to a representative menu of policy choices. These include:

- business as usual;
- a tax on carbon emissions starting at $1/tonne in 2000 and increasing at 5%/year;
- a tax on carbon emissions starting at $5/tonne in 2000 and increasing at 5%/year;
- stabilizing global CO_2 emissions at 1990 levels; and
- stabilizing concentrations of CO_2 in the atmosphere at near current levels.

The difference between the two tax policies is primarily one of timing. The tax beginning at one dollar per ton rises to five dollars per ton by about 2030. If we ignore the relatively small deadweight losses incurred prior to this date, it is equivalent to delaying the imposition of the five dollars per ton tax by 30 years. For short hand, we will refer to this as the delayed tax alternative.

There is no unique formula for achieving stabilization – either of emissions or concentrations. Each involves

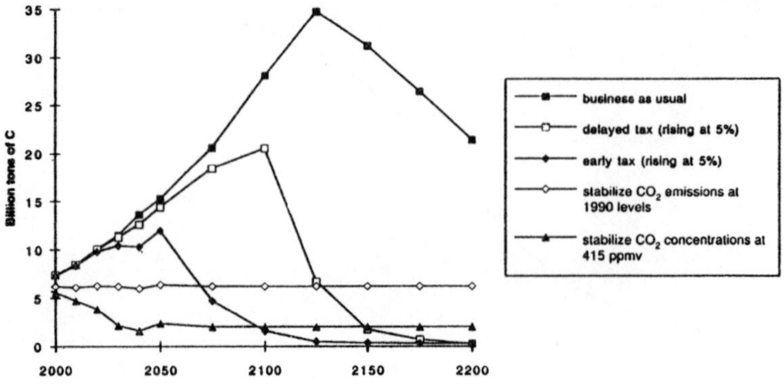

Figure 6 Global CO_2 emissions

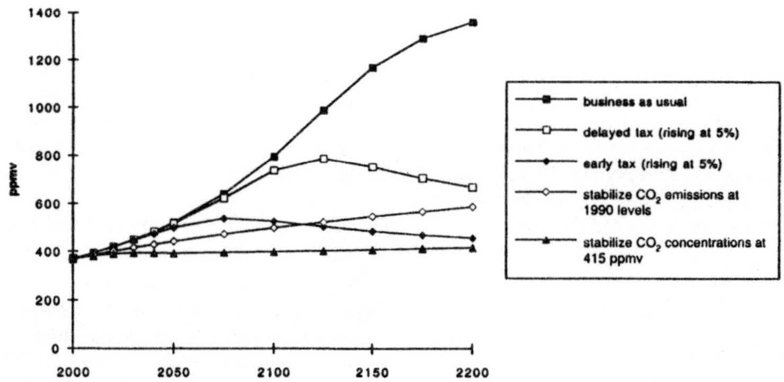

Figure 7 CO_2 concentrations

the difficult issue of the allocation of emission rights among regions. In this paper, we explore one widely discussed proposal, a gradual transition to equal per capita emission rights. We assume that carbon rights are initially distributed among regions in proportion to their 1990 level of emissions. Over time, the shares change gradually. By 2030, carbon rights are distributed in proportion to 1990 population levels. Note that the 1990 population base penalizes nations that fail to control their rate of population growth.

The IPCC (1990) estimates that emissions would have to be reduced immediately to 30% of 1990 levels in order to keep atmospheric CO_2 concentrations constant. Given that fossil fuels provide more than 90% of the world's commercial energy, this seems unrealistic. For our stabilization scenario, we assume that emissions are reduced gradually to 30% of 1990 levels by 2030. This results in stabilization of atmospheric concentrations at 415 ppmv – approximately 50% higher than pre-industrial levels.

Figure 6 compares carbon emissions for the five alternatives. Under business as usual, emissions continue to rise into the 22nd century. The growth reflects the increasing dependence on coal in the non-electric sector. With the eventual exhaustion of low-cost oil and gas resources, coal based synthetic fuels are apt to become the marginal source of non-electric energy supply. Unfortunately, the synthetics emit twice as much CO_2 per unit of energy as conventional oil.

In the longer term, CO_2 emissions will begin to decline – even under business as usual. Coal, like other fossil fuels, is in limited supply. As economically recoverable resources are exhausted, coal too will be replaced by other alternatives. Even in the absence of CO_2 considerations, the world's dependence on fossil fuels is likely to decline sharply towards the end of the 22nd century.

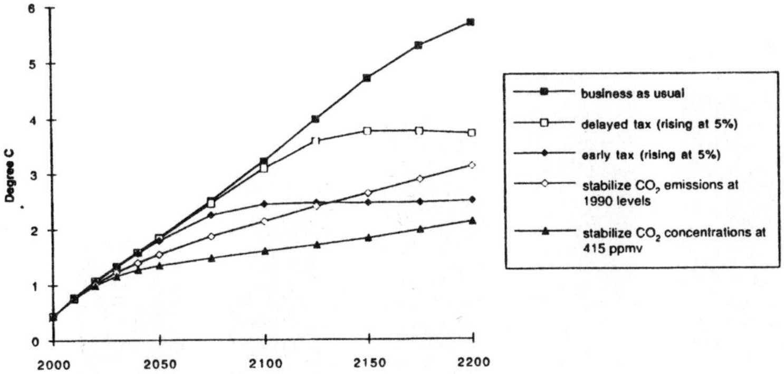

Figure 8 Mean global temperature change from 1990

The effect of the tax policies is to hasten the end of the synthetic fuels era. When the tax on carbon rises above $200 per tonne, the carbon free non-electric backstop becomes the technology of choice. With the delayed tax, this happens shortly after the year 2100. This explains the rapid fall off in carbon emissions as we move into the 22nd century. With the early tax, the $200 threshold is reached 30 years earlier. The synthetic fuels era is shortened accordingly.

Using the carbon cycle model, it is straightforward to convert emissions into atmospheric concentrations. Figure 7 compares results for the five alternatives. Under business as usual, concentrations are twice pre-industrial levels in the middle of the next century, and they double again by 2200.

The tax policies initially stabilize concentrations, and eventually lead to their decline. In the case of the delayed tax, concentrations approach 800 parts per million by volume (ppmv) and then return to 650 ppmv. For the early tax policy, they peak at 550 ppmv, then return to 450 ppmv.

By stabilizing emissions at 1990 levels, we do not halt the rise in atmospheric CO_2 concentrations. By 2200, concentrations approach 600 ppmv, and they continue to rise. The most stringent policy is designed specifically to achieve stabilization of concentrations. This happens at approximately 415 ppmv.

Figure 8 compares the impact on mean global temperature. Under business as usual, temperature increases by 2.5 °C (above the 1990 level) by 2075 and by nearly 6 °C by 2200. Not surprisingly, the delayed tax is the least effective in reducing temperature change. It does, however, serve to limit the long-term temperature rise to 4 °C.

By accelerating the imposition of the tax, there is a substantial impact on temperature change. The increase is now limited to 2.5 °C. Note also that over the longer term (post-2075), the early tax turns out to be more effective than the emissions stabilization policy (in terms of limiting temperature change).

Up to 2200, the policy with the greatest impact on global warming is one in which concentrations would be stabilized near current levels. Note, however, that stabilizing CO_2 concentrations does not cap the temperature rise. There are continuing emissions of the other greenhouse gases.

Benefits and costs

In the previous section, the effectiveness of the policy options were measured in physical terms – emissions, concentrations and mean temperature change. Although the exercise yielded some useful insights, it is a limited guide to greenhouse decision making. Sensible greenhouse policy requires a careful balancing of both benefits and costs. The next step in an integrated assessment is to translate temperature changes into market and non-market damages. To do this, we must first calibrate the regional damage functions.

In his original work on the impacts of climate change, Nordhaus (1991) focused primarily on market damages to the USA from a doubling of atmospheric CO_2 levels ($2 \times CO_2$). Noting that only 3% of US national output originates in climate sensitive sectors and another 10% in sectors that are modestly sensitive to climate change, he estimated that the net economic damages from a $2 \times CO_2$ scenario are likely to be of the order of 0.25% of national income.

Nordhaus acknowledged, however, that many valuable goods and services are excluded from conventional national income accounts. Correcting for these omissions and extending to the world, his best guess of the costs imposed by a doubling in atmospheric CO_2 concentrations was 1% of global GDP, with a range of 0.25–2%.

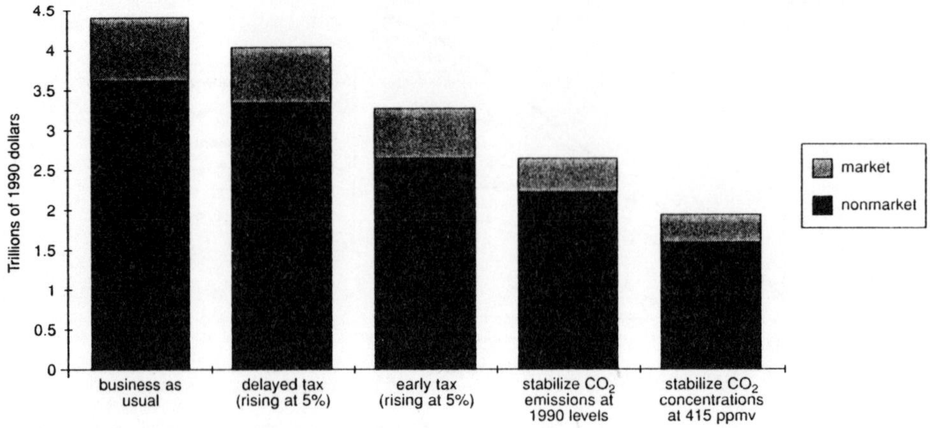

Figure 9 Global damages discounted to 1990 at 5%

Subsequent studies have attempted more comprehensive analyses for the USA with surprisingly similar results. Cline (1992) and Fankhauser (1993) estimate losses of the order of 1.1% and 1.3%, respectively, for a 2 × CO_2 scenario. When Cline adjusts for output composition in different countries, he projects total losses to the global economy of 1.3%. Similarly, Fankhauser suggests that the damage related to a doubling of atmospheric concentration may be of the order of 1–1.5% of GDP worldwide.

Although there appears to be some convergence in the GDP estimates, it is important to note that climate impact assessment is still in a rudimentary stage. Investigators are quick to stress the preliminary nature of their findings and the need for additional work. The present analysis makes no attempt to narrow the range of uncertainty surrounding damages, nor do we endorse a particular set of estimates. Rather, our goal is to demonstrate how damage projections should fit into the overall assessment of climate change. In doing so, we hope to shed some light on the sensitivity of greenhouse policy to alternative assumptions about the shape and nature of the damage function.

With these caveats in mind, we now describe the assumptions used to estimate the regional loss functions. If damages change quadratically with temperature, the calibration requires only a single point on each function.

For developed regions, we assume that a 2.5 °C increase in temperature (above 1990 levels) would result in market damages of 0.25% of GDP. This is consistent with Nordhaus' estimates of market losses to the USA for a 2 × CO_2 scenario.

For other regions, the assessment is more difficult. Relatively little attention has been devoted to estimating market damages outside of the OECD. Developing countries tend to have a larger share of their economies concentrated in climate sensitive sectors, particularly in agriculture. This vulnerability, however, is apt to shrink over time as these countries take on more of the characteristics of their industrialized counterparts. Moreover, they are more likely to be in equatorial regions where temperature changes are less severe than in polar latitudes. Nevertheless, many believe that market damages are apt to be a higher percentage of GDP for developing countries. For the present analysis, we assume a multiple of two.

For non-market damages, we begin with the willingness-to-pay function described in Figure 5. It is based on the assumption that consumers in high income countries would be willing to pay 2% of GDP to avoid the ecological damages associated with a 2.5 °C increase in temperature. As a point of reference, the USA currently devotes approximately 2% of its gross domestic product to all forms of environmental protection. This compares with 2.7% for research and development and 5.5% for national defense (Council of Economic Advisors, 1993). Given the highly subjective nature of the willingness to pay assumption, we will subject it to extensive sensitivity analysis.

We are now ready to conduct the next step in an integrated assessment – the estimation of market and non-market damages. Figure 9 summarizes the results in present value terms. For the business as usual alternative, the losses approach $4.4 trillion (discounted to 1990 at 5% per year). Consistent with the earlier studies,

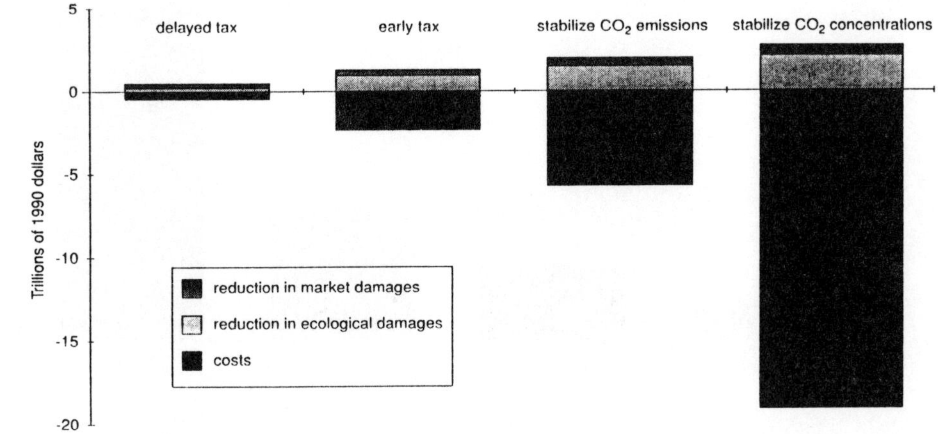

Figure 10 Global benefits and costs discounted to 1990 at 5%

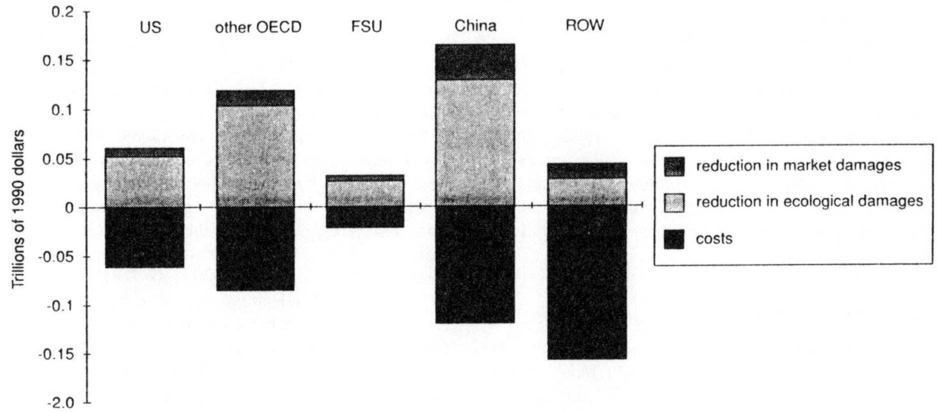

Figure 11 Regional benefits and costs from delayed tax policy (discounted to 1990 at 5%)

the bulk of the damages are in non-market categories. As a point of comparison, the projected losses for a 2.5 °C change in temperature amount to 1.4% of discounted worldwide GDP. This is also well within the range of the earlier estimates.

The economic benefits of an alternative policy may be measured in terms of the reduction in damages relative to business as usual. Total damages are reduced the most when CO_2 concentrations are stabilized at 415 ppmv. In this case, the benefits are nearly \$2.5 trillion. By contrast, the benefits from the delayed tax policy are less than \$0.5 trillion.

The desirability of a policy depends upon both benefits and costs. Using Global 2200, we calculate the economy-wide costs of each abatement alternative. The results are summarized in Figure 10. Although stabilizing concentrations buys the most – in terms of reduced market and non-market damages – it also carries the highest price tag. Costs outweigh benefits by almost eight to one. From a purely benefit-cost perspective, the only policy that appears attractive is the delayed tax. Both benefits and costs are of the order of \$0.4 trillion.

MERGE also allows us to examine the distribution of

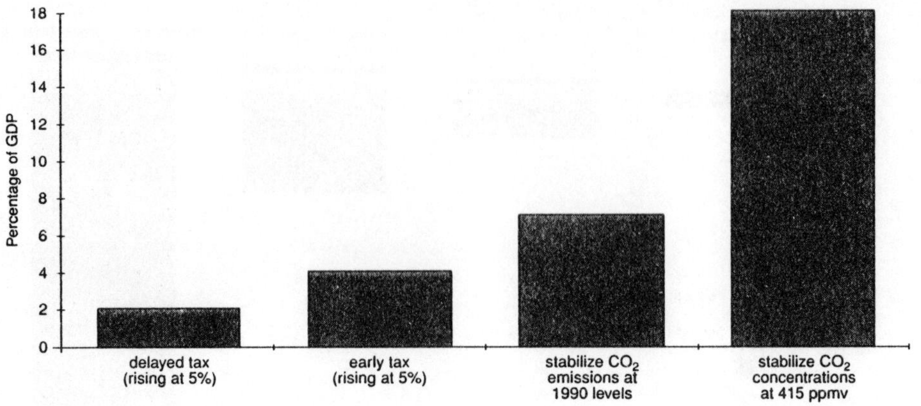

Figure 12 Break-even willingness to pay: 5% return on capital

benefits and costs across regions. Figure 11 shows such a breakdown for the delayed tax policy. Note that all regions are winners except ROW. There are two reasons for this. First, the temperature increase for the tropics is half of the global mean increase and most of ROW is in the tropics. Hence, it experiences less damage than the other regions. Secondly, because of its relatively low per capita income, it values nonmarket damages less than its more wealthy counterparts. In order to induce ROW to join into an agreement with the other four regions, there would have to be additional side payments.

Some sensitivity analyses

Like all quantitative analyses, the results are driven by the numerical inputs. In this section, we explore the sensitivity of the results to three key sets of assumptions: the willingness-to-pay to avoid ecological damages, the cost of alternative emission control policies, and the discount rate.

Given the highly subjective nature of the willingness to pay function, limited insight is gained from any particular set of estimates. It is more instructive to ask how much consumers must be willing to pay to avoid ecological damages in order for a particular policy to make sense from an economic cost–benefit perspective.

Figure 12 shows the break-even points for our four alternatives to business as usual. The delayed tax is justified when high-income consumers are willing to pay 2% of GDP to avoid a 2.5 °C increase in temperature. For a stringent emissions policy to be attractive, there must be a high willingness to pay. For the most costly policy, (stabilization of emissions at 415 ppmv), high-income consumers must be willing to pay approximately 17% of GDP to avoid a 2.5 °C increase in temperature. This is the point at which the benefits just break even with the costs.

Next, we look at how the break-even willingness to pay depends upon the costs of the emission control policies. Our reference case Global 2200 assumptions were described above. We now adopt more optimistic assumptions regarding two critical parameters: the elasticity of price induced conservation and the cost of the carbon free non-electric backstop (NE-BAK).

The degree of substitutability between the inputs of capital, labor and energy will affect the economic losses from energy price increases. Our earlier substitution elasticities were based on a backcasting exercise, and they best explain the US historical record (Manne and Richels, 1992). For the sensitivity analysis, we assume that the elasticities are 50% higher.

Similarly, we adopt considerably more optimistic assumptions about the cost of the non-electric backstop. For the sensitivity analysis, we assume that NE-BAK is only slightly more expensive than coal based synthetic fuels – \$55 as opposed to \$50 per barrel.

Figure 13 shows the impact of these optimistic abatement cost assumptions on the break-even willingness to pay. The effect is to lower the cost of the policy. This automatically lowers the value we must assign to ecological damages for control policies to be economically justifiable.

Finally, we turn to the issue of the discount rate ie the rate at which we discount goods and services from the

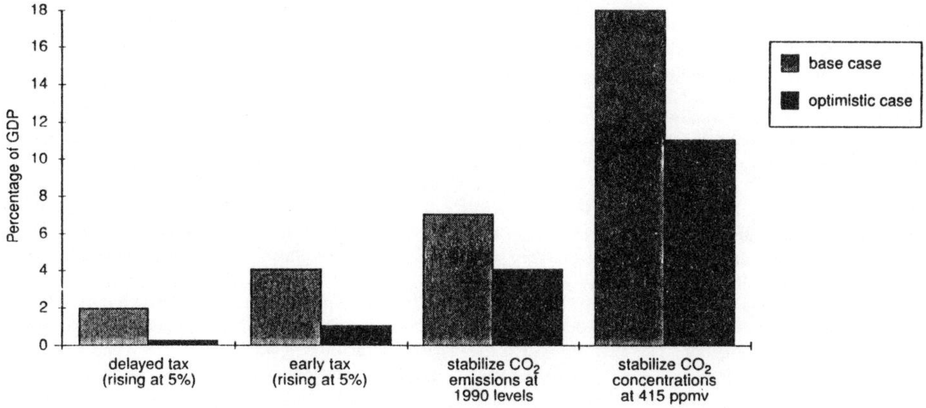

Figure 13 Break even willingness to pay: sensitivity to abatement costs

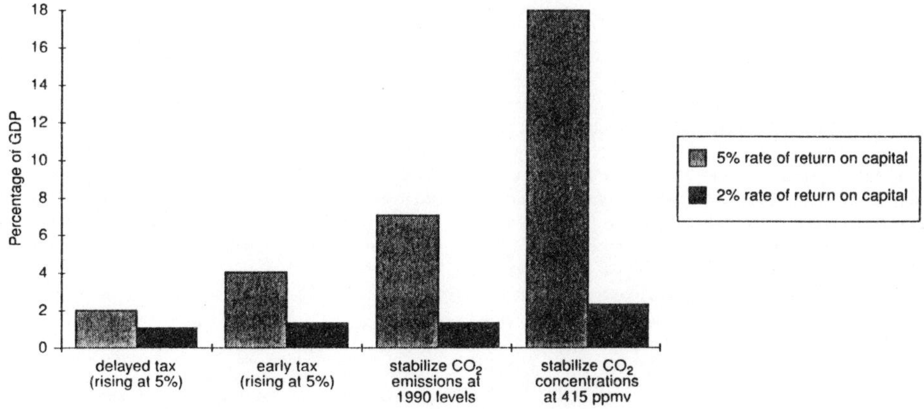

Figure 14 Break even willingness to pay: 5% versus 2% rate of retuun on capital

future to the present. Costs and benefits are both discounted at the same rate as investment opportunities elsewhere in the economy. From Equation (3), recall that in an economy growing at 2% per year, a 5% marginal productivity of capital is consistent with a 3% pure rate of utility time preference. If one adopts a prescriptive rather than a descriptive approach, one might assume a lower rate of utility time preference. This is a logically consistent scenario, but would require an unrealistically rapid increase in investment in the early years. The result would be to drive down the marginal productivity of capital to some much lower value than its current level of 5%. For example, Cline (1992) suggests that the appropriate number is closer to 2%. MERGE is designed so that such differences in viewpoints can be incorporated easily and their significance assessed.

Figure 14 shows the impact of the lower discount rate on the break-even willingness-to-pay. Increased weight is placed on future impacts. Hence, damages do not have to be as high in the future in order to justify action. As a result, the break-even point is lower.

Some final comments

MERGE provides a framework for thinking about climate change policy proposals. The model is designed to be sufficiently flexible and transparent so that it can be used to explore alternative views on a wide range of contentious issues eg costs, damages, non-market valuation and discounting.

Given the level of uncertainty which pervades the climate debate, it would be unrealistic to expect cost–benefit analysis to lead to consensus on a bottom line – at least in the immediate future. Rather, models such as MERGE should be viewed as research tools capable of providing insights into which aspects of the debate may be most important. In this way, they can help focus the discussion and identify the areas where additional research may have the highest pay off.

The present analysis has indeed identified a number of issues for further investigation. The evaluation of damages – particularly ecological damages – deserves a great deal more attention. The conventional wisdom, that losses will be higher for developing countries, needs to be re-examined. True, developing countries tend to derive a larger share of their economic output from climate sensitive sectors. But market damages may turn out to be a relatively minor part of the story.

The more important issue may be how we allocate and value ecological losses. Unfortunately, the accounting is far from straightforward. For the present analysis, we assumed that inhabitants of a particular region place the same value on global ecological losses regardless of their geographical location. The losses are then evaluated in terms of willingness to pay. This means that low income countries have lower ability to pay, and therefore lower losses than high income countries.

Given the importance of ecological factors to the overall climate cost–benefit calculus, more attention needs to be devoted to their valuation. 'Non'-market damages do not lend themselves easily to monetary valuation. Yet there is little alternative but to try. The best course of action may be to frame the issue in terms of a break-even willingness to pay. The various parties can then debate whether the benefits of a policy outweigh the costs.

The analytical framework itself needs further attention. Perhaps the most serious limitation is the treatment of uncertainty. The climate issue entails high stakes. Waiting leads to the risk of irreversible damages. Immediate action leads to the risk that large costs will be incurred in the near future, and there is considerable disagreement on how these actions will eventually affect the world's climate. Policy makers need help in identifying hedging strategies which balance the costs of delay against those of premature action.

To be a more effective policy tool, MERGE needs to deal more explicitly with critical climate-related risks. We need to provide the capability to replace point estimates with probability distributions reflecting the full range of possible outcomes. Much of the greenhouse debate centers around low probability but high consequence events. These concerns must be included in the analytical process.

We must also do a better job of capturing the true nature of policy making. Greenhouse decision making is apt to be a sequential process with many opportunities for learning and midcourse corrections. The issue is not deciding what to do about greenhouse emissions over the next century. It is deciding what to do over the next decade in the light of alternative consequences over the next century.

Policy frameworks are needed which reflect this 'act – then learn' character of the decision process. With the help of such tools, we will be better able to address two central issues. Which of the many climate related uncertainties are most important to today's decisions? And which strategies provide the best hedge against greenhouse related risks?

Acknowledgements

This research was funded by the Electric Power Research Institute (EPRI). The authors are much indebted to Tim Olsen and Susan Swinehart for research assistance. We have benefited from discussions with William Cline, Jae Edmonds, Sam Fankhauser, Larry Goulder, Erik Haites, William Nordhaus, Stephen Peck, Stephen Schneider and Thomas Teisberg. The views presented here are solely those of the individual authors, and do not necessarily represent the views of EPRI or its members.

References

Brooke, A, Kendrick, D and Meeraus, A (1988) *GAMS: A User's Guide* Scientific Press, Redwood City, CA

Chakravarty, S (1969) *Capital and Development Planning* MIT Press, Cambridge, MA

Cline, W (1992) *The Economics of Global Warming* Institute for International Economics, Washington, DC

Council of Economic Advisors (1993) *Economic Report of the President* US Government Printing Office, Washington, DC

Dorfman, R and Dorfman, N (1976) *Economics of the Environment* W W Norton, New York

Dowlatabati, H and Morgan, G (1993) 'A model framework for integrated studies of the climate problem' *Energy Policy* **21** (3)

Energy Modeling Forum (1993) *Reducing Global Carbon Emissions: Costs and Policy Options* EMF-12, Stanford University, Stanford, CA

Eykmans, J, Proost, S and Schokkaert, E (1992) *Efficiency and Distribution in Greenhouse Negotiations* Draft, Katholieke Universiteit Leuven, Leuven

Falk, I and Mendelsohn, R (1993) 'The economics of controlling stock pollution: an efficient strategy for greenhouse gases' *Journal of Environmental Economics and Management* **25** 76–88

Fankhauser, S (1993) 'The economic costs of global warming' in *Costs, Impacts, and Benefits of CO_2 Mitigation* International Institute for Applied Systems Analysis, Laxenburg, Austria

Grubb, M (1993) 'The costs of climate change: critical elements' in Kaya, Y, Nakicenovic, N and Toth, F (eds) *Costs, Impacts, and Benefits of CO_2 Mitigation* International Institute for Applied Systems Analysis, Laxenburg, Austria
Hope, C, Anderson, J and Wenman, P (1993) 'Policy analysis of the greenhouse effect' *Energy Policy* **21** (3)
IPCC (1990) *Climate Change: The IPCC Scientific Assessment* Intergovernmental Panel on Climate Change, Cambridge University Press, Cambridge
IPCC (1992) *Climate Change 1992: The Supplementary Report to the IPCC Scientific Assessment* Intergovernmental Panel on Climate Change, Cambridge University Press, Cambridge, UK
Maier-Reimer, E and Hasselmann, K (1982) *An Ocean Circulation Model for Climate Studies* Technical Report, Max-Planck-Institute, Hamburg
Maier-Reimer, E and Hasselmann, K (1987) 'Transport and storage of carbon dioxide in the ocean: an inorganic ocean-circulation carbon cycle model' *Climate Dynamics* **2**
Manne, A and Richels, R (1992) *Buying Greenhouse Insurance: The Economic Costs of Carbon Dioxide Emission Limits* MIT Press, Cambridge, MA
Manne, A (1992) *The Rate of Time Preference: Implications for the Greenhouse Debate* Working paper, Stanford University, CA
Nordhaus, W (1991) 'To slow or not to slow: the economics of the greenhouse effect' *Economic Journal* **101** 407
Peck, S and Teisberg, T (1992) 'CETA: a model for carbon emissions trajectory assessment' *Energy Journal* **13** (1)
Rutherford, T (1992) *Sequential Joint Maximization* Working paper, University of Colorado, CO
Working Group III (1990) *Formulation of Response Strategies* Intergovernmental Panel on Climate Change, Washington, DC
Zachariah, K and Vu, M (1988) *World Population Projections, 1987–1988 Edition* World Bank, Johns Hopkins University Press, Baltimore, MD

[14]

BUTTERWORTH HEINEMANN

Energy Policy, Vol. 23, No. 4/5, pp. 373–378, 1995

0301-4215/95 $10.00 + 0.00

The economics of stabilizing atmospheric CO_2 concentrations

Richard Richels

Energy Analysis and Planning Department, Electric Power Research Institute, 3412 Hillview Avenue, PO Box 10412, Palo Alto, CA 94303, USA

Jae Edmonds

Global Environmental Change Program, Pacific Northwest Laboratories, 901 D Street SW, Suite 900, Washington, DC 20024-2115, USA

This paper examines the economics of stabilizing atmospheric CO_2 concentrations. A particular concentrations target can be achieved in a variety of ways. It turns out that the choice of emissions time-path is as important as the concentrations level itself in determining the ultimate price-tag. Rather than choosing arbitrary emission trajectories, more attention needs to be devoted to identifying those paths that minimize the costs of achieving a specific target.

Keywords: Global warming; Energy modeling; Optimization

The United Nations Framework Convention on Climate Change has as its ultimate objective the 'stabilization of greenhouse gas concentrations in the atmosphere at a level that would prevent dangerous anthropogenic interference with the climate system'. Under the terms of the Convention, the costs of the policies and measures for achieving the target are to be given little weight in establishing the level at which concentrations are to be stabilized. The level is to be based upon our understanding of the greenhouse effect and its potential consequences – not upon a balancing of benefits and costs.

Whereas economic considerations play little role in establishing the target, they play an important role in determining how the target is to be achieved. The Convention states that 'policies and measures to deal with climate change should be cost-effective so as to ensure global benefits at the lowest possible cost'. A particular concentrations target can be achieved in a variety of ways. Some will be more costly than others. Economic analysis is needed to identify emission paths that minimize the costs of achieving the desired concentrations level.

This paper addresses the issue of cost-effectiveness in the selection of an emissions time-path. The focus is on carbon dioxide, the most important anthropogenic greenhouse gas. A reduced form carbon cycle model is used to identify alternative emission paths for achieving a given CO_2 concentrations target. Using two widely used energy–economy models, we then calculate how costs to the global economy vary with the choice of emissions trajectory. It turns out that the emissions time-path is as important as the concentrations level itself in determining the ultimate price-tag.

Carbon emissions and concentrations under business as usual

We begin by examining how CO_2 emissions and atmospheric concentrations are apt to evolve under business as usual. That is, in the absence of explicit abatement measures. For purposes of the present analysis, emissions are divided into three categories: (1) fossil fuels, (2) net deforestation and land-use change, and (3) others.

The fossil fuel emissions baseline is established with the Global 2100 model of Manne and Richels (1992).

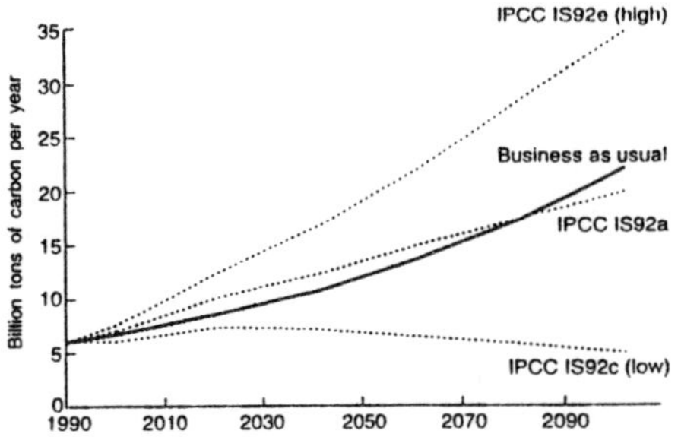

Figure 1 Projections of future carbon emissions

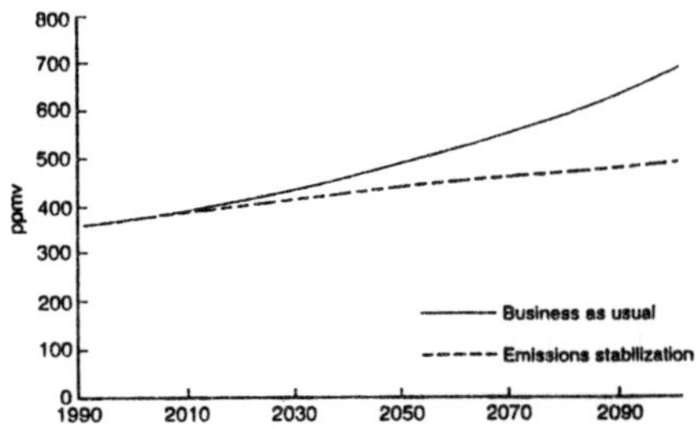

Figure 2 Atmospheric CO_2 concentrations – business as usual and emissions stabilization

Global 2100 is a dynamic non-linear optimization model which divides the globe into five geopolitical regions: the USA, the other OECD countries, the former USSR, China, and the rest of the world. The model is designed to evaluate the options that are realistically available to each region as the world moves away from its heavy dependence on fossil fuels and toward a more diversified energy economy.

The model's key parameters are calibrated using the results of a recent expert survey.[1] Figure 1 reports the business as usual emissions path. For purposes of comparison, we also show the range of emissions scenarios contained in the recent supplement to the IPCC scientific assessment (IPCC, 1992). Note that the results are quite similar to the IPCC's central case (IS92a).

For emissions from net deforestation and land-use change, we assume a 'neutral biosphere'. That is, the net biospheric input of CO_2 is exactly balanced by natural uptake mechanisms.[2] Further, we suppose that global cement production will contribute an additional 200 Mt of carbon to the atmosphere annually.

Ocean uptake of carbon is approximated by the impulse–response function of Maier-Reimer and Hasselmann.[3] Figure 2 shows atmospheric concentrations under business as usual. According to our calculations, concentrations will be double pre-industrial levels in the middle of the next century and exceed 700 ppmv by the year 2100.

Stabilizing atmospheric CO_2 concentrations

The level at which 'dangerous anthropogenic interference with the climate system' will occur has yet to be established. Indeed, the question of what constitutes an appropriate level is likely to remain the subject of intense discussion for some time. The present analysis makes no attempt to contribute to this debate. Nor does it endorse any particular set of limits. Rather, we explore the economic implications of holding concentrations to a range of alternative levels.

It is noteworthy that most proposals for dealing with climate change have focused on emissions and not concentrations. For example, the Toronto Conference on The Changing Atmosphere (June 1988) called for a 20% worldwide reduction in CO_2 emissions by the year 2005 and the Hamburg Conference (November of 1988) called for a 30% reduction by the year 2000.

More recently, the focus of National Action Plans mandated by the Framework Convention has been to return emissions to their 1990 levels by the year 2000. This is seen as the first step in achieving the Convention's ultimate objective of stabilizing atmospheric concentrations. The second step, however, remains unclear. As a point of departure, we explore the implications of a policy which would permanently hold global emissions to 1990 levels.

Figure 2 shows the impact of such a policy upon atmospheric concentrations. The rate of increase has been considerably reduced. Atmospheric concentrations in 2100 are now only 500 ppmv. Stabilizing emissions,

[1]Manne and Richels (1994). A group of knowledgeable individuals were polled on their beliefs about five factors that are important to future CO_2 emissions. These are: (1) potential GDP growth rates, (2) the elasticity of price-induced substitution between energy and capital–labor, (3) the rate of autonomous energy-efficiency improvements, (4) the availability of economically competitive carbon-free alternatives to coal-fired electricity and (5) the costs of a non-electric backstop alternative to liquid fuels. The Global 2100 model runs reported here are based on the poll averages.

[2]This hypothesis is consistent with the post-World War II historical record: Post *et al* (1990).

[3]Maier-Reimer and Hasselmann (1987). The model partitions carbon emissions into five classes, each with different atmospheric lifetimes. Parameters are least-square fitted to the computed response of a full ocean carbon cycle model.

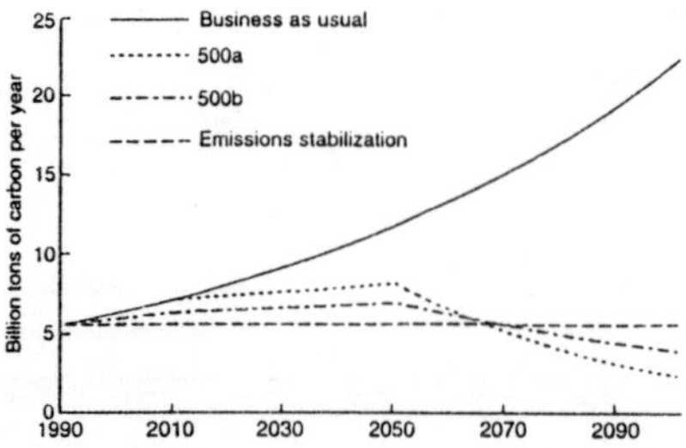

Figure 3 Alternative emission profiles for stabilization at 500 ppmv

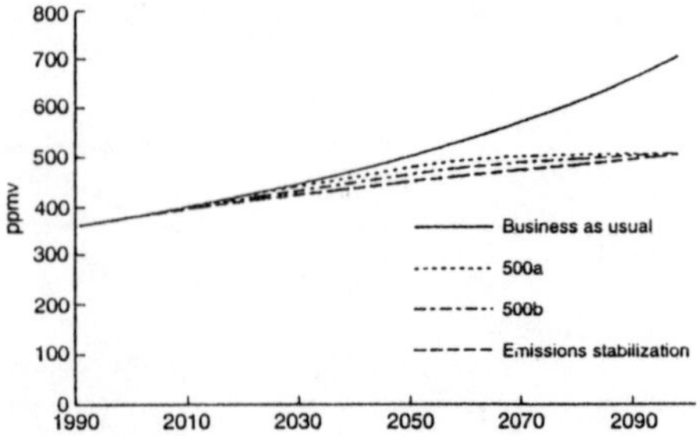

Figure 4 Alternative concentration profiles for stabilization at 500 ppmv

however, does not serve to stabilize concentrations. Although the rate of increase is slowing, the absolute level is still increasing.

A particular concentrations target can be achieved in any of a number of ways. Figure 3 shows three alternative emission trajectories that would lead to 500 ppmv in 2100. The line labeled 500a represents a scenario in which emissions are allowed to follow the business as usual path through 2010; are reduced gradually between 2010–50; and are reduced sharply thereafter. The line labeled 500b represents an emissions scenario which lies between the 500a and the emissions stabilization scenarios. Note that the areas under the lower three curves are approximately the same.

Figure 4 shows the corresponding curves for atmospheric concentrations. The 500a and 500b emission scenarios were chosen so that concentrations never exceed 500 ppmv. This is in contrast to the emissions stabilization scenario where concentrations continue to rise into the 22nd century.[4]

The alternative concentration profiles have implications for temperature change. Emissions stabilization results in lower atmospheric concentrations prior to 2100. Hence, it is apt to produce slightly less temperature change during the next century. Over the longer term, however, the scenarios which succeed in capping concentrations at 500 ppmv are likely to be more effective in limiting temperature change.

The economic costs of holding concentrations to 500 ppmv

Next, we turn to the economics of holding concentrations to 500 ppmv in 2100.[5] Two models are used to quantify economic losses: Global 2100 and the Edmonds–Reilly–Barns Model (ERB).[6] The models are alike in a number of respects. They are both long-term energy–economy models. They both provide relatively detailed representations of the energy system. And they both evaluate the losses from a carbon constraint through the impacts on the price of energy.

The models differ, however, in a number of important ways. ERB is a recursive rather than an intertemporal optimization model. It employs a so-called 'putty–putty' rather than a 'putty–clay' approach to the vintaging of capital stocks. And it provides more regional disaggregation – nine versus five regions.[7]

The two models were chosen more because of their differences than their similarities. The application of alternative approaches can provide valuable insights into the robustness of a set of the results. For the present analysis, the models were calibrated to the same emissions base line. Independent cost analyses were then conducted to identify the least-cost strategy. From Figure 5, we see that the results are quite similar. Costs are evaluated through 2100 and presented as a percentage of gross world product. Stabilizing emissions at 1990 levels is by far the most costly strategy. Shifting the emission reductions into the outer years reduces costs by as much as 50%.

[4]The principal conclusions of our paper turn out to be independent of the choice of carbon cycle model. We performed the same exercise with the model of Wigley (1993) using a balanced carbon cycle model with feedbacks and the IPCC IS92a land-use emissions trajectory. While the atmospheric concentrations associated with stable 1990 fossil fuel carbon emissions are somewhat lower, 461 ppmv in 2100, cases 500a and 500b cap concentrations at 462 and 463 ppmv, respectively. Thus, the characteristic relationship between the time path of emissions and atmospheric concentrations holds.

[5]Nordhaus (1979) was the first to couple a carbon cycle model and an energy-economy model in order to explore cost-effective strategies for stabilizing atmospheric concentrations.

[6]For a description of the ERB model, see Edmonds and Barns (1992).

[7]For a detailed model comparison, see Energy Modeling Forum (EMF), 1993.

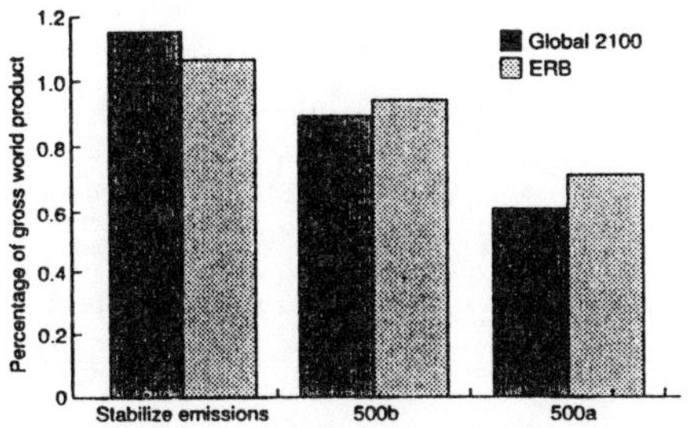

Figure 5 Costs of capping atmospheric CO_2 concentrations at 500 ppmv by 2100

There are several reasons why a less restrictive near-term emissions path may turn out to be less expensive. First, the shift to a less carbon intensive economy cannot happen overnight. The time scale for large-scale deployment of new supply technologies is typically measured in decades. Widespread adoption of highly efficient end-use technologies also takes time. Energy-efficient systems are often embedded in long lived durable goods (autos, housing, equipment, structures), and these will not be replaced instantaneously. The process can be accelerated, but at a cost.

Secondly, there is apt to be a shortage of low-cost substitutes during the early decades of the 21st century. There are constraints on the rate at which new supply and end-use technologies can enter the marketplace. Having more time to manage the transition away from fossil fuels will be worth a great deal. We can emit more in the early years when the marginal cost of emissions abatement is high. The 'pay back' can come in later years when low-cost technological alternatives are more plentiful.

Finally, scenarios 500a and 500b are less expensive because of the time value of money. With a positive discount rate, a dollar in the future is worth less than a dollar today. Even if the cost of reducing a tonne of carbon were the same in all years, we would still prefer to make the expenditures later on. Doing so will result in a lower discounted present value.[8]

[8]It is important to distinguish between the rate of time preference and the marginal productivity of capital. The above discussion relates to the latter. We have assumed a net real rate of return on capital of 5% per year. For purposes of illustration, suppose that it costs US\$100 to remove a tonne of carbon – regardless of the year in which the reduction occurs. If we were to remove a tonne today, it would cost US\$100. Alternatively, we could invest US\$25 today to have the resources to remove a tonne of carbon in 2020. Hence, if the focus is on the cumulative total rather than year by year emissions, we will prefer strategies which shift reductions into the outer years.

The economic costs of stabilizing concentrations at alternative levels

The focus thus far has been on capping concentrations at 500 ppmv. As noted earlier, our selection of a particular target was arbitrary. The goal was to explore the sensitivity of compliance costs to the choice of emissions trajectory. Costs are also a function of the target itself. To explore the nature of this relationship, we examine the costs of capping concentrations at alternative levels. For each target, we have attempted to identify an emissions path close to the least-cost solution: that is, we have selected paths which provide the greatest degree of near-term flexibility.

Figure 6 shows emission trajectories for capping concentrations at 400, 450 and 500 ppmv. Figure 7 presents the corresponding time profiles for atmospheric concentrations. In the case of the most stringent target, there is little alternative but to reduce near-term emissions. In 1990, concentrations were already at 353 ppmv. Limiting concentrations to 400 ppmv will require an early and rapid departure from the business as usual trajectory.

As the concentrations limit is relaxed, we gain in degrees of freedom. Capping concentrations at 450 ppmv allows for some growth in near-term emissions. Capping at 500 ppmv allows for more growth still.

Figure 8 shows abatement costs for the alternative targets. The figure highlights the importance of flexibility in the early years. This is when the marginal cost of emissions abatement is likely to be highest. Concentration targets requiring sharp reductions in near-term emissions will be particularly costly.[9]

Final comments

The economic costs of stabilizing concentrations will depend upon both the target and the manner in which the target is achieved. Under the terms of the UN Framework Convention on Climate Change, implementation costs are not a principal consideration in the choice of a concentrations limit. The decision is to be based on a scientific assessment of what constitutes 'dangerous anthropogenic interference with the climate system'. Nevertheless, international negotiators would do well to recognize the nonlinear relationship between concentration levels and implementation costs. The perceived price tag will undoubtedly influence the willingness of nations to abide by the treaty.

[9]The explanation for why the two cost estimates differ so markedly for the 400 ppmv case is straightforward. Unlike ERB, Global 2100 keeps track of the economic lifetime of the existing capital stock. As a result, its price responsiveness is lower in the short run than over the long run.

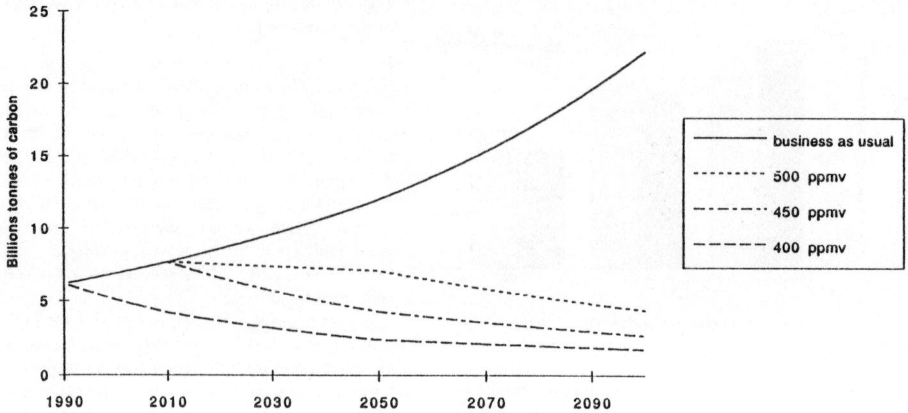

Figure 6 Global carbon emissions associated with alternative stabilization targets

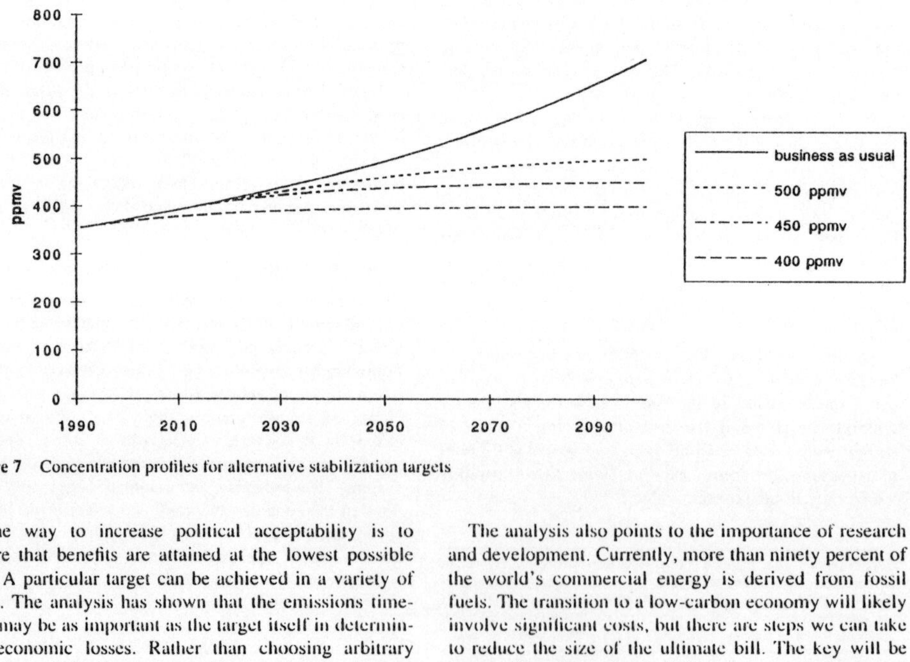

Figure 7 Concentration profiles for alternative stabilization targets

One way to increase political acceptability is to ensure that benefits are attained at the lowest possible cost. A particular target can be achieved in a variety of ways. The analysis has shown that the emissions time-path may be as important as the target itself in determining economic losses. Rather than choosing arbitrary emission trajectories, more attention needs to be devoted to identifying those paths that minimize the costs of achieving a particular target.

The analysis also points to the importance of research and development. Currently, more than ninety percent of the world's commercial energy is derived from fossil fuels. The transition to a low-carbon economy will likely involve significant costs, but there are steps we can take to reduce the size of the ultimate bill. The key will be timing. Time is needed both for an economical turnover of the existing capital stock and to develop ample supplies of low-cost substitutes.

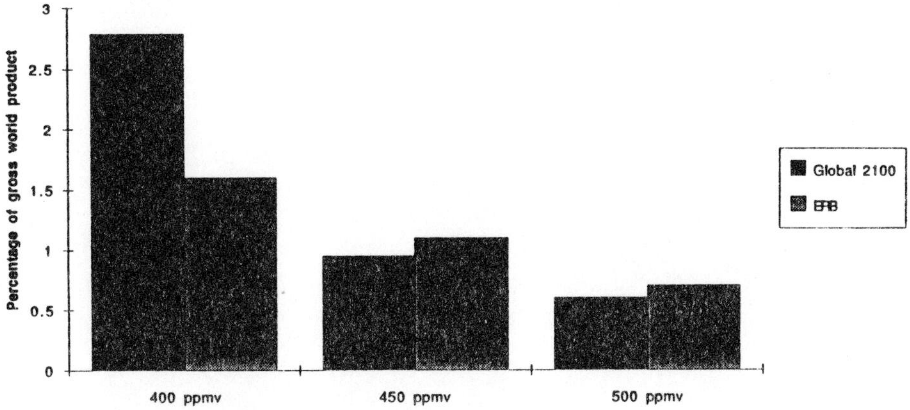

Figure 8 Costs of capping atmospheric CO_2 concentration at alternative levels

There are a number of promising carbon-free options, both on the supply and demand sides of the energy sector. New technologies, however, take many years for market penetration. A sustained research effort is needed to clarify their potential role and to ensure that cost-effective options are available to the greatest extent possible.

Acknowledgements

This research was funded by the Electric Power Research Institute and the US Department of Energy. The authors wish to thank Erik Haites, Michael Jefferson, Alan Manne, William Nordhaus, Paul Portney, Michael Schlesinger, and Tom Wigley for their comments on an earlier draft. The views presented here are solely those of the individual authors.

References

Edmonds, J and Barns, D (1992) 'Factors affecting the long-term cost of global fossil fuel CO_2 emissions reductions' *International Journal of Global Energy Issues* **4** (3) 140–166

Energy Modeling Forum (EMF) (1993) *Reducing Global Carbon Emissions: Costs and Policy Options* EMF-12, Stanford University, Stanford, CA

IPCC (Intergovernmental Panel on Climate Change) (1992) *Climate Change 1992: The Supplementary Report to the IPCC Scientific Assessment* Cambridge University Press, Cambridge

Maier-Reimer, E and Hasselmann, K (1987) 'Transport and storage of carbon dioxide in the ocean: an inorganic ocean-circulation carbon cycle model' *Climate Dynamics* **2** (2) 63–90

Manne, A and Richels, R (1992) *Buying Greenhouse Insurance: The Economic Costs of Carbon Dioxide Emission Limits* MIT Press, Cambridge, MA

Manne, A and Richels, R (1994) 'The costs of stabilizing global CO_2 emissions: a probabilistic analysis based on expert judgments' *The Energy Journal* **15** (1) 31–56

Nordhaus, W (1979) *The Efficient Use of Energy Resources* Yale University Press, New Haven, CT

Post, W, Peng, T, Emanuel, W, King, A, Dale, V and DeAngelis, D (1990) 'The global carbon cycle' *American Scientist* **78** (4) 310–326

Wigley, T (1993) 'Balancing the carbon budget: the implications for projections of future carbon dioxide concentration changes' *Tellus* **45B** (5) 409–425

Part III
The Choice of Instruments

A
Carbon Taxes

[15]

JOURNAL OF ENVIRONMENTAL ECONOMICS AND MANAGEMENT 27, 259–274 (1994)

Efficient Climate Policy in the Presence of Free Riders[1]

MICHAEL HOEL

Department of Economics, University of Oslo, P.O. Box 1095, Blindern, N-0317 Oslo, Norway

Received June 22, 1993; revised August 5, 1993

If all countries participate in a climate agreement, policies affecting the supply of fossil fuels have identical economic consequences as policies affecting the demand of fossil fuels. This is no longer true when there is limited participation in an international agreement. In the paper, it is assumed that a group of countries commit themselves to cooperating, while the remaining countries act independently and in pure self-interest. The paper compares supply and demand policies in this context and demonstrates that policies which affect only demand or only supply of carbon generally are inferior to policies affecting both demand and supply of carbon. The optimal mix of supply and demand policies is derived in the paper, with particular emphasis on carbon taxes. An important result is that if the marginal environmental costs of total emissions are sufficiently high (in the cooperating countries), then both production and consumption of carbon should be taxed at a positive rate. If on the other hand the marginal environmental costs of total emissions are sufficiently low, it may be optimal to subsidize production or consumption. In this case, production should be subsidized if the cooperating countries are net importers of carbon, while consumption should be subsidized if the cooperating countries are net exporters of carbon. Another important conclusion of the paper is that the cooperating countries should try to induce the non-cooperating countries, through appropriate transfers, to tax consumption and production of carbon at the same rate as in the cooperating countries. © 1994 Academic Press, Inc.

1. INTRODUCTION

As long as there is no international law to force countries to participate in an international climate agreement, each country may have an incentive to be a free rider, i.e., to stay outside the agreement instead of participating in it. This free rider incentive remains even if the agreement is such that all countries are better off with the agreement than without: A country may be better off participating in an agreement than it would be without any agreement. But it will usually be even better off if the other countries cooperate, while it itself stays outside the agreement and pursues its self-interest.

In spite of the free rider incentive, a stable coalition of cooperating countries may exist. The coalition is stable in the sense that it is not in the self-interest of any country to break out of the coalition. The size of such a stable coalition depends on what happens if a country chooses to be a free rider instead of cooperating. In, e.g., [1, 7] it is assumed that each country first decides (simultaneously) whether to cooperate, after which the cooperating and non-cooperating countries decide upon their emission levels. In a game of this type, the only

[1]Financial support from the Research Council of Norway, the Nordic Council of Ministers, and the European Communities' Environment program is gratefully acknowledged. Significant parts of the paper are based on work done by the author while employed at CICERO (Center for Climate and Energy Research, Oslo); see Hoel [8, 9]. A number of colleagues and participants at various seminars have given useful comments to these earlier papers. I am also grateful to Rolf Golombek and two anonymous referees for useful comments to an earlier version of the present paper.

0095-0696/94 $6.00

negative consequence for a country of breaking out of the coalition is that the optimal emission levels of the remaining cooperating countries may increase. Only if this increase in emission levels hurts the defector more than the costs it saves by defecting will it be optimal for a potential defector to cooperate. The studies mentioned above argue that for problems such as the climate problem, the number of countries in a stable coalition is likely to be very small. Moreover, total emissions from all countries will not be much lower than they are in the non-cooperative equilibrium.

The assumptions that the decisions of whether to cooperate are made simultaneously and once and for all are not particularly realistic. Bauer [2] has described a situation in which the decision of one country of whether to cooperate may affect the corresponding decisions of other countries. In this model, there exist equilibria with many countries cooperating. A similar approach is to extend the two stage games of the type described above to a repeated game. It is well known from the literature on game theory that it may be possible to sustain tacit cooperation as a perfect equilibrium of a non-cooperative (infinitely) repeated game, see, e.g., [11] for a discussion in the context of international environmental agreements. The fact that decisions about greenhouse gas emissions are frequently repeated may thus solve the free rider problem. However, as repeated games of this type have multiple equilibria, the coordination problems of reaching a Pareto optimal equilibrium are large. Obviously, these coordination problems are larger, the larger the number of countries involved. It therefore seems likely that only a subset of all countries will commit themselves to cooperation. Barrett [1] has also shown that if one restricts oneself to those equilibria of the repeated game which are renegotiation proof, the number of cooperating countries may be quite small.

The literature discussed above assumes that each country is only concerned with its own welfare level, defined (loosely) as income minus environmental costs. However, social norms and conventions may play an important role in sustaining international environmental agreements. At the individual level, there is a lot of behavior which is easiest explained by social norms and conventions; e.g., [5]. This may very well apply at the country level as well: The government will feel uncomfortable if it breaks the social norm of sticking to an agreement of reduced emissions, even if it, in strict economic terms, may benefit from being a free rider. The "cost" of breaking a social norm of this type is likely to increase as the number of cooperating countries increases. In this situation we may have multiple equilibria. In particular, many countries may wish to participate in an international climate agreement, as long as many other countries are participating in an agreement to reduce emissions. However, if a sufficient number of countries for some reason or other do not cooperate, then the cost of breaking the social norm of cooperating will be reduced for the remaining countries. In such a situation most countries may choose to be free riders, and we get an equilibrium with only a small number of cooperating countries.[2]

The Rio Convention on Climate Change does not give the signatories any explicit quantitative commitments for greenhouse gas emissions. It seems plausible, however, that some form of agreement will be reached during the next decade. At least initially, it seems unlikely that all countries will participate in such an

[2]An example of a game of this type is given in the Appendix, available upon request.

agreement. Nevertheless, the discussion above suggests that it may be possible to reach an agreement between a considerable number of countries.

The starting point of the present paper is therefore that some countries commit themselves to cooperating. These countries (henceforth called the cooperating countries) are assumed to coordinate their policies in a way which maximizes the sum of welfare for the whole of this group. The remaining countries (henceforth called the non-cooperating countries) act independently and in pure self-interest, taking greenhouse gas emissions from other countries, as well as international prices of fossil fuels, as given.

Limited participation in the agreement has important consequences. In a situation where all countries participate in an agreement to reduce CO_2 emissions, it can be shown that taxes on the consumption of fossil fuels (or other policies reducing consumption) and taxes on the production of fossil fuels (or other policies reducing production) have identical economic consequences, provided the tax revenues are used identically in the two cases (see, e.g., [12]).[3] This is no longer true when there is limited participation in the international agreement, as in the present context. In this case the international price of fossil fuels, and therefore consumption and production of fossil fuels in the non-cooperating countries, depends on the policies chosen in the cooperating countries. In particular, if the cooperating countries reduce their demand for fossil fuels, international fuel prices will decline. This gives the non-cooperating countries an incentive to increase their use of fossil fuels, thus partly offsetting the reduced emissions from the participating countries. Pezzey [10] has showed that this effect might be quite strong. It is therefore of interest to see if there are ways to avoid or reduce this offsetting emission increase from non-cooperating countries.

One way of avoiding increased emissions from non-cooperating countries is to reduce the supply of fossil fuels from the non-cooperating countries. This has been studied by Bohm [3], who assumes that the cooperating countries can buy or lease fuel deposits in the non-cooperating countries. This option is ignored in the present paper. Instead, it is assumed that the cooperating countries themselves have a positive supply of fossil fuels. This supply can therefore be affected by domestic policy, e.g., a production tax or some type of direct regulation of production of fossil fuels. In Section 3 this type of policy is compared with a policy which aims at reducing the demand for fossil fuels. One of the results is that without more detailed information on demand and supply functions, it is not possible to know whether a demand policy or a supply policy is best for the cooperating countries.

Section 4 considers the optimal choice of supply and demand policies in more detail. It is shown that there are plausible cases in which it is optimal to have policies which aim at reducing both the demand and the supply of fossil fuels. Obvious candidates for policy instruments are taxes (or subsidies) on consumption and production of fossil fuels. Throughout the paper, it is assumed that these are the policy instruments used to achieve desired levels of consumption and production. However, the focus of the analysis is on the optimal levels of consumption and production of fossil fuels, not on what instruments one uses to implement these quantities. The results of the paper are therefore valid (with appropriate reformulations) also for other policy instruments than taxes.

[3]This is also shown in more detail in the Appendix, available upon request.

In addition to choosing their own policies, the cooperating countries are in Section 5 assumed to be able to induce the non-cooperating countries to implement policies which affect consumption and/or production of fossil fuels in these countries.[4] This may be achieved through transfers from the cooperating to the non-cooperating countries, making each country of the latter group equally well off with the transfer and its policy toward fossil fuel consumption/production as it would have been without the transfer and without this policy.

Policy options in the non-cooperating countries are assumed to include policies directed both toward their consumption and their production of fossil fuels.[5] The policy instruments assumed in the analysis are consumption and production taxes (or subsidies). However, also for the non-cooperating countries it is the levels of consumption and production of fossil fuels which matters, not what instruments these countries use to reach these levels. With appropriate reformulations, the results are therefore valid also for policy instruments other than taxes in the non-cooperating countries.

2. NOTATION AND KEY ASSUMPTIONS

Throughout the paper, the markets for fossil fuels are treated as one aggregate, competitive market, called the carbon market. In reality, there are of course several (interrelated) markets for different types of fossil fuels, and all of these deviate more or less from perfect competition. However, these simplifying assumptions are not crucial for the intuitive discussion of the results given in Sections 3–5. It therefore seems plausible that similar results hold also under assumptions less restrictive than those used in the formal analysis. This is partially confirmed by Golombek *et al.* [6], who treat issues related to the present issues, and who extend the analysis to non-competitive supply conditions in the non-cooperating countries.

All consumers and producers (at home and abroad) are assumed to be price takers. In addition, the authorities of the non-cooperating countries regard the price of carbon as exogenous. The reason for this is that these countries act independently, and each of them is assumed to be so small that its influence on the price of carbon is negligible. The cooperating countries, on the other hand, are explicitly assumed to cooperate to maximize the total welfare of the whole group. When choosing their policies, the authorities of the cooperating countries are therefore assumed to take into consideration the effect their policies have on the international price of carbon.

Throughout the analysis, corner solutions are disregarded. In other words, for all relevant prices, some or all countries both among the cooperating and the non-cooperating groups of countries are assumed to have positive consumption and production of carbon. Most of the results would remain valid even if, e.g., production of carbon was zero in one of the two groups of countries.

[4]Using somewhat different assumptions than the present paper, Carraro and Siniscalco [4] have studied the possibility of a group of committed countries to expand the cooperating coalition through the use of transfers to the non-committed countries.

[5]Hoel [9] also considers the two cases in which the policies in the non-cooperative countries are (i) restricted to demand policies and (ii) restricted to supply policies.

TABLE I
Notation

	Cooperating countries	Non-cooperating countries
Consumption	y	Y
Production	x	X
Valuation of consumption	$u(y)$	$U(Y)$
Production cost	$c(x)$	$C(X)$
International price	p	p
Consumer price	q	Q
Producer price	r	R
Demand	$d(q)$	$D(Q)$
Supply	$s(r)$	$S(R)$
Consumer tax	$t^c = q - p$	$T^c = Q - p$
Producer tax	$t^p = p - r$	$T^p = p - R$
Consumer surplus plus producer surplus plus reimbursed tax	$w = u(y) - c(x) - p \cdot (y - x)$	$W = U(Y) - C(X) - p \cdot (Y - X)$
Total emissions	$e = y + Y = x + X$	
Environmental costs in cooperating countries	$n(e)$	
Transfer from cooperating to non-cooperating countries	I	

The analysis of the paper shares the standard limitations of all partial equilibrium analyses. Moreover, all problems related to aggregation and distribution within countries and across the cooperating countries are ignored.

The main notation of the present paper is shown in Table I (where all variables are scalars): The valuation functions $u(y)$ and $U(Y)$ measure the areas under the relevant demand functions $d(q)$ and $D(Q)$, i.e., the aggregate willingness to pay. Consumer surpluses are thus given by $u(y) - qy$ and $U(Y) - QY$. The demand functions $d(q)$ and $D(Q)$ are thus given by $u'(y) = q$ and $U'(Q) = Q$. Similarly, the supply functions $s(r)$ and $S(R)$ are given by $c'(x) = r$ and $C'(X) = R$. It is assumed that d' and D' are negative, and that s' and S' are positive. Throughout, it is assumed that tax revenues are reimbursed to consumers through lump-sum transfers. The sum of consumer surplus, producer surplus, and tax revenue for the cooperating countries (before any transfer to the non-cooperating countries) is given by

$$w = [u(y) - qy] + [rx - c(x)] + [t^c y + t^p x].$$

Inserting $t^c = q - p$ and $t^p = p - r$ and rearranging gives

$$w = u(y) - c(x) - p \cdot (y - x)$$

as in Table I. A similar derivation gives the expression for W in Table I.

3. DEMAND OR SUPPLY POLICY?

In this and the next section it is assumed that there is no climate policy in the non-cooperating countries. With the notation of the previous section, this means that $T^c = T^p = 0$ and $Q = R = p$, and that $I = 0$. If there is no climate policy in the cooperating countries either, we have $q = r = p$, and the equilibrium price and output level are given by the point (p^0, e^0) in Fig. 1, i.e., by $e^0 = d(p^0) + D(p^0) = s(p^0) + S(P^0)$.

Assume now that the cooperating countries implement policies with the aim of reducing *total* carbon emissions to a target level e^* $(< e^0)$. The determination of the optimal level e^* is treated in Section 4. In this section, we regard e^* as exogenous. The target level e^* may be achieved through policies directed toward either the demand or the supply of carbon in the cooperating countries (or a combination, as will be discussed in Section 4). Consider first a policy directed toward the demand of carbon (called a demand policy for short). More specifically, assume that the governments of the cooperating countries introduce a suitable consumption tax t^c in order to make their total use of carbon equal to y^*. In order to reach the target level e^* of total emissions, y^* must be given by

$$y^* + D(p^D) = s(p^D) + S(p^D) = e^* \tag{1}$$

where p^D denotes the international price of carbon under a demand policy.

Given the target e^* for total carbon emissions, the two equations given by (1) determine y^* and the equilibrium international price p^D. Figure 1 illustrates this new equilibrium. It is assumed that the target level e^* is chosen so that the value of y^* solving (1) satisfies $y^* \in (0, d(p^0))$.

As an alternative to reducing their demand to y^*, the cooperating countries could use a "supply policy," limiting their supply to x^*. In order to reach the target level e^* of total emissions, x^* must be given by

$$d(p^S) + D(p^S) = x^* + S(p^S) = e^*, \tag{2}$$

where p^S denotes the international price of carbon under a supply policy.

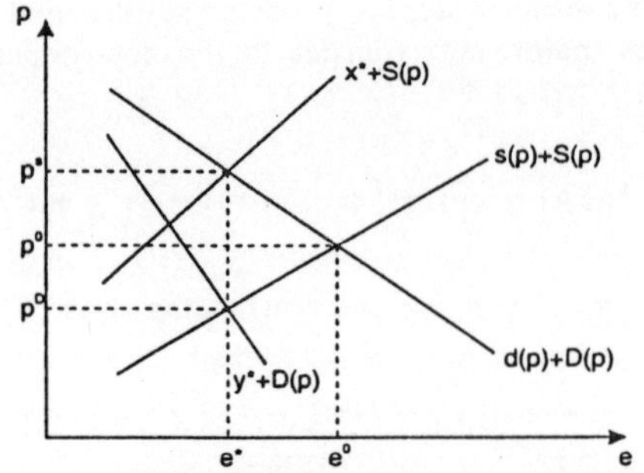

FIGURE 1

The equations given by (2) determine x^* and the equilibrium price p^S for the case of a supply policy. The equilibrium is illustrated in Fig. 1. Similarly to what was assumed above, it is assumed that the target level e^* is chosen so that the value of x^* solving (2) satisfies $x^* \in (0, s(p^0))$.

The prices p^D and p^S are the international market prices for carbon under demand and supply policies, respectively. From Fig. 1 we immediately have the following result:

RESULT 1. *The international price of carbon is higher under a supply policy than under a demand policy.*

Result 1 implies that a non-cooperating country, which is a net exporter of carbon in the case where the cooperating countries use a demand policy, must also be a net exporter of carbon in the case where the cooperating countries use a supply policy (since supply functions are increasing and demand functions are declining). Being a net exporter under both types of policies, the non-cooperating country is clearly best off under the policy type which gives the highest price, which is a supply policy (cf. Result 1). A similar argument holds for a non-cooperating country which is a net importer of carbon under the supply policy. We thus have our second result.

RESULT 2. *If a non-cooperating country is a net exporter of carbon whatever type of policy the cooperating countries use, it is better off when the cooperating countries use a supply policy than it is when the cooperating countries use a demand policy. If a non-cooperating country is a net importer of carbon whatever type of policy the cooperating countries use, it is better off when the cooperating countries use a demand policy than it is when the cooperating countries use a supply policy.*

Under a demand policy, the consumer price which makes demand in the cooperating countries equal to y^*, denoted by q^D, is given by $d(q^D) = y^*$. Similarly, the producer price under a supply policy is given by $s(r^S) = x^*$. The following result holds:

RESULT 3. *In the cooperating countries, both the consumer price and the producer price of carbon are higher under a demand policy than they are under a supply policy, i.e.,* $q^D > q^S = p^S$ *and* $p^D = r^D > r^S$.

Proof. Consumer prices for the two types of policies satisfy the two equations

$$\begin{aligned} d(q^D) &= y^* = e^* - D(p^D) \\ d(q^S) &= d(p^S) = e^* - D(p^S), \end{aligned} \tag{3}$$

which give

$$d(q^D) - d(q^S) = D(p^S) - D(p^D) < 0, \tag{4}$$

where the inequality follows from $p^S > p^D$ and $D' < 0$. Since $d' < 0$ it follows from this inequality that $q^D > q^S$. This completes the proof of the first part of Result 3. The second part of the Result ($r^D > r^S$) is proved in a similar way. ■

Without more detailed information on demand and supply functions, it is not possible to know whether a demand policy or a supply policy is best for the cooperating countries. Households in the cooperating countries pay for their use of carbon, they receive income from production of carbon, and they receive reimbursed revenue from the taxation of carbon consumption or carbon production. It follows from Result 3 that as users of carbon, households in the cooperating countries are better off with a supply policy than with a demand policy. On the other hand, Result 3 also implies that as recipients of income from carbon production, households in the cooperating countries are better off with a demand policy than with a supply policy. Moreover, in addition to the user and producer role, the households in the cooperating countries receive reimbursed tax revenue in both cases. The tax revenue generally differs in the two cases. Even if the distribution of tax revenue among the cooperating countries is the same under the two types of policies, the reimbursed tax revenue to households in each country will generally differ between the two types of policies. It follows from the discussion above that there is no unambiguous answer to the question of which type of policy is best for the households of the cooperating countries. However, other things equal, a move from a demand to a supply policy (with unchanged distribution of tax revenue among the cooperating countries) will give a relative increase of welfare to those of the cooperating countries which have high net imports (or low net exports) of carbon from those which have low net imports (or high net exports) of carbon.

While it is not obvious which of a demand and a supply policy is best for the cooperating countries, it intuitively seems likely that some combination of these two policy types might be better than any of the two. In the next section it is shown that this intuition is correct, and the optimal mix of demand and supply policies in the cooperating countries is derived and discussed.

4. THE OPTIMAL MIX OF DEMAND AND SUPPLY POLICIES

Using the notation from Section 2, we see that market equilibrium implies that

$$y + D(p) = x + S(p) \tag{5}$$

which gives

$$p = p(y - x), \tag{6}$$

where

$$p' = \frac{1}{S' - D'} > 0. \tag{7}$$

The cooperating countries are assumed to maximize the sum of consumer and producer surplus plus tax revenues (i.e., w) minus their environmental costs ($n(e)$). Since it is total emissions from all countries which affect the climate, it is these total emissions (e) which enter the environmental cost function of the cooperating countries.

The optimal choice of x and y for the cooperating countries is thus the solution to

$$\max_{x,y} u(y) - c(x) - p(y - x) \cdot (y - x) - n(e). \tag{8}$$

Inserting $e = y + D(p)$ and (6) into (8) and assuming an interior solution, we find

$$u' = p + (y - x)p' + \frac{S'}{S' - D'}n'$$
$$c' = p + (y - x)p' + \frac{D'}{S' - D'}n'. \tag{9}$$

Writing these equations in terms of the consumer and producer tax, we have

$$t^c \equiv u' - p = (y - x)p' + \frac{S'}{S' - D'}n'$$
$$t^p \equiv p - c' = -(y - x)p' + \frac{-D'}{S' - D'}n', \tag{10}$$

giving

$$t^c + t^p = n'(e). \tag{11}$$

Equation (11) may be expressed as the following result:

RESULT 4. *The* sum *of the optimal consumer and producer tax rates in the cooperating countries is positive and equal to the marginal environmental costs of total emissions (in the cooperating countries).*

Result 4 implies that there can never be a subsidy both on production and consumption. This is intuitively obvious, as such a combined subsidy would *increase* emissions compared with the case of no climate policy.

Result 4 says nothing about if and when one should subsidize either consumption or production. Our four next results deal with this question. We start by giving the results and proofs, after which an interpretation is offered.

RESULT 5. *If the aggregate consumption of carbon in the cooperating countries is sufficiently close to their aggregate production of carbon (i.e., if $|y - x|$ is sufficiently small), both consumers and producers in the cooperating countries should be taxed.*

Proof. This result follows directly from (10), since S', $-D'$, and n' are all positive. ■

RESULT 6. *For a sufficiently large equilibrium value of the marginal environmental costs of total emissions (in the cooperating countries), both consumers and producers in the cooperating countries should be taxed.*

Proof. From (10) it is easily seen that t^c can only be negative if $n' < -(1 + (-D')/S')(y - x)p'$, which cannot hold if n' is sufficiently large. Similarly, t^p can only be negative in $n' < (1 + S'/(-D'))(y - x)p'$, which cannot hold if n' is sufficiently large. ■

RESULT 7. *For a sufficiently low equilibrium value of the marginal environmental costs of total emissions (in the cooperating countries), there should almost always be either a consumer subsidy or a producer subsidy in the cooperating countries.*

Proof. For n' sufficiently low, $|y - x| \cdot p'$ will almost always be larger than n'. It follows from (10) that t^c or t^p in this case will be negative. ■

RESULT 8. *If either consumption or production is subsidized, then production is subsidized if the group of cooperating countries is a net importer of carbon (i.e., if* $y - x > 0$*), while consumption is subsidized if the group of cooperating countries is a net exporter of carbon (i.e., if* $y - x < 0$*).*

Proof. The result states that $y - x > 0$ is a necessary condition for $t^p < 0$, and that $y - x < 0$ is a necessary condition for $t^c < 0$. This follows directly from (10), since S', $-D'$, and n' are all positive. ■

To interpret the four previous results, it is helpful first to consider the case with no environmental concern. In this case n' is zero, and it is clear from (9) that consumers and producers in the cooperating countries should face the same price of carbon. This price is equal to the international price of carbon (p) plus an "optimal tariff" term, given by $(y - x)p'$. It is clear from (10) that t^c and t^p have opposite signs as long as $n' = 0$.

If the group of cooperating countries is a net importer of carbon in equilibrium (i.e., $y > x$), it wants the international price of carbon to be kept low. This can be achieved by raising the domestic price of carbon above the international price, which encourages domestic production and discourages domestic consumption. Net imports are thus discouraged, which presses down the price of carbon. The most obvious way to raise the domestic price from p to $p + (y - x)p'$ for both consumers and producers is to have an import tariff equal to $(y - x)p'$. Alternatively, one could tax consumption and subsidize production, both at the rate $(y - x)p'$, which is precisely what follows from (10) for $n' = 0$.

If on the other hand the group of cooperating countries is a net exporter of carbon in equilibrium (i.e., $y < x$), we get the opposite effect: In this case the cooperating countries want the international price of carbon to be kept high. This can be achieved by keeping the domestic price of carbon below the international price, which discourages domestic production and encourages domestic consumption. Net exports are thus discouraged, which presses up the price of carbon. The most obvious way to reduce the domestic price from p to $p + (y - x)p'$ for both consumers and producers is to have an export tariff equal to $-(y - x)p'$. Alternatively, one could tax production and subsidize consumption, both at the rate $-(y - x)p'$, which is precisely what follows from (10) for $n' = 0$.

We next turn to the case in which n' is positive. It is useful to first consider what implications a positive n' has for the optimal tax structure, ignoring the optimal tariff argument above. In this case, i.e., for $y = x$, it follows from (10) that t^c and t^p are both positive. From the discussion above we also know that the optimal tariff argument by itself requires a subsidy on either production or consumption of carbon and a tax of the same magnitude on the side of the market which is not subsidized. The isolated effect of the environmental cost, on the other hand, is that both production and consumption should be taxed. The message of Results 7 and 8 is that as long as the equilibrium marginal environmental cost is not too large, and domestic consumption of carbon is not too close to domestic production, the optimal tariff argument for subsidizing production or consumption of carbon is stronger than the argument for taxing both production and consumption of carbon as a consequence of the environmental cost. However, it follows from Results 5 and 6 that if the equilibrium marginal environmental cost is sufficiently large, or domestic consumption of carbon is sufficiently close to domestic production, then consumption and production should both be taxed. In other words, in these cases the argument for positive taxes on both production and consumption, due to the environmental cost, dominates the optimal tariff argument for subsidizing consumption or production.

5. EXTENDING THE POLICIES TO THE NON-COOPERATING COUNTRIES

In this section, it is assumed that the levels of both consumption and production in the non-cooperating countries, i.e., both Y and X, are determined by the cooperating countries. As mentioned in the introduction, each of the non-cooperating countries must be compensated for letting its consumption and production be determined by the cooperating countries, instead of choosing the consumption and production which is optimal at the international price p. For the whole group of non-cooperating countries, the optimal consumption and production levels are $D(p)$ and $S(p)$, which will generally differ from the imposed levels Y and X. The sum of consumer and producer surplus and tax revenues for the group of non-cooperating countries (excluding the transfer I) is denoted W^{opt} when consumption and production is chosen optimally and by W^{act} when consumption and production are given by Y and X. From the notation of Section 2 it is clear that

$$W^{\text{opt}} = W^{\text{opt}}(p) = \max_{\chi,\psi}\left[U(\psi) - C(\chi) - p\cdot(\psi-\chi)\right]$$

$$W^{\text{act}} = W^{\text{act}}(p,Y,X) = U(Y) - C(X) - p\cdot(Y-X). \tag{12}$$

The sum of transfers from the cooperating countries to the non-cooperating countries is given by

$$I = I(p,Y,X) = W^{\text{opt}}(p) - W^{\text{act}}(p,Y,X). \tag{13}$$

Inserting from (12) and using the envelope theorem, it follows from (13) that

$$I_p = Y - X + S(p) - D(p)$$

$$I_Y = -U'(Y) + p$$

$$I_X = -p + C'(X). \tag{14}$$

The group of cooperating countries chooses the levels of y, x, Y, and X, subject to the balance equation $y + Y = x + X$. This means that demand and supply of carbon are equal whatever the international price p is. The international price may therefore be treated as a choice variable for the cooperating countries. The interpretation of this is the following. The choice of the vector (y,x,Y,X) implicitly gives a vector of consumer and producer prices (q,r,Q,R), given by $d(q) = y$, $s(r) = x$, $D(Q) = Y$, and $S(R) = X$. For any price p, the corresponding consumer and producer taxes (which may be positive or negative) are $t^c = q - p$, $t^p = p - r$, $T^c = Q - p$ and $T^p = p - R$. To any vector (y, x, Y, X, p) there thus corresponds a tax vector (t^c, t^p, T^c, T^p). Given these taxes, the international price follows from

$$d(p + t^c) + D(p + T^c) = s(p - t^p) + S(p - T^p). \tag{15}$$

Choosing the vector (y, x, Y, X, p) is thus equivalent to choosing an appropriate tax vector (t^c, t^p, T^c, T^p), and letting the international price of carbon be determined by (15). The taxes t^c and t^p are implemented in the cooperating countries, while the non-cooperating countries must be given transfers which add up to $I(p, Y, X)$ (given by (13)) to be willing to impose the taxes T^c and T^p on their consumption and production of carbon.

The cooperating countries choose the vector (y, x, Y, X, p) so that the following maximization problem is solved:

$$\underset{x,y,X,Y,p}{\text{maximize}}\,[u(y) - c(x) - p\cdot(y - x) - I(p, X, Y) - n(y + Y)],$$
$$\text{s.t. } y + Y = x + X. \tag{16}$$

From (16), it is clear that whatever the values of (y, x, Y, X), the cooperating countries would like the international price p to minimizes their total payments to the non-cooperating countries. In other words, p is chosen to minimize $p\cdot(y - x) + I(p, X, Y)$, which gives[6] $y - x + I_p = 0$, or, using (14),

$$D(p) = S(p) \tag{17}$$

The following result follows directly from (17):

RESULT 9. *The optimal tax policy gives an equilibrium international price which is such that the group of non-cooperating countries in the absence of any taxes would want to consume and produce the same amount of carbon.*

Note that Result 8 does *not* say that net imports are zero in both groups of countries in equilibrium. It only says that the international price of carbon is such that *if* all taxes were removed *and* the international price of carbon nevertheless remained unchanged, then the sum of the optimal net imports of carbon to the non-cooperating countries would be zero.

The remaining first-order conditions for the maximization problem (16) are straightforward to calculate (using (14)) and are given by

$$u'(y) = U'(Y) = p + \mu + n'(e)$$
$$c'(x) = C'(X) = p + \mu, \tag{18}$$

where μ is the shadow prices of the constraint $y + Y = x + X$. The shadow price μ cannot in general be signed.

[6]Since $I_{pp} = S'(p) - D'(p) > 0$ (from (14)), the second-order condition for this minimization problem is satisfied.

From the notation of Section 2 it is clear that (18) may be written in terms of taxes as

$$t^{c} = T^{c} = \mu + n'(e)$$
$$t^{p} = T^{p} = -\mu. \tag{19}$$

From (18) or (19) we immediately get the following result:

RESULT 10. *Both the consumer price and the producer price of carbon should be the same in the non-cooperating as in the cooperating countries, i.e., the tax rates on consumption and production of carbon should both be the same in the non-cooperating and in the cooperating countries.*

Equation (19) implies that

$$t^{c} + t^{p} = T^{c} + T^{p} = n'(e). \tag{20}$$

It is clear from (20) that Result 4, which says that the sum of the consumer and producer tax rates is positive (equal to the marginal environmental costs of total emissions), is valid also in the present case.

To see whether Results 5–8 remain valid, it is useful to illustrate the equilibrium of the present situation. In Fig. 2, the equilibrium price p^* is given by the intersection of the demand and supply curve for the non-cooperating countries (17). The *aggregate* demand and supply curves must always lie outside (i.e., to the right) of the corresponding curves for the non-cooperating countries. In Fig. 2, the intersection between the aggregate demand and supply curves is denoted by A. This intersection point, with the corresponding price p^A, will generally lie above (as in Fig. 2) or below the p^* line; only by coincidence with $p^A = p^*$.

From Fig. 2, it is easily verified that Results 6 and 7 are valid also for the present case. Consider first the case in which the equilibrium value of n' is "large." More precisely, assume that this equilibrium value is equal to the distance of the line segment EF in Fig. 2. When the line segment EF is sufficiently long, it is clear that its intersection with the aggregate demand curve, denoted by q^0 in Fig. 2, must lie above p^*, and that its intersection with the aggregate supply curve,

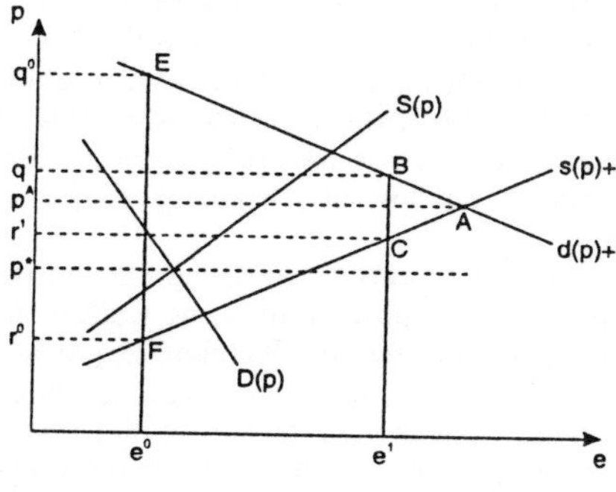

FIGURE 2

denoted by r^0 in Fig. 2, must lie below p^*. In other words, we get

$$t^c = T^c = q^0 - p^* > 0$$
$$t^p = T^p = p^* - r^0 > 0, \tag{21}$$

which proves Result 6.

Consider next the case in which the equilibrium value of n' is "small." More precisely, assume that this equilibrium value is equal to the distance of the line segment BC in Fig. 2. When the line segment BC is sufficiently short, it is clear from Fig. 2 that the line segment BC must lie entirely above (as in Fig. 2) or entirely below the p^* line, except for the special case in which $p^A = p^*$. If $p^A > p^*$, as in Fig. 2, we therefore get $q^1 > r^1 > p^*$, i.e., $t^p = T^p = p^* - r^1 < 0$, i.e., production is subsidized. If $p^A < p^*$, we would have $p^* > q^1 > r^1$, i.e., $t^c = T^c = q^1 - p^* < 0$, i.e., consumption is subsidized. This proves Result 7.

Finally, it is shown in the Appendix that also Results 5 and 8 remain valid in the present case.

6. CONCLUSIONS

An important conclusion from the analysis above is that with limited participation in a international climate agreement, policies should be directed toward both consumption and production of fossil fuels. The paper gives particular emphasis on carbon taxes as the policy instruments and describes some of the properties of the optimal mix of consumption and production taxes. Loosely speaking, the main results are that production should be subsidized if the marginal environmental costs of total emissions are sufficiently low (in the cooperating countries) and the cooperating countries have sufficiently large net imports of carbon. Consumption should be subsidized if the marginal environmental costs of total emissions are sufficiently low and the cooperating countries have sufficiently large net exports of carbon. In all other cases both production and consumption of carbon should be taxed at a positive rate.

Another important conclusion of the paper is that the cooperating countries should try to induce the non-cooperating countries to implement policies which affect consumption and/or production of fossil fuels in these countries. This may be achieved through transfers from the cooperating to the non-cooperating countries, making each country of the latter group equally well off with the transfer and its policy toward fossil fuel consumption/production as it would have been without the transfer and without this policy. It is shown that when this possibility exists, it is optimal for the cooperating countries to induce the non-cooperating countries to tax their consumption and production of carbon at the same rate as in the cooperating countries.

Note that the arrangement with a transfer-induced climate policy in the "non-cooperating" countries gives these countries the same welfare as they have in the absence of transfers and in the absence of any climate policy. The cooperating countries, on the other hand, are better off when they can induce the non-cooperating countries to have a climate policy than they are if there is no climate policy in the non-cooperating countries, in spite of the transfer the cooperating countries must pay in the former case. In the decision each country must make of whether to join the group of countries who commit themselves to cooperation, an important factor is the welfare of the countries within and outside the cooperating group. The incentive to cooperate is clearly higher as the short-run welfare gain of free-riding becomes smaller. It follows from the discussion above that the incentives to commit oneself to cooperation are higher in a situation with transfer-induced climate policies in the non-cooperating countries than they are in the case where the non-cooperating countries have no climate policy.

The analysis in this paper has focused on the greenhouse problem. However, several of the qualitative results are relevant also for other international environmental problems. For several such problems, policies in a group of countries which reduce their emissions will also affect international prices of traded goods. This may in turn affect production and consumption patterns, and thereby emissions, in other countries. Whenever there are international spillovers of harmful emissions, this must be taken into consideration by the countries which have an explicit environmental policy. One possibility is to combine emission taxes with imports and export tariffs (positive or negative) for those traded goods whose prices may affect harmful emissions from countries which have no explicit environmental policy. An alternative (or complementary) possibility would be to induce the latter countries, through appropriate transfers, to introduce policies which reduce their emissions.

APPENDIX

Proof of Results 5 and 8 for the Case in Which Policies are Extended to the Non-cooperating Countries

We start by proving Result 8. From the notation of Section 2 it is clear that

$$y - x = X - Y = S(R) - D(Q). \tag{A.1}$$

If production is subsidized, we know from Result 4 that consumption is taxed. This means that $Q > R > p^*$. Inserting this into (A.1) and using Result 9 gives

$$y - x = S(R) - D(Q) > S(p^*) - D(p^*) = 0. \tag{A.2}$$

If consumption is subsidized, we know from Result 4 that production is taxed. This means that $p^* > Q > R$. Inserting this into (A.1) and using Result 9 gives

$$y - x = S(R) - D(Q) < S(p^*) - D(p^*) = 0. \tag{A.3}$$

We have thus shown that a subsidy on production can occur only if $y - x > 0$, and that a subsidy on consumption can occur only if $y - x < 0$. This proves that Result 8 is valid for the present case.

We now turn to Result 5. From (18) we know that $Q = R + n'$; thus (A.1) may be written as

$$y - x = S(R) - D(R + n'). \tag{A.4}$$

For any given n'-value, the r.h.s. of (A.4) is increasing in R. If production is not taxed, $R \geq p^*$, i.e.,

$$y - x \geq S(p^*) - D(p^* + n') = \alpha(n'), \tag{A.5}$$

where $\alpha(n')$ is a positive number (increasing in n'), since $S(p^*) = D(p^*)$, $n' > 0$, and $D' < 0$.

From $Q = R + n'$ we could rewrite (A.1) as

$$y - x = S(Q - n') - D(Q). \tag{A.6}$$

If consumption is not taxed, $Q \leq p^*$, i.e.,

$$y - x \leq S(p^* - n') - D(p^*) = \beta(n'), \tag{A.7}$$

where $\beta(n')$ is a negative number (declining in n'), since $S(p^*) = D(p^*)$, $n' > 0$, and $S' > 0$.

From the discussion above, it is clear that if

$$\beta(n') < y - x < \alpha(n'), \tag{A.8}$$

then neither of the inequalities (A.5) or (A.7) are satisfied. In other words, both production and consumption must be taxed if (A.8) is valid. This proves Result 5 for the present case.

REFERENCES

1. S. Barrett, "Self-enforcing international environmental agreements." CSERGE Working Paper GEC 92-24, Centre for Social and Economic Research on the Global Environment, London (1992).
2. A. Bauer, International cooperation over environmental goods, mimeo, University of Munich (1993).
3. P. Bohm, Incomplete international cooperation to reduce CO_2 emissions: alternative policies, J. Environ. Econom. Manage. **24**, 258–271 (1993).
4. C. Carraro and D. Siniscalco, Strategies for the international protection of the environment, *Journal of Public Economics* **52**, 309–328 (1993).
5. J. Elster, Social norms and economic theory. J. Econom. Perspect. **3**, 99–118 (1989).
6. R. Golombek, C. Hagem, and M. Hoel, The design of a carbon tax in an incomplete international climate agreement, *in* "Trade, Innovation, Environment" (C. Carraro, Ed.), Milano, in press.
7. M. Hoel, International environment conventions: The case of uniform reductions of emissions. Environ. Resour. Econom. **2**, 141–159 (1992).
8. M. Hoel, Tradeable emission quotas for CO_2: Quotas on use of carbon or on production of carbon?, CICERO Working Paper 92:1, Center for Climate and Energy Research, Oslo (1992).
9. M. Hoel, Efficient international climate agreements in the presence of free riders, CICERO Working Paper 92:7, Center for Climate and Energy Research, Oslo (1992).
10. J. Pezzey, Analysis of unilateral CO_2 control in the European community, Energy J. **13**, 159–172 (1992).
11. A. Torvanger, Efficient contracts in a game of nations pursuing greenhouse gas emissions abatement, CICERO Working Paper 93:3, Center for Climate and Energy Research, Oslo (1993).
12. J. Whalley and R. Wigle, The international incidence of carbon taxes, *in* "Global Warming: Economic Policy Responses," (R. Dornbusch and J. M. Poterba, Eds.), MIT Press (1991).

[16]

European Economic Review 36 (1992) 400–406. North-Holland

Carbon taxes

An international tax or harmonized domestic taxes?*

Michael Hoel

Center for International Climate and Energy Research (CICERO), N-0316 Oslo, Norway

1. Introduction

At least since Pigou, it is well known that emission taxes are efficient instruments in environmental policy. It is therefore not surprising that some kind of carbon tax often is proposed as part of an international climate policy. It is, however, not always clear exactly what kind of carbon tax the proponents have in mind. It is important to distinguish between three types of carbon taxes:

- a domestic tax to achieve a domestic target,
- an international carbon tax,
- internationally harmonized domestic taxes.

After a brief discussion of the first of these three types of carbon taxes in section 2, sections 3 and 4 discuss the two latter types of carbon taxes. It is for these two latter types of carbon taxes that the tax rate(s) are determined as part of an international greenhouse gas agreement. It is shown that provided emissions of CO_2 can be monitored at negligible costs, an international carbon tax is superior to harmonized domestic taxes.

2. A domestic carbon tax

A domestic carbon tax is relevant no matter how a country's emission target is determined. A country may have a unilateral policy of reducing CO_2 emissions; it may have a target level imposed on it through some rigid

*The paper is based on research at CICERO and the Centre for Research in Economics and Business Administration (SNF), Oslo.

type of international agreement;[1] or it may have chosen its own emission level taking account of the price it has to pay for emission permits under an international agreement involving tradeable emission permits. No matter how the target level of emissions is determined, a domestic carbon tax is superior to most other policy instruments to achieve the target level of emissions, provided we are considering a country which has reasonably well functioning markets.

3. An international carbon tax

An international tax could be designed in the following way: The government of each country pays a tax, proportional to its CO_2 emissions, to an international agency. The tax revenue (minus administrative costs) is reimbursed to the governments of the participating countries according to some specified rules. Without loss of generality, we may think of these rules as a set of fixed reimbursement shares. Assuming that each country is so small that it considers both total CO_2 emissions and total tax revenue as (practically) independent of its own CO_2 emissions, it also considers its own reimbursement as given, since this reimbursement is a constant share of the total tax revenue.[2] A country which maximizes its national income therefore chooses its emission level so that the sum of CO_2 taxes paid to the international agency and its domestic costs of reducing CO_2 emissions are minimized. This cost minimization implies that CO_2 emissions should be chosen so that the marginal cost of reducing CO_2 emissions is equal to the international tax rate per unit of CO_2 emissions. Since this tax rate is the same for all countries, we thus get an outcome in which the marginal cost of reducing CO_2 emissions are equal for all countries. This is the condition for international cost efficiency, defined as an emission pattern which achieves a goal for world wide emissions at as low costs as possible. An international CO_2 tax of the above type thus gives a distribution of CO_2 emissions which minimizes the costs of reaching a specified level of total CO_2 emissions. Total CO_2 emissions follow from the choice of the tax rate: the higher the tax rate, the lower are total emissions.

An important feature of the scheme above is that no matter how the tax revenue is reimbursed, the end result is an efficient allocation of emissions between countries. The distribution of total costs (i.e., costs of reducing emissions plus net taxes) between the participating countries is determined by the reimbursement rules. These rules can thus be determined purely from

[1] E.g. a uniform percent reduction of emissions for all participating countries.

[2] See Hoel (1991, 1992b,c) for a discussion of the complications which arise if some countries are 'large', in the sense that they do not ignore their own influence on total emissions.

considerations of fairness. It is beyound the scope of the present paper to discuss various concepts of 'fairness'. However, in order to achieve broad voluntary participation in an agreement, a plausible *minimum* requirement is that each country regards itself better off under the agreement than without any international cooperation.[3]

In the discussion above, each country has been assumed to maximize its national income, defined as its GDP minus net tax payments to the international agency. This is obviously a drastic simplification compared to real world policy goals. An important question is therefore to what extent our conclusions from previous sections remain valid with more complex preferences and/or various market failures and constraints on economic policy. In particular: How robust is the conclusion that an international CO_2 tax (combined with some reimbursement rules) gives an efficient allocation of CO_2 emissions between countries?

To see whether or not a uniform international CO_2 tax is efficient, consider the welfare function $w_j(v_j, s_j)$, where v_j is emissions from country j, while s_j is net transfers to country j from other countries. This function is defined as follows:

$$w_j(v_j, s_j) = \max_{x_j} \{u_j(x_j) \quad \text{s.t.} \quad x_j \in X_j(s_j);\ v_j = f_j(x_j)\}. \tag{3.1}$$

Here x_j is a vector of economic variables, such as production and consumption of different goods, employment in particular sectors, various prices, policy instruments such as taxes and subsidies, etc. The objective function of the country is $u_j(x_j)$, and may depend in a complex way of all variables included in the vector x_j. The feasible set X_j of possible x_j-vectors may depend on the net transfers from abroad.[4] Finally, total emissions depend on the vector x_j, usually in a very simple way.[5]

Since $u_j(x_j)$ is a very general function, and the constraints implied by $X_j(s_j)$ may include all relevant institutional constraints, the welfare function $w_j(v_j, s_j)$ is very general. A reasonable assumption is that w_j is strictly increasing in s_j. Moreoever, we assume that w_j is increasing in v_j for v_j-values below some critical value.

Given a constraint on total emissions, denoted by V, an efficient $(v_1, \ldots, v_N, s_1, \ldots, s_N)$ vector is defined by the following maximization problem:

[3]For a further discussion of issues of fairness and of the decision of whether or not to participate in an international climate agreement, see Barrett (1989, 1990), Grubb (1990), Hoel (1992a), Pearce (1990) and Welsch (1991).

[4]In the optimization problem below, the sum of emissions from all countries are exogenous, and is therefore not explicitly included as a variable which will affect u_j and possibly X_j.

[5]If one only considers CO_2 emissions from fossil fuels, the function f_j will be a linear function of the use of different types of fossil fuels.

$$\text{Maximize } \Sigma_i \alpha_i w_i(v_i, s_i), \quad \text{s.t.} \quad \Sigma_i v_i \leqq V, \quad \text{and} \quad \Sigma_i s_i = 0, \tag{3.2}$$

where the α_i's are some non-negative weights. The solution to this maximization problem of course depends on the vector $\alpha = (\alpha_1, \ldots, \alpha_N)$. Whatever the α-vector is, the solution to (3.2) must satisfy the condition

$$w_{1v}/w_{1s} = \cdots = w_{Nv}/w_{Ns} \tag{3.3}$$

(where $w_{jv} = \partial w_j / \partial v_j$, etc.). All allocations of $(v_1, \ldots, v_N, s_1, \ldots, s_N)$ satisfying (3.3) are thus *efficient* in standard economic terminology. Which of these efficient allocations is regarded as *best* depends on the α-vector.

The special case in which the objective of each country is to maximize its national income is given by $w_j = r_j(v_j) - s_j$, where $r_j(v_j)$ is country j's GDP. In this case (3.3) implies that the marginal income of emissions, $r_j'(v_j)$, should be equalized across countries. For this special case there is thus a unique efficient emission vector $(v_1, \ldots, v_N)$, while the vector $(s_1, \ldots, s_N)$ will depend on the α-vector. For more general w_j-functions, however, the vector $(v_1, \ldots, v_N)$ will depend on the α-vector.

Consider an international tax on CO_2 emissions. Country j pays a tax equal to tv_j, and is given a reimbursement equal to $\beta_j tV$. Without loss of generality we may ignore other transfers between countries, so that $s_j = -tv_j + \beta_j tV$ and $w_j = w_j(v_j, -tv_j + \beta_j tV)$. By assumption, country j regards t, β_j and V as exogeneous, so that maximization of $w_j(\cdot)$ implies

$$w_{jv}/v_{js} = t. \tag{3.4}$$

It is immediately clear from (3.3) and (3.4) that as long as all countries face the same CO_2 tax t, an efficient allocation of emissions is reached.

Notice that *any* efficient allocation $(v_1^*, \ldots, v_N^*, s_1^*, \ldots, s_N^*)$, i.e. a solution to (3.2) for *any* non-negative α-vector, may be reached by an international CO_2 tax and a suitable β-vector. The appropriate CO_2 tax follows directly from (3.3) and (3.4):

$$t = w_{jv}(v_j^*, s_j^*)/w_{js}(v_j^*, s_j^*). \tag{3.5}$$

and the appropriate tax reimbursement vector β follows from the equality $s_j^* = -tv_j^* + \beta_j t \Sigma_i v_i^*$, giving

$$\beta_j = (s_j^* + tv_j^*)/t\Sigma_i v_i^*. \tag{3.6}$$

In particular, all efficient allocations satisfying the constraint

$$w_j(v_j^*, s_j^*) \geqq w_j^0 \ \forall j, \tag{3.7}$$

may be reached through a suitable choice of (t, β), where w_j^0 is country j's welfare level under the non-cooperative equilibrium.[6]

4. Harmonized domestic taxes

Consider next an international agreement which instead of an international tax requires each participating country to impose a specific domestic carbon tax on its CO_2 emissions. Wouldn't such an agreement have properties which are quite similar to those of the international carbon tax discussed above? The answer is no: The most important difference is that an agreement of harmonizing domestic carbon taxes would have no built in mechanism to make the distribution of burdens satisfactory. As a first approximation (see below for other differences), harmonized domestic taxes would be similar to an international tax where the reimbursement rules happened to be set so that each country *in equilibrium* pays zero net taxes. This gives a particular cost distribution between countries, which only by chance would be reasonably fair by any chosen criterium.

Distributional issues aside, there are other problems with an international agreement of harmonizing domestic CO_2 taxes: It is not clear what such an agreement implies, in particular if some of the participating countries are non-market economies. Even in market economies, it is not clear whether such a harmonized CO_2 should be in *addition* to various other domestic taxes on fossil fuels the countries might have, or *instead* of all existing taxes on fossil fuels. If existing taxes reflect domestic externalities in transportation etc., an internationally agreed upon CO_2 tax ought to be an addition to existing taxes, while one could argue that the CO_2 tax should take the place of other taxes on fossil fuels if these taxes are pure revenue raising taxes.

Another problem with an international agreement requiring equal domestic CO_2 taxes is the free rider problem: It is in each country's interest to have little or no restrictions on their own CO_2 emissions, given the CO_2 emissions from other countries. If a country is required to have a CO_2 tax through an international agreement, it is therefore in the interest of the country to try to make this tax be as ineffective as possible. One way to do this is to reduce other domestic taxes on fossil fuels, e.g. taxes on gasoline which several countries have for domestic purposes. Even if a country doesn't directy reduce such domestic taxes, it might raise them less than it would have, had it not been for the imposed CO_2 tax. Another way to reduce the effect of the imposed CO_2 tax is to manipulate prices of other domestic goods. Roughly speaking, a country should tax close substitutes to fossil fuels

[6]The results above should not come as any surprise: As John Moore pointed out to me in private conversation, they follow directly from the two fundamental theorems of welfare economics applied to the present context.

and subsidize complements. Obvious examples are taxes on other types of energy (e.g. hydroelectric power) and subsidies on automobiles and air conditioning. This type of price policy will reduce the effect of an imposed CO_2 tax on a country's consumption and production pattern, and thereby reduce the cost for the country, even though the country in a formal sense is sticking to the international agreement. In spite of the possibility of each country to reduce the effect of a CO_2 tax, any realistic goal of global CO_2 emissions could probably be achieved with a sufficiently high CO_2 tax. The point is, however, that each country's attempts to reduce the effect of the CO_2 tax leads to inefficiencies at the national level: Each country could have achieved whatever emission level it has in equilibrium at a lower cost if it was allowed to reduce the CO_2 tax and at the same time remove all counteracting policies of the type mentioned above.

5. Concluding comments

This analysis above suggests that an international tax has some advantages over internally harmonized domestic taxes. However, it may be politically more realistic to reach an agreement of harmonizing domestic taxes rather than setting up the necessary international institutional framework for an international carbon tax. In any case, harmonization of domestic taxes is likely to be preferable on efficiency grounds to rigid types of agreements of the type 'uniform percentage reductions of emissions'. It might also be easier to monitor compliance with an agreement of harmonizing domestic taxes than the necessary monitoring of emissions under an international carbon tax: Although CO_2 emissions are closely linked to the burning of fossil fuels, and one *today* has reasonably good statistics over uses of such fuels, this need no longer be the case if one introduces an international tax on the use of fossil fuels.

References

Barrett, S., 1989, On the nature and significance of international environmental agreements, Mimeo. (London Business School, London).

Barrett, S., 1990, International environmental agreements as games, Mimeo. (London Business School).

Grubb, M., 1989, The greenhouse effect: Negotiating targets (Royal Institute of International Affairs, London).

Hoel, M., 1991, Efficient international agreements for reducing emissions of CO_2, The Energy Journal 12, no. 2, 93–107.

Hoel, M., 1992a, International environment conventions: The case of uniform reductions of emissions, Environmental and Resource Economics, forthcoming.

Hoel, M., 1992b, Emission taxes in a dynamic game of CO_2 emissions, in: R. Pethig, ed., Conflict and cooperation in managing environmental resources (Springer Publishing House), forthcoming.

Hoel, M., 1992c, How should international greenhouse gas agreements be designed?, in: P. Dasgupta, K.-G. Maler and A. Vercelli, eds., The economics of transnational commons (Oxford University Press) forthcoming.

Pearce, D., 1990, Greenhouse gas agreements: Part 1 (University College, London).

Welsch, H., 1991, Equity and efficiency in international CO_2 agreements, Institute of energy economics, Mimeo. (University of Cologne, Cologne).

[17]

OXFORD REVIEW OF ECONOMIC POLICY, VOL. 7, NO. 2

EMPIRICAL MEASURES OF CARBON TAXES

ALAN INGHAM
JAMES MAW
ALISTAIR ULPH
University of Southampton

I. INTRODUCTION

The possibility that increases in the atmospheric concentration of carbon dioxide (CO_2), one of the 'greenhouse gases', as a result of emissions of CO_2 associated with the combustion of fossil fuels may contribute to the problem of global warming is one of the factors that policy makers now need to consider in shaping energy policies. (See Houghton, Jenkins, and Ephraums, 1990, for a useful summary of the scientific evidence about global warming.) The fact that there are still substantial uncertainties surrounding global warming, and that the problem is of a long-term and global nature make the analysis of such energy policies challenging (the papers by Pearce and Helm, Dasgupta, Barrett, and Mäler in a recent issue of the *Oxford Review of Economic Policy* (1990), give useful introductions to how economists address such problems, not just in the context of global warming).

In this paper we are concerned not with the general arguments about policy towards global warming and CO_2 emissions in particular (see Ingham and Ulph, 1990*a*, for a discussion of these general arguments), but rather with much narrower question of trying to provide an empirical assessment of the possible scale of the impact of any such policy on the UK energy sector. To assess this, we shall consider one widely discussed policy instrument—a tax on CO_2 emissions—and ask at what level would such a tax need to be set to achieve particular policy targets, and what would be the consequent costs of imposing such a tax. We shall not discuss the desirability of such targets, but just use some of the more commonly discussed possible targets, such as the UK Government's target of stabilizing CO_2 emissions at their 1990 level by the year 2005. The choice of a carbon tax as the policy instrument to calculate does not reflect a belief either that this is necessarily the best policy instrument or that the government would actually impose such a tax. It is chosen because the use of a tax ensures that costs of achieving the target are minimized (see Tietenberg, 1990, for a recent discussion of this familiar argument), and because it is a readily understood measure of policy.

In the next section we shall survey some of the previous empirical studies of carbon taxes before presenting our own results for the UK. In section III we outline the models and baseline assumptions used to calculate carbon taxes. In section IV we calculate the taxes needed to achieve targets for emissions in particular years in the future. Finally in section V we consider targets expressed in terms of cumulative emissions over the period 1990–2010, and show how the costs of achieving such targets might be affected by delaying the introduction of policy measures, perhaps because policy makers wish to obtain better information to reduce some of the uncertainties currently surrounding global warming.

II. SURVEY OF PREVIOUS EMPIRICAL STUDIES OF CARBON TAXES

In this section we shall review a number of empirical studies of carbon taxes (no claim is made for comprehensiveness, especially given the rapid growth of literature in this field; for other surveys see Cline, 1989; Barrett, 1990*a*; Boero, Clarke, and Winters, 1991). As will become clear there are a wide range of such estimates, and it is important to determine how far these differences are due to (i) differences in the methodologies of the studies; (ii) differences in estimates of crucial parameters, such as price elasticities of demand for fossil fuels; (iii) differences in assumptions about future values for key variables, such as the rate of economic growth; (iv) differences in targets set for CO_2 emissions. We shall group the studies in descending order of comprehensiveness, beginning with global models over long time-horizons and progressing to studies of sectors in individual countries over fairly short time-horizons.

(i) Global Models

One of the most important contributors to the study of carbon taxes has been Nordhaus, beginning with his seminal study (Nordhaus, 1977; 1979). This was part of a long-range study of the efficiency of world energy markets, focusing on the depletion of fossil fuels and the need to switch in the long run to non-exhaustible sources such as fast breeders, nuclear fusion, solar, and hydrogen. The methodology employed was a linear programming model which simulated competitive markets by maximizing the present value of world consumer and producer surplus in energy markets, subject to constraints such as reserves of fossil fuels, dates of availability of new technologies, etc. The model split the world into two regions—USA and ROW, and covered periods up to 160 years from 1975 in blocks of either 10 or 20 years. To introduce carbon taxes, Nordhaus began by modelling how the burning of fossil fuels was linked to emissions of CO_2 into seven different strata (two parts of the atmosphere, two layers of the oceans, and three parts of the biosphere). Concentrations of CO_2 in these strata were modelled by studying transfers of CO_2 between strata over time, in addition to new emissions. No direct attempt was made to assess damage costs; rather restrictions were imposed on atmospheric concentrations relative to pre-industrial levels (limits of 50 per cent, 100 per cent, and 200 per cent increases). There were four interesting conclusions from the study (we quote the fuller 1979 version). First the rates of tax on carbon emissions started off at between 0.07 and 12.90 (1975 $ per ton carbon) in 2000, rising to between $43 and $188 by 2140, depending on the severity of the concentration target sought. These are relative to prices of fossil fuels (in comparable units) of $25, $100, and $200 for coal, oil, and gas respectively. For no fuel would price be doubled before the middle of next century. Second, the tax rises steadily till 2100 and then falls slightly. Interestingly, in the most stringent scenario, the exhaustion rents for coal and shale oil in the US are driven to zero. Third, the predominant response in this model is fuel switching rather than demand reduction. Finally the overall cost to the world economy of imposing carbon taxes is equivalent to a reduction of GDP of less than 0.05 per cent. The overall message of this study was perhaps rather optimistic, reflecting assumptions about the relatively low cost and early availability of technologies for substituting away from the use of fossil fuels.

The above studies made no attempt to assess damage costs directly, but just imposed a limit on concentrations. In a later paper (Nordhaus, 1989), Nordhaus attempts a fairly heroic cost–benefit calculation. He looks at the costs of reducing greenhouse-gas emissions in 2050 and the benefits in terms of reduced concentrations in 2100, but evaluated in terms of impacts on world GDP in 2050. This is a highly simplified way of dealing with the dynamic structure of the problem. The world in 2050 is viewed as a scaled-up version of the US in 1981. He then constructs empirical estimates of marginal damage and marginal abatement costs. For abatement he considers four strategies—reducing CFC emissions, imposing a carbon tax, reforestation, and a gasoline tax (a special case of a carbon tax). For damage costs, he first points out that for a developed economy, 87 per cent of the activities generating GDP are insensitive to climate, 10 per cent would be moderately affected, and only 3 per cent severely affected (agriculture, forestry). The specific effects he evaluates are impacts on agriculture, effects of sea-level rises, the need for extra energy for space cooling (net of savings in space heating); he was unable to assess damages to non-marketed activities. The overall measured damage costs amounted to about 0.25 per cent of world GDP in 2050, with error bounds of plus or minus 2 per cent of GDP (a minus 2 per cent bound would make overall damage costs negative, corresponding to a situation in which agriculture benefits from global warming). From these calculations he derived central and high estimates of marginal damage costs of \$3 and \$37 per ton CO_2 equivalent respectively. Using these estimates he concluded that it would pay to eliminate between 90 per cent and 95 per cent of CFCs, and between 9 per cent and 28 per cent of CO_2 emissions, depending on the estimate of marginal damage cost; reforestation was not worthwhile, and obviously a carbon tax was preferable to the use of a gasoline tax. The appropriate carbon tax is given by the estimated marginal damage costs.

A rather more sophisticated version of this cost-benefit analysis which models the dynamic aspects more fully is in Nordhaus (1990), in which he constructs a multi-country intertemporal general equilibrium model. For each country there is a simple one-sector optimal growth model, in which output is produced from labour and capital using a Cobb–Douglas production function. To incorporate greenhouse gases, it is assumed that emissions of these gases are proportional to output, though the factor of proportionality can be reduced. These emissions add to atmospheric concentrations, (which have a natural decay rate), and concentrations in turn affect global mean surface temperature. Finally, increases in temperature and reductions in the factor of proportionality linking emissions to output both reduce the level of output according to a fairly complex cost function. This is parametrized to ensure that doubling CO_2 concentrations would reduce world GDP by 1 per cent, while reducing greenhouse-gas emissions by 50 per cent also reduces output by 1 per cent. Assumptions on technology and population growth ensure that the model reaches steady-state after the middle of next century, and parameters are selected for the model to fit observations for the years 1965, 1975, and 1985. The model has six regions—USA, Japan, rest of OECD, USSR, China, and ROW. The results of the model suggest that, for the USA, emissions should be cut by about 10 per cent now from their uncontrolled level, rising to about 12 per cent by 2030, and then falling slightly. Japan follows a similar, slightly lower path; for China the emissions reduction should only be between 2 per cent and 5 per cent. In terms of carbon taxes, these should start off at about \$4 per ton carbon now and rise at about \$1 per decade; these are rather lower than previous estimates. Nordhaus stresses the preliminary nature of these results; in particular there are difficulties in model solution.

A widely quoted study is that by Edmonds and Reilly (1983; 1985). Like the first Nordhaus study it is based on a model of equilibrium in world energy markets. Total energy demand in each country and time period is determined from GNP (which is driven by population growth and labour productivity growth) and price and income elasticities; total energy demand is then subdivided into individual fuels, taking account of relative fuel prices.

Demand functions use constant elasticities, derived from other studies. Energy supply is modelled in a rather *ad hoc* fashion (compared, say, to the analysis of exhaustible resources in Nordhaus, 1979). Production of commodities mainly uses fixed input–output co-efficients. Parameters are based either on other studies or calibration against past trends. Using this model, Edmonds and Reilly calculated that carbon taxes of 100 per cent, 78 per cent, and 56 per cent on coal, oil, and gas respectively, would reduce CO_2 emissions by 40 per cent from their unconstrained baselines (which were quite high), but this would only delay the date of doubling of atmospheric concentrations of CO_2 by 10 years (from 2049–2067 to 2058–2083). The cost of these measures was calculated to be 1 per cent of world GNP. A subsequent sensitivity analysis of their model (Reilly, Edmonds, *et al.*, 1987) which took account of correlations between uncertainties in parameter values, suggested that the median baseline emissions were much lower than in the original study, with unconstrained concentrations not doubling until 2100. The authors did not conduct an analysis of carbon taxes in this later study, but this was done by Cline (1989), who showed that taxes of 150 per cent, 100 per cent, and 50 per cent on coal, oil, and gas respectively, imposed from 2000 onwards, would cut emissions in 2050 by 57 per cent from their baseline level (and 21 per cent below their 1985 level). This would require a significant shift in fuel use towards nuclear power (21 per cent of total world primary energy), and a significant reduction in energy use. World GNP would fall by 7.4 per cent by 2075 (though no allowance is made for smoothing the impacts on GNP over time).

The last global study we shall report on is by Whalley and Wigle (1989) (see also Whalley, 1990). This employs a general equilibrium model of international trade in commodities and energy products between three blocs—developed countries, developing countries, and oil exporters. Unlike the earlier models surveyed, this is a purely static model. Nested CES functions are used to represent production and demand. The model parameters are calibrated to data for 1986. Taxes are imposed to reduce baseline emissions by 50 per cent, and three kinds of taxes are considered—national production-based taxes, national consumption-based taxes, and a global tax (in which case the distinction between consumption and production is irrelevant). Tax rates turn out to be fairly similar (no distinction is made between individual fossil fuels) at around 80 per cent on a tax inclusive basis (400 per cent on a net of tax basis). While tax rates are similar for the different tax schemes, the impacts on countries are very different; taxes on production favour oil exporters, taxes on consumption favour developed countries, global taxes (redistributed on a *per capita* basis), favour developing countries. The impacts on countries (measured as percentages of GDP) are quite large: (-2.36 per cent, –0.63 per cent) for developed countries; (–28.73 per cent, +10.8 per cent) for oil exporters; (–4.76 per cent, +2.96 per cent) for developing countries; with overall losses of around 2 per cent of world GDP.

(ii) Individual Country Studies

It is perhaps appropriate to begin with the widely cited study for the USA by Manne and Richels (1990*a* and *b*), since their model is actually designed as a global model, with parallel (i.e. non-interacting) models of five regions: USA, other OECD, USSR, China, and ROW (almost the same as Nordhaus, 1990, above). However at the time of writing we only have access to the version which gives results for the USA. The study exploits the now long-established ETA–MACRO model (Manne, 1976; 1981) which is designed to analyse long-run energy-market equilibria. The MACRO component is a simple macroeconomic model in which aggregate output is produced using a putty-clay vintage model in which the *ex-ante* production function is a nested CES/Cobb–Douglas production function, where there is an elasticity of substitution between a capital/labour input and an energy input (comprising electric and non-electric energy). This will allow new machines to change their energy intensity in response to changes in relative future prices; in addition there are autonomous improvements in energy efficiency. Two key parameters are the elasticity of substitution between energy and non-energy inputs and these exogenous improvements in technology. Output, net of energy costs, goes to consumption and investment according to utility-maximizing behaviour which, again, is very similar to Nordhaus (1990). The ETA component of the model is rather like the Nordhaus (1979) model in that it contains a detailed representation of depletion of non-renewable energy resources and the likely costs and availabilities of new technologies, in particular non-carbon-based electric and non-electric technologies which can be supplied with infinite elasticity at sufficiently high prices. This model is then used to simulate the effects of imposing restrictions on emissions of CO_2 (so there is no modelling of concentrations); the targets selected are to reduce CO_2 emissions to their 1990 level by 2000, 80 per cent of that level by 2020, and stabilizing at that level thereafter. This is done under different assumptions about the exogenous rate of improvement in energy efficiency (either 0.0 per cent or 1.0 per cent per annum) and the backstop technologies likely to be available. Under the most constrained case, the economy meets the limits on CO_2 emissions by a mixture of fuel-switching, particularly towards natural gas, which drives up its price and encourages the exploitation of new, non-carbon technologies, and a reduction in energy demand. If a carbon tax was used to bring about these changes, it would start off at around $29 per ton of carbon in 2000, climb sharply to nearly $600 per ton of carbon by 2020, and stabilize at around $250 per ton of carbon from 2050 onwards. The loss of GNP amounts to about 5 per cent from 2030 onwards, a present value cost, at 5 per cent discount rate, of $3.6 trillion; alternatively the losses amount to a slowing down of growth in GNP or consumers' expenditure from 2.52 per cent to 2.46 per cent per annum (Barker, 1990). Allowing for exogenous improvements in energy efficiency and the development of technologies such as integrated-gasification combined-cycle (IGCC) plant with CO_2 removal (and subsequent disposal in the oceans), would reduce costs to $0.8 trillions.

We turn now to the UK, beginning with economy-wide models. Barker and Lewney (1990) adapt the Cambridge Econometrics multi-sectoral dynamic model (MDM) of the UK economy to study carbon taxes. Although the MDM model includes an energy sub-model (based on translog models of energy demand—Peterson, 1987), the authors thought the estimates were too out-dated to be reliable, and so used estimates of energy-demand elasticities for non-household sectors derived from the Department of Energy together with assumptions about fuel-mix in electricity generation; the household sector of MDM was used to provide estimates of household energy demand elasticities. Carbon taxes were imposed (by the UK alone) on coal, oil, and gas in proportion to CO_2 content and rose over time in proportion to the number of years from 1990. VAT was assumed to be adjusted to hold inflation constant, and changes in net revenues of the public sector are assumed to be saved. World prices of oil and coal, and UK prices of gas and domestic coal are assumed unaffected by carbon taxes. The targets set were to reduce emissions by 2005 to either the 1990 level or 80 per cent of the 1988 level (the Toronto target). To reach the 1990 target would require a carbon tax of 3.1 per cent initially, rising at 3.1 per cent per year; to reach the Toronto target would require an initial tax in excess of 9.6 per cent (and rising at 9.6 per cent per annum). The tax rates in 2005 would be 132 per cent, 92 per cent, and 48 per cent on coal, oil, and gas respectively under the 1990 target (with electricity prices rising by 27 per cent), while under the Toronto target the tax rates would be 374 per cent, 327 per cent, and 141 per cent respectively (with electricity prices rising by 54 per cent). The impacts of these taxes on macroeconomic variables and the balance of payments were found to be negligible.

The UK's input to the IPCC Response Strategies Working Group is contained in a report compiled by the Department of Energy (1989). The report does not contain any assessment of policy options, so there are no calculations of carbon taxes. The report sets out possible levels of emissions of CO_2 for 2005 and 2020 under different scenarios about economic growth (1.25 per cent p.a., 2.25 per cent p.a., or 3.25 per cent p.a.), its sectoral decomposition, and energy prices. The report carries out detailed sectoral analyses of energy demand in the absence of any policy using a range of approaches, and from this draws conclusions about CO_2 emission levels; thus in the medium growth scenario with low prices, CO_2 emissions would rise from 158m tons of carbon in 1985 to 233m tons of carbon in 2020. The report then considers various strategies for reducing CO_2 emissions, such as increased use of non-fossil-fuelled methods of generating electricity, improvements in energy efficiency, increased use of combined-heat-and-power, or raising temperatures in fossil-fuelled power stations.

For each of these routes the report calculates the likely scope and cost per ton of carbon for reducing CO_2 emissions under different scenarios. In principle this could be used to build up the kind of marginal abatement cost curve constructed by Nordhaus (1989), but this is not done. No attempt is made to compute overall economic effects of reducing CO_2 emissions.

Although the Department of Energy did not compute carbon taxes, Barrett (1990*b*) used the demand elasticities estimated in the Department of Energy's demand models (in Helm *et al.*, 1989) to compute a set of carbon taxes. He showed that to reduce the CO_2 emissions by 20 per cent would require taxes on coal, oil, and gas of 67 per cent, 54 per cent, and 40 per cent respectively in the short run, and 24 per cent, 19 per cent, and 14 per cent in the long run.

Another study which draws in part on the work of the Department of Energy is by Skea (1990). The report considers four sectors—industry, services (commercial and public), transport, and household—and projects their demands for fuels for a base 'business as usual' policy stance; these demand projections draw on a range of other studies, rather than being derived from a consistent set of models. Using a model of electricity supply which takes into account assumptions about new capacity and restrictions on fuel use, estimates of CO_2 emissions are derived for this base case. Skea then considers two more positive policy cases; the first assumes that the full potential for energy efficiency is realized, while the second assumes additionally that there is enhanced use of gas, nuclear, and renewable sources of electricity generation. For improvements in energy efficiency the report draws on the work of the Energy Efficiency Office. The report largely favours non-price-based policy measures, particularly for energy efficiency, such as improved information, building regulations, and appliance standards. Much of the concern with energy efficiency derives from the calculation that such measures would be a cheaper way of reducing CO_2 emissions than the greater use of nuclear power (calculated to cost £110 per tonne of carbon reduced). Carbon taxes may encourage some fuel switching (at £70–£90 per tonne of carbon, gas would be significantly encouraged), but are not thought to have much impact on reducing energy demand (the household sector is explicitly assumed not to be responsive to energy prices).

Finally we turn to two studies of individual sectors. Pearson and Smith (1990) consider the household demand for energy for transport and non-transport purposes. While they discuss carbon taxes, the calculations they perform are related to more conventional taxes. Thus, they note that household consumption of coal, gas, electricity, and heating oil are zero-rated for VAT, and consider the impact on domestic fuel consumption of imposing VAT at 15 per cent (carbon taxes would differentiate taxes on fuel types by carbon content). Using the IFS model of consumer expenditure based on consumption data on 116,000 households (Baker *et al.* , 1990; Blundell *et al.*, 1989), they show that household energy consumption would fall by 5.5 per cent (contrary to the claim by Skea noted above), and the tax would generate total revenue of about £1.7 billion. However the effect is rather different across household groups. Weekly tax bills would rise by £1.02 for the lowest decile income group, and by £1.92 by the highest decile group, so the tax is fairly regressive; moreover the lowest decile group cut their energy consumption by 10 per cent, while the upper decile group scarcely cut theirs. Separate analysis was also carried out for pensioners. For transport, the authors consider two types of tax. First, an increase in petrol duty of 55p per gallon of 4-star (the amount needed to bring petrol prices in 1990 back to their 1975 peak) was calculated to reduce household petrol consumption by 8 per cent in the short run (ignoring changes in vehicle size), and to generate £2.1 billion in revenue. This tax is slightly progressive if we look at all households, but this is because wealthier households are more likely to own a car; looking only at car owners, the tax is slightly regressive. A second transport-related tax considered is a vehicle excise duty differentiated by engine size to encourage a switch to smaller cars. The authors note that this would be a blunter instrument than an increase in petrol duty (or a carbon tax), but do not carry out any calculations of the effect of such a tax. These studies provide some useful indicators of possible distributional impacts of a carbon tax, although, as they note, the study considers only the direct impact; there will be indirect effects through the change in relative prices of other goods depending on the energy intensity of their production.

A more detailed study of transport demand can be found in a report for the Energy Efficiency Office of the Department of Energy by Martin and Shock (1989), providing forecasts of energy use and fuel efficiency up to the year 2010. The report contains a mixture of econometric modelling, engineering studies (especially of possible future energy efficiencies), traffic surveys, and some judgemental assumptions (e.g. imposition of upper limits on demand to simulate saturation effects if the unconstrained forecast seems implausible), to produce forecasts of energy consumption for the UK transport sector (broken down into sub-sectors) for the year 2100, under three different scenarios about economic growth (1.25 per cent p.a. or 3.25 per cent p.a.), oil prices (high or low), and energy efficiency improvements. The forecasts range from 1,919 peta-joules (almost the same as the 1986 level) under the low growth/high price/high take-up of energy efficiency scenario, to 3,284 under the high growth/low price/no energy efficiency improvements scenario. However the study does not explicitly consider CO_2 emissions or the effect of carbon taxes to reduce them.

III. CARBON TAXES IN THE UK: BACKGROUND TO OUR OWN ESTIMATES

We now turn to our own work on carbon taxes for the UK. In this section we shall outline the background to our work, and in the next two sections we present the results. We have previously reported results for the UK manufacturing sector (Ingham and Ulph, 1990*b* and 1991). In this paper we shall summarize and slightly extend the results for the UK manufacturing sector reported in Ingham and Ulph, 1991, and present our findings for the transport and household sectors based on some very preliminary analysis of energy demand in those sectors.

The focus of our work is not on very long-term questions of optimal taxes, which would be quite inappropriate given the partial nature of our results, but on the taxes and resulting costs required to achieve a range of targets over the relatively short period of 10–20 years. We shall be interested in questions such as the sensitivity of tax rates and costs to differences in assumptions about the baseline scenarios, the nature of the cost curve for achieving different levels of reductions (to see whether there are levels of reductions at which costs rise sharply). A particular feature of the analysis is the emphasis on the dynamics of adjustment, and this leads us to study the question of the impact of variations in the timing of policy on tax rates and costs; we shall consider two aspects—variation in the date by which a particular target should be reached and variation in the date from which policy starts.

Three factors are important in determining the tax rates required to achieve particular targets for reducing CO_2 emissions: the models of how energy demand responds to changes in prices induced by carbon taxes, the baseline assumptions about what will happen to factors such as economic growth and energy prices in the absence of any policy, and finally the targets for policy and the way in which taxes are to be imposed. In the next three subsections we discuss each of these in turn.

(i) The Models Employed

(*a*) Manufacturing Sector

We employ a sophisticated vintage model of energy demand which has been developed over a number of years with financial support from the Department of Energy. A detailed description can be found in Ingham and Ulph (1990*b*). Like conventional vintage models, we incorporate the fact that producers can change the design of new machines in response to expected relative prices over the planning horizon, so that technical progress, in terms of improvements in energy efficiency, for example, is determined in part endogenously. But unlike conventional models we allow for the possibility that factor use on machines already installed can be changed in response to changes in current relative factor prices; moreover, the extent to which factors can be varied *ex post* (i.e. the flexibility of the structure of production), is also chosen by producers, with greater flexibility obviously coming at a cost. It should be noted that in the version of the model used in this paper the level of output is taken as exogenous, so that when we come to study carbon taxes we shall ignore the effect of such taxes on the demand for output of the manufacturing sector. We are only capturing the substitution response of producers in reducing their use of carbon-intensive inputs. The model is estimated over the period 1955Q1 to 1987Q4 (see Ingham, Maw, and Ulph, 1990, for parameter estimates).

(b) The Household Sector

This is based on a simple AIDS model of expenditure shares, in which we consider six categories of expenditure: food, clothing, energy, transport, durables, and other. As this implies we are not able to distinguish individual fuel use in this sector. To capture dynamics we have included lags in the expenditure shares, with the first and fourth lags proving significant. The model has been estimated on quarterly data from 1963Q1 to 1989Q3, and estimates are presented in Appendix A. Estimated short-run own-price and income elasticities of demand for energy are –0.258 and 0.911 while the corresponding long-run elasticities are –0.656 and 2.317 respectively. Although the model is very simple, we compared the results for carbon taxes derived from this model with the results from the IFS model which was estimated on micro data and was described in the previous section. The results are broadly comparable.

(c) Transport Sector

We do not cover the whole of the transport sector, but only the demand for gasoline. The model we have estimated is similar to that of Drollas (1984). Rather than trying to model the dynamics of gasoline demand explicitly by modelling changes in the stock of vehicles and the utilization of vehicles, we simply estimate a gasoline demand function with lagged values of the dependent and independent variables. Long-run gasoline demand is a log linear function of income, gasoline prices, car prices, seasonal dummies, and a time trend. A general-to-specific framework was used for estimating the lag structure beginning with four lags on all variables, and then eliminating insignificant variables. The model is estimated on quarterly data from 1975Q1 to 1989Q4, and details of the model and the estimated parameters can be found in Appendix B. The short-run own-price and income elasticities are –0.172 and 0.223 respectively, while the long-run elasticities are –0.306 and 0.360. These elasticities are very low (see, e.g., Hughes, 1990), and so the results for this sector need to be treated with considerable caution.

(d) Welfare Costs of Carbon Taxes

One consequence of the very partial analysis we have carried out for the three sectors is that this requires us to take a very simple view of the economic costs of imposing carbon taxes, or rather the economic costs of having to reduce CO_2 emissions. In the manufacturing sector, as already noted, output is taken as exogenous, so consumers are unaffected by the carbon tax. The impact is therefore borne by producers in higher costs and reduced profits, so we calculate the welfare cost as the increase in total present value costs following the imposition of a carbon tax less the present value of the tax revenues. In both the transport and domestic sectors the imposition of a carbon tax raises the prices faced by consumers of transport services and domestic energy, reducing consumption of these items, and hence consumer welfare. For these sectors we calculate the welfare cost as the present value of the reduction in consumer surplus less the present value of tax revenues.

(ii) The Background Assumptions

The assumptions we make about the future paths of output and factor prices are related to those used by the Department of Energy for the Intergovernmental Panel on Climate Change Response Strategies Working Group (Department of Energy, 1989). They are summarized in Table 1. The main difference is that the DoE were making forecasts from 1985. For energy prices we have used the same rates of growth in energy prices as were used by the Department of Energy but applied them to data for 1990. Given the much lower base for energy prices in 1990 relative to 1985 we are therefore assuming much lower energy prices throughout than in the DoE study. The high output growth is rather out of line with what would be derived by simply extrapolating historical experience, so that, combined with the low energy prices, we are choosing baselines which will imply high levels of CO_2 emissions in the absence of policy. This needs to be borne in mind in assessing the results that follow.

Table 1
Assumptions: Annual Rates of Growth (Real)

	Low	High
Output	1.25	3.25
Oil/gas/transport fuels	0.60	3.50
Coal	0.80	1.00
Household energy	0.60	3.50
Wages	1.04	3.04

Note: All other real prices and interest rates are assumed to stay constant at their 1990 levels.

(iii) Targets and the Structure of Carbon Taxes

In setting targets for policy for the imposition of carbon taxes we will use two variants. In the next section, we will consider targets for levels of emissions in a particular year; we take 2005 as the benchmark case, but consider other years, specifically 2000 and 2010 as sensitivities. The targets for emissions in the target year we consider are (i) stabilization of emissions at their 1990 level (the target currently set by the UK government); and (ii) cutting emissions to 75 per cent of their 1990 level. In section V we shall consider cumulative emissions over the period 1990–2010, the costs of achieving reductions in such cumulative emissions from their unconstrained value by percentages up to 60 per cent, and how such costs would be affected by delaying the implementation of policy till 1995 or 2000.

A carbon tax is a specific tax on emissions of CO_2, i.e. a tax of a fixed amount of money per tonne of CO_2 emitted at a particular period of time irrespective of the source of the emission. In practice, given the difficulties of measuring CO_2 emissions, the tax will be levied on the fossil fuels whose combustion emits CO_2, and since there is currently no technology for varying the amount of CO_2 emitted per unit of fossil fuel, there is no loss of efficiency in levying the tax on fossil fuels rather than emissions. In converting the specific tax on CO_2 emissions into a specific tax on fossil fuels we take account of the fact that coal, oil, and gas differ in the amount of CO_2 they emit per unit of energy. Although the tax is calculated as a specific tax, we report it as a percentage rate of tax, and differences in these percentage rates of tax across fuels reflect not only the different CO_2 levels just noted, but also differences in fossil fuel prices per unit of energy.

Although we shall be considering as possible targets specifying the level of emissions of CO_2 in a particular year in the future, we calculate the taxes that will be required for every year from the start of the policy until 2010. This requires that we make assumptions about how tax rates will be set in different years. What we assume is that the specific tax per unit of CO_2 rises over time at a specified percentage rate. One possible justification for having the tax rise over time can be derived from the analysis of optimal pollution taxes for pollutants where the damages are related to the stock of the pollutant in the environment rather than to the flow of emissions of the pollutant, which is appropriate for CO_2. That analysis shows that at each instant the tax should be set equal to the present value of the marginal costs of the damage that will arise from the stock at each future period; if marginal damage costs are increasing in the stock, and the stock is itself rising in the optimal policy, then the tax must be rising over time (see Dasgupta, 1982; Plourde, 1972; Ingham and Ulph, 1990*a*) for further discussion). A less formal argument might be that it would be politically more acceptable to introduce a carbon tax in a gradual manner. The former argument suggests that if the damage costs from global warming are all incurred after the period 1990–2010, then during that period a carbon tax should rise at a rate equal to the real interest rate plus the rate at which the stock of CO_2 in the atmosphere would decay naturally (this assumes that CO_2 concentrations would decay exponentially; this is not thought to be quite correct, for example CO_2 concentrations would not decay to zero; however it is approximately correct and provides a useful benchmark for deriving policy rules). The latter rate is estimated to be low (0.6 per cent a year, see Houghton *et al.*, 1990, pp. 7–18), and long-run real interest rates for the UK have been estimated to lie in the range 2–4 per cent (Dimson, 1988); so we have taken 3 per cent and 5 per cent as plausible estimates of the rate at which a carbon tax might rise over the period 1990–2010.

IV. TAXES REQUIRED TO ACHIEVE EMISSION TARGETS IN FUTURE YEARS

In this section we consider taxes to achieve targets of stabilizing emissions at their 1990 level by 2005 (the government's target) and cutting emissions to 75 per cent of their 1990 level by 2005, with sensitivities of achieving these targets in 2000 and 2010. We begin by considering the emissions that would arise in each of the three sectors in the absence of any policy. Then we calculate the taxes required in each sector to achieve these targets in each sector. Of course this is not an efficient policy; if the government wants to stabilize emissions at their 1990 level by 2005 it does not necessarily want to do that in each sector; what it should aim to do is to achieve an overall stabilization of emissions by 2005 at least cost, and that requires setting the same tax in each sector. So we then calculate the carbon taxes required to achieve the targets assuming the same tax rate is set throughout all sectors, and we compare the cost of this efficient policy with the cost of the inefficient policy of achieving the same target in each sector.

(i) Unconstrained Emissions

We begin by considering what would be the emissions levels in 2005 relative to 1990 under each of the scenarios about income and price growth. As Table 2 shows, the greatest relative uncontrolled increase in emissions is in the industrial sector, the least in the transport sector, with the domestic sector almost exactly midway between the two. This ranking reflects two factors—the elasticities of energy demand, and the assumption in the industrial sector of rising real wages, which acts to increase energy demand and hence CO_2 emissions.

Table 2
Unconstrained CO_2 Emissions

	Output growth	Energy price growth	CO_2 emissions 2005/1990
Manufacturing:			
	Low	High	1.275
	Low	Low	1.442
	High	High	2.010
	High	Low	2.250
Household:			
	Low	High	1.299
	Low	Low	1.306
	High	High	1.703
	High	Low	1.714
	Output	Price	CO_2 emissions 2005/1990
Transport:			
	Low	High	1.063
	Low	Low	1.208
	High	High	1.258
	High	Low	1.428

(ii) Tax Rates—Achieving Targets Sector-By-Sector

We now consider for each sector individually the tax rates that would be needed to achieve the two emission targets in 2005 under different assumptions about output, income, and price growth. Tables *3a* and *3b* show the results for manufacturing, 3*c* and 3*d* for the domestic sector, and 3*e* and 3*f* for the transport sector. In interpreting these tax rates note that for the transport sector the tax rate is essentially a tax on oil products, and so comparable with the tax rates for oil in the manufacturing sector, while the tax rate for the domestic sector is an average across all fuels used in the domestic sector.

In all sectors we get the expected results that tax rates are higher the higher is the rate of output growth, the lower is the rate of price growth, the lower is the rate of growth of tax rates and the lower is the target for emissions in 2005. Moreover these differences have large effects on tax rates, with the difference between the highest and lowest tax rates across scenarios being a factor of thirty or forty in the manufacturing and transport sectors. In the domestic sector the difference in tax rates across scenarios is rather smaller than in the other sectors. For example, the impact of different assumptions on income growth is rather less marked for the domestic sector than for the other two. In particular, with low growth and the stabilization target, tax rates are really rather modest in transport and manufacturing; raising the rate of growth causes tax rates to rise by factors of between three and ten in the transport and manufacturing sectors. By contrast, even with low growth, tax rates are quite significant in the domestic sector under stabilization, and at most double when the high growth scenario is considered. This difference between the domes-

Table 3*a*
Tax Rates (%) for Different Scenarios—Manufacturing
2005 Target: 1990 Level

Fuel	Growth	Low				High			
	Price	High		Low		High		Low	
	Tax growth	5%	3%	5%	3%	5%	3%	5%	3%
Coal	1990	19	22	24	27	249	379	270	407
Coal	1995	24	25	29	30	304	420	335	460
Coal	2000	29	28	36	34	370	466	417	520
Coal	2005	35	31	45	39	452	516	519	588
Coal	2010	43	34	56	44	551	573	645	665
Oil	1990	19	21	23	26	244	371	264	398
Oil	1995	20	21	29	30	262	363	330	453
Oil	2000	22	21	36	34	282	355	413	515
Oil	2005	24	21	45	39	303	347	517	586
Oil	2010	25	20	57	44	326	339	647	667
Gas	1990	11	12	13	15	135	205	146	221
Gas	1995	12	13	16	17	158	219	185	253
Gas	2000	14	14	20	19	185	233	233	290
Gas	2005	17	15	26	22	216	247	294	333
Gas	2010	20	16	32	25	253	263	371	382
Electricity	1990	9	10	11	12	118	180	128	194
Electricity	1995	9	10	12	12	129	186	143	204
Electricity	2000	9	9	13	12	142	192	160	214
Electricity	2000	10	8	14	12	156	198	179	225
Electricity	2010	11	8	17	12	172	205	200	237

OXFORD REVIEW OF ECONOMIC POLICY, VOL. 7, NO. 2

Table 3*b*
Tax Rates (%) for Different Scenarios—Manufacturing
2005 Target: 75% of 1990 Level

Fuel	Growth	Low				High			
	Price	High		Low		High		Low	
	Tax growth	5%	3%	5%	3%	5%	3%	5%	3%
Coal	1990	122	196	143	224	518	700	530	704
Coal	1995	149	217	178	253	632	777	659	796
Coal	2000	181	241	221	286	770	861	820	901
Coal	2005	221	267	275	323	939	951	1020	1018
Coal	2010	269	296	342	366	1145	1059	1268	1152
Oil	1990	119	192	140	219	507	686	519	690
Oil	1995	128	187	175	249	545	671	650	785
Oil	2000	138	183	220	283	587	656	813	893
Oil	2005	148	179	275	322	631	642	1017	1016
Oil	2010	160	175	344	367	678	628	1272	1156
Gas	1990	66	106	78	121	281	380	288	382
Gas	1995	77	113	98	139	329	404	363	438
Gas	2000	90	120	124	160	385	430	458	503
Gas	2005	106	128	156	183	450	458	578	577
Gas	2010	124	136	197	210	526	487	729	662
Electricity	1990	58	93	68	106	247	335	252	336
Electricity	1995	63	96	76	112	268	345	280	353
Electricity	2000	70	99	86	118	292	356	310	370
Electricity	2000	78	102	97	124	319	367	340	389
Electricity	2010	86	106	110	131	347	379	382	408

Table 3*c*
Tax Rates (%) for Different Scenarios—Domestic
2005 Target: 1990 Level

Growth	Low				High			
Prices	High		Low		High		Low	
Tax rates	5%	3%	5%	3%	5%	3%	5%	3%
1990	65	90	130	172	135	180	240	320
1995	83	104	166	200	172	209	306	371
2000	104	123	212	232	220	242	391	430
2005	133	140	270	268	280	280	499	499
2010	170	163	345	311	358	325	637	578

Table 3d
Tax Rates (%) for Different Scenarios—Domestic
2005 Target: 75% of 1990 Level

Growth	Low				High			
Prices	High		Low		High		Low	
Tax rates	5%	3%	5%	3%	5%	3%	5%	3%
1990	138	187	242	330	260	330	405	540
1995	176	217	309	383	332	383	517	623
2000	224	251	394	444	423	444	660	726
2005	286	291	503	514	540	514	843	842
2010	365	338	643	596	690	596	1075	976

Table 3e
Tax Rates (%) for Different Scenarios—Transport
2005 Target: 1990 Level

Growth	Low				High			
Prices	High		Low		High		Low	
Tax rates	5%	3%	5%	3%	5%	3%	5%	3%
1990	17	22	44	58	87	114	115	151
1995	18	21	54	65	93	111	141	169
2000	20	21	67	73	100	109	175	190
2005	21	20	83	82	108	106	216	214
2010	23	20	102	92	116	104	268	241

Table 3f
Tax Rates (%) for Different Scenarios—Transport
2005 Target: 75% of 1990 Level

Growth	Low				High			
Prices	High		Low		High		Low	
Tax rates	5%	3%	5%	3%	5%	3%	5%	3%
1990	166	218	195	256	343	450	373	490
1995	178	213	239	286	367	440	457	548
2000	191	208	296	322	395	429	566	617
2005	205	203	367	362	424	419	701	694
2010	221	198	454	408	456	409	869	781

tic and other sectors reflects the rather low price and income elasticities of demand in the domestic sector.

(iii) Efficient Tax Rates

The previous section presented tax rates for each sector based on achieving a stated target for emissions in 2005 for each sector in isolation. But an important rationale for using a carbon tax as a policy instrument is to ensure an efficient policy, i.e. achieving an overall target for emission reductions at minimum total cost, and that requires setting the same carbon tax across all sectors, and allowing sectors where it is relatively cheaper to abate emissions to bear a larger proportionate reduction in emissions than a sector where it is relatively more expensive to reduce emissions. To illustrate the difference between the two approaches, we have considered the scenario of low growth of output (income), high growth of prices, 5 per cent growth in the carbon tax over time, and a target of achieving 75 per cent of 1990 emissions by 2005. Table 4*a* reproduces from Table 3 the tax rates for each sector for this scenario when emissions in each sector are reduced to 75 per cent of their 1990 levels. This shows that carbon taxes are highest in the transport sector and lowest in the manufacturing sector, with domestic tax rates, initially, intermediate, suggesting that it is most expensive to reduce emissions in the transport sector, cheapest in the industrial sector. Table 4*b* shows the tax rates that would be set if total emissions across all three sectors are reduced to 75 per cent of their 1990 level

Table 4*a*
Tax Rates (%) to Cut Emissions to 75% of 1990 level by 2005, Sector by Sector

	Industrial				Domestic	Transport
	Coal	Oil	Gas	Elec.		
1990	122	119	66	58	138	166
1995	149	128	77	63	176	178
2000	181	138	90	70	224	191
2005	221	148	106	78	286	205
2010	269	160	124	86	365	221

Table 4*b*
Tax Rates (%) to Cut Emissions to 75% of 1990 level by 2005, Across All Sectors

	Industrial				Domestic	Transport
	Coal	Oil	Gas	Elec.		
1990	152	149	83	72	127	149
1995	178	160	89	81	152	160
2000	208	172	95	90	181	172
2005	243	185	103	101	218	185
2010	284	199	111	113	264	199

Background assumptions: Low growth of output (income), high growth of prices, carbon tax rising at 5 per cent p.a.

by 2005, but the same tax rate is set in each sector. As would be expected this involves raising the tax rate in the industrial sector, cutting it in transport, and, to a lesser extent, in the domestic sector. This efficient policy requires that by 2005 emissions relative to their 1990 levels are 86 per cent in transport, 78 per cent in the domestic sector, and 65 per cent in the manufacturing sector, confirming that efficiency requires greatest reductions in the industrial sector, least in the transport sector. The total present value cost of the efficient policy is 20 per cent less than for the inefficient policy. Figure 1 shows the comparison of total costs under the equal tax (efficient) policy and the equal per cent reductions policy for different levels of emissions in 2005 relative to their 1990 level. Figure 2 shows how emissions in the three sectors in 2005 fall in response to differences in the initial value of a carbon tax, set at the same level for all sectors. As this shows, the response of emissions in the industrial sector to the tax is initially quite great, but quickly fades; emissions in the domestic sector fall almost linearly with increases in the tax, as do emissions in the transport sector, but at a slower rate than in the domestic sector.

(iv) Variations in Target Dates

The scenarios above have all been based on the assumption of achieving a target by 2005. We now consider the sensitivity of tax rates to target dates. For this we have again chosen the scenario with low output (income) growth, high price growth, taxes rising at 5 per cent, and a target of a 25 per cent reduction on 1990 emission levels. Tables 5*a*, *b*, and *c* show the required tax rates for the manufacturing, domestic, and transport sectors respectively, of using target dates of 2000, 2005, and 2010. These tax rates are based on achieving the targets on a sector-by-sector basis, so the tax rates for 2005 are those shown in Fig 4*a*. As would be expected, tax rates are higher the closer is the date by which the target must be reached. However there are marked differences across the sectors. For the domestic and transport sectors, the effects of raising and lowering the target date by five years are more or less symmetric, although the impact is very much smaller in the transport sector than in the domestic sector. For the manufacturing sector, there is marked asymmetry, with the extension of the target date to 2010 causing a very sharp fall in the tax rates. We believe

Figure 1
Alternative Polices Across Sectors, Reducing the Rate of Omissions

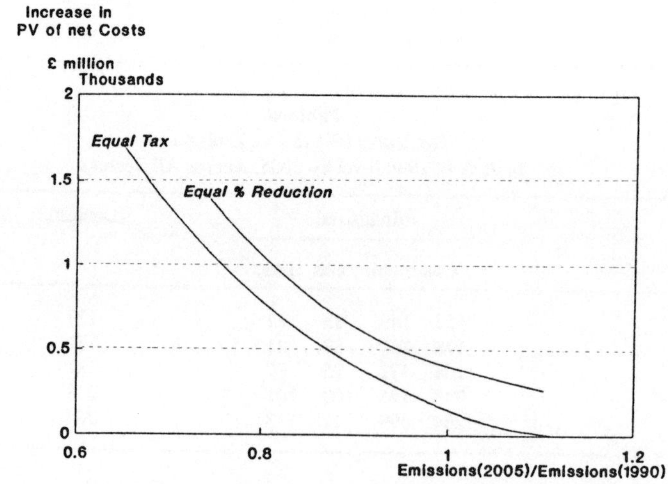

OXFORD REVIEW OF ECONOMIC POLICY, VOL. 7, NO. 2

Figure 2
Effects of Tax Level on Emissions of Sectors

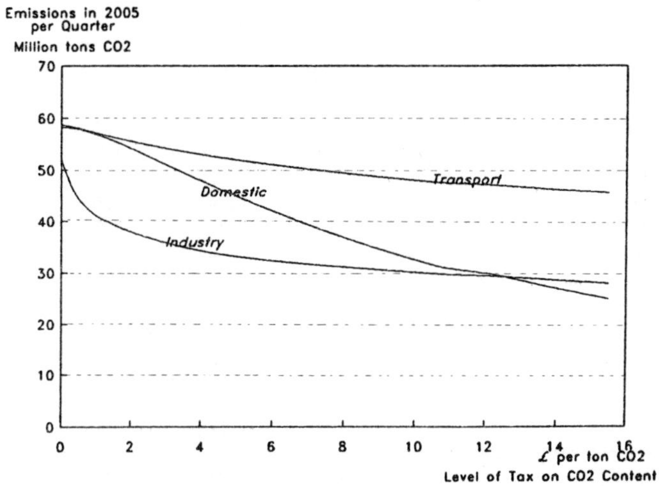

Table 5a
Manufacturing Sector: Tax Rates (%) for Different Target Dates

Target Date		Tax Rates %		
		2000	2005	2010
Coal	1990	180	122	32
Coal	1995	211	149	38
Coal	2000	246	181	44
Coal	2005	288	221	52
Coal	2010	336	269	60
Oil	1990	177	119	32
Oil	1995	190	128	34
Oil	2000	204	138	37
Oil	2005	220	148	39
Oil	2010	236	160	42
Gas	1990	98	66	18
Gas	1995	105	77	19
Gas	2000	113	90	20
Gas	2005	122	106	22
Gas	2010	131	124	24
Electricity	1990	85	58	15
Electricity	1995	91	63	15
Electricity	2000	98	70	16
Electricity	2005	105	78	17
Electricity	2010	113	86	18

Table 5b —Domestic Sector
Tax Rates (%) for Different Target Dates

Date of Tax	Target Year		
	2000	2005	2010
1990	194	138	96
1995	248	176	123
2000	316	224	156
2005	403	286	200
2010	515	365	255

Table 4c—Transport
Tax Rates (%) for Different Target Dates

Date of Tax	Target Year		
	2000	2005	2010
1990	177	166	156
1995	190	178	167
2000	204	191	179
2005	219	205	193
2010	235	221	207

that this reflects the more careful modelling of the dynamic process of adjustment that is captured by the vintage model for the industrial sector.

V. COSTS OF DELAY IN IMPLEMENTATION

In the previous section we have considered taxes required to achieve targets expressed in the form of a level of emissions at a future date relative to current emissions. Since any damages associated with CO_2 will be related to the atmospheric concentrations of CO_2, and hence to cumulative emissions of CO_2 rather than emissions in any one year, such targets are inherently unsatisfactory. For example, in the previous section, for any given set of assumptions about growth of output and prices, we have considered two scenarios designed to achieve the same target for emissions in 2005, one in which taxes rise at 5 per cent per annum and another in which they rise at 3 per cent per annum; these have different implications for cumulative CO_2 emissions.

This concern with the formulation of targets is particularly important for the issue we now wish to address, the cost of delay in implementing policy. In the previous section we assumed that policy was implemented in 1990, irrespective of target date. But an obvious response to the considerable uncertainties surrounding the potential damages from global warming and the costs imposed on the economy of reducing CO_2 emissions would be to delay for a number of years taking any policy initiative other than 'no regret' policies, i.e. policies which one would wish to undertake for other reasons, but have the additional side-effect of reducing CO_2 emissions. Examples of such 'no regret' policies could be measures to correct possible market failures in energy conservation, or policies to reduce coal burning in power stations to reduce acid rain. An argument against such a 'do nothing you might regret' policy is that when policy-mak-

OXFORD REVIEW OF ECONOMIC POLICY, VOL. 7, NO. 2

ers come to take action they will be doing so from a higher level of cumulative CO_2 emissions than they would otherwise, and these could impose higher costs on society if global warming proves to be serious. These higher costs arise from either the effect of the damages caused by the higher concentrations of CO_2, or the need to make a sharper reduction in CO_2 emissions once policy is undertaken in order to bring cumulative emissions back into line with what they would have been had policy started earlier.

A proper appraisal of whether a delay in implementing policy would be worthwhile requires a full cost–benefit study, including a modelling of damage costs and the likely acquisition of information by delaying, and such an analysis is beyond the scope of this paper. What we have done is to provide a calculation relating to only one aspect of such a full analysis, and even that in a rather dramatic form. The question we pose is to suppose that there is a target for the cumulative emissions of CO_2 over the period 1990–2010, and ask what are the present value (in 1990) costs of achieving that target if policy is implemented in 1990, 1995, or 2000. The advantage of delay is simply that it postpones having to incur costs of abatement; against that are the disadvantages that in each year in which policy operates, reductions in emissions will need to be greater the later is the start, first because there are fewer years in which to cut emissions, and second because it is necessary to offset the higher emissions in the years before policy was introduced; the compression of the time period for meeting the target will have the additional cost disadvantage that rather more of the reduction in emissions will come about through adjustments in factor use on existing capital stock rather than re-equipping with capital stock that is designed to reduce the use of carbon-intensive fuels. We emphasize that this falls far short of a full analysis of the costs of delay, and by requiring that a target for cumulative emissions be met by 2010 we are imposing a tight deadline which will exaggerate the disadvantages of delay. Finally we conduct this analysis on a sector-by-sector basis, rather than considering an efficient policy.

Figures 3*a*, *b*, and *c* show for the manufacturing sector, domestic sector, and transport sector respectively the percentage increase in total present value cost of meeting targets for reduction in cumulative

Figure 3*a*
Industrial Sector

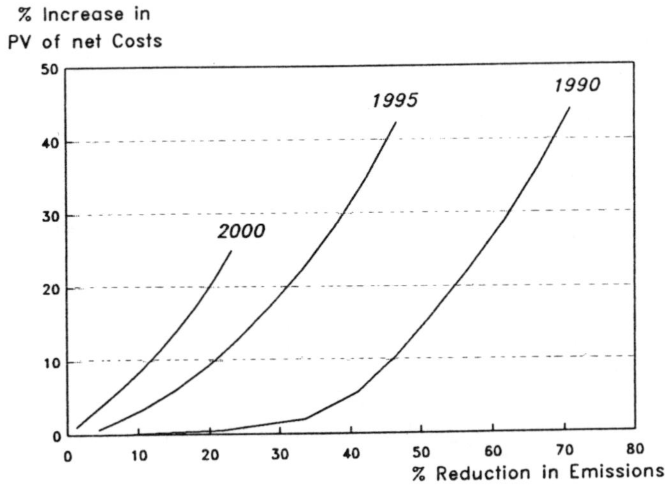

Figure 3b
Household Sector

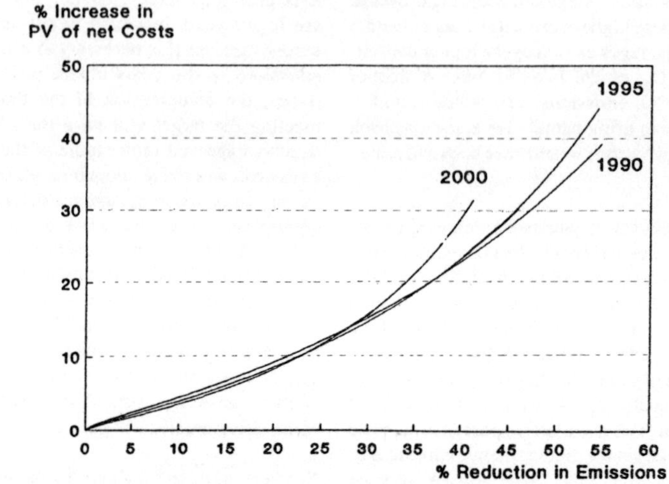

Figure 3c
Transport Sector

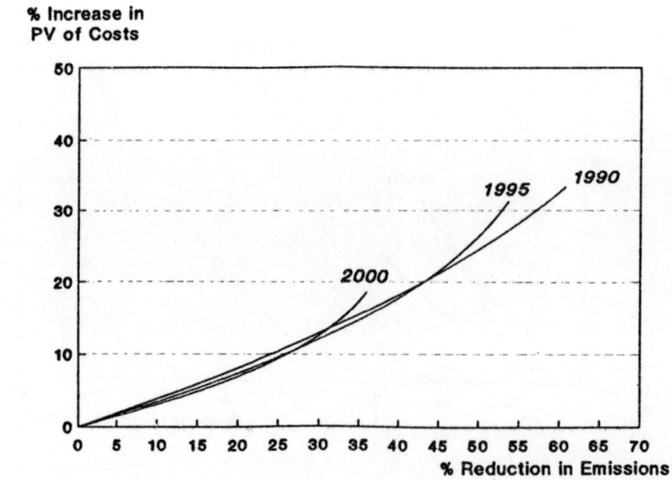

emissions over the period 1990–2010 of up to 60 per cent. There is a striking difference across sectors. For the domestic and transport sectors there is little difference in terms of percentage increase in total costs of variations in start date. For low levels of reduction (less than 33 per cent in the household sector, less than about 27 per cent in transport) delay to 2000 is worthwhile; for intermediate reductions in cumulative emissions (33–43 per cent in the household sector, 27–45 per cent in transport), delay to 1995 is worthwhile; for larger reductions in emissions, no delay is worthwhile. In the manufacturing sector, however, for all levels of reduction in cumulative emissions that we have considered delay is never worthwhile, and costs rise sharply as policy is delayed further; thus for reductions of 25 per cent in cumulative emissions, abatement of emissions would add about 1 per cent to manufacturers' costs; if policy were delayed till 1995, the cost increase would be over 10 per cent, while if the delay was to 2000 the increase in costs would be about 25 per cent. We believe that this marked difference across sectors again reflects the greater care with which costs of adjustment are modelled using the vintage model of the manufacturing sector.

In summary, where costs of adjustment are not very great, delay could reduce the present value cost of abating emissions while ensuring that a target for cumulative emissions by 2010 is attained, provided that target is moderate. Where costs of adjustment are significant, then delay in implementation could significantly raise the costs of attaining a target for cumulative emissions by 2010. We emphasize again that these calculations cannot be used to draw conclusions about the desirability or otherwise of delaying implementation of policy, because of all the caveats mentioned earlier, but they do throw some light on some of the factors that would need to be considered in assessing such a decision.

VI. CONCLUSIONS

In this paper we have presented estimates of the rates of carbon taxes that might need to be imposed in the UK to achieve particular targets, based on models of energy demand in the manufacturing, domestic and transport sectors. The emphasis in this work has not been on very long-term projections of tax rates, but rather with tax rates needed to achieve targets that might be set for the next ten to twenty years. In particular we have been concerned to explore the implications for carbon tax rates and costs of abating CO_2 emissions of the simple fact that energy demand in the short run is related to the currently available stock of energy using equipment, so that it becomes easier to reduce CO_2 emissions as newly designed capital stock is installed. While we capture these effects in the manufacturing sector through a relatively sophisticated vintage model of energy demand which has been developed over a number of years of research, the models for the other two sectors were developed for this paper, and are of a considerably more rudimentary nature, for example capturing the dynamics of adjustment of energy demand through simple lag structures.

We have used baseline scenarios for future growth of income and prices which are related to the work of IPCC, but are rather extreme projections by comparison with historical data. With these caveats, our results confirm the findings of our survey of other work that rates of carbon taxes are very sensitive to background assumptions about future economic growth and future changes in relative prices. We have also assessed the extent to which a policy of aiming for equal proportionate reductions in emissions in each of the three sectors differs from an efficient policy of imposing an equal tax on all sectors; our findings are that it would be efficient to encourage rather greater reductions in the manufacturing sector than in the domestic sector, and less again in the transport sector. This efficient policy would reduce costs by 20 per cent.

The importance of the dynamic process of adjusting energy demand emerges in the analysis of the effects on tax rates and costs of varying either the date at which a target for emissions reduction must be achieved or the date at which policy is implemented. We have shown that in the manufacturing sector the reduction in tax rates from lengthening the target date by five years is markedly greater than the increase in tax rates from shortening the target date by five years. There are also substantial costs of delaying the date of policy implementation given that a target for cumulative emissions over the period 1990–2010 needs to be met. These effects reflect both the advantage of being able to cut emissions by changing the capital stock rather than changing utilization on existing stock, and the long lags involved in changing the capital stock. Neither of these effects is as marked in the domestic and transport sectors, which we believe may reflect the rather crude modelling of dynamics of demand by simply incorporating some rather short lags. The important message that emerges for policy-makers is that for the medium term the level of taxes and costs of reducing CO_2 emissions depend importantly on the time-scale that economic agents are given to meet targets, while for researchers the message is the importance of carefully modelling the dynamic process of demand adjustment.

APPENDIX A:ESTIMATION OF THE DEMAND MODEL FOR THE DOMESTIC SECTOR

The model used is the well known Almost Ideal Demand System, extended to include quarterly dummies and dynamic adjustment. The latter was done by including up to the fifth lag on the commodity share, and then eliminating lags with insignificant coefficients. Thus general to specific modelling was done albeit in a very restricted way. The lagged dependent terms and the quarterly dummies were introduced in such a way that adding up was preserved. The model thus retains theoretical valid-

Table A1
Parameter Estimates for Demand Model

Parameter	Estimate	t-Statistic	Parameter	Estimate	t-Statistic
α_2	0.6808	5.189	γ_{22}	0.0735	4.4541
α_3	–0.0900	–1.214	γ_{23}	–0.0168	–1.8907
α_4	0.0983	1.065	γ_{24}	–0.0130	–1.3409
α_5	–0.2209	–2.430	γ_{25}	–0.0169	–2.3501
α_6	0.2909	1.647	γ_{26}	–0.0276	–1.7436
β_2	–0.0574	–4.959	γ_{33}	0.0233	3.5283
β_3	0.0109	1.621	γ_{34}	0.0012	0.2171
β_4	–0.0056	–0.678	γ_{35}	–0.00297	–0.63243
β_5	0.0209	2.508	γ_{36}	–0.00723	–0.97024
β_6	–0.0050	–0.313	γ_{44}	0.0459	5.0252
χ_1	0.1915	3.315	γ_{45}	–0.00324	–0.73233
χ_4	0.4153	6.905	γ_{46}	–0.0361	–3.2893
δ_{12}	–0.0015	–0.8524	γ_{55}	0.02229	2.8340
δ_{22}	–0.0015	–0.8571	γ_{56}	–0.0119	–1.1916
δ_{32}	–0.0012	–0.8926	γ_{66}	0.1253	4.5631
δ_{13}	–0.0139	–6.6613			
δ_{23}	–0.0077	–5.7170			
δ_{33}	–0.0096	–6.9581			
δ_{14}	0.0041	2.5130			
δ_{24}	–0.0092	–4.1146			
δ_{34}	–0.093	–5.2993			
δ_{15}	0.0196	5.0863			
δ_{25}	0.0118	4.5110			
δ_{35}	0.0124	4.7840			
δ_{16}	–0.0237	–3.8770			
δ_{26}	0.0011	0.2506			
δ_{36}	–0.0002	–0.0444			

OXFORD REVIEW OF ECONOMIC POLICY, VOL. 7, NO. 2

ity which is important for sensible results over relatively long time horizons, which for a model which included the explanatory variables in a very general form would be hard to ensure.

The selection of lags led to the first and fourth being selected. Thus the equations estimated were

$$w_{it} = \alpha_i + \Sigma_j \gamma_{ij} \log(p_{jt})$$
$$+\beta_i \log(Y_t / \Sigma_k w_{kt} \log(p_{kt}))$$
$$+\delta_{1i} Q_1 + \delta_{2i} Q_2 + \delta_{3i} Q_3 + \chi_1 w_{it-1} + \chi_4 w_{it-4} + u_{it}$$

$$i = 1,n \; ; \; t = 1,T$$

The data used was quarterly data on consumers' expenditure in constant and current price terms from *Economic Trends* from 1963Q1, aggregated into six groups: 1. durables, 2. food, 3. clothing, 4. energy, 5. transport, 6. other.

The model was estimated using FIML, and parameter estimates and diagnostic statistics are given in tables A–1 and A–2. Estimated short-run elasticities for income and price (evaluauted in the last year of the sample) are 0.911 and –0.258 respectively; long-run elasticities for income and price are 2.317 and –0.656.

Table A 2
Diagnostic Statistics for Individual Equations

	Food	Clothing	Energy	Transport	Other
SSR	0.00076	0.00029	0.00086	0.00325	0.0115
SER	0.00273	0.0017	0.00290	0.00564	0.0106
DW	1.6585	0.0170	1.7277	2.0593	1.7864

Notes: SSR = Sum of squared residuals
SER = Standard error of regression
DW = Durbin–Watson statistic

APPENDIX B: THE MODEL OF TRANSPORT DEMAND

The model is related to that of Drollas(1984) but estimated on more up-to-date data. The dynamic structure of petroleum demand is captured not by explicitly modelling changes in the vehicle stock, but by including lags on the dependent and independent variables in a single equation of gasoline demand. The independent variables are the price of petrol, the price of cars, and income, the dependent variable is petrol consumption (all measured in logs). Up to five lags were included for each variable and insignificant lags then eliminated. Quarterly dummies and a time trend were also included. The model was estimated on quarterly data from 1975Q1 to 1989Q4, using OLS. The estimates are contained in Table B1. The estimated short-run own-price and income elasticities are -0.172 and 0.223, while the corresponding long-run elasticities are -0.306 and 0.360 respectively.

Table B1
Estimates of Transport Demand Model

Variable	Co-efficient	t-Statistic
Constant	4.140	5.00
LG (–1)	0.381	3.46
LY	0.223	2.50
LPP	–0.172	–4.22
LPP(–1)	0.124	2.58
LPP(–3)	–0.141	–4.08
LPC(–4)	–0.081	–2.69
Q1	–0.045	–6.56
Q2	0.066	4.96
Q3	0.067	10.38
Time	0.002	4.74

Notes: R3 = 0.9899 S.E.R. = 0.0147 DW = 2.126
LG: log petrol consumption; LY: log income; LPP: log petrol price; LPC: log price of cars.

REFERENCES

Baker, P., Blundell, R., McKay, S., Symons, E., and Walker, I. (1990), 'A Simulation Programme of Consumer Expenditure', IFS Working Paper (forthcoming)

Barker, T. (1990), 'Measuring Economic Costs of CO_2 Emission Limits', Working Paper, Dept. of Applied Economics, Cambridge.

Barker, T., and Lewney, R. (1990), 'Macroeconomic Modelling of Environmental Policies: the Carbon Tax and Regulation of Water Quality', paper presented to conference, The UK Economy and the Green 1990s, Cambridge, 11–12 July.

Barrett, S. (1990*a*), 'Economic Instruments for Global Climate Change Policy', Report for OECD Economic Directorate, London Business School.

— (1990*b*), Memorandum for the House of Commons Energy Select Committee.

— (1990*c*), 'The Problem of Global Environmental Protection', *Oxford Review of Economic Policy*, 6(1), 68–79.

Blundell, R., Pashardes, P., and Weber, G. (1989), 'What Do We Learn About Consumer Demand Patterns from Micro-data', IFS Micro to Macro Paper, 3.

Boero, G., Clarke, R., and Winters, A. (1991), 'The Macroeconomic Consequences of Controlling Greenhouse Gases—A Survey', mimeo, University of Birmingham.

Cline, W. R. (1989), 'Political Economy of the Greenhouse Effect', mimeo, Institute for International Economics.

Dasgupta, P. (1982), *The Control of Resources*, Basil Blackwell, Oxford.

— (1990), 'The Environment as a Commodity', *Oxford Review of Economic Policy*, 6(1), 51–67.

Department of Energy (1989) 'An Evaluation of Energy Related Greenhouse Gas Emissions and Measures to Ameliorate Them', Energy Paper 58, HMSO, London.

Dimson, E. (1988), 'An Analysis of the Cost of Capital and Risk Premiums', evidence to the Hinkley Point 'C' Public Inquiry.

Drollas, L. (1984) 'The Demand for Gasoline: Further Evidence', *Energy Economics*, 6(1), 71–82.

Edmonds, J., and Reilly, J. (1983), 'Global Energy and CO_2 to the Year 2050', *The Energy Journal*, 4(3), 21–47.

— — (1985), *Global Energy: Assessing the Future*. New York, Oxford University Press.

Helm, D., Kay, J., and Thompson, D. (1989), *The Market for Energy*, Oxford, Clarendon Press.

Houghton, J. T., Jenkins, G. J., and Ephraums, J. J. (eds.) (1990), *Climate Change: The IPCC Scientific Assessment*, Cambridge, Cambridge University Press.

Hughes, G. (1990), 'On the Consistency of Short and Long Run Models of Gasoline Demand', in D. Hawdon (ed) *Energy Demand: Evidence and Expectations*, University of Surrey Press, Guildford (forthcoming).

Ingham, A., and Ulph A. (1990*a*), 'The Economics of Global Warming', in J. Bennett (ed.) *Economics and the Environment: A Reconciliation*, Australian Institute for Public Policy (forthcoming).

— — (1990*b*), 'Carbon Taxes and the UK Manufacturing Sector', in T.Dietz, R. van der Ploeg and J. van der Straten (eds), *Environmental Policy and the Economy*, North Holland, Amsterdam (forthcoming).

— (1991), 'Market-Based Instruments for Reducing CO_2 Emissions—the Case of UK Manufacturing', *Energy Policy*, **19**, 138–48.

Ingham, A., Maw, J., and Ulph, A. (1990), 'Energy Conservation in the UK Manufacturing Sector—a Vintage Approach' in D. Hawdon (ed.) *Energy Demand—Evidence and Expectations*, University of Surrey Press, Guildford (forthcoming).

Mäler, K-G. (1990), 'International Economic Problems', *Oxford Review of Economic Policy*, **6**(1), 80–108.

Manne, A. (1976), 'ETA: A Model for Energy Technology Assessment', *Bell Journal of Economics and Management*, **7**(2), 379–406.

— (1981), 'ETA-MACRO: A User's Guide', EA–1724, Electric Power Research Institute, Palo Alto.

Manne, A.S. and R.G. Richels (1990*a*), 'CO_2 Emission Limits: an Economic Cost Analysis for the USA', *The Energy Journal*, **11**(2), 51–74.

— — (1990*b*), 'Global 2100: Model Formulation', paper presented at Workshop on Economic/Energy/Environmental Modelling for Climate Policy Analysis, Washington.

Martin, D. J., and Shock, R. A. W. (1989), 'Energy Use and Energy Efficiency in UK Transport Up To The Year 2010', Energy Efficiency Series, 10, Department of Energy, London, HMSO.

Nordhaus, W. D. (1977), 'Economic Growth and Climate—The Carbon Dioxide Problem', *American Economic Review*, **67**(1), 341–6.

— (1979), *The Allocation of Energy Resources*, New Haven, Yale University Press.

— (1989), 'The Economics of the Greenhouse Effect', mimeo, Yale University.

— (1990), 'An Intertemporal General-Equilibrium Model of Economic Growth and Climate Change', paper presented to Workshop on Economic/Energy/Environmental Modelling for Climate Policy Analysis, Washington.

Pearce, D., and Helm, D. (1990), 'The Assessment: Economic Policy Towards the Environment', *Oxford Review of Economic Policy*, **6**(1), 1–16.

Pearson, M., and Smith, S. (1990), 'Taxation and Environmental Policy: Some Initial Evidence', IFS Commentary No. 19.

Peterson, A.W. A. (1987), 'The Demand for Energy', in T. S. Barker and A. W. A. Peterson (eds.) *The Cambridge Multisectoral Dynamic Model of the British Economy*, Cambrdige, Cambridge University Press.

Plourde, C. (1972), 'A Model of Waste Accumulation and Disposal', *Canadian Journal of Economics*, **5**, 119–25.

Reilly, J., Edmonds, J., Gardner, R., and Brenkert , A. (1987), 'Uncertainty Analysis of the IEA/ORAU CO_2 Emission Model', *The Energy Journal*, **8**, 1–29.

Skea, J. (1990), 'A Case Study of the Potential for Reducing Carbon Dioxide Emissions in the UK', SPRU Discussion Paper.

Tietenberg, T. (1990), 'Economic Instruments for Environmental Regulation', *Oxford Review of Economic Policy*, **6**(1),17–33.

Whalley, J. (1990), 'The Interface Between Environmental and Trade Policies', mimeo, University of Western Ontario.

Whalley, J., and Wigle, R. (1989), 'Cutting CO_2 Emissions: The Effects of Alternative Policy Approaches', mimeo, University of Western Ontario.

[18]

The International Incidence of Carbon Taxes

John Whalley and
Randall Wigle

This chapter discusses how different countries may fare (because of international incidence effects) under any carbon-tax scheme that may be adopted in the next few years as part of efforts to limit global buildup of carbon dioxide and other greenhouse gases.[1] As such, the chapter makes no claims to provide an evaluation of the severity or otherwise of the greenhouse effect,[2] but instead makes the assumption that there will be a policy response to the risk of significant global warming over the next few decades. A component of such a response may be a tax on carbon use (carbon embodied in fossil fuels); we evaluate what the intercountry effects of such a measure could be.

International incidence effects from carbon-tax schemes reflect a number of factors. One is the sharply differing configuration of production and consumption of carbon-based energy sources (oil, coal, natural gas) across countries. Depending upon whether carbon taxes are passed forward in higher prices to carbon-based energy users or passed backward in reduced rewards to producers, effects across countries will differ. If supply elasticities of carbon-based energy products are low, producers (including OPEC countries) will finish up largely bearing the burden of any carbon taxes. If supply elasticities are high, more of the burden of the tax will be borne on the demand side by consumers (including OECD countries).

Another factor is how the revenues raised by such taxes are treated. These are potentially very large, and how they are disbursed (i.e., whether there is any international redistribution of revenues) makes a large difference. Yet another factor is trade in energy-intensive manufactures (primary metals, fabricated metal products, chemicals), since countries may bear some of the carbon-tax burden because they are importers. Hence consumption, production, and trade in energy and in products embodying energy, as well as tax revenue distribution, all need to be considered in evaluating international incidence effects from carbon taxes.

International incidence effects also need to be evaluated relative to a particular reference point, and that may or may not involve the full participation of all countries in the tax. Developing countries have long claimed that global environmental problems largely reflect the cumulative growth process of the developed countries over a 200-year or even longer period.[3,4] In their view, only the developed and not the developing world should be responsible for limiting carbon emissions. In contrast, some developed countries point to the more rapid growth of developing countries as the source of potentially major global environmental difficulties in the decades ahead. Combined with their perception of inefficient energy-conversion technologies that needlessly generate carbon emissions in such countries, their view is that developing countries should be involved.

In this chapter we report results from a global numerical general-equilibrium model that we use to investigate international incidence effects quantitatively. The model captures trade, production, and consumption of carbon-based and non-carbon-based energy products, as well as energy-intensive manufactures, and other goods between six major world trading areas. Demand and production functions are used with specified parameter values, and the model is calibrated to a 1990–2030 base-case scenario. A series of counterfactual equilibrium analyses with the model then allow various carbon tax schemes to be analyzed and their effects to be compared to the base-case model solution (no carbon tax).

Despite a number of weaknesses in the model,[5] a series of striking features emerge from our results. First, we show that both the base and revenues generated by such a tax are likely to be large. We estimate that a global carbon tax base in 1990 present-value terms could be $43 trillion over the period 1990–2030. This is about 10 percent of the projected 1990–2030 global value of product (gross world product). To meet an emission reduction target set at 50 percent of base-case emissions over the period, carbon-tax rates would need to be around 85 percent of the consumer (gross of tax) price of carbon-based energy products, or around $460 per ton of carbon. The tax revenues involved would be in the $40–$50 trillion range; that is, around 10 percent of gross world product would be raised by such a tax.

Perhaps not surprisingly, model results also show that what happens to these tax revenues is crucial for international incidence. If a significant portion is recycled to lower-income countries, then their benefit from participation in a global carbon tax scheme will be large, probably more than outweighing benefits they now receive from aid flows.

Model results also confirm elasticity assumptions as important for international incidence effects. Low supply elasticities for carbon-based energy products imply more redistribution away from producers (owners) of carbon-based energy, as well as higher tax rates in order to achieve emission reduction targets. With higher supply elasticity values, the converse is true. We survey literature elasticities in an effort to produce a best-guess assessment of carbon-tax incidence effects using the model. Interestingly, however, even though we use a wide range of elasticity values in sensitivity calculations we later report, international incidence effects seem substantially more dependent on the treatment of revenues under such a tax than they are on elasticity values.

The form that any carbon tax might take also turns out to be important in determining international incidence effects. There are, for instance, sharp differences between a production-based and a consumption-based tax, simply because some countries are in net export and others in net import positions. In addition, such features as a common, cross-country, per-capita emission target for the tax, instead of a common equiproportional emission cut, have sharply differing incidence effects. The former, for example, may induce relocation of manufacturing activity from high-emission (developed-country) regions to low-emission (developing-country) regions. Thus international incidence effects can also work through induced-capital-flow and production-relocation effects.

The model also captures international incidence effects operating through trade in energy-intensive manufactures. In the model, trade is differentiated between these and other products, and takes place between a number of developed and developing countries. Model results show that direct incidence effects from trade in energy products (oil) dominate those from trade in energy-embodying products.

In summary, two central themes emerge from the reported model results. The first is that international incidence effects of carbon-tax schemes will likely be large. The second is that their precise form depends crucially both on the design features embodied in such a tax

Table 7.1
Summary features of current carbon-tax proposals

Objective
- To achieve reduction in global emissions of carbon relative to some specified baseline projection (say, 20 percent reduction by 2020, and stabilize emission levels thereafter).

Features
- Tax base to include carbon-based energy products (oil, coal, natural gas).
- Taxes apply to carbon content of fuel (rates highest on coal, then oil, then natural gas).
- Tax rates typically in 50–100 percent range (around $250 per ton of carbon).

Ambiguities
- Consumption- or production-based tax.
- New tax or replacement for existing taxes.
- Revenues collected nationally or through a supranational agency.
- Revenue from the tax to finance global alternative technologies fund or to be recycled.
- All countries or subset of countries to participate.
- Common rate across all countries or country-variable rates.
- With or without annual exemption set at specified emission level per capita, or some other more beneficial treatment for low emitters (developing countries).

(revenue disbursement, and whether a production or consumption basis is used for the tax) and on key model parameters (supply elasticities).

The Carbon-Tax Proposal

As table 7.1 indicates, the idea underlying most carbon-tax proposals is to reduce consumption of carbon-based energy products (oil, coal, natural gas) and slow (or even stabilize) the buildup of atmospheric carbon dioxide. Most involve taxing the carbon content of individual fuels; hence different tax rates would apply to coal, natural gas, and oil.[6]

The tax is seen as one that would meet a specified target in terms of reduction in emission levels. Targets commonly discussed are between 20 and 50 percent of projected baseline levels (i.e., with no policy intervention to slow CO_2 accumulation) by some specified date.[7] Taxes are usually proposed as a fixed dollar (or currency) amount on a physical unit basis—say, $250 per ton of carbon embodied in each fuel.[8] Ad valorem equivalent carbon tax rates (which vary by fuel) are in the region of 50–100 percent (higher in some cases). The rate needed in any given carbon tax is thus dependent on a variety of factors, including the target emission reduction and assumed elasticity values for demand and supply functions. The latter will, in part, determine how closely the emission reduction target will be met by any given tax rate.

While carbon-tax proposals have been discussed for some time at an academic level,[9] the idea of responding to global warming in this way was popularized by the World Conference on the Changing Atmosphere held in Toronto in June 1988 (see Climate Institute 1988). The concluding conference statement proposed a 20 percent cutback in global CO_2 emissions (from their then current level) by 2005, with a larger 50 percent cut in emissions beyond 2005 to stabilize atmospheric CO_2 content.[10] To achieve these objectives, the conference called for a World Atmospheric Fund to finance the development of new technologies to help reduce carbon emissions, with industrialized countries to contribute to the fund, and new technologies focused on the developing world. The statement proposed that the fund be financed by a tax on fossil-fuel consumption in industrialized countries. Since this conference took place, such taxes have also been proposed by the U.K. environment secretary (in 1989) and by the West German Social Democratic Party.[11]

As can be seen from table 7.2, carbon use around the globe is highly concentrated, suggesting that significant intercountry effects may accompany the introduction of such a tax. Emissions per capita in developed countries are on average around 20–30 times those of developing countries; three countries (United States, U.S.S.R., and China) account for more than 50 percent of emissions. On the other hand, carbon use per dollar of GNP is typically much higher in developing countries, reflecting the use of relatively inefficient energy-conversion technologies.

However, despite the calls for a carbon tax, and as table 7.1 indicates, there is a surprising ambiguity and lack of specificity in the proposals currently under discussion. There is ambiguity in whether their main purpose is to finance schemes to develop new energy-efficient technologies (as in the Toronto Conference statement), with incentive effects on energy consumption viewed as of secondary consequence, or whether their primary aim is to use the price mechanism to reduce energy consumption, and hence carbon emissions.

The lack of specificity in current proposals reflects the many design

Table 7.2
Country characteristics of fossil-fuel carbon emissions, 1987

	Emissions per capita (tons/year)	Country emissions (percent of world total)	Grams of carbon per dollar of GNP
United States	5.03	21.9	276
Canada	4.24	1.9	247
Australia	4.00	1.1	320
Soviet Union	3.68	18.5	436
Saudi Arabia	3.60	0.8	565
Poland	3.38	2.3	492
West Germany	2.98	3.3	223
United Kingdom	2.73	2.8	224
Japan	2.12	4.5	156
Italy	1.78	1.8	147
France	1.70	1.7	133
South Korea	1.14	0.8	374
Mexico	0.96	1.4	609
China	0.56	10.7	2,024
Egypt	0.41	0.4	801
Brazil	0.38	0.9	170
India	0.19	2.7	655
Indonesia	0.16	0.5	403
Nigeria	0.09	0.2	359
Zaire	0.03	0.01	183
Average for the world	1.08		327

SOURCE: Information reported in this table has been extracted from Flavin (1990, Table 2-1, p. 19), who, in turn, uses a variety of primary sources in his calculations.

and implementation issues with such a tax. These are not clarified as far as we are aware in any recent literature. First of all, the definition of liability for the tax is greatly complicated by the integrated nature of production of refined energy products. As currently proposed, the carbon tax is not a sales tax and is in no way related to conventional retail sales, value-added, turnover, or even excise taxes. Any attempt to tax the carbon content, say, of oil used by an integrated producer/refiner who extracts and refines crude oil and then sells refined products will, for tax purposes, become involved with separating out, in some way, transactions internal to the firm. The same is true of natural gas, where natural gas producers convert gas into feedstock or into subsequent petrochemical production.

In addition, it is unclear whether carbon taxes as proposed are to apply to the consumption or production of carbon. The vocabulary commonly used is "carbon use," seemingly implying a consumption-based tax. However, among other things, if the tax were administered akin to a value-added or retail sales tax, this approach would involve taxing carbon as it crosses national borders, requiring customs officials to infer the carbon content of various imports. Such imports might include not only crude oil, but also other energy products at various stages of refinement, and even nonenergy products at higher stages of fabrication. If "carbon use" is defined to include carbon taxed in the origin country at lower stages of fabrication before entry into the destination country, such difficulties may be potentially soluble, but enormous problems in administering such a tax likely remain. The alternative of a production-based tax seems simpler, but then becomes an extraction tax rather than a tax on "use."

As pointed out earlier, this distinction between a production- and a consumption-based carbon tax is extremely important in any discussion of international incidence, because individual countries differ sharply in their net trade position in carbon. Under a production-based tax, OPEC oil exporters keep the tax revenues; under a consumption-based tax, oil importers keep the revenues. Also, the carbon content of fuels even within broad fuel groups varies substantially, making any application of the tax, at best, inexact.

The definition of the base for any carbon tax is especially crucial in any evaluation of its international incidence effects. In table 7.3, we present some calculations of the potential global and regional tax bases in present-value terms over the forty-year period 1990–2030, using the same data base employed in the general-equilibrium-model calculations we subsequently report. The base for such a tax is large, around $43 trillion in 1990 present-value terms (approximately 10 percent of gross world product over the period). Also, for individual regions the difference between the consumption and production tax base is substantial (EC, Japan, oil exporters). These data clearly suggest that a carbon tax will have major consequences for global economic activity, with potentially large intercountry effects.

Tax rates are also important to international incidence effects, and a variety of issues arise here. One is whether rates should be uniform or nonuniform across countries. Uniform rates by country are viewed as desirable, since they are nondistorting of trade and production

Table 7.3
Estimated global and regional carbon-tax bases for the period 1990–2030 (billions of 1990 dollars)

	Value of carbon-based energy production	Value of carbon-based energy consumption
European Community	3,078	6,060
North America	10,195	11,231
Japan	92	1,842
Other developed countries	999	1,053
Oil exporters	8,666	2,290
Developing/centrally planned	19,900	20,454
World	42,929	42,929

SOURCE: Data base constructed by the authors for general-equilibrium model discussed later.

location decisions. On the other hand, if environmental quality is a luxury good and if taxes are related to the country-specific benefits from emission reductions, then arguments in favor of nonuniformity of tax rates emerge. Developing countries concerned not to truncate their development, as well as countries for whom the cost of global warming seems small (Canada and the U.S.S.R., for instance), may argue that they should face lower tax rates.

A further issue, which is surprisingly important, if conceptually pedantic, is whether the tax rates at issue are on a gross or net basis. A $250 per ton carbon tax may imply an effective 500 percent tax on coal calculated on the net of tax price, but only an 80 percent tax when calculated on the gross of tax price. There seems to be confusion in the literature over the difference in these two rates, with incomplete discussion of which base particular proposals relate to.

The uses to be made of tax revenues generated by any carbon-tax scheme actually implemented are also crucial to international incidence effects. If the base is of the order of magnitude that we have suggested and if ad valorem equivalent rates are large, who gets the revenues (i.e., to whom they are distributed) becomes very important. Particularly crucial is whether the carbon tax is agreed to globally but left to be collected by national government agencies (i.e., revenues redistributed within nation-states) or whether the tax is instead a truly global tax, collected by a supranational agency (the World Atmospheric Fund, or WAF, or the UN). Under a true global tax, revenues would presumably be recycled to countries on some mutually agreed basis, perhaps an equal amount per capita. As the results of the model we present later clearly indicate, such a redistribution scheme, not surprisingly, is extremely beneficial to developing countries.

Even the preceding does not exhaust the list of design and implementation issues with the tax. Whether carbon taxes replace existing taxes within countries or are new taxes, and whether exemptions of a per-capita amount of carbon emissions are to accompany the tax (as developing countries seek), will affect international incidence effects. Complex international negotiation issues that also seem likely to be involved in agreeing on rates as part of any global carbon-tax scheme will come into play. Small countries will generally have an incentive to get a free ride—levy no taxes themselves and let large countries reduce their carbon emissions. Whether large countries will be able to force small countries to participate in a global carbon-tax scheme at full tax rates, without using other non-carbon-tax sanctions, such as trade measures, is thus a question.[12]

Beyond all these considerations, there are also uncertainties as to how successful a carbon tax would be in practice in achieving its objectives. If the tax is not passed on in higher prices but merely redistributes wealth away from owners of energy sources, it could have less than the intended effect (or even, in the limit, no effect) on consumption, and hence on emissions. Moreover, with underlying real growth in both the global and in individual regional economies, unless carbon-tax rates increase over time, any carbon-tax scheme will only have the effect of delaying atmospheric buildup of carbon dioxide, not preventing it. Nordhaus and Yohe (1983), for instance, estimated that a 60 percent carbon tax might only delay a doubling of atmospheric carbon dioxide content by fifteen years, a period which Edmonds and Reilly (1983), using somewhat higher tax rates, suggested could be even shorter, perhaps no more than ten years.

Because of this wide range of options, uncertainties, and ambiguities in actually implementing any carbon-tax scheme, our discussion of international incidence effects of carbon taxes in the later sections of this chapter considers a number of different variants for such a tax. Our model results, in part, emphasize that, beyond the potentially large effects from such a tax, the form the tax takes can be the key factor in determining its international incidence effects.

Table 7.4
Regions in the global-equilibrium model used to evaluate international incidence effects of carbon taxes

1. European Community (of the twelve)
2. North America: United States, Canada
3. Japan
4. Other OECD: Austria, Switzerland, Finland, Iceland, Norway, Sweden, Australia, New Zealand
5. Oil exporters (OPEC countries, plus major non-OPEC exporters): Algeria, Libya, Nigeria, Tunisia, Mexico, Venezuela, Indonesia, Iran, Iraq, Kuwait, Saudi Arabia, United Arab Emirates
6. Rest of the world (developing countries and centrally planned economies): This is a residual category containing all other countries, including the U.S.S.R., Eastern Europe, China, Brazil, India, and other developing countries not in category 5.

A Numerical General-Equilibrium Model for Assessing the International Incidence Effects of Carbon Taxes

To help make discussion of international incidence effects of carbon taxes more concrete, we have used a global general-equilibrium model[13] to generate a number of counterfactual calculations that we use to provide an indication of what could happen to the global economy after the introduction of carbon taxes of various kinds. The model incorporates trade, production, and consumption of both energy and nonenergy products for a number of countries (or groups of countries) over a single forty-year projection period of 1990–2030.[14,15] To keep the model manageable, we do not identify fuel types within the broader category of carbon-based energy products, even though in practice the various elements within this category (oil, coal, natural gas) would be taxed at different rates. We also have not incorporated existing taxes on energy products. In practice, these vary by region, and could also affect results.

In the model, the world is divided into six regions, as indicated in table 7.4. The European Community, North America, and Japan are separately identified. Oil exporters include all OPEC countries and major non-OPEC energy exporters. Most developing countries (those who are not oil exporters) are included in a residual rest-of-the-world category, along with the centrally planned economies. Data difficulties preclude a separation between the two groups (i.e., between centrally planned and other developing countries).

Nested functional structures are used in the model to represent production and demand in each region. Each region is endowed with four nontraded primary factors: (1) primary factors, exclusive of energy resources; (2) carbon-based energy resources (deposits of oil, gas, and coal); (3) other energy resources (hydroelectric capacity and nuclear); and (4) sector-specific skills and equipment in the energy-intensive manufacturing sector. Both energy resources are converted into the relevant energy products through a refining/extraction process, which uses other resources (primary factors). There are three internationally traded commodities in the equilibrium system: carbon-based energy products, energy-intensive manufactures, and other goods (all other GNP). Energy-intensive manufactures, other goods, and the composite energy product (carbon-based and non-carbon-based energy) are the commodities that enter final demands.

For each of five produced goods in each region, production is represented by a constant elasticity of substitution (CES) function. Carbon-based and non-carbon-based energy products use the respective energy resources and primary factors. Non-carbon-based energy products are nontraded, since hydroelectric, solar, and nuclear power are not traded in significant quantities between the regions as defined here.[16] A domestic energy composite is produced by a third (energy-conversion) industry, using inputs of the two energy products. The two final goods (energy-intensive manufactures and other goods) use primary factors and the composite domestic energy product as inputs. Perfect competition is assumed throughout for all sectors and in all regions.

Equilibrium in this model involves full global market clearing in the single forty-year period in all three traded goods (carbon-based energy products, energy-intensive manufactures, and other goods). For the two nontraded goods (noncarbon energy products and composite energy) there is domestic market clearing within each economy. Since prices in this system are treated as completely flexible, they will adjust to the levels required to clear the relevant international and domestic markets.

The introduction of a carbon tax into this system will raise prices of carbon-based energy products, and hence the price of composite energy to users, and reduce prices to producers. These price changes will have the effect of cutting both consumption and production of carbon-based energy products, with prices adjusting until the global market in carbon-based energy products clears. If prices to consumers increase fully by the amount of the tax, the tax will have been fully passed on to users of carbon. If prices to consumers change little, then the tax will have been passed backward to producers of carbon-energy-based products. Substitution will also occur between carbon and noncarbon energy products, and between the energy composite product and other inputs. The strength of all of these effects depends on the elasticity values used at the respective nodes in the nesting of substitution possibilities.

To evaluate international incidence effects of alternative carbon-tax schemes on the six regions, we use the model in what we term counterfactual mode.[17] This involves constructing a base (or reference) case solution to which the equilibrium model is calibrated (see Mansur and Whalley 1984); that is, parameters are chosen such that the model will reproduce the base-case data as a full-equilibrium solution prior to the introduction of any policy change. The base-case data thus fully describe a no-policy-change model solution. In this case, this is a general-equilibrium model solution covering the forty-year period 1990–2030 with transactions represented in present-value terms. Different carbon-tax schemes can then be introduced, with a new equilibrium (counterfactual equilibrium) solution computed for each. Using a pair-wise comparison (base case versus counterfactual), international incidence effects over the period 1990–2030 are evaluated after each such calculation.

Data

The model is benchmarked to a base-case equilibrium solution that has been constructed to represent the future evolution of the global economy over the forty-year period 1990–2030. The model is solved to yield a forty-year baseline solution representing the world economy in the absence of any response to global warming over the period 1990–2030 (in discounted present-value terms at 1990 prices, and in billions of U.S. dollars). Policy experiments are then evaluated against this baseline, with a comparison of the base and counterfactual equilibria.

To construct this base-case solution, Hicksian neutral growth is assumed to occur in each of the regions in the model at average growth rates reported in the 1989 *World Development Report* of the International Bank for Reconstruction and Development (IBRD). These are assumed to apply over the entire period under consideration, including the interim period 1982–1989. The oil-exporting region is assumed to grow at 2.5 percent, the "rest of the world" at 2.7 percent, and the remaining regions at 2.3 percent. Each region's endowment of nonproduced factors during this period thus reflects the present value of its resources (at constant prices) over the entire period. We assume that a 5 percent real discount rate applies for all years in the period.

The structure of the regional economies in the base data used to calibrate the model corresponds largely to data available for 1982. Data for regional population and GNP in 1982 (the benchmark year) are obtained from the 1987 *World Tables* of the IBRD (1988). Value added, production, and trade in energy-intensive manufactures (primary metals, glass, ceramics, and other basic manufactured products) are obtained from Nguyen, Perroni, and Wigle (1990). These are identified as those industries having the highest energy input requirements. Input ratios used in Ross (1989) are used to infer energy input requirements for energy-intensive and other industries.

Production, consumption, and trade in carbon-based energy products and non-carbon-based energy (for 1982) come from United Nations (1986) energy statistics. Raw data are in (metric) kilotons of coal equivalent. These are converted to determine the carbon content of production and consumption for the regions in the model using conversion coefficients from the CDIAC CO_2 glossary (Milleman 1988). To convert the data into value terms, we use price information from World Resources Institute (1990).

Some of the main features of the base-case data we use are displayed in table 7.5. Consumption and production of carbon-based energy over the period 1990–2030 are each around \$43 trillion (in 1990 dollars), from a worldwide GDP figure over the same period of approximately \$450 trillion. A carbon tax, therefore, will apply to approximately 10 percent of world product and will produce revenues potentially of a comparable order of magnitude. Put another way, the tax may yield a global policy instrument with the potential to redistribute 10 percent of global income (and, effectively, over a forty-year period 10 percent of global wealth).

Table 7.5 also gives details of the concentration by region in both population and emission-generating activities, and reports net trade by region. Because the model considers only net trades (rather than gross) and uses a highly aggregated product classification, the volume of trade relative to GDP is substantially reduced compared to

Table 7.5
Global baseline data for 1990–2030 used in the general-equilibrium model
A. Global production and consumption of carbon-based energy products[a] and GDP (billions of 1990 dollars)

	Production	Consumption	GDP
European Community	3,078	6,060	96,739
North America	10,195	11,231	127,480
Japan	92	1,842	39,879
Other OECD	999	1,053	21,362
Oil exporters	8,666	2,290	26,242
Developing/centrally planned	19,899	20,454	132,472
World	42,929	42,929	444,174

B. Trade in carbon-based energy products, energy-intensive manufactures, and other goods by region (net exports [+] or imports [−] over the period 1990–2030 in billions of 1990 dollars)

	Carbon-based energy products	Energy-intensive manufactures	Other goods
European Community	−2,982	638	2,300
North America	−1,036	36	985
Japan	−1,751	551	1,174
Other OECD	−54	246	−193
Oil exporters	6,376	−234	−6,049
Developing/centrally planned	−554	−1,237	1,783

C. Shares of population, GDP, and global emissions by region, 1990–2030 (percentages)

	Population	GDP	Emissions
European Community	7.1	21.8	15.0
North America	6.0	28.7	27.3
Japan	2.8	9.0	4.8
Other OECD	1.2	4.8	2.6
Oil exporters	8.7	5.9	4.1
Developing/centrally planned	74.4	29.8	46.1
World	100.0	100.0	100.0

a. These data on consumption and production are identical to those in table 7.3, which gives the carbon-tax base over the period 1990–2030.

actual and likely future trade. As there is relatively limited cross-hauling of trade in carbon-based energy products, the amount of netting out that occurs in this category is small.

Elasticities

In this model, the implied demand and supply elasticities for carbon-based energy products are important parameters as far as international incidence effects are concerned. The values of these elasticities are not directly specified in the model formulation, but instead reflect preferences and intermediate production technology on the demand side, and the relative importance of fixed and variable factors in carbon-based energy production (oil in the ground versus extraction costs) and the marginal productivity of variable factors in this production process on the supply side.

We are able to solve the model under varying carbon-energy product-supply elasticity assumptions relatively easily, since in the special case where the production function for carbon-based energy products is Cobb-Douglas, the supply elasticity is given simply as $(1 - \theta)/\theta$, where θ is the share parameter on the carbon resource input in the carbon-based energy-product production function. We use a central-case carbon-based energy supply elasticity of 0.5, with sensitivity analysis over the wide range 0.1 to 1.5.

Surprisingly, there are relatively few estimates of energy supply elasticities in the literature; most estimation is concerned with income and own-price elasticities on the demand side, or interfuel substitution elasticities (see Kouris 1981 and Pindyck 1979). The supply elasticities that have been estimated are mainly for OPEC countries and focus on strategic supply response over relatively short periods of time; not the longer-run elasticities at issue here that reflect exploration and extraction cost functions.

The ease with which the composite energy output can be produced from varying mixes of energy inputs (carbon- and non-carbon-based energy) and the ease of substitution between composite energy and the two nonenergy products in consumption are the model features that affect the elasticity of demand for energy. In the absence of any literature estimates of the elasticity of input substitution in composite energy production, we set this equal to 1.0. This value reflects two competing factors. First, for many industrial and home purposes (such as heating), it is relatively easy to substitute between fossil fuels and electricity. At the same time, substitution between fossil and nonfossil fuels is not as easy in transportation. Literature on elasticity estimation for industrial electricity demand (see Nainar 1989) provides some support for a range around the value we use.

In the final demand function involving composite energy and the two nonenergy final products, the CES function used has an elasticity of substitution set equal to 0.5. A demand elasticity for energy in this range is consistent with that suggested by several studies of the long-run demand for energy (see Pindyck 1979; Kouris 1981; Nguyen 1986; Hunt and Manning 1989; Considine 1989; and Estrada and Fugleberg 1989).

Since all of the traded goods produced in each of the three regions are perfect substitutes, the model does not employ the Armington structure of product heterogeneity by region for traded goods common in other applied general-equilibrium models (see Shoven and Whalley 1984).

Counterfactual Policy Experiments Using the Model

We have used the model to evaluate possible international incidence effects that would follow the introduction of alternative carbon-tax schemes. The results that follow initially evaluate a target of a 50 percent cut in carbon emissions relative to the baseline model solution over the period 1990–2030. We later report model calculations with different reduction targets for one of our tax options. A 50 percent cut in emissions from the baseline is on the high side of the range of targets currently used by other modelers evaluating carbon-tax proposals, but corresponds approximately to that called for in the 1988 Toronto Conference statement.

In all our model runs, we endogenously determine the ad valorem carbon-tax rate, which, in this structure, applies to all fuels at the same rate. We also calculate the implied carbon tax in dollars per ton of carbon. As emphasized earlier, the model does not capture the benefit side of reduced atmospheric CO_2 concentrations—that is, the benefits of reduced or slowed global warming.

The carbon tax variants we consider are as follows:

1. Production-based carbon taxes collected by national governments. The emission target assumed is a 50 percent reduction in each region's production of carbon-based energy. Under this scheme, tax rates will vary by region.

2. Consumption taxes collected by national governments. The emission target assumed is a 50 percent reduction in each region's consumption of carbon-based energy. Under this scheme, tax rates will vary by region.

3. A global consumption (or production)[18] tax collected by an international body such as the UN, with revenues distributed to countries according to population, and achieving a global 50 percent emission reduction. Under this scheme, tax rates are identical across regions.

4. A series of national consumption taxes designed to reduce global emissions by 50 percent by achieving the same per-capita consumption of carbon-based energy. Under this scheme, tax rates vary sharply across regions.

These tax schemes are each evaluated using the general-equilibrium model in counterfactual mode. The model is first calibrated to reproduce the original (benchmark) data as a baseline equilibrium solution over the period 1990–2030, with no policy changes in effect. The model is then re-solved with a carbon tax in place.[19] The international incidence effects attributable to any of the tax schemes are determined by comparing the original (benchmark) and the new (or counterfactual) equilibria.

Table 7.6 reports the central-case results from our model experiments. We consider three alternative forms of carbon tax: a national production-based tax, a national consumption-based tax, and a global tax. In each case, the target emission reduction is 50 percent relative to the base-case model solution for the period 1990–2030.

As part A table 7.6 indicates, the carbon-tax rates involved in meeting such emission reduction targets are high, in the range of 85–90 percent. Importantly, these are calculated on a gross-of-tax basis, meaning that from each dollar spent by carbon users on carbon-based products, 85 cents would go in taxes and only 15 cents directly on the purchase of carbon-based energy products. Thus, if calculated on a net-of-tax basis, these rates would be considerably higher.

We report required carbon tax rates for these targets in the range of $450 per ton of carbon. These tax rates are higher than other estimates in the literature in part because our model assumes a finite supply elasticity for carbon-based fuels. With a given quantity reduction target, required tax rates increase as the supply elasticity of carbon-based energy products falls.

Part B of table 7.6 reports our central-case estimates of the gains or

Table 7.6
Central-case model analyses of 50 percent reduction in global carbon use
A. Carbon-tax rates needed to meet emission reduction targets

Option	Tax rate[a] (percent)	Carbon-tax rate[b] (dollars per ton of carbon)
National production-based tax	88.7–89.6	448.0
National consumption-based tax	89.3–89.7	448.0
Global tax	88.6	439.0

B. Gain or loss by region under each option (Hicksian EVs over period 1990–2030 in billions of 1990 dollars; figures in parentheses are percent of GDP in present-value terms, 1990–2030)

Region	Tax option		
	National production-based tax	National consumption-based tax	Global tax
European Community	−3,840.5	−1,006.1	−3,724.5
	(−4.0)	(−1.0)	(−3.8)
North America	−5,494.4	−4,576.5	−12,442.0
	(−4.3)	(−3.6)	(−9.8)
Japan	−1,459.6	215.3	−366.9
	(−3.7)	(0.5)	(−0.9)
Other OECD	−487.0	−440.1	−939.9
	(−2.3)	(−2.1)	(−4.4)
Oil exporters	1,191.4	−4,901.9	−3,416.7
	(+4.5)	(−18.7)	(−13.0)
Developing/ centrally planned	−9,392.3	−8,994.4	2,371.2
	(−7.1)	(−6.8)	(1.8)
World	−19,482.4	−19,704.1	−18,516.1
	(−4.4)	(−4.4)	(−4.2)

a. These are gross tax rates; that is, an 85 percent tax rate means that for each dollar spent by consumers, 85 cents is collected in tax. The equivalent net tax rate is 566 percent (i.e., 85/15%).
b. These are calculated by taking the present value of revenues raised by the tax and converting this to a dollar revenue flow for each year, including the 1990 base year. This is then divided by projected global carbon emissions for the year (5.038 billion tons of carbon in 1990), multiplied by one minus the emissions cut. We use data from Milleman (1988) and World Resources Institute (1990) to provide a base-year estimate of global carbon emissions from fossil fuel consumption of 5.038 billion tons.

losses for the regions identified in the model under each of the tax options. In making these estimates, we have calculated Hicksian equivalent variations (EVs) over the period 1990–2030 in billions of dollars (expressed in 1990 prices). These measures capture the combined gains or losses to regions from the production and sale of carbon-based products, as well as the consumption side gains associated with price changes. They also capture the economy-wide effects of energy price changes as they feed through the model.

As can be seen from table 7.6, the global implications of such taxes are large. For both a national production-based carbon tax and a global carbon tax, our results suggest the global economic cost to be in the region of $19 trillion over the period 1990–2030. With global economic product over this period of time valued at approximately $450 trillion, this represents a cost of approximately 4 percent of world product.

Effects by region differ dramatically across the various tax options. North America loses $12 trillion, or approximately 10 percent of the value of income, under a global tax, whereas the loss under a national consumption-based carbon tax is smaller: $4.5 trillion, or 3.6 percent of income. Developing and centrally planned countries lose more than $9 trillion (or 7 percent of income) under a national production-based carbon tax, whereas they gain $2.3 trillion (or 1.8 percent of national income) under a global tax. Oil exporters gain substantially from a national production-based tax, while they lose dramatically from either a global tax or a national consumption-based tax.

These results, therefore, strongly underline the point we have emphasized previously—namely, that the form any carbon tax takes will have major implications for the international incidence effects of the tax. Under a national production-based tax, energy producers collect tax revenues; hence oil exporters gain. Under a national consumption tax, oil importers collect revenues, and hence Europe gains, North America loses only marginally, and oil exporters lose. Under a global tax, the revenues from the tax are redistributed on a per-capita basis, and developing and centrally planned countries gain substantially.

Data reported in table 7.7 showing tax revenues collected and redistributed by region under each tax option reiterate the same theme. Under a global tax, the associated intercountry transfers are extremely large. The revenues raised are in the region of $46 trillion, slightly more than 10 percent of global income over the period 1990–2030. Under the first two tax options, taxes are both collected and

Table 7.7
Revenue redistribution under carbon-tax schemes in central-case model analyses
(taxes paid and revenues received by region under each tax option over the period 1990–2030; in billions of 1990 dollars)

Region	National production-based tax (taxes paid and received)	National consumption-based tax (taxes paid and received)	Global tax		
			Taxes paid	Transfers received	Net transfer
European Community	3,324.7	6,698.2	6,492.3	3,270.1	−3,222.2
North America	11,039.1	12,362.0	11,824.6	2,765.2	−9,059.4
Japan	98.7	1,998.2	1,957.6	1,273.8	−683.8
Other OECD	1,078.5	1,144.4	1,116.4	537.3	−579.1
Oil exporters	9,404.3	2,505.7	2,517.1	4,029.7	+1,512.6
Developing/centrally planned	21,681.2	21,950.1	22,410.8	34,442.6	+12,031.8
Global revenue raised	46,626.5	46,658.6	46,318.7	46,318.7	—

redistributed within countries with no international redistribution. Under a global tax, taxes are collected and redistributed to countries on an equal per-capita basis. In this case, developing countries receive $12 trillion over the period 1990–2030 (around 5 percent of income), with a majority of the transfer originating from North America, which loses $9 trillion. The revenue redistribution scheme accompanying any carbon-tax option is, therefore, crucial in determining international incidence results.

Table 7.8 provides more-detailed model results on the potential impacts of a global carbon tax. We report the percentage change in consumption and production of carbon-based energy products by region, consumption of non-carbon-based energy, and change in the consumption of the composite energy product. The impacts reported largely reflect the 50 percent emission target and the use of similar demand and supply side elasticity values by region in the central-case specification. Consumption falls approximately 50 percent in most regions, as does the production of carbon-based energy, with a much smaller response in consumption of non-carbon-based energy (solar and nuclear).[20] The percentage change in consumption of total energy is reported in the final column.

There are large changes in the prices of carbon-based energy products in this case. For the global tax scheme that we consider, the price of carbon-based energy for users increases by 140 percent, whereas the price for sellers falls by 70 percent, reflecting the substantial price wedge associated with approximately 85 percent tax rates (calculated on a gross-of-tax basis).

We have also examined the impacts of global carbon taxes on trade patterns between regions in the model, emphasizing those cases where there is a change in trade pattern as a result of the imposition of the carbon tax. For the six regions and the two products involved, changes in trade patterns occur in four out of the twelve possibilities. Both Europe and Japan become net importers of energy-intensive products, where previously they were net exporters. A change also occurs for oil exporters from a net import to a net export position for energy-intensive manufactures. Developing and centrally planned countries change from a net export to a net import position in other goods.

Table 7.9 presents results that show the impact of using a ceiling-emissions per-capita approach to implementation of a carbon tax. In this set of model calculations, emission levels within countries are

Table 7.8
More detailed model results for a global carbon tax

A. Results by region

Region	Change in consumption of carbon-based energy (percent)	Change in production of carbon-based energy (percent)	Change in consumption of non-carbon-based energy (percent)	Change in consumption of total energy (percent)
European Community	−48.6	−49.8	+3.1	−46.6
North America	−49.5	−49.1	+3.2	−52.9
Japan	−49.0	−50.3	+2.6	−46.1
Other OECD	−49.1	−50.0	+2.7	−41.1
Oil exporters	−47.3	−48.5	+4.7	−46.5
Developing/centrally planned	−47.4	−47.3	+5.5	−46.1

B. Global results

Change in the users' price of carbon-based energy	+145.4%
Change in the sellers' price of carbon-based energy	−72.7%
Change in the price of non-carbon-based energy (effects vary by region)	+17% to +25%

Table 7.9
Impact of a per-capita emissions ceiling implemented through national carbon taxes

A. Gain or loss by region (Hicksian EVs over period 1990–2030 in billions of 1990 dollars; figures in parentheses are percent of GDP in present-value terms, 1990–2030)

	National consumption-based carbon tax	
Region	Central-case results	Per-capita emission ceiling
European Community	1,354.3 (1.4)	−6,220.3 (−6.4)
North America	−1,529.8 (−1.2)	−23,698.5 (−18.6)
Japan	1,208.3 (3.0)	−985.0 (−2.5)
Other OECD	76.9 (+0.4)	−1,431.2 (−6.7)
Oil exporters	−4,374.5 (−16.7)	−3,959.8 (−15.1)
Developing/centrally planned	−5,948.0 (−4.5)	−1,549.9 (−1.2)
World	−9,212.8 (−2.1)	−37,844.7 (−8.5)

B. Carbon-tax rates by region (ad valorem, gross basis)

European Community	88.6	95.0
North America	88.6	98.0
Japan	88.6	93.2
Other OECD	88.6	95.1
Oil exporters	88.6	81.1
Developing/centrally planned	88.6	81.4

C. Changes in trade patterns involving energy-intensive manufactures[a]

European Community	NE → NI	NE → NI
North America	NC	NE → NI
Japan	NE → NI	NE → NI
Other OECD	NC	NE → NI
Oil exporters	NI → NE	NI → NE
Developing/centrally planned	NC	NI → NE

a. NC denotes no change in the direction of trade; NE → NI denotes net exporter changed to a net importer; NI → NE denotes net importer changed to a net exporter.

reduced to an equal per-capita level across countries. As might be expected under such a scheme, results are more dramatic than for the carbon-tax case presented in table 7.6. The global economic costs are four times larger. The losses in North America are approximately fifteen times larger; while losses in developing and centrally planned countries are nearly four times smaller. Carbon-tax rates by region are substantially higher, particularly if these are recalculated on a net-of-tax basis. Once again, the form of the tax makes a major difference to international incidence.

In table 7.10 we report model results showing the impacts of alternative supply elasticity assumptions for carbon-based energy products. We run our model with alternatively low (0.1) and high (1.5) carbon-based energy-product supply elasticities, examining a global carbon tax in both cases. Lower elasticities imply higher carbon tax rates, both in ad valorem terms and in dollars per ton of carbon. Gains and losses by region vary with these elasticity parameters, but not by as much as with changes in the form of the tax. These results also have the feature that the lower the carbon-based supply elasticity, the larger the welfare cost. At first sight, this result may appear counterintuitive, but with the constant emission (output) reduction target across all these cases, a higher tax rate is required in low-elasticity cases, implying larger costs in terms of economic dislocation for the global economy. Effects by region vary across cases in a similar way.

Finally, in table 7.11 we report model calculations of the tax rates and associated welfare costs involved in meeting alternative emission reduction targets for a global carbon tax. We consider emission reduction targets from 5 to 50 percent, with the welfare cost involved in meeting each increasing more than proportionally. This pattern of cost estimates by size of emission-reduction target reflects the well-known Harberger rule that welfare costs of taxes vary (approximately, depending on cross-price effects) with the square of the tax rate. The rapid increase then slowing rise in the ad valorem equivalent rate, in part, reflect our use of gross tax rates (rates are bounded from above by 100 percent), although a similar, if less pronounced pattern, comes through for carbon-tax rates.

These results suggest that a 50 percent cut, as we consider here (taking our lead from the Toronto conference), is a severe cut, and milder cuts would be less costly with less pronounced international incidence effects. Nonetheless, it is also worth reiterating that any

Table 7.10
Impacts of alternative supply elasticity assumptions on global carbon-tax model results

	Central-case model specification with carbon-based energy supply elasticities of 0.1	Central-case model specification	Central-case model specification with carbon-based energy supply elasticities of 1.5
A. Gain or loss by region[a]			
European Community	−4,331.7 (−4.5)	−3,724.5 (−3.8)	−3,394.5 (−3.5)
North America	−14,845.4 (−11.6)	−12,442.0 (−9.8)	−10,261.3 (−8.1)
Japan	−335.8 (−0.8)	−366.9 (−0.9)	−526.2 (−1.3)
Other OECD	−1,158.0 (−5.4)	−939.9 (−4.4)	−754.0 (−3.5)
Oil exporters	−5,080.2 (−19.6)	−3,416.7 (−13.0)	−1,744.0 (−6.6)
Developing/centrally planned	304.8 (0.2)	2,371.2 (1.8)	2,715.6 (2.0)
B. Carbon-tax rate ad valorem in dollars per ton of carbon	99.3% 484.0	88.6% 460.0	74.2% 380.0
C. Revenues raised globally from carbon tax (billions of dollars, 1990, over period 1990–2030)	50,399.7	46,318.7	39,529.3

a. Hicksian EVs over period 1990–2030 in billions of 1990 dollars; figures in parentheses are percent of GDP in present-value terms, 1990–2030.

Table 7.11
Global welfare cost estimates and tax rates for alternative emission reduction targets (global carbon tax)

Emission reduction target (percent)	Global welfare cost (percent of global income)	Ad valorem carbon-tax rate (gross basis)	Carbon-tax rate (dollars per ton)
5	0.0	7.9	34
10	0.1	25.9	113
15	0.3	40.4	180
20	0.6	52.1	237
25	0.9	61.7	286
30	1.4	69.5	328
35	1.9	75.8	366
40	2.6	81.0	400
45	3.3	85.2	431
50	4.2	88.6	460

attempt to stabilize carbon emissions in the face of both population and real GDP per capita growth would likely require ever-increasing tax rates.

Concluding Remarks

This chapter discusses what the possible international incidence effects (which countries gain and which lose) from global carbon taxes could be if they are used in the future to reduce carbon emissions because of concerns over global warming.

We first emphasize that carbon-tax proposals, while discussed extensively at popular level in recent years, are lacking in specificity. The treatment of the revenues raised by such a tax can make a major difference to international incidence. In addition, the difference between production- and consumption-based taxes is of substantial consequence.

Second, we try to help place future discussion of the international incidence effects of such taxes in a quantitative context by using a global general-equilibrium model covering six countries or regions and capturing production, consumption, and trade in energy and nonenergy products to evaluate a range of carbon-tax options. The model is calibrated to a base-case scenario for the global economy projected over the period 1990–2030 using current growth rates. Using this model, we introduce alternative carbon taxes, compute a new model solution, and compare each of these to the base-case model solution.

A number of themes emerge strongly from our results. The first is that the effects of carbon taxes are potentially very large. The base and revenues involved could well account for 10 percent of global income over the period we consider (1990–2030). Tax rates needed to meet the form of emission targets that we consider will typically be high. The introduction of such a tax, therefore, has to be treated as one that potentially is of major consequence for global economic activity.

At the same time, our results show that the precise form of the tax will largely determine what international incidence effects will likely occur. Under global taxes collected supranationally (say, via the UN) and redistributed to countries on a per-capita basis, developing countries benefit. Under national production-based taxes, energy exporters (OPEC) do well. Under national consumption-based taxes, energy importers (OECD) do well.

Notes

An earlier version of this paper was presented to the conference Economic Policy Responses to Global Warming organized by the Istituto Bancario Sao Paolo di Torino, held at Palazzo Colonna, Rome, October 4–6, 1990. We are grateful to Ngee Choon Chia for comments and help, and to David Newbery, William Nordhaus, and Carlo Perroni for comments.

1. An earlier and more preliminary discussion of some of the issues taken up here can be found in Whalley and Wigle (1989), where calculations related to those presented here also appear.

2. See the discussion of the possible scientific dimensions of the greenhouse effect in Schneider (1989). A range of economic consequences are set out in Cline (1989).

3. See also the discussion of linkages between development and the greenhouse effect in Arrhenius and Waltz (1990).

4. We do not capture potential international incidence effects from carbon taxes via country growth performance in the calculations we report later. In many developing countries the growth process has involved moving resources first from agriculture to labor-intensive manufactures, then to heavy manufacturing, before moving on into consumer electronics and other higher-technology manufactured products. The argument is that carbon taxes may make the energy-intensive heavy industry phase of such development more difficult, potentially truncating the growth of the least-developed countries.

5. To simplify the model, we work with a single carbon-based energy product, and hence different tax rates by type of fuel are not explicitly considered. Also, existing taxes, and particularly excise taxes on gasoline and other energy products, are not explicitly captured. The time profile that carbon-tax rates may be expected to follow can also complicate the analysis in ways we do not capture. If carbon-tax rates are expected to increase through time, then the expectation of future tax changes will change the time profile of supply, and an analysis of international incidence effects of carbon taxes based on an assumption of fixed tax rates can be misleading. A final omission is macro dislocation and adjustment costs associated with carbon taxes. These may be large and asymmetric across countries, and contribute to shorter-run international incidence effects.

6. Edmonds and Reilly (1983), for instance, consider taxes at rates of 100 percent on coal, 78 percent on oil, 56 percent on gas, and 115 percent on shale oil. Flavin (1990, fn. 45) suggests tax rates proportional to the carbon content of fuels; with coal averaging 24.12 kilograms of carbon per gigajoule, oil 19.94 (82 percent of coal), and natural gas 13.78 (57 percent of coal).

7. Manne and Richels (1990), for instance, consider a target of a 20 percent reduction from baseline emission levels by 2020, with a stabilized level of emissions thereafter, until 2100. See also IEA (1990) and Nordhaus (1977, 1990).

8. Tax rates of this order are considered by Manne and Richels (1990) and Nordhaus (1990). They imply more than a fivefold increase in the price of coal and a doubling or more in gasoline prices.

9. See Nordhaus and Yohe (1983) and Edmonds and Reilly (1983), for instance.

10. Subsequent conferences at The Hague and in Cairo have focused on similar proposals.

11. See the discussion in Flavin (1990, 28), who also notes the West German SDP endorsement of the carbon tax as an "ecological redistribution of taxes," on the grounds that revenues could finance development of alternative renewable resources and energy conversion and efficiency improvements.

12. See also the early discussion of this issue in Kosobud and Daly (1984), who highlight the free-rider problems with reaching a global cooperative outcome on a carbon-emission-reduction scheme.

13. See also the presentation in Whalley and Wigle (1989) of earlier, preliminary calculations made using a simplified version of this model.

14. This has been chosen somewhat arbitrarily to capture the initial period and subsequent intermediate term during which a carbon tax would have its largest effects, since with discounting the significance in present-value terms of later-year effects recedes. It is relatively easy to run the model for a longer projection period (say, eighty or 100 years), and were this done we believe that the main themes of our results would remain.

15. A weakness of this forty-year projection period approach is that in the base-year data used for these projections, most carbon energy trade takes place in oil rather than in other carbon-based fuels. If, as some expect, trade in oil is slowly replaced by trade in coal into the next century, the data used here may be misleading because the countries that are potential future coal exporters (United States, U.S.S.R., Australia, China) are quite different from current oil exporters.

16. There is trade in radionuclides, but it is ignored.

17. This is a technical term used in the applied general-equilibrium modeling literature; see Shoven and Whalley (1984, in press) for more details.

18. In the case of a global tax, production and consumption taxes are equivalent. Revenues are distributed in the same way, and the divergence between consumer and producer prices is the same.

19. Each counterfactual equilibrium solution is computed using the MPS/GE code developed by Thomas Rutherford of the University of Western Ontario (see Rutherford 1988).

20. This result reflects their small share in production of the composite energy product, and the substitution elasticity used in the composite energy production function.

References

Arrhenius, E., and Waltz, T. W. 1990. The greenhouse effect—Implications for economic development. World Bank Discussion Paper No. 78, Washington, DC.

Climate Institute. 1988. Toronto conference calls for sharp cuts in carbon dioxide emissions. *Climate Alert* 1, no. 3 (Fall).

Cline, W. R. 1989. Political economy of the greenhouse effect. Mimeo. Institute for International Economics, Washington, DC, August.

Considine, T. J. 1989. Estimating the demand for energy and natural resource inputs: Trade-offs in global properties. *Applied Economics* 21 (7): 931–945.

Edmonds, J., and Reilly, J. 1983. Global energy and CO_2 to the year 2050. *Energy Journal* 4, no. 3 (July): 21–47.

Estrada, J., and Fugleberg, O. 1989. Price elasticities of natural gas demand in France and West Germany. *Energy Journal* 10, no. 3 (July): 77–90.

Flavin, C. 1990. Slowing global warming. In Lester R. Brown (ed.), *State of the World*. New York: W. W. Norton.

Hunt, L., and Manning, N. 1989. Energy price and income elasticities of demand: Some estimates for the U.K. using the cointegration procedure. *Scottish Journal of Political Economy* 36, no. 2 (May 1989): 183–193.

IEA. 1990. Follow-up to Noordwijk Ministerial Conference on Atmospheric Pollution and Climate Change. IEA/SLT(90)2, Paris.

International Bank for Reconstruction and Development. 1988. *The World Tables, 1988*. Washington, DC: International Bank for Reconstruction and Development.

International Bank for Reconstruction and Development. 1989. *World Development Report*. Oxford University Press.

Kosobud, R. F., and Daly, T. A. 1984. Global conflict or cooperation over the CO_2 climate impact. *Kyklos* 37: 637–659.

Kouris, G. 1981. Elasticities—science or fiction? *Energy Economics*, April 1981, pp. 66–69.

Manne, A. S., and Richels, R. G. 1990. Global CO_2 emission reductions—The impacts of rising energy costs. Mimeo. Revised version of a paper presented to the International Association of Energy Economics, New Delhi, February.

Mansur, A., and Whalley, J. 1984. Numerical specification of applied general equilibrium models: Estimation, calibration and data. In J. B. Shoven and H. Scarf (eds.), *Applied General Equilibrium Analysis*, 69–127. Cambridge University Press.

Milleman, R. E. (ed.). 1988. *A glossary for carbon dioxide and climate*. No. ORNL/CDIAC-22. Washington, DC: U.S. Department of Energy.

Nainar, S.M.K. 1989. Bootstrapping for consistent standard errors for translog price elasticities. *Energy Economics*, October 1989, pp. 319–322.

Nguyen, H. V. 1986. Energy elasticities under Divisia and Btu aggregation. *Energy Economics*, October 1987, pp. 210–213.

Nguyen, T. T., Perroni, C., and Wigle, R. M. 1990. A microconsistent data set for the analysis of world trade: Sources and methods. Working paper, Wilfrid Laurier University, Waterloo, Ontario, August.

Nordhaus, W. D. 1977. Economic growth and climate: The carbon dioxide problem. *American Economic Review*, Papers and Proceedings, February, pp. 341–346.

Nordhaus, W. D. 1990. A survey of estimates of the cost of reduction of greenhouse gas emissions. Mimeo. New Haven: Yale University, Department of Economics.

Nordhaus, W. D., and Yohe, G. W. 1983. Future paths of energy and carbon dioxide emissions. In National Research Council, *Changing Climate: Report of the Carbon Dioxide Assessment Committee*. Washington, DC: National Academy Press.

Pindyck, R. S. 1979. Interfuel substitution and the industrial demand for energy: An international comparison. *Review of Economics and Statistics* 61: 169–179.

Ross, Marc. 1989. Improving the efficiency of electricity use in manufacturing. *Science* 244 (April): 311–317.

Rutherford, T. F. 1988. General equilibrium modelling with MPS/GE. Department of Economics, The University of Western Ontario.

Schneider, Stephen. 1989. The greenhouse effect: Science and policy. *Science* 243 (February): 771–781.

Shoven, J., and Whalley, J. 1984. Applied general equilibrium models of taxation and international trade: An introduction and survey. *Journal of Economic Literature*, September, pp. 1007–1051.

Shoven, J., and Whalley, J. In press. *Applying General Equilibrium*. Cambridge University Press.

United Nations. 1986. *UN Energy Statistics Yearbook, 1984*. New York: United Nations.

Whalley, J., and Wigle, R. 1989. Cutting CO_2 emissions: The effects of alternative policy approaches. Paper presented to an NBER conference on Applied General Equilibrium Modeling, San Diego, September.

World Resources Institute. 1990. *World Resources, 1990–1991*. Oxford University Press.

B
Tradeable Permits

[19]

Tradeable Emission Permits and the Control of Greenhouse Gases

GEOFFREY BERTRAM

This article reiterates the case for tradeable permits as a global policy option for limiting greenhouse gas emissions, and considers the detailed design of a global tradeable-permit regime, emphasising the importance of the initial assignment of property rights, and arguing that the relevant property rights in this case are the rights of every member of the world community to share in a sustainable global atmosphere and climate. The allocation of permits should therefore be done on a per capita basis across the world community, with the result that rents generated by the process of reducing carbon emissions would accrue to non-polluters, most of whom live in the 'South'. The international transfers of income and wealth implied by the proposed scheme are large but feasible. There is therefore a real prospect that an international convention on carbon dioxide emissions could end the debt crisis and finance sustainable development in the South.

I. INTRODUCTION

The past two years have brought a flurry of scientific and diplomatic activity in the quest for an international convention to limit global emissions of the gases which are responsible for global warming. Attention has focused especially upon carbon dioxide as the most important of those gases. While there appears to be widespread agreement on the need for policy initiatives, no consensus has yet emerged regarding the best mechanism through which the international community could implement a programme of cutting back global emissions. The three leading contenders are direct quantitative emission restrictions, carbon taxes, and tradeable emission permits, with most attention in the economic debate having been given to carbon taxes.

The author is at the Economics Department, Victoria University of Wellington, New Zealand.

The Journal of Development Studies, Vol. 28, No. 3, April 1992, pp. 423–446
PUBLISHED BY FRANK CASS, LONDON

Bertram, Stephens and Wallace [*1989*] argued that a world-wide tradeable-permits system could be an effective way of advancing the interests of developing countries in harmony with the global community's interest in protecting the atmosphere, provided that permit entitlements were allocated in such a way that the resulting rents accrued mainly to poor countries. This theme was taken up by Grubb [*1989: Ch. 5*] and has been the subject of some discussion since [*e.g. Hoel, 1991*]. This article develops the suggestion with particular emphasis on the need for any international convention to be based upon an appropriate set of property rights giving priority to the world's non-polluting poor.

II. THE CONCEPT OF TRADEABLE PERMITS

The idea of using tradeable permits to control pollution emerged in the 1960s and 1970s from a long theoretical debate over the economics of externalities [*Dales, 1968; Baumol and Oates, 1988: Chs. 5 and 12; Pearce and Turner, 1990: Ch. 8*]. Externalities are benefits or costs, arising from economic activities, which are experienced by third parties without their having participated in any market transaction. Global atmospheric pollution is the classic example of an externality which has all the characteristics of a public good (or rather, bad). No human being can escape the effects, short of leaving the planet; the effects are shared by the entire global population, albeit in uneven ways; and there is, as yet, no market to which people can go to purchase an alternative state of the world's atmosphere and climate. (It is taken for granted here that the world does indeed face a real present choice amongst alternative future climatic states, as outlined in, for example, Rosenberg *et al.* [*1989*]; Leggett *et al.* [*1990*], despite the scientific uncertainty about what actual outcomes will flow from choices made now – see [Solow [*1990*]. It is assumed also that there exist technologically-feasible means of stabilising and eventually restoring the gaseous composition of the earth's atmosphere, especially by the adoption of energy-efficient techniques of the sort outlined in [Goldemberg *et al.* [*1988*].)

The idea that the problem may best be thought of in terms of a missing market harks back to two famous economics papers: Scott Gordon's [*1954*] study of the tragedy of the commons in the marine fishing industry, and Ronald Coase's [*1960*] paper on social cost. Both suggested that if an appropriate set of enforceable property rights could be established, together with a competitive low-cost market, it should be possible for individuals to trade their way out of problems such as overfishing and pollution. Government could thus address such problems without embarking on detailed direct intervention through devices such as taxes on polluters, or regulations to control the quantity and quality of emissions according to a uniform standard. Such tax and regulatory instruments, however, could still appropriately be used alongside a tradeable-permits regime if the benefits of doing so could be established. (The general rule recognised by Coase is that the cost-effectiveness of government intervention can be determined only by analysis, not as an a priori matter of principle.)

The advantages of allowing individual agents to move freely towards pollution abatement, in a framework which makes it worth their while to do so, are considerable. Those who face the highest costs of reducing their emissions will tend to be buyers of permits which enable them to continue existing operations. Those whose clean-up costs are less will be able to sell off permits and use part of the proceeds to pay for pollution reduction, while still being left with a profit. The result of such trading on a world scale would be to concentrate global cleanup investment in those locations and industries where cost effective abatement options are available, so that the pollution which remains is from 'hard-core' (high net abatement cost) sources.

An important advantage of concentrating abatement effort according to sensible economic criteria, rather than relying upon rigid country-by-country targets, is that industrialisation in the South need not be compromised, since countries with high marginal returns on carbon-emitting activities will be able to reap those returns, so long as permits are held and the technology used is energy-efficient and not unnecessarily pollution-intensive. The idea becomes more attractive the more efficiently the market mechanism works, relative to the efficiency of a governmental regulator. Two important issues here are the transactions costs involved in the working of the hypothetical market, and the possible possession of market power by some individual or group which might enable them to skew the outcome to their own advantage.

Consideration of the real world suggests that in practice, both of these problems are likely to be greatest in markets of medium size. In a very small market involving transactions between two individuals, with no other parties involved, face-to-face dealing minimises transactions cost, while market power reduces to the personal characteristics of the two parties. (This was the base case for Coase's [*1960*] analysis.) In very large (say, global) markets, economies of scale permit the profitable operation of institutions to facilitate individual trading – stock markets, publicly posted prices, networks of brokers – while the very large number of participants makes it difficult to establish uncontested market power. It is when we turn to markets in the middle – for example, small national economies – that transaction costs and market power pose the greatest obstacles. For this reason, a tradeable-permit regime to combat atmospheric pollution is likely to work better at global than at national level, whereas direct regulations and carbon taxes are likely to succeed at national level if anywhere.

In confronting the problem of greenhouse gas emissions, a tradeable-permit regime has several major advantages over alternatives such as carbon taxes or country-by-country quotas. It provides the incentive for industry worldwide to cut back emissions, without placing on national governments the heavy burden of negotiating and enforcing specific country-by-country limits. Besides being an allocatively efficient means of addressing the problem of global pollution abatement, it should also be easier to negotiate than country-by-country quotas, while avoiding the severe moral-hazard problems of a global carbon tax (which would result in the administering agency gaining control over annual revenues of staggering magnitude). It avoids also the problems of harmonising a system of nationally-administered carbon taxes, and the enormous bureaucratic effort required to fix and enforce detailed case-by-case restrictions. All that a tradeable-permit scheme requires of national governments is their agreement to global emission limitation, and their adherence to the international legal convention required for enforcement. There need be no fiscal burden placed on national governments, and only limited surrender of sovereignty if international agencies and other non-nationals are given access to the courts of any country to pursue offenders.

III. SOME RECENT DEVELOPMENTS

The concept of tradeable permits was cautiously supported by Working Group III of the Intergovernmental Panel on Climate Change in October 1989:

> of all the instruments examined, the system of tradeable emission rights came in for the most attention and was considered to be most promising. It offers the advantages of flexibility, efficiency in pollution abatement, direct control of total emission levels, a mechanism for trading reduction in different gases, and incentives for research into pollution abatement technology (IPCC WG3-II/Doc. 3, cited in Grubb [*1990a: 3*]).

Unfortunately, the subsequent development of discussion in Working

Group III of the IPCC was increasingly dominated by diplomats from major industrial powers such as the USA and Japan, anxious to protect their national's vested interests as the world's major emitters of greenhouse gases. In its June 1990 report [*Intergovernmental Panel on Climate Change, 1990: 37*] the discussion of a possible tradeable-permit system was given a single half-hearted paragraph in the 46-page paper, with the sole sentence on permit allocation eliminating at a stroke any prospect of support from developing countries for this option. The sentence reads: 'Once an overall limit on emissions has been set, emissions entitlements amounting to that limit could be provided *to emitting sources* and free trading of such entitlements allowed' [*emphasis added*]. By this preemptively slipping in the US practice of 'grandfathering' permits (that is, giving them to established polluters) [*Baumol and Oates, 1988: 179*] as though it were the only allocation option available, the June 1990 document sought to remove the choice of property rights assignment from the negotiating agenda. But this is the central issue which could render a tradeable-permit system attractive to the world community at large. In effect, the IPCC committee proposed that the property right of all citizens of the world in a sustainable atmosphere and climate should give way to the right of existing polluters to have their activity legitimated and rewarded. On the face of it, this violates Coase's [*1960*] principle that initial property rights should be allocated on the basis of which party would suffer 'greater harm' from having to bear the cost of the transactions needed to secure the socially-optimal outcome. This theme is developed below.

IV. A SUGGESTED FRAMEWORK

The main advantage of tradeable-permit regimes is their ability to achieve environmental aims with a minimum of bureaucratic apparatus. The central problem with most such schemes to date has been that the 'licence to pollute' has been granted to firms which were already major polluters, with the result that the rents associated with a growing scarcity of pollution entitlements fell into the hands of those firms – a result with obvious equity problems, which provides the wrong incentives both to polluters and to others.

Bertram, Stephens and Wallace [*1989*] argued that that a global system to regulate greenhouse gas emissions over time spans of several decades should start from a strong presumption in favour of the long-run property right of the world's population to inhabit a sustainable global ecosystem. This means that any tradeable-permit system should be designed in such a way as to incorporate the principle that polluters should pay, through the mechanism of being obliged to buy at least part of their right to pollute. A second basic principle was that the large flows of rents generated by those payments should be channelled directly from polluters to the owners of the property right, without passing through the hands of large international agencies on the way. This would mean an entirely new and massive channel of North–South transfer payments, mediated through a newly-created world market in pollution rights. The possibilities which this opens up both for development and for solving debt-servicing problems are obvious.

Suppose that an international consensus in favour of action on the greenhouse problem emerges. What institutions might then be established, and how would they work? The remainder of this section outlines the tradeable-permits system proposed in Bertram, Stephens and Wallace [*1989*]. Some alternative options are then reviewed, some lessons are drawn from recent real-world experience with tradeable permit systems, and the possible orders of magnitude of North–South redistribution are indicated.

First, a global emissions budget for each greenhouse gas would be formulated and announced by an international agency relying on the best obtainable scientific advice. This budget would be specified in terms of annual permitted emissions over a time period – say, the next ten years – but would be subject to annual revision as scientific information accumulated, and in the light of the development of carbon sequestration projects such as reafforestation. Operators of polluting activities in all countries would be required, under an international convention recognised by their national governments, to hold a portfolio of permits corresponding to the composition of their emission streams. The discussion which follows abstracts from this diversity of greenhouse gases and assumes, for simplicity, that we can treat carbon emissions as homogeneous. In practice the composition of emissions would of course have to be confronted in the design of the range of permits.

Around five billion 'shares' in this emission budget would be notionally issued on a one-share-per-head basis to the world population. In the first instance, national governments would be allocated shares corresponding to their populations, but there is no reason in principle why allocations could not be made, with the national government's agreement or by its own initiative, to sub-national groups - for example tribe of indigenous people, or local-government agencies, or even individuals or families. (Transactions costs should in theory be the main determinant of how widely entitlements can be distributed. In practice, few national governments are likely to surrender the opportunity to capture the rent revenues attached to share allocations.)

Each share would entitle the holder to emit a specified proportion of the global budget for each year of the permit's life. In the case of global carbon emissions, the numbers are convenient: the present level of gross emissions is six billion tonnes per year [*Grubb, 1989: 28 Table 3*] and the world's population is somewhat over five billion people. The human race is therefore emitting (that is, utilising the atmosphere and associated ecosystems as a dump for) roughly one tonne of carbon per year per person. In 1990, thus, each share would equate to roughly one tonne of carbon - that is, one-five-billionth of global emissions.

A typical scientists' target for the required reduction in emissions is 60 per cent globally in the next twenty years, which would imply a fall in per capita emissions to about one-third of a tonne by 2010, with nearly eight billion shares on issue and a global budget of 2.4 billion tonnes. (This assumes a 1.5 per cent p.a growth of eligible world population. A permit system relying on a per-head-of-population share issue would probably have to include some ceiling on growth of eligible population, in order not to be exposed to criticism for giving an incentive to some countries to promote accelerated population growth.)

Each share would be valid for a specified time period, after which it would be replaced by a newly-issued share, again allocated on the per-capita rule. Once the system had reached a steady state, a good rule of thumb might be that all shares should be ten-year entitlements, with one-tenth of the total outstanding stock expiring in each year. Thus in each year a new issue of one-tenth of a share per capita would be made on the same basis as the original issue. This principle of continually turning-over the stock of permits has two rationales: first, to keep the market liquid so that barriers to entry cannot be erected by cartels of industrial interests; and second, to ensure the availability at any time of entitlements lasting ten years into the future, which would make it possible for a firm contemplating an investment in a carbon-emitting plant to secure at the outset a portfolio of ten-year entitlements for its expected emission stream. (Ten years is a rough estimate of the typical planning horizon for industrial investment decisions.)

In the first year of the system, only one-tenth of the issued permits would be for ten years; the remainder would be spread evenly down the spectrum from nine years to one year. By the end of the ninth year of operation of the system, these shorter-term permits would all have been retired and replaced by ten-year permits. The secondary market would thus from the start be setting spot prices for permits with different periods to expiry, and the outstanding stock of permits would be comprised, in approximately equal proportions, of ten different maturities.

Secondary trading in permits would be entirely unrestricted, and any barriers to the emergence of a free world market would be removed. The emission permits would then be internationally-traded paper assets comparable to shares and bonds, and would probably be traded mainly through the world's sharemarkets, using existing networks of brokers and with no restrictions on the identity of market participants or the currency of trading. A central computerised register would be maintained in order to monitor the global distribution of permitted emissions, and this register would be open to public scrutiny to enable national governments, commercial competitors and NGOs to carry much of the monitoring workload.

Enforcement of the permits would be designed on a cost-effective basis. Ideally, each national government would sign a convention binding its citizens to acquire permits for all carbon-emitting activities, and opening channels through which legal action against offenders could be initiated by non-nationals, with uniform sanctions agreed under international law. It would then be straightforward both for the international budgeting agency to initiate cases against major offenders, and for governments, environmental organisations or individuals to mount test cases in any country whose nationals offended. Hopefully in addition, national governments would be willing to use their own policing mechanisms to ensure compliance.

If fines are to be levied on offenders, the issue of who captures the resulting revenues is a touchy one. Grubb [*1990a: 4*] proposes that fines should go into an international fund which could be used to buy-in permits on the open market, thus contributing to the downward pressure on global emissions through a mechanism closely akin to open-market operations by a central bank. This would be a superficially attractive idea for a tradeable-permit system based on specified *quantities* of permitted emissions per permit, because as outstanding permits were bought-in the budget would automatically be tightened. However, this would not work in the Bertram–Stephens–Wallace proposed system. There, the predetermined global pollution budget is proportionally split among outstanding permits, so that each permit would represent a fractional share of the total, not a right to a specific tonnage of pollution. Thus a programme of open-market buying, withdrawing permits from circulation, would simply raise the quantity of pollution allowed to each remaining permit, and so would give capital gains to the holders of the remaining permits without affecting total permitted global pollution. In this case it seems preferable that fines or other levies be devoted to the promotion of carbon sinks to directly improve the state of the atmosphere.

The key goal, however, is to enforce compliance rather than raise revenue via fines. If in the process national governments fine their own locals, it is not essential that the resulting revenues go into an international fund. Indeed, this is one area where the incentives facing national government would be fully compatible with the operation of the Bertram–Stephens–Wallace regime. If budgetary goals lead national governments to fine offending polluters, or to levy domestic carbon taxes, such initiatives would serve not as alternatives to the global permit system, but as reinforcements to it.

The suggested procedure of leaving cases to be heard, and determined, by the existing judicial systems of specific nations within the framework of an international legal convention, involves inevitable compromises of national sovereignty. It would be feasible only if the tradeable-permit regime offered strong incentives favouring participation by otherwise-recalcitrant national governments. The proposed allocation system, with its deliberate bias towards the poor countries which contain a majority of the world's population, should provide such an incentive for many governments in the South. Governments in the North, whose powerful industrial interests will be faced with the bill for buying-in permits from offshore, are more likely to be a problem. The USA, in particular, might well seek to reverse the property-rights basis of the permit allocation, claiming privilege for its industries' right to pollute over the world population's right to a sustainable atmosphere. The assumption behind this paper is that the rich-country governments have a sufficient stake in the international rule of law, and a sufficiently informed and vocal environmental constituency at home, to persuade them to withstand pressure from industrial interests and thus commit themselves to a policy package which solves the pressing economic development problems of the South in tandem with the global problem of ecological sustainability.

The technology for monitoring emissions is here assumed to be either available or likely to be profitably developed once the permit system became operational. Poor-country governments, once in control of substantial portfolios of marketable permits with an annual top-up through reissues, would have a clear and direct incentive to ensure that the system was enforced, and it could be anticipated that a profitable industry providing the technology and services required would quickly emerge.

V. OTHER OPTIONS

Tradeable permits are only one of four broad classes of policy instruments which can be used to reduce a pollution externality. The four are direct regulation, taxation, provision of a framework for private litigation, and the creation of new markets in which rights to pollute can be traded. These options are not necessarily mutually exclusive, but each has a distinctive strategic character.

Direct regulation has traditionally been the main policy instrument used by national governments to deal with pollution inside their own territories. Its advantages lie in its transparency and certainty. Transparency occurs because the authorities can explicitly set out in written form the limits within which private economic agents must constrain their behaviour, and individual offenders can be prosecuted on the basis of concrete evidence regarding that behaviour. Certainty is gained, to the extent that the authorities put resources into policing the regulations, because enforcement is direct, in contrast to the indirect reliance on impersonal market mechanisms which occurs in some of the alternative systems. There are, however, three important drawbacks to direct regulation as the basis for an international policy regime.

The first problem lies in the cost of administering the regulations, including the task of designing and enforcing the small print. Even at national level, detailed environmental regulations absorb large amounts of time, create large bureaucracies, and tie up resources in hearings, enquiries, investigations and the like. At international level this problem is compounded.

The second problem lies in the difficulty of designing effective regulations which can apply across all countries, given the very diverse characteristics of different societies, economies and legal systems. In practice, it is not really conceivable that an international organisation could undertake detailed regulation of individual polluters in all countries. Consequently the sort of regulations most likely to emerge in practice would be ones which placed more-or-less binding obligations on governments to regulate their own nationals. Examples of such obligations are the Montreal protocol on restricting emissions of CFCs, and the recent proposal for 20 per cent target reductions in carbon dioxide emissions by each country over the next 15 years or so. Such country-by-country quantitative targets are grossly inefficient, unfair and unsustainable. They serve only two clear purposes: first to enable politicians and diplomats to indulge in the pretence that they are responding to the greenhouse problem; and second to reinforce the dominance of the industrial great powers in the world political and economic order.

The third problem is the difficulty of enforcement, given that a world government does not yet exist, and is not likely to be allowed to develop much beyond the present embryonic institutions of the United Nations. International enforcement of global regulations would probably take the form of the imposition of sanctions (economic, diplomatic or military) on offending countries, in effect holding national governments accountable for the behaviour of their polluters. It is not difficult to envisage the successful imposition of such sanctions on a small country by an international coalition led by large powers. It is less easy to see how such enforcement mechanisms would be brought to bear on the USA, the USSR, Japan, or other major economic and military powers. Because enforcement would be so crucially subject to the asymmetry of the global distribution of power and wealth, the moral legitimacy of the regulations would tend to be undermined in the long run, and there would be an incentive for 'underdog' countries to defect as a coalition.

Carbon taxes provide the second broad option for policy. Such taxes are the familiar textbook response to pollution, learned by every first-year economics student. The advantages of the tax approach, again, are familiar. First, it allows some scope for economic agents (both producers and consumers) to make optimising adjustments in their behaviour. Second, governments are experienced in implementing such taxes. Third, taxes raise revenue, which puts resources in the hands of the authorities, enabling them to finance both the administration of the tax itself, and further measures to reduce pollution or to promote research into pollution-related problems.

The carbon tax is the focus of most of the recent economics literature on the greenhouse problem [*e.g. Whalley, 1990; Schelling, 1990; Nordhaus, 1990a; 1990b; 1990c; Common, 1990; Hoel, 1991; Pearce, 1991*] and its relevance as an instrument which national governments might use to control domestic polluters is not in doubt. At the international level, however, it is an unattractive option. In the first place, there would be the problem of the currency in which the tax would be paid; exchange-rate issues would render it extremely difficult to achieve a uniform tax incidence across polluters in different countries, and would open up opportunities for manipulation. Second, and more important, any carbon tax on the scale required to make major inroads into world fossil fuel consumption would generate revenues on an unprecedented scale (several percent of world GDP) which would have to be collected and administered by someone. It is simply not conceivable that the world community would tolerate the emergence of an international agency with revenue-raising powers on this scale – nor should they, given the large element of moral hazard involved.

It is sometimes argued that carbon taxes have superior targeting properties to tradeable permits in cases where the marginal cost curve for pollution abatement is not known with certainty but is more steeply-sloping than the marginal benefit curve (for a summary of the static theory see Baumol and Oates [*1988: 68–9*]. Intuitively, one expects the greenhouse problem to be characterised by a marginal cost curve which is convex downward and a marginal benefit curve which is concave, which makes it extremely difficult to apply this principle in practice. Interestingly, however, a recent guess at the total greenhouse abatement cost curve by Nordhaus [*1990a: 20*] appears to imply a roughly linear marginal cost curve up to 90 per cent abatement, by which stage the marginal-benefit curve could be argued to have steepened significantly. Nordhaus' data thus appear to be consistent with advocacy of a permit system in preference to a carbon tax. As Baumol and Oates note [*1988: 74*], 'where the marginal benefits function is quite steep, close control over quantity becomes important'.

Recent work by Hoel [*1991*] and Pearce [*1991*] has suggested that by imposing a uniform carbon tax and then distributing targeted reimbursements, an international government could mimic the effects of the tradeable-permits option outlined earlier. Quite apart from the underlying assumption that an international agency exists which could be relied upon to do this efficiently, the suggestion seems unnecessarily cumbersome and difficult to negotiate. If the aim is simply to imitate the operation of a tradeable-permit system whose allocative design targets the financial benefits to the poor, then it is probably better simply to implement the tradeable-permits system itself. This argument is reinforced when it is recalled that a key reason for allocating a large share of the benefits to developing countries is to provide 'side-payments' sufficient to persuade the South to join the system. The permits arrangement locks-in the side-payments up front, before polluting activity takes place and associated payments are made. The alternative of charging a carbon tax first, and making side-payments later, is obviously open to greater credibility problems, and would be less attractive to the South for any given level of side-payments.

Private litigation is the third option. If this approach were taken, an international convention would specify the standards within which all polluting activities would be legally obliged to operate, and would establish the right of any member of the world community to sue offending polluters before some appropriate court or tribunal. The simplest way to do this would be for each national government to write appropriate emission standards into its domestic law, together with a provision granting non-nationals the legal standing to sue in the courts of that country, possibly with a right of appeal to a world court. The great advantage would be the low administrative cost, compared to the regulatory approach. The work of identifying and investigating individual offenders would be left to third parties with an interest in doing it – for example, environmental organisations such as Greenpeace and Friends of the Earth; international agencies such as UNEP; commercial competitors anxious not to be undercut by pollution-intensive countries; and ordinary citizens affected or offended by some polluter's activity. Precedents are provided by the international law relating to oil spills at sea, and by transboundary pollution treaties in Europe. The obvious disadvantages would lie first in the ability of large wealthy polluters to stall legal proceedings, possibly indefinitely; second in the question of what sanctions the courts could impose and how these would be enforced; and third in the natural reluctance of national authorities to see their courts invaded by litigants from outside their sovereign jurisdiction.

Tradeable permits are the fourth strategic approach. They should provide incentives for optimising adjustment by all economic agents to bring their collective emissions within a clearly-defined global pollution budget; and to these strong efficiency properties could be added the very desirable equity effect of transferring income and wealth from North to South, without any need for large flows of funds to pass through the hands of an international bureaucracy, provided that the initial allocation is along the lines suggested in this paper. If implemented together with the private litigation option outlined above, the scheme could be largely self-policing at very low fiscal cost to the world's governments; and it should be politically sustainable in the long run because of the strong incentives for a majority of countries to participate. Most important of all, it would spread the burden of adjustment across the world economy in such a way as to minimise the cost of abatement, which means that the poor countries would not see their economic growth blocked by the imposition of rigid and irrational country-by-country emission-reduction targets.

There is one obvious political problem. The large industrial countries would have to shoulder an adjustment burden proportional to the scale of their existing polluting activity, since the scheme would oblige the polluters to pay the rest of the world community for their right to pollute. The leading polluters would naturally be reluctant, and whether the opposition of the United States, Japan and some key European countries could be overcome by persuasion or pressure cannot be forecast with any certainty. However, the peoples of the rich countries have a large stake in protecting the global environment, which might well outweigh political pressures from powerful industry lobby groups.

One big problem is the possibility that an otherwise desirable tradeable-permits programme might be 'captured' by the large industrial powers as a means of advancing their own interests against the rest of the world community, Whether tradeable permits at the global level are a workable idea or not hinges critically upon the property-right assumptions which underly any programme. The choice of property-rights assumption should be guided by the principle of seeking the highest possible degree of incentive compatibility to induce national governments to co-operate. One key requirement for a workable international scheme would be that permits *not* be allocated preferentially to existing polluters.

VI. WHOSE PROPERTY RIGHTS? THREE LESSONS FROM THE UNITED STATES

Coase [*1960*] offered the theoretical conjecture that, provided transaction costs and wealth effects were zero, establishment of a property right was the sufficient condition for emergence of an allocatively efficient outcome, and that it made no difference which property right was chosen. If individuals were free to trade, he suggested, the same result should be reached regardless of whether polluters had to compensate the victims of pollution for damage suffered, or the victims had to bribe the polluters to stop. Coase offered this conjecture in the context of an attack on the tax-based policy prescription of Pigou [*1938*], and many readers have misinterpreted him as arguing that the 'polluter-pays principle' was wrong – that is, that there is some presumption that the victims, rather than the polluters, should pay. What Coase actually argued was that pure (neoclassical) economic theory should leave one indifferent between the two options in the zero-cost, full-information, no-wealth-effects case, so that the choice would rest upon other criteria – including a consideration of real-world transaction costs, wealth effects, and legal precedents. As many subsequent commentators have pointed out, 'when wealth effects are admitted it is easy to see that the initial assignment of rights is everything ... To assign or allocate rights is also to assign or allocate power and the control over future benefit streams. A property right is, above all, the ability to hold something off the market until a possible buyer meets the price that the owner is free to set' [*Bromley, 1988: 54*].

In conservative hands, the Coase proposition had major policy ramifications in the 1980s. In particular, it provided vested interests in large industry with an apparent argument against the polluter-pays-principle. Rather than polluters having to pay for the damage they caused, so the argument ran, why should the victims not pay for the cleanup by bribing the polluters? This implied that the property right to be legitimated should be the polluter's right to pollute, rather than the victim's right to enjoy a pollution-free environment. Applied to a tradeable-permit system, this meant that the permits were allocated to the existing polluters on a pro-rata basis – that is, the biggest polluters received the largest share of the right to pollute. This approach was applied by the Environmental Protection Agency (EPA) during the Reagan era in the United States to a number of experiments with tradeable permits as a means of regulating air pollution within 'hotspots' such as major urban centres.

As Cook [*1986: 221*] has pointed out, the EPA programme has been open to valid criticism on the grounds that it legitimates and rewards the activities of large corporate polluters: 'Beyond the concerns for cost-effectiveness, does society really want to grant full property rights in air resources to large corporations? What would the economic and political (both present and future) impact be if one, or even a few, large pollution sources got control of the available air resources in a region?' The US experience draws attention to three lessons for the design of tradeable emission permits. First, there are important distributional implications flowing from the property-rights decision, and these in turn have incentive effects from which Coase simply abstracted in his original article. Second, the choice of a property-rights framework that legitimates the position of existing polluters enables those polluters to erect barriers to entry which serve to obstruct the desired process of abatement [*Misiolek and Elder, 1989*]. Third, the Coase conjecture is vitiated if the allocation of property rights reinforces, rather than countervails, any asymmetry in budget-constraint regimes between polluters as a group and victims as a group.

To illuminate these three issues, consider the contrast amongst three possible ways of establishing a tradeable-permit regime to limit air pollution.

(1) In one approach, the starting point is the assumption that existing polluters have acquired, by their actions, a presumptive property right to emit pollution, and that other parties who have not caused pollution in the past do not have any share in that property right. Permits adding-up to the total pollution budget for the period are then allocated in accordance with this rule. Assuming that the pollution budget is tight enough to impose a binding constraint on emissions, there will be scarcity rents attached to the possession of a permit, until sufficient technological progress has been achieved to enable industry to work within the budget. These rents accrue to the holders of permits, either as excess profits on their pollution-causing activities, or as revenue from the sale of some or all of their permits to third parties wishing to enter the polluting activity.

The rents in effect constitute a reward to polluters for their past behaviour, while at the same time providing the means and the incentive to adopt pollution-abating techniques in future. The obvious advantage of this property-rights allocation is that it avoids what would otherwise be a bruising political confrontation between the regulator and industrial vested interests. The equally obvious disadvantage is that the costs of the policy fall upon the rest of the population, who find themselves paying for high profits in the polluting industries, while at the same time they continue to experience the detrimental effects of pollution during the transitional period until abatement takes effect. Under these circumstances the political legitimacy of the policy regime is difficult to sustain, if governments are to any extent accountable to the wider public rather than merely to industrial vested interests (the latter, of course, including workers as well as capitalists in the industries concerned).

(2) A second approach would be for a governmental agency to print a fixed number of permits and put them up for open auction. The property-rights assumption here would be that while the commons are the property of the community at large, the government acts as agent for the community in regulating access. The revenues from the auction would then be captured by the government, providing resources which could be directed to other budgetary needs (presumed to be ultimately of benefit to the community whose property had been auctioned by the government agency). Would-be polluters would then have to pay for permits, either by acquiring them at the initial auction, or by secondary trading. At the initial auction, bidders would offer prices reflecting their estimate of the value to them of permits to emit pollutants, subject to their budget constraints. Given that budget constraints are the key to bidding strength, access to finance would be a powerful factor in determining the identities of successful bidders, and those bidders who could anticipate early cash returns from their possession of permits would obviously be in a preferred position relative to those not intending to make quick commercial use of permits. A bidding ring or rings with financial advantage could potentially sew up a large share of the initially-auctioned allocation, and by subsequently restricting the supply of permits to the secondary market, could drive up the price for speculative advantage, while at the same time erecting a significant barrier to the entry of new competitors (who, having to pay a premium to acquire the necessary permits, would begin at a competitive disadvantage).

(3) A third possibility is that the property right of all community members in the commons could be recognised by the issuing of a large number of individual shares in the resource or in some aspect of it. The precise

institutional detail would be a matter for negotiation and debate, since individual shareholders would obviously have the right to nominate agents to receive and manage the shares on their behalf. (The view in (2) above that government acts as the agent of citizens at large is merely an extreme version of such delegation.) Would-be polluters would now have to enter the open secondary market to acquire permits from the initial recipients. Assuming that the spot price quickly adjusted to reflect the rental value of permits, then the rents associated with permitted pollution would come into the hands of the initial holders of share-permits as they opted to sell. The more open and efficient the market, the less would be the possibility of speculative or 'hold-out' behaviour by any group of initial permit recipients. The distinctive feature of this decentralised regime is that it removes from the hands of central government the control of the permit market, and with it the ability to realise the seigniourage rents which would otherwise come into the hands of the permit-issuing agency.

In the case of a global tradeable-permit regime, this property-rights issue is a far more serious problem than is the case with national or local programmes to check air pollution. This is because the political legitimacy of any international convention will rest heavily upon unanimous agreement amongst governments. If the US model of handing out permits to existing polluters were to be adopted for a global system, this would correctly be seen as a device to reinforce the existing global distribution of power and wealth, leaving poor countries to bear the main burden of financing the rich countries' clean-up operations, while having to pay twice-over to secure development opportunities for themselves. Rich-country governments would have a strong constituency of winners from the regime. Poor-country governments would have to represent, however imperfectly, large constituencies which would clearly stand to lose. The payoff matrix thus would look very unfavourable for the achievement of international agreement; and even if achieved, an agreement would probably not be politically sustainable, since the benefits to individual poor countries from free-riding on the world programme would be too great to ignore.

The world community has a strong interest in promoting greater equality of economic welfare by means of the economic development of poor countries. Such development will imply rising pressure on energy supplies, and poses the prospect of rising greenhouse gas emissions from at least some countries. If industrialisation is to proceed in the South in tandem with a global transition to a sustainable world, then that industrialisation must utilise state-of-the-art (and better) energy-saving technologies [*Goldemberg et al., 1988*]. One way to make this feasible is to enable the economies of the South to capture the rents from a tradeable-permit regime, while providing them with the incentive to plow those rents into energy-efficient development. (Equally, it will be essential that established industries in the North face as much pressure to abate their emissions as do new industries in the South.)

If the opportunity is lost to tackle development and sustainability as simultaneous parts of a joint problem, then the global outlook darkens seriously. Either the greenhouse effect could be held at bay by condemning the poor countries to long-term underdevelopment; or the South might grow for a generation or two without regard to the environmental consequences, exposing the entire global community to the risk of catastrophic climate change.

The fact that a joint solution implies large transfers of wealth (side-payments) from North to South means that the proposals outlined here must be expected to encounter some degree of self-interested obstruction from the rich, particularly the United States and Japan, but it would be wrong to fudge the issue simply on that account. The world community faces an historic chance actually to achieve the development goals to which so much lip service is paid on the diplomatic circuit, as a by-product of that community's willingness jointly to confront the greenhouse issue. The developing countries deserve no less than full partnership in this process. If full partnership is denied them, they have the ability credibly to threaten ecological disaster. Prudence, as well as benevolence, should prompt the rich to tolerate economic redistribution on a very considerable scale.

VII. NEW ZEALAND FISHERY REGULATION: THREE LESSONS

In his discussion of the fishery case, Gordon [*1954*] identified the absence of a full set of markets, which resulted in unregulated private exploitation leading to overfishing. Over the following 30 years the world has moved to 200-mile exclusive economic zones, within which a number of governments have experimented with quasi-market mechanisms. New Zealand, for example, now operates a system of Individual Transferable Quotas - entitlements to a certain volume of fish catch of named species, which can be bought and sold by private individuals without government involvement. The New Zealand experience highlights three key problems to be overcome if a tradeable-permit system is to deliver the required environmental payoffs.

First, the decision to use the initial quota auction as a revenue-raising device exposed the scheme to severe moral hazard from the outset. The

New Zealand government declared itself to be in effect the monopoly possessor of fishing rights, and invited would-be fishery operators to bid for them. There was no reason why the short run revenue-maximising supply of quotas to this auction market should coincide with the sustainable yield of the fishery. (In theory, with full information, perfect foresight, and perfect competition, the two might coincide. In the real world, with very incomplete information, great uncertainty about the ecology of the fishery and the likely future trend of policy, and strong incentives for strategic behaviour by private-sector participants, matters were otherwise.) In the event, the urgent desire for revenue to help balance the budget in the short term led the government to oversell the fisheries, with the result that the quota system's implementation led to serious depletion of several fish species around the New Zealand coastline. The government then had to tighten the catch limits. The lesson is that the agency which fixes the volume of quotas to be issued should not itself stand to profit from an increase in their number. This principal-agent problem is the dominant reason why a tradeable-permit system should not be introduced by means of a revenue-raising auction. (As already noted, the risk of placing enormous revenue-raising power in the hands of any agency constitutes, equally, a strong reason for arguing that a carbon tax should not be contemplated on a global scale.)

The second problem with the New Zealand fisheries experiment was the mismatch between the species mix encountered in actual fishing, and the species mix in any given firm's quota porfolio. The result was that 'unwanted' species (those not included in the firm's quotas), when caught, are often dumped at sea – a wasteful practice that accelerates the damage to fish stocks and is impossible for government to monitor. The lesson is that economists' tendency to assume that resources are 'homogeneous' needs to be tempered by strong input from the scientific community about the structure of the real world.

The third problem with the New Zealand fishery quotas was the tendency for quotas to be aggregated into the hands of a few powerful companies, leading to a situation where government regulators faced a concentrated and influential lobby group. The lesson for a global tradeable-permit system is that there must be built-in safeguards to keep the market liquid and competitive.

VIII. NORTH–SOUTH REDISTRIBUTIVE IMPLICATIONS AND INCENTIVE EFFECTS

Having outlined a pollution-abatement regime which provides the vehicle for very large North-South transfers, we may close with some remarks on the orders of magnitude involved. There are some economists who feel that the whole exercise is out of the question. Mishan [*1990: 42n*], for instance, remarks:

> The notion, occasionally broached by Western politicians, of annual tribute to poorer countries in order to encourage them to finance ambitious programmes of reafforestation, and of changeover to less toxic technologies, cannot be taken seriously. The astronomical cost of attempting so grandiose a global economy strategy takes it out of the range of the politically feasible.

Read [*1990b*] has estimated (probably rather optimistically) that a programme to grow enough trees to sequester six billion tonnes of carbon annually would cost $400 billion per year. Nordhaus [*1990a: 20*] estimates that a 60 per cent cut in carbon emissions would cost the world over $300 billion per year. As Leggett [*1990: 3*] remarks 'The implications of concerted action to cut global emissions of greenhouse gases are not for the politically faint-of-heart.'

Before reaching the view that such numbers are too big to contemplate seriously, one should place them into perspective. The outstanding debt of developing countries as a whole in 1990 was US $1,265.2 billion, with annual debt-service payments of $162.7 billion [*IMF, 1990: 184 Table A45*]. Grubb [*1989: 2.8 Table 3*] estimates that a one per cent ad valorem tax on world carbon consumption would net $0.4 per tonne, or a total revenue of $2.4 billion annually; so a tax (or comparable permit system) which doubled world energy prices would thus raise $240 billion at current energy consumption rates. Official development assistance flows from OECD and OPEC countries are between $40 billion and $50 billion per year [*World Bank, 1990: Table 19*]. Total Gross Domestic Product of developing countries in 1988 was $3,061 billion, and that of the industrial market economies was $13,868 billion [*World Bank, 1990: Table 3*]. Annual arms spending by the OECD countries is of the order of $500 billion, and world arms spending over $800 billion [*Deger, 1990*].

Table 1 below shows the orders of magnitude involved in the tradeable-permit scheme outlined above. The estimates are based on a world population of 4.9 billion with one ton per head of carbon entitlements, in a world with the 4.9 billion tons of annual emissions identified by Bolin [*1989*]. (The population figure and the carbon emissions figure are low relative to 1990 estimates; both represent the situation in the mid-late 1980s.) With permits allocated on the basis of one ton per person, with no age or gender restriction, the mismatch between initial permit entitlement and actual emission performance is as shown in Column 4, if we leave out the effects of rainforest burning. (Inclusion of deforestation-related

TABLE 1
HYPOTHETICAL PERMIT-TRADING OUTCOME FOR MID-LATE 1980S DATA

Country	*(1) Per capita carbon emission (tons)*	*(2) Population millions*	*(3) Total carbon emissions excluding rainforest burning million tons*	*(4) Net buy-in required million tons (3) minus (2)*
USA	4.9	241.6	1183.8	942.2
Democratic Republic of Germany	4.9	16.6	81.3	64.7
Canada	4.4	25.6	112.6	87.0
Czechoslovakia	4.1	15.5	63.6	48.1
Australia	3.9	16.0	62.4	46.4
USSR	3.3	281.1	927.6	646.5
Poland	3.0	37.5	112.5	75.0
West Germany	2.9	60.9	176.6	115.7
UK	2.5	56.7	141.8	85.1
Netherlands	2.5	14.6	36.5	21.9
Netherlands	2.5	8.4	18.5	10.1
France	2.0	55.4	110.8	55.4
Japan	1.9	121.5	230.9	109.4
Italy	1.5	57.2	85.8	28.6
Spain	1.4	38.7	54.2	15.5
People's Republic of China	0.5	1054.0	527.0	−527.0
Brazil	0.3	138.4	41.5	−96.9
India	0.1	781.4	78.1	−703.3
Other	0.5	1868.3	843.9	−1024.4
World Total	1.0	4889.4	4889.4	0.0
Total net purchases by deficit countries				2351.6
Total net sales by holders of surplus permits				2351.6

Figures from Bertram, Stephens and Wallace [*1989: Appendix 3*], based on Bolin [*1989: Table 4*] estimates of actual carbon emissions and World Bank [1988: Table 1] estimates of population.

emissions could eliminate much of Brazil's gain from the system – a powerful potential incentive to secure Brazilian efforts to halt forest destruction.) The resulting estimate is that at least half the stock of permits would have to be traded internationally in order to legitimate the prevailing distribution of emission sources; in practice the figure would be somewhat higher because of the diversity of the countries included in the 'other' category, which includes New Zealand and several small European industrial economies as well as a mass of poor countries.

The implied gross transfer of purchasing power in the first round of trading would then be determined by the dollar value of permits per ton of permit entitlement. Each dollar on the carbon permit value, in the Table 1 example, would add over $2.4 billion to the North–South transfer. A $20 per ton permit valuation would thus transfer roughly $50 billion per year, assuming no abatement in rich-country pollution and no rise in poor-country utilisation of permits. This would be equivalent to a doubling of current development assistance, to a level which would still be below 1 per cent of the GDP of rich countries. A $40 per ton valuation would transfer $100 billion. Numbers of this order of magnitude lie well within the world community's grasp.

final version received November 1991

REFERENCES

Baumol, W. and Oates, W., 1988, *The Theory of Environmental Policy*, 2nd edition, Cambridge: Cambridge University Press.

Bertram, I.G., Stephens, R.J. and C.C. Wallace, 1989, Economic Instruments and the Greenhouse Effect, paper prepared for New Zealand Ministry for the Environment, Wellington, September. Reprinted as Working Paper 3/90, Graduate School of Business and Government Management, Victoria University of Wellington.

Bolin, D., 1989, 'Preventive Measures to Slow Down Man-Induced Climate Change', mimeo, University of Stockholm.

Bromley, D.W., 1988, *Property Rights and the Environment: Natural Resource Policy in Transition*, Wellington, Ministry for the Environment.

Coase, R.J., 1960, 'The Problem of Social Cost', *Journal of Law and Economics*, Oct.

Common, M.S., 1990, 'Policy Instrument Choice', paper presented to Workshop on Moving Toward Global Sustainability – Options and Implications for Australia, ANU, Canberra, 9–10 Aug.

Cook, B.J., 'Review of T.H. Tietenberg, 1986, *Emissions Trading: An Exercise in Reforming Pollution Policy*', *The Journal of Politics*, Vol.48, pp.220–22.

Dales, J.H., 1968, *Pollution, Property and Prices*, Toronto: University of Toronto Press.

Deger S., 1990, 'World Military Expenditure,' Ch. 5 in Stockholm International Peace Research Institute, *SIPRI Yearbook 1990: World Armaments and Disarmament*, Oxford: Oxford University Press.

Goldemberg, J., Johansson, T.B., Reddy, A.K.N. and R.H., Williams, 1988, *Energy for a Sustainable World*, New York: Wiley & Sons.

Gordon, H.S., 1954, 'The Economic Theory of a Common-Property Resource: The Fishery', *Journal of Political Economy*, Vol.62, April, pp.124–42.

Grubb, M.J, 1989, *The Greenhouse Effect: Negotiating Targets*, London: Royal Institute of International Affairs.

Grubb, M.J., 1990a, 'International Marketable Emission Permits: Key Issues', paper prepared for IPCC/OECD workshop on Financial and Economic Measures as a Response to Climate Change, Paris, April.

Grubb, M.J., 1990b, 'Issues and Options in Implementing Greenhouse Gas Control Agreements', paper for IPCC Energy and Industries Subgroup meeting on greenhouse gas emissions, London, June.

Hoel, M., 1991, 'Efficient International Agreements for Reducing Emissions of CO2', *The Energy Journal*, Vol.12,No.2,pp.93–107.

Intergovernmental Panel on Climate Change, 1990, *Policymakers Summary of the Formulation of Response Strategies: Report prepared for IPCC by Working Group III*, WMO/UNEP, June.

International Monetary Fund (IMF), 1990, *World Economic Outlook*, Washington, DC, May.

Leggett, J. (ed.), 1990, *Global Warming: The Greenpeace Report*, Oxford: Oxford University Press.

McKinsey and Company, 1989a, 'Protecting the Global Atmostphere: Funding Mechanisms' interim report to Steering Committee for Ministerial Conference on Atmospheric Pollution and Climatic Change, Netherlands, June.

McKinsey and Company, 1989b, 'Background Paper on Funding Mechanisms for the Ministerial Conference on Atmospheric Pollution and Climatic Change, with Particular Attention to Global Warming, 6th and 7th November 1989', Oct.

Marks, R.E., 1990, 'Reducing Carbon Dioxide Emissions from Australian Energy Use', Working Paper 90–027, Australian Graduate School of Management, University of New South Wales, July.

Mishan, E.J., 1990, 'Economic and Political Obstacles to Environmental Sanity', *National Westminster Bank Quarterly Review*, May, pp.25–42.

Misiolek, W.S. and H.W. Elder, 1989, 'Exclusionary Manipulation of Markets for Pollution Rights', *Journal of Environmental Economics and Management* 16, pp.156–66.

Nicolaisen, J. and P. Hoeller, 1990, 'Economics and the Environment: A Survey of Issues and Policy Options', OECD Department of Economics and Statistics Working Paper No. 82, Paris, July.

Nordhaus, W.D., 1990a, 'Greenhouse Economics: Count Before You Leap', *The Economist*, 7 July, pp.19–22.

Nordhaus, W.D., 1990b, 'A Survey of the Costs of Reduction of Greenhouse Gases', *The Energy Journal*, Fall.

Nordhaus, W.D., 1990c, 'Economic Approaches to Greenhouse Warming', paper presented to Conference on Economic Responses to Global Warming, Istituto San Paolo di Torino, Rome, 5–6 Oct.

Nordhaus, W.D., 1991, 'To Slow or Not to Slow: The Economics of the Greenhouse Effect', *Economic Journal*, Vol. 101, July, pp. 920–37.

Pearce, D., 1991, 'The Role of a Carbon Tax in Adjusting to Global Warming', *Economic Journal*, Vol. 101, July, pp. 938–48.

Pearce, D.W. and R.K. Turner, 1990, *Economics of Natural Resources and the Environment*, Hemel Hempstead: Harvester Press.

Pigou A.C., 1938, *Economics of Welfare*, 4th ed., London: Macmillan.

Read, P., 1990a, 'Global Warming: Why Mrs Thatcher Should Be More Ambitious', *National Westminster Bank Quarterly Review*, Nov.

Read, P., 1990b, 'Tradeable Obligations on Energy Firms and Global Redevelopment with Energy Environment Sustainability (GREENS): A Policy Response to Global Warming', Massey University; New Zealand, unpublished.

Rosenberg, N.J., Easterling, W.E. Jr, Crosson, P.R. and J. Darmstadter, (eds), 1989, *Greenhouse Warming: Abatement and Adaptation*, Washington, DC: Resources for the Future.

Schelling, T.C., 1990, 'International Burden Sharing and Coordination: Prospects for Cooperative Approaches to Global Warming', paper presented to Conference on Economic Responses to Global Warming, Istituto San Paolo di Torino, Rome, 5–6 Oct.

Solow, A.R., 1990, 'Is There a Global Warming Problem?', paper presented to Conference on Economic Responses to Global Warming, Istituto San Paolo di Torino, Rome, 5–6 Oct.

Tietenberg, T.H., 1985, *Emissions Trading: An Exercise in Reforming Pollution Policy*, Washington, DC: Resources for the Future.

Whalley, J., 1990, 'The International Incidence of Carbon Taxes', paper presented to

Conference on Economic Responses to Global Warming, Istituto San Paolo di Torino, Rome, 5–6 Oct.
Whalley, J. and Wigle, R., 1989, 'Cutting CO2 Emissions: The Effects of Alternative Policy Approaches', paper presented to NBER Conference on Applied General Equilibrium Modelling, San Diego, Sept.
World Bank, 1990, *World Development Indicators 1990*, Washington, DC.

[20]

Oxford Economic Papers 46 (1994), 841–856

GLOBAL TRADEABLE CARBON PERMITS, PARTICIPATION INCENTIVES, AND TRANSFERS

By BJORN LARSEN *and* ANWAR SHAH

The World Bank, 1818 H Street NW, Washington DC 20433, USA

1. Introduction

THIS PAPER examines the feasibility of alternative tradeable permits regimes to stabilize global carbon dioxide emissions at their 1987 level by the year 2000.[1] An optimal policy response to global climate change would be a cooperative agreement in which each country reduces emissions until the marginal costs of reductions for each country are equal to the sum of marginal benefits of reductions across all countries (Hoel, 1990; 1991a, b). However, reliable estimates of marginal benefits of greenhouse gas reductions are simply not available, and therefore, one can only strive for cost efficiency in meeting emissions targets. Some countries may consider the costs of greenhouse gas reductions to exceed potential benefits and therefore may not be willing to participate in a treaty unless suitably compensated. Furthermore, a treaty must also address the free rider problem.

Most OECD countries are committed to stabilizing their carbon emissions at 1990 levels by the year 2000, and some to reduce emissions to 80–60% of 1990 levels by the year 2005 or 2010. However, most non-OECD countries are reluctant to reduce emissions to combat global climate change, arguing instead that such policies would forestall their development and that the stock of greenhouse gases in the atmosphere is primarily due to historical emissions from OECD countries and the former Soviet Union, and it is they who should be made to bear the costs of such abatement policies.

This paper evaluates alternative permit allocations for a global tradeable permit regime that minimizes costs of stabilizing world carbon emissions from fossil fuel combustion at 1987 levels by the year 2000. Particular attention is paid to the issue of inducing broad participation in a treaty. Alternative permit allocations are evaluated in terms of costs or benefits of participating in a global permit system as well as efficiency gains of inducing broader participation. The paper is organized as follows. Section 2 is concerned with current and projected carbon emissions. Section 3 details marginal cost functions and determination of equilibrium permit prices. Section 4 evaluates the costs and benefits of alternative permit allocations and incentives to participate in an international treaty. A final section summarizes main conclusions.

[1] According to Nordhaus (1991a), carbon dioxide emissions are responsible for about 80% of global warming potential associated with the accumulation of all greenhouse gases in the atmosphere.

2. Current and projected carbon emissions

The sample countries were selected on the basis of their share of world emissions, *per capita* emissions, and/or their emissions per dollar of GDP in order to have a diverse representation. Nine lower income, five middle income, four emerging market economies, all 24 OECD countries are included (see Table 1).[2] The sample countries accounted for almost 90% of world carbon emissions from fossil fuel combustion in 1987. The shares of world emissions from the OECD countries and the non-OECD sample countries were about 46% and 43% respectively. The remaining non-OECD countries are aggregated and presented as the 'rest of world'. The share of world emissions is as high as 10% for China, 18% for the former Soviet republics (12% for Russia), and 22% for the United States. *Per capita* emissions vary from 30 kg in Bangladesh to 5,930 kg in Luxembourg. Emissions per dollar of GDP vary from 0.06 kg in Switzerland to 2.06 in North Korea.

Projections of carbon emissions to the year 2000 for the same sample countries are presented in Table 2. Total world carbon emissions from fossil fuel combustion are projected to be 36% higher in year 2000 relative to the 1987 emission level. This implies an average annual growth rate of 2.4%. The share of world carbon emissions for the OECD countries are relatively unchanged from 1987. However, China's share of world carbon emissions are projected to increase from 10% to 14%. Emissions shares in most of the other low and middle income countries are also projected to increase significantly. Former Soviet republics' share is projected to decline from 18% to 13% as a result of economic restructuring. Emissions shares are also projected to decline in the eastern european countries. Carbon emissions *per capita* are projected to increase from 1,112 kg to 1,254 kg for the world as a whole, while emissions per dollar of GDP are projected to decline from 0.34 to 0.31 as a result of increased efficiency in energy use.

3. Marginal cost functions and equilibrium permit price

The purpose of a tradeable emissions permit regime is to limit aggregate emissions to a targeted level. We consider the year 2000 to which carbon emissions are projected as a baseline scenario of business as usual. The equilibrium permit price in the year 2000 is determined by the aggregate permit level (world emissions in 1987) and the marginal cost function of emission reductions in each country.[3] At the present time, country specific marginal cost functions are not available for most non-OECD countries. Even for OECD countries estimated cost functions are mostly based on different models and

[2] Some of these countries no longer exist, i.e. the Soviet Union, Yugoslavia and Czechoslovakia.

[3] See Koseobud *et al.* (1991), Quinn *et al.* (1992), and South *et al.* (1991) for a discussion and estimates of walfare gains of initial allocations of multiperiod undated permits to allow intertemporal efficiency.

TABLE 1
Carbon emissions and GDP data for selected countries—1987

	GDP per capita (US $)	*Carbon emissions (000 tons)*	*Percent of world emissions*	*Carbon emissions to GDP (kg/US $)*	*Carbon emissions per capita (kg/cap)*
Bangladesh	166	3,155	0.06	0.18	30
Nigeria	229	8,962	0.16	0.37	84
China	286	569,900	10.38	1.87	533
India	322	145,400	2.65	0.57	182
Pakistan	325	13,154	0.24	0.39	128
Indonesia	443	26,258	0.48	0.35	153
Zimbabwe	598	4,160	0.08	0.77	463
Egype	709	19,037	0.35	0.54	380
North Korea	889	39,203	0.71	2.06	1,834
Mexico	1,715	77,200	1.41	0.55	943
Brazil	2,145	50,300	0.92	0.17	356
South Africa	2,493	75,900	1.38	0.92	2,292
Venezuele	2,629	23,307	042	0.49	1,276
South Korea	3,121	44,900	0.82	0.34	1,067
Poland	1,700	125,700	2.29	1.96	3,338
Yugoslavia	2,700	32,839	0.60	0.52	1,403
USSR	2,900	1,012,900	18.45	1.23	3,578
Czechoslovakia	2,400	64,000	1.17	1.71	4,110
Switzerland	26,115	10,330	0.19	0.06	1,580
Iceland	21,873	481	0.01	0.09	1,955
Norway	19,963	8,580	0.16	0.10	2,048
Denmark	19,830	16,610	0.30	0.16	3,238
Sweden	19,257	15,150	0.28	0.09	1,812
Germany	16,754	266,700	4.86	0.20	3,427
Finland	18,070	14,420	0.26	0.16	2,925
Luxembough	16,331	2,200	0.04	0.36	5,930
France	15,913	91,000	1.66	0.10	1,636
Austria	15,441	13,000	0.24	0.11	1,717
The Nethererlands	14,521	35,600	0.65	0.17	2,428
Belgium	14,457	26,000	0.47	0.18	2,637
Italy	13,176	97,000	1.77	0.13	1,691
United Kingdom	12,024	154,000	2.80	0.23	2,707
Ireland	8,353	7,540	0.14	0.25	2,120
Spain	7,452	43,600	0.79	0.15	1,123
Greece	4,619	14,370	0.26	0.31	1,437
Portugal	3,612	7,700	0.14	0.21	758
Turkey	1,293	34,150	0.62	0.50	649
Japan	19,437	237,100	4.32	0.10	1,942
United States	18,434	1,246,100	22.69	0.28	5,112
Canada	16,056	109,100	1.99	0.26	4,221
Australia	11,364	63,900	1.16	0.35	3,932
New Zealand	10,749	5,700	0.10	0.16	1,709
OECD		2,520,331	45.90	0.20	3,015
Rest of World		634,394	11.55	0.55	639
World		5,491,000	100.00	0.34	1,112

Sources: World Resource Institute; World Bank.

TABLE 2
Projections of carbon emissions for selected countries—2000

	Carbon emissions (000 tons)	*Percent of world emissions*	*Carbon emissions per capita (kg/cap)*	*Carbon emissions to GDP (kg/US $)*
Bangladesh	8,078	0.11	58	0.29
Nigeria	13,161	0.17	82	0.44
China	1,074,631	14.28	839	1.46
India	291,640	3.88	294	0.60
Pakistan	24,804	0.33	161	0.39
Indonesia	41,066	0.55	195	0.29
Zimbabwe	5,048	0.07	397	0.66
Egypt	40,605	0.54	642	0.61
North Korea	44,617	0.59	1,614	2.06
Mexico	113,371	1.51	1,098	0.55
Brazil	89,141	1.18	506	0.18
South Africa	104,629	1.39	2,351	0.98
Venezuela	30,150	0.40	1,244	0.52
South Korea	90,059	1.20	1,905	0.28
Poland	125,700	1.67	3,210	1.73
Yugoslavia	32,839	0.44	1,298	0.46
USSR	1,012,900	13.46	3,441	1.08
Czechoslovakia	64,000	0.85	3,953	1.50
Switzerland	13,363	0.18	1,866	0.05
Iceland	622	0.01	2,401	0.08
Norway	11,099	0.15	2,515	0.09
Denmark	21,487	0.29	4,081	0.14
Sweden	19,598	0.26	2,169	0.08
Germany	345,005	4.59	4,376	0.18
Finland	18,654	0.25	3,639	0.14
Luxembourg	2,846	0.04	7,474	0.32
France	117,718	1.56	2,009	0.09
Austria	16,817	0.22	2,109	0.10
The Netherlands	46,052	0.61	2,832	0.15
Belgium	33,634	0.45	3,324	0.16
Italy	125,480	1.67	2,159	0.11
United Kingdom	199,215	2.65	3,412	0.20
Ireland	9,754	0.13	2,638	0.22
Spain	56,401	0.75	1,434	0.13
Greece	18,589	0.25	1,835	0.27
Portugal	9,961	0.13	968	0.18
Turkey	44,177	0.59	657	0.44
Japan	306,714	4.08	2,416	0.09
United States	1,611,963	21.43	5,886	0.24
Canada	141,132	1.88	4,922	0.23
Australia	82,661	1.10	4,301	0.30
New Zealand	7,374	0.10	1,993	0.14
Total OECD	3,260,317	43.34	3,611	0.17
ROW	1,056,313	14.04	823	0.55
World	7,523,069	100.00	1,254	0.31

Source: Model Calculations.

assumptions. We have therefore chosen to apply the following marginal cost function (assuming zero intercept) derived by Nordhaus (1991a, b) based on a survey of cost studies in OECD countries.

$$R = 1 - e^{-0.0054MC} \tag{1}$$

where R is percentage emissions reductions and MC is marginal costs of reductions (US \$/ton).[4] It is plausible that costs of reductions are lower in countries that have done very little to improve energy efficiency. In particular, there may be net economic gains from initial carbon reductions in countries with significant fossil fuel subsidies (Churchill and Saunders, 1991; Larsen, 1993; Larsen and Shah, 1994; Larsen and Shah, 1992a, b; Shah and Larsen, 1992a) such as the former Soviet Union, China, and Venezuela. We assume that all countries by year 2000 would have reached the point at which marginal costs are non-negative, and therefore marginal costs of reductions are zero for the first unit of carbon emission. Each country is assumed to reduce emissions until the marginal costs of reductions are equal to the emission permit price.[5] The equilibrium permit price, given the cost function in (1), that clears the permit market and thus limits total world emissions to the 1987 level, is estimated to equal US \$58 per ton of carbon and is independent of any permit allocation.

4. Costs of alternative permit allocations

An important aspects of alternative permit allocaton schemes is the international support they may expect to receive. It is reasonable to assume that a treaty will be signed primarily by nations that expect positive net benefits from participation. In view of the scientific uncertainty at present, it is extremely difficult to incorporate benefits from greenhouse gas emission reductions in terms of damages averted by delaying the threshold of global warming. Thus the approach taken in this paper is to assume that some countries have a positive willingness to pay for emissions reductions while other countries have small or zero willingness to pay. Permit allocations must then be such that the countries with negligible or zero willingness to pay are induced to participate in a treaty. Furthermore, countries with a positive willingness to pay must also gain from such an allocation compared to the case with emissions reductions exclusively in the latter countries.[6] In particular, it is assumed that OECD countries have positive willingness to pay for emission reductions, while non-OECDs have negligible or zero willingness to pay.

[4] The studies upon which this marginal cost function is based ignore possible benefits of local pollution reductions associated with reductions in carbon intensive energies such as fossil fuels. For further discussion and empirical evidence see Glomsrod *et al.* (1992), Shah and Larson (1922a, b) and Summers (1991).

[5] For cost efficient national policy responses, see Pearce (1991) and Poterba (1991).

[6] Bohm (1992) considers the impact on negotiated allocations in the case of no permit trade versus the case of allowing permit trade. Other effects of unilateral reductions such as increased emissions in countries not participating in a treay due to shifts in the marginal benefit curve (Hoel, 1991b) or changes in comparative advantage are also ignored.

Each country will reduce emissions until marginal costs of reductions are equal to the permit price of US $58 per ton of carbon. If the number of emissions permits is less (greater) than baseline emissions minus reductions, the country will purchase (sell) permits for US $58 per ton of carbon. The net costs or benefits in US dollars of alternative permit allocations, with linear approximation of the marginal cost function over the relevant range, are given by

$$C_j = P^*[E_j^P - (E_j - E_j^R)] - 0.5P^*E_j^R \qquad (2)$$

where P^* is the permit price, E_j^P is the allocated permit volume, E_j is the baseline (year 2000) emissions, and E_j^R is emissions reductions in country j. The first term on the right-hand side of (2) is the costs (−) or revenues (+) from permit purchases or sales, and the second term is the cost of emissions reductions with E_j^R given by (1) multiplied by the baseline emissions. Thus given the permit price, baseline emissions and cost function, the net costs or benefits are determined by the permit allocation scheme.

The permit allocations considered are: (i) by population; (ii) by GDP; (iii) by a combination of population and GDP; (iv) to each non-OECD country equivalent to its projected baseline emissions and to OECD countries as a group equivalent to the world emissions target minus non-OECD permits; and (v) unilateral reductions by the OECD countries.

4.1. *Permit allocation by population*

The permit allocation by population is calculated for the year 2000 by

$$E_j^P = E^P\,(POP_j/POP_w) \qquad (3)$$

where E_j^P is emissions permits allocated to country j (units of thousand tons), E^P is world permit level ($E^P = \Sigma_j E_j^P$), POP_j is the projected population of country j, and POP_w is the projected world population. E^P is in this case equivalent to 1987 world emissions. Permit allocation per capita is the same in all countries and is given by $E^P/POP_w = 0.915$ tons of carbon.

Permit allocations relative to projected baseline emissions in the year 2000 are presented in Table 3. The ratio is greater than one, i.e., allocations are larger than projected emissions, for all the low-income sample countries expect for North Korea. The ratio is less than one for all the middle-income, emerging market economies and OECD countries, except for Brazil and Turkey. OECD countries as a group would only receive permits equivalent to 25% of their projected emissions.

From (2) and (3) the net costs or benefits relative to GDP are given by

$$C_j/GDP_j = P^*\,1/gdp_j[E^P/POP_w - (1 - 0.5R)E_j/POP_j]$$

where gdp_j is *per capita* GDP for country j. Given the permit allocation *per capita* and (1), we have

$$C_j/GDP_j = P^*\,1/gdp_j[0.915 - 0.865E_j/POP_j] \qquad (4)$$

TABLE 3
Ratio of permit allocation to carbon emissions (year 2000)

	Allocation by population	*Allocation by GDP*	*Non-OECD fully covered**
Bangladesh	15.75	0.77	1.00
Nigeria	11.17	0.51	1.00
China	1.09	0.15	1.00
India	3.12	0.37	1.00
Pakistan	5.70	0.57	1.00
Indonesia	4.70	0.78	1.00
Zimbabwe	2.30	0.34	1.00
Egypt	1.43	0.37	1.00
North Korea	0.57	0.11	1.00
Mexico	0.83	0.41	1.00
Brazil	1.81	1.27	1.00
South Africa	0.39	0.23	1.00
Venezuela	0.74	0.43	1.00
South Korea	0.48	0.79	1.00
Poland	0.29	0.13	1.00
Yugoslavia	0.71	0.49	1.00
USSR	0.27	0.21	1.00
Czechoslovakia	0.23	0.15	1.00
Switzerland	0.49	4.22	0.38
Iceland	0.38	2.85	0.38
Norway	0.36	2.49	0.38
Denmark	0.22	1.56	0.38
Sweden	0.42	2.71	0.38
Germany	0.21	1.25	0.38
Finland	0.25	1.58	0.38
Luxembourg	0.12	0.70	0.38
France	0.46	2.48	0.38
Austria	0.43	2.29	0.38
The Netherlands	0.32	1.52	0.38
Belgium	0.28	1.40	0.38
Italy	0.42	1.99	0.38
United Kingdom	0.27	1.13	0.38
Ireland	0.35	1.00	0.38
Spain	0.64	1.69	0.38
Greece	0.50	0.82	0.38
Portugal	0.95	1.22	0.38
Turkey	1.39	0.51	0.38
Japan	0.38	2.55	0.38
United States	0.16	0.92	0.38
Canada	0.19	0.97	0.38
Australia	0.21	0.74	0.38
New Zealand	0.46	1.60	0.38
Total OECD	0.25	1.28	0.38
ROW	1.11	0.41	1.00
World	0.73	0.73	0.73

Note: * Non-OECD countries are allocated emissions permits equivalent to projected baseline emissions. OECD countries as a group are allocated permits equivalent to the targeted world emissions less permits allocated to non-OECD countries. Each OECD country recives permits in proportion to their projected baseline emissions.
Source: Authors' calculations.

From (4), we observe that the break-even point is 1.058 tones of carbon baseline emissions *per capita*. The net benefits (net costs) are decreasing (increasing) in *per capita* baseline emissions and decreasing (decreasing) in *per capita* GDP. The net benefits and costs of this allocation in the year 2000 are presented in Table 4. Brazil, Portugal, Turkey, and all the low-income sample countries except for North Korea, would benefit from this permit allocation scheme. In particular, the net benefits for Bangladesh and Nigeria would be more than 25% of.GDP in year 2000, and more than 11% for Pakistan. From (4) and Table 2 we see that this is because of the low *per capita* baseline emissions and the low *per capita* GDP. The net costs would be as large as more than 5% of GDP for Poland and Czechoslovakia as a result of high *per capita* baseline emissions and relatively low *per capita* GDP. The net costs to the OECD as a group would be more than 0.6% of GDP. The net costs are high for OECD countries such as Luxembourg, Australia, the United States, and Canada because their baseline *per capita* emissions are signficantly higher than in most other OECD countries.

Given the net costs involved for most of the middle-income countries and particularly the emerging market economies, these countries are unlikely to participate in a global treaty based on a permit allocation by population. The sample countries in these groups are projected to contribute more than 20% of world emissions in the year 2000, and therefore participation by these groups is vital for an efficient global treaty. Furthermore, several large OECD countries may also be reluctant to participate given that they experience net costs of about 1% of GDP.

4.2. *Permit allocation by GDP*

The permit allocation by GDP is calculated for the year 2000 by using the following expression

$$E_j^P = E^P(GDP_j/GDP_w) \tag{5}$$

where GDP_j is country j's GDP in millions of US dollars, GDP_w is aggregate world GDP in the same units in year 2000, and E^P is equivalent to 1987 world emissions. Permit allocation per dollar of GDP is given by $E^P/GDP_w = 0.23$ kg of carbon.

Permit allocation relative to projected baseline emissions in the year 2000 are presented in Table 3. The ratio is less than one for all the low-income sample countries in contrast to the case of permit allocaton by population. It is also less than one for the middle-income (except for Brazil) and the emerging market economies as in the case of allocation by population. Thus, the high income countries of the OECD end up with more permits than the baseline emissions, except for Luxembourg, Australia, the United States, and Canada, and relatively lower income OECD countries such as Greece and Turkey.

From (1), (2), (5) and permit allocation per dollar of GDP, we have

$$C_j/GDP_j = P^*[0.23 - 0.865E_j/GDP_j] \tag{6}$$

TABLE 4

Net costs (−) or benefits (+) as a percentage of GDP of alternative permit allocations (year 2000)

	Allocation by population	*Allocation by GDP*	*Non-OECD fully covered*	*OECD unilateral reductions*
Bangladesh	25.38	−0.16	0.23	
Nigeria	26.63	−0.93	0.35	
China	1.92	−6.06	1.15	
India	7.92	−1.73	0.47	
Pakistan	11.11	−0.68	0.31	
Indonesia	6.41	−0.14	0.23	
Zimbabwe	5.58	−2.04	0.52	
Egypt	1.98	−1.75	0.48	
North Korea	−3.58	−9.10	1.62	
Mexico	−0.10	−1.46	0.43	
Brazil	0.97	0.42	0.14	
South Africa	−2.72	−3.63	0.77	
Venezuela	−0.39	−1.30	0.41	
South Korea	−0.64	−0.13	0.22	
Poland	−5.83	−7.39	1.36	
Yugoslavia	−0.42	−0.99	0.36	
USSR	−3.79	−4.16	0.85	
Czechoslovakia	−5.56	−6.28	1.18	
Switzerland	−0.12	1.04	−0.15	−0.30
Iceland	−0.22	0.91	−0.22	−0.44
Norway	−0.26	0.85	−0.26	−0.51
Denmark	−0.54	0.58	−0.41	−0.81
Sweden	−0.21	0.89	−0.24	−0.47
Germany	−0.69	0.40	−0.51	−1.02
Finland	−0.51	0.59	−0.41	−0.80
Luxembourg	−1.38	−0.30	−0.91	−1.81
France	−0.22	0.85	−0.26	−0.51
Austria	−0.25	0.82	−0.28	−0.55
The Netherlands	−0.47	0.57	−0.42	−0.83
Belgium	−0.55	0.50	−0.46	−0.91
Italy	−0.29	0.74	−0.32	−0.64
United Kingdom	−0.69	0.31	−0.56	−1.12
Ireland	−0.68	0.18	−0.64	−1.26
Spain	−0.18	0.64	−0.38	−0.75
Greece	−0.58	−0.07	−0.78	−1.55
Portugal	0.09	0.38	−0.53	−1.04
Turkey	1.36	−0.92	−1.26	−2.50
Japan	−0.25	0.87	−0.25	−0.50
United States	−1.01	0.08	−0.70	−1.38
Canada	−0.92	0.14	−0.66	−1.31
Australia	−1.16	−0.23	−0.87	−1.72
New Zealand	−0.33	0.60	−0.40	−0.79
Total OECD	−0.62	0.43	−0.50	−0.99
ROW	0.80	−1.47	0.43	
World	−0.24	−0.24	−0.24	

Source: Authors' calculations.

From (6) we see that the break-even point is 0.26 kg of carbon baseline emissions per dollar of GDP, and the net benefits (net costs) are decreasing (increasing) in baseline emissions per dollar of GDP. The net benefits of this allocation in the year 2000 are presented in Table 4. All the OECD countries, except for Turkey, Luxembourg, Australia, and Greece, would benefit under this permit allocation. Even though the United States and Canada would receive less permits than their baseline emissions, they would still benefit from this allocation because their baseline emissions per dollar of GDP are less than 0.26 kg (see Table 2). All the low-income, middle-income countries (except Brazil) and the emerging market economies would experience net costs from this allocation. In particular, the net costs would be large for North Korea, China, Pland, Czechoslovakia, and USSR because of their high baseline emissions per dollar of GDP.

Given the estimated high net costs, most non-OECD countries are unlikely to participate in a global treaty based on a permit allocation by GDP. Non-OECD countries as a group are projected to contribute more than 56% of world carbon emissions in the year 2000 and therefore represent a group whose participation is vital for an efficient global treaty. Thus a permit allocation by GDP appears even less likely to induce broad participation than an allocation by population. Furthermore, negotiations regarding an allocation by GDP will also have to resolve the problematic issue of determining the level of a common currency GDP.

4.3. *A permit allocation by a combination of population and GDP*

A permit allocation that is based on a combination of population and GDP is unlikely to induce any broader participation than an allocation by population because the middle-income countries and the emerging market economies will end up with net costs under both allocations and therefore under any allocation that is also a combination of the two (Table 4).

4.4. *A permit allocation that fully covers non-OECD projected emissions*

An alternative allocation, given that allocations by population or GDP are unlikely to induce paritipation from all important groups, is identified: non-OECD countries are allocated emissions permits equivalent to projected baseline emissions in year 2000

$$E_k^P = E_k \tag{7}$$

for each country k of the non-OECDs. The OECD countries as a group are allocated permits equal to the targeted world emissions minus permit allocations to the non-OECDs

$$E_{OECD}^P = E^P - E_{non-OECD}^P \tag{8}$$

The permits to the OECDS as a group can be distributed to each OECD country by several allocation schemes. We have arbitrarily chosen that each

OECD country is allocated permits in proportion to their projected baseline emissions

$$E_l^P = E_l E_{OECD}^P / E_{OECD} \tag{9}$$

where E_l is baseline emissions for country l and E_{OECD} baseline emissions.[7] Permit allocations relative to projected baseline carbon emissions for the year 2000 are presented in Table 3. The ratio is equal to one for all the non-OECD countries, and 0.38 for all the OECD countries.

From (1), (2) and (7), we have that the net benefits relative to GDP for the non-OECD countries are

$$C_k/GDP_k = 0.135P^* E_k/GDP_k \tag{10}$$

and from (1), (2) and (9), that net costs relative to GDP for the OECD countries are

$$C_l/GDP_l = -0.485P^* E_l/GDP_l \tag{11}$$

These net benefits and costs for the year 2000 are presented in Table 4. Net benefits relative to GDP are highest (more than 1% of GDP) for North Korea, Poland, Czechoslovakia, and China, and the net costs are highest (more than 0.85% of GDP) for Turkey, Luxembourg, Australia, and Greece. The average costs to the OECD countries are estimated at 0.50% of GDP.

All the non-OECD countries would benefit from this allocation,[8] and are therefore far more likely to participate in a global treaty under this permit allocation compared to the allocations by population, GDP, or some combination of the two.

4.5. *Unilateral reductions by the OECD countries*

As an alternative to allocations that potentially could induce participation from the non-OECD countries, the OECD countries could unilaterally reduce emissions so as to stabilize world emissions at 1987 levels. In this case the same quantity of permits would be allocated to the OECDs as under the allocation in which the projected emissions of the non-OECDs are fully covered, and we assume that permits to each of the OECD countries continue to be allocated in proportion to their baseline projected emissions. However, the OECDs can now only trade permits among themselves, and not with the non-OECDs. In this case the equilibrium permit price is estimated at US $181 per ton of carbon as compared to US $58 if all countries participate. The costs relative to GDP

[7] An interesting alternative allocation is one in which permits are allocated so as to equalize costs relative to GDP for each OECD country. Given relatively large differences in *per capita* income, and perhaps marginal cost functions, this can give rise to a very different permit allocation than an allocation that is proportional to projected emissions (see Bohm and Larsen, 1994).

[8] See Bohm and Larsen (1994) for an allocation under which non-OECD countries (Eastern European countries and states of the former Soviet Union in their case) receive permits so as to cover the costs of emissions reductions, i.e., without net benefits.

for the year 2000 are presented in Table 4. The costs are on average 0.99% of GDP, or almost twice as high as under the allocation in which the projected emissions of the non-OECDs are fully covered. The cost efficiency gains to the world as a whole is about 68%, which is significantly higher than the cost reductions to the OECD countries under the permit allocation in which the non-OECDs receive permits equivalent to their projected emissions. This is because a proportion of efficiency gains accrue to the non-OECD countries.

4.6. *A comparison of costs and benefits of alternative permit allocations*

We have evaluated five alternative permit allocations in terms of their costs and benefits to participating countries and their likely merits to induce broad participation from OECD and non-OECD countries. An allocation by GDP is rather unlikely to induce participation from the non-OECD countries given the high net costs involved, although this allocaton would result in net benefits to most of the OECD countries. An allocation by population would be less costly for the OECD countries than unilateral reductions. However, most of the middle income and emerging market economies would be unlikely to participate if permits are allocated relative to population. We observed that the allocation in which the projected baseline emissions were fully covered for the non-OECDs would involve less average net costs for the OECDs than if permits were allocated by population, or OECDs were to undertake unilateral reductions. The non-OECD countries would realize net benefits from this allocation. This leads us to conclude that the allocation in which non-OECDs are fully covered in terms of their projected baseline emissions could be attractive to both OECD and non-OECD countries. However, a potential problem with the latter allocation scheme is that net benefits relative to GDP for non-OECDs would be larger for non-OECDs with higher emissions per dollar of GDP, such as China, North Korea, and Poland, and therefore countries that have achieved higher energy or carbon emission efficiency at their own cost prior to an international treaty would not be rewarded. One may argue that this consideration is not important for non-OECD countries because those that have achieved higher energy or carbon emission efficiency have done so on national interest grounds. OECDs with higher energy efficiency would nevertheless be rewarded. For instance, the costs relative to GDP for the United States would be more than twice those for France if the marginal cost function in (1) is reasonably correct for both countries. However, if marginal costs are lower in OECDs with relatively low carbon or energy efficiency, the cost difference between low and high carbon efficiency OECD countries would be less than those reported in Table 4.

The allocation scheme in which non-OECDs are fully covered is noteworthy as developing countries are not forced to reduce emissions that may slow their development efforts and OECD countries do not assume an unrealistic cost burden of emissions reductions. An additional merit of this allocation scheme is that it recognizes that OECDs had the advantage of unrestricted emissions during their own development path, while non-OECDs are far from having used their 'quota'. Smith (1991) developed a 'natural debt index' to illustrate this point by examining data on accumulated historical emissions. Furthermore, an international treaty on greehouse gas emissions reductions needs to be such that countries have an incentive to partipate. A simple but useful perspective may be to have an allocation scheme at relatively low cost to OECDs, but which allows large enough transfers to induce participation from non-OECD countries. The allocation scheme presented here, in which projected emissions of non-OECD countries are fully covered, may satisfy such a condition.

4.7. *The free rider problem*

An optimal policy response to global warming would be a cooperative agreement in which each country reduces emissions until the marginal costs of reductions for each country are equal to the sum of marginal benefits of reductions across all countries. In this case the marginal costs of reductions for any given country are generally larger than the marginal benefits to that country alone. Thus the country is generally better off by not partipating in the treaty, i.e. by being a free rider, if a similar treaty is established among the remaining countries.

The free rider problem is therefore an issue with respect to the permit allocations evaluated in this paper. We have estimated the costs and benefits of participating in a global treaty based on marginal costs of emissions reductions and permit allocations, ignoring benefits (or possibly costs for some countries) of greenhouse gas reductions in terms of climate changes. In this case, countries for which participation involves net costs after emissions reductions and permit trade, would in general be better off by not participating. However, countries for which participation involves net benefits would be worse off by not participating. In fact these countries would have zero benefits if they do not participate.

Of those we have evaluated, the permit allocaton that we found most likely to induce the broadest participation, and at significantly lower costs for the OECDs than unilateral reductions, was the one in which each non-OECD country is allocated permits equivalent to its projected baseline emissions and the OECDs are allocated permits equivalent to the targeted world emissions minus permits to the non-OECDs. Under this permit allocation, all the non-OECD countries would be worse off if they do not participate in the treaty. However, the free rider problem would remain among the OECD countries because their participation would involve net costs. In particular, this could be a problem among the lower income OECD countries. These countries might demand a larger volume of permits to participate.

4.8. *Variations in marginal cost functions*

A final consideration is possible deviations from the assumption of uniform marginal cost functions. We will consider one special case: suppose that the

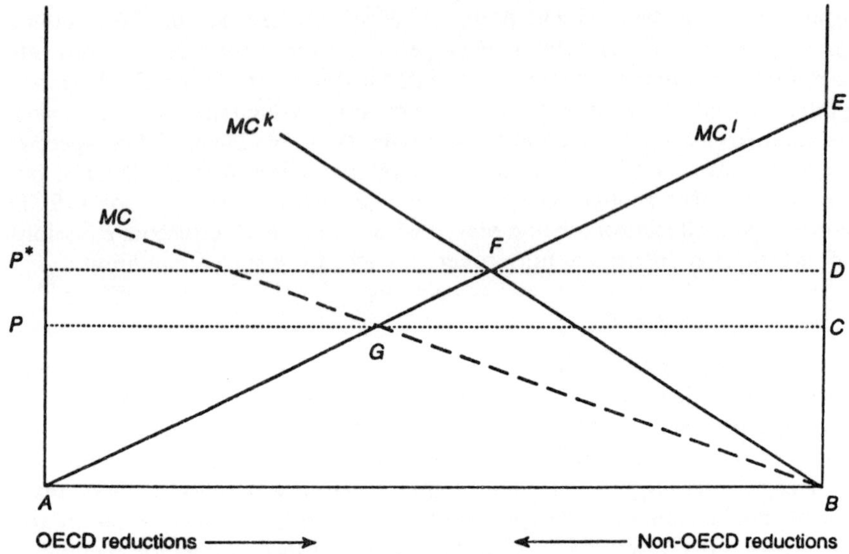

FIG. 1. Costs/benefits of emissions reductions and permit trade

OECD countries have the marginal cost functions as in (1), but that the non-OECD countries have lower marginal costs of emissions reductions than OECD countries because, on average, OECDs have already exploited the lower cost options of improving their energy efficiency and reducing reliance on fossil fuels. With lower marginal costs in non-OECDs the equilibrium permit price will be lower, given the same world emissions target and baseline emissions. Emissions reductions would then partially shift from OECDs to non-OECDs where costs are lower. Thus the total cost to the OECDs of achieving the world emissions target will be lower. This implies that the costs to OECDs, as presented in Table 4, are overstated if the marginal costs in OECDs are no higher and the marginal costs in the non-OECDs are lower. This is illustrated in Fig. 1 with a linear approximation of the marginal cost function in (1). For the equilibrium permit price P^*, and marginal cost functions MC^l and MC^k for the OECDs and non-OECDs respectively, the total costs to the OECDs are the area *ABDF*. In case the marginal costs in the non-OECDs are lower, say *MC*, the costs reductions for the OECDs are the area *CDFG*.

Lower marginal costs in the non-OECD countries would also impact on their net benefits or costs. Under the allocation in which the non-OECDs receive permits equivalent to their projected emissions, and their marginal cost functions are given by MC^k, their net benefits are the area *BDF*. If marginal costs are lower, say *MC*, the net benefits are *BCG*. This area could be larger or smaller than *BDF*, depending on the decline in permit price relative to the increase in carbon reductions in the non-OECDs. Net benefits increase with an increase in the elasticity of the MC' curve (OECD). The MC curve in Fig. 1, i.e. eq. (1) as estimated by Nordhaus (1991a), is inelastic in the relevant range, and thus net benefits of the non-OECDs would be lower than presented in Table 4 if marginal costs of emissions reductions for the non-OECDs are lower than assumed.

5. Conclusions

This paper has evaluated alternative tradeable permit allocations in a global permit regime for stabilization of world emissions at 1987 levels by the year 2000. An allocation by population is likely to be unacceptable to most middle income countries as well as to the emerging market economies of Easten Europe and the former Soviet Union because of the large net costs involved. An allocation by GDP will most likely be unacceptable to almost all non-OECD countries because they would end up bearing a large proportion of the costs of emissions reductions. Furthermore, an allocation based on a combination of population and GDP is unlikely to induce partipation from most middle income and emerging market economics because of the net costs involved under allocations by both population and GDP.

An alternative allocation scheme, proposed here, entails that non-OECD countries are allocated permits equal to their projected emissions, and the OECD countries and allocated permits equal to the world emissions target minus the permit allocations to the non-OECD countries. Such a scheme may be able to generate sufficient support for a viable international agreement on greenhouse gas emission reductions as it represents a reasonable allocation scheme given that the OECDs have a higher willingness to pay for emissions reductions, and the fact that non-OECDs are only responsible for a significantly smaller part of the stock of greenhouse gases in the atmosphere. Under this permit allocation, non-OECD countries would benefit from participation, and the net costs to the OECD countries would be only one-half of the costs of unilateral OECD reductions. The cost savings would be even larger if marginal costs of reductions in the non-OECD countries are lower than in the OECD countries.

ACKNOWLEDGEMENTS

The authors are grateful to an anonymous reviewer and the editors for helpful comments on an earlier version of this paper. The findings, interpretations and conclusions expressed in this paper are entirely those of the authors. They do not necessarily represent the views of the World Bank, its affilitiates or its member countries.

REFERENCES

BOHN, P. (1992). 'Distributional Implications of Allowing International Trade in CO2 Emission Quotas', *The World Economy*, 15, 107–14.

BOHM, P. and LARSEN, B. (1994). 'Fairness in a Tradeable-Permit Treaty for Carbon Emissions Reductions in Europe and the former Soviet Union', forthcoming in *Environmental and Resource Economics*.

CHURCHILL, A. A. and SAUNDERS, R. J. (1991). 'Global Warming and The Developing World', *Finance and Development*, **28**, 28–31.

GLOMSROD, S., VENNEMO, H., and JOHNSON, T. (1992). 'Stabilization of Emissions of CO2: A Computable General Equilibrium Assessment', *Scandinavian Journal of Economics*, **94**, 53–69.

HOEL, M. (1990). 'CO2 and the Greenhouse Effect: A Game Theoretic Exploration', Univeristy of Oslo.

HOEL, M. (1991a). 'Efficient International Agreements for Reducing Emissions of CO2', *Energy Journal*, **12**, 93–107.

HOEL, M. (1991b), 'Global Environmental Problems: The Effects of Unilateral Actions Taken by One Country', *Journal of Environmental Economics and Management*, **20** 55–70.

KOSOBUD, R. F., SOUTH, D. W., DALY, T. A., and QUINN, K. G. (1991). 'Tradeable CO2 Emission Permits for Cost-Effective Control of Global Warming', presented at the 13th Annual North American Conference of International Association for Energeny Eonomics, Chicago, Illinois, November 18–20.

LARSEN, B. (1993). 'Word Fossil Fuel Subsidies and Global Carbon Emissions in a Model with Interfuel Substitution', World Bank Policy Research Working Paper Series, No. 1256, Washington, DC.

LARSEN, B. and SHAH, A. (1994). 'Energy Pricing and Taxation Options for Combatting the 'Greenhouse Effect', in A. Amano, (ed.), *Climate Change: Policy Instruments and their Implications*, Intergovernmental Panel on Climate Change, Tsukuba, Japan, 34–35.

LARSEN, B. and SHAH, A. (1992a). 'World Energy Subsidies and Global Carbon Emissions', World Bank Policy Research Working Paper Series No. 1002, Washington, D.C.

LARSEN, B. and SHAH, A. (1992b). 'Combatting the Greenhouse Effect', *Finance and Development*, **29**, 37–64.

NORDHAUS, W. D. (1991b). 'To Slow Or Not To Slow: The Economics of The Greenhouse Effect', *Economic Journal*, **101**, 920–37.

PEARCE, D. (1991). 'The Role of Carbon Taxes in Adjusting to Global Warming', *The Economic Journal*, **101**, 938–48.

POTERBA, J. M. (1991). 'Tax Policy to Combat Global Warming: On Designing a Carbon Tax', in R. Dornbusch and J. M. Poterba (eds), *Global Warming: Economic Policy Response*, MIT Press, Cambridge, MA.

QUINN, K., MCGRATH, D. T., and KOSOBUD, R. (1992). 'A Cost-Effective CO2 Emission Permit System for GlobalClimate Change Mitigation', presented at the 56th Annual Meeting of the Midwest Economics Association, Chicago, Illinois, March 26–8.

SHAH, A. and LARSEN, B. (1992a). 'Carbon Taxes, the Greenhouse Effect and Developing Countries', World Bank Policy Research Working Paper Series No. 957, the World Bank, Washington, DC.

SHAH, A. and LARSEN, B. (1992b). 'Global Warming, Carbon Taxes and Developing Countries', paper presented at the 1992 Annual Meetings of the American Economic Association, New Orleans, LA, January 3.

SMITH, K. R. (1991). 'Allocating Responsibility for Global Warming: The Natural Debt Index', *Ambio*, **20**, 95–6.

SOUTH, D., KOSOBUD, R. F., and QUINN, K. G. (1991). 'Greenhouse Gas Emissions Control by Economic Incentives: Survey and Analysis', presented at the 13th Annual North American Conference of International Association for Energy Economics, Chicago, Illinois, November 18–20.

SUMMERS, L. H. (1991). 'The Case For Corrective Taxation', *National Tax Journal*, **XLIV**, 289–92.

WORLD BANK (1992). *World Development Report 1992*, Washington, DC.

WORLD RESOURCES INSTITUTE (1990). *World Resources 1990–91*, Washington, DC.

[21]

Resource and Energy Economics 15 (1993) 117–146. North-Holland

The efficiency and equity of marketable permits for CO_2 emissions*

Adam Rose

Department of Mineral Economics, The Pennsylvania State University, University Park, PA 16802, USA

Brandt Stevens

Fuels Planning Office, California Energy Commission, Sacramento, CA 95814, USA

Final version received December 1992

This paper examines the efficiency and equity implications of alternative assignments of marketable permits for carbon dioxide. A non-linear programming model is used to estimate the net welfare changes of permit allocations based on Sovereignty and Rawlsian equity criteria for 8 countries/regions covering the spectrum of economic development levels.

The net welfare gains associated with an overall 20% reduction in CO_2 emissions are estimated to be nearly \$20 billion, an increase of several billion dollars over a system of inflexible emission quotas requiring 20% abatement in each country. Also, although the welfare changes implied by alternative permit assignments may vary greatly between countries before trading, the trading process significantly reduces the disparities. This result stems from the Coase Theorem, which implies a uniquely efficient outcome. That is, individual country abatement levels and, hence, costs, are the same under all permit assignments after trading, and net welfare for a given nation differs only by the amount of permit revenues/expenditures associated with the application of alternative equity criteria.

Foremost among the paper's policy implications is that although equity criteria may differ significantly in principle, their welfare implications in practice may be very similar for various subsets of these criteria. This should reduce tensions at the bargaining table and facilitate the negotiation of greenhouse gas agreements.

1. Introduction

Several policy instruments have been proposed to address the acceleration of global warming caused by CO_2 emissions [e.g., Flavin (1990), Grubb and Sebenius (1991)]. Although significant progress has been made in analyzing the theoretical aspects of these instruments [e.g., Poterba (1991) on taxes, and Task Force (1991) on marketable permits], there are still some

Correspondence to: Professor Adam Rose, Department of Mineral Economics, The Pennsylvania State University, 221 Eric A. Walker Building, University Park, PA 16802, USA.

*The views expressed in this paper are those of the authors and do not represent any official position.

unresolved issues. These include the free-rider problem and the optimal distribution of tax revenues or permit assignments. Furthermore, empirical assessments of the performance of these instruments are needed for policy-makers to reach an informed decision on the control of greenhouse gases. Most analyses thus far have focused on the direct costs [e.g., Darmstadter and Plantinga (1991)] and some on macroeconomic impacts [e.g., Manne and Richels (1991)] of various policies, but few have examined both benefits and costs [Nordhaus (1991b) is a notable exception].

The purpose of this paper is to perform an economic analysis of a tradeable permits approach to global warming policy. Section 2 contains a theoretical analysis of efficiency implications of this instrument. Also presented are alternative welfare measures, distinguished according to assumptions regarding opportunity costs and free-riders. Because there is no consensus on a single best definition of equity, section 3 presents several international equity criteria for consideration. Each alternative is then translated into an operational rule for assigning permits among the nations of the world.

Section 4 presents the details of a mathematical programming model we construct to perform our empirical analysis, including a discussion of cost and benefit data. In section 5, the model is used to estimate the welfare implications of utilizing various equity criteria for the initial assignment of permits. Pre- and post-trading outcomes are compared for eight countries/regions: United States, Canada, Western Europe, Commonwealth of Independent States, Brazil, Central Africa, Indonesia, and China.

Our empirical results indicate that permit trading can lower mitigation costs by tens of billions of dollars in the aggregate vis-à-vis an inflexible system of emission quotas. For the countries/regions examined, aggregate and individual net benefits of moderate CO_2 cutbacks are positive. At the same time, we found that the popular policy proposal of a 20% reduction in current CO_2 emission levels (considered necessary to stabilize atmospheric concentrations) may be higher than the economically efficient level. Our analysis also shows that initial welfare changes vary considerably between assignments of permits according to various equity criteria, but that the trading process helps reduce these disparities significantly. This result should be of great value to policy-makers. It helps narrow the distinctions between equity criteria, thereby enabling policy-makers to focus on those most likely to bring about an agreement on CO_2 emission reductions.

2. Analytical framework

The rationale for the marketable permits approach stems from the work of Coase (1960). What has become known as the Coase Theorem states: externalities can be eliminated effectively through market transactions if

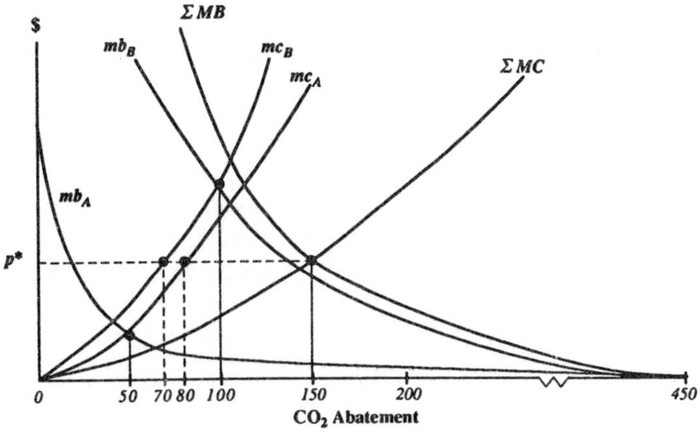

Fig. 1. Globally efficient permit trading.

property rights can be assigned, regardless of who receives the rights (assuming no significant transactions costs or income effects). This approach has been applied to pollution problems in the United States since 1975, and has also been implemented in a number of other countries [see Tietenberg (1985) and Hahn and Hester (1989)].[1]

The workings of marketable permits in the context of CO_2 abatement can be illustrated by a two-country example. In fig. 1, the global marginal benefit function is denoted $\sum$MB. It is the vertical summation of marginal benefits for the two countries (MB_A and MB_B), due to the public good nature of CO_2 abatement. The individual country marginal benefit functions exhibit the usual downward slope because the avoided damages they represent decrease with lower emissions of CO_2. The total amount of CO_2 emitted is indicated by where the global marginal benefit curve intersects the horizontal axis. To the extent there is some immediate CO_2 absorption by environmental media, the marginal benefit function would intersect the axis at a lower level. Fig. 1, however, ignores this possibility with no loss of generality in the analysis.

The global marginal cost function, $\sum$MC, is the horizontal summation of marginal costs for the two countries (MC_A and MC_B). The individual country cost functions exhibit the typical upward slope reflecting diminishing returns to pollution abatement. If some conservation is costless [e.g.,

[1]For a discussion of the merits of this approach for the control of GHGs, see Grubb and Sebenius (1991). For a comparison with other policy instruments, see Epstein and Gupta (1991). These topics and the implementation of the permits approach are also discussed in Barrett et al. (1992). For an excellent introduction to some of the theoretical issues involved, especially with regard to the game-theoretical aspects not addressed here, the reader is referred to Barrett (1990).

Chandler et al. (1990), Cline (1992)], the marginal cost curves would have a negative vertical intercept. Again, there is no loss of generality from omitting this consideration in our diagrams.[2]

The intersection of the $\sum$MB and $\sum$MC curves is the optimum level of CO_2 abatement for our two country example from the standpoint of efficiency. The intersection occurs at 150 million tons of CO_2, which is 33% of the 450 million tons (170 + 280) emitted. Subtracted from total emissions of the two countries, this represents the overall optimal quantity of permits. The intersection also reflects the social value of CO_2 abatement and therefore the market clearing price for permits.

Of course, this optimum will be attained only if both countries are signatories to an agreement. If Country A and B independently mitigate GHGs, in the absence of a treaty, each would control emissions where its own marginal benefits equal its own marginal costs. These independent country actions would, on the surface, imply control of 50 and 100 units for countries A and B, respectively. However, independent action by Country B might undercut the formulation of a treaty because Country A can enjoy benefits in excess of its own private optimal level (50 million tons) without undertaking any abatement of its own. This is of course the basis of the free-rider problem.

The prospects for permit trading are evident from fig. 1. The trading optimum equalizes the marginal cost curves for both countries, which calls for country A to abate 80 million tons of CO_2 and Country B to abate 70 million tons. Country A, which has the relatively lower abatement cost curve, would wind up selling 30 million permits to Country B.

The welfare implications of a tradeable permit scheme depend on various assumptions regarding free-riding, trading, and opportunity costs. These will be discussed as a prelude to our subsequent empirical analysis.

Case I – Voluntary agreement with no free-riding and no trading

This case would correspond to a global pact to limit CO_2 emissions without the trading feature (as in establishing emission quotas corresponding to the global optimal level of 33% in fig. 1). The implications for Countries A and B are illustrated in figs. 2A and 2B, respectively, and are summarized in line 1 of table 1. For Country A, the total benefit would be the area under its MB curve from 0 to 150 million tons of CO_2 abatement; this would exclude only the small area under the MB curve to the right of the 150 unit level. The total cost would be the much smaller area under Country A's marginal cost curve up to a level of 50 million tons of CO_2 abatement. The assessment

[2]The differences in costs and benefits between countries reflect different technologies and vulnerabilities to global warming, respectively. This is consistent with the empirical evidence to be discussed below.

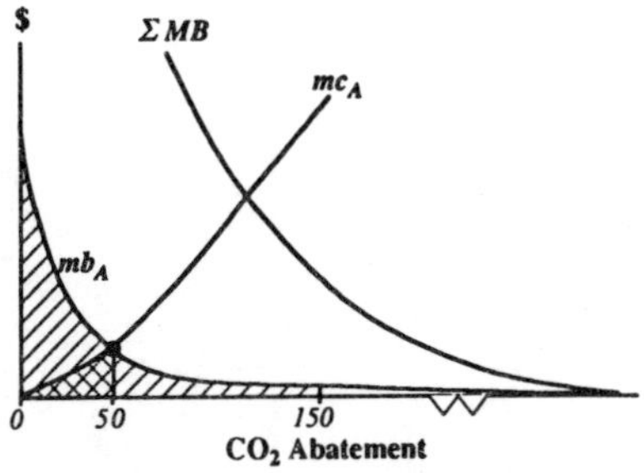

Fig. 2a. Country A net benefits (no trading).

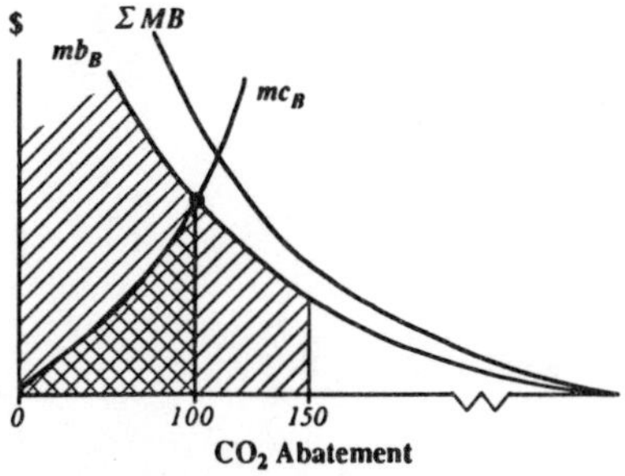

Fig. 2b. Country B net benefits (no trading).

of the total benefit for Country B would be analogous. Clearly net benefits (the difference of the areas under the marginal benefit and marginal cost curves) for both countries are positive.

Case II – Voluntary agreement with no free-riding but with trading.

As noted in the context of fig. 1, the optimal amount of permit trading would equalize the marginal costs of abatement for both countries. The welfare implications are presented in figs. 3A and 3b. For Country A, the initial benefit from damages avoided is identical to Case I. Under conditions of global efficiency, the total cost expands to the area under the marginal cost curve up to 80 million tons of CO_2 abated. However, benefits to Country A are also greater by the amount of revenues from permit sales.[3] This result depends on how many permits each country is initially allocated. Let us assume that Country A is issued permits to emit 160 millions tons

[3]The revenue gains and losses from permit trading are of course merely transfers from a social valuation standpoint and cancel out in the last column of table 1. However, trading does result in some real resource cost savings.

Table 1

Welfare analysis under various assumptions about permit trading.

Case	Assumptions	Welfare gain Country A	Welfare gain Country B	Welfare gain Global
I	Voluntary and no trading (individual efficiency)	$\int_0^{150} mb_A - \int_0^{50} mc_A$	$\int_0^{150} mb_B - \int_0^{100} mc_B$	$\int_A^B \int_0^{150} mb - \left[\int_0^{50} mc_A + \int_0^{100} mc_B\right]$
II	Voluntary with trading (global efficiency) $TP_A = 160$; $TP_B = 140$	$\int_0^{150} mb_A - \int_0^{80} mc_A + \int_{10}^{80} p^*$	$\int_0^{150} mb_B - \int_0^{70} mc_B - \int_{70}^{140} p^*$	$\int_A^B \int_0^{150} mb - \left[\int_0^{80} mc_A + \int_0^{70} mc_B\right]$ where $mc_A^{80} = mc_B^{70}$
II–I	Net gains from trade (global efficiency) $TP_A = 160$; $TP_B = 140$	$\int_{50}^{80} mc_A + \int_{10}^{80} p^*$	$\int_{70}^{100} mc_B - \int_{70}^{140} p^*$	$\int_{70}^{100} mc_B - \int_{50}^{80} mc_A$
III	Voluntary with trading (cost-effectiveness) $TP_A = 150$; $TP_B = 120$	$\int_0^{180} mb_A - \int_0^{105} mc_A + \int_{20}^{105} p^*$	$\int_0^{180} mb_B - \int_0^{75} mc_B - \int_{75}^{160} p^*$	$\int_A^B \int_0^{180} mb - \left[\int_0^{105} mc_A + \int_0^{75} mc_B\right]$ where $mc_A^{105} = mc_B^{75}$
IV	Free-rider and no trading	$\int_0^{100} mb_A$	$\int_0^{100} mb_B - \int_0^{100} mc_B$	$\int_A^B \int_0^{100} mb - \int_0^{100} mc_B$
V	Free-rider with trading	$\int_0^{x} mb_A - \int_0^{y} mc_A + \int_0^{y} p^*$	$\int_0^{x} mb_B - \int_0^{z} mc_B - \int_z^{x} p^*$ where $\int_z^{x} mc_B = \int_z^{x} p^*$	$\int_A^B \int_0^{x} mb - \left[\int_0^{y} mc_A + \int_0^{z} mc_B\right]$
V–IV	Net gains from trade (free-rider case)	$\int_{100}^{x} mb_A - \int_0^{y} mc_A + \int_0^{y} p^*$	$\int_{100}^{x} mb_B - \int_z^{100} mc_B - \int_z^{x} p^*$	$\int_A^B \int_0^{x} mb - \left[\int_0^{y} mc_A + \int_z^{100} mc_B\right]$

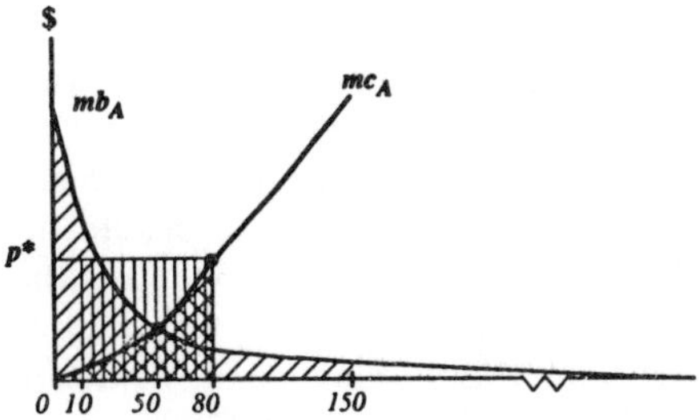

Fig. 3a. Country A net benefits (after trading).

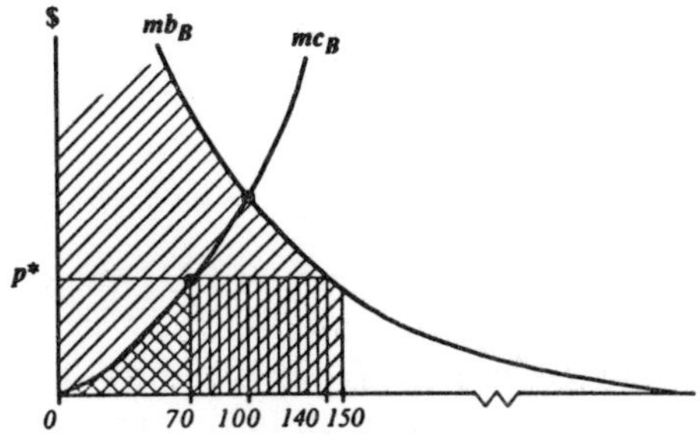

Fig. 3b. Country B net benefits (after trading).

(thereby requiring it to abate only 10 million tons), and Country B is allocated permits for 140 million tons (with abatement at 140 million tons). The optimal permit exchange is 70 million tons. This is illustrated as a rectangular revenue gain to country A, equivalent to $P^* \times 70$ million (see fig. 3a). It is also a revenue loss to Country B of an equivalent amount (see fig. 3b). This amount is still much less than the associated total cost of pollution abatement that Country B would incur if it abated 140 million tons as opposed to 70 million tons of CO_2.

The net gains from permit trading can be determined by simply subtracting the results of Case I from Case II (see row 3 of table 1). One interesting result is that since the overall control level is the same in both cases, so are total benefits, and hence they cancel out. The net gains from trade can be analyzed *from cost data alone* even in the case of optimality. This is an important practical result, given the difficulty of estimating benefits.

Case III – Cost-effectiveness analysis

For a variety of reasons, including uncertain or incomplete information, an optimal level of CO_2 abatement may not be the basis for policy. Instead, some global target or quota for emissions may be set and abatement requirements or permits assigned to each country. This policy objective should be met at the least cost, thereby attaining efficiency in the production of CO_2 abatement. The trading of permits can ensure that this cost-effective outcome will be forthcoming. For purposes of illustration, let us assume a global abatement requirement of 40% of the 450 million tons of CO_2 from our previous examples, with this burden being distributed on the order of permits to emit 150 million tons of CO_2 for Country A and permits for 120 million tons for Country B. Trading will result in abatement by Country A equal to 105 million tons, with 85 million tons being traded between the two countries. The global welfare gain is not as large as in Case II, since a non-optimal target level of abatement has been chosen. Since the abatement level is higher than the optimal level, the total benefits will be greater but so will the total cost to even a greater extent (see row 4 of table 1).[4]

Case IV – Free-rider and no trading

As noted earlier, the public good nature of CO_2 abatement makes it prone to the free-rider problem. Let us now illustrate its welfare implications. Assume Country A is the free-rider and Country B undertakes its own optimal level of abatement. Carrying through our previous examples, this means that Country A receives its individual optimal level (50 million tons of CO_2 abated) without undertaking any action. Of course, the global optimum is not reached. The welfare gains for Countries A and B are shown in the fifth row of table 1. Both countries receive a gain, though the global welfare gain is short of that attained by the global optimum case.

Case V – Free-rider with trading

In this case, we are really not referring to permit trading in the strict sense since free-riding implies no comprehensive CO_2 agreement has been reached, but to the revenue transfer as essentially a side-payment. The situation is such that Country B, because of its higher abatement costs, can improve its welfare if it pays Country A to undertake some abatement. Country B's total cost saving represents a boundary on how much would be paid to Country A. If Country B has relatively higher costs, this amount can pay for more abatement in Country A than Country B, thereby raising the overall abatement level between the two countries (this is indicated by the X-value in the domain of the integrals in row 5 of table 1, with Y and Z levels

[4]The area between the MC and MB curves to the right of the 150 million ton optimum represents a negative increment in producer and consumer surpluses.

representing the revised abatement levels of Countries A and B, respectively). It can readily be shown that the welfare gains from trade in this case fall short of the optimal outcome (Case II). The calculation of the net gains from permit trading under the free-rider cases can be computed in a manner similar to those of the voluntary agreement case. Again, there are gains from trade; however, in this case some knowledge of the benefit function is required.[5]

3. International equity criteria

Equity implications of environmental policy are often ignored. There are several reasons, including the irreconcilable fact that equity involves value judgments and, until recently, an absence of operational models to estimate distributional impacts [e.g., Rose et al. (1988) and Whalley and Wiggle (1991)]. Perhaps most important has been the concern that the promotion of equity would undercut efficiency. This neglect has been reinforced by the Coase Theorem (see section 2), which is operative irrespective of the distribution of property rights. However, the obverse of this part of the Coase Theorem is that there are no efficiency–equity tradeoffs in the case of marketable permits. Moreover, if principles of fairness can help bring about a global agreement, there is even less justification to ignore equity [Rose (1990), Burtraw and Toman (1991)].

Equity represents a normative evaluation of the social desirability of economic and non-economic disbursements, both positive and negative. The vast majority of attention has been devoted to the distribution of well-being among individuals in a given jurisdiction, but when gains and losses transcend political boundaries, the unit of analysis enlarges accordingly.

Below, we offer several equity principles applicable to global warming policy. The equity principles, a general operational rule emanating from each, and a rule applicable to the allocation of tradeable permits for CO_2

[5]There are some subtleties of the analysis not readily apparent from the diagrams. First, the MB function is a vertical summation of individual country gains because of the public good nature of GHG abatement. Second, the MC curve is horizontally summed, reflecting the private good nature of abatement from the production side. The results of these conditions translate into an elevating and steepening of the benefits function in relation to the cost function, *ceteris paribus*, hence raising the optimal abatement requirement at the global level over and above that which is indicated by any one country. Also, GHGs have characteristics of a *fund* pollutant [see Tietenberg (1992)], i.e., they exceed the assimilative capacity of the environment to some extent and hence build up over time. The static approximation of the figures will suffice for our purposes. We can conceive of the MB function reflecting the net present value of benefits over time, arising from a reduction in current levels of emissions. Thus, the annual damage estimates often presented in the literature understate the real values, since they are presented as avoided damages only for a given year, as opposed to perpetual damages, minus any absorption or decay function [Nordhaus (1991b) being a notable exception].

Table 2

Alternative equity criteria for global warming policy.

Criterion	General Operational Rule	Operational Rule for CO_2 Permits
Horizontal	Equalize net welfare change across nations (net cost of abatement as proportion of GDP equal for each nation)[a]	Allocate permits to equalize net welfare change (net cost of abatement as proportion of GDP equal for each nation)[a]
Vertical	Progressively share net welfare change (net cost proportions inversely correlated with per capita GDP)[a]	Progressively distribute permits (net cost proportions inversely correlated with per capita GDP)[a]
Ability to pay	Equalize abatement costs across nations (gross cost of abatement as proportion of GDP equal for each nation)[b]	Allocate permits to equalize abatement costs (gross cost of abatement as proportion of GDP equal for each nation)[b]
Sovereignty	Cut back emissions proportionally across all nations	Allocate permits in proportion to emissions
Egalitarian	Cut back emissions in proportion to population	Allocate permits in proportion to population
Market justice	Make greater use of markets	Auction permits to highest bidder
Consensus	Seek a political solution promoting stability	Distribute permits so (power weighted) majority of nations are satisfied
Compensation	Compensate net losing nations	Distribute permits so no nation suffers a net loss of welfare
Rawls' maximin	Maximize the net benefit to the poorest nations	Distribute large proportion of permits to poorest nations
Environmental	Cut back emissions to maximize environmental values	Limit permits associated with vulnerable ecosystems (e.g., forests)

[a]Net cost equal to the sum of mitigation benefits-abatement costs +permit sales revenues-permit purchase costs.

[b]Gross cost refers to abatement cost only.

emissions are summarized in table 2. A discussion of the economic, legal, and philosophical foundations of these criteria is contained in Rose (1992).[6]

Before proceeding, we point out a confusion in the literature arising from the failure to distinguish between fundamental equity criteria and what we

[6]Other recent work explicitly dealing with equity aspects of global warming policy includes: d'Arge (1989), Chapman and Drennen (1990), Rose (1990), Bergesen (1991), Burtraw and Toman (1991), Solomon and Ahuja (1991), and Young (1991). There are also major discussions of equity criteria in Grubb and Sebenius (1991) and Schelling (1991). We have not considered all possible equity criteria. Moreover, our equity criteria are somewhat ethnocentric, i.e., have a Western bias. Criteria based on precepts of Middle-eastern and Eastern religions should also be considered given the global scale of the issue [see, e.g., d'Arge (1989)]. Finally, there are several non-equity bases for permit allocations, primarily in game-theoretical approaches such as that of Barrett [see Barrett et al. (1992)].

will term 'reference bases.' The former refer to general concepts of distributive justice or fairness and the latter to specific measures or indices by which to gauge the equity principles. It is not surprising that many analysts emphasize the latter, since reference bases are more concrete and less controversial than ethical principles.[7] On the other hand, their equity implications are often ambiguous or erroneous. Examples are population (assigning permits on a per capita basis), which is consistent with several conflicting equity criteria (e.g., egalitarian, sovereignty, consensus), and land area, which is at best only weakly related to some criteria.

There is also the question of the object of equity analysis, or what exactly is being distributed. The most straightforward consideration is the cost of mitigation. From a comprehensive perspective, however, we should consider the net welfare change, or net benefit, which is equal to the sum of mitigation benefits and permit sales revenues minus abatement costs and minus permit purchase costs (as discussed in the previous section).[8] There is a temptation to simplify the analysis and merely confine it to abatement costs, but for the purpose of achieving an agreement one cannot ignore the magnitude and variations across nations of benefits from avoiding global warming damages. Even if the scientific community feels these benefits can never be accurately measured, nations are likely to have strong perceptions of their magnitude, a factor that will influence their negotiating position. Thus, while we can simplify the analysis of tradeable permits on the efficiency side by resorting to cost-effectiveness analysis (i.e., simply assuming a given CO_2 reduction level and examining whether it can be attained at least cost), we do not necessarily have this luxury from an equity standpoint.

The fact that GHGs are fund pollutants has important implications for the intergenerational allocation of resources forward in time, in terms of the usual issues relating to discounting [see, e.g., Nordhaus (1989) and d'Arge (1989)]. Going backward in time, in the sense of cumulative emissions, also has important international equity implications. In terms of efficiency, policy instruments should be applied only to current and future emissions, since built-up GHGs are analogous to sunk costs. From an equity standpoint, however, the fund nature of GHGs (recall footnote 5) does matter. Industrialized nations have developed by exploiting the global commons at little or no cost. Ignoring the past build-up and simply basing reduction requirements on subsequent emissions would be equivalent to penalizing developing countries

[7]Reference bases have the desirable properties of focal points, a term Schelling (1960) coined to represent a key, facilitating feature of negotiation processes. We will give primacy to equity principles in the presentation to follow. If global warming were an uncomplex issue that could be settled in a short time, we would endorse the more pragmatic 'rule of thumb' approach [see also Toman and Burtraw (1991)].

[8]It would be preferable to use general equilibrium effects (interactive impacts on the entire economy), as opposed to partial equilibrium measures (direct costs), but this is beyond the scope of our study.

for their progress, when no such penalty was imposed on industrialized countries [see Agarwal and Narain (1991)]. Fortunately, since the efficiency and equity aspects of marketable permits are generally separable, the latter objective can be pursued in this context without misallocating resources.

The equity criteria in table 2 can be reinterpreted in terms of cumulative emissions. Cumulative emissions can, for example, be thought of as affecting a country's GHG 'rights' or 'responsibilities,' where these are relevant. The consideration of rights arises in the context of sovereignty and the egalitarian ethic. Responsibility is not explicitly related to any of the criteria except sovereignty and the environmental ethic, and possibly rather weakly to consensus. Otherwise, it is simply compatible with all other criteria except market justice and one rigid form of egalitarianism.[9]

The effect of incorporating cumulative emission considerations into the reference base is to alter abatement requirements or permit allocations. Surprisingly, the cumulative emissions reference base would not change the distribution of the CO_2 cost burden dramatically. Smith et al. (1991) estimate the contribution to cumulative emissions between industrialized and developing nations as 60:40 on a current basis and 70:30 on a cumulative one.

The fund nature of GHGs, as well as the potential for them to be emitted at significantly higher rates as population grows and economic development spreads, justifies special concern over future emissions. For example, the contribution of developing countries to CO_2 emissions is likely to rise from 30–40% of the world total currently to 60+% in perhaps the next 30–40 years. As such, the pressure will mount to raise emission limits (reduce permits) for these countries. While there has been a big emphasis on the low cost of achieving emission reductions technologically in the Third World, other mitigation policies might be costly. One example is a policy to discourage or ban the use of fossil fuels. In countries like China, these costs could be huge. Under these circumstances, it may be legitimate to include the opportunity cost of devalued indigenous resources. Their inclusion into the net welfare calculus would place such countries in a position to receive extra permits under criteria like vertical equity and Rawls' maximin principle.

In the 'future reference' case, it must be emphasized that currently undertaking CO_2 abatement diverts resources from investment needed to stimulate economic development. The opportunity cost extends beyond the partial equilibrium aspects associated with fossil fuel reserves, and includes the general equilibrium losses from foregone economic growth and development. The result is problematic because developing nations are especially

[9]In addition, there are some equity criteria that are only relevant in the context of historical emissions and very explicitly so. A good example is Young's (1990) 'status quo' or 'adverse possession' criterion, which stipulates that the current rate of GHG emissions constitutes a 'right' legitimized by usage and custom. The latter has been proposed by Fujii (1990) in terms of every person having an equal emissions quota regardless of his/her location or *generation*.

needy and, according to some specifications of welfare functions and views of equity, suffer more from a comparable decrease in the rate of economic growth than industrialized nations. The effect of including future dynamic elements would therefore shift the allocation of entitlements in favor of developing countries.

4. The empirical model

To assess the efficiency and equity implications of greenhouse gas controls, we constructed a non-linear programming model that incorporates estimates of the costs of CO_2 control and the benefits of CO_2 reduction. The model is used to analyze individual and global net benefits before and after trading for eight countries or regions: the US, Canada, Western Europe, the Commonwealth of Independent States, Brazil, Central Africa, Indonesia, and China.

Sections 2 and 3 demonstrated that global net benefits and the equity consequences of greenhouse gas controls are sensitive to: the number of countries that are signatories to a greenhouse gas treaty or protocol, the extent that permit trading is allowed, and the initial distribution of emission control responsibilities between nations. If all countries are signatories and permit trading is allowed, the global efficiency version of the model, in accordance with Cases II and III of section 2, can be stated as:

$$\max\left[\sum_{i=1}^{8} \mathrm{TB}_i - \sum_{i=1}^{8} \mathrm{TC}_i - \sum_{i=1}^{8} (\mathrm{TE}_i \cdot P^*)\right], \tag{1}$$

where

TB_i = the ith country's total benefits (or avoided damages) from global control of greenhouse gases,
TC_i = the ith country's total emission control costs,
TE_i = the ith country's total revenues or payments for permits,
P^* = the equilibrium permit price.

Given the public good nature of greenhouse gas controls, the benefit for each country is a function of global emission reductions, while costs are determined by each country's abatement level. The third term of expression (1) represents the sum of individual country expenditures and revenues on permits, which cancel out.

The model's static formulation has two important implications. First, we assume that each nation can look forward in time and assess the implications of a greenhouse gas protocol on its own economic well-being, as well as how a protocol would improve or worsen its economic standing vis-à-vis other nations. There is considerable scientific uncertainty about the contribution of greenhouse gases to global warming, the rate of temperature increase, and ecosystem impacts, so negotiators are likely to have an imprecise estimate of the economic implications of an agreement. The model implicitly assumes risk-neutral behavior by each nation, though, clearly, attitudes towards risk will affect negotiating positions. It would be inordinately difficult to characterize each nation's risk attitude, but it is not necessary for our purposes because the model is designed to analyze efficiency and equity implications, not to predict the outcome of treaty negotiations.

A second limitation of the static model is it ignores issues regarding the rate of accumulation of greenhouse gases and the timing of emission reductions. Clearly, these issues are important, e.g., the timing of emission reductions will affect costs. We assume that a 'crash' program will not be implemented via a greenhouse protocol, and that in the absence of controls, a 3°C warming effect will occur near the middle of the next century. No other assumptions about the dynamics of global warming are made, partly because the dynamics are extremely complex and would rightly be the subject of another paper. In addition, there is little guidance in the literature on the rate and timing of emission controls. Our cost and benefit estimates are based on the assumption employed by Nordhaus (1991b) and others regarding the amount of warming and when it occurs.[10]

4.a. Cost estimates

Several studies estimate the costs of greenhouse gas controls, with estimates for the US being of better quality than estimates for developing nations [see the summaries in Nordhaus (1991a) and Cline (1992)]. We employ a logarithmic functional form for CO_2 reduction costs, similar to Nordhaus (1991a, 1991b) and Barrett (1991, 1992):

$$\ln(1-R_i)=\alpha_i \mathrm{MC}_i, \tag{2}$$

where

R_i = the ith country's percentage emission reduction,
α_i = a slope parameter,
MC_i = the ith country's marginal cost, in dollars per ton carbon, of CO_2 reduction.

[10]There are considerable differences of opinion over key aspects of the global warming problem and its resolution. For example, the 3°C warming projection we cite has recently been revised downward by the IPCC (1991) to best estimates of 1°C by 2025 and 3°C not until 2100 [see also the discussion of various estimates by Cline (1992)]. The range of estimates of the optimal level of abatement are also broad (see below). For example, Nordhaus (1991b) concludes that for his medium estimate of damages, greenhouse gas reductions of 11% would be economical, whereas Peck and Teisberg (1992) conclude that almost no control of greenhouse gases would be sensible if damages are linear (even though their damage function is based on Nordhaus' initial damage estimates). The difference lies primarily in the size of the gap in cost estimates between the two studies.

Table 3

Cost estimates, 10% reduction in CO_2.

Region	Marginal cost ($/ton CO_2)	Total cost (billions)	α_i Estimate
US	29.51	2.20	−0.00357
Canada	32.42	0.22	−0.00325
Western Europe	33.55	1.61	−0.00314
CIS	10.54	0.70	−0.01
Brazil	7.02	0.02	−0.015
Central Africa	16.21	a	−0.0065
Indonesia	7.02	0.01	−0.015
China	16.21	0.73	−0.0065

[a]Less than $0.005 billion.

This functional form has the desired property of positive and increasing marginal cost (for $\alpha_i > 0$). The parameter also captures technological and other distinctions that cause control costs to differ between countries. While statistical estimates of α_i are preferable, for those countries for which such estimates do not exist, an algebraic proxy from eq. (2) can be computed if the marginal control cost associated with a specific CO_2 percentage reduction is available.

Barrett (1991) concludes that Nordhaus' estimate of the cost of CO_2 reduction implies an alpha parameter of approximately −0.00357, or that the US could reduce emissions by 30% at a marginal cost of approximately $100/ton.[11] In addition, we use the average of Barrett's (1992) statistical estimates of α_i for several European countries to represent the European community as a whole.

Studies used for other regions include: Chandler (1990), Manne and Richels (1991), and Houghton et al. (1987).[12] Note the estimates in table 3 are consistent with the general consensus finding that mitigation costs are relatively higher for industrialized nations.

4.b. Benefit estimates

Nordhaus (1991b) estimates the damages from a 3°C warming in the latter

[11]Nordhaus estimates a global marginal cost function by fitting a curve through point estimates of costs from studies that differ by methods (econometric and mathematical programming models) and regions (i.e., the US, Sweden and the European Community, for example). Note that most of these estimates do not suggest a significant range of 'costless' abatement (due to conservation), as in several studies cited by Cline (1992).

[12]The estimates for α_i and corresponding marginal costs for each region are shown in table 3. The elasticity of marginal cost is $dR_i/dMC_i \cdot R_i/MC_i = -a_i\exp(a_iMC_i) \cdot R_i/MC_i$ so that the elasticity is identical for all regions. The elasticity ranges from 0.948 at a 10% emission reduction to 0.766 at 40%.

part of the next century at approximately 1% of global output.[13] This estimate serves as the basis for constructing a global marginal benefit function, though it requires several modifications. First, the economic value of damages from global warming will differ between nations according to their vulnerability to crop losses, sea level rise, loss of amenities, etc. Utilizing a recent IPCC study (1991) that qualitatively assesses these differences, we assume that, relative to the US, each qualitative adjustment implies a 10% decrease or increase in damages from the 1% reference point.[14] Under the assumption that the 3°C global temperature increase will occur in the middle of the next century, nations considering agreeing to a greenhouse protocol need to know their domestic output roughly 60 years hence and also need to know the present discounted value of damages. Using each nation's current GDP, and assuming annual growth rates of 3% for the US, Canada, and Western Europe, 5% for China, 4% for Indonesia and Central Africa, and 3.5% for Brazil and the CIS, as well as a 4% real interest rate, and the adjusted percentage of output that represents damages, we compute the present discounted value of damages[15] (discounted from 2050 to 2000)[16] for

[13]Nordhaus bases his estimate on an examination of various sectors of the US economy in 1981 that would be affected by a change in global climate. As Nordhaus notes, his medium estimate of 1% of global output includes an ad hoc estimate of damages to nonmarket sectors. Implicit in Nordhaus' extension of this estimate to the globe is that other nations' economies are structurally similar to the US, at least with respect to those sectors where damages occur and that the magnitude of damages in nonmarket sectors is similarly valued. For less-developed nations, agriculture is currently a much larger share of the economy, though in 40 or 50 years, that may no longer be true. In addition, the value of nonmarket damages may be lower in less-developed countries (consistent with the hypothesis that environmental quality has an income elasticity greater than one). Thus, while this assumption may appear to be quite strong, the disparities between currently industrialized nations and less-developed nations will not be as large in 50 years, and, as damages to the market sector decline (i.e., agriculture), the value of damages in nonmarket sectors will increase. For a critical discussion of Nordhaus' estimates, see Cline (1992).

[14]The adjustment factors results in damages as a percent of gross domestic product as follows: US, 1%; Canada, 0.7%; Western Europe, 0.7%; CIS, 0.5%; Brazil, 1%; Central Africa, 0.7%; Indonesia, 0.9%; and China, 1.3%. Note also that the benefit estimates employed here differ significantly from those in Rose (1992), because of greater reliance on Nordhaus' estimates and those of the IPCC in this paper.

[15]The real output discount rate employed here is consistent with the assumption in Nordhaus (1991b) that the global real output discount rate is in the range of 0% to 1%. In more recent work, Nordhaus (1993) uses a global real output discount rate of 3%, arguing that this is in line with real output growth rates in industrialized economies over the last twenty years. However, such a real output discount rate may not be realistic over the next sixty years, and real output discount rates may differ substantially between industrialized and developing economies. We acknowledge the importance of the issue since higher growth rates would increase the optimal amount of greenhouse gas control, perhaps substantially. We do not explore the issue further, partially due to the static nature of our model.

[16]We have chosen the year 2000 for our example for several reasons. First, it is reasonable to assume that it will take several years for an agreement to be ratified and for an abatement program to be initiated. Second, choosing an earlier, rather than later, year for illustrative purposes is preferable because projections are less prone to errors that could arise from changes in emission rates or shifts in benefit and cost functions (e.g., there is less of an opportunity for

each region in table 4.[17]

Much of the empirical evidence from the benefit estimation literature shows that marginal benefits decline as environmental quality improves. We assume this principle applies to global climate change as well, i.e., marginal benefits from reducing global temperature increases will be higher at higher levels of temperature change.[18] A negatively sloped marginal benefit function could be represented by several functional forms. We selected a logarithmic functional form to represent marginal benefits, using the adjusted 1% gross global output criterion to determine its parameters, thus:

$$\ln R_i = \beta_i \mathrm{MB}_i, \quad (3)$$

though given the public good nature of the benefits (each nation benefits from CO_2 abatement without detracting from the benefits received by other nations), we use the global, G, version of this function:

$$\ln R_G = \beta_G \mathrm{MB}_G. \quad (3a)$$

This functional form is similar to that we employed for marginal costs and is consistent with theoretical expectations for lower marginal benefits as global emission reductions increase. Nordhaus assumes marginal damages per ton of CO_2 are constant, while we assume a logarithmic marginal damage function calibrated to his 10% optimal mitigation level.

Thus, an estimate of the global marginal benefit slope parameter, β_G, is possible if we have an estimate of marginal benefits for a given emission reduction. To estimate the global slope parameter, we convert our global damage estimate into damages per ton of carbon. For this purpose, we assume that without any controls on greenhouse gases, each country's CO_2 emissions would grow, over the 60-year time horizon, at the growth rate of regional output (see above) minus an annual autonomous improvement in

---

technological change to affect the results in the short run). If such factors are stable over the policy horizon, then the year 2000 will be reasonably representative of all of the other years in any case. Note that the benefit function is such that the effects of warming still do not take place until the middle of the next century. These are simply discounted back to the year 2000 (though all figures are expressed in 1989 constant dollars).

[17] Note that the 'totals' in table 4 represent vertical (public good) summations of the individual country benefit functions (recall the discussion in section 2). No totals were presented for table 3 because of the private good nature of the entries in that table (i.e., they are horizontally summed to yield an aggregate marginal cost curve). In the analysis below, efficiency is achieved by the equality of individual country marginal costs and aggregate marginal benefits.

[18] The total benefits function may be nonconvex, due to the physical parameters governing global warming. Although further scientific inquiry would provide insights about the shape of the total benefits function, nonconvexity may not be troublesome. If the marginal benefits function is initially positively sloped and then negatively sloped, the optimum emission reduction would occur where marginal cost intersects the negatively sloped portion of marginal benefits.

Table 4

Benefit estimates, 10% reduction in CO_2.

Region	Marginal benefit ($/ton CO_2)	Total benefit (billions of $)	β_i Estimate
US	14.96	10.70	−0.1539
Canada	0.94	0.68	−2.4451
Western Europe	9.04	6.50	−0.2545
CIS	2.01	1.44	−1.1466
Brazil	1.34	0.96	−1.7240
Central Africa	0.12	0.09	−18.8105
Indonesia	0.41	0.30	−5.5783
China	4.74	3.40	−0.4861
Total	33.57	24.11	−0.0686

energy efficiency equal to 1%.[19] This yields total CO_2 emissions of 18.1 billion tons in 2050 for our eight countries. Nordhaus concludes the optimal level of greenhouse gas control is 11%. Following that result, we estimate damages per ton of CO_2 at a 10% reduction in CO_2, i.e., adjusted damages (not shown) are divided by a global emission reduction of 1.81 billion tons.[20] The resulting global and regional estimates of marginal damages per ton are shown in column 2 of table 4. The global slope parameter can be computed from eq. (2), since R_G is 0.1 and global marginal benefits are known. Column 4 of table 4 shows the global marginal benefit function slope parameter (as well as slope parameters for each region's marginal benefit function),[21] i.e., $\beta_G = -0.0686$.

This global marginal benefit function predicts global damages of approximately 1.3% of the present discounted value of global output for a 10% reduction in CO_2, in line with Nordhaus' estimate. However, the per ton estimate of damages at 10%, $33.57, is more than four times higher than Nordhaus' estimate. The difference in per ton damages is due to a difference in methodological approaches and assumptions. Nordhaus uses an intertemporal growth model to derive his estimate, while our approach is much less elaborate. More importantly, while our analysis is based on Nordhaus' initial estimate of damages equal to 1% of gross output, our estimates of the present discounted value of global output and total CO_2 emissions may

[19] This is consistent with Manne and Richel's (1991) recent mid-range estimate.

[20] Our method assumes that the global marginal benefit function passes through the optimal emission reduction level found by Nordhaus but for a downward-sloped (rather than horizontal) marginal benefit function. Note also that we use 10% as a reference point instead of Nordhaus' 11% in order to adjust for the contribution of CFCs.

[21] Since reductions in CO_2 are a public good, vertically summing each region's marginal benefit function to obtain an estimate of global marginal benefits requires summing the dollar value of damages at a given level of emission reductions, exactly as we have done, to estimate the global benefit function.

differ from his because we differentiate regions in terms of the likely global warming damages and the growth rate of output and CO_2 emissions.

5. Welfare analysis

In this section, we use our empirical model to estimate the welfare effects of using two equity criteria – sovereignty and the Rawlsian maximin principle – to distribute marketable permits for CO_2 emissions. Both will be examined in terms of current and cumulative reference bases. Also, we will examine both a cost-effectiveness policy relating to the oft-espoused 20% CO_2 reduction target[22] and a global efficiency policy determined by the intersection of aggregate marginal benefits and aggregate marginal costs. Insights to the utilization of other equity criteria for assigning permits will be noted as well.

5.a. Empirical results

Our analysis is undertaken for eight countries or country groupings, listed in column 1 of table 5, representative of various levels of development as denoted by their per capita Gross Domestic Product (GDP) in column 2. Columns 3 and 4 of table 5 list estimates of benefits (environmental damage avoided) from mitigation of 20% of CO_2 emissions in the year 2000 and the associated costs, respectively.[23] At this stage of scientific inquiry, the numbers are of course crude, but still are able to indicate the relative policy impacts. For example, benefits are relatively small for Canada and the Commonwealth of Independent States (CIS), since a warming of their climate would increase agricultural productivity and would offset considerably the damage due to sea level rise and loss of recreational benefits and tourist revenues. At the same time, some of the more populated areas of Europe and Indonesia could be inundated by a fairly modest melting of the polar ice caps.

Mitigation costs are evaluated in terms of actual direct expenditures rather than general equilibrium or macroeconomic effects [cf. Manne and Richels (1991)]. The reliance of China on its coal resources, and the cost of fuel

[22]A 20% reduction in baseline emissions has been estimated as necessary to avoid a doubling of CO_2 in the atmosphere by the middle of the 21st century (even taking into account atmospheric removal rates). Most serious policy pronouncements have not been so rigid as to require a 20% reduction across the board; most policy-makers are cognizant of the status of LDCs, and simply require some capping of emissions or allowing for modest growth to avoid stifling development. This dichotomy is actually more consistent with vertical equity or the Rawlsian principle. Still, we choose to use the stylized version of the actual policy pronouncement in order to simplify the analysis and to begin with the depiction of the pure sovereignty case.

[23]Note that the carbon dioxide emission estimates we used do not include those associated with deforestation; thus, the relatively low cost estimates for Brazil.

Table 5

Changes in national welfare due to static sovereignty permit assignments[a] (cost-effectiveness case: 20% abatement).

Country (area)	Per capita	(All cases)	(Without trading)		(With trading)		
	GDP[b]	Benefit	Cost	Net change	Cost	Trading	Net change
US	20,725	16.98	9.13	7.85	3.64	4.19	9.15
Canada	18,648	1.07	0.89	0.18	0.30	0.43	0.34
Western Europe[c]	14,322	10.27	6.67	3.60	2.08	3.24	4.95
CIS	3,491[d]	2.28	2.90	−0.62	7.72	−6.14	0.70
Brazil	2,167	1.52	0.10	1.42	0.53	−0.63	1.62
Central Africa[e]	475	0.14	0.02	0.12	0.02	−0.01[f]	0.12
Indonesia	527	0.47	0.06	0.41	0.30	−0.36	0.53
China	375	5.38	3.03	2.35	3.72	−0.73	2.39
Total	–	38.11	22.80	15.31	18.31	0.00	19.80

[a]All figures in billions of 1990 dollars except per capita GDP.
[b]Data for 1989 from World Bank (1991).
[c]Western Europe includes the European Community countries: Belgium, Denmark, France, Germany, Greece, Ireland, Italy, Luxembourg, Netherlands, Portugal, Spain, and the United Kingdom.
[d]Data for 1988 for Soviet Union from Euromonitor (1991).
[e]Central Africa includes: Burundi, Cameroon, Central African Republic, Congo, Equatorial Guinea, Gabon, Rwanda, and Zaire.
[f]Less than 0.005 billion dollars.

switching, for example, are the main reasons for its relatively high mitigation costs in relation to other developing countries.

Sovereignty criterion – Cost-effectiveness

The net change in national welfare from an application of the (20%) cost-effectiveness version of the sovereignty criterion, but without permit trading, is presented in column 5 of table 5. All eight countries/regions would stand to reap positive net benefits except the CIS. The sum of net benefits is $15.31 billion.

Our analysis of the sovereignty principle has not yet included the feature of permit trading, which will take place as long as some countries have marginal abatement costs greater than or equal to the world market price of permits and some countries have abatement costs less than or equal to this price. It is usually assumed, and generally borne out by the cost studies, that abatement costs are relatively higher in developed countries than developing ones. Trading will thus lower the total costs to the former and provide a transfer to the latter, which in effect also lowers the net abatement costs of developing countries.

The effect of trading on the welfare of the eight world regions is presented in the last three columns of table 5. The results emanate from our model's determination of optimal individual abatement levels by equalizing marginal costs across countries. They indicate that the US, Canada, and Western Europe would be permit buyers and the other five countries/regions would be permit sellers.[24] Nearly 205 million permits would be traded with a total value of $7.86 billion (see table 6) at the equilibrium price of $38.35 per ton of CO_2. The revenues from the sale of permits more than offset the increased costs of abatement incurred by the sellers.

Permit trading has several positive implications. First, the equalization of marginal costs of abatement saves this global group $4.49 billion, or almost 20% of the $22.80 billion of abatement costs associated with the uniform emission reduction requirement. The net cost savings (including expenditures on permits) to the three Western industrialized countries/regions is $2.80 billion, or about 17% of their costs at uniform abatement levels. In addition, permit trading is estimated to be a source of revenue for developing and transitional economies. In fact, for Brazil and Indonesia, the permit revenues more than offset their total abatement costs. Finally, all of the countries examined receive positive net benefits from a 20% CO_2 emission reduction

[24]Positive entries in the column marked 'Trading' in table 5 (and tables 7 and 8) refer to expenditures on permits, which are added to abatement costs to arrive at net 'costs' and net benefits estimates. Negative entries in the 'Trading' column refer to revenues from permit sales (cost offsets).

Table 6

Permit trades from alternative CO_2 abatement levels.

Country (area)	Cost-effectiveness case		Global efficiency case	
	Permits (thousands)[a]	Value (billions of $)[a]	Permits (thousands)[a]	Value (billions of $)[a]
US	109,267	4.19	84,623	2.34
Canada	11,193	0.43	8,630	0.24
Western Europe	84,399	3.24	64,986	1.80
CIS	−160,044	−6.14	−124,024	−3.43
Brazil	−16,512	−0.65	−13,201	−0.37
Central Africa	−114	[b]	−82	[b]
Indonesia	−9,269	−0.36	−7,410	−0.21
China	−18,918	−0.73	−13,520	−0.37
Total	204,859[c]	7.86[c]	158,239[c]	4.38[c]

[a]Positive entries represent permit purchases; negative entries represent permit sales (cost offsets).
[b]Less than 0.005 billion dollars.
[c]Sum of absolute value of either permit purchases or sales.

Table 7

Changes in national welfare due to static sovereignty permit assignments[a] (global efficiency case: 14.9% abatement).

Country (area)	(All cases)	(Without trading)		(With trading)		
	Benefit	Cost	Net change	Cost	Trading	Net change
US	14.13	5.03	9.10	1.94	2.34	9.85
Canada	0.89	0.49	0.40	0.16	0.24	0.49
Western Europe	8.54	3.68	4.86	1.11	1.80	5.63
CIS	1.90	1.60	0.30	4.31	−3.43	1.02
Brazil	1.26	0.05	1.21	0.30	−0.37	1.33
Central Africa	0.12	0.01	0.11	0.01	−[b]	0.11
Indonesia	0.39	0.03	0.36	0.17	−0.21	0.43
China	4.47	1.67	2.80	2.02	−0.37	2.82
Total	31.70	12.56	19.14	10.02	0.00	21.68

[a]All figures in billions of 1990 dollars.
[b]Less than 0.005 billion.

policy, though it is important to note that the US and Western Europe received the highest net change overall and the greatest gains from permit trading. There is, of course, the possibility that some countries not in our sample would have negative net benefits, but they would still be better off than if permit trading were not allowed.

Sovereignty criterion – Global efficiency

The global efficiency case is summarized in table 7. First, our model estimates that the economically efficient abatement level for our eight countries/regions is 14.9%; thereby implying a 20% reduction target overshoots the mark.[25] Individual country benefits are itemized in column 2 of table 7 and the associated mitigation costs in column 3. The estimates in these columns are all smaller than their counterparts in table 5; however, the net benefits in column 4 are not. This is because the case presented in table 7 reflects the globally efficient abatement requirement. Note also that no country suffers negative net benefits, even before trading, though there are significant disparities between countries.

The effects of trading are depicted in the last three columns of table 7. Again the US, Canada, and Western Europe would be permit buyers and the other five countries/regions would be permit sellers. Over 158 million permits would be traded with a total value of $4.38 billion (see table 6) at the price of $27.67 per ton of CO_2. The lower permit price than in the 20% cost-effectiveness case reflects the lower equilibrium abatement level. Again, permit trading lowers global abatement costs, in this case from $12.56 billion to $10.02 billion, or slightly more than 20%.

Sovereignty criterion – Cumulative reference base

The results of an application of the sovereignty criterion under a cumulative reference basis are presented in the first part of table 8. Recall that this reference base corresponds to the accumulation of CO_2 in the atmosphere, and thus the overall 20% CO_2 abatement requirement is apportioned according to each country's/region's contribution to the build-up over time.[26] The higher abatement cost estimates in column 2 of table 8 for the US and Western Europe, relative to their costs in column 4 of table 5, reflect the increased abatement requirements for these two areas.

Overall, pre-trading CO_2 abatement costs are higher by almost $5 billion because of the exponential nature of the abatement cost functions. The associated net benefits for our eight region globe are thus also lower by almost $5 billion relative to the example in table 5.

However, individual and aggregate abatement costs following trading are identical for the two sovereignty cases, as are the aggregate net benefits. This

[25]Our benefit estimates rely on Nordhaus' midrange estimate of damages equal to 1% of global GDP. Nordhaus believes that the upper bound of damages is equal to 2% of GDP. Sensitivity analysis of the global efficiency results when permit trading is allowed by the global warming protocol, using 0.5% and 2% of GDP as the raw damage estimates, yielded the following results: at 0.5% of GDP, global net benefits would be equal to $7.7 billion, the efficient percentage reduction in CO_2 would be about 9.6%, and permit prices would be $17.08 per ton; while at 2% of GDP, global net benefits would be $58.4 billion, the efficient percentage CO_2 reduction would be 22.4% and permit prices would be $43.68 per ton.

[26]Estimation of CO_2 emissions far back in time is tenuous at best, so we utilized the data series compiled by Smith et al. (1991), which covers the period 1950 to 1986, and then added our own projections to the year 2000.

Table 8

Changes in national welfare due to dynamic sovereignty and Rawlsian permit assignments[a] (cost-effectiveness case: 20% abatement).

Country (area)	Sovereignty cumulative reference case					Rawlsian cumulative reference case				
	Without trading		With trading			Without trading		With trading		
	Cost	Net	Cost	Trading	Net	Cost	Net	Cost	Trading	Net
US	12.76	4.22	3.64	6.22	7.12	17.10	−0.12	3.64	8.26	5.08
Canada	1.04	0.03	0.30	0.51	0.26	1.40	−0.33	0.30	0.68	0.09
Western Europe	10.08	0.19	2.08	4.87	3.32	13.51	−3.25	2.08	6.23	1.96
CIS	2.33	−0.05	7.72	−7.17	1.73	2.33	−0.05	7.72	−7.17	1.73
Brazil	0.07	1.45	0.53	−0.73	1.72	0.07	1.45	0.53	−0.73	1.72
Central Africa	0.01	0.13	0.02	−0.02	0.14	0.001	0.14	0.02	−0.04	0.16
Indonesia	0.03	0.44	0.30	−0.44	0.61	0.002	0.47	0.30	−0.60	0.77
China	1.22	4.16	3.72	−3.24	4.90	0.07	5.30	3.72	−6.63	8.29
Total	27.54	10.57	18.31	0.00	19.80	34.48	3.61	18.31	0.00	19.80

[a]All figures in billions of 1990 dollars.

is caused by the uniqueness of the efficient trading outcome, despite a different number of permits traded – 306 million in the cumulative reference case vs. 205 million in the current reference case (see again table 6). This means that aggregate gains from trading are significantly higher in the cumulative case, amounting to more than 33% of the $27.54 billion initial level.

It is also important to look at the distribution of gross and net costs, permit revenues and expenditures, and net benefits between the current and cumulative reference cases. Before trading, the CIS suffers net losses, but trading helps it achieve positive net benefits. The CIS and China gain the most over the current reference case, through receipt of permit revenues, while the US and Europe have relatively lower net benefits because of higher permit expenditures.

Rawlsian criterion – Cumulative reference base

This case accentuates the shift of the initial abatement burden to industrialized nations by way of assigning more permits to developing countries, in this case the three lesser-developed countries (LDCs) in our sample. The Rawlsian rule favors the most disadvantaged group, in principle, though there is no specific guidance on exactly how much. Our approximation is to increase permit assignments from 80% of projected year 2000 emissions to 95% for LDCs. The permits assigned to the three Western industrialized nations decrease in number so as to still attain an exact 20% global reduction. We then make the same adjustment as in the previous case for cumulative emissions. Interestingly, the initial control requirements for industrialized countries do not increase much (they are: 23.8% for cumulative sovereignty, 25.8% for the current Rawlsian, and 27.4% for the cumulative Rawlsian). The initial control requirements for LDCs do drop sharply, however (13% to 5% to 3.2% for China for the three cases, respectively).

The initial results are presented in the second part of table 8. Control costs do go up dramatically to nearly $35 billion, with global net benefits less than $4 billion. Moreover, all four industrialized nations stand to incur negative net benefits.

As in the other cases we examined, trading improves the situation greatly. Permits for over 400 million tons of CO_2 would be traded with a value of $15.2 billion. Post-trading abatement costs are the same as for the other two cost-effectiveness cases examined (and for current and cumulative emission reference bases for all ten equity criteria). Global net benefits increase dramatically to nearly $20 billion, and all countries receive positive net benefits. China is clearly the largest net winner in this case and the US and Western Europe do not fare as well as in the other cases examined. Note that the net benefits for Brazil and the CIS are the same in both cases in table 8, since the Rawlsian adjustment to permit assignments does not apply to them.

5.b. Interpretation of the results

We have obtained several important findings. First, gains from permit trading are likely to be substantial, especially for the cost-effectiveness case. Abatement cost savings are on the order of 25% for the current reference base case (the sovereignty criterion) and 33% for its cumulative counterpart. They are nearly 50% in the Rawlsian cumulative reference case. While there is sizable uncertainty in the data used, the results represent relative improvements and are therefore less susceptible to variations in the data than are absolute levels of cost savings. Our more tenuous estimates of these levels are that post-trading CO_2 abatement yields positive aggregate net benefits on the order of $20 billion for the cost-effectiveness case and $22 billion for the global efficiency case. Also, permit trading will result in transfers in the tens of billions of dollars per year to developing countries in the cost-effectiveness cases.

A definitive finding is that trading substantially erodes the differences in real world implications of various equity criteria. In a comparison of the four cases we examined (and several others we could not present for lack of space), there is a striking uniformity in the final outcomes of individual country abatement levels and a strong similarity in net welfare changes. This is due to the uniqueness of the Coase Theorem equilibrium. That is, no matter how permits are assigned there is a single, least-cost mix of abatement levels between countries. Put another way, this means that post-trading abatement costs for any country will always be the same for a given CO_2 reduction target no matter what the initial allocation of permits. Since these abatement costs will be a major proportion of the net costs (including permit transactions) this feature will significantly reduce the variance between nations. Only those countries engaging in large permit transactions will have noticeably varying net welfare changes. In addition, we have found that because the global efficiency and cost-effectiveness abatement levels are fairly similar, the welfare changes associated with these two options will be similar as well. All of this has very important policy implications. It, in effect, narrows the distinct number of equity criteria that may need to be examined, and hence could speed the negotiation process. Two countries may differ in their first choice of equity principles to apply to permit assignments, but if the two principles have the same welfare implications, a deadlock can be avoided.

We emphasize that these results do *not* mean that initial permit assignments do not matter or that equity is not an actionable policy criterion. In our empirical analyses, the outcomes for some countries likely to engage in permit trading on a large scale, such as the US and China, were significantly different across criteria and reference bases. Moreover, relatively high initial permit assignments may give some countries a greater sense of security about this policy instrument, thereby making their participation in an international agreement more likely. What the results do mean are that marketable permits have some inherent limitations, if redistribution is a major policy objective. No matter how large a revenue gain is forthcoming from the sale of a sizable allocation of permits, there is a partially offsetting cost incurred in undertaking the abatement necessary to make the sale legal; moreover, this offset level is constant over all permit assignments.

There are several other important findings. None of the eight countries/regions that we examined would appear to be a net loser. In fact, Brazil and Indonesia receive permit revenues in excess of their abatement costs. Even if we have grossly overestimated the benefits for these countries, there is no question that their welfare will be improved as a result of global warming policies like those we examined.

Our efficiency analysis identified 14.9% as the global optimum abatement level in our eight country/region world. Net welfare would be $2 billion, or 10%, higher than with a 20% target. Thus it could be that the oft-espoused 20% reduction target may be too high, though we caution that this result could change if all countries were considered.

Of course, some of our empirical results might change significantly if we were to have explicitly addressed some other critical issues, such as the free-rider problem. We plan to investigate considerations such as this in future work.[27]

6. Conclusion

We have performed a conceptual and empirical analysis of a tradeable permits approach to global warming policy. We first derived some welfare measures associated with different policy cases and then laid out operational

[27]There are some other considerations briefly worth noting. First, we have assumed that the caveats to the Coase Theorem are not significant in the cases examined. The transactions costs associated with negotiating a greenhouse gas treaty alone will not undercut the efficacy of the marketable permit approach, and, although monitoring and enforcement portions of such costs are non-trivial, they would not affect our results either. Also, welfare changes of less than 3% of GDP will not generate any significant income effects at the national level. At the same time, we acknowledge that the use of a monetary common denominator appears to give billion dollar welfare changes equal weight in, say, the US and China. We have, however, stated several of our equity criteria in terms of proportions (of GDP), though this still does not incorporate any considerations of differing marginal utilities of income between countries. Finally, we note that our results, which consistently show transfers from industrialized nations to transitional and developing countries, are highly dependent on the relative shapes of abatement cost curves. Although studies generally find marginal cost curves to be steeper in industrialized nations than in others, there may be some important exceptions outside the range of the 8 countries/regions we examined.

rules for the assignment of permits under a diversity of equity criteria. Our empirical analysis involved the application of a non-linear programming model to analyze the costs and benefits before and after permit trading.

We found that net benefits are positive and on the order of $20 billion for the eight countries/regions we examined. Trading could involve transfers of tens of billions of dollars towards developing and transitional economies. Trading also helps arrive at a much more efficient outcome.

Our most important finding is that the post-trading outcome is rather similar between different initial assignments of permits. The reason is the uniqueness of the efficient equilibrium. Post-trading abatement levels and abatement costs are the same in each case, with only the level of transfers being different. This means that, although various equity criteria would appear to have significant philosophical and political differences, their implications for the welfare of nations may be very similar in many instances. This fact should facilitate negotiations by minimizing tensions over alternative initial permit assignments.

References

Agarwal, A. and S. Narain, 1991, Global warming in an unequal world: A case of environmental colonialism (Centre for Science and Environment, New Delhi).

Barrett, S., 1990, The problem of global environmental protection, Oxford Review of Economic Policy 6, 68–79.

Barrett, S., 1991, Reaching a CO_2 emission limitation agreement for the community: Implications for equity and cost-effectiveness, Report to OECD Environment Directorate (London Business School).

Barrett, S., 1992, Acceptable allocations of tradeable carbon emission entitlements in a global warming treaty, in: S. Barrett et al., Combating global warming.

Barrett, S., M. Grubb, K. Roland, A. Rose, R. Sandor and T. Tietenberg, 1992, Combating global warming: A global system of tradeable carbon emission entitlements (UNCTAD, Geneva, Switzerland).

Bergesen, H., 1991, A legitimate social order in a 'greenhouse' world: Some basic requirements, International Challenges 11, 21–30.

British Petroleum, 1988, BP statistical review of world energy (BP, London, UK).

Burtraw, D. and M. Toman, 1991, Equity and international agreements for CO_2 containment, Discussion Paper ENR 91-07 (Resources for the Future, Washington, DC).

Chandler, W. (ed.), 1990, Carbon emissions control strategies (World Wildlife Fund and The Conservation Foundation, Washington, DC).

Chapman, D. and T. Drennen, 1990, Economic dimensions of CO_2 treaty proposals, Contemporary Policy Issues 8, 16–28.

Cline, W., 1992, The economics of global warming (Institute for International Economics, Washington, DC).

Coase, R., 1960, The problem of social cost, Journal of Law and Economics 3, 1–44.

d'Arge, R., 1989, Ethical and economic systems for managing the global commons, in: D. Botkin et al., eds., Changing the world environment (Academic Press, New York, NY).

Darmstadter, J. and A. Plantinga, 1991, The economics of CO_2 mitigation: A review of estimates for selected world regions, Discussion Paper ENR91-06 (Resources for the Future, Washington, DC).

Energy Information Administration, 1988, International energy annual (EIA, Washington, DC).

Epstein, J. and R. Gupta, 1991, Controlling the greenhouse effect: Five global regimes compared (The Brookings Institution, Washington, DC).

Euromonitor, 1991, European marketing data and statistics 1991 (Euromonitor Publications, London).

European Community, 1990, Cost effectiveness analysis of CO_2 reduction options: Report for the commission of the European Community, Paper presented at a Workshop on Economic/Energy/Environmental Modeling for Climate Policy Analysis (Washington, DC).

Flavin, C., 1990, Slowing global warming: A worldwide strategy, Worldwatch Paper 91 (Worldwatch Institute, Washington, DC).

Fujii, Y., 1990, An assessment of the responsibility for the increase in the CO_2 concentrations and inter-generational carbon accounts, IIASA Working paper WP-05-55.

Grubb, M. and J. Sebenius, 1991, Participation, allocation and adaptability in international tradeable emission permit systems for greenhouse gas control, Paper prepared for OECD Workshop on Tradeable Greenhouse Gas Permits (Paris, France).

Hahn, R. and G. Hester, 1989, Marketable permits: Lessons for theory and practice, Ecology Law Quarterly 16, 361–406.

Hoeller, P., A. Dean and J. Nicolaisen, 1990, A survey of studies of the costs of reducing greenhouse gas emissions, Organization for Economic Cooperation and Development, Department of Economics and Statistics Working Paper No. 89.

Houghton, R., R. Boon, J. Fruci, J. Hobbie, J. Melillo, C. Palmer, B. Peterson, G. Shaver and G. Woodwell, 1987, The flux of carbon from terrestrial ecosystems to the atmosphere in 1980 due to changes in land use: Geographic distribution of the global flux, Tellus 39B, 122–139.

Intergovernmental Panel on Climate Change, 1991, Climate change: The IPCC scientific assessment (revised) (Cambridge University Press, New York, NY).

Kosobud, R. and T. Daly, 1984, Global conflict or cooperation over the CO_2 climate impact, Kyklos 37, 638–659.

Manne, A. and R. Richels, 1991, Global CO_2 emission reductions: The impacts of rising energy costs, Energy Journal 12, 87–107.

Morris, S., B. Solomon, D. Hill, J. Lee and G. Goldstein, 1990, A least-cost energy analysis of U.S. CO_2 reduction options, in: J. Tester et al., eds., Energy and the environment in the 21st century (MIT Press, Cambridge, MA) .

Morrisette, P. and A. Plantinga, 1991, The global warming issue: Viewpoints of different countries, Resources 103, 2–6 (Resources for the Future, Washington, DC).

Nordhaus, W., 1989, The economics of the greenhouse effect, Paper presented for the Workshop on Energy and Environmental Modeling and Policy Analysis, MIT Center for Energy Policy Research, July 31–August 1.

Nordhaus, W., 1991a, The cost of slowing climate change: A survey, Energy Journal 12, 35–67.

Nordhaus, W., 1991b, To slow or not to slow: The economics of the greenhouse effect, Economic Journal 101, 920–37.

Nordhaus, W., 1993, Rolling the DICE: The optimal transition path for controlling greenhouse gases, Resource and Energy Economics 15, 27–50.

Parry, M., 1990, Climate change and world agriculture (Earthscan Publications, London, UK).

Peck, S. and T. Teisberg, 1992, CETA: A model for carbon emissions trajectory assessment, Energy Journal 13, 55–77.

Poterba, J., 1991, Tax policy to combat global warming, in: R. Dornbusch and J. Poterba, eds., Global warming: Economic policy responses (MIT Press, Cambridge, MA).

Rose, A., 1990, Reducing conflict in global warming policy: Equity as a unifying principle, Energy Policy 18, 927–935.

Rose, A., 1992, Equity considerations of tradeable carbon entitlements, in: S. Barrett et al., Combatting global warming.

Rose, A. and D. Webber, 1993, Interregional equity and acid rain policy, Regional Studies (forthcoming).

Rose, A., B. Stevens and G. Davis, 1988, Natural resource policy and income distribution (Johns Hopkins University Press, Baltimore, MD).

Schelling, T., 1991, Economic responses to global warming: Prospects for cooperative approaches, in: R. Dornbusch and J. Poterba, eds., Global warming: Economic policy responses (MIT Press, Cambridge, MA).

Smith, K., J. Swisher, R. Kanter and D. Ahuja, 1991, Greenhouse gas indices: Crucial link between science and policy (East-West Center, Honolulu, HI).

Solomon, B. and D. Ahuja, 1991, An equitable approach to international reductions of greenhouse gas emissions, Global Environment Change 1, 343–350.

Spash, C. and R. d'Arge, 1989, The greenhouse effect and intergenerational transfers, Energy Policy 17, 88–96.

Stevens, B. and H. Vaux, Jr., 1990, Assessing strategies for contol of irrigation-induced salinity in the upper Colorado River Basin: Comment, American Journal of Agricultural Economics 72, 493–496.

Task Force on the Comprehensive Approach to Climate Change, 1991, A comprehensive approach to addressing potential climate change, (U.S. Department of Justice, Environment and Natural Resources Division, Washington, DC).

Tietenberg, T., 1985, Emissions trading: An exercise in reforming pollution policy (Resources for the Future, Washington, DC).

Tietenberg, T., 1992, Environmental and natural resource economics, 3rd edition (Harper Collins, New York, NY).

U.S. EPA, 1990, Policy options for stabilizing global climate, Draft report to Congress, Office of Policy, Planning, and Evaluation (U.S. Environmental Protection Agency, Washington, DC).

Whalley, J. and R. Wiggle, 1991, Cutting CO_2 emissions: The effects of alternative policy approaches, Energy Journal 12, 109–24.

World Bank, 1991, World development report 1991 (Oxford University Press, New York, NY).

World Resources Institute, 1990, World resources 1990/91 (World Resources Institute, Washington, DC).

Young, H., 1991, Sharing the burden of global warming, Equity and Global Climate Change Project, School of Public Affairs, University of Maryland.

[22]

RELEVANT EXPERIENCE WITH TRADEABLE ENTITLEMENTS

Tom Tietenberg*

Introduction

One useful point of departure for crafting a tradeable entitlement approach to controlling greenhouse gases is provided by the lessons derived from existing applications of this economic incentive approach. Though the translation of these experiences from their purely national settings to the international setting will not be simple, the lessons are instructive. They help to clarify the key design issues and to avoid potential pitfalls by anticipating and resolving them.

The tradeable entitlement or permit approach to pollution control was first introduced in the United States where it has received its most vigorous application up to date. Beginning in 1975, burgeoning costs associated with the rigidities inherent in its traditional, predominantly legal, approach to controlling air pollution led the United States Environmental Protection Agency (EPA) to begin experimenting with a transferable permit approach now known as the Emissions Trading Program. Since that time the tradeable entitlement (permit) concept has been applied to several new areas of environmental policy and is currently being proposed for several more.

Support for the use of this market approach to environmental control has clearly grown in the United States, as reflected in the favourable treatment it has received both in the popular business[1] and environmental[2] press. Some public interest environmental organizations have even adopted tradeable entitlement approaches as a core part of their strategy for protecting the environment.[3] As described below this entitlement approach is also spreading to other nations.

Our knowledge about tradeable entitlements has grown rapidly during the last two decades in which it has received serious analytical attention. Not only have the theoretical models become more focused and the empirical work more detailed, but we have now had over a decade of experience with several limited but operational versions of this approach. It seems a propitious time to stand back and to organize what we have learned about this practical and promising approach to pollution control.

This chapter draws upon economic theory, empirical studies, and actual experience with implementing the various versions of tradeable entitlement programmes to provide

* Department of Economics, Colby College, Waterville, Maine.

1 See, for example, Main (1988).

2 See, for example, Stavins (1989).

3 See the various issues in volume XX of the *EDF Letter*, a report to members of the Environmental Defense Fund.

a brief overview of some of the major lessons that are instructive in crafting a similar approach tailored to the control of greenhouse gases.[4] Some of the proposed extensions to new settings will also be discussed in order to suggest the range of applicability.

A. Current experience

1. *Transboundary application*

The ozone shield in the stratosphere protects humans from harmful ultraviolet radiation. Current evidence suggests that this shield is being depleted by man-made chemicals such as chlorofluorocarbons and halogens, and the resulting increase in ultraviolet rays raises the incidence of both skin cancer and damage to eyesight. These gases also contribute to global warming.

Responding to this threat, 24 nations signed the Montreal Protocol in September 1988. According to this agreement signatory nations have to restrict their production and consumption of the chief gases responsible to 50 per cent of 1986 levels by 30 June 1998. In one of the provisions of the agreement, production quotas are transferable both within and between nations. Since transferable production quotas have some similarities in design and concept to transferable carbon dioxide entitlements, this is an important international precedent.

2. *Applications in the United States*

(a) *Conventional air pollutants*

Stripped to its essentials, the foundation of the United States approach to pollution control is based upon a traditional command-and-control regulatory system. Ambient standards, which establish the highest allowable concentration of each conventional pollutant in the ambient air or water, represent the targets of this approach. To reach these targets, emission or effluent standards (legal discharge ceilings) are imposed on a large number of specific discharge points such as stacks, vents, outfalls or storage tanks. Following a survey of the technological options of control, the control authority selects a favoured control technology and calculates the amount of discharge reduction achievable by that technology as the basis for setting the emission or effluent standard. Technologies yielding larger amounts of control (and, hence, supporting more stringent standards) are selected for emitters in areas where it is very difficult to meet the ambient standard and also for new emitters. The responsibility for defining and enforcing these standards is shared in legislatively specified ways between the national government and the various state governments.

In an attempt to inject more flexibility into the manner in which the objectives of the Clean Air Act were met during the last half of the 1970s, the United States EPA created what has now become known as the Emissions Trading Program. The programme attempts to facilitate compliance by allowing sources a much wider range of choice in how they satisfy their legal pollution control responsibilities than is possible under the command-and-control approach. Any source choosing to reduce emissions at any discharge point to a greater extent than required by its emission standard can apply to the control authority for certification of the excess control as an "emission reduction credit" (ERC). Defined in terms of a specific amount of a particular pollutant, the certified emission reduction credit can be used either to satisfy emission standards at other discharge points (presumably more expensive to monitor) controlled by the creating source, or it can be sold to other sources. By making these

[4] In the limited space permitted by this chapter only a few highlights can be illustrated. All the details of the proofs and the empirical work can be found in the references listed at the end of the chapter. For a comprehensive summary of this work see Tietenberg (1980), Tietenberg (1985), Dudek and Palmisano (1988) and Hahn and Hester (1989).

credits transferable, the EPA has allowed sources to find the cheapest means of satisfying their requirements, even if the cheapest means are under the control of another firm. The ERC is the currency used in emissions trading, while the offset, bubble, emissions banking and netting policies govern how this currency can be stored and spent.[5]

The offset policy requires major new or expanding sources in "nonattainment" areas (those areas with air quality worse than the ambient standards) to secure sufficient offsetting emission reductions (through the acquisition of ERCs) from existing firms to make the air cleaner after their entry or expansion.[6] Prior to this policy no new firms were allowed to enter nonattainment areas on the grounds that they would interfere with the attainment of the ambient standards. By introducing the offset policy, EPA allowed economic growth to continue while ensuring progress towards attainment.

The bubble policy receives its unusual name from the fact that it treats multiple emission points controlled by existing emitters (as opposed to those expanding or entering an area for the first time) as if they were enclosed in a bubble. Under this policy only the total emissions of each pollutant leaving the bubble are regulated. While the total leaving the bubble must be no larger than the total permitted by adding up all the corresponding emission standards within the bubble (and, as a result of recent changes in the rules, must be 20 per cent lower), emitters are free to control some discharge points less than dictated by the corresponding emission standard as long as sufficient compensating ERCs are obtained from other discharge points within the bubble. In essence, sources are free to choose the mix of control among the discharge points as long as the overall emission reduction requirements are satisfied. Multi-plant bubbles are allowed, opening the possibility for trading ERCs among very different kinds of emitters.

Netting allows modifying or expanding sources (but not new sources) to escape from the need to meet the requirements of the rather stringent new source review process (including the need to acquire offsets), so long as any net increase in emissions (counting any ERCs earned elsewhere in the plant) is below an established threshold. In so far as netting allows firms to escape particular regulatory requirements by using ERCs to remain under the threshold that triggers applicability, it is more properly considered regulatory relief than regulatory reform.

Emissions banking, the final component, allows firms to store certified ERCs for subsequent use in the offset, bubble or netting programmes or for sale to others. The banking component provides key institutional support for the programme by serving as an easily accessible storehouse of credits available for sale, and as a means of facilitating trades when demand and supply in respect of credits are not perfectly synchronized.

(b) *Lead in gasoline*

Since the inception of the Emissions Trading Program, the marketable entitlement concept has been applied in a number of new and rather different areas, demonstrating its versatility. One of those applications involved a programme to facilitate a rapid transition to the widespread availability of unleaded gasoline.

In the month prior to the issuance of new and more stringent regulations on lead in gasoline, EPA announced the results of a cost-benefit analysis of their expected impact. The analysis concluded that the proposed standard of 0.01 grams per leaded gallon (gplg) would result in $36 billion (1983 dollars) in benefits (from reduced adverse health effects), at an estimated cost to the refining industry of $2.6 billion. The regulation was unquestionably justified on efficiency grounds.

A main concern of EPA in issuing these regulations was the rigidity of the intermediate deadlines used to implement the phase-down. While some refiners could meet early deadlines with ease, others could do so only at a significant increase in cost. Recognizing that meeting the goal did not require every refiner to meet each deadline, EPA initiated the lead banking programme to provide additional flexibility in complying with the regulations.

While not a part of the Emissions Trading Program discussed above, the lead bank-

5 The details of this policy can be found in "Emissions trading policy statement" 51 *Federal Register* (4 December 1986):43829. This document is based upon a decade of experience with the programme. For a description of its evolution see Tietenberg (1986).

6 Offsets are also required for major modifications in areas that have attained the standards if the modifications jeopardize attainment.

ing programme did share some design similarities with it. Under this programme, refiners reducing lead more than required by the applicable standard in each quarter of the year could bank the credits for use or sale in some subsequent quarter. Banked credits were fully transferable among refiners.

The lead banking programme, though plagued by less-than-perfect implementation procedures, eased the transition to the new and more stringent regulatory regime.[7] Refiners had an incentive to respond quickly because reductions in lead undertaken prior to the deadlines became valuable under the new programme. Acquiring these credits made it possible for other refiners to comply with the deadlines, even in the face of equipment failures or Acts of God; fighting the deadlines in court - the traditional response - was no longer necessary. Designed purely as a means of facilitating the transition to this new regime, the lead banking programme ended as scheduled on 31 December 1987.

The lead banking programme shows how marketable entitlement systems can be used to facilitate the transition to more strict regulatory regimes. By providing more flexibility coupled with incentives for meeting the deadlines early, it respected differences in the capabilities of polluters to meet stringent deadlines without sacrificing the overall environmental objectives. Since deadlines will be an essential part of any international approach to controlling global warming, this is a useful attribute.

(c) Ozone-depleting chemicals

On 12 August 1988 the United States Environmental Protection Agency issued regulations implementing a marketable permit system to achieve the United States targeted reductions.[8] According to these regulations all major United States producers and consumers of the controlled substances were allocated baseline allowances for which 1986 levels were used as the basis for the pro rata calculations. Each producer and consumer is allowed 100 per cent of this baseline allowance for each year before 1 July 1993, 80 per cent between 1 July 1993 and 30 June 1998 and 50 per cent after 30 June 1998. These allowances are transferable within producer and consumer categories. Production allowances can be transferred across international borders to producers in other signatory nations provided that the transaction is approved by EPA and results in the appropriate changes in the buyer's and seller's production allowances in their respective countries. Production allowances can be augmented by demonstrating the safe destruction of an equivalent amount of controlled substances by approved means, but so far no means have been approved.

Though the EPA chose to allocate the production allowances on the basis of historical levels of use, it was clearly uneasy about this choice.[9] Since the demand for these allowances is quite inelastic, supply restrictions increase revenue. By allocating allowances to the seven major domestic producers of CFCs and halons, EPA was concerned that its regulatory action would result in sizeable windfall profits (estimated to be in the region of billions of dollars) for those producers. It was further concerned that the existence of these windfalls would create an economic incentive for producers to delay the introduction of more environmentally benign chemical substitutes.

Confronted with scientific evidence that the reduction in CFC use, mandated by the Montreal Protocol, might not be sufficiently stringent, concerned about the excess profits to which the allowance holders would be entitled and seeking new sources of revenue to counteract a growing budget deficit, the United States Congress complemented this marketable permit or entitlement approach with a special tax on ozone-depleting chemicals.[10] The amount of the tax per pound of the chemical sold or used is equal to a base tax multiplied by an "ozone-depleting factor" specific to each chemical. These factors range from a low of 0.8 for CFC-113 to 10.0 for

---

7 For some of the details of the programme as well as a description of some of the implementation difficulties, see Nussbaum (1991).

8 53 *Federal Register* 30598 (12 August 1988). The final rules governing transfer of import and production allowances are in 54 *Federal Register* 6376 (9 February 1989). These rules have been consolidated in 40 *Code of Federal Regulations* 82.

9 When issuing the rules for the ozone depletion gas permit system, EPA sought public comment on the possibility of ultimately replacing its free distribution of permits on the basis of historical use by an auction market. Though EPA never followed up on this possibility (a tax was imposed instead), it demonstrates a recognition of, and concern for, the wealth effects inherent in a restrictive, cost-effective system of pollution control.

10 Ozone-depleting chemicals are defined in the Act as CFC-11, CFC-12, CFC-113, CFC-114, CFC-115, Halon-1211, Halon-1301 and Halon-2402. See 103 Stat 2365.

Halon-1301. The base tax rates rise from \$1.37 in 1991 to \$2.65 in 1994, with further annual increases of 45 cents each year thereafter.

One of the useful lessons for controlling global warming that is to be drawn from this application is that it demonstrates the complementarity of using transferable entitlements in conjunction with fees. When used together, both are allowed to fulfil the roles for which they are best suited - entitlements produce the desired control of quantities emitted while fees raise revenue.

(d) Water pollution

Virtually all water pollution control policy in the United States has been focused on point sources: large identifiable polluters emitting effluents from specific discharge points. As the controls on point sources have been tightened, nonpoint sources (such as the overflow from urban storm sewers or the runoff from farms) become responsible for a larger share of the remaining pollution. Although nonpoint sources are now responsible for the bulk of the problem that still subsists, they are notoriously difficult to regulate by traditional means.

Because the marginal cost of controlling nonpoint sources is now considerably lower than the marginal cost of securing further reductions from point sources, ways of bringing nonpoint sources under control are of great interest. One result of the attention paid to them has been the introduction of an option to trade waste reduction entitlements between point and nonpoint sources for bodies of water where further control is required.

One example of this approach has occurred in the Lake Dillon Watershed, a primary source of water for Denver, Colorado.[11] Changing land uses and development have produced near-eutrophic algae in the lake, despite the fact that all waste water treatment plants on the lake currently employ advanced methods for phosphorus removal. Even with zero phosphorus discharge from treatment plants in the future, nonpoint phosphorus sources from growth areas would make the lake eutrophic.

Transferable permits (entitlements) were introduced as part of an overall management plan for the lake. According to this plan all waste treatment plants would continue to apply advanced waste water treatment, and state-of-the-art nonpoint source controls would be applied by local governments to all new development projects in the basin. An effluent trading system was developed to accommodate expected source growth beyond 1990 by encouraging reductions for nonpoint sources that existed prior to 1984 and therefore were not subject to the new regulations. Authorized by the State Commission, the trading rules require a 2:1 ratio to account for uncertainties in the process. One pound of waste reduction credit is certified for every two pounds of phosphorus eliminated from pre-existing nonpoint sources. This approach is expected to keep the phosphorus loading at 1982 levels while accommodating expected source growth. It allows local governments to control phosphorus in the most cost-effective manner without being limited to further reductions in point sources.

The nonpoint source application demonstrates how a marketable entitlement system can reduce costs by bringing previously unregulated sources under control. By coupling economic incentives for controlling new sources with procedures forcing traders to bear a rather heavy burden of proof in demonstrating the magnitude and enforceability of the proposed reductions, regulators can spread the control burden over a larger number of sources while assuring the integrity of the process.[12]

(e) Acid rain

A version of the emissions trading concept has also been incorporated into the United States approach for achieving further reductions in pollutants contributing to acid rain from electric utilities.[13] Under this inno-

---

[11] See Northwest Colorado Council of Governments (1984).

[12] As the political forces are gearing up for the periodic reauthorization of the United States Clean Water Act, transferable permits (entitlements) are being discussed in three different contexts. First, significant interest has materialized in generalizing and expanding the point-nonpoint trading approach as illustrated by the Dillon Reservoir case. Second, the use of transferable development rights as a means of protecting wetlands is being discussed. Finally, the possibility of trading responsibilities for pretreating waste entering into publicly owned sewage treatment plans is also being explored. At this stage it is impossible to know which, if any, of these ideas will find their way into the revised law. (This information was gathered by means of a telephone conversation with Dr. William O'Neil, United States Environmental Protection Agency, 7 June 1991.)

[13] This title was included in the Clean Air Act Amendments of 1990, which became law on 15 November 1990. See

vative approach allowances to emit sulphur oxides will be allocated to older plants, and the number of allowances will be restricted to assure a reduction of 10 million tons in emissions from 1980 levels. Although these allowances are defined for a specific calendar year and cannot be used prior to that year, the Act contains provisions for carrying unused allowances forward into the next year. Unused allowances are transferable among the affected sources. Any plant reducing emissions more than required by its initial allocation of allowances could transfer the unused allowances to other plants. These acquired allowances can be used by the owners of the purchasing source to ensure that its emissions do not exceed the levels permitted by the allowances (allocated plus acquired) that it holds. The Act even anticipates the possibility of interpollutant trading between sulphur and nitrogen oxides by requesting a study of its feasibility.[14]

Adopting this approach resulted from the Congressional belief that the acid rain reduction goals could be achieved at a much lower cost than would otherwise be possible. Utilities that could afford to reduce their emissions relatively cheaply would do so, selling or leasing any excess reduction to those for whom further reductions would be much more expensive. Both the plants and their customers could be made better off by the allowance transfer. The allowances would go where they could be best used, minimizing the overall cost of control in the process, while the legislatively mandated total emission reduction would be held constant.

The acid rain application is particularly interesting for this study because the acid rain tradeable allowances treat sulphur oxides in the same manner that greenhouse gases would be treated in a global transferable entitlement market designed to control global warming, i.e. tradeable on a one-for-one basis. The expectations of a vigorous market in these acid rain allowances are sufficiently high for a spot market in these allowances to exist already and the Chicago Board of Trade expects to open a forward market for allowances in 1992-1993.[15]

3. *Applications in Germany*[16]

Germany has used a limited form of transferable emission entitlements for a number of years. Under this programme plant operators are allowed to deviate from strict standards, providing that ambient air concentrations are reduced. The first programme (governed by the plant renewal clause) allows new plants seeking to locate in areas where allowable ambient concentrations have been exceeded to acquire emission offsets from the renovation or renewal of old plants; without these compensating reductions the establishment of new plants in these areas would not be allowed. In the second programme authorities can deviate from the new licensing procedure required for plants undergoing renewal, providing greater reductions are achieved from the proposed deviation (including the acquisition of offsets) than would have been achieved from the follow-up order. A unique characteristic of the German programme is that it does not limit the offsetting reductions to the same pollutant; reductions of other pollutants can serve as offsets providing they have comparable effects on the environment.

4. *Applications in Canada*[17]

An intra-utility form of trading (similar to the United States bubble concept) is already allowed in Ontario. The utility involved is required to meet an aggregate emissions cap, but is given great flexibility in how it chooses to fulfil that responsibility.

Canada is currently investigating the use of transferable permits (entitlements) in two new contexts. The first application would

104 Stat 2584. For an analysis of the application of the emissions trading concept to acid rain, see Atkinson (1983); Raufer and Feldman (1984); Oates and McGartland (1985); Feldman and Raufer (1987); and Tietenberg (1989).

14 104 Stat 2591.

15 Remarks by Mr. Richard Sandor, Chairman of the Clean Air Committee of the Chicago Board of Trade at the UNCTAD Expert Workshop on a Global Scheme for Tradeable Carbon Emission Entitlements, Geneva, Switzerland, 31 October-1 November 1991.

16 Opschoor and Vos (1989).

17 The information in this section was derived from a telephone conversation with Erik Haites, a principal with Barakat & Chamberlin, the consulting firm involved in examining the feasibility of these proposals.

allow the trading of NO_X and VOC credits within each of Canada's three ozone nonattainment areas (Vancouver, Ontario and Quebec, and southern New Brunswick). The second application would involve SO_2 trading.

5. Applications in Eastern Europe and Asia

Interestingly the transferable entitlement approach has spread to the less industrialized countries. In crafting its environmental policy for the future, Poland is committed to the introduction of a form of transferable entitlements patterned on the emissions trading system in the United States.[18] This choice was justified on the grounds that transferable entitlements provide a degree of environmental security possible only with a quantity-based system of controls, while offering polluters the flexibility for cost-effective achievement of those goals. Interestingly this choice followed an investigation (and rejection) of emission charges.

According to press reports China has also recently adopted a transferable entitlement system of pollution control to complement its programme of fines for violations of the entitlements, though the details of the system remain sketchy.[19]

6. A proposed application in the United Kingdom[20]

The United Kingdom is considering applying a variant of a marketable entitlements approach to promote energy efficiency and pollution control in the transport sector.[21] The basic proposal involves the setting of a fuel efficiency target for manufacturers and importers to achieve in the aggregate from sales of new vehicles. As each new car is registered, it would qualify for a credit if it performed better than a prespecified baseline standard. The credit would be issued to the new car manufacturer or importer by the regulatory authority. Conversely manufacturers or importers would have to deposit sufficient credits to offset the amounts by which new vehicles exceeded the baseline. Credits could be acquired within the company from the rest of the fleet whose performance improved upon the baseline. They could also be purchased from other manufacturers or importers, or acquired from banks where excess credits had been deposited by those who generated them.

One of the interesting characteristics of this system is that it demonstrates how a transferable entitlement market can be implemented even for pollution sources that are small but numerous (such as automobiles). In this case the suggested solution is to apply the programme at the point of production rather than at the point of use. Because the number of producers and importers is small, transactions costs can be kept acceptably small.

7. Applications other than pollution control

(a) Fisheries in New Zealand[22]

A quite different application of the tradeable entitlement concept has been used in New Zealand to reduce the stress on an already overexploited renewable resource. Some particularly desirable fish species in New Zealand were being seriously overfished. While the need to reduce the amount of pressure on the fish population was obvious, how to accomplish that aim was not at all clear. Although it was relatively easy to prevent new fishermen from entering the fisheries, it was harder to decide how to induce persons who had been fishing in the area for many years to limit their catch. As fishing is characterized by economies of scale, reducing each person's catch proportionately would not make much sense. It would simply raise everyone's costs and waste a great deal of fishing capacity as boats remained idle for a significant amount of time.

---

18 Zylicz (1991).

19 Nadelson (1991).

20 Taylor (1991). The author is a member of the United Kingdom Department of Transport.

21 A similar scheme is now being considered in California. See Stevens (1991). The author of the California report is on the staff of the California Energy Commission.

22 See Campbell (1989) and Pearse (1991).

A better solution would clearly be to have fewer boats harvesting the stock. Each boat could then be used closer to its full capacity without depleting the fish population. The question then is which fishermen should be asked to give up their livelihood and leave the industry?

The tradeable entitlement approach addressed this problem by imposing catch quotas on fishermen for which they had to pay an annual fee. The revenues derived from this fee were used to buy out fishermen who were willing to forgo future fishing for the species in jeopardy. Essentially each one stated the lowest price that would be acceptable for leaving the industry; the regulators selected those who could be induced to leave at the lowest price, paid the stipulated amount from the tax revenues, and withdrew their licence to fish the species concerned. It was not long before a sufficient number of licences had been withdrawn and the population was protected. Since the programme was voluntary, those who left the industry did so only when they felt they had been adequately compensated. Meanwhile, those who paid the fee realized that this small investment would benefit them greatly in the future as the population recovered. Thus, a difficult and potentially dangerous pressure on a valuable natural resource was avoided by the creative use of a tradeable entitlement system.

(b) Milk production quotas in the EC[23]

To reduce the milk surplus produced as a result of price supports, in the mid-1980s every member state in the EC was assigned a production quota based on 1983 milk production minus 8.65 per cent. The milk quotas were tradeable, but the EC Council ultimately decided to restrict their transferability. The quotas were linked to the amount of land that a farmer used for the raising of dairy cattle in the reference year. By acquiring new land or selling currently owned land, the owner buys or sells quotas as well. With some anomalies apparently attributable to the restrictions on transfers,[24] the scheme has reportedly worked well.

B. Lessons

1. *Lessons from theory*

Theory can help us to understand the characteristics of this economic approach in the most favourable circumstances and assist in the designing of the instruments for maximum effectiveness.

(a) Assuming all participants are cost minimizers, a "well-defined" transferable entitlement market could cost-effectively allocate the control responsibility for meeting a predefined pollution target among the various sources of pollution despite incomplete information on the control possibilities by the regulatory authorities.

The reasoning behind this theorem is not difficult to grasp. Cost-minimizing firms seek to minimize total entitlement and control costs. Minimization will occur when the marginal cost of control is set equal to the entitlement price. Since all sources will see that their marginal control costs are equal to the same price, an entitlement market will equalize marginal control costs across all discharge points, precisely the condition required for cost effectiveness.[25] In essence,

---

23 Information supplied by Paul Koutstaal, Faculty of Law, Department of Economics, University of Groningen.

24 Because of the linkage between land and the quota, the system has had some peculiar side-effects that were never intended. For example, farmers in the southern part of the Netherlands, who wanted to enlarge their herds but could not do so because of the quotas, bought land in the northern provinces which they left idle. The quotas linked to this newly acquired land were then used to justify higher milk production in the south.

25 It should be noted that, while the allocation is cost effective, it is not necessarily efficient. It would only be efficient if the predetermined target happened to coincide with the efficient amount of pollution. Nothing guarantees this outcome.

sources needing extra emission reduction, either as a precondition to entering an area or to meet emission standards, can purchase transferable entitlements from sources with the lowest marginal costs of control.

Though derived from the rarified world of theory, the practical importance of this theorem should not be underestimated. Transferable entitlements offer a unique opportunity for regulators to solve a fundamental dilemma. Control authorities usually want to allocate the responsibility for control cost effectively, but they rarely have the information required to achieve this objective. Transferable entitlements create a system of incentives in which those who have the best knowledge about control opportunities - the environmental managers for the industries - are encouraged to use that knowledge to achieve environmental objectives at minimum cost. Information barriers do not preclude effective regulation.

(b) *Transferable permits encourage more technological progress in pollution control than the command-and-control system.*[26]

Traditional command-and-control policies usually base standards on technologies known to the regulators. Meeting those standards can be accomplished simply by adopting the identified technology. Adopters have little incentive to search for superior technologies.

With transferable entitlements, adopters have an incentive not only to search for new technologies that reduce the cost of compliance with the mandated standard, but also to search for technologies that can reduce emissions more than required by the standard. Selling the emission reduction credits (or unused allowances) produces revenue which can be used to finance the new technologies. Entitlements allocated on the basis of past emissions provide fewer incentives for technological progress than either emissions charges or auctioned entitlements, but more incentives than a command-and-control system.

(c) *What constitutes a "well-defined" transferable entitlement system depends crucially on the attributes of the pollutant being controlled.*[27] *Fortunately, the category containing greenhouse gases is the simplest to design and implement.*

To be consistent with a cost-effective allocation of the control responsibility, the entitlements would have to be defined in different ways for different types of pollutants. Three differentiating characteristics have been identified in the literature so far. First, entitlements designed to control pollutants that are uniformly mixed in the atmosphere (such as CFCs) can be defined simply in terms of a rate of emissions flow per unit of time. Secondly, entitlements to control pollutants (such as heavy metals) that build up in the environment over time would be defined in terms of a stock of emissions rather than an emissions flow. Lastly, entitlement design becomes quite complicated when the pollution target pursued is defined in terms of concentrations (such as sulphur oxides or particulates) measured at a number of specific receptor locations. In this case it would be necessary for the cost-effective design to take into account the location of the emissions (including injection height) as well as the amount of emissions. As greenhouse gases can be treated as falling within the first category, designing the entitlement is easiest for this type of pollutant.

(d) *As long as markets are competitive and transactions costs are low, the trading benchmark in an emissions trading approach does not affect the ultimate cost-effective allocation of control responsibility. When markets are non-competitive or transactions costs are high, however, the final allocation of control responsibility is affected.*[28]

Once the control authority has decided how much pollution of each type will be allowed, it must then decide how to allocate the entitlements among the sources. In theory, entitlements can either be auctioned off, with the sources purchasing them from the control authority at the market-clearing price, or allocated among existing sources free of charge based on some regulatory distribution rule. The premise suggests that either approach will ultimately result in a cost-effective allocation of the control responsibility among the various polluters as long as they are all price-takers, transactions costs are low, and the entitlements are freely transferable. Even when they are randomly distributed without cost among existing sources, the after-market

26 Milliman and Prince (1989).

27 For the technical details supporting this proposition, see Montgomery (1972), Palmer *et al.* (1980) and Tietenberg (1985).

28 See Hahn (1984) for the mathematical treatment of this point. Further discussions can be found in Tietenberg (1985) and Misiolek and Elder (1989).

in which firms can sell excess entitlements and purchase additional ones corrects any problems with the initial allocation. This is a significant finding because it implies that under the right conditions the initial allocation can be used to pursue distributional goals without interfering with cost effectiveness. This is a particularly noteworthy characteristic in view of the possibility of incorporating transfers to developing countries in any permanent solution to the global warming problem.

When firms are price-setters rather than price-takers, however, cost effectiveness will only be achieved if the control authority initially allocates the entitlements in such a way that a cost-effective allocation would be achieved even in the absence of any trading. For all other initial entitlement distributions an active market would exist, offering the opportunity for price-setting behaviour. The larger the deviation of the initial allocation of entitlements to a single price-setting source the greater the likelihood that the source in question can exercise market power.[29]

Similar problems exist when transactions costs are high, since these can preclude or reduce trading by diminishing the gains from trade. When the costs of concluding the transaction exceed the potential gains, the incentive to participate in the market is lost.

2. *Lessons from empirical research*

(a) *A vast majority, though not all, of the relevant empirical studies have found that control costs are substantially higher with the regulatory command-and-control system than the least cost means of allocating the control responsibility.*[30]

While theory tells us unambiguously that the command-and-control system will not be cost effective except by coincidence, it cannot tell us the magnitude of the excess costs. The empirical work adds the important information that the excess costs have typically been very large. This is an important finding because it provides the motivation for introducing a reform programme; the potential social gains (in terms of reduced control cost) from breaking away from the status quo are sufficient to justify the trouble. Although the typical estimates of the excess costs attributable to a command-and-control system overstate the cost savings that would be achieved by even a completely unrestricted entitlement market (a point discussed in more detail below), the general conclusion that the potential cost savings from adopting market-based approaches are large remains accurate even after correcting for overstatement.

(b) *Only a transferable entitlement system that allocates entitlements free of charge to sources on the basis of their historical emission rate would guarantee that existing sources would be no worse off than they would under a command-and-control system imposing the same degree of control. The financial outlays associated with acquiring allowances or ERCs in an auction market (or a comparable emissions charge) would be large enough for sources typically to have lower financial burdens with the traditional command-and-control approach than with these particular economic incentive approaches.*[31]

From the point of view of the source required to control its emissions, two components of the financial burden are significant: (1) control costs and (2) expenditures on entitlements. While only the former represent real resource costs to society as a whole (the latter are merely transferred from one group in society to another), to the source both represent a financial burden. The empirical evidence suggests that when an auction market is used to distribute entitlements (or, equivalently, when all uncontrolled emissions are subject to an emissions tax), the entitlement expenditures (tax revenue) would frequently be larger in magnitude than the control costs. The sources would spend more on entitlements (or pay more in taxes) than they would on the control equipment.

Under the traditional command-and-control system firms make no financial out-

---

29 When a price-setting source is initially allocated more than its cost-effective allocation, it can inflict higher control costs on others by withholding some entitlements from the market. However, when a price-setting source is initially allocated fewer entitlements than would be justified by its cost-effective responsibility, it can exercise power on the buyer side of the market.

30 These studies are surveyed and compared in Tietenberg (1985).

31 See Atkinson and Tietenberg (1982), Atkinson and Tietenberg (1984), Hahn (1984), Harrison (1983), Krupnick (1986), Lyon (1982), Palmer *et al.* (1980), Roach *et al.* (1981), Seskin *et al.* (1983) and Shapiro and Warhit (1983) for the individual studies, and Tietenberg (1985) for a summary of the evidence.

lays to the government. Although control costs are necessarily higher with that system than with a marketable entitlement system, they are not so high as to outweigh the additional financial outlays required in an auction market entitlement system (or an emissions tax system). For this reason existing sources understandably oppose the distribution of entitlements by an auction market despite its social appeal, unless the revenue derived is used in a manner that is approved by the sources, and the sources with which they compete are required to absorb similar expenses.

(c) While it is clear in theory that larger trading areas and larger numbers of traders offer opportunities for larger potential cost savings in a transferable entitlement programme, some empirical work suggests that substantial savings can be achieved even when the trading areas and the number of potential trades are rather small. This is an important point because it suggests that even very limited steps towards a transferable entitlement system could be beneficial.

Sometimes political considerations mandate a limited transferable entitlement programme. Whether large trading areas and large numbers of traders are essential for the effective use of this policy is, therefore, of some relevance. The amount of cost savings achieved by any entitlement market would be a function of the size of the area within which the entitlements may be traded and of the number of traders. In general, larger trading areas and larger numbers of traders increase the potential cost savings owing to the wider set of cost-reduction opportunities made available. The empirical question is how sensitive the cost estimates are to these components.

One study of utilities in the United States has found that even allowing a plant to trade among discharge points within that plant could save 30 to 60 per cent of the costs of complying with new sulphur oxide reduction regulations, compared to a situation in which no trading whatsoever was permitted.[32] Expanding the trading possibilities to other utilities within the same state permitted a further reduction of 20 per cent, while allowing interstate trading permitted another 15 per cent reduction in costs. If this study is replicated in other circumstances, it would appear that even small trading areas offer an opportunity for significant cost reduction.[33]

(d) Although only a few studies of the empirical impact of market power on the cost effectiveness of transferable entitlement markets have been carried out, their results are consistent with a finding that market power does not seem to have a great effect on regional control costs in most realistic situations.[34]

Under favourable circumstances for the exploitation of market power, the available evidence suggests that, while price manipulation in auction entitlement markets could have a fairly large impact on the regional financial burden, under normal circumstances its effect on control costs would be rather small. Estimates typically suggest that control costs would rise by less than 1 per cent if market power were exercised by one or more firms.

The effects of market power depend on whether the entitlements are distributed by auction or are free of charge. In an auction market the price-setting source reduces its financial burden by purchasing fewer entitlements than otherwise in order to drive the price down. To compensate for the smaller number purchased, the price-setting source must spend more on controlling its own pollution, thereby limiting the gains from price manipulation.

With regard to free distribution rules, some rules create a larger potential for strategic price behaviour than others. In general the larger the divergence between the number of entitlements received by the price-searching source and the cost-effective number of entitlements, the greater the potential for market power. When allocated an excess of entitlements by the control authority, price-searching firms can exercise power on the selling side of the market, and when allocated too few entitlements, they can exercise power on the buying side.

According to the existing studies it takes a rather considerable divergence from the cost-effective allocation of entitlements to produce a significant difference in regional control costs. Since most realistic rules used to distribute entitlements are estimated to affect control costs to such a small degree, the

32 ICF, Inc. (1989).

33 As indicated below, the fact that so many emissions trades have actually taken place within the same plant or among contiguous plants provides some confirmation of this result.

34 For individual studies see de Lucia (1974), Hahn (1984) and Maloney and Yandle (1984). For a survey of the evidence see Tietenberg (1985).

deviations from the least cost allocation caused by market power pale in comparison to the much larger potential cost reductions achievable by implementing a transferable entitlements system.[35]

3. Lessons from implementation

Although comprehensive data on the effects of the various transferable entitlement programmes do not exist because the relevant information has not yet been collected in a systematic way, some of the major aspects of the experience are clear.

(a) The transferable entitlement programmes in the United States have evolved considerably since their inception in the 1970s. Starting with a very limited system which only remotely resembled the theoretical concept, the programmes have evolved to the point where the newest programme - acid rain allowances - is expected to approximate a transferable entitlement market rather closely.

Rather than establishing a vigorous market with well-defined prices as envisioned by early proponents, the early emissions trading programme involved negotiated trades between sources, individually supervised and approved by regulatory agencies on a case-by-case basis. This programme was a complement to, rather than a substitute for, the traditional command-and-control approach.

Experience with the programme has instilled confidence in it among the major participants. Air pollution regulators, who were once very opposed to transferable entitlements, have become a major source of support for the sulphur allowance programme and were instrumental in its design. Environmental groups have also begun to lend their support.

(b) The existing programmes in the United States have lowered compliance costs while improving air quality.

The programme has unquestionably and substantially reduced the costs of complying with the requirements of the Clean Air Act. Most estimates place the accumulated capital savings for all components of the programme at over $10 billion. This does not include the recurring savings in operating costs.[36]

Somewhere between 7,000 and 12,000 trading transactions have been completed. Each of these transactions was voluntary and for the participants represented an improvement over the traditional regulatory approach. Several of the transactions involved the introduction of innovative control technologies.

Though designed primarily to reduce the compliance costs of achieving predefined environmental goals, the emissions trading programme has also brought about an improvement in air quality. Facilitating compliance with the pollution control acts by making compliance cheaper and easier has been one avenue for improving air quality. Trading has increased the possible means for compliance and sources have responded. Other avenues by which emissions trading has improved air quality include: (1) providing a basis for more stringent standards, (2) encouraging banked emission reduction credits, and (3) developing trading rules that extract an "improvement premium".

In the negotiations over the sulphur allowance programme it became clear that industrial representatives were more willing to accept a larger reduction of sulphur oxides once it became clear that allowance trading was going to be possible. Since introducing trading lowers the cost of compliance (making it easier to reach any given target reduction), larger target reductions could be achieved at a lower financial cost than would have been possible with a pure command-and-control approach. Allowing trading increased the political acceptability of larger reductions.

Banking not only provides incentives to create credits for future use, it also offers an incentive for some credits to remain unused. While normal credits are used to justify emissions, the pollutant flows justified by banked credits are not being emitted. This reduction in actual emissions represents an improvement in air quality for as long as the credits remain banked.

---

[35] Strategic price behaviour is not the only potential source of market power problems. Firms could conceivably use entitlement markets to drive competitors out of business. See Misiolek and Elder (1989). For an analysis which concludes that this problem is relatively rare and can be dealt with on a case-by-case basis should it arise, see Tietenberg (1985).

[36] For detailed evaluations see Dudek and Palmisano (1988), Hahn (1989), Hahn and Hester (1989a) and Tietenberg (1985) and (1990).

Finally, the emissions trading programme has opted to improve air quality by requiring trading ratios in excess of 1:1. For example, buyers must typically acquire 120 per cent of the entitlements they actually intend to use. If a facility wishes to use credits to justify a 10 ton reduction, they must actually purchase 12 tons. The extra two tons represent an improvement in air quality.[37]

(c) Transferable entitlements seem to have worked particularly well for trades involving uniformly mixed pollutants and for trades of non-uniformly mixed pollutants involving contiguous discharge points.

It is not surprising that most trades concluded have been internal (with the buyer and seller either forming part of the same corporation or sharing a common corporate parent) rather than external. Not only are the uncertainties associated with interfirm transfers avoided, but most internal trades involve contiguous facilities that are not required to supply dispersion model results. The fact that so many trades have taken place between contiguous discharge points serves as confirmation that substantial savings can be achieved even if the geographic boundaries of the trading area are quite restricted.

It is also not surprising that the majority of trades realized involve volatile organic compounds, which are uniformly mixed pollutants. Since dispersion modeling is not required for uniformly mixed pollutants even when the trading sources are somewhat distant from one another, trades involving these pollutants are cheaper to realize. In addition, trades involving uniformly mixed pollutants need not be concerned about local air quality deterioration since the location of the emissions is not a matter of policy consequence.

(d) Transferable entitlements have encouraged technological progress in pollution control. Although, in general, the degree of progress has been modest, it has been more dramatic in areas where emission reductions have been sufficiently stringent to restrict the availability of emission reduction credits created by more traditional means.[38]

Theory would lead us to expect more technological progress with emissions trading than with a command-and-control policy because it changes the incentives so drastically. Under a command-and-control approach technological changes discovered by the control authority typically lead to more stringent standards (and higher costs) for the sources. Therefore, sources have little incentive to innovate and a good deal of incentive to hide innovations from the control authority. With emissions trading, on the other hand, innovations allowing excess reductions create saleable emission reduction credits.

The evidence suggests that the expectations based on this theory have been borne out, but to a limited degree in the operating programme. The most prominent example of technological change has been the substitution of water-based solvents for solvents containing volatile organic compounds. Though somewhat more expensive, this substitution made economic sense once the programme was introduced.

The number of new innovations stimulated by the programme has been rather small. As long as cheaper ways of creating credits within existing processes are available, it would be unreasonable to expect large investments in new technologies with unproven reliabilities. On the other hand, as the degree of control rises and the supply of readily available credits dries up, the demand for new technologies can be expected to rise as well. This expectation seems to be borne out in those areas where unusually low air quality or stringent regulatory rules have served to limit the available credits.[39] This point is highly relevant to the control of greenhouse gases since control costs are ultimately expected to be very high.[40]

---

37 Note that unlike the introduction of banking and encouragement of greater compliance, this source of improved air quality extracts a cost-effectiveness penalty. The use of trading ratios to produce better air quality leads to fewer trades and lower cost savings.

38 For more details see Tietenberg (1985) and Dudek and Palmisano (1988).

39 For the experience in California see Dudek and Palmisano (1988).

40 See, for example, Blitzer *et al.* (1991), Hogan (1990), Jorgenson and Wilcoxen (1991) and Manne and Richels (1990).

C. Conclusions

Key lessons

The experience with market-based programmes has demonstrated that they have had a positive role in environmental policy and can continue to have this in the future. The issue is no longer ***whether*** they have a role to play, but rather ***what kind*** of role they should play. The available experience with operating versions of these programmes allows us to draw some specific conclusions which facilitate defining the boundaries for the optimal use of market-based approaches.

(a) *Transferable entitlement systems offer the possibility of achieving greater environmental improvement at a lower cost than the conventional approach of allocating nontransferable quotas.*

Allowing reductions above the required amount to be transferable increases the opportunities for compliance. Enhanced opportunities for compliance in turn reduce the costs of compliance and lead to greater willingness to accept more stringent reduction goals. Transferable entitlements also encourage the development and installation of new and more environmentally benign technologies.

(b) *Full-blown entitlement markets are not necessary to achieve environmental and cost-effectiveness goals. Experience in the United States has shown that even hesitant initial steps towards a fully transferable entitlement system can provide the foundation for the eventual evolution of a rather sophisticated system engendering significant political support.*

Even a limited transferable entitlement market characterized by bilateral trades can provide a viable alternative to traditional regulatory approaches. Every trade realized reduces compliance costs. The experience of participants with limited trading lays the groundwork for the evolution of more sophisticated approaches which circumvent the pitfalls uncovered by the early experience. Supporting institutions (such as brokers and exchange markets) begin to evolve along with the entitlement system and help to make the system operate more smoothly and successfully. As the key participants gain familiarity with (and confidence in) the approach, its continued evolution is facilitated.

(c) *As greenhouse gases are appropriately treated as uniformly mixed, they fall within the category of pollutants for which transferable entitlement systems work best.*

For uniformly mixed pollutants only the amount emitted is a policy concern; the location of the emissions does not matter. The transferable entitlements system is particularly effective for this category of pollutants because trades can be on a one-for-one basis, no diffusion modeling is required (reducing transactions costs), and local "hot spots" are not a problem.

(d) *Compared to local or regional applications, the global nature of a transferable entitlement market in greenhouse gases offers the maximum potential for cost reduction and the minimum potential for problems such as the rise of market power.*

In general, the larger the number of potential participants in a transferable entitlement market the greater the number of possibilities for lowering the cost of compliance. Global markets offer the maximum opportunities for participation.

More participants not only open up more potential sources of emission reduction, they also diminish the likelihood of market power. Large numbers of buyers and sellers make the cartelization of any market difficult if not impossible.

(e) *As transferable entitlement systems allow the issue of who will pay for control to be separated from that of who will undertake control, they allow distributional and cost-effectiveness goals to be pursued simultaneously. They also facilitate technology transfer by providing a means for cost-sharing and risk-sharing.*

Regardless of the initial allocation of entitlements, the trading that would subsequently take place would provide the means for greenhouse gases to be controlled cost effectively. This is a particularly important feature when, as is the case with control of the greenhouse gases contributing to global warming, concerns about fairness and affordability preclude simple solutions such as equal proportional reductions. Even very complex international allocations of the control responsibilities, which are sensitive to a host of individual country concerns, can be fully compatible with achieving the desired emission target at the lowest possible cost.

Cost-sharing and risk-sharing can be achieved even for very limited versions of transferable entitlement systems. In the United States now, for example, some manufacturers of sulphur oxide control equipment have indicated a willingness to install pollution control equipment free of charge, taking only the sulphur oxide allowances in return.[41] In this way the recipient utility incurs neither a financial burden nor a financial risk; the equipment supplier is willing to assume both by accepting the allowances as payment. It is not difficult to imagine equipment manufacturers striking a similar deal with developing countries should a transferable carbon dioxide entitlement system be implemented.

(f) Implementing a free distribution of entitlements would allow global warming goals to be achieved with a lesser financial burden for polluters than either a traditional regulatory approach or an emissions charge approach which achieves comparable reductions.

Traditional regulatory approaches are not cost effective; they impose extra control costs which can be eliminated, or at least reduced, by a transferable entitlement system. Those with high marginal control costs can secure entitlements from others with lower control costs. Entitlement buyers gain because they can fulfil their responsibilities at a lower cost than without trading. Sellers gain because they receive revenue that can be used not only to meet the costs of undertaking the greater-than-required reductions, but for other worthy projects as well.

While emissions charges can support cost-effective pollution control, they impose an extra financial burden. All uncontrolled emissions are subject to a fee and result in a flow of revenue to the government. With an emission fee set high enough to produce the desired reductions in greenhouse gas emissions, this revenue flow (and the financial burden associated with it) would be very large indeed. Under a free distribution entitlement system no fee would be automatically assessed on uncontrolled emissions.

(g) If raising revenue becomes an important component in the coalition building strategy followed by negotiators, revenue could be raised while using a transferable entitlement system by levying a low annual fee on each entitlement. This revenue could be used for financing technology transfers or for other worthy purposes without jeopardizing the cost effectiveness of the system.

Although an annual fee is not a necessary component of a transferable entitlement system, it can be added, if so desired. This combined system would leave the control of emissions to the quantity-based entitlements, while using the fee to raise revenue.

Coupling a low annual fee with a free distribution of entitlements provides an attractive alternative to both auctioned entitlements and an emissions charge. Owing to the political reluctance to establish rates as high as would be required to achieve the kinds of targets that are being discussed, emission charges have traditionally not been very effective in producing the desired level of emission reduction but they have been effective in raising revenue for environmental purposes. Conversely, while entitlements distributed free of charge have been effective in producing the desired level of control, they produce no revenue. The combined system provides complete assurance of sufficient emission reduction while raising revenue and encouraging even greater reductions.

Since uniform annual fees applied to the quotas do not affect the cost effectiveness of the system, the fee can be as low or as high as necessary to achieve the desired revenue result. While the fees do lower the entitlement price, they do not affect the incentive to trade.[42]

---

41 I am indebted to Dan Dudek of the Environmental Defense Fund for pointing this out.

42 Although the entitlement price is lowered, the incentive to trade is preserved by the fact that a seller not only receives the price for the entitlement, but is also able to avoid paying the annual fee any longer on it.

References

Atkinson, Scott E. (1983), "Marketable pollution permits and acid rain externalities", *Canadian Journal of Economics*, 16(4) (November): 704-722.

Atkinson, Scott E. and T.H. Tietenberg (1982), "The empirical properties of two classes of designs for transferable discharge permit markets", *Journal of Environmental Economics and Management*, 9(2) (June): 101-121.

----- (1984), "Approaches for reaching ambient standards in non-attainment areas: Financial burden and efficiency considerations", *Land Economics*, 60(2) (May): 148-159.

----- (1987), "Economic implications of emission trading rules for local and regional pollutants", *Canadian Journal of Economics*, 20(2) (May): 370-386.

----- (1991), "Market failure in incentive-based regulation: The case of emissions trading", *Journal of Environmental Economics and Management*, 21(1) (July): 17-31.

Blitzer, C.R., R.S. Eckaus *et al.* (1991), "A general equilibrium analysis of the effects of carbon emission restrictions on economic growth in a developing country", *Proceedings of the Workshop on Economic/Energy/Environmental Modeling for Climate Policy Analysis* (Cambridge, MA: MIT Center for Energy Policy Research): 434-480.

Campbell, H.F. (1989), "Fishery buy-back programmes and economic welfare", *Australian Journal of Agricultural Economics*, 33(1): 20-31.

Congressional Budget Office (1986), *Curbing Acid Rain: Cost, Budget, and Coal-Market Effects* (Washington, D.C.: US Government Printing Office).

de Lucia, Russell J. (1974), *An Evaluation of Marketable Effluent Permit Systems.* Report No. EPA-600/5-74-030 to the United States Environmental Protection Agency (September).

Dudek, Daniel J. and John Palmisano (1988), "Emissions trading: Why is this thoroughbred hobbled?", *Columbia Journal of Environmental Law*, 13(2): 217-256.

Feldman, Stephen L. and Robert K. Raufer (1987), *Emissions Trading and Acid Rain. Implementing a Market Approach to Pollution Control* (Totowa, NJ: Rowman & Littlefield).

Hahn, Robert W. (1984), "Market power and transferable property rights", *Quarterly Journal of Economics*, 99 (November): 753-765.

----- (1989), "Economic prescriptions for environmental problems: How the patient followed the doctor's orders", *The Journal of Economic Perspectives*, 3(2) (Spring): 95-114.

Hahn, Robert W. and Gordon L. Hester (1989), "Marketable permits: Lessons from theory and practice", *Ecology Law Quarterly*, 16: 361-406.

----- (1989a), "Where did all the markets go? An analysis of EPA's Emission Trading Program", *Yale Journal of Regulation*, 6(1) (Winter): 109-153.

Harrison, David, Jr. (1983), "Case study 1: The regulation of aircraft noise", in Thomas C. Schelling (ed.), *Incentives for Environmental Protection* (Cambridge, MA: MIT Press): 41-143.

Hogan, W.W. (1990), "CO_2 emission limits: An economic analysis for the USA: Comments", *Energy Journal*, 11(2): 75-85.

ICF Resources, Inc. (1989), "Economic, environmental, and coal market impacts of SO_2 emissions trading under alternative acid rain control proposals". Report prepared for the Regulatory Innovations Staff, USEPA (March).

Jorgenson, D. and P. Wilcoxen (1991), "The cost of controlling U.S. carbon dioxide emissions", *Proceedings of the Workshop on Economic/Energy/Environmental Modeling ..., op. cit.*: 262-376.

Krupnick, Alan J. (1986), "Costs of alternative policies for the control of nitrogen dioxide in Baltimore", *Journal of Environmental Economics and Management*, 13(2) (June): 189-197.

Lyon, Randolph M. (1982), "Auctions and alternative procedures for allocating pollution rights", *Land Economics*, 58(1) (February): 16-32.

Main, Jeremy (1988), "Here comes the big cleanup", *Fortune* (21 November): 102.

Maloney, Michael T. and Gordon L. Brady (1988), "Capital turnover and marketable property rights", *Journal of Law and Economics*, 31 (April): 203-226.

Maloney, Michael T. and Bruce Yandle (1984), "Estimation of the cost of air pollution control regulation", *Journal of Environmental Economics and Management*, 11(3) (November): 244-263.

Manne, A.S. and R.G. Richels (1990), "CO_2 emission limits: An economic cost analysis for the USA", *Energy Journal*, 11(2): 51-74.

Milliman, Scott R. and Raymond Prince (1989), "Firm incentives to promote technological change in pollution control", *Journal of Environmental Economics and Management*, 17(3) (November): 247-265.

Misiolek, Walter S. and Harold W. Elder (1989), "Exclusionary manipulation of markets for pollution rights", *loc. cit.*, 16(2) (March): 156-166.

Montgomery, W. David (1972), "Markets in licences and efficient pollution control programs", *Journal of Economic Theory*, 5(3): 395-418.

Nadelson, Robert (1991), "The ruined earth", *Far Eastern Economic Review*, 153(38) (19 September): 39.

Nelson, R. and T.H. Tietenberg (1990), "Differential environmental regulation: The effects on electric utility capital turnover and emissions". Paper presented at the American Economics Association Meetings, Washington, D.C. (29 December).

Noll, R.G. (1982), "Implementing marketable emissions permits", *American Economics Review*, 72(2) (May): 120-124.

Northwest Colorado Council of Governments (1984), "Point sources-nonpoint sources trading in the Lake Dillon watershed", unpublished final report (September).

Nussbaum, B. (1991), "Phasing down lead in gasoline in the U.S.: Mandates, incentives, trading and banking". Paper presented at the OECD Workshop on Tradeable Permits to Reduce Greenhouse Gas Emissions, 26-27 June (Paris).

Oates, W.E. and A.M. McGartland (1985), "Marketable pollution permits and acid rain externalities: A comment and some further evidence", *Canadian Journal of Economics*, 18(3) (August): 668-675.

Oates, W.E. *et al.* (1989), "The net benefits of incentive-based regulation: The case of environmental standard setting", *American Economic Review*, 79(5) (December): 1233-1242.

Opschoor, J.B. and H.B. Vos (1989), *The Application of Economic Instruments for Environmental Protection in OECD Countries*, (OECD: Paris).

Palmer, Adele R., William E. Mooz, Timothy H. Quinn and Kathleen A. Wolf (1980), *Economic Implications of Regulating Chlorofluorocarbon Emissions from Nonaerosol Applications*. Report #R-2524-EPA prepared for the United States Environmental Protection Agency by the Rand Corporation (June).

Pearse, P. (1991), "Developing property rights as instruments of natural resources policy: The case of the fisheries", Paper presented at the OECD Workshop on Tradeable Permits ..., *op. cit.*

Raufer, Roger K. and Stephen L. Feldman (1984), "Emissions trading and what it may mean for acid deposition", *Public Utilities Fortnightly*, 114(4) (16 August): 17-25.

Roach, Fred, Charles Kolstad, Allen V. Kneese, Richard Tobin and Michael Williams (1981), "Alternative air quality policy options in the four corners region", *Southwestern Review*, 1(2) (Summer): 29-58.

Seskin, Eugene P., Robert J. Anderson, Jr. and Robert O. Reid (1983), "An empirical analysis of economic strategies for controlling air pollution", *Journal of Environmental Economics and Management*, 10(2) (June): 112-124.

Shapiro, M. and E. Warhit (1983), "Marketable permits: The case of chlorofluorocarbons", *Natural Resource Journal*, 23(3) (July): 577-591.

Stavins, Robert N. (1989), "Harnessing market forces to protect the environment", *Environment*, 31(1) (January/February): 4-7, 28-35.

Stevens, Brandt K. (1991), " A tradeable vehicle and fuel credits program". Paper presented at the Western Economics Association International Conference, Seattle, Washington (29 June).

Taylor, Stuart (1991), "Marketable permits: Variants for the transport sector". Paper presented at the OECD Workshop on Tradeable Permits ..., *op. cit.*

Tietenberg, T.H. (1980), "Transferable discharge permits and the control of stationary source air pollution: A survey and synthesis", *Land Economics*, 56(4) (November): 391-416.

----- (1985), *Emissions Trading: An Exercise in Reforming Pollution Policy* (Washington, D.C.: Resources for the Future, Inc.).

----- (1986), "Uncommon sense: The program to reform pollution control policy", in *Regulatory Reform: What Actually Happened* (Boston: Little Brown and Company): 269-303.

----- (1989), "Acid rain reduction credits", *Challenge*, 32(2) (March/April): 25-29.

----- (1990), "Economic instruments for environmental regulation", *Oxford Review of Economic Policy*, 6 (Spring): 17-33.

United States General Accounting Office (1986), *EPA Program to Assist Leaded-Gasoline Producers Needs Prompt Improvement*. Report CED-86-182.

Zylicz, Tomasz (1991), "The greening of the post-communist Europe?". Paper presented at the Second Annual Meeting of the European Association of Environmental and Resource Economists, Stockholm, Sweden (14 June).

[23]

Policy Sciences **24**: 199–222, 1991.

Limits of market-based strategies for slowing global warming: The case of tradeable permits

DAVID G. VICTOR
Department of Political Science. Massachusetts Institute of Technology, Cambridge, MA 02139, U.S.A.

Abstract. Concern over an enhanced greenhouse effect has yielded interest in an international effort to control rising concentrations of greenhouse gases. Among the possible mechanisms for doing this is a comprehensive marked-based approach. This approach would establish a global system of tradeable permits for the sources and sinks of the major greenhouse gases: carbon dioxide (CO_2), nitrous oxide (N_2O), methane (CH_4), and tropospheric ozone (O_3). (Chlorofluorocarbons (CFCs) are not closely considered here because they are already controlled under the Montreal Protocol on Substances that Deplete the Ozone Layer.)

The main finding is that – with the possible exception of CO_2 – not enough is currently known about these gases to accurately monitor such a system of tradeable permits. The main argument is made on the basis that the anthropogenic sources and sinks of non-CO_2 gases are quite uncertain, making nation-by-nation monitoring very difficult. Also, monitoring many of the sources and sinks for these non-CO_2 gases will be very complicated – the administrative costs of the endeavor may be much larger than benefits of including these gases in a global system of tradeable permits. One alternative to the proposed comprehensive marked-based approach is a more limited system that includes only CO_2. This finding has broader application to all greenhouse control mechanisms, not just tradeable permits.

In addition to these natural science-related issues, the paper also briefly raises several other issues such as fear that markets will allow firms to 'buy up the environment.'

Introduction

Over the last few years attention has turned to the prospects for controlling global warming due to an enhanced greenhouse effect. At present, the majority of scientific opinion supports the theory of global warming with its associated effects on public welfare due to increasing temperatures and changes in sea level,[1] agriculture,[2] hydrology,[3] and other integral components of society.[4] Highly accurate space-based sensors now validate many features of the computerized global circulation models, lending credibility to current projections of global warming.[5] However, an unambiguous 'signal' of anthropogenic global warming is yet to be detected,[6] and many of the important consequences of global warming remain uncertain and highly dependent upon the modelling assumptions used.[7] Further, viable arguments dissenting from the scientific 'consensus' have underscored the complexities and uncertainties in climate projections.[8] In addition to natural science uncertainties, there are important social science uncertainties regarding the structure and values of the future society that will have to adapt to climate change.[9]

The uncertainties of climate change contribute to difficult policy choices. They present challenges to decision analysis,[10] and make it difficult to set 'optimal' policy.[11] Even prudential strategies – those that reduce egregious risks even when uncertainties are accounted for[12] – are problematic because it is extremely difficult to identify the likelihood of such egregious outcomes.[13]

Another contributor to difficult policy choices is the balance of the present and future costs and consequences of greenhouse controls.[14] Stringent measures to control global warming could 'blacken' the economy,[15] depriving present and future generations of the wealth needed for 'quality of human life and enterprise.'[16] One strategy to control global warming would employ a nontrivial annual energy tax of $ 350 billion on the U.S. alone,[17] nearly four times the current public and private spending on all forms of pollution abatement in the United States.[18] Alternatively, if different modelling assumptions are used, policies for slowing global warming may be much less costly,[19] especially if policies such as improving energy efficiency that help cool the greenhouse effect also achieve other social objectives such as improving energy security or lowering other energy-related environmental problems.[20] Some have even suggested that modest greenhouse policies could reap an economic windfall because they would yield the benefits of energy conservation.[21]

Beyond assessments that worry about uncertainties and costs, the policy process also responds to many other signals.[22] For example, there are ethical questions of whether tampering with the atmosphere and climate of the whole planet – the 'end of nature' – through endless industrial growth should be allowed.[23] Fundamental questions on the role of man in nature – and the former's manipulation of the latter – are raised by 'deep ecologists.' And, there are questions of duty to future generations – under the concept of intergenerational equity – that may compel controlling greenhouse gases.[24]

In sum, what has emerged as the 'Great Climate Debate' is a highly complex topic for which there are persuasive arguments on many sides.[25] Over the last two years, the arguments in favor of controlling greenhouse gases have gained ground as the greenhouse effect becomes the preeminent environmental issue in the industrialized world. In 1988, the Toronto Conference on the Changing Atmosphere set a benchmark for controlling CO_2 emissions (the largest contributor to global warming): a 20% reduction below 1987 levels by 2005, and a 50% reduction by 2015. The Prime Ministers of Canada and Norway have endorsed that goal, and Chancellor Kohl of (West) Germany seeks a 25% reduction by 2005. In the U.S., proposed legislation in the House[26] and the Senate[27] have framed a 20% reduction by the year 2000.[28] Environmental groups have called for a longer term vision with even more stringent reductions.[29] The exact ends of greenhouse policy are still highly debated, but many have called for a climate convention to be signed by 1992.[30] President Bush has offered to host the first negotiating session for that convention in February, 1991.[31]

Ends and means

Given the rising interest in achieving certain ends of greenhouse policy – slowing, stopping, or even reversing the global warming – it would be wise to consider carefully the means of achieving those ends.[32] As illustrated in figure 1, the means of greenhouse policy are controlling the sources and sinks of greenhouse gases so that, in turn, the atmospheric concentrations of those gases are controlled.[33]

One possible mechanism for controlling global warming would be to follow the structure of the Montreal Protocol on Substances that Deplete the Ozone Layer and to set targets for future emissions (based on past emissions) that each country must meet.[34] This would leave individual nations to find the means within their own borders to meet those targets at the specified times. As recently negotiated for the Montreal Protocol,[35] signatories could renegotiate the terms of the agreement as necessary.

Another possibility would be to address the greenhouse problem with a global mechanism rather than strictly on the terms of individual nations. This could take several forms, but one interesting proposal is to use a comprehensive, market-based approach.[36] This proposal is reflected in the declaration from the recent Houston G-7 Economic Summit.[37]

The proposal consists of two distinct components. First, the approach is termed 'market-based' because it would utilize a global system of tradeable greenhouse gas emissions permits.[38] Under such a system, each nation would be allocated a target level of greenhouse gas emissions, and the nation would then distribute permits to emitters such that the total annual emissions allowed by the permits would not exceed the nation's target. Emitters could use those permits on their own, sell or lease them to other emitters, and/or augment their permits with purchases of additional permits. National as well as international trading on global 'greenhouse markets' would be permitted.[39] Emissions of greenhouse gases would be debited against that emitter's permits. Conversely, absorption of greenhouse gases from the atmosphere – e.g.,

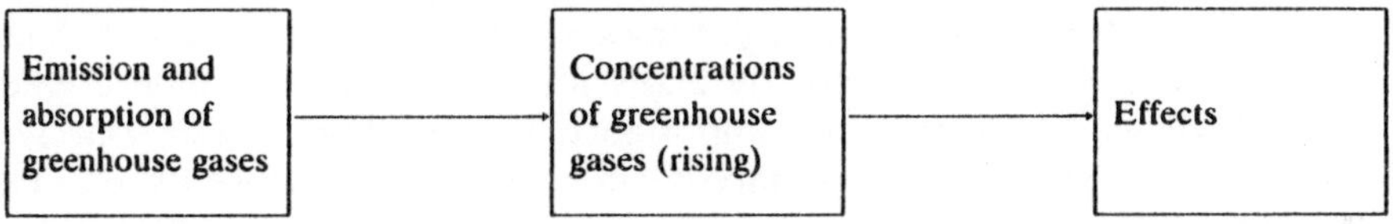

Fig. 1. Illustration of the links between sources and sinks of greenhouse gases, concentrations of greenhouse gases, and the effects of global warming. Much of the debate over the ends of greenhouse policy – whether or not and to what degree we should combat global warming – has focussed on the link between rising concentrations of greenhouse gases and the effects. As attention turns to the means of greenhouse policy we must focus earlier in the chain – on the sources and sinks of greenhouse gases – and the implications of uncertainty there for designing greenhouse control mechanisms.

by afforestation, which removes atmospheric CO_2 – would be credited to the permits ledger. Each year, the books must balance so that the total net greenhouse emissions from an emitter do not exceed the number of permits held. One could imagine other features of the tradeable permits system such as requirements that excessive net emissions in one year be compensated for in subsequent years; also, there might be large fines, trade sanctions, or other enforcement mechanisms for excessive net emissions. Presumably any international mechanism would also include a fund to assist developing nations with the costs of controlling greenhouse gases. Other structures for a global system of tradeable permits are also possible, but the principles remain the same as outlined above.

Second, the approach is termed 'comprehensive' because it would include all the main greenhouse gases: carbon dioxide (CO_2), nitrous oxide (N_2O), methane (CH_4), and tropospheric ozone (O_3). In contrast, others have proposed marked-based systems using tradeable permits[40] or emissions taxes,[17] but these have been limited to only carbon dioxide. Since the comprehensive approach would include all sources and all sinks of all greenhouse gases, there would be maximum opportunity for trading to find those sources or sinks of those gases where the costs of greenhouse abatement are lowest. (CFCs are also strong greenhouse gases, but they are not closely considered here because they are already being phased out under the Montreal Protocol. Also, including CFCs in a greenhouse agreement would greatly complicate the existing, complicated and fragile agreements reached with the Montreal Protocol and its recent amendments and additions.)[41]

A considerable literature has emerged on the economic advantages of tradeable permits and on a wide variety of technical issues related to tradeable permits systems.[42] In the U.S., experience with tradeable permits for air and water pollution is limited and mixed but generally supportive of theoretical claims that such market approaches can reduce the costs of controlling pollution.[43] After years of non-market strategies – broadly known as command-and-control regulation – there is increasing interest in marked-based approaches because experience with command-and-control has not been satisfactory.[44] Some studies have noted the extreme political difficulties of moving away from existing strategies towards market-based systems.[45] Since no international strategy currently exists for controlling greenhouse gases, these findings underscore the importance of installing a well-conceived mechanism from the beginning since it will be difficult to change the mechanism in the future.[46]

Criteria: is the system feasible?

In contrast to command-and-control, a tradeable permits system (and other types of marked-based approaches) directly links economic incentives to the sources and sinks of greenhouse gases. However, the system only works if the administrator of the permits system can quantify those emissions or absorptions.[47] Without such quantitative information it is impossible to identify violators – except possibly in egregious cases – and the permits system will fail to optimally achieve its economic and environmental objectives. Indeed, without adequate information the permits system may even be counterproductive: the world may think it has tackled global warming when the problem actually remains unimproved.

The purpose of this essay is to assess whether or not the natural science understanding of the sources and sinks of these greenhouse gases is sufficiently advanced to provide the needed information to an administrator of a comprehensive, global system of tradeable permits for greenhouse gases. Two main criteria are used in assessing the feasibility of the system.

First, most of the focus in this essay is on uncertainty in the estimates of the sources and sinks of these greenhouse gases. For the ends of greenhouse policy, it was suggested above that scientific uncertainty has not and should not necessarily prohibit actions to achieve the ends of greenhouse policy. However, uncertainty is much more problematic for employing the marked-based means. For those sources or sinks which are highly uncertain it is impossible to accurately assess credits and debits as required to administer the permits system. The main case against the comprehensive approach is made on the basis that for all the greenhouse gases considered here – except CO_2 – the sources and sinks are sufficiently uncertain that an all-inclusive marked-based mechanism for controlling global warming is not feasible.

The second criterion is one of administrative complexity. A system whose administration is very complicated is undesirable because administrative costs will consume some (or even all) of the economic benefits of using a marked-based approach.

A brief review of the methodology used to compute sources and sinks reveals the monitoring-related issues of uncertainty and administrative complexity. For all the gases considered here, the most promising means of monitoring sources and sinks are not by directly measuring the fluxes of gases into and out of the atmosphere. Instead, the science of these sources and sinks is constructed on 1) a global accounting of the location and duration of greenhouse gas-emitting activities, 2) emissions per unit of those greenhouse gas-emitting activities (the 'emission factor' or 'emission intensity'), and 3) adjustments for complicating factors. As shown schematically in figure 2, these fac-

$$\boxed{\begin{array}{l}\text{Teragrams}\\ \text{greenhouse gas}\\ \text{emitted (or absorbed)}\end{array}} = \left(\begin{array}{l}\text{Amount of gas-}\\ \text{emitting (or}\\ \text{absorbing) activity}\end{array}\right)\left(\begin{array}{l}\text{Emission (or}\\ \text{absorption) per}\\ \text{unit of activity}\end{array}\right)\left(\text{Complications}\right)$$

Fig. 2. Schematic presentation of a method used to compute fluxes of greenhouse gases. As discussed in the text, the complications are sometimes numerous and are usually adjustments to the 'emissions per unit of activity' term to account for variations in soil types, fertilization, industrial practices, etc.

tors are multiplied to compute the greenhouse emissions.[48] Uncertainties arise when any one of the factors is not known. Complexities arise when the source (or sink) cannot be adequately described with only the first two factors. All of these factors must be accounted for at least with nation-by-nation spatial resolution. Depending on the design of the permits system, the global administrator may also have jurisdiction at the subnation level, in which case the required spatial resolution must be even finer.

Comparing greenhouse gases

Before evaluating the sources and sinks of the various greenhouse gases, in passing it must be noted that a system of tradeable permits which includes more than one greenhouse gas rests heavily on conversion factor(s) to translate sources (and sinks) of different gases into common greenhouse units. This was done in the Montreal Protocol with an Ozone Depletion Potential (ODP) which was used to translate emissions of different CFCs and Halons into a common unit of ozone depletion. Rather than controlling each different CFC and Halon independently, the ODP allows collective control measures and gives signatories flexibility to choose *which* CFCs and Halons to control in order to meet the stated ozone depletion targets.[49]

The analogous index that must be developed for the greenhouse case would be a global warming potential (GWP) and would account for two important factors. First, different concentrations of greenhouse gases yield different magnitudes of global warming. A 1 part per million (ppm, a unit of concentration) increase in the methane concentration has 26 times the contribution to global warming as a 1 ppm increase in the abundance of CO_2. Second, different gases have different atmospheric lifetimes, so their integrated contribution to global warming will vary. The atmospheric lifetime of methane is about 10 years whereas the lifetime of CO_2 is much longer, perhaps 100 or 200 years. Thus, emitting (or absorbing) one mole of carbon dioxide has a longer effect on global warming.[50] The second factor is especially problematic because although the lifetimes are well known for the non-CO_2 gases, it is not clear what the atmospheric lifetime of CO_2 is. Since CO_2 is the most important greenhouse gas, it is essential that a GWP adequately address this.

Several attempts have been made to compute GWPs,[51] but uncertainties remain in the treatment of CO_2.[52] Additionally, for gases which are chemically active in the atmosphere, emissions may also affect the concentrations of other greenhouse gases, enhancing or dampening the initial global warming. All of these must be adequately included. This issue is not discussed further here, but it is essential to resolve these issues before installing a comprehensive system of tradeable permits. A global market in tradeable permits could amount to hundreds of billions or even trillions of dollars, so small changes in the GWPs could have large monetary effects and large effects on the strategic planning and relative costs of greenhouse gas emitting (and absorbing) activities.[53]

Global budgets for greenhouse gases

This section reviews the budgets for the greenhouse gases with particular attention to the feasibility of accurately measuring sources and sinks. As argued above, monitoring obstacles to a tradeable permits system arise either when the anthropogenic sources and sinks are either uncertain (i.e. difficult to quantify) or complicated (i.e. expensive and difficult to monitor). In addition to evaluating the existing uncertainties and complexities related to these greenhouse gases, attention is also given to the prospects of monitoring those sources and sinks in the future. The distinction is important because if a comprehensive, market-based approach is employed then there will be incentives to reduce emissions (or enhance absorption) of these gases. If, in doing this, monitoring of the sources and sinks becomes even more uncertain and/or complicated then the case against the comprehensive market-based approach is more compelling. It will be shown that for many of the sources and links (especially for non-CO_2 greenhouse gases), promising emission reduction strategies consist of a variety of add-on technologies or modifications of industrial or agricultural practices. Accurately accounting for the existence and efficacy of these modifications will further enhance the complexity of the system and may yield a monitoring nightmare.

Carbon dioxide (CO_2)

Increasing concentrations of CO_2 are the largest contributor to global warming and thus the greatest anthropogenic leverage point on the greenhouse effect. Table 1 shows the net sources and sinks for CO_2.[54] As currently understood, there are no large net natural sources of CO_2, and about 1/2 or more of the anthropogenic CO_2 source is taken up by natural processes (presumably the oceans) while the rest stays airborne. Since the deforestation source of CO_2 is uncertain, it is not clear exactly how much of the total CO_2 source remains airborne. Overall, that airborne fraction is dependent upon the pattern of emissions over time; in the future, that airborne fraction will not necessarily remain at past levels.[55]

By far, the largest anthropogenic source of CO_2 is the combustion of fossil fuels. There is good agreement on fossil fuel consumption data, and the CO_2 emission factors are fairly well worked out for the different fossil fuels.[56] There already exists a United Nations (U.N.) energy data system which collects and disseminates data on fossil fuel production and trade;[57] furthermore, individual governments and independent organizations also collect such data and can provide important verification of the U.N. system.[58] Over-

206

Table 1. Net sources and sinks of carbon dioxide.

	Tg (10^{12}g) C yr^{-1}
Anthropogenic sources	
Fossil fuel combustion:	
Coal	2400 ± 240
Oil	2400 ± 240
Natural gas	930 ± 100
Gas flaring	45 ± 5
Cement production	136 ± 15
Deforestation	1300 +1600, −500
Total sources:	7200
Natural sinks	
Net uptake (by oceans and land)	3600 ± 1000
Total sinks:	3600 ± 1000

all, uncertainty in the CO_2 estimates is probably no more than about 10% although estimates of fossil fuel CO_2 for some countries are probably higher since the quality of fossil fuel consumption data is uneven. Indeed, economists generally have low regard for the U.N. energy database, so there will probably have to be significant improvements in the data quality. Regarding complicating factors, once the consumption of fossil fuels (by type) is relatively well known then additional complications in computing CO_2 emissions to the atmosphere are relatively few.

The second largest anthropogenic source of CO_2 is deforestation. There is considerable controversy over the rate of deforestation, emission intensity, and the variety of complicating factors. However, the overall estimate of CO_2 flux from deforestation has narrowed considerably in the last eight years.[59] About half of the current uncertainty is due to uncertain estimates of deforestation rates, which should improve with improved satellite-based estimates.[60] The other main source of uncertainty is in estimates of the emission intensity (i.e. tons of CO_2 emitted per hectare of deforestation). This remains problematic although improved satellite-based monitoring along with *in situ* assessment of the carbon content of soils and vegetation can help lower uncertainties. Overall, greater international cooperation in monitoring-related activities can reduce these uncertainties.

In practice, the uncertainty in deforestation sources of CO_2 can be reduced with a scientific agreement on a single methodology, but it is not clear that such an agreement can be reached at the present time.[61] Even with such an agreement, the number of complicating factors will remain large. Notably, deforestation CO_2 estimates must account for the different carbon contents of various forest stands and soils; also, the net change in carbon content (i.e. the net release of CO_2) is affected by the land use after deforestation. Certainly it is technically feasible to establish nation-by-nation estimates of deforestation CO_2 emissions since that has already been done with the 1980 data set.[62] However, the problems of uncertainty remain, administrative costs for such an endeavor will be very high since forests would have to be surveyed for changes in carbon content every year or so (however, some of this would be done anyway as part of international scientific research on deforestation).[63]

Related to the problem of deforestation is the general question of whether CO_2 sink credits for afforestation and reforestation will work. As with deforestation, there are problems of verification and enforcement. Furthermore, a sink credit creates a fundamentally new set of constraints on land ownership and land use when the biomass on a particular plot of land has been used for such a credit. There also remains a question of liability for biomass lost from land used for a CO_2 credit.[64] These problems aside, the greenhouse sink credit for CO_2 is technically feasible to implement; a piece of land would be nominated for sink credit, surveyed for carbon content by the administrator and then monitored for carbon content changes during the reforestation or other carbon-increasing operation. Monitoring of reforestation and afforestation would presumably be done in conjunction with the broader effort to monitor deforestation. More work is needed to estimate administrative costs.

Less discussed are the options for extracting CO_2 from the atmosphere or from smoke stacks which are technically feasible but far from cost effective at present.[65] As currently understood, these systems would present no major verification problems, but most likely there will be some debate on their efficacy in removing CO_2 from the atmosphere. Do proposed systems which pump CO_2 into the deep ocean get a 100% credit, or does some of the CO_2 resurface and/or reduce the capacity of the oceans to absorb excess CO_2 elsewhere?[66] Most of these questions can be answered when they appear although there may be some large uncertainties and complicated monitoring requirements.

In sum, a treaty which controls anthropogenic sources and sinks of CO_2 has no insurmountable technical problems in either verification or the issuance and administration of permits. If forest sources and sinks (i.e. deforestation and afforestation/reforestation) are included then administrative costs will be much higher, but it is technically feasible to include deforestation sources and reforestation sinks (and this problem of administrative costs exists for all mechanisms, not just tradeable permits). Doing so will require some methodological agreements and a great deal of international cooperation. Complicating factors will remain, but many of them can be included in the estimates if there is increased monitoring of the forests. Although estimates of deforestation CO_2 will not be perfect, it is probably essential that these sources (and sinks) of CO_2 be included since deforestation is a highly visible environmental issue (which is also linked to other issues such as loss of biodiversity), and controlling deforestation probably offers a relatively inexpensive way to reduce CO_2 emissions.[67]

Nitrous oxide (N_2O)

Global budgets for nitrous oxide are less well understood than for CO_2. Over the last 10 years the magnitude of the global N_2O budget and the distribution of sources have changed dramatically in the light of new scientific evidence. The overall budget (table 2) is probably known within about 20%, but uncertainties on all the individual sources are much larger.[68] Those anthropogenic sources which are known seem to be very small or highly uncertain. Further, growth in the atmospheric concentration of N_2O will contribute relatively little (5%) to global warming over the next 50 years,[69] so even if all anthropogenic sources of N_2O were included there would be little additional leverage on the global warming problem.

There is a small opportunity to control anthropogenic N_2O emissions through limiting nitrogen fertilizer use, but the evidence shows that the rate of N_2O emissions is dependent upon fertilizer types, application rates, and other complications. On the clay soils of Amazonia, for example, the conversion of nitrogen in fertilizers to N_2O is significant but variable with the type of fertilizer applied. Under different conditions, however, conversion rates vary

Table 2. Sources and sinks of nitrous oxide.

	Tg (10^{12}g) N yr^{-2}
Natural sources	
Oceans	2 ± 1
Grasslands	<0.1
Soils:	
boreal and temperate	0.1 –0.5
tropical and subtropical forests	3.7
Anthropogenic sources	
Combustion:	
coal and oil	<0.5
biomass	0.02–0.11
Agricultural fertilizers (artif. + manure)	
direct conversion to N_2O	0.01–1.4
from leaching into groundwater	0.5 –1.1
Tropical deforestation to pastures	0.8 –1.3
Dry and wet deposition	0.13–5.0
Total sources:	13.4 ± 3
Natural sinks	
Reactions in the stratosphere:	
$N_2O + hv \rightarrow N_2 + O$	
$N_2O + O(^1D) \rightarrow 2NO$	10.5 ± 3

Notes: Annual source computed from observed rise in atmospheric N_2O and estimates of the annual sink. The range of individual sources do not add up to the range of the total global source.

by more than a factor of 10.[70] Thus, while the administrator can monitor global use of nitrogen fertilizers,[71] the N_2O emissions from a particular area are subject to many complications which vary significantly around the world.

Regarding measures that may be used to reduce fertilizer-deprived N_2O emissions, Elkins notes that soil management practices could substantially decrease fertilizer sources of N_2O.[72] For the tradeable permits system, this adds further complications since the administrator would have to monitor where, when, and with what efficacy such soil management practices are pursued.

Beyond direct conversion of nitrogenous fertilizers (and nitrogenous wastes) to N_2O, leaching of nitrogen fertilizer into groundwater may also be a significant source of N_2O (about 1 Tg N per year). However, assessing permit credits and debits is even more difficult than with the direct fertilizer N_2O source because the primary cause (nitrogenous fertilizer application) is further removed from the ultimate release of N_2O to the atmosphere (and the emission intensity is more variable, uncertain, and complicated).

Until recently, it was thought that fossil fuel combustion was a large N_2O source, but it is evident that those results were based on experimental error,

Table 3. Sources and sinks of methane.

	Tg (10^{12}) CH_4 yr^{-1}
Natural sources	
Animal fermentation	4 ±2
Natural wetlands	115 +85, −15
Termites	40 +60, −30
Oceans	10 + 10, −5
Freshwaters	5 +20, −4
Methane hydrates	5?
Anthropogenic sources	
Domestic animals	74 + 11
Coal mining	35 ± 10
Natural gas operations	45 +5, −20
Rice paddies	110 ±50
Biomass burning	55 +45, −5
Landfills	40 +30, −10
Total sources:	533 + 100, −140
Natural sinks	
Tropospheric reaction:	
$CH_4 + OH^{\cdot} \rightarrow CH3^{\cdot} + H_2O$	425
Stratospheric reactions:	
$CH_4 + OH^{\cdot} \rightarrow CH3^{\cdot} + H_2O$	
$CH_4 + Cl \rightarrow CH3^{\cdot} + HCl$	
$CH_4 + O(^1D) \rightarrow CH3^{\cdot} + OH^{\cdot}$	75
Total sinks:	500

and the direct N_2O source of fossil fuels is probably insignificant.[73] However, N_2O emissions due to acid deposition may be large and, therefore, controls on acid-causing emissions might help (although such acid-causing, fossil fuel-related emissions would also be controlled to some degree under a CO_2-only greenhouse treaty since acid rain is caused by fossil fuel combustion, especially coal). At present, however, the uncertainties on acid deposition-related N_2O emissions are quite large (they vary by a factor of 40).

The sinks for N_2O are probably unchangeable because they are stratospheric reactions dependent primarily upon the abundance of sunlight and oxygen (i.e. the prospects for human intervention are essentially nil.)[74]

In sum, there are some opportunities for reducing anthropogenic N_2O emissions, but all of them present serious problems for accurately monitoring a system of tradeable permits. Potentially the largest anthropogenic N_2O sources – from fertilizer leaching into groundwater and from acid precipitation – are very difficult to monitor because the eventual emission of N_2O is far removed from the initial activity thought to cause those emissions. Uncertainties are large and there are many complicating factors. Finally, monitoring problems will increase if emitters take steps to decrease emissions such as by changing soil management practices to reduce fertilizer N_2O because the practices will have to be monitored in addition to fertilizer use itself.

Methane (CH_4)

As shown in table 3,[75] the prospects for methane reductions appear larger than those for N_2O. Over half of the annual methane source is anthropogenic. Under closer scrutiny, methane reductions within a system of tradeable permits are problematic. In part, this is because the scientific uncertainties are so large. Globally, biomass burning, rice paddles, and landfill sources of methane have uncertainties up to 50% or more; the uncertainty for domestic animal sources of methane is approximately 25%. At higher spatial resolution (e.g. on a per-nation or per-permit basis), the uncertainties for many of these sources are probably much larger since monitoring in developing countries is poor and there are so many complicating factors. Based on rice paddy measurements in Spain, for example, a research team computed the average annual global source of methane from rice paddies at 35 to 59 Tg.[76] Using rice paddy emission factors next door, in Italy, two scientists from the same research team computed the average global source at 70 to 170 Tg.[77] Methane emissions from rice paddies vary significantly, depending on factors such as the amount of organic matter in the soils, fertilization, light and temperature.[78] World rice production, however, is well known.[79]

If these anthropogenic sources of methane are included in a tradeable permits system, complicating factors will compound when permit holders seek the lowest abatement cost. For example, there are significant opportunities to reduce methane emitted from livestock, but it appears that the most effective manner is to improve the quality and quantity of feeds.[80] This will increase the productivity per cow and decrease both the overall methane emissions and the emissions per cow for a given level of beef and/or milk production. But verifying claims that permit holders have actually reduced their methane emissions by changing feeds will require monitoring the diets of individual cows (or, at least, with crude estimates for herds). With the current cattle population at 1.3 billion worldwide, the task seems large.[81]

Likewise, preliminary work has shown that relatively inexpensive measures to reduce methane emissions may exist for landfills, rice paddies, and coal mines.[82] But as with the feed adjustments for livestock, the scientific evidence shows that CH_4 emissions can be reduced at lowest cost through a variety of add-on technologies (e.g. CH_4 traps on landfills) or through changes in agricultural methods (e.g. altering the rice growing cycle, using different rice strains, or using different fertilizers) which will add complications to monitoring efforts. All of these problems related to monitoring will be more serious in developing countries where necessary data-collecting infrastructure does not currently exist.[83]

Roughly one-fifth of the annual methane source is leakage from coal (35 ± 10 Tg) and natural gas operations (45 +5, −20 Tg).[84] World consumption of these fossil fuels is well known, but the methane emission intensities of the various fossil fuels is highly debated because there are so many complicating factors. For coal, it is known that methane emission intensities vary with coal location – deep coal emits more methane than shallow coals – but this finding has not been rigorously tested outside the industrialized world. For natural gas, the sources – somewhere between the wellhead and the burner tip – are not well known since it is extremely difficult to measure leaks in the highly dispersed natural gas recovery and supply system. Natural gas leakage from some segments of the natural gas supply system is relatively well known in western industrialized nations,[85] but there are extremely large uncertainties in the leakage elsewhere, especially in the former Soviet Bloc. Monitoring will be extremely complicated.

Preliminary calculations suggest that the effects of leaking methane, if left uncontrolled while CO_2 is controlled in a greenhouse treaty, may offset the greenhouse benefits of lowering CO_2 emissions in some cases.[86] If so, this creates a nontrivial policy dilemma. This is because a policy which attempts to control global warming should, presumably, give incentives to switch *away* from leaky natural gas to other fuels. However, a CO_2 reduction policy would favor switching *to* natural gas and away from oil and coal since natural gas is the least CO_2-intensive of the fossil fuels (see note 56). This problem exists for any greenhouse control mechanism. Still, it probably makes sense to explore a variety of possible treaty incentives to reduce such leaks.

In sum, all the anthropogenic sources of methane are both highly complicated and uncertain. Computing accurate nation-by-nation estimates will be very difficult.[87] The problem of methane leakage from natural gas systems is the only case where failure to pursue the comprehensive approach may result in emitters reducing emissions of controlled substances (i.e. reducing CO_2 by switching from coal to gas as an energy source) while, as an unintended byproduct of the policy, they are increasing an uncontrolled substance (i.e. CH_4 released from leaky natural gas systems). Ausubel *et al.* have persuasively argued that natural gas will be an important energy source in the future,[88] so it would be important to closely consider the options for including incentives to build tight natural gas systems. Beyond that, the uncertainties and complexities of anthropogenic methane sources make it difficult to monitor them for inclusion in the tradeable permits system.

Tropospheric ozone (O_3)

At present, it is impossible to establish the global budget for tropospheric ozone with any degree of certainty. Sources and sinks vary annually, seasonally, diurnally, and spatially. Measured at various locations around the world, it is generally found that tropospheric ozone concentrations are rising, but it is not clear precisely which and whose emissions are at fault (this is especially true in geographically compact areas such as Europe). This is because anthropogenic sources of tropospheric ozone are not direct emissions of ozone into the atmosphere. Instead, ozone is a secondary pollutant formed with the complex concurrence of three factors: reduced carbon (methane, other hydrocarbone, and/or carbon monoxide), oxides of nitrogen (NO_x), and sunlight.[89] The relative abundance of all these precursors directly affects ozone formation. As part of a global warming control mechanism it would be possible to measure tropospheric ozone at hundreds or thousands of locations around the world and estimate greenhouse debits based on excessive levels of ozone (e.g. above some agreed 'natural' level). For a tradeable permits system, however, the administrator would have to accurately estimate debits based on the actual sources of ozone precursors; this is extremely difficult even in theory, and doing it worldwide would be extremely costly.[90]

In sum, the schematic process of ozone formation is relatively well known, but compared with the level of quantified information needed to administer a system of tradeable permits, the sources, sinks and atmospheric dynamics of ozone precursors are poorly known.[91]

Conclusion

Although the uncertainties of climate change do not preclude taking steps towards certain ends of greenhouse policy, this paper has argued that uncertainties in the means of achieving those ends are much more problematic. The case analyzed here is a comprehensive system of tradeable permits for greenhouse gases. The argument developed is that not enough is known about the sources and sinks of the non-CO_2 gases to adequately include them in such a system. While the comprehensive approach may be theoretically desireable, it suffers from insufficient information.

It is also suggested that the cost of developing a monitoring body to collect the needed information for the comprehensive approach would be sufficiently costly that it would significantly offset the supposed economic benefits of using a marked-based mechanism. However, this argument of administrative costs is made without any estimates of actual costs, so a logical point for further research would be estimate the administrative costs. Such an estimate might be the basis of a broader assessment of the arguments raised in this paper, which are related to the *optimality* of the comprehensive market-based approach. In practice, no system will provide optimal results, but the question here is whether a tradeable permits system that includes all the sources and sinks of all the greenhouse gases (i.e. the comprehensive approach) is markedly better than alternative strategies (e.g. CO_2-only or command-and-control) that might be administratively easier.

Beyond the conclusions specific to a comprehensive system of tradeable permits for greenhouse gases, two issues emerge that have broader application. First, for mechanisms to control global warming, systems linked to highly quantified estimates of sources and sinks are problematic unless they are limited to CO_2 (even that will be difficult for forest-related sources and sinks). Enough is also known about the sources of CFCs (not considered here) to include them in such systems,[92] but the other gases present serious monitoring obstacles. Although this essay has made the argument using the case of a marked-based approach, most of the conclusions on the feasibility of monitoring of permits also apply to other market-based means and to monitoring compliance with command-and-control provisions.

Second, it is instructive to separate the ends and means of pollution control. The ends may be agreed upon even in the face of great uncertainty, but in designing a *mechanism* for achieving those ends uncertainty and complexity can prove to be extreme obstacles. This is especially true when highly quantified strategies such as markets are employed. Additionally, in evaluating whether the means are technically feasible it is useful to consider not only the prospects for monitoring the existing sources and sinks but also the possibility of monitoring sources and sinks once emitters (and absorbers) modify their practices in response to the policy.

To close, it is noted that there are other obstacles to marked-based approaches for greenhouse gases. For example, the arguments presented here support the technical feasibility of a CO_2-only market-based approach. The CO_2-only approach is also advocated by other authors on ground of technical feasibility.[93] Additionally, controlling CO_2 gives considerable leverage over global warming both because CO_2 emissions are the largest anthropogenic contributor to global warming and because emissions of other greenhouse gases (e.g. N_2O from acid deposition, and O_3 precursors) are linked to activities which are also linked to CO_2 emissions. However, starting with a CO_2-only approach may foreclose future expansion of the system to include other

214

gases if the monitoring problems are solved. This is because the permits system works by creating a property right in the permit, but in doing so makes the system inflexible to future changes since some property holders will object to future redistributions of that property. Such redistributions would result either from: 1) changing the permit system from a CO_2-only approach to one that also requires permits for emissions of other gases; or, 2) renegotiating the nation-by-nation targets. Starting with a CO_2-only system may foreclose these two important options that are needed for long term, global management of the greenhouse effect and its related economic and environmental issues.[94] Additionally, there are important ethical issues – such as the fear that a system of tradeable permits will allow the rich to 'buy up the environment' – which will also factor into whether this market-based approach is accepted as the mechanism to control global warming.[95]

Acknowledgements

The author is grateful to J. H. Ausubel and A. D. Hecht for initiating this research and for sage advice. Thanks also to D. Ahuja, R. U. Ayres, W. C. Clark, K. Heaton, A. Hecht, C. Hogan, H. Lee, J. A. Logan, E. Matthews, M. B. McElroy, H. F. Murray, E. J. Ozer, S. Seidel, E. B. Skolnikoff, A. Turner, S. C. Wofsy and two anonymous reviewers for helpful conversations, critiques, and data. The initial research for this paper was funded by the U.S. Environmental Protection Agency. The views expressed there and in this paper are the author's own and not necessarily those of the Environmental Protection Agency or the United States Government. For the longer paper see D. G. Victor, 1990. 'Tradeable permits and greenhouse gas reductions: some issues for U.S. negotiators,' Global Environmental Policy Project Discussion Paper Series, Energy and Environmental Policy Center, John F. Kennedy School of Government.

Notes

1. e.g. J. Oerlemans, 1989. 'A projection of future sea level,' *Climatic Change* 15: 151–174. And, National Research Council, 1987. *Responding to Changes in Sea Level: Engineering Implications* (Washington: National Academy Press).
2. M. L. Parry, T. R. Carter, and N. T. Konijn (eds.), 1988. *The Impact of Climatic Variations on Agriculture*, 2 vols. (Dordrecht: Kluwer).
3. See generally, P. E. Waggoner (ed.) 1990. *Climate Change and U.S. Water Resources* (New York: Wiley).
4. For reviews of effects see: J. B. Smith and D. A. Tirpak (eds.) 1989. *The Potential Effects of Global Climate Change on the United States* EPA Draft report to Congress. B. Bolin, B. R. Doos, J. Jager, and R. A. Warrick (eds.) 1986. *The Greenhouse Effect, Climatic Change, and Ecosystems* (New York: Wiley). National Research Council, 1983. *Changing Climate* (Washington: National Academy Press).

5. A. Raval and V. Ramanathan, 1989. 'Observational determination of the greenhouse effect,' *Nature* 342: 758–761.
6. R. W. Spencer and J. R. Christy, 1990. 'Precise monitoring of global temperature trends from satellites,' *Science* 247: 1558–1562. And, A. R. Solow and J. M. Broadus, 1989. 'On the detection of greenhouse warming,' *Climatic Change* 15: 449–453.
7. e.g. R. M. Adams et al., 1990. 'Global climate change and US agriculture,' *Nature* 345: 219–224.
8. e.g. Marshall Institute, 1989. *Scientific Perspectives on the Greenhouse Problem* (Washington: Marshall Institute), 2 vols. Also, R. S. Lindzen, 1990. 'Some coolness concerning global warming,' *Bulletin of the American Meteorological Society* 71: 288–299. For a response to some of these arguments see J. E. Hansen and A. A. Lacis, 1990. 'Sun and dust versus greenhouse gases: an assessment of their relative roles in global climate change', *Nature* 346: 713–719.
9. T. C. Schelling, 1983. 'Climate Change: Implications for Welfare and Policy,' in National Research Council, *Changing Climate* (Washington: National Academy Press).
10. J. H. Ausubel, 1980. 'Economics in the Air – An introduction to economic issues of the atmosphere and climate,' in J. H. Ausubel and A. K. Biswas (eds.), *Climatic Constraints and Human Activities* (Oxford: Pergamon), pp. 13–59. In addition to the general problem of analysis with uncertainty, Ausubel notes that many of the critical uncertainties related to climate change may not improve in the near future.
11. The effect of uncertainty on the outcome of risk/benefit assessments is noted, for example, in several case studies reviewed in R. Wilson and E. Crouch, 1982. *Risk/Benefit Analysis* (Cambridge, MA: Ballinger), esp. 25, 135, 142–144, 190–191.
12. J. G. Morone and E. . Woodhouse, 1986. *Averting Catastrophe: Strategies for Regulating Risk Technologies* (Berkeley: University of California Press), 160–163.
13. And, making conservative assumptions in the face of uncertainty systematically overstates the real risks. See C. G. Whipple, 1986. 'Dealing with uncertainty about risk in risk management,' in National Research Council, *Hazards: Technology and Fairness* (Washington: National Academy Press), pp. 44–59.
14. For the economic principles of combatting global warming (both through reducing emissions and through adaptation) see W. D. Nordhaus, 1990. 'Economic policy in the face of global warming,' in H. Aaron (ed.) *Setting National Priorities, 1990* (Washington, Brookings).
15. R. W. Gale, 1990. 'A Carbon Tax Could Blacken Economy,' *Wall Street Journal*, 6 July.
16. G. H. W. Bush, 1990. 'Remarks by the President at the Intergovernmental Panel on Climate Change,' 5 February.
17. A. S. Manne and R. G. Richels, 1989. 'CO_2 Emission Limits: An economic Cost Analysis for the USA,' *The Energy Journal* 11: 51–74. Initially, the tax is lower then rises steeply to $ 600 per ton of carbon then stabilizes at $ 250 per ton by about the year 2050. The authors argue that this schedule of taxes is needed to stabilize U.S. carbon emissions at 1.37 billion tons of carbon per year (the 1990 rate of emissions). The discounted present value of the losses from 1990 through 2100 due to this carbon tax would be about $ 3.6 trillion.
18. 1987 data in: K. D. Fraber and G. L. Rutledge, 1989. 'Pollution abatement and control expenditures, 1984–87,' *Survey of Current Business*, table 6, pp. 24–25 (June). U.S. Department of Commerce, Bureau of Economic Analysis. Washington. Reproduced in Council on Environmental Quality, 1990. *Environmental Quality* (Washington: Government Printing Office), p. 430.
19. J. Goldemberg, T. B. Johansson, A. K. N. Reddy, and R. H. Williams, 1988. *Energy for a sustainable world* (New Delhi, Wiley Eastern).
20. S. H. Schneider, 1989. 'The Greenhouse Effect: Science and Policy,' *Science* 243: 771–781.
21. McKinsey and Company, 1989. 'Background paper on funding mechanisms,' Prepared for the Ministerial Conference on Atmospheric Pollution and Climatic Change, with Particular Attention to Global Warming. See especially exhibit 1–4.

22. D. A. Stone, 1988. *Policy Paradox and Political Reason* (Glenview, IL: Scott, Foresman and Co.), esp. chapter 1.
23. B. McKibben, 1989. *The End of Nature* (New York: Random House).
24. E. B. Weiss, 1989. *In Fairness to future generations: International law, common patrimony, and intergenerational equity* (Dobbs Ferry, NY: Transnational Publishers). Also see E. B. Weiss, 1989. 'Climate Change, Intergenerational Equity and International Law: An Introductory note,' *Climatic Change* 15: 327–335.
25. R. M. White, 1990. 'The Great Climate Debate,' *Scientific American* 263 (1): 36–43.
26. 'Global Warming Prevention Act,' U.S. House of Representatives, H.R. 1078, 101st Congress (1989).
27. 'National Energy Policy Act of 1989,' U.S. Senate, S.324, 101st Congress (1989).
28. For a summary of the logic behind these two proposed bills see C. Schneider, 1989. 'Preventing Climatic Change,' *Issues in Science and Technology* 5 (4): 55–62.
29. L. R. Brown et al (eds.), 1990. *State of the World* (New York: Norton), pp. 173–190. See also their summary of enacted and proposed global warming strategies on page 33.
30. e.g. see the Declaration from the G-7 economic summit in Houston. 'Excerpts from the Economic Declarations by Chiefs of Seven Industrial Nations,' *New York Times*, p. A15 (12 July 1990).
31. Executive Council, World Meteorological Organization,,1990. 'Framework convention on climate change,' Resolution 4.1/1 (EC-XLII).
32. A review of several possible mechanisms is found in M. Grubb, 1989. *The Greenhouse Effect: Negotiating Targets* (London: Royal Institute of International Affairs).
33. Another end of greenhouse policy is adaption to climate change. This essay is limited to policy responses aimed at reducing the concentrations of greenhouse gases, but close consideration of adaptive responses is equally important.
34. In the Montreal Protocol this is done with two schedules for cutting the consumption of ozone-depleting substances: one for developed countries and one for developing countries. The difference between the two classes of country is established according to the per-capita consumption of chlorofluorocarbons (developing countries are below 0.3 kg per capita). See United Nations Environment Programme, 1987. *Montreal Protocol on Substances that Deplete the Ozone Layer.* Final Act done 16 September. Both the Montreal Protocol and the Vienna Convention (to which the Montreal agreement is a Protocol) are reprinted in *International Legal Materials* 26 (Nov. 1987), 1516–1561.
35. M. W. Browne, 1990. '93 Nations move to ban chemicals that harm ozone,' *New York Times*, 30 June, p. A1.
36. See 'U.S. Concept Paper: Comprehensive greenhouse gas approach to addressing climate change,' filed with the Intergovernmental Panel on Climate Change on 29 December 1989 (Response strategies working group). Also see 'Discussion paper: A "comprehensive" approach to addressing potential global climate change,' presented at an informal seminar, 3 February 1990, U.S. Dept. of State. The intellectual background for this proposal is persuasively developed in B. A. Ackerman and R. B. Stewart, 1988. 'Reforming environmental law: The democratic case for market incentives,' *Columbia Journal of Environmental Law* 13: 171–199. Also see the preceding article: R. B. Stewart, 1988. 'Controlling environmental risks through economic incentives,' *Columbia Journal of Environmental Law* 13: 143–169.
37. See paragraph 63: 'Work on appropriate implementing protocols [to a climate convention] should be undertaken as expeditiously as possible and should consider all sources.' Reprinted in *supra* note 30.
38. In addition to the advantages of using the market in pollution abatement, a tradeable permits system has certain advantages over other market approaches such as greenhouse gas taxes since the system of tradeable permits would ensure that certain environmental goals are achieved. With tradeable permits, the total emissions of greenhouse gases would be limited by the number of permits issued. Assuming all nations abide by the agreement, a

fixed number of permits would effectively cap the net emissions of greenhouse gases to the atmosphere. In contrast, with a tax based system there is no inherent cap. Further, a tax-based system presents the sticky problem of what level the tax should be set at; if the tax is too low then emissions will be too high and vice-versa.

39. It is even possible that a whole range of financial services and secondary markets, akin to those now connected to capital markets, might emerge.

40. J. N. Swisher and G. M. Masters, 1989. 'International Carbon Emission Offsets: A tradeable currency for climate protection services,' Technical Report # 309, Civil Engineering Department, Stanford University.

41. Also, not discussed here is the gas carbon monoxide (CO) which is not, itself, a greenhouse gas. However, CO reacts with OH˙ radicals in the atmosphere. CH_4 is also removed from the atmosphere by reaction with OH˙ radicals. It is theorized (and generally seems true) that increasing concentrations of CO will deplete OH˙. Such OH˙ depletion will extend the atmospheric lifetime of CH_4 because the rate at which CH_4 is removed from the atmosphere is directly dependent upon the availability of OH˙ with which it reacts. Indirectly, then, CO is a greenhouse gas since CO and CH_4 are coupled through OH˙. The exact extent of this OH˙-coupling is highly debated. CO emissions also contribute to the formation of tropospheric O_3, another strong greenhouse gas. For one calculation see A. M. Thompson, R. W. Stewart, M. A. Owens, and J. A. Herwhe, 1989. 'Sensitivity of tropospheric oxidants to global chemical and climate change,' *Atmospheric Environment* 23: 519–532. For several future scenarios see A. M. Thompson, M. A. Huntley, and R. W. Stewart, 1990. 'Perturbations to Tropospheric Oxidants, 1985–2035; 1. Calculations of Ozone and OH in Chemically Coherent Regions,' *Journal of Geophysical Research* 95: 9829–9844.
CO is also a greenhouse gas because CO oxidizes to form CO_2 (the chemical pathway starts with the OH˙ reaction described above). This is a nontrivial source of CO_2, but one can account for this CO_2 source by including CO emissions in the CO_2 emission factors. At least for fossil fuels (about 20% of the total CO source), this is feasible. Additionally over 20% of the CO source is from CH_4 oxidation (a byproduct of the chemical pathway which starts when CH_4 reacts with OH˙) and will be controlled if CH_4 is controlled in a greenhouse treaty. Another 20% is due to forest clearing and can be accounted for if deforestation CO_2 is included in a greenhouse treaty. In sum, there is no compelling reason why CO must be separately included in a greenhouse treaty. This is especially true since it is quite unclear where the CO is coming from. CO source data used here are as estimated in World Meteorological Organization, 1986. *Atmospheric Ozone: 1985*, Global ozone research and monitoring project report # 16, p. 106.

42. T. H. Tietenberg, 1985. *Emissions Trading: an exercise in reforming pollution policy* (Washington: Resources for the Future). For current possible applications of market incentives, including tradeable permits, see: *Project 88: Harnessing Market Forces to protect our environment*, A public policy study sponsored by Senators T. E. Wirth and J. Heinz. The Project 88 ideas are also described in R. N. Stavins, 1989. 'Harnessing Market forces to protect the environment," *Environment* 31: 4ff. Also see R. K. Raufer and S. L. Feldman, 1987. *Acid rain and emissions trading: Implementing a market approach to pollution control* (Totowa: Rowman and Littlefield).

43. For a review of past uses of marketable permits in the U.S. see 'Using incentives for environmental protection: An overview,' Congressional Research Service, 89—360 ENR.

44. This is evident not only in the U.S. but also in other industrialized countries. See J. B. Opschoor and H. B. Vos, 1989. *Economic instruments for environmental protection* (Paris: OECD), especially chapter 3.

45. R. W. Wahl, 1989. *Markets for federal water: Subsidies, Property Rights, and the Bureau of Reclamation* (Washington: Resources for the Future), esp. pp. 293–295.

46. Indeed, the first large-scale U.S. experiment with tradeable permits will be for acid rain abatement, an area where there has been no fully articulated U.S. abatement strategy. See

'Clean Air Act Amendments of 1990,' S.1630, as passed by the Senate (3 April) and the House of Representatives (23 May), 101st Congress, 2nd session.

47. This constraint is well discussed in the emissions trading literature. See, for example, T. H. Tietenberg *supra* note 42 at page 169.
48. This is not the only way to compute fluxes of gases into and out of the atmosphere. One can, for example, construct a global or regional budget of sources and sinks through a series of constraints such as the observed atmospheric concentrations, the distribution of various isotopes, etc. However, for the purposes of monitoring and verifying relatively small sources at high resolution (i.e. on the geographical scale of nations), the system shown schematically in figure 1 is probably the most viable.
49. See annex A of the Montreal Protocol *supra* note 34.
50. There is also a question of discount rates since the global warming of long-lived gases is largely in the future.
51. D. A. Lashof and D. R. Ahuja, 1990. *Nature* 344: 529–531. And, H. Rodhe, 1990. 'A comparison of the contribution of various gases to the greenhouse effect,' *Science* 248: 1217–1219.
52. It is not clear what the 'correct' atmospheric lifetime of CO_2 is because the effective atmospheric lifetime of CO_2 emissions is highly dependent upon the CO_2 emissions scenario. D. G. Victor, 1990. 'Calculating greenhouse budgets,' *Nature* 347: 431.
53. It must be noted that the relevant uncertainties in formulating the ODP were much lower and less fundamental (i.e. the scientific disputes were not as large) as they are for a Global Warming Potential. The one area where there were larger disputes – converting Halon ODPs into the same units as chlorofluorocarbon ODPs – was somewhat avoided since the Montreal Protocol does not permit trading between Halons and chlorofluorocarbons. Thus success in formulating the ODP for the Montreal agreement does not prescribe success in formulating a GWP for a greenhouse agreement. For a recent assessment of ODPs see D. A. Fisher et al., 1990. 'Model calculations of the relative effects of CFCs and their replacements on stratospheric ozone,' *Nature* 344: 508–512. Note that the range of ODP values is relatively small, indicating the extent to which there is scientific agreement.
54. Sources for table 1: fossil fuel data from W. C. Clark (ed.), 1982. *Carbon Dioxide Review: 1982* (New York: Oxford), pp. 458–460 (data scaled to 1988 using fossil fuel consumption figures in *BP Statistical review of world energy*, June 1989 (London: BP). See also G. Marland et al., 1989. *Estimates of CO_2 Emissions from fossil fuel burning and cement manufacturing, based on the United Nations Energy Statistics and the U.S. Bureau of Mines Cement Manufacturing Data*, ORNL/CDIAC-25, NDP-030, available from DOE/Oak Ridge National laboratory. Net uptake data based on an estimated airborne fraction of 50% as discussed in B. Bolin, 1986. 'How much CO_2 will remain in the atmosphere,' in B. Bolin, B. R. Doos, J. Jager, and R. A. Warrick (eds.). *The greenhouse effect, climatic change, and ecosystems: A synthesis of present knowledge* (Chichester: Wiley), pp. 157–203. However, there is considerable uncertainty here since it is not clear exactly where the CO_2 is going. See P. P. Tans, I. Y. Fung, and T. Takahashi, 1990. 'Observational constraints on the global atmospheric CO_2 budget,' *Science* 247: 1431–1438. The deforestation number is an estimate based on a review of the literature; however, it is about 30% lower than the number given by Houghton et al., 1987 (see note 62). The numbers shown in this table are net sources and sinks and do not include the entire carbon cycle.
55. This balance between sources and sinks is an important factor for setting global anthropogenic CO_2 controls since it *may* be possible to 'manage' atmospheric CO_2 by reducing emissions of CO_2 to the rate at which the oceans (and the terrestrial biota) absorb excess CO_2. The overall 'allowable' emissions (i.e. those emissions that would not increase atmospheric concentrations) could be calculated with a model of the global carbon cycle. L. D. D. Harvey, 1989. 'Managing Atmospheric CO_2,' *Climatic Change* 15: 343–381. This argument is based on a computation of net oceanic uptake of CO_2 and is somewhat challenged by new results on the role of the terrestrial biota (see note 54).

56. The emission factors are as follows: Coal, 0.683; Oil, 0.52; and Gas, 0.411. Units are in tons of carbon emitted per ton of coal-equivalents burned. The figures are computed based on the carbon content of the fuels and on different assumptions about the % oxidation of the fuels (99% for coal, 91.8% for crude petroleum, and 98% for gas). Lack of 100% oxidation is due in part to incomplete burning, in part to fuel leaks, and in part to uses of fuel for non-combustion purposes (e.g. petrochemicals). Computations in J. H. Ausubel, A. Grubler, and N. Nakicenovic, 1988. 'Carbon Dioxide Emissions in a Methane Economy,' *Climatic Change* 12: 245–263. Data based primarily on R. M. Rotty and C. D. Masters, 1985. 'Carbon Dioxide from fossil fuel combustion: Trends, resources, and technological implications,' in J. R. Trabalka (ed.) *Atmospheric Carbon Dioxide and the Global Carbon Cycle*, DOE/ER-0239.
57. The U.N. and other data sources also include trade in fossil fuels. It is assumed that debits for CO_2 emissions are attached to the fossil fuels themselves and not to the country where they are recovered from the ground. The question of how to manage CO_2 credits and debits due to trade in fossil fuels is probably easy to solve. One can model the system after the Montreal Protocol, for example, which controls 'calculated consumption' defined in articles 1 & 3 as production plus imports minus exports. The Montreal Protocol also controls production of halocarbons at slightly less stringent levels than for consumption. This allows exporting nations to slightly expand production to service the halocarbon needs of nations which have little or no domestic CFC production facilities. Such exports have certain constraints regarding the Montreal Protocol participation of countries to which the exports are made; most notably, the increased exports are only allowed for satisfying the needs of developing countries operating under article 5 of the protocol.
A combination of consumption *and* production restraints might be a good way to structure a greenhouse treaty. However, such a treaty will lower the value of fossil fuel reserves still in the ground. Presumably, resolution of this operational equity problem would be a function of the bargaining process and would be reflected in different control levels for consumption or production such as was done the Montreal Protocol.
58. e.g. private corporations such as British Petroleum collect world energy data; regional economic organizations such as the Organization of Economic Cooperation and Development (OECD) compile energy statistics on their member nations; nations, themselves, also compile national statistics (e.g. the Energy Information Administration of the U.S. Government).
59. This does not necessarily mean that the estimates are narrowing on the 'right' number but just that there is greater agreement among scientists.
60. e.g. Y. J. Kaufman, C. J. Tucker and I. Fung, 1990. 'Remote sensing of biomass burning in the tropics' *Journal of Geophysical Research* 95: 9927–9939.
61. Deforestation *rates* could be measured *in situ* or by remote sensing using a Landsat-type system (presumably both approaches would be used). However, there still exists an expensive and sticky (but solvable) problem of funding and flying an international array of satellites or other remote sensing devices. Carbon intensity (i.e. how much CO_2 is released per hectare deforested) would have to be established through an agreed methodology and would require a great deal of *in situ* data (thus requiring an array of data collectors and substantial international cooperation). This could be very tricky since there are many important factors which affect the CO_2 emission intensity of deforestation.
62. R. A. Houghton et al., 1987. 'The flux of carbon from terrestrial ecosystems to the atmosphere in 1980 due to changes in land use: geographic distribution of the global flux,' *Tellus* 39B: 122–139. See, especially, figure 2 and pages 128–129 for tabulation of the net carbon fluxes by nation.
63. Perhaps the greatest benefit of post cold war thaw is that nations will be more willing to release high resolution satellite technologies for use in deforestation monitoring. This 'peace dividend' could help reduce costs since the research and development costs for such satellites have already been incurred.

64. This problem becomes especially difficult when the loss is due, for example, to an unforeseeable natural disaster which, itself, may have been due to global warming. For example, is a nation responsible for reforesting an area after a big forest fire if the fire happened during a dry season which might have been caused by climatic change or by natural variations in climate? Are the CO_2 emissions from the fire included in the nation's total permit allocation; if so, are they included immediately or phased in over time? What if the fire was due to dryness of the forests but the nation could have taken measures such as increased patrols of the forest which would have prevented or limited the scope of the fire? Is it irrelevant that after such a fire the clear land may be useful for crops or pastures? How can one distinguish between accidental fires and ones which only appear as accidents?
65. For a number of technological options to reduce emissions of greenhouse gases see generally: OECD, 1989. *Energy Technologies for reducing emissions of greenhouse gases, proceedings of an experts' seminar*, Paris, 12–14 April, 2 vols.
66. One system which proposes disposing of CO_2 into the oceans is C. Marchetti, 1977. 'On Geoengineering and the CO_2 problem,' *Climatic Change*: 59–68.
67. It is unclear whether high deforestation CO_2 estimates will work to the benefit or detriment of tropical nations. They may benefit such nations since high *present* deforestation will make the baseline from which CO_2 is controlled higher than it would be if deforestation were low or excluded. Also, in a 'debt for nature swap" or other enticement to reduce deforestation, the country might receive both debt relief *and* the CO_2 credit for not deforesting. That would presumably work to the benefit of the country. Alternatively, including deforestation in a greenhouse treaty may force countries with high deforestation rates to make sweeping changes in domestic policy. Presumably that would work to the detriment of the country (assuming those nations find it expensive or undesirable to make such sweeping changes and to reduce such deforestation).
68. Sources for table 2: primarily from J. W. Elkins, 1989. 'State of the research for atmospheric nitrous oxide (N_2O) in 1989,' contribution for Intergovernmental Panel on Climate Change. See also the budget in World Meteorological Organization, 1986. *supra* note 41 at page 81.
69. V. Ramanathan, R. J. Cicerone, H. B. Singh, and J. T. Kiehl, 1985. 'Trace gas trends and their potential role in climate change,' *Journal of Geophysical Research* 90: 5547–5566, from figure 3.
70. M. Keller, W. A. Kaplan, S. C. Wofsy, and J. M. Da Costa, 1988. 'Emissions of N_2O from tropical forest soils: Response to Fertilization with NH_4^+, NO_3^-, and PO_4^{3-},' *Journal of Geophysical Research* 93: 1600–1604.
71. These data are already complied by the United Nations. See Food and Agriculture Organization, *Fertilizer Yearbook* published annually. Data for nitrogen and other fertilizers also summarized annually in the *United Nations Statistical Yearbook*.
72. J. W. Elkins, 1989 *supra* note 68, citing J. M. Bremner, G. A. Breitenbeck, and A. M. Blackmer, 1981. 'Effect of nitropyrin on emission of nitrous oxide from soil fertilized with anhydrous ammonia,' *Geophysical Research Letters* 8 (4): 353–356.
73. W. P. Linak et al., 1990. 'Nitrous oxide emissions from fossil fuel combustion,' *Journal of Geophysical Research* 95: 7533–7541.
74. In theory, however, distant technological developments may enable a cost effective device that removes N_2O or other greenhouse gases from the atmosphere.
75. Compiled primarily from R. J. Cicerone and R. S. Oremland, 1988. 'Biogeochemical aspects of atmospheric methane,' *Global Biogeochemical Cycles* 2: 299–327. Further data resolution on livestock from P. J. Crutzen, I. Aselmann, and W. Seiler, 1986. 'Methane production by domestic animals, wild ruminants, other herbivorous fauna, and humans,' *Tellus* 38B: 271–284. Wetlands data from E. Matthews and I. Fung, 1987. 'Methane emission from natural wetlands: global distribution, area, and environmental characteristics of sources,' *Global Biogeochemical Cycles* 1: 61–86. Landfill data from H. G. Bingemer and P. J. Crutzen, 1987. 'The production of methane from solid wastes,' *Journal of Geophysical Research* 92: 2181–2187.

76. W. Seiler, A. Holzapfel-Pschorn, R. Conrad, and D. Scharffe, 1984. 'Methane emission from rice paddies,' *Journal of Atmospheric Chemistry*: 241–268.
77. A. Holzapfel-Pschorn and W. Seiler, 1986. 'Methane emission during a cultivation period from an Italian Rice Paddy,' *Journal of Geophysical Research* 91: 11803–11814.
78. Even if these technical problems are solved, there are still many political problems. Notably, controlling methane production from rice paddies is politically difficult given the staple role of rice in some cultures. (There may, however, be opportunities to reduce methane emissions by using new varieties of rice.) The same is true for livestock emissions which might be even more problematic than rice since countries raise cattle for local consumption and for lucrative export contracts. Thus controls will be difficult because they affect both local diets and the current account.
79. Annual statistics are collected by the Food and Agriculture Organization of the United Nations. As with the U.N. energy statistics, there are questions of data quality (but many of these can be resolved fairly easily).
80. M. J. Gibbs, L. Lewis, and J. S. Hoffman, 1989. *Reducing methane emissions from livestock: Opportunities and issues*, EPA 400/1-89/002.
81. An equity issue may also emerge here since the solution to lower methane emissions is better feed. Poor nations will see this as a tactic by rich nations which can feed their livestock better food. The EPA work has shown that better food might pay for itself in higher production, but the issues are not always seen that way by poor subsistence herders.
82. Intergovernmental Panel on Climate Change, Response Strategies Working Group, 1990. *Methane Emissions and Opportunities for Control*, available from the U.S. Environmental Protection Agency.
83. For example, see the assumptions used to calculate emissions of CH_4 from landfills and solid waste in developing countries: H. G. Bingemer and P. J. Crutzen, 1987 *supra* note 75, esp. pp. 2183–2186.
84. Perhaps leakage is much higher than presented here. See D. C. Lowe, C. A. M. Brenninkmeijer, M. R. Manning, R. Sparks and G. Wallace, 1988. 'Radiocarbon determination of atmospheric methane at Baring Head, New Zealand,' *Nature* 332: 522–525. One way to determine how much fossil fuel-related methane is emitted into the atmosphere is to measure the isotopic distribution of the carbon in the methane. Fossil methane will have an 'old' signature, but this is complicated because methane locked in tundra and methane from clathrates will have the same isotopic signature as fossil methane. The paper cited above claims that 32% of atmospheric methane is 'old;' other papers find lower numbers, about 20% or even lower. See Cicerone and Oremland, 1988 *supra* note 75.
85. However, the total leak rate is highly debated. Some believe the leakage in the U.S., for example, is less than 1% whereas others believe a higher number (2% or even higher) is the real rate. Very little work has been done on the total system leak from wellhead to end user.
86. D. G. Victor, 1990. 'Leaking methane from natural gas vehicles: Implications for transportation policy in the greenhouse era,' *Climatic Change* (forthcoming). See also H. Rodhe *supra* note 51.
87. Recently, the World Resources Institute published nation-by-nation estimates. However, the data they used to compute their estimates are much softer than their results (reported to two significant digits for each country) have implied. See *World Resources 1990–91* (New York: Oxford), pp. 346–349. For another global budget, including past contributions, see S. Subak, 1990. 'Accountability for Climate Change,' Research Memorandum (Stockholm Environment Institute).
88. J. H. Ausubel, A. Grubler, and N. Nakicenovic, 1988 *supra* note 56.
89. A review of the chemical processes of ozone formation is found in: J. A. Logan, 1985. 'Tropospheric Ozone: Seasonal behavior, trends, and anthropogenic influence,' *Journal of Geophysical Research* 90: 10463–10482.
90. As part of national and international efforts to monitor urban air pollution, some of the needed monitoring is already being done. However, the monitoring is highly incomplete, especially in developing countries. Also, using data from a network of urban monitoring

stations would only provide information on urban contributions of ozone to global warming. Adequately including rural sources of ozone and/or ozone precursors would require even a larger array of monitoring activities.

91. The effects of ozone on the climate are somewhat complicated. At low altitudes (e.g. the urban ozone problem) the net effect of ozone is as a greenhouse gas: increasing ozone leads to warming of the Earth's surface. In the upper troposphere and in the stratosphere, increasing ozone will decrease temperatures at the Earth's surface. The difference is because the role of ozone in the stratospheric heat budget is quite different from the role of ozone in the lower troposphere. The ozone issues raised in this memo primarily affect ozone concentrations in the lower troposphere and, thus, rising concentrations contribute to the greenhouse effect. However, ozone depletion from chlorofluorocarbons will affect (increase) the abundance of tropospheric ozone, so the initiatives to phase out chlorofluorocarbons are important in addition to the direct role of CFCs as greenhouse gases.
92. If CFCs were included in a system of tradeable permits they would not present any serious monitoring problems. World production and trade of CFCs is reasonably well monitored, and under the Montreal Protocol those monitoring activities have been expanded. Under the recent revisions of the Montreal Protocol, CFCs are being phased out by the year 2000 (see notes 34 and 35). CFC alternatives that deplete ozone are being phased out by 2030 (those alternatives are also greenhouse gases, so phasing them out, although over a longer time period, is an important action for controlling global warming). Of greatest interest, however, are the non-ozone-depleting CFC alternatives (e.g. CFC-134a). These gases are not controlled under the Montreal Protocol but are strong greenhouse gases, so they are candidates for inclusion in mechanisms to control global warming. As with CFCs, there are no serious technical obstacles to monitoring production, trade, and consumption of these CFC alternatives. For more on the global warming potentials of the CFC alternatives see D. A. Fisher et al., 1990. 'Model calculations of the relative effects of CFCs and their replacements on global warming,' *Nature* 344: 513–516.
93. See especially Grubb *supra* note 32. Also see Swisher and Masters *supra* note 40 and G. Marchant, 1990. 'Freezing CO_2 emissions: an offset policy for slowing global warming,' forthcoming in the Discussion Paper Series, Energy and Environmental Policy Center, John F. Kennedy School of Government.
94. These points were raised by T. C. Schelling and other participants in a 27 February 1990 seminar at the John F. Kennedy School of Government at which an earlier version of this paper was presented. Foreclosing the first option is unique to the CO_2-only permits system; foreclosing the second option is a characteristic of any tradeable permits system (i.e. whether or not the system includes all sources and all sinks of all greenhouse gases).
95. Several of these ethical and structural issues are discussed in greater detail in D. G. Victor, 1990. 'Tradeable permits and greenhouse gas reductions: some issues for U.S. negotiators,' Global Environmental Policy Project Discussion Paper Series, Energy and Environmental Policy Center, John F. Kennedy School of Government, Harvard University.

C
Hedging Strategies

[24]

THE TIMING OF CO_2 CONTROL IN THE FACE OF UNCERTAINTY AND LEARNING

Charles D. Kolstad[1]

ABSTRACT

This chapter concerns the optimal regulation of greenhouse gases that lead to global climate change. In particular, the focus is on uncertainty and learning (which, over time, resolves uncertainty). An empirical stochastic model of climate-economy interactions is presented, along with results on the tension between postponing control until more is known vs. acting now before irreversible climate change takes place.

4.1 INTRODUCTION

Uncertainty is a dominant characteristic of environmental externalities, including the accumulation of greenhouse gases leading to climate change. We understand well neither the effects of climate change nor the costs of controlling greenhouse gases. This is one reason considerable sums are expended in trying to better understand this problem. An additional factor frequently comes into play having to do with the cumulative or stock effects of greenhouse gases. It is not the *emissions* of greenhouse gases that *directly* cause adverse effects; rather it is the stock of these gases that may lead to climate change and these stocks change slowly with a great deal of momentum. These two aspects of the problem--stock effects and uncertainty--lead to a tension between instituting control and delaying control.[2] Some in society will desire control of greenhouse gases before climate change is well understood. Others in society may urge delaying control until the problem is clearly delineated. If, *ex post*, the problem turns out to be less severe than expected then those urging delay will

[1] Research supported by a grant from the Research Board of the University of Illinois, by NSF Grant SES-92-09943 and by USAID cooperative agreement DHR-5555-A-00-1086. Superb research assistance from Ying Zhang is gratefully acknowledged. An earlier version of this paper was presented at the International Workshop on Costs, Impacts and Possible Benefits of CO_2 Mitigation at IIASA in Austria in September 1992 (see Kolstad, 1993). Thanks to Prof. William Nordhaus of Yale for providing an early version of his 'DICE' model.

[2] There are other examples with these basic characteristics: hazardous wastes and groundwater, acid rain, species extinction, pesticide accumulation.

have been proved correct (*ex post*). If on the other hand, the problem turns out to be more severe than expected, then delay can be very costly indeed.

This chapter concerns one of the most fundamental questions in the climate change/greenhouse gas control policy arena: when and to what extent to regulate the generation of greenhouse gases when uncertainty exists and learning is taking place. Thus this chapter seeks to determine how the *fact* that we are learning about climate change influences our actions *today* to control greenhouse gases. In our application, there is uncertainty on the damage from climate change. While there has been some work related to this question (Manne and Richels, 1992; Peck and Teisberg, 1992, 1993; Nordhaus, 1991), explicit treatment of the learning process has yet to appear in the empirical literature on climate change.[3] Our approach to the problem is to adapt a simple optimal growth economy-climate model (Nordhaus, 1992a, 1993) to include uncertainty and learning.

Two primary results emerge from our analysis. If emission control is perfectly reversible (no sunk capital), then the fact that one is learning does not appreciably affect today's emission control policies. This is because of the long lags in emissions contributing to temperature change. In the case where emission control investments, once made, become sunk costs, then rapid learning does tend to *modestly* reduce optimal current period emissions.

The next section of the chapter reviews some important contributions to the theoretical literature on learning, as well as existing empirical analyses of climate change and learning. The subsequent section presents the model of optimal regulation used in this study. The case of uncertainty in the disutility of pollution is then considered, followed by results.

4.2 BACKGROUND

4.2.1 Irreversibilities and Stock Externalities

A major literature has developed in the area of investment under uncertainty in the presence of externalities. Arrow and Fisher (1974) initiated much of the work in this area by focusing on a two period model with uncertainty about the benefits of an environmental asset that is to be exploited (eg, a canyon flooded to make electricity). With some uncertainty resolved between the two periods and the impossibility of undoing development of the environmental asset, it turns out to be optimal to bias

[3] See Kolstad (1993) and Cunha-e-sa and Kolstad (1992) for a discussion of theoretical issues surrounding learning and stock externalities.

development in favor of preservation of the environmental asset. Henry (1974) published similar results. In essence, taking an irreversible action has a cost in terms of reducing the value of information. Arrow and Fisher (1974) introduced the notion of quasi-option value, the value of the information gained by waiting before exploiting the environmental asset. Since then, there has been a considerable literature on irreversibilities and on quasi-option value (eg, see Fisher and Hanemann, 1987, 1990; Freeman, 1984; Olson, 1990; Conrad 1980; and Miller and Lad, 1984). Of course there is also a large literature on option value in the field of finance. In particular, a number of recent papers concern the optimal timing of capital investments (eg, oil field development) when learning is taking place (eg, oil field exploration); see Paddock et al. (1988).

Another related literature, primarily from the early 1970's concerns optimal growth in the presence of environmental externalities, particularly stock externalities. This was a natural extension of the optimal growth models that were popular in the 1960's and early 1970's. An important and characteristic paper in this genre is that of Keeler et al (1971). In that paper a simple optimal growth model is posited where utility is a function of consumption and a stock of pollution. Optimal paths for accumulation of capital and pollution are developed for several different types of pollution control. Other papers of this type include Plourde (1972), d'Arge and Kogiku (1973), Smith (1972), Plourde and Yeung (1989) and Forster (1973). Cropper (1976) also considers such a model of optimal growth but focuses on catastrophic environmental effects--the ultimate in irreversibilities.

4.2.2 Learning

There are three basic types of learning which are potentially applicable to global warming. One is active learning whereby observations on the state of the economy and the climate system convey information about uncertainty. Thus by perturbing emissions, one can obtain information about uncertain parameters (eg. see Balvers and Cosimano, 1990). A second type of learning is purchased learning whereby knowledge is purchased and the amount of knowledge purchased (R&D expenditures) depends on its cost and benefits (see Grossman and Shapiro, 1980). A third type of learning can be called autonomous learning where the mere passage of time reduces uncertainty. It is this third type of learning that we examine in this chapter.

4.2.2.1 Information Structures

The typical approach to including autonomous learning in models of irreversibility is to posit a two or three period model where uncertainty changes from one period to the next. Miller and Lad (1984) use a two period model with an *ex ante* probability distribution on period i benefits (b_i) of $f(b_1,b_2)$. After observing period one benefits, the *ex post* marginal distribution is obtained: $f(b_1,b_2 \mid b_1)$. While this is clearly learning, we need a way to parameterize the rate of learning so that the effects of the *rate* of learning can be deduced. Jones and Ostroy (1984), Olson (1990) and Marshak and Miyasawa (1968) provide such a framework through the concept of an ordering on information structures. Starting with a set of states of nature and an informative message, an information structure consists of a prior on the probabilities of receiving specific messages, along with a conditional probability on states of nature, given a specific message. Of two information structures with the same prior on states of nature, the one that has the greater variability in terms of possible posteriors is viewed as being "more informative." This is equivalent to the more informative structure yielding a higher attainable expected utility when the consumption bundle depends on the state of nature (see Jones and Ostroy, 1984, who point out that more flexibility can only be advantageous). Thus if two learning processes yield two comparable information structures, then the structure that is more informative corresponds to greater learning.

To quantify this concept of learning further, suppose there is a set of possible states of nature, indexed by $s=1,...,S$. Furthermore, suppose there is a finite set, Y, of possible "messages" containing information on the state of nature. Suppose the prior on receiving particular messages is q (dimension equal to the size of Y) and the conditional probability on states of nature (after the message $y \epsilon Y$ has been received) is $\pi(y)$. We use the term "prior" to refer to a probability distribution on state, before the message is received and posterior to refer to distributions assuming a message has been received. Let Π be a matrix with columns consisting of $\pi(y)$ with a different column for each y. Thus Π has S rows and the same number of columns as members of Y. (Π,q) is an information structure. A first goal is to develop an economically relevant ordering on information structures. A standard definition of the comparative value of information is provided by Jones and Ostroy (1984) (see also Laffont, 1989).

4.2.2.2 A Special Parameterization of Learning

We consider a special restriction on the set of comparable information structures. In particular, if there are S possible states of nature, we assume a message consists of a noisy signal as to the true state of nature and there are S possible noisy signals. Let

$\lambda \epsilon [0,1]$ reflect the level of information in the signal with 0 being no information and 1 being perfect information. Thus given a prior $\bar{\pi}$ we define the *star-shaped* information structure (Π, q) where $q \equiv \bar{\pi}$ and the sth column of Π is

$$\pi^s = (1-\lambda)\bar{\pi} + \lambda e_s \quad (4.1)$$

where e_s consists of all zeros except with a one in the sth position. Clearly $\pi_q = \Pi\bar{\pi} = \bar{\pi}$. Furthermore if $\lambda = 0$, each column of Π is $\bar{\pi}$ and if $\lambda = 1$, $\Pi = I$.

As an example, suppose you can receive one of three messages indicating whether the state of nature is 1, 2 or 3. We thus assume that number of possible messages equals the number of possible states-of-nature, which need not be the case. A message that conveyed the maximum amount of information would resolve all uncertainty on the state of nature. If the message is too noisy to contain any information, then the posterior on states of nature is the same as the prior. This is illustrated in Figure 4.1 where the simplex of probabilities on states of nature is shown. The prior is $\bar{\pi}$. The set of posteriors associated with a star-shaped spreading of beliefs, spread all the way out to the vertices, is shown by the three lines radiating out from $\bar{\pi}$. Perfect learning would move you to one of the three vertices following receipt of the message. Less learning would move you to one of the three points marked with circles. Even less learning would move you to one of the three points marked with x's after receiving the message.

The advantage of representing learning by this star-shaped spreading of beliefs is that the process can be parameterized by the λ in equation (4.1). The disadvantage is that we have eliminated perfectly legitimate and orderable learning processes (emanating from $\bar{\pi}$ in Figure 4.1).

4.2.3 Economy-Climate Models

Economic models have played a critical role in the formulation of environmental policy in the US over the past three decades. The main function of these models has been to simulate the economy's response to particular environmental regulations. Before putting a regulation in place, Congress or regulatory bodies desire to estimate the economic effects as well as the environmental effects of there regulations. That is precisely what an economic model can do, at least in theory.

One of the first economics papers in the global warming area is by Nordhaus (1977) and one of the earliest models is due to Nordhaus and Yohe (1983). They utilize a highly aggregated model, specifying in a single equation the relationship between world GNP and inputs of non-energy factors (such as labor), fossil fuels and nonfossil fuels. Technical change is explicitly represented. Using this highly simplified representation of the world economy, the authors focus on the effect of uncertainty in the underlying parameters on levels of CO_2 over the next 125 years. They are also able to infer which aspects of their model most affect atmospheric CO_2 levels.

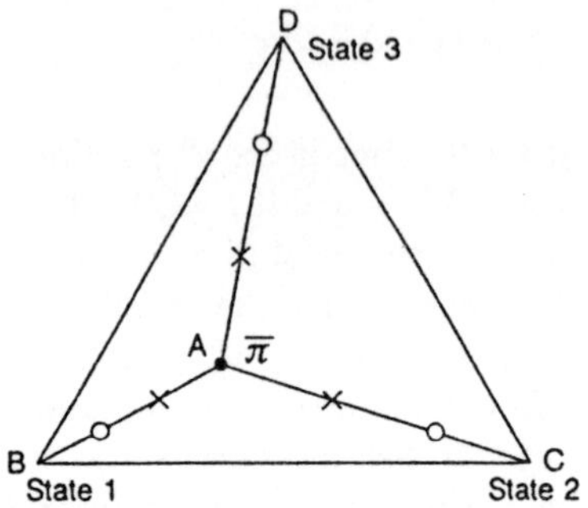

Figure 4.1 Star-shaped spreading of beliefs from $\bar{\pi}$.

William Nordhaus' work has also evolved considerably since the early 1980's. In recent papers (Nordhaus, 1992ab, 1993) he augments his economic model by incorporating equations representing the evolution of the atmosphere in response to greenhouse gas emissions. He has also conducted a useful and thorough review of the costs of control of greenhouse gas emissions (Nordhaus, 1991).

Also in the early 1980's a much more detailed model of the relationship between CO_2 and world energy demand was developed at Oak Ridge Associated Universities (Edmonds and Reilly, 1983). The model divides the world into nine regions. In each region aggregate energy demand is a function of prices and income. Supply of energy is represented in some detail, with various technological options represented separately. Using the model, they demonstrate the effect of carbon taxes, either on a world-wide basis or just for the US. Thus they are able to demonstrate the effect of regulation as well as the difficulty in controlling the problem at the sub-global level.

Edmonds and Reilly have developed their model further in recent years. In a recent paper (Darmstadter and Edmonds, 1989), they demonstrate the dramatic effect uncertainty can have on future CO_2 emissions. For example, they show that there is a 5% probability that CO_2 emissions will actually be substantially lower in 2050 than at present, at least given the probability distributions they assume for their exogenous parameters.

Recent entrants into the greenhouse gas analysis arena, though not newcomers to energy modeling, are Alan Manne and Richard Richels. They have extended Manne's ETA-Macro model (Manne, 1981) to include the generation of greenhouse gases (Manne and Richels, 1990). The model is used to determine the level of a carbon tax that would be necessary to support particular CO_2 emission goals for the US. Related to this question, they have used their model to look at the value of R&D in climate change by looking at the payoff from resolving uncertainty (Manne and Richels, 1991b, 1992).

The Manne and Richels framework has proved to be very popular in greenhouse policy circles. Peck and Teisberg (1992, 1993) have introduced a damage function into the Manne and Richels model to examine the influence of the curvature of the greenhouse gas damage function on optimal control policies. Manne and Richels (1991a) have substantially extended their model (referred to as the "Global 2100" Model) by considering several distinct regions of the world.

There are several other models of the economics of CO_2 generation that should be mentioned. Marks et al, (1989) and Dixon and Johnson (1989) have examined the effect of CO_2 emission controls on the Australian economy using the ORANI general equilibrium model. Richard Kosobud has developed a series of models, sometimes alone, sometimes in conjunction with others, to examine specific greenhouse gas issues (Kosobud, 1989, 1990).

4.3 A STOCHASTIC MODEL WITH LEARNING

4.3.1 The Stochastic Growth Model

In this section we present a general model of the dynamic evolution of an economy, corporating emission control, pollution accumulation and pollution damage. To a large extent it is a standard optimal growth model, although some aspects having to do with the climate are nonstandard. It is based on the climate-economy model of William Nordhaus (Nordhaus, 1992b). His model is deterministic however, and our model is stochastic.

The model is not regionally differentiated and involves the maximization of the net present value of expected utility. Utility is enhanced by consumption and depressed by pollution damage. Output can be channeled to consumption, emission control or investment. Uncertainty enters in that several states of the world (s) are possible and one wishes to maximize expected utility. The following represents such a model. The parameter h_t can be ignored (assumed constant) for the time being; it represents the history of learning and will be used later when we consider learning.

$$\max_{I,E} \sum_t (1+\rho)^{-t} \sum_{h_t} \mu(h_t, t) \sum_s \pi_s(h_t, t)\, u\left[c(h_t, t), d(h_t, t), s\right] L(t) \tag{4.2}$$

$$s.t.\ I(t,h_t), E(t,h_t) \geq 0 \tag{4.3}$$

$$Y(t,h_t) = f[K(t,h_t), L(t), E(t,h_t), t] \tag{4.4}$$

$$c(t,h_t) = [Y(t,h_t) - I(t,h_t)] / L(t) \tag{4.4a}$$

$$d(t,h_t) = g[T(t,h_t), Y(t,h_t)] / L(t) \tag{4.4b}$$

$$E(t+1,h_{t+1}) \leq e[E(t,h_t), Y(t,h_t),\ Y(t+1,h_{t+1})] \tag{4.4c}$$

$$K(t+1,h_{t+1}) = (1-\delta_K)K(t,h_t) + I(t,h_t) \tag{4.4d}$$

$$M(t+1,h_{t+1}) = (1-\delta_M)M(t,h_t) + \beta E(t,h_t) \tag{4.5}$$

$$T(t+1,h_{t+1}) = s[T(t,h_t), M(t,h_t), 0(t,h_t)] \tag{4.5a}$$

$$0(t+1,h_{t+1}) = r[T(t,h_t), 0(t,h_t)] \tag{4.5b}$$

where

I	= investment	(control)
E	= emissions of greenhouse gases	(control)
K	= capital stock	(state)
M	= stock of greenhouse gases	(state)
T	= mean atmospheric temperature	(state)
O	= mean deep ocean temperature	(state)
c	= per-capita consumption	
d	= per-capita climate damage	
Y	= gross output of goods and services	
ρ	= pure social rate of time preference	

δ_K	= capital depreciation rate
δ_M	= greenhouse gas decay rate
L	= population/labour supply
t	= time
s	= state-of-the world
β	= greenhouse gas emission factor
π_s	= probability of state s given learning history h_t: $\Sigma\pi_s = 1$
μ	= probability of learning history h_t

Equations (4.2 - 4.4) constitute the basic economic model and equations (4.5) describe the evolution of the climate. Equation 4.4c is used to constrain the emission rate of greenhouse gases based on previous control decisions; i.e. investment in emission control constrains future emission choices. The links between the economic model and the climate are E and T. Emissions of CO_2 (E) increase the stock of carbon dioxide (M) which increase average global temperature (T) which causes damage per capita (d) which, finally, yields disutility. The goal is to choose the investment path and emission path that maximize expected utility.

4.3.2 Learning

Introducing learning into this model involves introducing a second set of states corresponding to different messages that might be received (see discussion of eqn. 4.1). Each message yields a different outcome of the learning process where an outcome is a new probability on states of nature, $\pi_s(t)$. Let Y_t be the set of possible single period outcomes of the learning process at time t. One can think of these as messages as was previously discussed (or see Laffont, 1989). For instance, Y_t could contain three elements, $Y_t = \{y_1, y_2, y_3\}$ where y_1 = learning which increases the likelihood that global warming is serious, y_2 = learning which indicates global warming is a modest problem. y_3 = learning which indicates that global warming is not serious. Before learning occurs we do not know whether y_1, y_2 or y_3 will be realized although we do know the probabilities of elements of Y_t occurring.

While Y_t indicates the possible outcomes of the learning process at t, to know the current state of knowledge, it is also important to know the learning that has preceded t. We call this the history of learning, $H_t = \{(y_o, \ldots, y_t) \mid y_i \in Y_i \; \forall \; 0 \leq i \leq t\}$. Notationally, H_t contains the learning that occurred up to time period t. An element of H_t is a particular history.

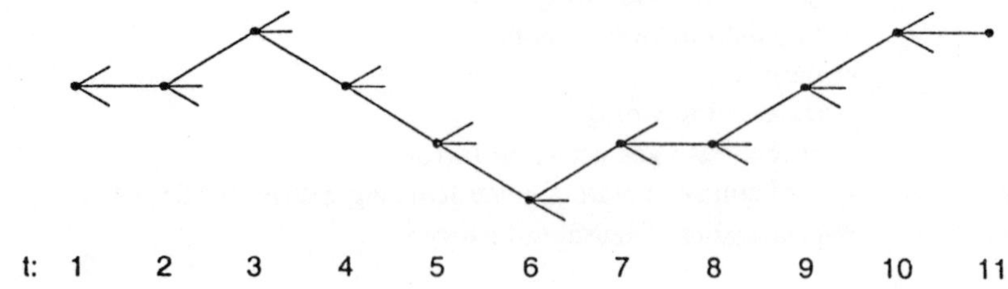

Figure 4.2 The Learning History: (0,1,-1,-1,-1,1,0,1,1,0,1,1,0).

For instance, consider a ten period world in which learning can proceed in three directions {-1, 0, 1} at any point in time. Figure 4.2 illustrates the history (0, 1, -1, -1, -1, 1, 0, 1, 1, 0). If we partition any h_t as (h_{t-1}, y_t), then we define the "predecessor" and "most recent" functions, φ: $H_t \rightarrow H_{t-1}$ and ψ: $H_t \rightarrow Y_t$ as $\varphi(h_t) = h_{t-1}$, $\psi(h_t) = y_t$. The function ψ indicates the most recent learning whereas φ indicates learning that occurred earlier. This allows one to functionally represent the learning path and to compute the probability vector on states of the world, $\pi(h_t,t)$. Define the transition matrix $\Pi_t(h_t)$ such that each column is a posterior probability vector corresponding to a different element of Y_{t+1}. Thus $\pi(h_t,t)$ is the column of $\Pi_{t-1}(\varphi(h_t))$ corresponding to $\psi(h_t)$. Furthermore if $q_t(h_t)$ is the probability vector associated with different elements of Y_{t+1}, then

$$\pi(h_t,t) = \Pi_t(h_t) q_t(h_t) \tag{4.6}$$

$(\Pi_t(h_t),q_t(h_t))$ is a learning structure as described earlier.

It is "easy" to modify model (4.2 - 4.5) to incorporate this learning. All of the variables in the model are already indexed by h_t we need only define how π (knowledge) and μ (probability of a history) evolve:

$$\pi(h_{t+1}, t+1) = \begin{cases} (1-\lambda)\pi(h_t,t) + \lambda e_s \\ (1-\lambda)\pi(h_t,t) \end{cases}$$

$$\text{if } s = \hat{s} \quad \text{otherwise} \tag{4.7a}$$

$$\mu(h_{t+1}, t+1) = \mu(h_t, t)\,\pi(h_t, t) \tag{4.7b}$$

where

$$h_t = \varphi(h_{t+1}) \tag{4.7c}$$

$$\hat{s} = \psi(h_{t+1}) \tag{4.7d}$$

where e_s is a vector of 0's and 1's with a 1 in the sth position and, 0 in the rest of the positions and λ is the rate of learning. The message space is assumed to be the same as the space of possible states of the world. Thus messages are noisy indications of the true state of the world.

There are many ways uncertainty can enter a model such as this. We assume uncertainty in the damage from global warming. Specifically, we write utility as

$$u[c(t), d(t), s] = \log[c(t) - \Delta_s d(t)]. \tag{4.8}$$

We further assume there are two states of the world, B and L, corresponding to global warming being a big problem (B) vs. global warming being a little problem (L). We assume $\pi_B(t=0) = .2$ and $\pi_L(t=0) = .8$ with $\Delta_B = 5$ and $\Delta_L = 0$. Thus damage could be five times as high as assumed by Nordhaus ($\Delta_B = 5$). This is somewhat arbitrary but yields an expected value of delta of 1 and reflects the fact that damage could be serious. The variance of Δ is 4. Figure 4.3 shows how the expected variance of Δ changes with learning at various learning rates.

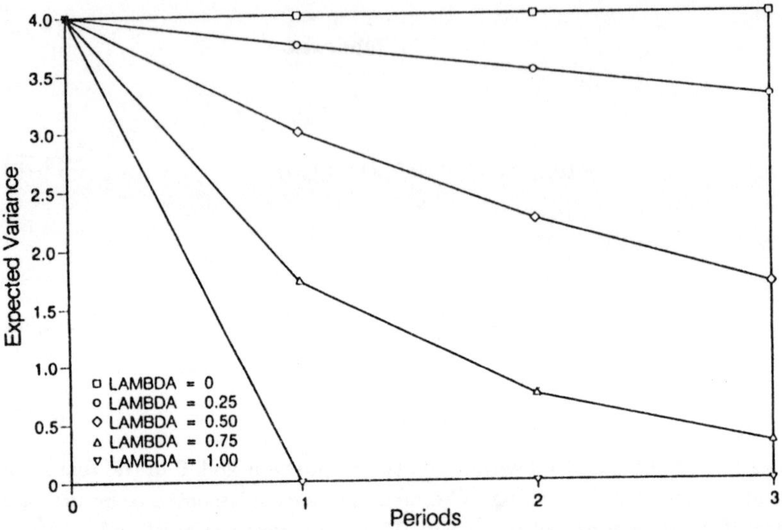

Figure 4.3 Evolution of Expected Variance of $[\pi_0(\Delta=5)=.2;\ \pi_0(\Delta=0)=.8]$.

4.4 RESULTS

The model described by eqn (4.2 - 4.8) has been implemented using time points at 10-year intervals beginning in 1965.[4] See the appendix for details on the implementation. The first three points (1965-85) are used as calibration and control of emissions is fixed at zero. Optimal emission control levels are computed beginning in 1995. Learning has occurred in 2005, 2015 and 2025. No learning occurs thereafter. Learning can be of two types, B or L, corresponding to suggesting global warming is a big problem (B) vs. suggesting global warming is a little problem (L). Figure 4.4 illustrates the possible paths learning can take. If the probability of B in 1965 is $\pi(0)$, it will stay at that value through 1995. At that point the probability change depends on how learning progresses. In 2025 and thereafter there will be eight possible values

[4] The model is solved using GAMS/MINOS. I am grateful to Lars Mathiesen for suggesting a stochastic programming approach.

of this probability, depending on the learning history. All of the variables in the model must be indexed on the path learning takes.

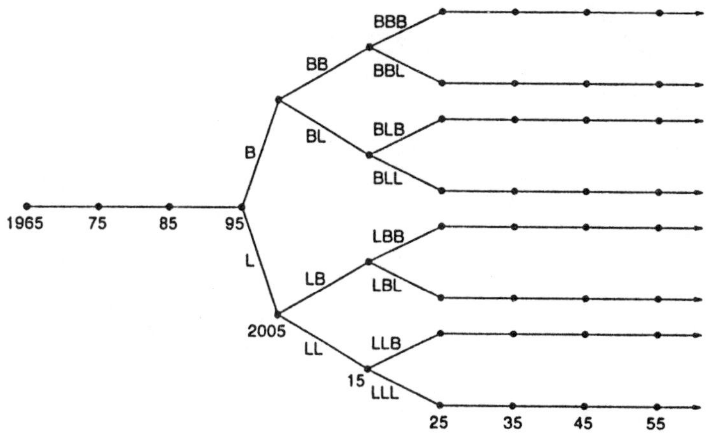

Figure 4.4 Three-period Learning.

The model described in the previous section was an infinite horizon model. Such a model takes a fair amount of computer time to solve so we have chosen to approximate this with a 20 period/200 year finite horizon model. Figure 4.5 shows, for the case of the deterministic model (4.2 - 4.5), the effect of the horizon on optimal emission control rates.[5] Clearly the control level in 1995 is largely unaffected for horizons in excess of 20 periods.

[5] The figure shows the solution to the deterministic model using eqn (8) as an objective with $\Delta_s = 1$ for S = B,L. This is the expected value of delta, and no learning occurs.

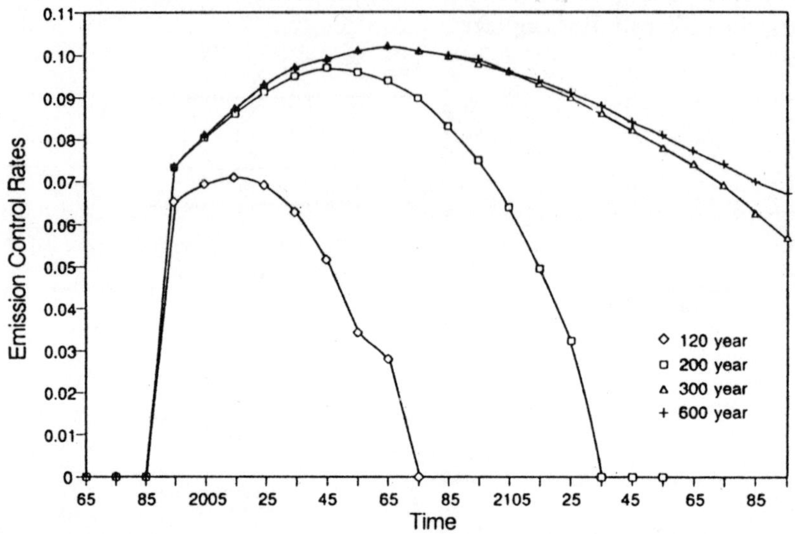

Figure 4.5 Optimal Emission Control Rates.

Model (4.2 - 4.8) was solved under two conditions. In one, emissions were allowed to take any positive value. In the other, the emission control rate, χ, was restricted to be monotonically non-decreasing, reflecting the fact that emission control investments tend to involve sunk costs: once a control level is implemented, it is unlikely to be decreased at a later date.

Focusing on the year 1995, with reversible emission control rates (i.e., eqn 4.4c is relaxed), we find that optimal control levels for greenhouse gases are virtually unaffected by the rate of learning. If one over controls today, then that error can be corrected in the future. Thus the fact that learning is taking place does not impact current decisions to control emission.

It is for this reason that we focus on emission control rates that are monotonically non-decreasing. In this case, once an emission control level is implemented it cannot later be reduced, even if climate change turns out to be less significant. Figure 4.6 shows how the level of emission control in 1995 is affected by the rate of learning, λ. Recall that $\lambda=0$ corresponds to no learning whereas $\lambda=1$ corresponds to resolution of all uncertainty in one period. Thus when learning is occurring rapidly, it pays to reduce greenhouse gas control rates from approximately 7% to as low as 3%. More modest learning rates involve more modest reductions in control rates, but reductions nevertheless.

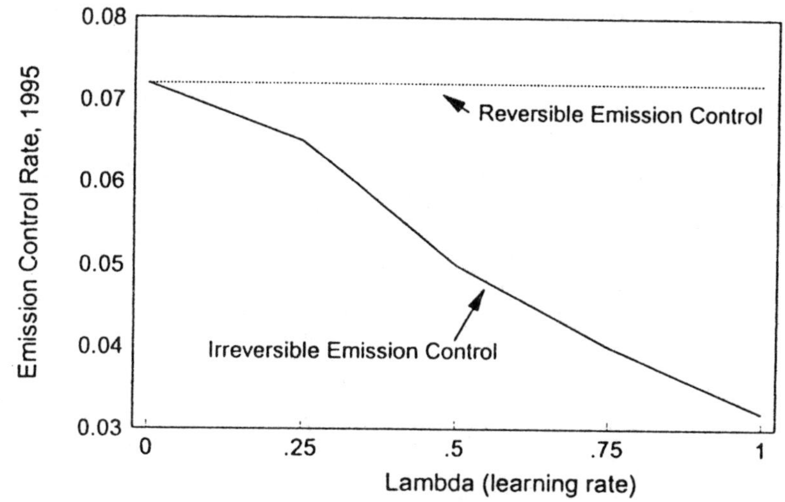

Figure 4.6 1995 GHG Control Rates vs Learning Rates.

The reason for lower emission control levels with more rapid learning is that it is better to defer irreversible control decisions until more is known. It is interesting that the irreversibility in control capital dominates the climate irreversibility. This is no doubt because of the long lags involved in climate change. Figure 4.7 shows for $\lambda=0.5$, how atmospheric temperature increases vary with learning histories.[6] Atmospheric temperature changes come long after the emissions occur. This suggests the potential payoff from control measures that are reversible.

It is interesting to compare this result to that of Manne and Richels(1992). While their model is substantially different than ours, they show that immediate resolution of uncertainty (very rapid learning) results in lower emission control rates (higher emissions) compared to the case when uncertainty is not resolved. This is qualitatively the same as our result. Figure 4.8 shows how emission control evolves with learning over time for $\lambda=0.5$. In 1995, control goes to 5%, before learning occurs. In 2005 one period of learning occurs. If that learning is B, control goes to 11%; otherwise it stays

[6] The learning histories are shown in the figure. The path BBL, for instance, corresponds to learning in the direction B in the first two periods, followed by learning in the direction L.

at 5%. After another period of learning, the range of control levels grows even further.

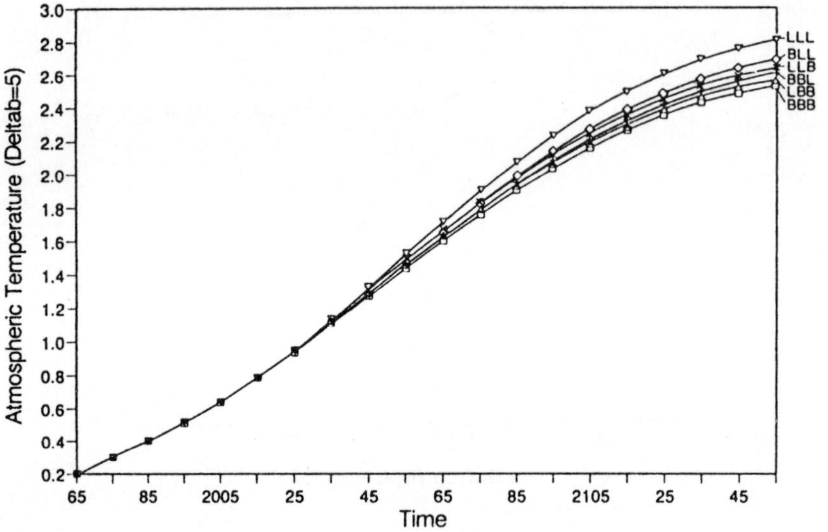

Figure 4.7 The Evolution of Atmospheric Temperature.

Figure 4.9 shows the value of information as a function of λ. The expected value of perfect information is the difference between the net present value of expected consumption less damage if emission control can be made completely contingent on the state of nature and the same figure with uncertainty and non-state-dependent controls. The expected value of perfect information (EVPI) shows what it could be worth to resolve all uncertainty instantaneously. Even modest learning rates substantially reduce the EVPI. This is consistent with Peck and Teisberg (1993) who show only modest value to resolving uncertainty.

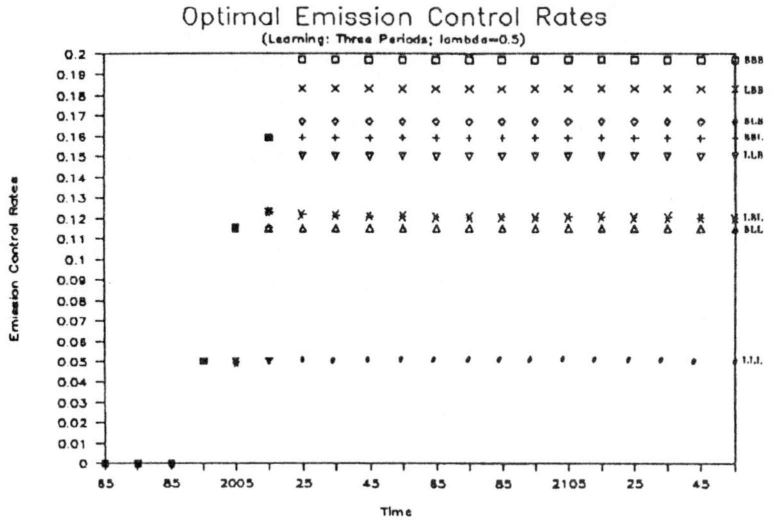

Figure 4.8 Emission Control Paths.

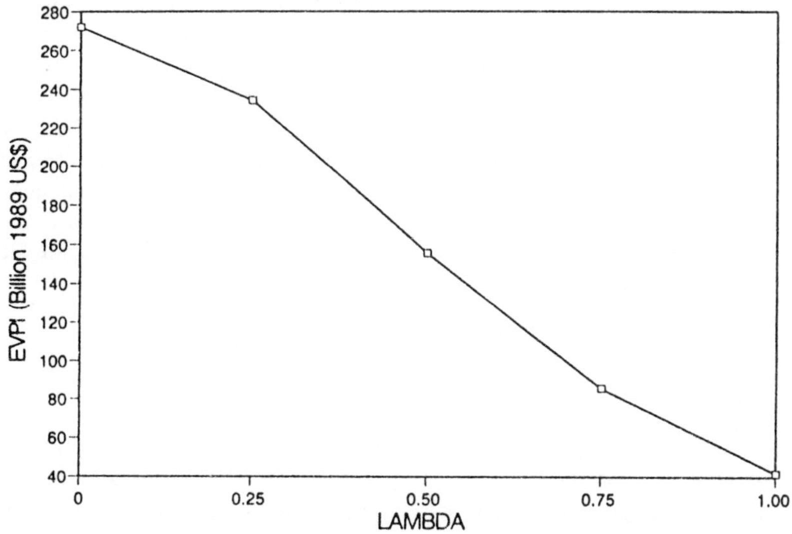

Figure 4.9 Expected Value of Perfect Information.

4.5 POLICY IMPLICATIONS

The results of this simple model do suggest that rapid learning biases current period CO_2 control levels downward. Thus these results tend to support those that argue that policy-makers should take a wait-and-see approach to CO_2 mitigation. There are two important qualifiers to this result, however. One is that learning only biases control downward and does not eliminate the desirability of some control. Secondly, the assumption of complete irreversibility of emission control investment may be extreme and certainty suggests the development of control strategies which are reversible; this is related to but not quite the same as a no-regrets strategy where actions are taken that we will never wish to reverse.

The most important caveat of all, of course, is that the model presented here is highly stylistic and is really only intended as an illustration or research tool; it is certainly not intended to be used to develop policy.

4.6 CONCLUSIONS

In this chapter we have presented an empirical model of learning with uncertainty in the context of the control of greenhouse gases. The particular focus of the model is on the conflicting irreversibilities of emission control investment and the stock effect of greenhouse gas accumulation. The issue is which of these irreversibilities dominates in determining today's emission control level.

Our conclusion is that there is no identifiable irreversibility effect associated with greenhouse gas accumulation. Furthermore, there is an effect of irreversibility in control investment. The net result is that with irreversibilities in control capital and significant learning, today's emission control levels should be reduced relative to either the no-learning case or the case of no control capital irreversibilities. Another implication is that a high priority should be placed on identifying reversible emission control.

APPENDIX: MODEL IMPLEMENTATION

In implementing the model (4.2 - 4.8), we have followed closely the deterministic DICE model (Nordhaus, 1993). Time units are ten year intervals; thus, suppressing h_t for the moment, the functions f,c,g,s and r in equations (4.4 - 4.5) are defined as f: $$Y(t) = [1 - 0.0686\chi(t)^{2.887}]A(t)K(t)^{0.25}L(t)^{0.75} \quad (4A.1)$$

where $$\chi(t) = 1 - \frac{E(t)}{\sigma(t)Y(t)} \quad (0 \leq \chi \leq 1) \quad (4A.2)$$

e: $$\sigma(t+1)\,Y(t+1) - E(t+1)) \geq (1-\delta_E)\{\sigma(t)\,Y(t) - E(t)\} \quad (4A.3)$$

g: $$d(t) = \frac{0.00148\,Y(t)\,T(t)^2}{L(t)[1 + 0.00148\,T(t)^2]} \quad (4A.4)$$

s: $$T(t+1) = T(t) + 0.226\{F(t) - 1.41\,T(t) - 0.440[T(t) - 0(t)]\} \quad (4A.5)$$

where $$F(t) = 4.1 \log [M(t)/590]/\log(2) \quad (4A.6)$$

r: $$0(t+1) = 0(t) + 0.02\ [T(t) - 0(t)] \quad (4A.7)$$

where A(t) and σ(t) are exogenous technology change parameters (see Nordhaus, 1993). The damage equation (4A.4) appears to be different than used by Nordhaus but some manipulation demonstrates the form is exactly as in Nordhaus (1992b). Starting values for state variables (in 1965) are T_o = 0.2(°C), O_o = 0.1(°C), M_o = 677 (gigatons) and K_o = 18.11 (trillion US$). In addition δ_M = 0.0833 and β = 0.64 (see eqn 4.5). Furthermore, δ_K = 0.35 and δ_E = 0 or 1. Remember that δ_M, δ_E, and δ_K are per decade. The pure rate of time preference is 3% per annum.

REFERENCES

Arrow, K.J. and A.C. Fisher (1974) *Environmental Preservation, Uncertainty, and Irreversibility*. Quarterly J. of Economics, 88:312-19.

Balvers, R. and T. Cosimano (1990) *Actively Learning about Demand and the Dynamics of Price Adjustment*. Economic J., 100:882-898.

Conrad, John (1980) *Quasi-Option Value and the Expected Value of Information*. Quarterly J. Economics, 92:813-19.

Cropper, M.L. (1976) *Regulating Activities with Catastrophic Environmental Effects*. J. Env. Econ. Mgmt., 3:1-15.

Cunha-e-sa, M.A. and C.D. Kolstad (1992) *Controlling a Stock Externality in the Presence of Active Learning*. unpublished manuscript, University of Illinois.

D'Arge, R.C. and K.C. Kogiku (1973) *Economic Growth and the Environment*. Rev. Econ. Stud., 40:61-77.

Darmstadtor, J. and J. Edmonds (1989) *Human Development and Carbon Dioxide Emissions: The Current Picture and Long-term Prospects*. Chapter 3 in N.J. Rosenberg et al. (eds) Greenhouse Warming: Abatement and Adaptation (Resources for the Future, Washington, DC.

Dixon, Peter B. and David T. Johnson *Estimation of the Macroeconomic Effects on Australia of Attempting to Reduce CO_2 Emissions by 20 percent by 2005*. University of Melbourne draft manuscript.

Edmonds, J. and J. Reilly (1983) *Global Energy and CO_2 to the Year 2050*. Energy Journal, 4(3):21-47.

Fisher, A.C. and W.M. Hanemann (1987) *Quasi-Option Value: Some Misconceptions Dispelled*. J. Env. Econ. & Mgmt, 14:183-90.

Fisher, A.C. and W.M. Hanemann (1990) *Information and the Dynamics of Environmental Protection: The Concept of the Critical Period*. Scand. J. of Economics, 92:399-414.

Forster, B.A. (1973) *Optimal Consumption Planning in a Polluted Environment*. Econ. Record, 49:534-545.

Freeman, A. Myrick (1984) *The Quasi-Option Value of Irreversible Development*. J. Env. Econ. & Mgmt., 11:292-94.

Grossman, S. and C. Shapiro (1986) *Optimal Dynamic R&D Programs*. Rand J., 17:581-593.

Henry, Claude (1974) *Option Values in the Economics of Irreplaceable Assets*. Rev. Econ. Studies Symposium, pp. 89-104.

Jones, R.A. and J.M. Ostroy (1984) *Flexibility and Uncertainty*. Rev. Economic Studies, 51:13-32.

Keeler, E., M. Spence and R. Zeckhauser (1971) *The Optimal Control of Pollution*. J. Economic Theory, 4:19-34.

Kolstad, C.D. (1993) *Looking vs. Leaping: The Timing of CO_2 Control in the Face of Uncertainty and Learning*. pp. 63-82 in Y. Kaya, N. Nakicenovic, W.D. Nordhaus and F.L. Toth (eds), *Costs, Impacts and Benefits of CO_2 Mitigation*. IIASA, Austria.

Kosobud, R.F. (1989) *Economic Analysis of Greenhouse Effect Policy Issues: East and West*. University of Illinois Institute for Government and Public Affairs Working Paper #10, Urbana, IL.

Kosobud, R.F.(1990) *How Much Greenhouse Policy Intervention, and When? A Great Lakes Regional Study*. paper presented at AAAS meeting, New Orleans.

Laffont, Jean-Jacques (1989) *The Economics of Uncertainty and Information*. MIT Press, Cambridge, Mass..

Manne, A.S. and R.G. Richels (1990) *CO_2 Emission Limits: An Economic Cost Analysis for the USA*. Energy Journal, 11:51-74.

Manne, A.S. and R.G. Richels (1991a) *Global CO_2 Emission Reductions--The Impacts of Rising Energy Costs*. Energy Journal, 12:87-107.

Manne, A.S. and R.G. Richels (1991b) *Buying Greenhouse Insurance*. Energy Policy, 19:543-552.

Manne, A.S. and R.G. Richels (1992) *Buying Greenhouse Insurance: The Economic Costs of CO_2 Emission Limits*. MIT Press, Cambridge, MA.

Manne, A.S. (1981) *ETA-MACRO: A User's Guide*. Electric Power Research Institute Report EA-1724, Palo Alto, California.

Marks, R.E., D.L. Swan, P.B. Dixon, D.T. Johnson, P. McLennan and R. Schodde (1989) *The Feasibility and Implications for Australia of the Adoption of the Toronto Proposal for CArbon Dioxide Emissions*. Report to CRA Ltd.

Marshak, J. and K. Miyasawa (1968) *Economic Comparability of Information Systems*. Int'l Econ. Rev., 9:137-74.

Miller, J.R. and F. Lad (1984) *Flexibility, Learning and Irreversibility in Environmental Decisions: A Bayesian Approach*. J. Env. Econ. & Mgmt., 11:161-72.

Nordhaus, W.D., (1977) *Economic Growth and Climate: The Case of Carbon Dioxide*. Amer. Econ. Rev.

Nordhaus, W.D. and G.W. Yohe (1983) *Future Paths of Energy and Carbon Dioxide Emissions*. in National Research Council Carbon Dioxide Assessment Committee, Changing Climate (National Academy Press, Washington, D.C..

Nordhaus, W.D. (1991) *To Slow or Not to Slow: The Economics of the Greenhouse Effect*. Economic Journal.

Nordhaus, W.D. (1992a) *An 'Optimal Transition Path for Controlling Greenhouse Gases*. Science, 258:1315-1319.

Nordhaus, W.D. (1992b) *The 'DICE' Model: Background and Structure of a Dynamic Integrated Climate-Economy Model of the Economics of Global Warming*. Cowles Foundation Discussion Paper 1009, New Haven, CT.

Nordhaus, W.D. (1993) *Rolling the 'DICE': An Optimal Transition Path for Controlling Greenhouse Gases*. Resource and Energy Economics, 15:27-50.

Olson, Lars J. (1990) *Environmental Preservation with Production*. J. Env. Econ. & Mgmt., 18:88-96.

Paddock, J.L., D.R. Siegel and J.L. Smith (1988) *Option Valuation of Claims on Real Assets: The Case of Offshore Petroleum Leases*. Quart. J. Econ., 103:497-508.

96 *The timing of CO_2 control in the face of uncertainty and learning*

Peck, S.C. and T.J. Teisberg (1992) *CETA: A Model for Carbon Emissions Trajectory Assessment*. The Energy Journal, 13:55-77.

Peck, S.C. and T.J. Teisberg (1993) *Global Warming Uncertainties and the Value of Information: An Analysis Using CETA*. Resource and Energy Economics, 1571-97.

Plourde, C.G. (1972) *A Model of Waste Accumulation and Disposal*. Canad. J. Econ., 5:119-125.

Plourde, C. and D. Yeung (1989) *A Model of Industrial Pollution in a Stochastic Environment*. J. Env. Econ. & Mgt., 16:97-105.

Smith, V.L. (1972) *Dynamics of Waste Accumulation: Disposal Versus Recycling.* Quart. J. Econ., 86:600-616.

[25]

Buying greenhouse insurance

Alan S. Manne and Richard G. Richels

There have been numerous proposals for immediate cutbacks in CO_2 emissions. Proponents argue that sizable reductions are necessary as a hedge against unacceptably rapid changes in climate. This paper provides a decision tree analysis of the problem. We examine how the optimal hedging strategy might vary with: a) the damage potential associated with the continued buildup of greenhouse gases; b) the accuracy and timing of climate research; and, c) the prospects for new supply and conservation technologies. The analysis is from the perspective of a single country, the USA. Optimal hedging strategies need to be explored on a region-by-region basis.

Keywords: Global warming; Decision analysis; Value of information

The greenhouse debate is enlivened but not necessarily illuminated by rhetoric on 'an irreversible ecological catastrophe' *v* 'the staggering costs of reducing emissions'. Suppose that instead we take the view of a thrifty purchaser of insurance who knows that the climate experts are currently deadlocked on the chances of a global calamity if we follow a business-as-usual policy. What steps should we take today to reduce the risks to future generations?

What are our options? Three forms of greenhouse insurance dominate current discussions: continued intensive science research to reduce climate and impact uncertainties, development of new supply and conservation technologies to reduce abatement costs, and immediate reductions in emissions in order to slow down climate changes.

Although these options define the rough outlines of the greenhouse policy debate, there is little agreement on their relative merit. Some argue that it is premature to think about doing anything other than intensive research. Others claim that scientific certainty on the greenhouse issue is an elusive goal, and that the risks of waiting are simply too great.

The problem with such perspectives is that they tend to over simplify the choices. The issue is not one of 'either/or', but one of finding the right blend of options. Uncertainty need not lead to paralysis. Policymakers must decide how to divide our greenhouse insurance dollars among competing needs. What portion goes to resolve climate uncertainties? What portion goes to technology development? And what portion goes to immediate abatement of emissions?

As is usually the case with insurance, the key to selecting an optimal policy is to understand how the options interact. For example, better information about greenhouse mechanisms and impacts is apt to influence technology development and emission reduction decisions. Similarly, the prospect of low-cost carbon-free supply alternatives and highly efficient end-use of technologies will affect both the need for better scientific information and the allocation of emission rights among generations.

In this paper, we analyse some of these interactions. A logical starting point is the issue of near-term emission cuts. There have been a number of recent proposals to reduce emissions in the industrialized nations. Proponents argue that such reductions are necessary as a hedge against unacceptably large greenhouse warming. It is important to understand how the optimal hedging strategy varies with the accuracy and timing of climate research, and with the prospects for new supply and conservation technologies.

We also look at the value of improved information. The USA and other Intergovernmental Panel on Climate Change (IPCC) participants have undertaken a substantial research programme in order to resolve climate-related uncertainties. Ideally, improved information will lead to better decisions. What is the value of reducing scientific uncertainty? How much accuracy is needed in order to improve our decisions on carbon limitations? How do the prospects for new supply and conservation technologies affect the value of information?

Alan Manne is Professor of Operations Research, Stanford University, Stanford, CA, USA; and Richard Richels is Director, Energy Analysis and Planning Department, EPRI, 3412 Hillview Avenue, PO Box 10412, Palo Alto, CA 94303, USA.

0301-4215/91/060543-10

Buying greenhouse insurance

As with our previous studies on CO_2-related issues, we employ the Global 2100 model.[1] Our model takes account of realistic rigidities in adapting both supplies and demands to changes in the value of carbon emission rights. We allow for the role of both price-induced and non-price energy conservation. The present analysis is restricted to the USA. Uncertainty is a major factor in issues involving insurance. A probabilistic version of Global 2100 is used to provide for the explicit consideration of uncertainty. In the next section, we employ a 'back-of-envelope' calculation to introduce some of the basic concepts that underlie an analysis of this type. We then provide a more rigorous treatment based on Global 2100.

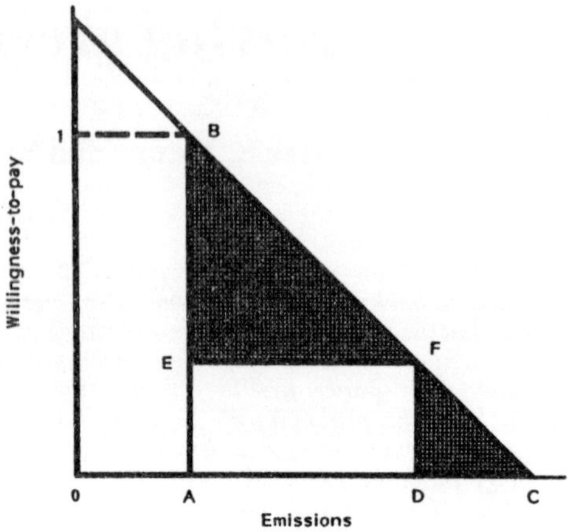

Figure 1. Willingness-to-pay for greenhouse emissions.

The value of information – an illustrative calculation

Our first calculations are intended to be illustrative but far from definitive. We begin by sketching out a decision-theoretic framework for handling the uncertainties inherent in the global greenhouse controversy. The basic idea is to avoid the controversies inherent in the debate over whether the benefits of control outweigh their costs. Instead, the analysis is focused upon identifying the value of information in making near-term decisions on emission reductions. In this first example, it is assumed that the ultimate benefits of control are either very large, or that they are negligible. Explicit probabilities are assigned to the two alternative long-term scenarios.

Consider the single period demand curve for emission rights shown in Figure 1. For simplicity, the willingness-to-pay for such rights is shown as a *linear* function of emissions. The *identical* function is used in order to characterize demands during the two 30-year periods before and after the resolution of climate uncertainties in 2020.

In the absence of a greenhouse tax, emissions during each period would be OC. With a sufficiently high tax (AB), emissions could be reduced to OA. Suppose that AC is the total *cumulative* amount of reductions required to avert a global catastrophe if the greenhouse pessimists turn out to be correct. One possibility might be to hedge by undertaking a small reduction (DC) during the first period and then a large reduction (AD) during the second period – if the pessimists are indeed correct. By assumption, the greenhouse danger will be identified early enough so that there is no long-term irreversible damage from a hedging policy. In the worst case (the business-as-usual scenario with no reductions in the first period, ie $DC = 0$), all that is required is a rapid replacement of capital stocks (housing, power plants, etc) during the years 2020–2050.

Without loss of generality, we may choose the units of emissions and of willingness-to-pay so that the distances $AC = AB = 1$. Let x be the decision variable that denotes the fraction of the total reductions to be undertaken during the first period, and let $1-x$ represent the fraction of the reduction during the second period. That is, x corresponds to the distance DC, and the lower shaded triangle DCF corresponds to the immediate costs of this decision. The second period's emissions are $AD = (1-x)$. The upper left shaded triangle BEF corresponds to the second period's costs. The costs during the first and second period are, respectively $.5\,x^2$ and $.5(1-x)^2$.

Let p be the probability that the greenhouse effect would be catastrophic, with this uncertainty resolved in 2020. *Without* time discounting, the expected money value of the costs incurred during the two periods together would depend as follows upon the probability p and the near-term decision variable x: $EMV(x) = .5x^2 + .5p\,(1-x)^2$.

With this cost function, it is straightforward to show that the optimal policy is to set $x = p/(1+p)$. That is, if we were certain that the greenhouse phenomenon will prove to be catastrophic ($p = 1$), the optimal policy would be to undertake half the cumulative reduction during the first period, and half during the second. Similarly, if there is only a 50-50 chance of a catastrophe, the optimal hedging strategy would be to commit to a 33% reduction in the near term, and a 67% reduction subsequently –

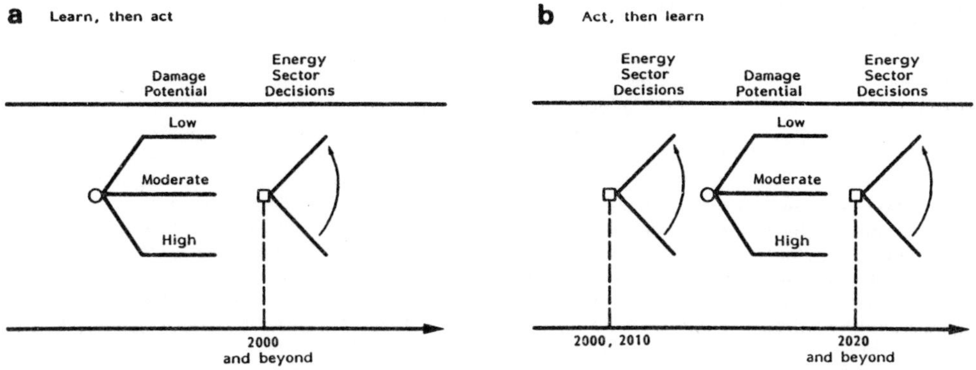

Figure 2. a 'Learn, then act'; b 'Act, then learn'.

should this prove to be necessary. This is a 'low regrets' policy. In this example (with $0 < p < 1$), there is no way to construct a 'no regrets' policy.

If $p = .5$, and therefore the optimal $x = .33$, the reader may verify that $EMV(.33) = .167$. These are the expected costs associated with decisions under uncertainty ('act, then learn'). By contrast, under a scenario approach ('learn, then act'), all uncertainties are resolved prior to the date at which decisions are taken. The expected value of costs would then drop to .125. This can be seen as follows: (a) If we know that there is no need to act, the cost is zero in both time periods; (b) If we know that we do need to act, the optimal policy is to set $x = .5$. The total cost in the two periods will then be: $.5(.25 + .25) = .25$. Since outcomes (a) and (b) are equally likely, the expected costs are .125. Therefore the expected value of perfect information, $EVPI = .167 - .125 = .042$. Note that the $EVPI = .042/.167 = 25\%$ of the expected costs under an 'act, then learn' policy. In plain English, this means that there is a large payoff to the resolution of uncertainties prior to 2020.

This example is suggestive but it needs to be extended to include more realistic features. The following analysis includes *non*-linear demand functions that shift from one period to the next. It includes time discounting and economic growth. Each of these ideas may be incorporated within the Global 2100 model, but we retain the basic ideas underlying the two-period sample. That is, if there is a positive probability that cumulative carbon emissions will lead to a global catastrophe, how much greenhouse insurance should we buy during the near future? Is it worthwhile to delay taking action until improved climate modeling information becomes available?

A framework for dealing with uncertainty

In previous applications of Global 2100, there was only an indirect treatment of uncertainty. For a given scenario, all uncertainties were resolved *prior* to decisionmaking. The analysis proceeded *as if* we had the opportunity to learn the state-of-world before taking action. Scenario analysis is one way to look at the here-and-now significance of alternative futures. Figure 2a shows the decision tree for this 'learn, then act' characterization of the decision problem. A circle denotes a chance node – a point at which an uncertainty is resolved. A square denotes a decision node – a point at which actions are required.

Figure 2b indicates the basic structure that underlies our introduction of uncertainty into Global 2100. It is characterized by the phrase 'act, then learn'. Decisions are taken at discrete points of time separated by intervals of one decade. The energy sector's supply and demand decisions for 2000 and 2010 must be made under certainty about the needs for carbon limits.

From 2020 onwards, decisions will be made *after* the resolution of uncertainty. In order to introduce uncertainty into the model, a state-of-world (*SOW*) subscript is attached to each of the original decision variables and constraints. During the two initial periods (2000 and 2010), the decision variables must take on identical values for all three *SOW*, but during the subsequent periods (from 2020 on), they may take on different values, depending on the magnitude of the carbon constraint.

Aside from the *SOW* subscripts, the only departure from the previous formulation is that we introduce one idea that is central to the back-of-envelope

Buying greenhouse insurance

Table 1. Emissions level.

Damage potential	**Carbon emissions**
Low	Unlimited
Moderate	20% below 1990 level
High	50% below 1990 level

calculation. We now allow for the possibility that carbon emission rights may remain unused during one time period, but may then be carried over for use at a later point in time. In effect, this means that the carbon tax cannot rise more rapidly than the real rate of return on capital – approximately 5%/year.

Just as in the two-stage model suggested by Figure 1, the cumulative allowable carbon limit is determined *exogenously* by a probabilistic mechanism. However, the period-by-period allocation of carbon emissions is determined *endogenuously* so as to maximize the expected discounted utility of consumption in the USA.

In order to introduce uncertainty into Global 2100, two additional sets of parameters are needed: (a) alternative carbon emissions scenarios; and, (b) the probabilities of realization of these scenarios. Figures 2(a) and 2(b) show three long-term possibilities regarding the damage potential associated with the continued buildup of greenhouse gases. For each of these possible outcomes, there is an appropriate constraint on carbon emissions. The more dire the outcome, the more drastic becomes the cutback in emissions.

Table 1 shows the emissions constraint ascribed to each of the three states-of-the-world under a 'learn, then act' procedure. If the damage potential is low, carbon emissions remain unconstrained. Otherwise, emissions are held at their 1990 level through the turn of the century and then reduced to the desired target by 2020. With a moderate damage potential, the 2020 target is 20% below 1990 levels. With a high damage potential, the 2020 target becomes a 50% reduction.

Under 'act, then learn', we make the decisions for 2000 and 2010 without knowing which of these three states will occur. Accordingly, we must select an optimal hedging policy for the emission levels during these years. This policy will depend upon the probabilities of the alternative outcomes. For illustrative purposes, suppose that the probability of low damage is 60%, and that the probability of a 20% reduction is 1.5 times that of a 50% reduction. That is, if controls are required, they are more likely to be moderate than high.

With these probabilities, Figure 3 provides a comparison between the optimal carbon emissions paths under 'learn, then act' and 'act, then learn'. Along the dashed lines, we take our decisions for 2000 and 2010 in full knowledge of whatever turns out to be the potential damages associated with global warming. Either the emissions are unlimited, or we begin to impose carbon constraints. The three paths diverge immediately after the 1990 base year.

By contrast, the solid lines show what happens when the damage potential remains unknown during 2000 and 2010. In this case, note that the carbon emission paths do not diverge until after 2010. The hedging strategy consists of adopting an emissions level that lies between the extreme cases shown along the dashed lines. Note that the optimal initial decisions (the emissions in 2000 and 2010) are quite different between the dashed and solid lines. There are losses associated with uncertainty. The probabilistic version of Global 2100 identifies a strategy which minimizes the *expected* value of these losses.

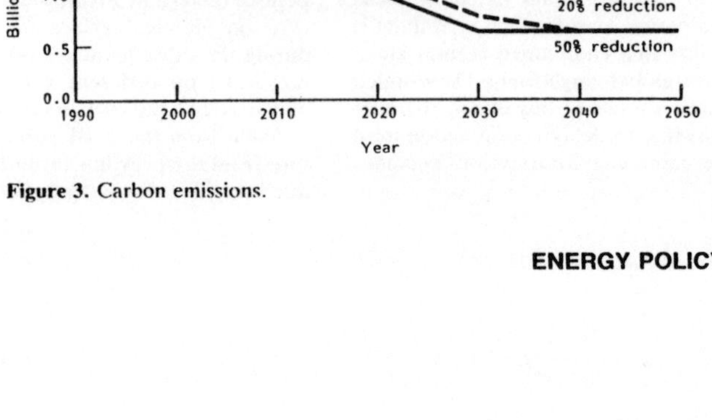

Figure 3. Carbon emissions.

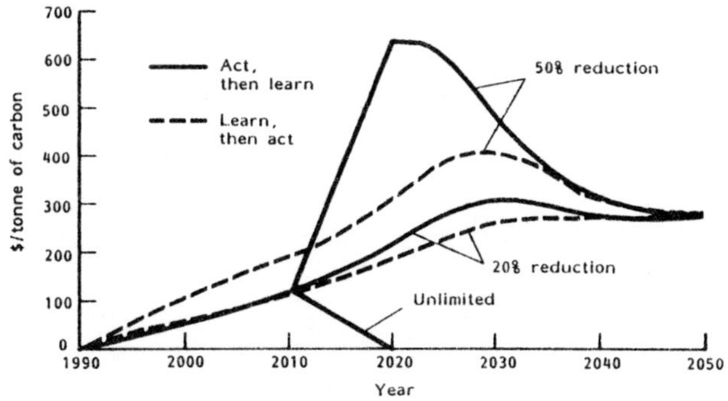

Figure 4. Carbon taxes.

Clearly the size of the hedging strategy will depend upon the probabilities assigned to alternative scenarios.

Associated with each emissions scenario, there is an appropriate carbon tax (Figure 4). This is sometimes known as 'state-contingent' pricing. With 'learn, then act', it is possible to adapt smoothly to whichever of the three states-of-world eventually occurs. With 'act, then learn', we do not know the state-of-world during the initial years (2000 and 2010). The tax rate must be identical until the certainties are resolved. There is an abrupt transition at the point when we learn which of the three carbon limitations must be adopted. If carbon limitations are required, the long-term equilibrium tax rate turns out to be $250/tonne. Because of the costs of rapid transition, there is a substantial overshoot above this level when a 50% carbon reduction is imposed, but only a modest overshoot in the case of a 20% reduction.

Perfect information

A substantial research programme is underway to reduce climate-related uncertainties. What is the value of such a programme? First, consider the case of perfect information. All uncertainties are resolved prior to the energy sector's decisions for the year 2000. Although this is straightforward from an analytical perspective, it is also quite unrealistic. Nonetheless, it represent a useful point of departure. It places an upper bound on the value of a climate research programme. In Figure 5, the value of perfect information is shown as a function of the subjective probability that emissions will remain unconstrained. (Throughout the following computations, the probability of a 20% reduction is 1.5 times that of a 50% cutback – regardless of the probability that emissions will remain unconstrained.)

The value of perfect information is the value of being able to chart the right course from the outset. In the face of uncertainty, it is prudent to hedge our bets. 'Act, then learn' leads to initial decisions which lie between the extreme values shown under 'learn, then act'. With a hedging strategy, we will always be somewhat off-course. Near-term actions *are* affected by the distant-future uncertainties. If we find out that the damage potential is low, we will have overly constrained carbon emissions during the near term. If the true damage potential turns out to be moderate or high, our near-term emissions will have been excessive. In either case, getting back onto the long-term track will involve costs. These costs may be eliminated by knowing the true damage potential in advance.

According to Figure 5, perfect information may be worth tens of billions of dollars to the USA in terms of its impact on macroeconomic consumption. Discounted to 1990 at 5%, the EVPI (expected value of perfect information) is at its maximum ($90 billion)

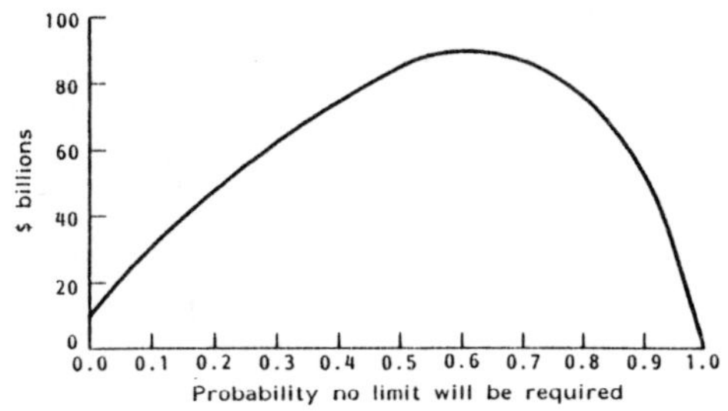

Figure 5. Expected value of perfect information.

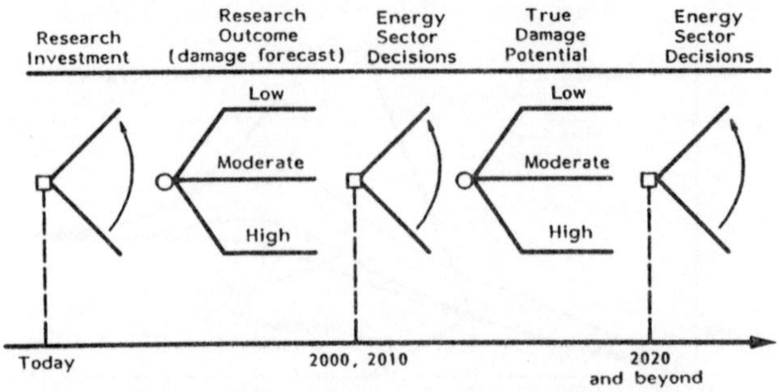

Figure 6. Decision tree for value of information analysis.

when the probability of unlimited emissions is .60. The EVPI falls to zero as the probability approaches one. Note, however, that the curve is not symmetric. A zero probability of unlimited emissions does not imply a zero value of information. We would still not know whether a 20% or a 50% emissions cut is going to be required. The greater the uncertainty on the need for an emissions constraint, the higher becomes the value of information. On the other hand, if one already knows the true outcome, further information will be of little value. It will only confirm that which is already believed.

The value of improved information

Perfect information is an elusive goal. What is the value of improved information? How much accuracy in climate modeling is needed in order to improve our decisions on carbon limitations? What if uncertainties are resolved in stages spanning a period of several decades?

Figure 6 shows how the original decision tree may be modified to address these questions. Our initial decisions relate to how much we are willing to pay to reduce uncertainties about the damage potential. Assume that the research programme will yield improved information soon enough so that we can revise our supply and conservation decisions for the year 2000. Regardless of the magnitude of research expenditures, it is unlikely that uncertainty can be completely resolved by 2000. Accordingly, we assume that the damage potential will remain unknown well into the second decade of the 21st century. It is only then that we will know with certainty what emissions level is appropriate.

The decision tree structure reflects the fact that the energy sector's near-term supply and conservation decisions must be made in the face of uncertainty regarding the ultimate damage potential. The degree of uncertainty can be represented by a likelihood table indicating the probability that the damage forecast will be low, moderate, or high – given that the true damage potential is low, moderate, or high. This representation allows the use of standard probabilistic techniques to incorporate climate research outcomes into the carbon decision model.

Table 2 illustrates the conditional probabilities for the research outcomes. The information would be perfect if the diagonal entries in Table 2 were equal to unity and all other cells were zero. In this case, there could be no error associated with any forecast resulting from the climate research programme. What if the research programme turned out to be useless, and the forecast were independent of the

Table 2. Conditional damage forecasts.

Research outcome (damage forecast)	True damage potential Low	Moderate	High
Low	Pr(Low forecast/low damage)	Pr(Low forecast/moderate damage)	Pr(Low forecast/high damage)
Moderate	Pr(Moderate forecast/low damage)	Pr(Moderate forecast/moderate damage)	Pr(Moderate forecast/high damage)
High	Pr(High forecast/ low damage)	Pr(High forecast/ moderate damage)	Pr(High forecast/ high damage)

Table 3. Assumed likelihood matrix.

Research outcome (damage forecast)	True damage potential Low	Moderate	High
Low	σ	(1−σ)/2	(1−σ)/2
Moderate	(1−σ)/2	σ	(1−σ)/2
High	(1−σ)/2	(1−σ)/2	σ

actual damage potential? This case would be presented by equal entries in all cells of the likelihood table.

A reasonable research programme should produce something less than perfect and something better than no information. The accuracy of the information depends on the particular characteristics of the likelihood table. In Table 3, we assume that the errors are symmetrically distributed. If a forecast is wrong, it is equally likely to be wrong on the low side as on the high side. The parameter σ is a measure of the accuracy of the forecast.

Consider two extreme cases. If σ = 1/3, the research programme is useless. All forecasts would be equally probable regardless of the damage potential. Conversely, if σ = 1.0, this would be the case of perfect information. Accordingly, the accuracy of climate forecasts will increase with σ over the interval between 1/3 and 1. We now show how Global 2100 may be employed to estimate the economic returns from a research programme with different values of the accuracy parameter σ.

Hedging strategies under imperfect information

We now turn to the issue of information accuracy – the case in which σ is less than unity. When σ = 1, the climate research programme will unequivocally reveal the true damage potential soon enough so that we can revise our supply and conservation decisions for the year 2000. When $\sigma < 1$, the climate research results will be imperfect. Just as in weather forecasting, there will be a margin for error.

Now continue to assume that the probability of a low damage potential is .60. Figure 7 illustrates the relationship between information value and information accuracy. Note how quickly the value of information drops off as we adopt lower values for σ. For example, if there is one chance in five that the research results are misleading (σ = .80), the value of information falls to $40 billion. This is still significant but less than half the value of perfect information. If there are two chances in five of the results being correct (σ = .60), the value drops to approximately $10 billion. In other words, it is worth $30 billion to increase research accuracy from .60 to .80.

The value of the research programme is closely tied to our confidence in the results. Perfect information eliminates the need to hedge against surprises. The more faith we have in the forecast, the less need to hedge. For example, Figure 8(a) shows the optimal hedging policy when the research programme provides no new information (σ = .33). The dashed lines display the paths under 'learn, then act'. These are the policies that would be chosen if we knew the true damage potential in advance. By contrast, the solid line indicates the optimal hedge that is associated with 'act, then learn'.

Suppose that before committing to an emissions strategy we are able to observe the results of a research programme, and that our confidence in the findings is expressed by setting σ = .75. Further, suppose that the research indicates a low value of the potential damage. Figure 9 is a standard application of Bayesian analysis. It shows how the research outcome leads to a revision of our beliefs. Prior to receiving the information, the odds of a low damage potential are six chances in 10. They now rise to nine in 10.

Returning to Figure 8(b) we see the impact on the optimal hedging policy. As one would expect, results pointing to a low damage potential move us closer to the unlimited emissions strategy. The research programme has strengthened our belief that the true damage potential is low. In doing so, it has reduced the need to hedge against large future emission cuts.

Better information need not result in business-as-usual – that is, in an unconstrained emissions policy. If the research results indicate a significant damage potential, rational decisionmakers would modify their beliefs accordingly and adopt a more restrictive emissions policy. The value of a climate research programme stems from its potential to modify our near-term decisions. The more confidence we attach to its results, the less we need to hedge.

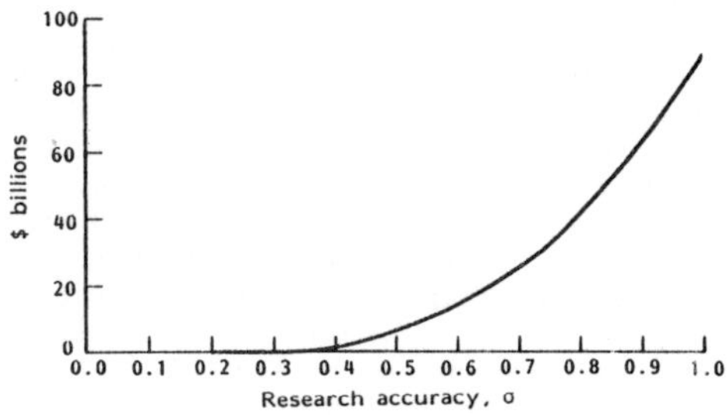

Figure 7. Expected value of imperfect information.

Buying greenhouse insurance

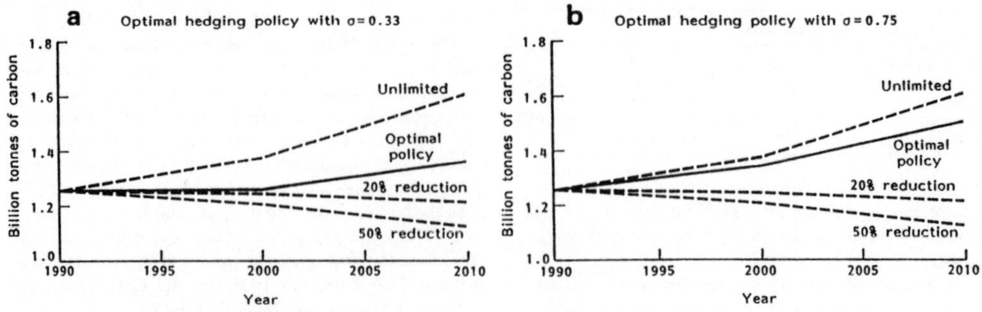

Figure 8. Optimal hedging policies.
Note: Research indicates low damage potential.

The impact of supply and conservation estimates

The value of information is sensitive to one's degree of optimism or pessimism concerning new supply and conservation technologies. This can be demonstrated through an illustration involving two key parameters: (a) the availability of low-cost carbon-free substitutes for conventional crude oil; and, (b) the rate of autonomous energy-efficiency improvements.

In Global 2100, carbon-free non-electric energy sources are grouped into two broad categories: low-cost renewables such as ethanol from biomass (RNEW) and high-cost backstops such as hydrogen produced through photovoltaics and electrolysis (NE-BAK). The key distinction is that RNEW is in limited supply, but that NE-BAK is available in unlimited quantities at a constant but considerably higher marginal cost. The following calculations will focus on RNEW. Up to this point, we have adopted annual limits of 5 quads for this technology. For illustrative purposes, we now investigate limits of 0 and 10 quads. These might reflect alternative views on the availability of land and water for cultivating biomass.[2]

On the demand-side of the model, there are two mechanisms for decoupling energy consumption from GDP growth. There is conservation induced by an increase in the price of conventional crude oil, and there is also the impact of *costless* autonomous energy-efficiency improvements (AEEI). Such non-price induced efficiency improvements may be brought about by deliberate changes in public policy, eg mandatory efficiency labeling standards. Hitherto, we have assumed an AEEI of 0.5%/year. We now investigate alternative values of 0.0% and 1.0%.[3]

Three possibilities are summarized in Table 4. The optimistic scenario describes an energy future that is considerably more attractive than the base case. Low-cost carbon-free supply alternatives are in greater supply and there is a lower overall demand for energy. The pessimistic scenario describes a very different future – one with few options for energy supplies and conservation.

The top three dashed lines in Figure 10 show the implications for carbon emissions when no limits are imposed. Under the optimistic scenario, emissions rise more slowly than in the base case. There are low-cost carbon-free substitutes for crude oil, and there are lower energy demands. Conversely, for the pessimistic scenario, there is greater dependence on coal-based alternatives to petroleum, there are higher demands for primary energy, and there are higher carbon emissions.

These differing perspectives have a direct impact upon the EVPI. (See Figure 11.) Technological optimism lowers the value of information. Why? From Figure 10, recall that optimism leads to low carbon emissions – even under a business-as-usual

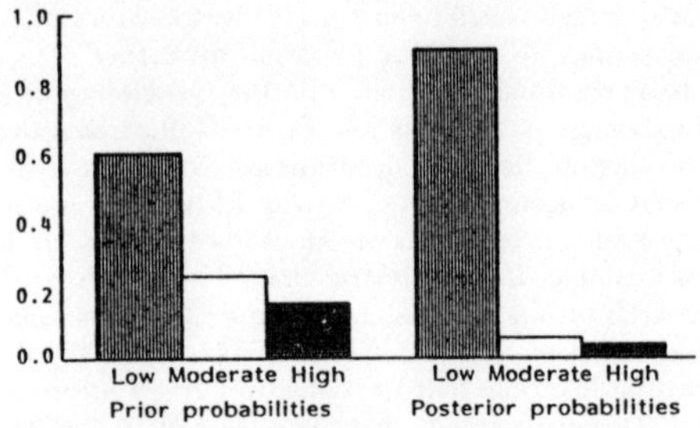

Figure 9. Effect of low damage forecast on subjective probabilities.

Table 4. Alternative supply and demand scenarios

Scenario	RNEW (quads/year	AEEI (%/year)
Pessimistic	0	0.0
Base	5	0.5
Optimistic	10	1.0

scenario. There is a smaller *difference* between the constrained and unconstrained paths than in the base case. When the constrained and unconstrained scenarios lie close together, there are lower costs associated with being off-course. This reduces the value of knowing the desired path in advance. Conversely, when the constrained and unconstrained scenarios lie further apart, there are higher costs of being off the right track. Technological pessimism tends to raise the value of information.

Technological optimism reduces the size of the optimal hedge. (See Figure 12 – based on 'act, then learn'.) In this case, the constrained and unconstrained emission paths lie close together. Future generations will have less difficulty in adapting to a carbon limit. This considerably reduces the incentive for immediate abatement measures. By contrast, technological pessimism increases the size of the optimal hedge.

Conclusions

There have been numerous proposals for immediate cutbacks in CO_2 emissions. Proponents argue that sizeable reductions are necessary as a hedge against unacceptably rapid changes in climate. This paper examines how the optimal hedging strategy might vary with: (a) the probability of eventual carbon limitations; (b) the accuracy and timing of the climate research programme; and, (c) the prospects for new supply and conservation technologies.

Our analysis yields several useful insights. We find that the size of the hedge is quite sensitive to the

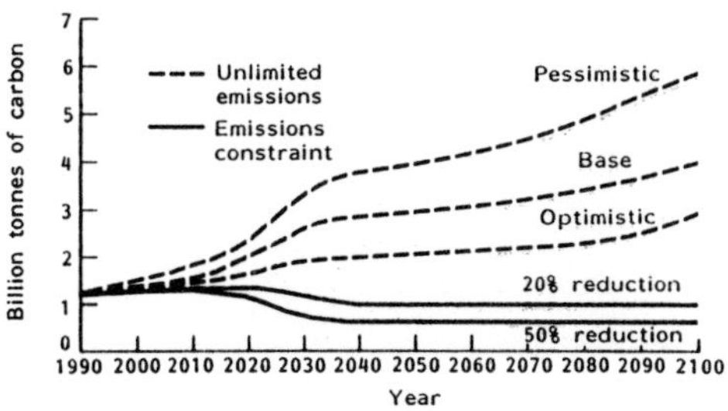

Figure 10. Carbon emissions for three supply/demand scenarios.

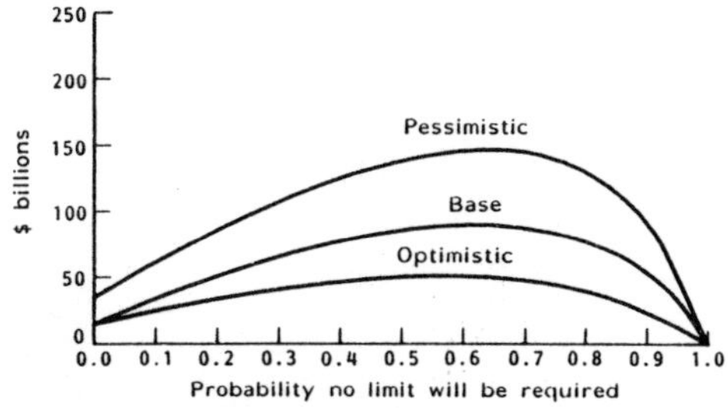

Figure 11. Expected value of perfect information.

quality and timing of the climate research results. In the best of all possible worlds, uncertainties would be completely resolved prior to energy sector decisionmaking. With perfect information today, we could chart the best course of immediate action. There would be no need to hedge our bets. By contrast, the less faith we have in the possibility of near-term research accuracy, the greater the need for precautionary actions.

It is straightforward to calculate the benefits from reducing scientific uncertainty. Better information leads to better decisions. Global 2100 allows us to calculate the aggregate economic impacts. The analysis leaves little doubt that there can be a big payoff to reducing climate-related uncertainties – and that it could range upwards of $100 billions for the USA alone. At the same time, we find that the value of information is quite sensitive to one's confidence in its timeliness and reliability. Relatively small increases in accuracy can yield substantial dividends.

The optimal hedging strategy is also sensitive to the prospects for new supply and conservation technologies. Consider two alternative energy futures. In the optimistic scenario, there are abundant low-cost carbon-free supply alternatives and highly efficient end-use technologies. In the pessimistic scenario, there is heavy dependence on carbon-intensive synthetic fuels, and there is less autonomous conservation. Clearly, with the latter scenario, it is more difficult to adapt to a carbon constraint. If one is a technology pessimist, it may make sense to shift some of the burden of emissions abatement to the present generation. For the optimist, the story is reversed. Future generations will need little help in adapting to a carbon constraint. There is less pressure for immediate cutbacks.

The near-term policy implications are clear. There is less need for precautionary emission cutbacks if we undertake a sustained commitment to reducing climate uncertainty and to developing new supply

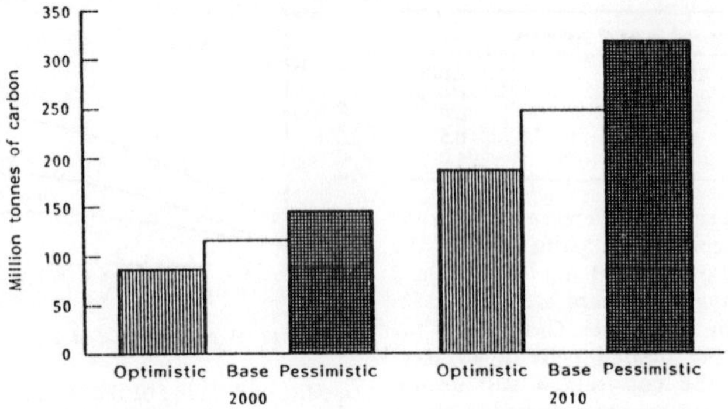

Figure 12. Optimal emission reductions for three supply/demand scenarios.

and conservation options. Better climate information reduces the need to hedge against a potentially hostile future. Improved supply and conservation technologies will enhance our ability to deal with such a future if it should occur.

The research reported in this paper was funded by the Electric Power Research Institute (EPRI). The views presented here are solely those of the individual authors, and do not necessarily represent the views of EPRI or its members. The authors are much indebted to Robert Luenberger for research assistance. We have benefitted from discussions with: Hung-po Chao, Lawrence Goulder, James Hammitt, William Hogan, David Montgomery, John Rowse, Stephen Peck and Leo Schrattenholzer.

[1]A. Manne and R. Richels, 'CO_2 emission limits: an economic cost analysis for the USA', *The Energy Journal*, Vol 11, No 2, April 1990.
[2]National Research Council, *Confronting Climate Change*, National Academy Press, Washington, DC, USA, 1990.
[3]For arguments supporting lower AEEI values, see W.W. Hogan and D.W. Jorgenson, *Productivity Trends and the Cost of Reducing CO_2 Emissions*, Energy and Environmental Policy Center, Harvard University, MA, USA, 1990; for a contrary view, see R.H. Williams, 'Low cost strategies for coping with CO_2 emission limits', *The Energy Jounal*, Vol 11, No 4, October 1990, pp 35–59.

Resource and Energy Economics 15 (1993) 71–97. North-Holland

Global warming uncertainties and the value of information: An analysis using CETA*

Stephen C. Peck

Electric Power Research Institute, Palo Alto, CA 94303, USA

Thomas J. Teisberg

Teisberg Associates, Weston, MA 02193, USA

Final version received November 1992

In this paper, we investigate the sensitivity of optimal carbon control strategies to parameters of the Carbon Emissions Trajectory Assessment (CETA) Model, and we use CETA in a simple decision tree framework to estimate the value of information about global warming uncertainties. We find that if an optimal control policy is used under uncertainty, the eventual resolution of uncertainty has high value relative to current research budgets, and resolving uncertainty about the costs of warming is nearly as important as resolving uncertainty about the extent of warming. In addition, we find that there is not a high premium on immediate resolution of uncertainty, if resolution would otherwise occur within, say, twenty years; this implies that time is available to plan and execute a carefully designed research program. On the other hand, we find that if the real world political process would result in a suboptimal control policy being chosen under uncertainty, and this choice could be prevented by early resolution of uncertainty, the benefit of early resolution may be as much as three orders of magnitude greater.

1. Introduction and summary

While there is great disagreement about what policies, if any, are appropriate responses to the threat of global warming, there is fairly general agreement on the need for research to improve our understanding of the warming problem. Nevertheless, there has been relatively little quantitative analysis of the value of information about global warming.[1]

Correspondence to: S.C. Peck, Electric Power Research Institute, 3412 Hillview Avenue, P.O. Box 10412, Palo Alto, CA 94303, USA.

*This paper does not represent the views of EPRI or of it members. We are grateful to the following persons for advice: Hung-Po Chao, Jae Edmonds, Robert Hahn, Charles Kolstad, Michael MacCracken, Alan Manne, Richard Richels, Michael Schlesinger, John Tukey, Franz Wirl, and Arnold Zellner, as well as to two anonymous referees. We are responsible for any remaining errors.

[1]Two exceptions are Manne and Richels (1991b) and Nordhaus (1991).

In this paper, we use the Carbon Emissions Trajectory Assessment (CETA) Model to investigate the value of information about global warming. CETA is an economic growth and energy use model that also incorporates representations of greenhouse gas accumulation, global mean temperature rise, and the adaptation/damage cost associated with this temperature rise.[2]

We begin our investigation of the value of information with a brief presentation of optimal policy for the central case parameter assumptions of the model. This is followed by results indicating the sensitivity of optimal policies to variations in key parameters. If optimal policy changes significantly as a result of variation in a particular parameter, the value of information about that parameter is likely to be high.

Continuing our investigation of the value of information, we use CETA in a simple decision tree framework to estimate the value of information about a few key model parameters. Based on the sensitivity results, the key parameters on which we focus in this paper are those relating to the extent of warming and the costs of warming.

A central issue concerns the choice of emissions control policy under uncertainty. In the standard decision analysis approach under uncertainty, the policy is set so as to maximize expected net benefits. However, the global warming problem is being considered in a highly political context involving governments of many countries with differing perspectives and interests. In this context, emissions control policies chosen in the absence of good information may be far from the optimal policy. Thus we present results assuming both that the policy under uncertainty is optimal, and that the policy is arbitrarily chosen in the political process.

If policy under uncertainty is chosen to maximize expected net benefits, our results suggest that the value of information can be up to hundreds of billions of dollars; numbers of this magnitude justify devoting serious attention to resolving uncertainties. While the value of information is greatest for information regarding the potential warming anticipated from a given increase in CO_2 concentration, the value of information regarding the future damage costs of warming and of adaptation to warming is nearly as great. Since research budgets are now directed primarily at resolving scientific uncertainties like that about the extent of potential warming, it would appear that budgets for research on impacts and adaptation are relatively under-funded.[3]

Although resolving uncertainty produces a large benefit relative to not resolving uncertainty, the benefit of resolving uncertainty quickly is surprisingly low. Specifically, we find that the benefit of resolving uncertainty now

[2]For a detailed description of CETA, see Peck and Teisberg (1992a).

[3]We say 'appear' because our analysis does not directly assess the *incremental* reduction of uncertainty obtainable for an *incremental* dollar spent on climate research vs. impacts research.

instead of 20 years from now is roughly 2% of the overall benefit of resolving uncertainty. This result is due to the fact that the optimal energy use policy in our model would be about the same over the next couple of decades, for any resolution of uncertainty about the key model parameters. However, by the middle of the next century, optimal energy use policies will become more sensitive to the key model parameters. Consequently, the benefits of accelerating uncertainty resolution by 20 years would be much higher later on.

If the emissions control policy under uncertainty is assumed to be arbitrarily chosen in a real world political process, there is obviously a wide variety of possible policies that might result. We consider two such policies: no emissions reductions and an emissions limit at the 1990 level. If these emissions control policies would be chosen under uncertainty, while optimal policies would be used once uncertainty is resolved, the value of resolving uncertainty now instead of twenty years from now is much greater – specifically, if the arbitrary policy under uncertainty were to be no emissions reduction, the benefit of resolving uncertainty is an order of magnitude greater; and if the arbitrary policy were to be an emissions limit at the 1990 level, the benefit is three orders of magnitude greater.

2. CETA model results and sensitivities

As a prelude to using the CETA model to estimate the value of information, this section briefly describes the model, discusses our Low, Central, and High case values for key model parameters, shows model results for our 'Central case' parameter value assumptions, and presents a brief summary of the sensitivity of the model's results to variations in key parameters.

2.a. The CETA model

The CETA model represents world-wide economic growth, energy consumption, energy technology choice, global warming, and global warming costs over a time horizon of more than 200 years.[4] Fig. 1 presents a schematic overview of the key relationships in the model. Energy technologies and the oil, gas, and coal resource bases are inputs to an energy submodel, which supplies energy inputs to a production submodel, and the

[4]See Peck and Teisberg (1992a) for a detailed description of CETA. The results in this paper are obtained using an updated version of CETA. In the new version, the utility function is population times the logarithm of consumption per capita, instead of simply the logarithm of consumption. The new utility function tends to increase marginal utility of consumption in the future relative to marginal utility of consumption now, because population grows over time. In addition, we now use a carbon cycle model adopted from Maier-Reimer and Hasselmann (1987), in which carbon emissions fall into five classes with differing atmospheric lifetimes of two years to infinity; before we had used a single class with a lifetime of 233 years.

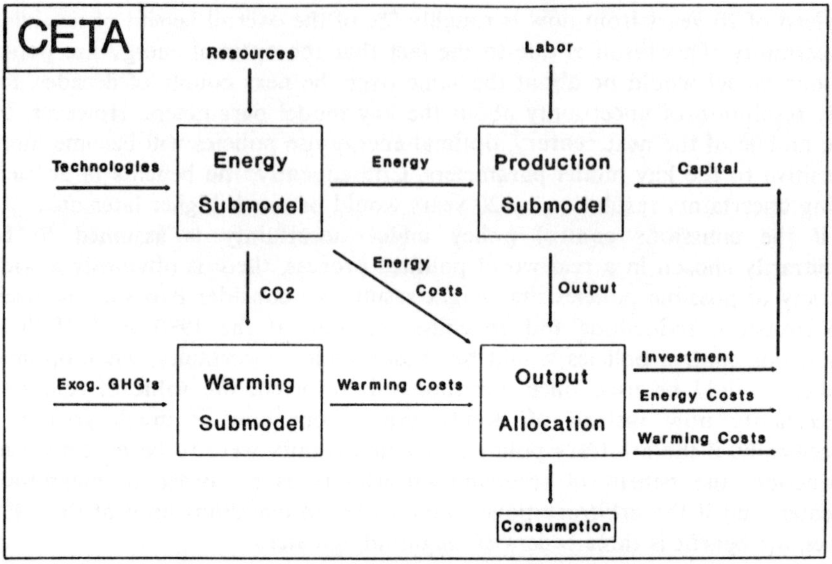

Fig. 1. CETA model overview.

CO_2 by-product to the warming submodel. In the production submodel, energy, labor, and capital inputs are used to produce output which is then allocated to consumption, investment, energy costs, and damage costs of warming. Because energy costs (and energy technology choices) are considered together with warming damage costs, the time paths of CO_2 emissions and carbon taxes in our model reflect an optimal balancing of the cost of emission reduction and the benefit of reduced global warming.

2.b. Low, Central, and High case assumptions for key parameters

To define our Central case and sensitivity cases, we identify a set of potentially important model parameters and define Low, Central, and High case values for each. We associate the label 'Low' with the parameter value that implies warming is less of a problem, and 'High' with the value that implies warming is more of a problem. While we frequently are able to rely on some authority in setting the Central case value, we typically can only offer our own rough estimates for the Low and High case values; we think of the latter as 5 and 95 percent points on subjective cumulative probability distributions. Table 1 lists the parameters we identify as potentially important, and gives the Low, Central, and High case values we have chosen for each.

The parameters in the table are grouped according to whether they effect

Table 1

CETA parameter values.

Parameter	Case Low	Central	High
1. *Cost of control related*			
AEEI (%/year)	0.5	0.25	0
Oil and gas resources (exajoules)	37110	24740	12370
Coal resources (exajoules)	75000	150000	225000
Non-electric backstop cost ($/GJ)	10	15	20
Electric backstop cost (mills/kWh)	37.5	50	62.5
2. *Benefit of control related*			
Warming per $2\times CO_2$ (°C)	1	3	5
Adaptation/damage function percent(% gross prod)	0.5	2	3.5
Adaptation/damage function power	1	2	3
Annual temp. adj. rate (%/yr)	1	2	10
Exogenous GHGs (see text)			
Carbon cycle (see text)			
3. *Cost and benefit related*			
Labor growth rate (% of EMF12)	75	100	125
Discount rate (%)	4	3	2

the cost of emission control, or the benefits of emission control, or both. Discussions of parameters in each of these categories follow.

2.b.1. Cost of control related parameters

Cost of control related parameters include those affecting future uncontrolled emissions and the costs of reducing those emissions. Future uncontrolled emissions depend on future gross production, energy use per unit of gross production, and the mix of energy sources used. Thus, many parameters in CETA play a role in determining future emissions. In the following paragraphs we identify those we believe to be most important.[5]

The Autonomous Energy Efficiency Improvement rate (AEEI) is the most important determinant of energy use per unit of gross production. Our Central case value is the one we use for the Stanford Energy Modeling Forum Global Climate Change study (EMF-12). The High case is twice this, and the Low case is zero.

The availability of various energy types, in particular the size of the Oil and Gas Resource Base and the size of the Coal Resource Base, are important parameters affecting (uncontrolled) carbon production per unit of energy used. Our Central case resource bases are those used in the EMF-12 study, and the Low and High cases are 50% and 150% of the Central case, respectively.

[5]The growth rate of the labor force is perhaps the most important determinant of future emissions. However, this parameter affects both the costs and benefits of emissions reductions and so appears in the third category of parameters.

Other cost of control related parameters include those affecting the cost of reducing future emissions. There are two main ways in which future emissions may be reduced. These are through energy conservation and through substitution of carbon-conserving energy for carbon-intensive energy. The more important mechanism in CETA is substitution, particularly of the Non-Electric Backstop technology (which is carbon-free) for synthetic fuels made from coal (which are very carbon intensive). Thus the Price of the Non-Electric Backstop technology is a key parameter affecting the cost of reducing future emissions. We take our Low, Central, and High cases to be \$10/GJ, \$15/GJ, and \$20/GJ, respectively; these compare to the EMF-12 base case assumption of \$16.67/GJ.

A second kind of substitution occurs in the electric sector, where the (carbon-free) Electric Backstop may be used in place of fossil based electric power generation. Although the Electric Backstop is quantitatively less important than the Non-Electric Backstop, it remains a significant parameter affecting the cost of reducing future emissions. For the Electric Backstop price, we use Low, Central, and High cases of 37.5, 50, and 62.5 mills per kilowatthour.

2.b.2. Benefit of control related parameters

This class includes parameters relating to the carbon cycle, the potential warming rate, exogenous greenhouse gas emissions, the lag between potential and actual temperature change, and the adaptation/damage cost function.

The carbon cycle is a complicated and poorly understood system by which CO_2 emissions produce elevated atmospheric CO_2 concentrations over decades or centuries following the emissions. In CETA, we represent the carbon cycle using the simple response function estimated by Maier-Reimer and Hasselmann (1987). This response function, which summarizes the behavior of a coupled ocean–atmosphere carbon cycle model, effectively divides new carbon emissions into five classes, each with a different atmospheric lifetime ranging from infinity to two years. Total atmospheric CO_2 over time is then the sum of atmospheric CO_2 across these five classes.

Our Central case uses the Maier-Reimer and Hasselman CO_2 class fractions (0.13, 0.20, 0.32, 0.25, 0.10), associated with lifetimes of infinity, 363, 74, 17, and 2 years, respectively. For our High case, we retain the same lifetimes, but change the fractions to: 0.18, 0.30, 0.32, 0.15, 0.05. For our Low case, we change the fractions to: 0.08, 0.10, 0.32, 0.35, 0.15.

Associated with a given concentration of CO_2 in the atmosphere is an equilibrium global mean surface temperature rise, or 'potential warming' from CO_2. Commonly, predictions of warming are presented in terms of the potential warming expected from a doubled atmospheric CO_2 concentration – e.g. a warming of 3°C may result from doubled CO_2 concentration. In CETA, a prediction of this nature is used to benchmark a logarithmic potential warming function of CO_2 concentration. Thus, we represent uncertainty about potential warming in terms of the potential warming function benchmark. We use 3°C as our Central case, 1°C as our Low case, and 5°C as our High case.[6]

In addition to CO_2, there are other greenhouse gases that contribute to global warming. Among the more important are methane (CH_4), nitrous oxide (N_2O), and chlorofluorocarbons (CFCs). Also, there is a certain amount of CO_2 emissions which is due to deforestation and is independent of the energy sector. In CETA, we treat the emissions of these greenhouse gases as exogenous.

Our Central case for exogenous gas emissions in CETA is roughly consistent with IPCC Scenario B.[7] To explore the implications of uncertainty about warming attributable to exogenous gases, we simply define Low and High emissions cases which are 50% and 150%, respectively, of the Central case.

Actual temperature will rise slowly to the equilibrium level implied by the concentration of CO_2 (and other greenhouse gases) in the atmosphere. In CETA, this is represented by a temperature rise in each time period that is proportional to the difference between the potential temperature in that time period and the actual temperature in the preceding time period. We refer to the proportionality factor as the annual temperature adjustment rate.

In our Central case, we assume the annual temperature adjustment rate is 2%.[8] For the Low and High cases we use 1% and 10%, respectively.

Great uncertainty surrounds the adaptation/damage cost function. In CETA, this function can be represented in a variety of ways. For example, it can depend on the level of temperature, or its rate of change, or both. Also, there is wide latitude to choose its functional form.

For purposes of exploring the implications of uncertainty, we use an adaptation/damage function that is defined on the level of temperature rise, rather than on its rate of change.[9] The damage function we use is of the form:

$$C_t = \alpha \cdot L_t \cdot T^{\lambda}_t, \tag{1}$$

where C_t = annual warming cost, time t,
α = a scaling constant,

[6]The range here is suggested by National Academy of Sciences (1991), based on the range of results obtained from climate models.

[7]See IPCC (1990, p.14).

[8]See Schlesinger and Jiang (1990).

[9]This choice is made for simplicity – a large number of model runs is required to analyze uncertainty, and these runs generally solve more easily for damage functions defined on the level of temperature rise. See Peck and Teisberg (1992b) for results from CETA using rate of temperature change damage functions.

L_t = labor input index, time t (1990 = 1.0),
T_t = temperature rise (above pre-industrial),
λ = power of the damage function.

Note that the damage function above shifts upward over time with growth in the labor force, L_t (which is expressed in efficiency units to reflect future productivity growth); this specification reflects the idea that willingness-to-pay to avoid warming effects will grow with population and income.

In the above function, the scaling parameter, α, is set so that damages are a certain percentage of gross production, at 3°C of temperature rise. We refer to this damage percentage at 3°C as the Adaptation/Damage Percent, and use it as one parameter representing uncertainty about the damage cost of temperature rise. For our Central case, we assume that the Adaptation/Damage Percent is 2.[10] For our Low and High cases we use 0.5% and 3.5%, respectively.

We also identify the Adaptation/Damage Power parameter, λ, as important.[11] For our Central case, we choose a power of 2, and for our Low and High cases we choose powers of 1 and 3, respectively.

2.b.3. Cost and benefit related parameters

The Labor Force Growth Rate (in efficiency units) is a key parameter which drives future potential production as well as affecting damage costs; thus this parameter affects both the costs and benefits of emission reductions.

Our Central case labor growth rates use the EMF-12 assumptions to the year 2100. After 2100, we assume that growth slows from roughly 2% per year in 2100 to 0.25% per year starting in 2150. Our Low case is 75% of the Central case growth rates, and our High case is 125% of the Central case rates.

A second important parameter, the utility discount rate, also affects both the cost and benefit of emission reduction. The discount rate affects the cost of reduction in our model because current production decisions affect future production possibilities. This dependence is due to the existence of exhaustible resources in the model, and to assumptions that prevent excessively rapid changes up or down in technology use rates over time. The discount rate affects the benefit of emission reduction because of the long time delay between emission reduction and the effect of that reduction on temperature.

[10] This is based on the work of Nordhaus (1990a), who presents an estimate of damage for the US (one-quarter of one percent of GDP), and offers a guess that the inclusion of non-quantified costs and extension of the estimate to the world as a whole might increase it to one or two percent of world gross production.

[11] In earlier work, we identified the power of the damage function as a key parameter affecting the optimal CO_2 emissions reduction policy – see Peck and Teisberg (1992a).

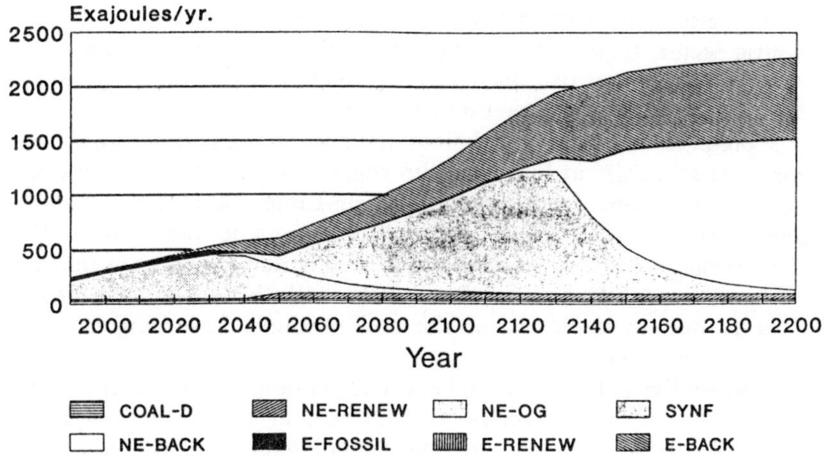

Fig. 2. Energy use, Central case.

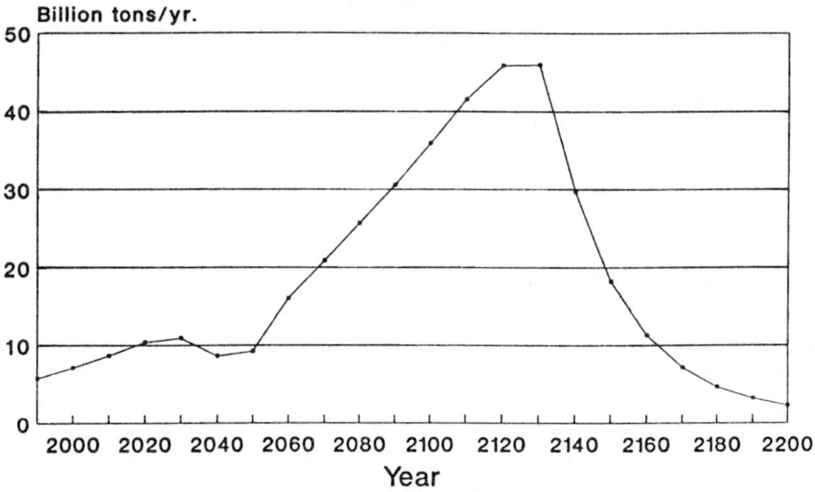

Fig. 3. Carbon emissions, Central case.

For our Low, Central, and High cases we use discount rates of 4%, 3%, and 2%, respectively.

2.c. Central case model results

Fig. 2 presents energy use patterns for our Central case parameter values. In the electric sector, the major energy use transition is the switch from fossil fuel technologies to the carbon-free backstop technology around 2030. In the non-electric sector, there are two major transitions: the switch from oil and gas to coal-derived synthetic fuels around 2050, and the switch from coal-derived synthetic fuels to the backstop around 2150.

Fig. 3 presents the carbon emissions path for the Central case. Carbon emissions rise gradually at first, then dip slightly around 2040 as the electric backstop comes on line. Beginning in 2060, emissions rise to a peak of about 45 billion tons per year before ultimately returning to pre-1990 levels by 2200. The surge in emissions beginning in 2060 is due to the transition from oil and gas to coal-derived synthetic fuels, and the subsequent drop in emissions is due to depletion of the coal resource base supporting the synthetic fuels industry.

Fig. 4 shows the path of atmospheric CO_2 concentration for the Central case. Concentration rises steadily until 2150, and then begins a gradual decline to the end of the time horizon shown. The initial rise in concentration occurs because new carbon emissions are being produced faster than the natural processes of the carbon cycle can remove carbon from the atmosphere. However, as the carbon emissions rate starts to fall beginning in 2140, the rate of carbon removal begins to exceed the rate of new emissions, causing the atmospheric concentration to begin a slow descent.

Fig. 5 shows the path of temperature for the Central case. Temperature rises steadily throughout the time horizon shown, reaching a level of about 6.5°C (over pre-industrial) by the year 2200. It is noteworthy that temperature rises steadily even though CO_2 concentration is declining by the latter part of this time period. Actual temperature responds with a long lag to concentration, and this accounts for the failure of temperature to turn down before the last year shown in the figure.

Fig. 6 presents the carbon tax path for the central case. The carbon tax begins at about $10 per ton and rises monotonically to about $160 per ton by 2200. The rise in the tax rate is attributable to rising temperature and the quadratic relationship between temperature and damage assumed as our Central case. However, the carbon tax does not rise high enough to make the carbon-free non-electric backstop technology competitive with synthetic fuels, and the backstop takes over only as a result of exhaustion of the coal resource base.[12]

2.d. Sensitivity results

To reduce the volume of information we need to report for the sensitivity cases, we present sensitivities in terms of the carbon emissions and carbon tax at two points in time. The first point in time is 2030, which is before the

[12] The tax would have to reach about $208 per ton to make the backstop competitive.

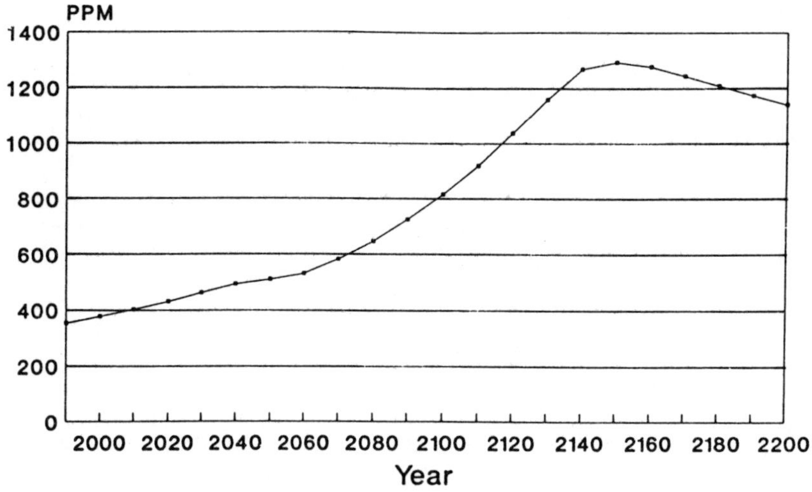

Fig. 4. CO_2 concentration, Central case.

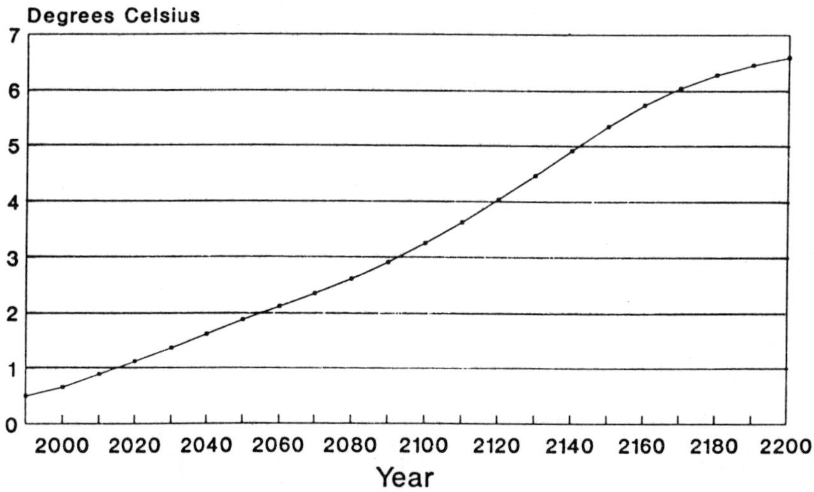

Fig. 5. Temperature rise, Central case.

major transition to synthetic fuels begins; and the second point is 2100, which is well into this transition. Tables 2 and 3 present these sensitivity results for the carbon emissions and the carbon tax, respectively.

Consider first the cost of control related parameters. Varying these parameters shifts the marginal cost of control function, while leaving the marginal benefit of control function undisturbed. Evidently, variations in these parameters induce only small shifts in the cost of control function in

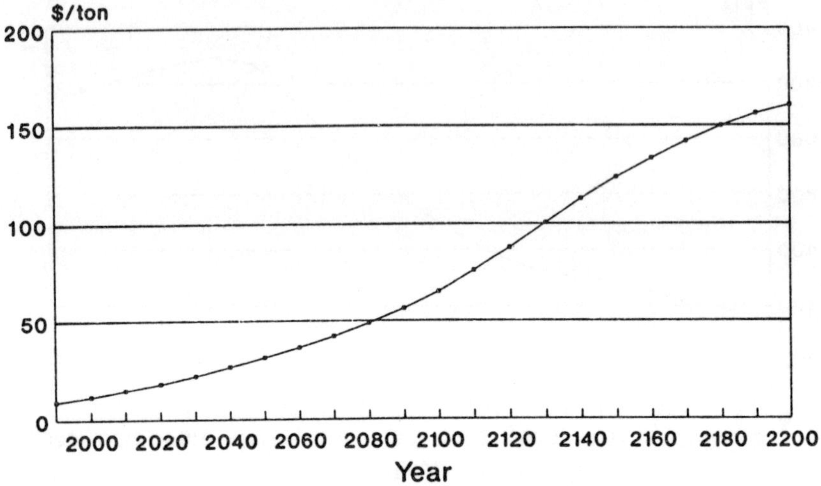

Fig. 6. Carbon tax, Central case.

Table 2

Sensitivity of optimal emission rates to parameter values.

Parameter	Optimal emissions rates 2030 Low	2030 High	2100 Low	2100 High
1. *Cost of control related*				
AEEI	10.0	11.8	29.3	42.4
Oil and gas resources	11.1	11.5	34.8	35.5
Coal resources	10.8	10.9	27.7	36.0
Non-electric backstop cost	10.9	10.9	2.5	34.3
Electric backstop cost	10.9	13.3	33.2	37.0
2. *Benefit of control related*				
Warming per $2 \times CO_2$	11.3	10.2	38.9	7.1
Adaptation/damage function percent	11.2	10.5	38.5	31.0
Adaptation/damage function power	11.1	10.7	38.3	21.0
Annual temp. adj. rate	11.1	10.3	37.7	23.7
Exogenous GHGs	11.0	10.8	35.9	34.9
Carbon cycle	10.9	10.8	35.8	34.9
3. *Cost and benefit related*				
Labor growth rate	8.5	13.9	20.9	54.2
Discount rate	10.4	11.2	35.5	28.0

2030, but larger shifts in 2100. Optimal carbon taxes, however, are relatively unaffected in either 2030 or 2100.

We can draw three conclusions from the above observations. First, the insensitivity of optimal emissions in 2030 implies that resolving uncertainty

Table 3

Sensitivity of optimal carbon taxes to parameter values.

	Optimal carbon taxes			
	2030		2100	
Parameter	Low	High	Low	High
1. *Cost of control related*				
AEEI	21.8	22.8	66.2	65.4
Oil and gas resources	22.1	21.8	65.8	65.6
Coal resources	22.9	22.3	70.3	64.1
Non-electric backstop cost	21.8	22.3	76.8	66.4
Electric backstop cost	22.1	22.4	66.1	65.3
2. *Benefit of control related*				
Warming per $2 \times CO_2$	3.4	57.6	8.7	197.6
Adaptation/damage function percent	5.5	39.6	15.6	122.6
Adaptation/damage function power	12.0	32.5	17.8	162.3
Annual temp. adj. rate	10.7	51.9	30.4	162.5
Exogenous GHGs	19.1	24.8	58.1	71.3
Carbon cycle	19.2	25.2	59.3	71.0
3. *Cost and benefit related*				
Labor growth rate	17.4	28.3	40.3	106.8
Discount rate	12.6	44.9	37.9	136.4

about these parameters before 2030 is not likely to have high value – if roughly the same policy is optimal in this time frame regardless of parameter values, little is gained by resolving uncertainty about these values. Second, and conversely, the sensitivity of optimal emissions in 2100 implies that resolving uncertainty well before 2100 is likely to have relatively high value. Third, the insensitivity of optimal carbon taxes in both 2030 and 2100 implies that the marginal benefit of emission reduction function is relatively flat across the range of optimal emissions changes induced by variation in the cost of control parameters.

Next consider the benefit of control related parameters. As before, optimal emissions are insensitive to parameter variation in 2030, but not in 2100.[13] Here, however, optimal carbon taxes are sensitive to variation in the benefit of control parameters, both in 2030 and in 2100. The latter result is not surprising since benefit of control parameter variations directly change the marginal cost of emissions.

Briefly, we note that the above results have implications for the relative merits of tax vs. quantity instruments for controlling emissions under

[13]There are two reasons for the difference in sensitivity of optimal emissions between 2030 and 2100. In the early time frame, there is heavy reliance on relatively clean and inexpensive oil and gas and it makes sense to use these resources regardless of how uncertainties may be resolved. In the later time frame, oil and gas is exhausted, but the non-electric backstop technology is available on a large scale and this technology allows carbon emissions to be reduced at constant unit cost by displacing carbon intensive synthetic fuels.

uncertainty.[14] Specifically, the insensitivity of optimal carbon taxes to variations in cost of control parameters indicates that the marginal benefit of control function is fairly flat; and the insensitivity of optimal emissions in 2030 to variations in benefit of control parameters indicates that the marginal cost of control function is quite steep, at least in this early time frame. If the benefit of control function is flat relative to the cost of control function, a tax policy will generally be the preferable control instrument, when there is uncertainty about cost of control related parameters. On the other hand, when there is uncertainty about benefit of control related parameters (but the cost of control is known), tax and quantity instruments are equivalent, since either can be set so as to produce exactly the same quantity of emissions.

In exploring the value of resolving uncertainty, we focus on uncertainty in benefit of control related parameters. We do this because our model is well suited to determining optimal quantity policies (i.e. carbon limits) under uncertainty, but not well suited to determining optimal carbon taxes under uncertainty. By restricting our analysis to benefit of control parameter uncertainty, we know that the optimal quantity policies we find are equivalent to the corresponding optimal tax policies. Whereas, if we applied our analysis to cost of control parameter uncertainty, we could find 'optimal' carbon limits, even though these represent a policy which is inferior to some other policy employing carbon taxes.

There is an additional reason for restricting our attention to uncertainty about benefit of control related parameters. Our analysis assumes that an uncertain parameter is initially unknown, and then becomes known all at once at a specified later date. However, the cost of control parameters by nature become known in other ways that are not well represented by this analytical approach. Specifically, the AEEI and the labor growth rate both are perfectly known for past time periods, and uncertain for future periods, within the model time horizon; resource base estimates are progressively improved over time as a result of production and new exploration; and the two backstop costs may be uncertain until the first date the backstop technologies become available, but after this date these costs are perfectly known for the rest of the time horizon (unless one assumes that these costs continue to change over time according to some stochastic process – but such a specification would also be outside the bounds of what we can handle using CETA in a simple decision tree framework).[15]

[14]These observations are based on results in Weitzman (1974).

[15]Various forms of progressive resolution of uncertainty could be modeled using stochastic dynamic programming. However, such an approach would be impractical unless CETA were simplified to reduce the number of state variables now represented in the model. See Peck et al. (1989) for an application of stochastic dynamic programming to determine jointly optimal emission control and learning policies when damages depend on a discharge stock.

The results in table 2 also suggest the subset of benefit related parameters that are of most interest, and which we choose to focus on in the next sections. Looking at 2100, the parameter with the greatest effect on optimal emissions is the warming rate per CO_2 doubling, and the parameter with the second greatest effect is the damage function power. The damage function percent parameter also induces a significant change in optimal emissions. Thus we choose to focus in the next sections on the warming rate (a parameter related to climate) and on the two parameters that together define the damage function (parameters related to impacts and adaptation costs).

3. Estimating value of information if optimal policies are used

Decision analysis provides a general paradigm for calculating the economic value of information.[16] In this paradigm, information is valued as the difference between (1) the expected value obtained if the state of the world is known before a policy must be adopted, thereby allowing a potentially different policy to be applied in every possible state of the world, and (2) the expected value obtained if a single policy must be adopted (without knowledge of the state of the world) and then applied across all possible states of the world. The first expected value is generally higher than the second, since the policy in the first case may be tailored to the actual state of the world. Mathematically, the expected value of perfect information (EVPI) can be expressed as follows:

$$\mathrm{EVPI}=\sum_i \rho_i \cdot \left[\max_{P_i} V(P_i, S_i)\right] - \max_P \left[\sum_i \rho_i \cdot V(P, S_i)\right], \tag{2}$$

where S^i = state of world i,
ρ^i = probability of state i,
P = uniform policy across all states,
P^i = tailored policy for each state i,
V = value function of policy, state.

Using decision tree diagrams, the expected value of perfect information can be represented as the difference in the value of two decision trees. These trees are illustrated in fig. 7. In the upper tree, uncertainty about the state of the world is resolved first, and then a potentially different policy is chosen depending on the state of the world. In the second tree, the policy is chosen first, and then uncertainty about the state of the world is resolved. Manne and Richels (1991b) characterize this difference using the phrases 'Learn, then act' and 'Act, then learn.'

While the above approach to valuing information is conceptually simple, there are numerous challenges in applying it to the global warming problem.

[16] See, for example, Raiffa (1968).

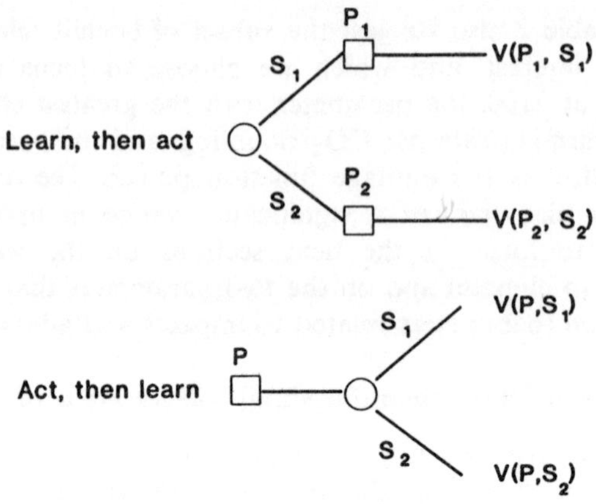

Fig. 7. Tree diagrams: Value of information.

First, there is a very large number of uncertainties involved in global warming. Second, available assessments of parameter uncertainties are typically limited to possible ranges at most, while information on distributional shapes and possible correlations among uncertain parameters is not available. Third, perfect information rarely becomes available all at once – instead, there is a continuing process of updating 'best estimates' over time as information is developed.[17] Finally, even without uncertainty, modeling of global warming is computationally demanding, since warming involves complex natural and human systems over a time scale measured in centuries.

In the face of these difficulties, we adopt certain simplifications in this paper. First, based on results in the previous section, we limit our consideration to three key parameters affecting the benefits of emission reductions. These are the extent of warming per CO_2 doubling and the level and power parameters of the warming damage function. Second, for most of our analysis we treat each parameter in turn as the only uncertain parameter, and represent its probability distribution using three points, with probabilities 1/6, 2/3, 1/6 for the Low case, Central case, and High case, respectively.[18] However, we do conduct an experiment to explore the implications of joint uncertainty about more than one parameter. For this experiment, we simplify our problem even further by assuming that the two parameters are

[17]See Peck et al. (1989) for an analysis of optimal emission control with progressive updating of information.

[18]This is consistent with the approach suggested by Miller and Rice (1983) to best represent a normally distributed random variable, using a three point distribution. In the context of the Miller and Rice approach, the Low and High cases to which 1/6 probability is assigned correspond to roughly the 5% and 95% probability points on the underlying normal being approximated; the Central case assigned 2/3 probability corresponds to the mean of the normal distribution.

independently distributed and that these parameters can take on only a High or Low value, each with probability 1/2.[19] Finally, in all the cases we consider, we assume that information perfectly reveals parameter values.

The basic CETA model is specified as a constrained non-linear optimization problem with both the objective function and some constraints non-linear. To determine optimal policies under uncertainty, this basic problem is augmented to solve a set of parallel problems, each representing a different state of the world (e.g. the High, Central, and Low parameter value outcomes, for uncertainty about only one parameter). Each parallel problem has its own complete set of variables and constraints, indexed by the state of the world. In addition, the objective function is re-specified as a probability weighted sum of the present value utilities obtained in each state of the world. Finally, additional constraints are added to the problem to require that solution variables be the same across all states of the world, for as long as the state of the world is assumed to be unknown. After this time, these constraints are eliminated and the model solution variables are allowed to differ across states of the world. With this specification, the solution before resolution of uncertainty is a single policy which maximizes expected utility across possible states of the world.

4. Value of information results if optimal policies are used

4.a. One variable at a time uncertain

To illustrate the nature of the model solutions underlying our value of information estimates, we initially present some detailed results for uncertainty about the warming rate per CO_2 doubling. First, in fig. 8, we show the optimal paths of emissions if the warming rate is known now. Notice that optimal emission paths for alternative values of the warming rate begin to diverge immediately, although the divergence is quite small until the oil and gas resource base becomes depleted around 2050. After oil and gas is depleted, a choice must be made between using synthetic fuels derived from coal or the carbon-free non-electric backstop technology. If the warming rate is 1°C, synthetic fuels are chosen, and the coal resource base is exhausted over the subsequent 150 years. If the warming rate is 3°C rather than 1°C, synthetic fuels are still chosen, but the rate of depletion of the coal resource base is modestly slowed, resulting in emissions that are delayed, but otherwise similar to those for the 1°C warming case. If the warming rate is 5°C, the use of synthetic fuels is curtailed, particularly after 2080; as a result, optimal CO_2 emissions are significantly lower in this state of the world.

[19]Here too we use the Miller and Rice (1983) approach. We choose the two point marginal distributions to be roughly equivalent to the three point distributions described earlier. The two points are the Central case plus or minus one standard deviation, each with probability 1/2.

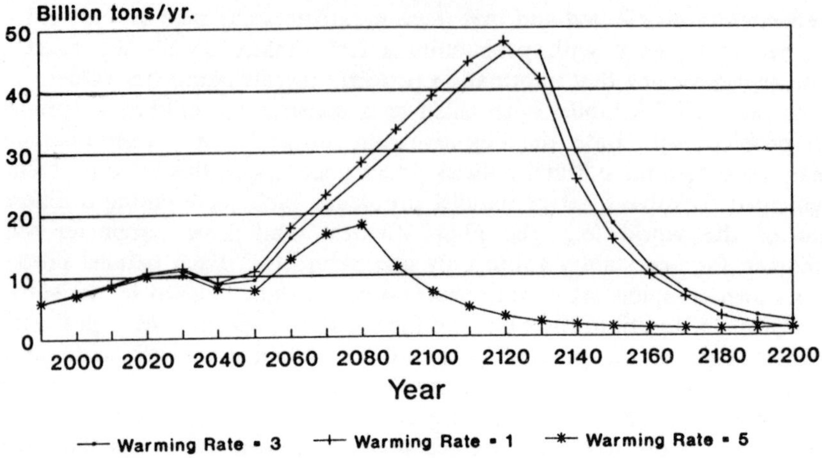

Fig. 8. Carbon emissions: Learn, then act.

If the true warming rate is assumed to be unknown for some period of time, the optimal carbon emissions paths will be the same initially, but later diverge once the warming rate becomes known. For example, suppose that the true warming rate becomes known after eighty years. In this case, the optimal emissions paths are as illustrated in fig. 9. Notice that the paths are identical over all states of the world until 2070, after which the optimal paths branch out to become roughly similar to the optimal paths in fig. 8.[20]

In the extreme, we might assume the warming rate remains unknown for the entire time horizon of the model. In this case, the optimal emissions paths will be the same across all states of the world for the entire time horizon. Fig. 10 illustrates this case.

The model solutions illustrated in figs. 8 through 10 may be used to calculate a variety of different values of information. First, by subtracting the objective function value for the solution in fig. 9 from the objective function value for the solution in fig. 8, we would get the value of knowing the warming rate now rather than having it be revealed in eighty years. This value is $56 billion.[21] Alternatively, by subtracting the objective function value for the solution in fig. 10 from the objective function value for the solution in fig. 9, we would obtain the value of knowing the warming rate in eighty years rather than never knowing it. This present value of information

[20]These paths are not identical to those in fig. 8 starting in 2080, because the optimal path under uncertainty prior to 2080 has affected important variables (e.g. the CO_2 concentration, the coal resource base, etc.).

[21]Actually, the difference in objective function values is denominated in expected "utils. Thus the difference in objective function values is converted to a dollar value using the dollar value of utility generated by the model solutions.

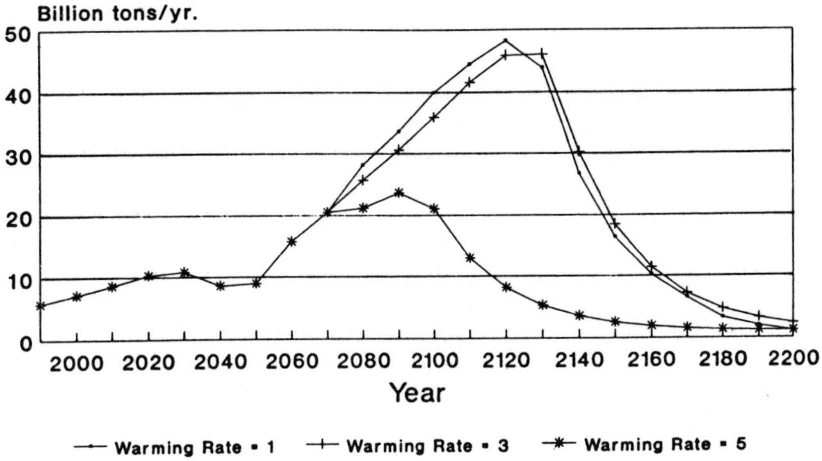

Fig. 9. Carbon emissions: Act, then learn.

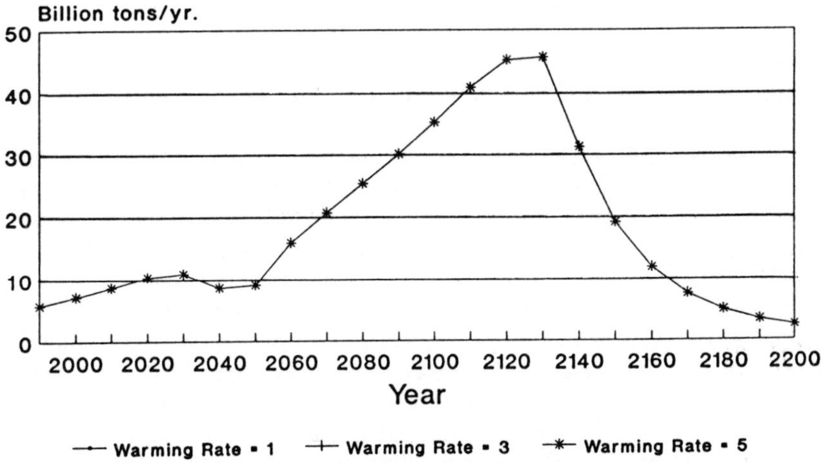

Fig. 10. Carbon emissions: Never learn.

is \$92 billion. Lastly, by subtracting the objective function value for the solution in fig. 10 from that for the solution in fig. 8, we would get the value of knowing the warming rate now rather than never. This value is \$148 billion.

The solutions illustrated by figs. 8 through 10 are actually three members of a whole family of solutions produced by assuming that perfect information is revealed at various dates between 1990 and the time horizon of the model. Obviously, there is a large number of reasonable values of information that

Table 4

Information now vs. never: Optimal policy under uncertainty.

Parameter	Value of information (billion $)
Warming rate per $2 \times CO_2$	148
Adaptation/damage function percent	27
Adaptation/damage function power	98
Total	273

Table 5

Information now vs. twenty years later: Optimal policy under uncertainty.

Parameter	Value of information (billion $)
Warming rate per $2 \times CO_2$	6
Adaptation/damage function percent	0.2
Adaptation/damage function power	0
Total	6.2

could be calculated from the solutions in this family. From this large number of values, we select a subset of values that are of particular interest. First is the value of information now vs. never; this is 'the value of information' in a gross sense. Second is the value of information now vs. in twenty years, which is a measure of the value of resolving uncertainty sooner rather than later. We also generalize this second measure by calculating the value of accelerating uncertainty resolution by twenty years, as a function of the date to which the acceleration occurs.

Table 4 shows the gross value of information now vs. never for the three parameters we consider in this paper. Notice that the value of information is greatest for the warming rate parameter. However, the adaptation/damage function parameters have a total value of information which is nearly as high. Thus resolving uncertainty about the impacts of warming appears to be about as important as resolving uncertainty about the extent of warming.

Yet, 95% of the US global warming research budget is presently devoted to resolving physical uncertainties such as that surrounding the potential warming rate, and only 5% is devoted to resolving uncertainties about warming impacts, mitigation strategies, and adaptation costs.[22] This relative imbalance may not be justified.[23]

[22]See National Academy of Sciences (1991, p.70.)

[23]A definitive conclusion in this regard would have to be based on an assessment of the incremental resolution of uncertainty per incremental dollar spent on alternative lines of research. Since our analysis does not address this issue, our results are only suggestive of a misallocation of research money.

Table 4 also shows a 'Total,' which needs to be interpreted cautiously. We believe that it significantly understates the actual total value of information about the warming problem. This is the case, most obviously, because there is a large number of other uncertain parameters whose values of information are not included in the total reported in table 4. In addition, results we report in the next section suggest that when two or more parameters are simultaneously treated as uncertain, the value of resolving uncertainty about all of them is greater than the sum of the values of resolving uncertainty about each of them treated in turn as the only uncertainty.

Table 5 shows the value of perfect information now, assuming that the same information would otherwise become available in twenty years. While the total of $6.2 billion dollars is not a trivial sum, it is not particularly large in relation to an annual US global warming research budget of $1 billion, and it is small relative to the gross values of information reported in table 4. The results in table 5 thus suggest that there is not a high premium attached to immediate resolution of global warming uncertainties.

We believe that the low premium on immediate resolution of global warming uncertainties is due to the short run insensitivity of optimal decisions to alternative parameter values. In general, when optimal policies are roughly the same regardless of how uncertainties are resolved, there is little economic benefit derived from resolving uncertainties. In the global warming problem, policies in the short run (20 or even 40 years) are not sensitive to parameter values. We believe that this insensitivity in the near term is the result of reliance during this period on inexpensive and relatively clean oil and gas – it makes sense to use this resource regardless of whether uncertain global warming parameters take on high or low values.

Although optimal emissions policies are relatively insensitive to parameter values in the short run, this is not true indefinitely. By the year 2100, for example, optimal policies are significantly different for different parameter values. Thus we should anticipate that the premium on early resolution of uncertainty would tend to increase over time, if uncertainty remains unresolved. To confirm this hypothesis, we consider another measure of the value of information: the value of accelerating uncertainty resolution by twenty years, as a function of the date to which the acceleration occurs.

Fig. 11 shows the value of resolving uncertainty twenty years earlier, as a function of the date at which this acceleration might occur. While not monotonically increasing, the lines in the figure are clearly upward trending. In a sense, the upward trends in the figure are understated, since the numbers graphed are the present values as of 1990 of these accelerations. If these values were re-expressed as contemporaneous present values, the upward trend would be more pronounced.

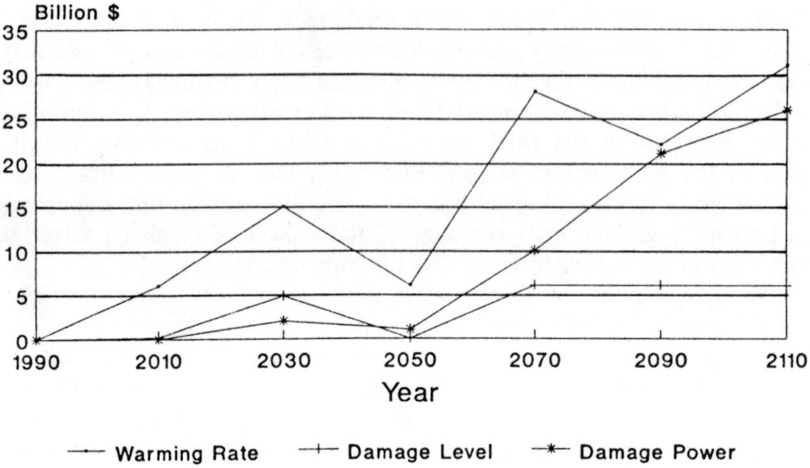

Fig. 11. Value of learning warming rate 20 years sooner vs. time.

4.b. Value of information if two parameters are uncertain

In this section we present results of an experiment intended to explore the implications of joint uncertainty about two or more parameters. Our results suggest that the value of resolving uncertainty about two or more parameters treated as jointly uncertain is likely to be greater than the sum of the values of resolving uncertainty about each parameter treated in turn as the only uncertainty.

Because of the computational difficulty involved in solving the CETA model over a large number of states of the world, we further simplify the problem we consider in this section.[24] First, we shorten the time horizon over which the model is solved to 2170 from 2250. Second, because of the shortened time horizon, we consider a relatively near term value of information question – specifically, what is the value of having perfect information now instead of forty years from now? Third, we substitute two point distributions for the three point distributions we used in the preceding section.

In shifting from three point distributions to two point distributions we again use the Miller and Rice (1983) approach. Under this approach, the two points to be used to approximate a Normal distribution are the mean plus or minus one standard deviation; each of these points is then assigned

[24]Analyzing one uncertain parameter with three possible values involves the simultaneous solution of three parallel CETA specifications. Two parameter uncertainty with three possible values for each uncertain parameter would involve simultaneous solution of nine parallel CETA specifications. Not only did the run-time become excessive in this case, but we also encountered numerical problems in converging to an optimal solution.

Table 6

Information now vs. in forty years: One parameter uncertain.

Parameter	Value of information (billion \$)
Warming rate per $2 \times CO_2$	24
Adaptation/damage function percent	9
Total	33

Table 7

Information now vs. in forty years: Two parameters uncertain.

Parameter	Value of information (billion \$)
Warming rate per $2 \times CO_2$	33
Adaptation/damage function percent	22
Both parameters together	47

probability one-half. Since we earlier identified the Low and High case parameter values with the mean plus or minus 1.732 standard deviations, it is a simple matter to calculate the implied standard deviation for each parameter, and from this to determine the parameter values associated with the Low and High points in the two point distribution.

Since we are using a near term value of information (perfect information now vs. forty years from now), we choose to consider uncertainty about the two parameters that have the highest value of information in the short term. As fig. 11 indicates, these are the warming rate and the adaptation/damage function percentage.

To create a reference point for comparison, we first calculate the value of information assuming single parameter uncertainty as was done in the preceding section; however, we here use the shortened time horizon and the two point distributions. Table 6 below shows our results; these are in fact quite similar to those we would have obtained (for learning now vs. forty years from now) using the longer horizon and the three point distributions of the preceding section.[25]

Table 7 shows the comparable results for value of information when two parameters are treated as simultaneously uncertain. While the general magnitude and pattern of results are similar, there are some noteworthy differences. First, the value of information about either parameter is higher if the other parameter is treated as uncertain (table 7) than if the other parameter is treated as known and equal to its Central case value (table 6).

[25]With the three point distributions and the longer time horizon, the values of information now vs. forty years from now are 21 and 5 billion for the warming rate and damage function percentage, respectively.

Second, the value of resolving uncertainty about both parameters simultaneously ($47 billion) is well in excess of the sum ($33 billion) of the values of resolving information for each of the two parameters treated as the only uncertain parameters. This result suggests that the sum of the values of information for two or more parameters each treated as the only uncertain parameter understates the value of resolving uncertainty about all those parameters at once.

5. Value of information results if suboptimal policies are used

In the previous section, we presented results assuming that emissions control policy under uncertainty would be based on a balancing of the expected costs and benefits of emissions reduction. In this section, we assume that policy under uncertainty will be arbitrarily determined by a real world political process involving the governments of many countries with differing perspectives and interests.

While it is difficult to forecast what kind of emissions reduction policy might emerge from the political process, the policy that emerges is unlikely to be the optimal one.[26] We consider two possible suboptimal policies that might emerge: one is a policy of no emissions reduction before uncertainty is resolved, and the other is a policy of limiting emissions to the 1990 level until uncertainty is resolved. In either case, we assume that when uncertainty is resolved, the policy will revert to the optimal one for whatever state of the world is revealed.[27]

As in section 4, we consider each parameter in turn, assuming the values of the other parameters are known and equal to their Central case values. As we discuss below, the bulk of the value of information we calculate in this section is attributable to the shift from a sub-optimal to an optimal policy, which we assume accompanies the resolution of uncertainty. Consequently, the sum of the values of information for each parameter is in this case likely to overstate the value of resolving all uncertainties together. We believe it may be most reasonable to interpret these numbers as follows: the largest value of information estimate for any single parameter may approximate the

[26]It is not plausible to believe that the policy process might first reach consensus on probabilities of alternative states of the world, and then agree on an emissions control policy that is optimal given this uncertainty. Rather, when there is uncertainty about the nature of a problem and there are multiple bargaining parties, there is a strong tendency for policy to gravitate toward readily understood 'focal points,' such as limiting emissions to their 1990 level; such focal point policies would be optimal only by chance.

[27]There is an argument, of course, that a suboptimal policy may also be chosen after uncertainty is resolved. However, this is less likely if uncertainty is resolved quickly, since in this case there is less time for vested interests to develop as a result of previous policy. If an optimal policy is more likely with quick resolution of uncertainty than with slow resolution, the benefit of resolving uncertainty quickly is even higher than our numbers suggest.

Table 8

Value of information (billion \$): Information now vs. twenty years later.

Parameter	Policy under uncertainty		
	No control	1990 Limit	Optimal
Warming rate per $2 \times CO_2$	90	2828	6
Adaptation/damage function percent	22	2911	0.2
Adaptation/damage function power	65	2897	0

value of resolving all uncertainties together, and thereby moving to the optimal policy for the state of the world that is revealed.

Table 8 presents estimates of the value of information now vs. twenty years later assuming no emissions control or a 1990 emissions limit in the interim before uncertainty is resolved. For comparison, the table also shows the comparable values of information from table 5; the latter, of course, are based on the assumption that an optimal policy is used under uncertainty.

Not surprisingly the new numbers in table 8 are much larger than the comparable numbers from table 5. One can interpret the new values of information in table 8 as being the sum of (1) the conventional value of information, and (2) the additional value from adopting a best policy under uncertainty instead of using one of the two sub-optimal policies. The previous value of information numbers from table 5 correspond to term (1) in the sum above, while the new numbers in table 8 correspond to the sum of (1) and (2). Evidently the value from adopting a best policy under uncertainty vs. no emissions control is high; and the value from adopting a best policy under uncertainty vs. the 1990 limit is huge.

Among our earlier results is the rather comforting one that information in twenty years is almost as good as information now, implying that we can proceed in a relatively unhurried fashion to design and implement a careful research program. However, the results in table 8 present a very different picture – if the emissions control policy under uncertainty will be some arbitrary policy like those considered here, it is critical to resolve uncertainty quickly.

6. Conclusion

The global warming problem is a complicated one, and placing a value on resolution of global warming uncertainties is a difficult task. In this paper, we have presented a first effort at such an analysis. Obviously, there are important caveats that should be attached to our analysis.

First, the CETA model cannot perfectly represent the future for the next 200 years, even if the key parameters of the model are completely known.

Both the climate model and the economic growth model in CETA are very simple representations of extremely complex systems over a very long period of time; these simple representations necessarily omit many real world factors that bear on the warming problem.

Second, our representation of uncertainty and learning is both limited and simplified. We limit the number of parameters that we treat as uncertain at a given time, we limit the number of possible values that each may take, and we assume that these parameter values are either completely unknown or perfectly known. In addition, the possible values that parameters may take are in most cases just our own estimates of 5 and 95% probability points for these parameters.

However, we have conducted a self-consistent exercise to identify important driving variables and to estimate the value of information for a selected subset of these variables. Caveats notwithstanding, we believe that our results support the following tentative conclusions:

(1) If an optimal policy is used under uncertainty, the value of information is large enough to justify current research efforts, and perhaps to justify increased emphasis on research into the impacts of warming and cost of adaptation to warming.
(2) If an optimal policy is used under uncertainty, ample time is available to plan and execute a well-designed research program to resolve uncertainties.
(3) However, if the political process will choose suboptimal policies and this choice could be prevented by early resolution of uncertainty, the urgency of resolving uncertainty is dramatically increased.

References

Baumol, W.J. and W.E. Oates, 1988, The theory of environmental policy (Cambridge University Press, Cambridge, MA).

Dickinson, R.E. and R.J. Cicerone, 1986, Future global warming from atmospheric trace gases, Nature 319, no. 9 (Jan.), 109–115.

Energy Modeling Forum, 1991, Study design for EMF12 global climate change: Energy sector impacts of greenhouse gas emission control strategies, Draft (Energy Modeling Forum, Stanford University, May).

Fulkerson, W., R.R. Judkins and M.K. Sanghvi, 1990, Energy from fossil fuels, Scientific American (Sep.), 128–135.

Hidy, G.M. and S.C. Peck, 1991, Organizing for risk oriented climate alteration research, Journal of Air and Waste Management Association 41, no. 12 (Dec.), 1570–1578.

IPCC, 1990, (Intergovernmental Panel on Climate Change) Climate change: The IPCC scientific assessment, Report prepared for IPCC by Working Group 1, Meteorological Office, Bracknell, UK.

Maier-Reimer, E. and K. Hasselmann, 1987, Transport and storage of CO_2 in the ocean – An inorganic ocean–circulation carbon cycle model, Climate Dynamics no. 2, 63–90.

Manne, A.S, 1986, GAMS/MINOS: Three examples, Draft (Department of Operations Research, Stanford University, Dec.).

Manne, A.S. and R.G. Richels, 1990, CO_2 emission limits: An economic cost analysis for the USA, The Energy Journal 11, no. 2, 51–74.

Manne, A.S. and R.G. Richels, 1991a, Global CO_2 emission reductions – The impacts of rising energy costs, The Energy Journal 12, no. 1, 87–107.

Manne, A.S. and R.G. Richels, 1991b, Buying greenhouse insurance, Energy Policy 19, 543–552.

Manne, A.S. and R.G. Richels, 1992. Buying greenhouse insurance: The economic costs of CO_2 emission limits (MIT Press, Cambridge, MA).

Miller, A.C. III and T.R. Rice, 1983, Discrete approximations of probability distributions, Management Science 29, no. 3, 352–362.

National Academy of Sciences, 1991, Policy implications of greenhouse warming (National Academy Press, Washington, DC).

Nordhaus, W.D, 1982, How fast should we graze the global commons?, American Economic Review (May), 242–246.

Nordhaus, W.D, 1990a, To slow or not to slow: The economics of the greenhouse effect, Draft (Department of Economics, Yale University, New Haven, Feb. 5).

Nordhaus, W.D, 1990b, Contribution of different greenhouse gases to global warming: A new technique for measuring impact, Draft (Department of Economics, Yale University, New Haven, Feb. 11).

Nordhaus, W.D, 1991, Uncertainties and research priorities in global environmental issues, Testimony before the House Committee on Science, Space, and Technology: Subcommittee on Science, Oct. 8.

Peck, S.C., Hung-Po Chao and T.J. Teisberg, 1989, Optimal control and learning strategies when environmental costs are cumulative, Proceedings of the IFAC/IFORS/IAEE International Symposium on Energy Systems, Management, and Economics, Oct. 25–27, Tokyo, Japan.

Peck, S.C. and T.J. Teisberg, 1992a, CETA: A model for carbon emissions trajectory assessment, The Energy Journal 13, no. 1, 55–77.

Peck, S.C. and T.J. Teisberg, 1992b, Temperature change related damage functions: A further analysis with CETA, Draft (Electric Power Research Institute, Palo Alto, CA, May 6).

Raiffa, H., 1968, Decision analysis: Introductory lectures on choices under uncertainty (Addison-Wesley, Reading, MA).

Schlesinger, M.E. and Xingjian Jiang, 1990, Simple model representation of atmosphere–ocean GCMs and estimation of the time scale of CO_2-induced climate change, Journal of Climate 3, (Dec.), 1297–1315.

U.S. Environmental Protection Agency, 1989, The potential effects of global climate change on the United States: Report to congress, Executive Summary, Draft (U.S. EPA, Washington, DC, Dec.).

Weitzman, M.L, 1974, Prices vs. quantities, Review of Economic Studies (Oct.), 477–449.

Wuebbles, D.J. and J. Edmonds, 1988, A primer on greenhouse gases, DOE/NBB-0083 (US Department of Energy, Washington, DC, Mar.).

Part IV
Ethics and Intergenerational Discounting

A
Defining Appropriate Discount Rates

BUTTERWORTH HEINEMANN

Energy Policy, Vol. 23, No. 4/5, pp. 379–389, 1995

0301–4215/95 $10.00 + 0.00

Intergenerational equity, discounting, and the role of cost–benefit analysis in evaluating global climate policy

Robert C Lind
Johnson Graduate School of Management, Cornell University, 562 Malot Hall, Ithaca, NY 14853, USA

When public policies with impacts far into the future are being debated, the question inevitably is raised whether cost–benefit analysis which discounts future costs and benefits is not biased against future generations and whether, if such discounting is appropriate at all, a lower rate should be used to avoid such bias. The debate on global climate change is no exception. This paper sketches and analyses the welfare foundations of cost–benefit analysis and from this perspective analyses the role of cost-benefit analysis in the climate policy debate, particularly with reference to intergenerational effects. The paper concludes that the cost–benefit criterion cannot provide a definitive basis for deciding whether we should commit to a longer-term programme to moderate climate change; the issues of intergenerational equity are not that global climate change will significantly lower the GNP of future generations, but relate to the possibility of science fiction-like changes in the planet that will produce catastrophic effects in the future; and the typical way in which the cost–benefit problem is posed obscures the basic choices that we should be evaluating.

Keywords: Global warming; Intergenerational equity; Discounting

Whenever cost–benefit analysis is used to evaluate projects or policies where benefits and costs extend over long time horizons, questions inevitably are raised with regard to the validity of the cost–benefit methodology in general and more particularly to whether the procedure of discounting future benefits and costs does not bias the analysis against the interest of future generations. Some contend that a procedure which discounts future costs and benefits using a positive rate necessarily gives less weight to the well-being of future generations; others contend that because the role of the discount rate becomes so dominant in the decision process when evaluating programmes or policies with long time horizons, perhaps a lower discount rate should be used; others question whether cost–benefit analysis is sufficiently robust to form the basis for the decisions regarding global warming which may dramatically affect the future of life on the planet.

This is vividly illustrated in an editorial 'Discounting our Descendants?' by Laura Wallace in the March 1993, issue of *Finance and Development*. This editorial is followed by two concise comments by William R Cline, the author of *The Economics of Global Warming* and Nancy Birdsall and Andrew Steer of the World Bank. These three pieces highlight almost all of these issues and demonstrate how tricky they can be even in the hands of the most competent economists. In addition, the piece by Cline provides actual estimates of benefits and costs over time which we use as a basis for discussing some of the intergenerational issues.

To set the stage for the analysis to follow, it is useful to quote liberally from 'Discounting our Descendants?' (Wallace, 1993, p 2):

> at the June 1992 Earth Summit in Rio de Janeiro, more than 150 countries signed a convention aimed at stabilizing concentrations of these gases . . . a mechanism for

> deciding on stronger measures, if warranted, was set up. But few specific measures or targets were included, reflecting a widespread disagreement on just how activist nations should be. The stumbling block . . . was the paucity of studies that vigorously set out both the costs and benefits.
>
> For that reason, *The Economics of Global Warming* . . . has drawn much attention. Using the technique of cost–benefit analysis, it concludes primarily on economic grounds – with many ecological effects omitted – an 'aggressive' action program is warranted . . . Cline advocates the use of a low (about 2%) discount rate . . . a controversial proposition to put it mildly.

The editorial goes on to observe that:

> even a small change in the discount rate used in a cost–benefit analysis has a tremendous impact on the bottom line . . . some people – especially environmentalists – insist that when it comes to setting the level of the discount rate, long-term investment to protect the environment must be treated differently. They advocate a special low discount rate.
>
> Cline, too, advocates a low rate, but . . . He argues environmental projects should be treated no differently from other projects . . .
>
> Others, such as the World Bank, worry that an especially low discount rate would actually reduce the wealth passed on to the next generation by financing projects whose rates of return are lower than those for other available investments.

These comments and concerns as expressed in the context of global warming have been expressed in previous policy debates, notably the debate over energy policy. Some of these concerns are based on misunderstandings, such as the concern that using a positive discount rate applied to benefits and costs necessarily places a lower value on the welfare of future generations. Other concerns about what we can or cannot conclude on the basis of cost–benefit analysis require a deeper look at the cost–benefit criterion itself and how the application of that criterion is limited in cases where payment of compensation would require intergenerational transfers of resources.

To address these issues we first set out the basic paradigm of cost–benefit analysis and its analytical foundation in welfare economics, including the well understood, but often neglected, fact that the basic cost–benefit criterion rests on the presumption that net gainers can at least in principle compensate net losers from a given policy action. We will demonstrate that it is not possible, as a practical matter, for governments to design and commit to the intergenerational transfer programmes across many generations. This means that even if the cost–benefit criterion is satisfied in the case of global warming, there are potentially important issues of intergenerational welfare that have to be considered in addition to the cost–benefit estimate. We cannot, for example, reject programmes to reduce global warming, simply because the present value of the costs exceeds the present value of the benefits. In addition, where there are significant issues of intergenerational equity, we cannot appropriately address these issues by simply adjusting the discount rate used in cost–benefit analysis. Further analysis is required on a case by case basis.

The next section analyses whether there is, in fact, a significant issue of intergenerational fairness relating to global warming using cost–benefit estimates provided by Cline (Cline, 1993, pp 3–5). The conclusion that emerges is that if the costs and benefits in Cline's analysis are captured within even a very wide margin of error, and if his assumptions about world economic growth are at all reasonable, the issue of intergenerational equity is not one of the current generation shortchanging future generations by not investing in greenhouse abatement, but rather one of whether the current generation should be making very large investments in abatement to reduce the costs of global warming for generations, who even after paying the costs of Cline's high damage scenario will be many times richer than we are today. If there is a significant issue of fairness to future generations, it is associated with the potentially catastrophic outcomes of global warming. The fundamental question of intergenerational equity hinges on the rate of per capita income growth and on the magnitude of the effect of global warming.

The next section discusses several different conceptual bases for the discount rate. It then demonstrates that using a positive rate of discount does not necessarily mean putting less value on the well-being of future generations and discusses the rationales for using a rate of discount in public policy analysis that is below the marginal rate of return on capital which in a competitive economy will equal the cost of capital.

The final section of the paper argues that in investment or policy environments where there is a high degree of uncertainty and where relevant information becomes available or can be produced over time, the typical formulation of the cost–benefit problem is not very helpful to decision making. Today's decision or policy problem related to global warming is not whether we should commit to and begin an aggressive programme of abatement or whether we should do nothing and wait and see. The important question is, given what we know today, what should we do to position ourselves so to respond effectively to future events? Such a programme may call for immediate aggressive abatement or it may call for waiting or taking some other action. The important point is that the decision framework should be sequential, capable of responding to new information, and should take into account the option value of alternative courses of action.

The foundations of cost–benefit analysis and intergenerational analysis

Cost–benefit analysis can be separated into two parts. The first is the measurement and display of costs and benefits over time. Costs and benefits are typically measured in terms of willingness to pay. For a benefit, it would mean estimating the maximum amount the beneficiary would be willing to pay for that benefit. For a cost, it would mean estimating the minimum amount that the person incurring the cost would be willing to accept as full compensation. To the extent that benefits and costs are represented by goods and services that are traded in competitive markets they can be measured by their market prices. The difficulty comes when benefits and costs, such as those pertaining to environmental changes, are not traded and valued in the market. Despite the fact that these benefits and costs are a more difficult measure, the willingness to pay criterion is the guiding principle of measurement.

Once benefits and costs over time are measured, they are then discounted to their present value using some rate of interest or discount. If we assume a perfectly competitive market, free of distortionary taxes and capital market imperfections, then there will be a single market rate of interest, r, at which businesses and individuals can borrow and lend. In this case, r should be used to discount benefits and costs over time to their present value, and the net present value of the project or policy is simply the present value of benefits minus the present value of costs. Note in this ideal case, r would equal both the marginal rate of return on investment and the consumer's rate of interest or marginal rate of time preference.[1]

The second component of cost–benefit analysis is the cost–benefit criterion which states that if the net present value of benefits is positive then the project should be adopted; if not, it should be rejected. In cases where there are several mutually exclusive alternatives, the cost–benefit criterion states that the project with the highest net present value should be chosen. This implies that in cases where there is a budget constraint, the cost–benefit criterion states that if there are more projects with positive net benefits than will fit into the budget, then one should choose that portfolio of projects which maximizes the net present value of the total feasible portfolio. Therefore, having positive net benefits is necessary but not sufficient to justify a project's adoption. This turns out to be a critical point in many cases where a below market rate of discount is used in a cost–benefit evaluation.

The welfare basis for the cost–benefit criterion is the Kaldor–Hicks compensation principle. If the net present value of benefits is positive, then it follows that if one adopts the project, the people receiving the benefits could compensate the people incurring the costs and still have something left over. Therefore, if the cost–benefit criterion is satisfied everyone can be made better off with the appropriate compensation of those incurring costs by those receiving the benefits. Because everyone can be made better off if the cost–benefit criterion is satisfied and appropriate compensation is paid, we can conclude that there is an unambiguous increase in welfare. The problem is that in most cases the appropriate compensation is not paid. Then even where a programme or policy meets the cost–benefit criterion, there will in practice be some people who enjoy positive net benefits and others who incur net costs. Unless we give equal weight to the net benefits or costs of each individual, we cannot conclude that there has been an unambiguous welfare gain.

Nevertheless, the cost–benefit criterion is quite powerful. Suppose, for example, that we wanted to subsidize a given group and the choice was between spending US$1000 000 on a project that would benefit that group or giving the group the US$1000 000 in cash. If the present value of the net benefits of the project is positive (ie the present value of future benefits exceeds the initial cost of US$1000 000), then it is better to adopt the project than to give them the cash. If the benefits are less than US$1000 000, it would be better to give them the cash. Therefore, whether or not we choose to require the full compensation of costs or to subsidize the beneficiaries of a project by not requiring them to pay the costs, the project should be adopted only if it meets the cost–benefit criterion. For this reason, we can say that whatever your policy is with regard to redistributing income between the beneficiaries of a project and those incurring the costs, the project should only be adopted if it meets the cost–benefit test.

In the case of global warming policy where the benefits and costs will be distributed over may generations, any compensation scheme will potentially involve intergenerational transfers. Therefore, the question of whether we can make intergenerational transfers in either direction is at the heart of the matter with regard to the applicability of the cost–benefit criterion to the analysis of global warming policy. To see this, suppose that we were in a perfectly competitive world and that all present and future generations could invest or borrow at the competitive market rate of interest, r. Then it can be correctly argued that a discount rate of r% should be applied in the cost–benefit analysis of global warming abatement. This argument, however, hinges on our ability to make intergenerational transfers over many generations.

Suppose we were considering a US$1 billion investment to reduce global warming that would provide a benefit in 100 years of US$7.24 billion which is

[1]For a complete development of this line of argument see Lind (1982).

equivalent to 2% rate of return for 100 years. Suppose that $r = 8\%$. The cost–benefit logic for rejecting that investment is that we could invest and reinvest the US$1 billion at 8% so that in 100 years we could give that generation US$2 199.76 billion or almost 304 times as much value as they would have received from the investment in abatement. This line of argument supports the requirement that global warming investments be evaluated using an 8% rate in the cost–benefit calculations. As Cline correctly points out, this is the essence of the argument made by Summers *et al* in arguing against his use of 2% (Cline, 1993, p 5):

> Instead, Summers, Birdsall, and Steer seem to argue much more broadly that society should do little to limit greenhouse gas emissions. They implicitly counsel that the public would be better off to set aside money in a 'Fund for Future Greenhouse Victims.' The money would be invested at 8% ... and the income would compensate future generations for global warming damage at a far smaller cost than through limiting emissions.

This is the 'give them cash not the project' line of argument that follows straight from the cost–benefit criterion. It is a powerful argument if there is a means of implementing the 'give them cash' alternative. Therefore, before adopting the cost–benefit criterion for evaluating global warming policy, we have to investigate the feasibility of implementing such an intergenerational compensation scheme as a practical matter. In many ways, the most interesting intergenerational issue is whether we can structure intergenerational transfers or trades so that we can then pursue efficient investment strategies to maximize the consumption stream across all generations and then take care of any inequities through intergenerational transfers.

There are a number of problems that make the concept of a Fund for Future Greenhouse Victims difficult to implement.[2] First, there is the problem that Cline alludes to of how we would set up and manage such a fund so that, say, 100 years in the future a vast stock of capital goods would be bestowed upon that generation for the production of consumption goods. What we are really talking about is increasing the total savings rate, private savings plus government savings, by increasing taxes and running a government surplus (or a reduced deficit) which will increase the rate of economic growth. However, in addition to the increased rate of investment, we would have to find a way, say through a trust fund, to guarantee that the returns on this additional investment were fully reinvested and not returned to the income stream where they would mostly go to consumption with only a relatively small part being reinvested. Without this continual reinvestment, the arithmetic of compounding which is fundamental to the 'give them the cash' alternative does not work. Furthermore, this policy would have to be sustained by successive governments for a hundred years. Even if this generation were to set up such a fund, it could not commit the governments of future intervening generations to do so. There would be every incentive for some future generation to break the chain of intergenerational transfers and consume all or part of the resources in the trust fund. Therefore, in the absence of a potentially viable plan for transferring resources to future generations, the powerful logic behind the cost–benefit criterion breaks down if there are potentially serious issues of generational fairness.

A case can therefore be made that we should not automatically apply the standard cost–benefit criterion if it can be demonstrated that future generations would be significantly worse off than the present one as a result of global warming. Whether future generations will be better or worse off will depend primarily on the rate of economic growth. Empirical evidence suggests economic growth has produced an increase in per capital income in the past and this process is continuing today. In fact, economic growth is such a fact of life that every one of the integrated economic models presented in this special issue has the assumption of positive economic growth built into their simulations. Therefore, in all likelihood future generations will be much richer than the present one and if they (future generations) want lower levels of greenhouse gases and lower temperature levels they should pay for them. The problem is that possibly the most cost-effective, or perhaps the only way of achieving this, is to begin abatement programmes today because of the irreversible nature of the process.[3] Therefore, to implement such a programme of control where future generations would pay the cost, we would need to implement a scheme whereby the future generations would compensate the present and near-term generations for their investments in emission abatement programmes. We know that in private markets such trades can be made through overlapping generations.[4]

To see how this would work through market transactions, suppose that epidemiologists were to predict that in 100 years a virus would return which would wipe out 50% of the world's population. Further, assume that to produce a known vaccine for this virus would take 100 years. In a competitive market, drug companies would at

[2]For a discussion of this and some other aspects of the intergenerational transfer problem in the context of global warming see Lind, (1990, pp S20–1).

[3]We typically treat global warming as an irreversible process; however, greenhouse gas concentrations will decay over time if not replenished, and it is possible that technologies can be developed which will reduce greenhouse gas concentrations or modify their effects. These possibilities should be explored carefully as they would be a significant factor in any decision to commit immediately to an aggressive abatement programme.

[4]The seminal paper of this literature is Samuelson (1958).

least in theory begin to produce the drug if they could profitably sell it in 100 years and earn a market rate of return even though the officers of these companies and the present shareholders would all be dead when the investment paid off. This can take place because present shareholders can sell their shares to the next generation and this process will repeat itself until the vaccine is sold and paid for by the generation that secures the benefits.

The value of each generation's share will go up by the increase in the present value of the investment as time passes. Since each generation has other alternative investments, the vaccine investment will have to earn a market rate of return, say 8%. Furthermore, if the generation facing the epidemic is assumed to be at least as wealthy as the ones before it, even after paying for the vaccine, there is no reason for prior generations to subsidize it. Therefore, the appropriate rate of return on the investment in the vaccine is 8%, the market rate, because the only reason for requiring investors in previous generations to accept a lower return, say through some government programme supported by taxes, would be if it were appropriate to subsidize the generation living 100 years from now.

The problem with this line of argument for global warming is that the output of investments in abatement are cooler temperatures in the future which are public goods and therefore cannot be provided profitably in the market. They will have to be provided by governments acting over time. Theoretically, governments could replicate the private market by appropriately transferring resources from younger to older generations. Each generation would make transfers to the previous one equal to the accumulated value of the investment in abatement and then receive later in life a similar transfer from the next generation. Clearly, this does not appear to be feasible in practice.

It follows from the previous example that if there is no reason to subsidize the generation 100 years from now, and if that generation would not be willing to pay for the cost of a century of investment fully compensating previous generations, which means returning their capital and paying a market rate of interest, ie the present value of benefits discounted at the market rate of 8% does not exceed the present value of costs, then the programme should be rejected both on the grounds of intertemporal efficiency and equity. Therefore, even if intergovernmental transfers could be carried out and guaranteed across generations so that compensation could be paid, it follows from the cost–benefit criterion and consideration of intergenerational equity that the discount rate should equal the market rate of return.

However, suppose that in the previous example the facts were altered to make the cost of the vaccine totally unaffordable to the generation living 100 years from now. More precisely, assume that the amount that this future generation would be willing to pay for the vaccine was less than the accumulated cost of producing it, not including any accumulated interest. Then the investment in the vaccine would have a negative rate of return and would be rejected on the cost–benefit criterion even if a zero rate of discount were used in the cost–benefit calculation. At the same time, current and future generations might well decide to fund the vaccine development on the grounds that it would be morally unacceptable not to develop the vaccine and subject a future generation to the ravages of such an epidemic.

Note that the solution to this particular case is not to lower the discount rate for public projects. We have to evaluate the issue of intergenerational equity separately and make a political decision about how we should deal with this ethical issue. The cost–benefit criterion does not help us with the decision as to whether current generations should subsidise future generations. What the cost–benefit format can do is help us determine the most cost–effective way of implementing that subsidy. This is no different for the case of intergenerational income transfers than for the case of intragenerational transfers. Cost–benefit analysis which uses the willingness-to-pay concept as the basis for measuring benefits and costs, which in turn depends on the intergenerational and intragenerational distributions of income as a starting point, does not provide a basis for determining whether that distribution of income is equitable or whether redistribution should occur. Furthermore, this situation cannot be solved by tinkering with the discount rate.

Whether we are dealing with public projects which span many generations or ones which occur within one generation, politically, it is often very difficult to design and implement mechanisms through which the beneficiaries of a project will compensate those that incur the costs. There is an extra set of problems when intergenerational transfers are involved because of our inability to commit intervening generations to the plan. This is not only true for the scenario that would give future generations the cash, but also for any continued programme of CO_2 emissions abatement. The present generation may start a programme based on the assumption that it will be continued by future generations, but there can be no guarantees that this will occur.

There is one kind of investment or programme that avoids this commitment problem. Investments in technology development that will provide future generations with the means of coping cost-effectively with greenhouse gas emissions or of reversing global warming have the property that they will be available to and benefit future generations regardless of what intervening generations choose to do. This characteristic means that such investments and programmes merit special attention.

Global warming and fairness to future

From the previous analysis it is clear that, because it is not possible to commit to and carry out the appropriate intergenerational transfers in either direction that would be required for the beneficiaries of a policy to compensate those incurring the costs or to remedy intergenerational inequities, the cost-benefit criterion is not a complete criterion for evaluating global warming policy. We must also ask whether there are issues of intergenerational equity that might require us to modify conclusions with regard to global warming policy reached on the basis of cost–benefit analysis alone. This question can be subjected to analysis.

To begin this analysis, it is useful to start with the cost and benefit estimates provided by Cline. He assumes (what he refers to as a 'modest') 1% rate of growth in per capita gross world product (GWP) (1993, p 5). This would result in per capita GWP levels which in 100 years would be 2.7 times what they are today and in 300 years be 19.8 times today's level. He also estimates potential damages for the year 2050 and the year 2300 for several scenarios. The highest damage estimate is 20% of GWP by the year 2300 (1993, p 4). For the purposes of this discussion we will use a damage estimate of 20% of GWP. This means that even if by 2093 people were 2.7 times as rich as people today, and if global warming were to reduce their wealth by 20%, they would still be over twice as rich as the present generation. By the year 2293, after deducting 20% for the cost of global warming, people would be over 15 times as rich as current generations. Can we justify current generations sacrificing 2–3% of GWP to increase the wealth of future generations who even after deduction for the high damage scenario are 2-15 times richer than the present generation? The answer is clearly no on the basis of intergenerational equity, which must weigh in favour of the current generation.

The crux of the question on intergenerational equity is whether per capita world income will continue to grow and at what rate, and what the magnitude of the costs imposed by global warming will be. If we take the most extreme cost and benefit estimates generated by Cline, the answer has to be that if there is an issue of intergenerational equity, it is whether the present generation should be transferring income through global warming programmes to future generations who will be much better off. The real disagreement between the environmentalists who advocate an all out programme to reduce greenhouse gas emissions, and economists and others who may be more sceptical, is a disagreement over whether future generations will be better off even with global warming than the present one. This in turn hinges on what the rate of per capita income growth will be and how severe the consequences of global warming will be. These two factors are interrelated. All things being equal, the greater the rate of economic growth, the greater will be greenhouse gas emissions and the severity of the greenhouse effect. Therefore, if we assume slower growth we probably should also assume a lower level of damage from global climate change. This is an issue that deserves further investigation using the integrated assessment models.

It appears from Cline's numbers and other numbers generated from integrated model runs that only if there were catastrophic greenhouse effects, outcomes that are not even close to the mean of what we would predict, is there possibly a crisis and a case for present generations subsidizing future generations. This position is well stated by William Nordhous, a noted contributor to the economic analysis of global warming (1993).

> Most economic studies of the impacts and policies concerning climate change are based on scenarios like the smooth and gradual warming. . . . The conclusion that emerges from most economic studies is to impose modest restraints, pack up our tools, and concentrate on more pressing problems. . . . Yet, even for those who downplay the urgency of most likely scenarios for climate change, a deeper anxiety remains about future uncertainties. . . . Given the potential of catastrophic surprises, perhaps we should conclude that the major concern lies in the uncertainties and imponderable impacts of climate change rather than in the smooth changes foreseen by the global models.

One might criticize this line of analysis on the grounds that GDP is not a particularly good measure of well-being. Perhaps even a tenfold increase in GDP would not be sufficient to compensate for a significant, but not catastrophic, reduction in non-market goods such as environmental quality. Therefore, it is possible that even with significantly higher measured per capita GDP, future generations could be less well off than the present one. What are the implications of this for both the economic analysis of climate change and for policy?

First, if we perform a full-blown cost–benefit analysis where we not only measure changes in GDP but quantify the environmental benefits and costs as well, then there is no problem because these costs and benefits have been measured by the same metric as GDP and can be compared with GDP. If a US$100 000 increase in per capita GDP is sufficient to compensate for the accompanying reduction in environment quality, then that tells us that the environmental costs are greater than the US$100 000 per person benefit from enhanced GDP.

A full cost–benefit analysis would pick this up. The problem is that few economic studies explicitly model environmental effects and quantify their costs and benefits. More work needs to be done along these lines and the resulting cost–benefit estimates need to be incorporated into the integrated assessment models.

However, it is unlikely that great progress along these lines will be made soon. While progress has been made in the science of climate change, the experience of the past suggests that we are not likely to get reliable, generally accepted projections of what climate changes will occur under various scenarios any time soon. Further, even if we knew with precision what climate changes would take place, it is difficult to predict their impacts on either man, his environment, or on human well-being generally. Estimating future benefits and costs associated with climate change is therefore virtually impossible. For these reasons, we will have to formulate policy given considerable uncertainty.

However, if people in the future will have much higher per capita incomes because of capital accumulation and technological advances, then they will have vastly more resources than we do today, greater technical capability and much more information about the true nature of the problems they face. Therefore, except for situations where we see an overwhelming case for believing that significant irreversibilities will make the position of future generations untenable, the best strategy is to leave the potential environmental impacts on future generations for those generations to deal with, given their superior resources, technology and information. This is not without risks, but it is superior to committing large quantities of current resources in a world with vast poverty to solve problems of future generations which are highly uncertain and may be more amenable to solution by those future generations.

What discount rate should we use?

Before discussing what rate should be used, we need to distinguish between several different concepts of the discount rate and how they may or may not fit into the standard cost–benefit model. In discussing the traditional cost–benefit paradigm in a competitive world, the rate of discount r was equal to the market rate of interest and, in turn, was equal to the marginal rate of return on capital and to every individual's marginal rate of time preference which determines the rate at which individuals would be willing to trade their own current consumption for their own future consumption. It is this own-consumption rate of discount which would be relevant to the cost–benefit model with overlapping generations described in this paper. Only if we accept the logic of the Kaldor–Hicks compensation scheme in the cost–benefit model is this rate relevant to the analysis of benefits and costs across many generations.

There is a second concept of discounting that has recently entered the literature and that relates to how we would value something received by someone else now as opposed to sometime in the future. An example of this kind of analysis is the study by Cropper and Portney (1992) which attempts to determine how people would value saving a life (not their own) now as opposed to saving a life in the future. Clearly, since people in general do not make this kind of trade off in their normal day to day transactions, we cannot look to a market rate of return or interest to determine their time preference in this sense. Such studies of time preference rely on questionnaires or experimental data. We can imagine asking people similar questions about their willingness to sacrifice their own consumption today for a reduction in the cost of global warming in the future and from that impute an implicit discount rate. This concept of the discount rate is noted for completeness and clarity, but will not be discussed further as it is discussed in another paper by Thomas Schelling (in this issue).

The third concept of the discount rate is the rate of discount applied to utility in the optimal growth models. In these models, the problem is to maximize social welfare over time by optimally allocating production at each point in time between consumption and investment. More specifically, the problem is to maximize

$$W = \int_0^{\infty} e^{-\delta t} U(C(t)) dt \qquad (1)$$

where $C(t)$ is consumption at time t, $U(C(t))$ is the utility of consumption at time t, and δ is the discount rate applied to utility, subject to the constraints of technology, the labour force and resource availability. There are several ways to interpret this formulation of the problem. One would be to consider this to be the optimization problem of a single or representative individual who lived for ever. A second interpretation, which is more useful for analysing intergenerational issues, would be to assume that there is a constant population of identical individuals with the same utility function ie $U(C(t))$ is the same for all individuals regardless of when their consumption takes place. Therefore, $U(C(t))$ can be taken to represent the utility of the population at time t. In addition, we will assume that generations are overlapping so that a competitive economy could operate through time.

In the optimal programme, one of the conditions of optimality is that

$$r_t = \rho_t = \delta + \eta(C_t)\frac{\dot{C}_t}{C_t} \qquad (2)$$

where r_t is the marginal rate of return on investment at time t, ρ_t is the individual consumer's rate of interest or time preference for consumption, δ is the utility rate of discount, $\eta(C_t) = -\frac{U''(C(t))}{U'(C(t))} C_t$ is the elasticity of the marginal utility of consumption, and $-\frac{\dot{C}_t}{C^t}$ is the rate of growth of consumption.[5]

If we look at the left side of Equation (2), we see that, on the optimal path, the marginal rate of return on investment, r, equals the consumer's rate of interest or time preference which is identical to the result for the multiperiod perfectly competitive economy. It would follow from this that the correct rate of discount should be the equilibrium value of r which would equal the market rate of interest and the consumer's rate of time preference.

How, then, can we justify a lower rate of discount than the market-determined cost of capital for use in cost–benefit analysis? This can be justified if the rate of discount that we should apply to consumption over time should for some reason be lower than the marginal rate of return on capital. There are two lines of argument that have been advanced to reach this conclusion.

The first line of argument for a lower discount rate for consumption begins with Equation (2). Instead of equating ρ, which we cannot observe directly, to the marginal rate of return on capital, r, which we can observe, some economists including Cline attempt to determine r from the right-hand side of Equation (2). Generally, the process begins by setting $\delta = O$ which is justified on the grounds that to have a positive δ would give less weight to the utility of future generations. The second step is to estimate the elasticity of the marginal utility of consumption η_t and the long-run growth rate of consumption $\frac{\dot{C}_t}{C_t}$, and the product of these two terms gives one the rate of time preference for use in discounting long-term costs and benefits. For example, Cline makes exactly this line of argument (1993, p 4). He states that

> To estimate the SRTP (social rate of time preference) – which I calculate to be 1.5% – I appeal to basic economic theory – which states that this rate equals the sum of two components.

In the notation used here, the social rate of time preference is ρ and its two components are δ and $\eta_t \frac{\dot{C}_t}{C_t}$. With regard to δ, Cline states,

> There is a strong tradition among economists to set this preference at zero, especially for comparisons between the present generations and future generations, who cannot participate in today's decisions

He goes on to reason as follows about the second component $\eta_t \frac{\dot{C}_t}{C_t}$:

> This component, in turn, equals the product of (1) the growth rate of per capita income and (2) the responsiveness ('elasticity') of marginal utility with respect to consumption ... For my study, I set the utility based rate at 1.5%. This estimate applies an 'elasticity of marginal utility' of 1.5 based on independent calculations by William Fellner and Maurice Scott. It also assumes an average per capita growth rate of 1% over the next three centuries.

It is important to note that even if the utility rate of discount is zero, the social rate of time preference will be positive as long as consumption is growing over time. Therefore, discounting benefits and costs at a positive rate does not necessarily give less weight to the welfare of future generations.

The second line of argument that has been used to justify discounting at a rate below the marginal rate of return on capital is based on the fact that because of personal and corporate income taxes, the after tax rate of return on investment to the individual investors is much lower than the before tax rate on investment so that individuals do not earn the market rate of return on investment on after tax basis. If we equate this after tax rate of return that an individual can earn to that individual's rate of time preference for consumption, then it can be argued that we should discount future consumption streams, ie future benefits and costs, at this lower rate. In other words, if we could organize the appropriate system of interegenerational transfers, then hypothetically future generations could fully compensate the previous ones by paying them a rate of return equal to their marginal rate of time preference which will be lower than the market rate and the marginal rate of return on capital.

The problem with this position is that it is not clear that individuals have a well defined rate of time preference which is lower than the marginal rate of return on capital. For example, many individuals borrow on credit cards and pay rates of interest of 15–25%. Should we be using these rates to evaluate global warming investments? Many economists say yes. More troubling still, from a theoretical point of view, is that we observe some people borrowing on credit cards at 15–25% and simultaneously investing at after tax rates of return in the 1–3% range. Which, if either, of these rates reflects these individual's rates of time preference?

Further, we know from experimental evidence, that

[5] For an excellent discussion of these conditions and terminology see Dasgupta (1982).

people exhibit different rates of time preference in different situations and for different goods. The question is how we would know what the appropriate rate is from this conflicting data. Put differently, do individuals have well defined rates of time preference, and if they do, do they equal any particular rate of interest or return?

Further, there are a number of serious problems with using a discount rate below the marginal rate of return on capital which is generally estimated to be in the range of 5–20% in real terms regardless of the logic behind it. The first problem is that if we were to believe that the rate of time preference were below the rate of return on capital then we are not on the optimal growth path as Equation (2) is not satisfied. This means that we should be saving more, which Cline fully understands and subscribes to (1993, p 5). However, Alan Manne uses a standard formulation of the neoclassical growth model to explore the implications for the level of saving and investment given the discount rate that Cline proposes, and assuming a conservative marginal rate of return on capital of 5%.[6] What he finds is that the optimal programme, given these assumptions, would require a large discontinuous increase in savings and investment. Savings rates in the transitional period would be much higher than any historical experience and much higher than rates that we would expect generations living in the near term to accept.

Furthermore, if we were going to increase savings and investment, for example, by running a government surplus which would create a larger pool of total savings in the economy or by raising taxes and making direct public investments, there are many private and public investments that we would want to make before making investments with a 2.0% rate of return. Given that we are not likely to be willing to increase the savings rate to the level implied by Cline's assumptions, it follows that there are many alternative investments to global warming that we should be making before we would invest in climate change abatement. The second-best argument frequently used by economists, including Cline, simply does not apply here, because there are potentially many investments in both the private and public sectors that could be made with the appropriate public policies given the savings that private and public sectors are willing to generate.

Using a 2% rate of discount, Cline calculates that an aggressive abatement programme would produce a cost–benefit ratio of 1.3. These programmes are projected to use up to 2 to 3% of GNP (1993, pp 1, 4). If we were to evaluate all potential private and public projects that could be implemented through public action using a 2% rate of discount, we would almost certainly find projects with substantially higher cost–benefit ratios than 1.3, the cost of which will exceed the budget for global warming. When there are constrained budgets, the cost–benefit criterion requires that we maximize the present value of net benefits of the total portfolio given the budget constraint. While we cannot prove that most or all of the money in the optimal portfolio would not be invested in CO_2 emissions abatement, I believe that this would probably be the case. In any case, to prove Cline's assertion that an aggressive programme of abatement is justified, even in a second-best world, we would have to demonstrate that there are no better alternative investments that we could make with this capital.

The problem of adjusting the discount rate for evaluating one particular investment or policy is that we run the risk of adopting an investment that would be outperformed by many other alternative investments if they were evaluated on the same basis. Even in a second-best world where it would not be possible to put into effect the more efficient investments in the private sector, there are many government investments in very worthwhile projects which would almost certainly outperform CO_2 abatement, even using Cline's numbers.

In addition, given the fact that economic growth will almost certainly make future generations better off than the present one, concerns that public decision criteria that use a market rate of discount discriminate against future generations are misplaced. Given any level of saving the interest of future generations is best served by making the highest return on investments in both the public and private sectors. Using a competitive market based rate in evaluating public policy decisions will support that objective. This is the response to Cline by Birdsall and Steer (1993, p 6).

> we feel that meeting the needs of future generations will only be possible if investible resources are channeled to projects and programs with the highest environmental, social, and economic rates of return. This is much less likely to happen if the discount rate is set significantly lower than the cost of capital.

We might argue that economic growth will not continue and future generations will not be better off, but in that case the greenhouse problem will be vastly diminished as well. However, the scenario of zero per capital income growth or of declining per capita incomes with global warming should be explored using the integrated modelling capability that we have developed.

Cost–benefit analysis in general, and Cline's study in particular, are of great value even though they may not provide definitive answers to difficult policy questions. For every public policy decision, it is critical that we try to identify and measure benefits and costs as accurately as possible. This exercise sharpens our understanding of the problem and informs the decision making process.

[6] See Manne (this volume). Similar results are reported in King and Rebelo (1993).

This is true even if we cannot measure all the benefits and costs and if, for one reason or another, we reject the cost–benefit criterion. It is also true regardless of the discounting procedure we chose and regardless of whether we draw the same inferences from the cost–benefit data or not. For example, the arguments in this paper using Cline's cost–benefit estimates do not support his conclusion that we should commit to and immediately begin an expensive, long-term programme of abatement. Nevertheless, the cost–benefit analysis that he has done is an important contribution to the global policy debate.

The public process as a sequential decision process

The typical cost–benefit analysis sets out an alternative to taking no action and then estimates benefits and costs over time. In general, mean value estimates are used and risk is supposedly accounted for in the discount rate. The entire analysis is structured as if the policy choice is to do nothing or to commit to some long-term course of action. For most large questions of public policy, when there is a high degree of uncertainty and where new information is becoming available and can be produced, the last thing we should do is commit to a long-term course of action. This is an inappropriate, although traditional, way of formulating the problem of investment under uncertainty.[7]

Many policy issues, including global warming policy, evolve incorrectly into polarized debates over whether we commit to all-out action now or no action at all. This polarization, which occurs in both the analytical and political communities, results from a mischaracterization of the policy issues which obscures the choices that we need to evaluate. This is reflected in the following statement with regard to the development of synthetic fuels (Weitzman *et al*, 1981):

> In economics as elsewhere, great confusion can result from posing issues of widely acknowledged significance in the wrong way. The wrong, but unfortunately common question to ask about large development projects like the government subsidization of synthetic fuel is: 'Should we launch a crash program to build a large synfuel industry, say one or two million bbls/day by 1990?
>
> This all-or-nothing 'Manhattan project mentality' is the wrong way to pose the issue. It misses the point that substantial uncertainties about the net benefits of large-scale synfuel production ... make a sequential, information-yielding strategy the relevant policy.
>
> The right question to ask is 'Do synfuels look good enough now to make taking a further look worthwhile?' It is quite possible we might reject synfuels development when it is posed as a false all-or-nothing dilemma, yet be in favor of financing one more step to find out more information when the issue is correctly posed as a problem in sequential decision making.

This statement characterizes the problem with much of the debate over what our response should be to the potential of global warming. The relevant options are not whether we should commit to a long-term programme of aggressive abatement or not. The more tractable question is: given our concern, what should we be doing over the next ten years to position ourselves to act on new information and new technological developments? We should be asking: do we need to begin an aggressive programme of abatement right now because of irreversibility or can we wait? If we invest successfully in new technologies that will substantially reduce the cost of abatement in ten years, can we wait and more cost-effectively reach the same level of total abatement by starting later but doing it more efficiently?

This type of analysis requires a sequential planning process. It may lead to a significant programme focused on research and development, planning, and information generation in the early years, with aggressive abatement being postponed until we understand better what we are facing and have better means of coping with it. In this type of analysis, timing becomes a critical factor to be analysed. It would be better to focus our attention on the short-run choices that we need to make. This approach would also avoid the paralysis that often develops when a long-run commitment to a policy such as aggressive abatement may be too expensive to obtain political support, so that the do nothing alternative gets adopted, when in fact, there are many things we should be doing to position ourselves to deal with the possibility of future climate change. As part of the analysis of our short-run strategy, we must explicitly consider the option values of the components of such a strategy as the value of most of what we will do today, including the possibility of aggressive CO_2 abatement, is related to how it will position us for the long-run decisions we will have to make when we have more information.

Acknowledgements

I am grateful to Dr Cline for supplying me with both a copy of his book (W R Cline, *The Economics of Global Warming*, Institute for International Economics, Washington, DC, USA, 1992) and the editorial section from *Finance and Development*, March 1993.

References

Birdsall, N and Steer, A (1993) 'Act now on global warming: but don't cook the books' *Finance and Development* **30**(1)

[7] For an excellent discussion of the 'real options' approach to the analysis of how we should analyse investment under uncertainty see Dixit and Pyndyck (1994).

Cline, W R 'Give greenhouse abatement a fair chance' *Finance and Development* **30** (1) 3–5

Cropper, M and Portney, P (1992) *Discounting Human Lives*, Resources for the Future, Washington, DC

Dasgupta, P (1982) 'Resource depletion, R&D, and the social rate of discount' in Lind, R C, Arrow, K L and Corey, G R (eds) *Discounting for Time and Risk in Energy Policy*, Johns Hopkins University Press, Baltimore, MD

Dixit, A K and Pyndyck, R S (1994) *Investment Under Uncertainty*, Princeton University Press, Princeton, NJ

Lind, R C (1982) 'A primer on the major issues relating to the discount rate for evaluating national energy options' in Lind, R C, Arrow, K L and Corey, G R (eds,) *Discounting for Time and Risk in Energy Policy* Johns Hopkins University Press, Baltimore, MD

Lind, R C (1990) 'Reassessing the government's discount rate policy in light of new theory and data in a world economy with a high degree of capital mobility' *Journal of Environmental Economics and Management*, **18** (2) S20–21

Nordaus, W D (1993) 'Reflections on the economics of climate change' *Journal of Economic Perspectives* **7** (4) 11–25

Rebelo, S T (1993) 'Transitional dynamics and economic growth in the neoclassical model' *American Economic Review* **83** (4) 908–931

Samuelson, A R (1958) 'An exact consumption-loan model of interest without the social convenience of money' *The Journal of Political Economy* **36** (4) 467–482

Wallace, L (1993) 'Discounting our Descendants' *Finance and Development* **30** (1) p 2

Weitzman, M L, Newey, W and Rabin, M (1981) 'Sequential R&D strategy for synfuels' *The Bell Journal of Economics* **12** (2) 574–590

[28]

"Give Greenhouse Abatement a Fair Chance"

WILLIAM R. CLINE
Senior Fellow, Institute for International Economics

Global warming from the buildup of carbon dioxide, methane, and other greenhouse gases is nearly unique as an environmental problem, because it involves global rather than local effects and is irreversible on a time scale of three centuries or more. In my recent book, I have estimated that scientists' central value of 2½°C for expected global warming by 2050 would imply warming of 10°C by 2300, and their upper-bound value of 4½°C by 2050 would mean 18°C by 2300. Economic damages under even a moderate-central estimate would be at least 1 percent of gross world product (GWP) by 2050 and 6 percent by 2300, and upper-damage cases reach over 4 percent of GWP by the first date and 20 percent by the second. Worse "catastrophic" consequences cannot be ruled out.

Reducing carbon emissions by about one third and holding them constant indefinitely thereafter would avoid the great bulk of this warming and damage. Costs to achieve this outcome could be held low at first, through a move to more efficient energy use and through low-cost carbon savings from afforestation and reduced deforestation. However, by about 2020, the abatement costs could reach some 3 percent of GWP as industrial emissions are curbed, based on several energy-economic-carbon models. The costs would decline thereafter to perhaps 2 percent of GWP, thanks to the advent of new energy technologies.

Cost-benefit analysis provides a basis for economic evaluation of policy toward global warming. This requires two key methodological decisions: what discount rate to use to compare effects over time and how to take risk into account. The discount rate has an unusually powerful influence because of the extremely long time horizon of global warming, and because abatement costs occur early, whereas greenhouse damages avoided (the "benefits" of abatement) show up only after several decades (see chart). I argue that the appropriate overall discount rate should be about 2 percent a year in real terms.

Choosing the discount rate

The debate over what discount rate to use for public policy purposes certainly is not a new one. Even within the United States, a variety of levels can be found (the Office of Management and Budget favors an inflation-adjusted rate of 10 percent a year, whereas the US Congressional Budget Office and General Accounting Office prefer a rate close to the government's borrowing costs—about 2.5 percent real, historically, for long-term government bonds and close to the rate I apply). For a project with a life of five to ten years, the difference is not necessarily large; but for a horizon of 200 years or more, the difference is extreme (see table on page 2).

Conceptually, there are two rates policymakers can call upon: the rate of return on private capital investments (opportunity cost of capital); and the social rate of time preference (SRTP)—or the extra value people place on consumption today rather than in the future. The former tends to be higher (say, 8 percent real) because of project risk, taxation of capital income, and capital market imperfections. The latter tends to be lower. Households are net savers, and the risk-free real rate of interest they can earn on savings is the Treasury bill rate (historically around 0.5 percent real a year).

"Conservative" economists argue that the rate of return on capital is the only one that matters, as the resources devoted to any undertaking could alternatively be placed into private projects where they would earn this rate. However, over the past two decades, mainstream cost-benefit analysis has moved to take both rates into account. In the 1960s, Arnold Harberger, Otto Eckstein, and William Baumol first proposed a weighted average of the two rates, with the weights reflecting how much of a project's resources were drawn from displaced capital investment as opposed to reduced consumption.

Today's state of the art discounting methodology—which I draw upon for my global warming cost-benefit analysis—was developed in the 1970s by Kenneth Arrow, David Bradford, Martin Feldstein, and Mordechai Kurz (the ABFK method). Their approach uses a "shadow price on capital" to translate all capital investment effects (such as the withdrawal of resources from alternative investments) into "consumption equivalents." The ABFK method then discounts consumption over time (including consumption-equivalents of capital effects) at the SRTP for consumption.

Social rate of time preference. To estimate the SRTP—which I calculate to be 1.5 percent—I appeal to basic economic theory, which states that this rate equals the sum of two components:

• The first is "pure" or "myopic" preference for consuming a good sooner rather than later—myopic, because it implies an inability to envision future pain imposed by robbing the future to increase today's pleasure. Ants have a low pure time preference rate; grasshoppers, a high one. There is a strong tradition among economists to set this preference at zero, especially for comparisons between the present generation and future generations, who cannot participate in today's decisions. "Emotional distance" does not justify pure time preference because it would invite imposing damage on others just because they are not ourselves. Nor do higher consumer borrowing rates constitute empirical evidence on pure time preference; typically, net borrowers expect their income to rise, so their discount rate is based on lower expected future marginal utility of consumption.

• The second is the "utility-based" discount rate, which takes account of declining "marginal utility" as income rises. Just as the third doughnut adds less satisfaction ("utility") than the second, the utility from an extra $1,000 is smaller for an individual at a $20,000 income than for the same individual at a $10,000 income. This component, in turn, equals the product of (1) the growth rate of per capita income and (2) the responsiveness ("elasticity") of marginal utility with respect to consumption—how fast the consumption value of an extra dollar drops off as the individual attains higher consumption levels. For my study, I set the utility-based rate at 1.5 percent. This estimate applies an "elasticity of marginal utility" of 1.5, based on independent calculations by William Fellner and Maurice Scott. It also assumes an average per capita growth rate of 1 percent over the next three centuries. Combined with rising population, even this seemingly modest rate multiplies GWP 25-fold over the horizon, which some would consider incompatible with global resources.

Pay now, enjoy later

Costs and benefits of aggressive abatement of greenhouse warming

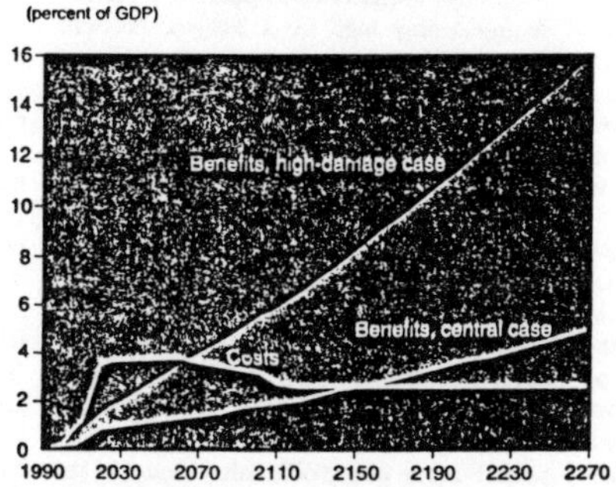

Source: *The Economics of Global Warming, 1992.*

Weights. The economic cost of carbon reduction is the sacrifice in output imposed by constraining fossil-fuel energy. This production loss should affect capital investment and consumption in proportion to their shares in the economy at large. I thus assume that the capital investment versus consumption origins of resources diverted to greenhouse abatement equal their economy-wide shares, or 20 percent and 80 percent, respectively.

Shadow price of capital. This I set at approximately 2 (one unit of capital is worth two units of consumption)—consistent with 1.5 percent for the SRTP, 8 percent for capital's rate of return, and a 15-year capital life.

The bottom line. The overall effect is approximately comparable to discounting at 2 percent real (i.e., 1.5 percent x 0.8 for consumption share in resources displaced, plus 1.5 percent x 2 capital shadow price x 0.2 for capital share).

Where does this method lead for greenhouse policy? My central scenario shows that the discounted benefits of limiting global warming would cover only about three quarters of the discounted costs. However, if risk aversion is incorporated by adding high-damage and low-damage cases and attributing greater weight to the former, benefits comfortably cover costs (with a benefit-cost ratio of about 1.3 to 1). Aggressive abatement is worthwhile even though the future is much richer, because the potentially massive damages warrant the costs. This conclusion takes on even more strength if the future is not richer, because then the SRTP should be set at zero.

In contrast, at a discount rate of 10 percent, in the central case the benefit-cost ratio falls from 0.74 to 0.33; and for the high-damage case, from 2.99 to 1.07. Not even risk-averse policymakers would adopt abatement if they discounted the future at 10 percent (unless they added "catastrophic" outcomes).

Broader implications

Certain questions naturally arise in interpreting the method I suggest. First, should the same method apply to both environmental and nonenvironmental projects? Because the ABFK method is mainstream cost-benefit analysis, the answer is yes. My approach is not based on the use of a different underlying methodology for environmental projects and does not mean that these projects should automatically have a lower discount rate than other projects. What matters most is where the resources come from: capital or consumption.

However, under some circumstances, the same method will generate a significantly higher discount rate. Thus, if there is severe crowding out of private investment by additional government borrowing, then a project may primarily displace investment, meaning the effective discount rate would be substantially higher than the SRTP. Moreover, the SRTP itself may be higher if the country's growth rate is high, or its populace is near starvation and has an unusually steep rate of drop-off in marginal utility—both conditions that may be likely in low-income countries.

Another question is: Will not low discount rates cause more environmental damage by justifying more projects? This argument is a red herring. If project analysis is done properly, it will incorporate the cost of adverse environmental side effects ("full-cost pricing"), and this will tend to rule out environmentally damaging projects.

A third question is: Does not this method imply that society undersaves and underinvests today, considering that the SRTP is well below the rate of return on capital? The answer is yes. However, where society is prepared to undertake more investment, as is arguably the case in global warming, second-best strategy recommends taking action if the cost-benefit results meet the criteria outlined here, rather than opposing action on grounds that something even better might be done with the money (but will not be). Moreover, this approach tends to increase—not reduce—total productive capacity passed on to the next generation.

Finally, does an extremely long horizon mean a lower discount rate? The answer is not necessarily, although it is likely to do so in practice. Per capita growth (and thus the SRTP) over the very long term is likely to be lower than that in the near term. Even the capital opportunity cost approach should apply a declining discount rate over a horizon of centuries, as capital accumulation relative to other production factors will reduce the rate of return on capital.

The World Bank's approach

Former Bank Chief Economist Lawrence Summers (*The Economist*, May 30, 1992) argues for a real discount rate of at least 8 percent in evaluating global warming. There are three possible reasons for this recommendation. First, the mainstream discount methodology yields a rate of 8 percent. Second, the SRTP is irrelevant; what matters is the opportunity cost of capital. Third, because alternative Bank projects achieve such rates, so should environmental projects.

The first of these views would be extremely difficult to defend, because it would require an implausibly high SRTP. Consider 8 percent discounting. Over 200 years, this discount factor reaches 4.8 million to one. Surely, Summers, along with Bank economists Nancy Birdsall and Andrew Steer (in this issue), do not believe society would have been better off over time if today's chairman of the Council of Economic Advisers could have been compelled to give up $4.8 million of consumption to make her bicentennial predecessor (Alexander Hamilton) better off by just one dollar of consumption (at constant prices). Even my SRTP of 1.5 percent means that this tradeoff is $20 to $1, already steep.

As for the second view, for narrow purposes of comparison among Bank projects, this approach may be adequate (with the qualification below). But for society's broad policy toward the greenhouse problem, it seems seriously misleading and imposes a far more stringent test than the mainstream cost-benefit analysis outlined above (e.g., ABFK). Even within this "conservative" position, 8 percent is implausibly high for a 300-year horizon. The rate of return on capital will decline as capital increases relative to labor and natural resources. Today, the Bank invests $15 billion per year. Compounded at 8 percent, just one year's investment of this amount would grow to 100,000 times my estimate of GWP at the end of the horizon! Birdsall and Steer make a modest bow in the direction of recognizing that an 8 percent rate is implausibly high by invoking a possible 5 percent rate to reflect lower investment return in industrial countries, at least for abatement by these countries. However, they do not accept the central point that the rate should take account of resource sourcing out of consumption as opposed to displaced private investment. They thus adhere conceptually to the "conservative" position.

If the Bank authors, however, are merely arguing the third view—that greenhouse and other environmental projects must achieve a return competitive with other Bank projects, then I agree (if the Bank really achieves 8 percent return). The Bank has a limited pool of capital. But in taking the greenhouse problem into account in all its projects, there should be a shadow price penalty on carbon emissions (e.g., from coal-fueled power plants) and a shadow price benefit added for projects that reduce these emissions (e.g., afforestation). This shadow price on emissions should be based on the society-wide cost-benefit analysis of abatement, and thus on the ABFK method with discounting values similar to those outlined above. This analysis would tend to place a price of $10-$20 per ton of carbon initially (the opportunity cost in afforestation or reduced deforestation), rising to $100-$200 per ton of carbon sometime after the turn of the century (and subject to further scientific confirmation). Once the greenhouse gas externalities are counted, it makes sense to require that the project compete with alternative uses of the Bank's scarce funds at the going rate of return. The only exception might be for funds specifically provided by donors for environmental projects, as in the Global Environment Facility.

Instead, Summers, Birdsall, and Steer seem to argue much more broadly that society should do little to limit greenhouse gas emissions. They implicitly counsel that the public would be better off to set aside money in a "Fund for Future Greenhouse Victims." The money would be invested at 8 percent (or, conceivably, at a still high 5 percent), and the income would compensate future generations for global warming damage at a far smaller cost than through limiting emissions.

There are several problems with this notion. First, as just discussed, real rates of return are unlikely to stay as high as 8 percent. Second, we cannot identify producer goods that yield a steady chain of producer goods that at the end of two centuries disgorge consumer goods of relevance to the population at that time (the "intertemporal transfer problem"). Third, we are not sure the damage estimates for global warming truly capture the change in the relative price between goods and environment. How many video cassette recorders will the future generations really consider an adequate compensation for 10°C or more global warming—especially if there is catastrophic risk? Fourth, it is not enough to talk about the possibility of such a Fund; tax revenue would actually have to be collected to implement it. It is unclear that the public will be willing to pay such taxes, whereas considerable public support for stemming global warming was already evident at the 1992 Rio Earth Summit.

Conclusion

So far, the message sent by the World Bank, at least in its *World Development Report 1992*, is that only minimal action on global warming is likely to be justifiable. An improved discounting methodology could change this diagnosis. Environmentalists have come a long way toward accepting economists' arguments, for example, by beginning to embrace tradable permits for emissions. Economists should not discourage this convergence by imposing a more stringent hurdle to action on global warming than would be required by mainstream cost-benefit analysis.

■

"Act Now on Global Warming—But Don't Cook the Books"

NANCY BIRDSALL AND ANDREW STEER
Director, Policy Research Department *Deputy Director, Environment Department and Director,* World Development Report 1992
The World Bank

William Cline's new book provides a thoroughly careful and fully transparent analysis of the economics of global warming—an excellent demonstration of the application of the economist's tool-kit to a pressing policy issue. But his conclusion—that an "aggressive" program of abating greenhouse gas emissions is warranted—rests heavily on lowering the discount rate used in cost-benefit analysis to around 2 percent.

We, too, believe that active policies to address greenhouse warming should be put in place now, but our recommendations do not hinge on such a low discount rate. Moreover, we believe strongly that it is wrong to assume that those who argue for relatively high rates are somehow less interested in the welfare of future generations than those arguing for relatively low ones. On the contrary, we feel that meeting the needs of future generations will only be possible if investible resources are channeled to projects and programs with the highest environmental, social, and economic rates of return. This is much less likely to happen if the discount rate is set significantly lower than the opportunity cost of capital.

We are concerned that the wealth passed on to future generations—in the form of clean air and water, and productive soil and forests, as well as the stock of technological knowledge, educated workers, and physical infrastructure—will be reduced if policymakers routinely accept investments that offer less than the best return. We are also concerned that too low a rate—by changing the ranking of projects—will induce a capital intensive pattern of development, and promote investments with high up-front costs, such as dams, that could be harmful to the environment. The fact that environmental investment has often been woefully inadequate is due to the failure to account for the costs of environmental damage in cost-benefit calculations, not because discount rates have been too high.

Why then do Cline and others argue that we should lower the discount rate when dealing with a subject like global warming? In this article, we will first address the general arguments commonly put forward: that the environment is a special case, and that long-gestating projects demand lower discount rates. We will then address the two specific technical arguments raised by Cline—actually two assumptions that underlie his calculation of the discount rate: that resources required for investing in preventing global warming will come from consumption rather than from existing savings, and that the individual components of the social rate of time preference (SRTP) are such that the SRTP is very low.

The "special case" argument

Should the discount rate for investments in environmental protection be lower than for, say, those on health, education, or food production? Such special treatment is sometimes defended on grounds of uncertainty, potentially large and irreversible impacts, and the "intrinsic" values associated with the environment. We believe that such factors deserve special treatment, but are too important to be handled indirectly through manipulating the discount rate.

Lowering the discount rate is an imperfect and often misleading tool for capturing uncertainty and for dealing with large irreversible impacts—as economists like Partha Dasgupta, Joseph Stiglitz, and J.V. Krutilla have pointed out. In the case of uncertainty, a lower discount rate simply increases the weight we put on risks in the distant future compared to the near future. And, of course, uncertainty can go both ways; unknown technological breakthroughs could greatly decrease the benefits of current investments to reduce later global warming. The problems of uncertainty and irreversibility should be addressed head on through appropriate valuation techniques, and on this score, Cline agrees (as his chart shows, he incorporates uncertainty by adding a "high-damage" scenario).

The "intrinsic" or "spiritual" values associated with the environment also deserve explicit consideration by policymakers. Cline would be the first to admit that his analysis fails to capture these values. The costs of global warming he calculates are economic values (loss of agricultural production, increased need for air conditioning, reduced revenues from ski lifts, etc.). These are valuable calculations, but clearly partial. It would surely be fooling ourselves to believe that by imputing higher values to these future economic costs (by lowering discount rates) we are somehow taking account of the future noneconomic costs that were ignored in the initial calculations. Economists need to recognize their own limits!

Cline is right in noting that many people today appear to be willing to make sacrifices to prevent global warming that they are not willing to make for child nutrition or soil protection in Africa—seeming to demonstrate a concern for future generations greater than implied by conventional discounting. Part of this apparent "willingness to pay" would disappear if citizens actually had to pay (which they have not yet), and priorities might shift if all available facts were known. But a strong concern about global warming nonetheless remains. It is hard to argue that this stems from a concern about the economic loss of a few percentage points of GDP two centuries from now (when by Cline's calculations, individuals will be seven times richer than we are today). If this were the case, citizens should be equally concerned about policies that promote economic growth over the long term, which would swamp any potential economic impacts of global warming.

The public's present concern is clearly more deep rooted. It may stem from a basic unwillingness to hand over to future generations a greatly altered, or "polluted," natural world, an unwillingness that transcends the economic impacts of such pollution. Policymakers can get some help from economic analysts in assessing the importance of such values; carefully designed contingent valuation (questionnaire) exercises and a recognition that these are values to people alive today (and thus do not need to be discounted) can help disentangle economic from noneconomic values. A complementary approach is to incorporate noneconomic values through participatory decisionmaking after the strictly economic costs and benefits have been calculated and made public. But in neither case is it necessary or helpful to lower discount rates.

The "long horizon" argument

Cline does embrace another common argument for low discount rates—that when, as with global warming and other environmental projects, an investment yields returns far into the future, a high discount rate makes no sense. By using a zero rate of pure time preference and assuming low long-term world growth rates in his calculation of the SRTP, he in effect argues that the discount rate should be adjusted to take into account the long gestation of global warming projects.

In reality, however, the line between quick- and slow-return projects is impossible to draw. All good projects benefit future generations—either directly or through reinvestment. Educate a girl in Africa today; the yield will begin immediately, but the full benefits will not yet be experienced one hundred years from now. Similarly, investment in technology in the eighteenth century enabled Europe's prosperity today, and investment in US railways and agricultural universities a hundred years ago made possible America's agricultural productivity today. We would thus argue for the same discount rate for a long gestating project as for a series of shorter gestating projects. Cline caricatures our view as arguing for a "Fund for Greenhouse Victims," implying an all-or-nothing decision at the outset between global warming avoidance or other types of investment. In the real world, the menu of options is richer—and each option deserves careful and equal treatment.

"Consumption matters" argument

How does Cline arrive at his 2 percent discount rate for global warming? He begins noting that the discount rate is a weighted average of the "capital share displaced," which incorporates the opportunity cost of capital and the "consumption share displaced," with the weights depending on whether resources come from consumption or other investment. (Cline's assumed SRTP is an input to *both* of these shares.) He then argues that the capital share deserves little weight since most (80 percent) of the resources society would allocate to reduce global warming would come from consumption not investment. His thinking runs as follows: people will not agree to pay taxes to finance more education or better roads—in other words, the overall level of investment is socially suboptimal. But because they are nervous about global warming, they might agree to some sort of controls on emissions, or even to taxes on emissions (which, in fact, by reducing GDP growth would ultimately tax them and their descendants anyway—and with a lower expected benefit than the tax on education). He thus argues for a *lower* discount rate for global warming than for, say, schools or roads.

We have no problem with the formula—that is, with the concept of incorporating the SRTP into the consumption share and capital share, which as Cline notes is now in the "mainstream" of economic theory. But we disagree with the weights used for the two shares, because we believe it is the capital share displaced that matters for purposes of allocating scarce investment resources—and allocating scarce resources is always the central problem. (What this means is that the discount rate calculated from the formula, with full weight to the capital share displaced, is conceptually equivalent to the opportunity cost of capital.) Another way to put the same point: it is the economist's job to assess the net benefits of an investment, telling policymakers where investment resources will yield the highest return—not to assume *a priori* how society will finance investment choices.

This is why at the Bank, the starting point for the choice of discount rates used in project analysis is the opportunity cost of capital in borrowing countries, a criterion that does not vary depending on the investment financed. Since a dollar used to finance investment "A" cannot also finance investment "B," the rate of return to a potential investment must be compared with alternative high return uses (taking into account environmental and social, as well as economic, factors)—or limited investment resources will be used poorly.

As a result, Bank-financed investments in roads, sanitation, education, environment, agriculture, and energy must pass a cost-benefit test applying discount rates in the order of 8–10 percent. We do accept that the opportunity cost of capital (and thus the discount rate for investment decisions) in industrial countries may be lower than the rates common in developing countries. Thus to the extent that investments in preventing global warming substitute for consumption or investment benefiting rich countries, a lower rate should be used (e.g., the 5 percent "test discount rate" used by the British Government).

A final point: although Cline concedes that in assessing how to allocate investment among competing uses it is the opportunity cost of capital that matters, he goes on to argue that in calculating the stream of costs and benefits from greenhouse projects, the shadow price on carbon emissions should be computed using his 2 percent discount rate. But it is logically inconsistent to calculate the shadow price at one discount rate and then calculate the present value of the project at another. Why should the monetary value of the externality used in calculating the shadow price be discounted at a different rate than other resources generated or consumed by the project?

"Low time preference" argument

The second factor underlying Cline's 2 percent discount rate is his use of an SRTP of 1.5 percent—which is the summation of a "pure" rate of time preference (which he sets at zero) and a discount of 1.5 percent based on the assumption that incomes will increase over time and our descendants will be better off than we are. If the only thing that really matters is the opportunity cost of not investing in higher return projects, the calculation of the SRTP is irrelevant. But for the sake of argument, let us examine this figure.

First, the "pure" rate. Cline's zero level is attractive, since it recognizes that we should take great care in making decisions that will affect future generations, who cannot participate in those decisions. Indeed, it builds in directly the notion of "stewardship"; we do not want to discount posterity's interests at all—not even to the extent we might discount our own future consumption (e.g., anticipating some probability of our own death cutting off future consumption benefits). But a zero rate implies I care as much about my and your descendants in the year 2300—about ten generations from now (within the time horizon Cline says should be considered)—as I care about the more than one billion people living today in poverty in developing countries. In terms of "emotional" proximity, are we sure that we would not rank higher an improvement in the welfare of children in Somalia today than an increase in the welfare of our descendants who might be living in the year 2300?

Moreover, there is another way to look at the issue. Consumers in developing countries are willing to borrow at high real rates of interest, higher than 10 percent, reflecting the tradeoff they see between current and future consumption for themselves. Cline maintains that the interest rate on consumer saving, not borrowing, better reflects time preferences, because the former is not affected by taxes and other distortions that drive up the cost of capital. But the fact remains that many of the poor in developing countries value current consumption highly—perhaps because they are poor and the marginal utility of additional current consumption is very high. Should we ignore this evidence?

Second, what about Cline's discount for future income growth? He assumes long-run future income growth per capita of 1 percent. But in the past four decades, such growth in the industrial and developing worlds has been 2 percent a year per capita. Furthermore, in Germany, Japan, and the Republic of Korea, per capita growth has been far greater, averaging more than 5 percent a year in the latter two over more than two decades. Predicting long-term growth rates is a tricky business; the World Bank has been overoptimistic in projecting income growth in Africa. However, in choosing between projects that would benefit Africans (e.g., education versus reduced global warming), does Cline really want us to assume that income growth will only be enough to provide the average Malian with an income of $2 a day in 2050?

What then is the right SRTP? We do not know, but we think it could be higher than 1.5 percent. If it is, then the appropriate discount rate would be greater than 2 percent. In fact, if we assume per capita income growth of 2 percent, a pure rate of time preference of 1 percent, and all of a project's resources potentially being used for other investments, the discount rate, using Cline's method, would be 8 percent. (As the reader should see, we thus have no argument with the formula for the SRTP itself—only with the weights Cline uses and his numerical assumptions in applying the formula to global warming.)

A strategy for global warming

We agree with Cline that active steps to address the possible costs of global warming should be put in place now. In fact, Cline's "aggressive" program for the next decade is not much different from that spelled out in the *World Development Report 1992*. The threefold strategy over the short term includes:

- adopt policies that can be justified for reasons in addition to their beneficial impact on global warming: eliminate energy subsidies, adopt modest carbon taxes (especially in countries with low energy taxes), and invest in afforestation and agroforestry;
- embark upon an aggressive program of research to reduce the uncertainty surrounding the problem and find solutions; in rich countries, of all public energy research funds, only 4 percent go to renewable energy;
- support the search for solutions in developing countries; rich countries should help poorer ones enjoy rapid economic growth while keeping greenhouse emissions low (e.g., through the Global Environment Facility).

Apparent differences with Cline arise mainly over his specific proposals on what to do starting a decade from now. For example, Cline recommends a tax of $100–$200 per ton of carbon "sometime after the turn of the century," but he qualifies this by saying the recommendation should be "subject to further scientific confirmation." This puts him in a similar position to the Bank, which, as the *WDR 1992* makes clear, also believes that over the longer term, as evidence accumulates, a stronger response may be warranted (e.g., international tradable permits for carbon emissions). We thus think Cline exaggerates the contrast between his and the Bank's position; his argument contrasting an "aggressive" policy to what the Bank now advocates is a "tempest in a teapot."

In sum, our difference with Cline arises over a principle rather than specific measures: while awaiting scientific confirmation, we think it will harm future generations to use low discount rates to justify locking in low-return investments. Indeed, if we are to make genuine progress in addressing the huge challenges of sustainable development, every investment will need to earn its way. As the Bank's former Chief Economist has noted in the article to which Cline refers: once costs and benefits are properly measured, it cannot be in posterity's interest for us to undertake investments that yield less than the best return. There is no need to cook the books. ■

B
The Suitability of the Intergenerational Discounting Framework

[30]

BUTTERWORTH HEINEMANN

Energy Policy, Vol. 23, No. 4/5, pp. 395–401, 1995

0301–4215/95 $10.00 + 0.00

Intergenerational discounting

Thomas C Schelling
School of Public Affairs, University of Maryland, College Park, MD 20742, USA

A 'discount rate' for the consumption of future generations from current investments for their benefit is typically composed of two parts: 'time preference' and an allowance for the lower marginal utility of consumption due to higher average levels of consumption in the future. Time preference would be involved if one were postponing one's own consumption; it has little or nothing to do with income redistribution, which is what greenhouse abatement is about. A lower marginal utility of consumption is an anomaly in income redistribution: we rarely deliberately transfer consumption from the less to the more well-to-do. Time may serve as a kind of measure of distance; we may prefer beneficiaries who are closer in time, in geographical distance, in culture, surely in kinship. Perhaps to keep our thinking straight we should use a term like 'depreciation', rather than 'discounting'.

Keywords: Global warming; Intergenerational equity; Discounting

Economists who deal with very long-term policy issues, like greenhouse gas emissions over the next century or two, are nearly unanimous that future benefits that take the form of additions to future consumption need to be discounted to be commensurable with the consumption earlier forgone to produce those benefits. And there is a near consensus that the appropriate discount rate should be conceptualized as consisting of two components (Cline, 1992; Manne, this issue; Nordhaus, 1992; Fankhauser, 1993).

One is pure time preference and 'deals with the impatience of consumers and reflects their inborn preference of immediate over postponed consumption' (Fankhauser, 1993, p 13). The second reflects the changing marginal utility of consumption with the passage of time, and is decomposed into a rate of growth of consumption per capita and an elasticity of utility with respect to consumption. The two components, pure preference for early over later utility and declining utility with growing per capita consumption, are used to compare not only utility increments in the year, say, 2050 with costs incurred in 2000, but to compare utility increments in the year 2150 with increments in the year 2050.

I first discuss 'pure time preference' and then the relevance of the elasticity of utility with respect to income or consumption. But first, since I am going to argue that 'discounting' is not the appropriate concept for dealing with the benefits of reduced greenhouse gas emissions in the distant future, I should clarify that I find traditional discounting perfectly appropriate for comparing costs and benefits of, say, hazardous-waste cleanup, as in the United States 'Superfund' program. In that kind of program discounting with appropriate rates of interest is crucial to determining which sites are worth cleaning up, how much they should be cleaned up, and when or in what order of priority they should be cleaned up. In that kind of program 'we' who pay the costs are saving and investing – foregoing some current consumption – in order to reap future benefits, along with our children and grandchildren. It makes sense to 'optimize' our investment portfolio by reference to appropriate discount rates. Global greenhouse gas abatement, I shall argue, is not like cleaning up our own land for our own benefit. *Costs* we incur for greenhouse gas abatement need to be discounted; *benefits* need an altogether different treatment.

Pure time preference

Alan Manne introduces his discussion of time preference with a quotation from Roy Harrod characterizing time preference as 'a polite expression for rapacity and the conquest of reason by passion'. I quoted Fankhauser above about 'impatience' and an inborn preference for immediate over postponed consumption. I am dubious about the ubiquity of that inborn impatience of consumers, at least for adults with decent levels of income, but my argument is that no kind of time preference pertinent to discounting the long-term costs and benefits of greenhouse gas abatement can have anything to do with the 'pure rate of time preference' defined in this fashion. That is because the alleged inborn preference for earlier rather than later consumption is exclusively concerned with the consumer's impatience with respect to his or her *own* consumption.

Alan Manne begins by asking us to 'consider an economy in which there is a single agent acting as producer, consumer, investor, and saver' (in this issue). An agent, it should be noticed, that is immortal. I suppose such an agent could have an inborn impatience about consumption. But greenhouse policy is not about saving for later consumption. It is about forgoing consumption in order that *somebody else* at a later time enjoy more consumption than would otherwise be available.

This assumption of the immortal agent, explicit in Manne, makes the issue one of 'optimization over time', one of maximizing the utilities of myriad heterogeneous peoples spread over continents and centuries as if they were all one family, all one 'agent'. It supposes that, whoever they are who pay for the investments that lead to increments in future consumption, they value increments in other people's utility *as if they were increments in their own utility*. It is this willingness to model all humankind as a single agent that makes optimization models attractive, feasible, and inappropriate.

The optimization models have no provision for redistributing current income. They redistribute only forward in time; contemporary Chinese get nothing from us, but future Chinese we treat as part of the family.

Introspectively I can find no impatience about an increment of consumption that may accrue to people whom I shall never know and who do not now exist, in the year 2150, compared with an increment closer in time, accruing to the people whom I shall never know, and who do not now exist, who might enjoy it instead in the year 2100, or closer still to the people in the year 2050.

I can imagine reasons – some of them may even appeal to me – for preferring a boost to consumption in 2025 to the same boost of consumption in 2075, but it is hard to see that it has anything to do with impatience and the inborn preference for immediate over postponed consumption. In 2025, my oldest son will be the age I am today and his brothers a little younger; with a little luck they will be alive and healthy and my grandchildren will be the ages that my children are today, and my great-grandchildren (whom I do not yet know) will have most of their lives ahead of them. Seventy-five years later they will all be strangers to me. My genes may be as plentiful in the population at the later date but they will be spread thinner. I probably would prefer the benefits to accrue to my own grandchildren rather than to their grandchildren, but I must remind myself that my grandchildren's happiness may depend on their perceived prospects for their own grandchildren, and my 'time preference' becomes attenuated.

The point of all this is that we may have grounds for preferring utility increments to occur earlier rather than later to the descendants of people now alive, but this cannot have anything to do with the kind of time preference that Roy Harrod or Samuel Fankhauser were talking about, or Alan Manne or William Nordhaus.

Actually, time may serve as a kind of measure of 'distance'. The people who are going to be living in 2150 I may consider 'farther away' than the people who will be living in 2050. They will also be different in racial composition and geographical distribution from the people I most identify with. I observe that in redistributing income via transfer payments, in providing foreign aid, in contributing to charity etc, people are expected to differentiate, and do differentiate, among recipient peoples according to several kinds of distance or proximity. One is geographical: Americans are expected to be more interested in their own cities than in distant cities, their own country than distant countries. Another is political: East Coast Americans are more interested in the people of Los Angeles than in the people of Quebec. Another is cultural: some people are closer in language, religion, and other kinds of heritage. Sheer familiarity seems to matter, and of course kinship does. (Kinship distance has both horizontal and vertical dimensions; just as children are closer than grandchildren, children are closer than nieces and nephews. Time just happens to correlate with vertical distance.)

To be less interested in the welfare of East Africans than former Yugoslavians is less like 'discounting' than, perhaps, 'depreciating'. When we count future welfare less than our own we are depreciating generations that are distant in time, in familiarity, in culture, in kinship, and along other dimensions. (There is no reason to suppose that the depreciation would be exponential. Beyond certain distances there may be no further depreciation for time, culture, geography, race, or kinship.)

The crucial point is that these are not 'saving' decisions we are talking about, ie not decisions about postponing our consumption, but decisions about redistributing income – our income. To invest resources now in reduced greenhouse emissions is to transfer consumption from ourselves – whoever 'we' are who are making these sacrifices – for the benefit of people distant in the future. It is very much like making sacrifices now for people who are distant geographically or distant culturally. Deciding whether I care more about the people who will be alive in 2150 than the people who will be alive in 2050 is a little like deciding whether I care more about people in one continent than in another, or about English speaking people more than people who speak other languages, or about people who share my history and my culture more than people who do not. People do have preferences about whom to help; the preferences show up in charitable giving, in foreign aid, in immigration policy, in military intervention.

What we are talking about is very much like a foreign aid program, with some of the foreigners being our own descendants who live not on another continent but in another century.

William Cline half agrees with me. He, too, argues that impatience or 'myopia' 'may be a legitimate basis for a single individual's preferring consumption earlier rather than later in his lifetime' but is 'hardly a justifiable basis for making intergenerational comparisons' (Cline, 1992). He disagrees in believing that we should not prefer – except on marginal utility grounds, which I am about to discuss – our own consumption to the consumption of future people. I expect that, whether or not we should, we all do. If we do not, there is a most extraordinary anomaly: we greatly prefer our own consumption to that of distant contemporaries, or even quite close ones, but not to that of people distant in future time. It would be strange to forgo a per cent or two of GNP for 50 years for the benefit of Indians, Chinese, Indonesians and others who will be living 50 to 100 years from now – and probably much better off than today's Indians, Chinese, and Indonesians – and not a tenth of that amount to increase the consumption of contemporary Indians, Chinese, and Indonesians. At its peak, the Marshall Plan took about 1.5% of US GNP; it went to the foreigners 'closest' to the Americans in most respects; and it was recognized as a short-run emergency. Americans do nothing like that now for anybody alive, except other living Americans. Whether that is good or bad, I do not see why we should expect them so much to prefer to help the unborn.

Marginal utility

The other component of the proposed discount rate is the rate of change over time of the marginal utility of consumption. The argument for including that component must be that in transferring income, or redistributing income, an important goal is to maximize the integral over time of the aggregate utility of consumption. The expectation is that on average the marginal utility of global consumption will decline over time as a result of rising consumption per capita. Resources invested now out of our own incomes will benefit people in the future who are expected to be better off than we are – an unaccustomed direction for redistributing income!

Both within countries and among countries, we expect civilized governments to redistribute toward the poorer countries and toward the poorer elements of their own populations. Doing it that way probably, as Abba Lerner argued in *The Economics of Control* (1944) 50 years ago, increases total utility. But I doubt whether that is the only reason why people prefer to see income redistributed from rich to poor rather than the other way around.

The argument for transferring consumption from the poor to the rich, or from the decently well off to the much better off, would be that the resources transferred grow in the process, and grow so much that though the marginal utility of the recipient is lower than that of the donor, the magnitude that the gift achieves in transit more than compensates. (That the resources so invested 'grow' by forestalling decrements does not affect the argument.)

There is not much room for this idea in *contemporary* transfers. If a poor farmer has some poor soil and a richer farmer has rich soil somebody could argue that extracting seed from the poor farmer and giving it to the rich farmer will so enhance the resulting crop that the somewhat utility-satiated rich farmer will gain more utility than the poor farmer loses. But ordinarily that is just an argument for trade: the poor farmer is better off selling the seed to the rich farmer, and their joint utility is even higher. The ethical interest arises only if trade is not possible, as when we outfit somebody who will emigrate to the new world, become rich, and never be heard from again, or as we contemplate transferring consumption forward in time to people who have no way to reciprocate.

Arthur Okun introduced the 'leaky bucket' in his 1974 Godkin Lecture (1975). Transferring consumption from those who have plenty to those who do not typically entails inefficiency – some administrative costs, some deadweight losses due to tax avoidance or transfer seeking, some transfers going to unintended and undeserving recipients. His analogy was carrying water from where it was plentiful to where it was scarce in a leaky bucket. The 'big trade-off' was deciding how leaky the bucket can be before we judge the effort not worthwhile. Clearly if the bucket arrives dry the effort was a mistake; if the bucket arrives three-quarters full or one quarter, or even a sixteenth, somebody in charge has to consider what discount ratio or augmentation ratio is acceptable.

Somebody might – not many people will – use some elasticity to calculate the marginal utility of consumption (Okun's water) where it is scarce and where it is not and decide whether total utility goes up with the leaky-bucket transfer. Marginal utility is clearly pertinent, for those who understand it, but probably rarely decisive even for those who do understand it. Enough attention was paid to John Rawls's *Theory of Justice* (1971) by quite sophisticated people to somewhat dethrone utility maximization, at least to deny it exclusive status. And Rawls was talking about transferring from rich to poor.

Okun never got round to talking about the other bucket, the 'incubation bucket' in which the good things multiply in transit so that more arrives at the destination than was removed from the origin. The trade-off question here would be, what sacrifice of food where it is scarce would be worthwhile if, in being transported to where it was abundant, it grew handsomely. An alternative phrasing would be, recognizing the advantage of moving resources from where they are less fruitful to where they are more fruitful, how much do we want to 'discount' the greater fruitfulness when it accrues to people who already enjoy bountiful supply? And of course Okun, concerned with contemporary transfers, could not be interested: the market would take care of the efficiency problem, and nobody is interested in helping the rich at the expense of the poor.

The conclusion I reach is that 'optimization' models are inappropriate for dealing with very long-term public investments, especially when the beneficiaries will be spread over the planet. Optimization models imply that it is our own consumption that we are promoting in the future, and it is not. Not only do they incorporate an irrelevant 'time preference', they imply that we want to treat increments in other people's utility from consumption as if they were our own. I can see no acceptance of the principle that consumption should be so distributed (redistributed) as to maximize utility. No such contemporary redistribution has ever been witnessed; and it would be strange to feel a strong obligation to redistribute income from ourselves to others in the future, *valuing their utility from consumption as if it were our own*, and no corresponding obligation to contemporaries whose marginal utilities exceed our own and probably exceed those of their own descendants.

Depreciating the consumption of high-income future people makes sense; but the 'optimization' approach is based on the principle that if the material benefits we procure for those future high-income people are large enough to offset their reduced marginal utilities, we should procure those future utilities just as if those utilities were our own. Few citizens who understood this principle would ever vote for it.

I conclude that most of us will want to discount or depreciate heavily the extra consumption provided for (or conserved for) descendants of the current population, because they are likely to be better off as well as because they are distant and it is *our* hard-earned consumption that somebody is proposing we transfer forward in time.

The analogy, or at least a better analogy, than 'optimization of consumption over time', is transferring resources from North America and Western Europe to Africa or the Middle East, South or South-east Asia, China, Russia, or deserving peoples anywhere. It is an aid program, not a savings program. There may be some reason for some people to prefer consumption increments that occur in 2075 over consumption increments that occur in 2125, just as there may be people who prefer consumption increments to occur in West Africa rather than East Africa, Russia than in Ukraine, or Boston rather than Los Angeles. And I have no quarrel with people who, when they are prepared to contribute large amounts to charity, indulge their own preferences about who the beneficiaries should be.

What should matter is your expectation about the course of per capita consumption over the next century or two. If both the developed and the developing worlds continue to grow in per capita consumption as they have done for the past 40 years, people in most countries are likely to be much better off in material welfare 50, 75 or 100 years from now than they are now. What we ought to feel we owe them is not the kind of ethical issue we have much practice with, because we are not used to thinking about making our own sacrifices, or imposing sacrifices on our contemporaries, for the benefit of people who are substantially better off.

The need to disaggregate

We must avoid a fallacy of composition here. If average per capita income rises in every country for the next 100 years, and if the poorer populations grow more rapidly than the wealthier populations, and if most of the economic sacrifices in the interest of carbon abatement are borne by the countries that can best afford it, the transfers will tend to be from the well-to-do people of Western Europe, North America, and Japan to the residents of what we now call the 'developing' countries, who should be far better off a century from now than they are now, but may not yet be as well off, during most of the intervening century, as we are now in Western Europe, North America, and Japan.

The significance of that point to me is that in deciding how to value consumption increments over the coming century or two, we need to disaggregate consumption according to the levels of per capita consumption at which they accrue. The optimization models err, on their own terms, in aggregating all future consumption and applying a uniform discount rate for declining marginal utility. Correctly, all increments in consumption should be valued at their own marginal utilities. In the optimization models, increments for poor people are discounted equally with increments for the rich; there is no adjustment for the fact that when Chinese per capita income has doubled, and Chinese marginal utility may have been halved – using the popular, but arbitrary, logarithmic utility function – Chinese marginal utility will still be many times that of the current populations most likely to pay for greenhouse abatement.

In neglecting to disaggregate, the optimization models make three assumptions in deriving a discount rate that are either dubious or wrong.

- They assume that elasticities of utility with respect to consumption are the same at all levels of consumption. (A logarithmic or power function meets that condition.) I count this one dubious.
- They assume that growth rates are uniform. Actually the optimization models themselves do not; on the contrary they recognize that China may develop much more rapidly than the currently advanced countries. But the discount-rate calculations use a single average growth rate.
- They assume that those who pay for abatement and those who benefit – or whose descendants benefit – are the same. Because all populations – all nations or regions – are assumed to enjoy increasing consumption per capita, and because investments in abatement precede benefits, the benefiting populations are assumed to have higher consumption levels and lower marginal utilities than the populations that finance the abatement.

I start from the premise that investments in greenhouse abatement, for the first 50 years, will be paid for by the countries that can afford it, ie the developed countries of Western Europe, North America, and Japan, and a few others. The beneficiaries of abatement will mainly be the descendants of those now living in the undeveloped countries, for several reasons explained below. Thus the consumption transfers will be from well-to-do countries that will mainly pay for abatement over the coming 50 years to the developing countries that, though probably better off progressively over the next 50 years, will have lower consumption levels 50 years from now than the current consumption levels of the developed countries. (I assume that benefits from abatement during the first 50 years will be negligible compared with benefits during the second 50 years.) Thus the consumption transfers, despite the hoped-for uniformly (not uniform) positive growth in GDP per capita everywhere, will be generally from rich to poor. That is, from lower marginal utility to higher marginal utility. The implications are startling, but first I should explain why the beneficiaries will be mainly the descendants of the populations now poor.

First, if the benefits of abatement were shared uniformly over the global population, they would accrue about 90% to the countries now considered undeveloped. The well-to-do are now about a fifth of the world's population; in 2075 the populations in countries now undeveloped are expected to be somewhere between seven-eighths and eleven-twelfths of the global population. So those populations will comprise most of the beneficiaries. Second, the material productivity of the developed countries currently appears to be substantially immune to weather and climate; productivity of the outdoor-dependent or agriculture-dependent less-developed countries is potentially much more susceptible to adverse effects of climate change. So besides outnumbering the descendants of the currently developed countries, they can suffer greater greenhouse damage per capita. (Absolutely, the more developed could suffer more lost GDP per capita even though the damage might not be noticeable.) Finally, the currently developed countries enjoy GDP per capita 10 times or more that of the undeveloped; during the second half of the coming century they will probably still be ahead by a factor of four or more. So the marginal utility of consumption of the nine-tenths of the population will be several times that of the richer tenth. Thus the benefits in utility increments from material consumption will be overwhelmingly inherited by the descendants of those that are currently poor. (If Chinese per capita income increased at 4% per year for the next 50 years and 2% for the following 50 years, and US per capita income increased at 1% over the 100 years, Chinese per capita income would still be less than half the US level at the end of the century and the Chinese population will be many times larger than that of the USA. At those rates of improvement, the Chinese will be about up to the present US level, and we may be expected to lose interest in further increments.)

I said the implications are startling. One has already been mentioned: virtually all the benefits from enhanced consumption will accrue to countries that will not participate much in financing the abatement. The transfers will be from the currently rich to the descendants of the currently poor, who will, when the benefits begin to be felt, be much less poor than they are now but still poorer than the descendants of the currently rich and probably still significantly poorer than the abatement-financing countries are now.

Second, the implicit 'discount rate' based on marginal utility comparisons will be negative. The currently popular optimization models cannot show this negative rate, because GDP per capita is assumed to rise everywhere. But even if it does, disaggregating shows that the beneficiaries will be both poorer and more numerous than those who finance the increments in consumption. The assumed positive discount due to diminishing marginal utility is based on a 'fallacy of composition'.

Third, if GDP per capita continues to increase in most of the developing world, as I expect and as the optimization models assume, marginal utilities of the beneficiaries will be much higher during the first 50 years – before abatement benefits become significant – than in the second 50 years. This factor substantially tilts the advantage toward any direct investments in development that can raise living standards in the first and second generations (while reducing dependence on climate-susceptible economic activities) compared with investment in climate stabilization.

Even more drastic: if marginal utilities will be higher in the fifth decade than in the sixth, in the third than in the fourth, and in the first than in the second, today's undeveloped populations have stronger claims, on the basis of marginal utility, than the populations two or four generations in the future. Once we disaggregate the world's population by income level, it becomes logically absurd to ignore present needs and concentrate on the later decades of the coming century.

And that means that no framework for considering the benefits and costs of greenhouse abatement can isolate itself from the opportunity cost: direct investment in the economic improvement of the undeveloped countries. Abatement expenditures should have to compete with alternative ways of raising consumption utility in the developing world.

A policy approach

Once we abandon the immortal-agent (optimization) approach, the two pertinent policy questions are:

- Do we want to help the future populations of China, Bangladesh, and Nigeria at our expense, because they are poor and will still be poor for some time?
- Assuming that we who pay for greenhouse abatement do not want to time-discount the consumption utility of poor countries, what mix of programs maximizes the integral of their consumption utility over time: greenhouse abatement, direct investment in their economic improvement, or direct subsidies to their consumption?

Nothing in this formulation can directly answer the question of *how much* to help the Chinese and the others. That remains a choice, not an analytical result. Just as there is no accepted formula for how much people want to help the Somalis, the Indians, the Chinese, or anybody else with current consumption, there is no formula that tells us how much people want to help with future consumption, whether with development aid or with greenhouse abatement. (I emphasize helping the poor; but there is likewise no analytical formula that could tell us how much to expect people to want to redistribute toward future Americans and Europeans.)

A different question is how to distribute Chinese and others' consumption over time when the marginal utility of their consumption is changing. The Chinese themselves may have a strong preference for current and near-term consumption over the distant, more than diminishing marginal utility would prescribe. We then have to decide whether to depreciate future Chinese less than today's Chinese would depreciate them.

If we abandon optimization based on the immortal-agent image, what analytical procedure do we replace it with? A first approximation may be:

- Assess benefits as consumption increments country by country, or region by region (homogeneous with respect to per capita income), for *alternative levels* of greenhouse abatement.
- Identify a utility function (not chosen for its mathematical simplicity) by which to convert the consumption increments into commensurable 'utility' increments.
- Treat the results as a *menu* of what we get for our money when we invest *alternative* amounts of greenhouse abatement in the enhanced welfare of currently poor countries, not as an optimization that tells us how much to invest.
- Estimate the increments in consumption, and corresponding 'utilities', that could be procured in a sample of different countries with direct investment in infrastructure, industry, public health, education, research and development etc, to identify – for any level of aggregate contribution to future welfare – the efficient mix of investments in abatement and economic development. (We cannot allocate resources between greenhouse abatement and economic development country by country because greenhouse abatement is necessarily uniform in CO_2 concentration; we cannot give India public health and education, and Indonesia greenhouse abatement, on the basis of comparative advantage.)

The rate of interest

Optimization models often look to the market rate of interest – some rate, somewhere – to get a handle on an appropriate discount rate. What role does my analysis ascribe to the market rate of interest?

- The market rate of interest tells us nothing about how much people would like to contribute now to help others in the future.
- It tells us something about when those people would like to contribute: the earlier, the more it costs.
- It tells us little or nothing about what return we would get on public investment in poor countries that may be alternatives to greenhouse abatement.
- It tells us about an opportunity cost: instead of greenhouse abatement, or investment in development, we could invest commercially and dedicate the proceeds to somebody's future consumption.

The utility function approach

The discount-rate question should disappear. In its place is what utility function to use in valuing future increments in other people's consumption. This is a real question, not a matter of mathematical convenience. It is not a question economists have much practice with. The policy implications of such an approach are troublesome

The discounting procedure has a pseudo-familiarity; the analogy with saving-for-future-consumption has appeal. Everybody knows about interest rates, and they are genuinely pertinent to much national public investment. Marginal utilities from consumption are alien to most policy thinking. People usually understand the argument for redistributing from rich to poor, but no legislator ever thinks about whether the utility function is a power function, a logarithmic function, or a polynomial; whether there should be a universal utility function for all cultures and all times; whether material consumption is the only argument of the function, or anything of the sort.

My procedure (properly) gives no answer to the question, how much? An optimization model answers that question – incorrectly, but it gives an answer. Facing an optimization, a policy maker can say, 'I will do half that, no more', but at least the policy maker thinks he or she is doing half of something objectively arrived at. My procedure requires a *choice* of how much, with no benchmark.

It is hard to think about the future with my procedure and not be forced to think about the present. If marginal utilities of consumption are the correct target for investing in future welfare, they must be pertinent to today's welfare. But today's welfare vanishes from the greenhouse-optimization models because there is no greenhouse-abatement increment in current or near-term consumption. Hence the next point: greenhouse abatement, largely identified with energy policy, is insulated in optimization models from economic development. An answer to how much (and when) to abate is given in optimization models independently of what else is going on. When greenhouse abatement is identified as a mechanism for making income transfers to future generations, especially to those whose consumption levels are still comparatively low, it has to compete with transfers for investment in economic development. So there is not only an unanchored choice of how much to help, but a choice of what proportions through greenhouse abatement and direct assistance to development.

Carbon dioxide abatement is probably 'target efficient'. Poorer countries are probably more vulnerable to climate change than wealthier countries. But direct investments in public health, birth control, training and education, research, physical infrastructure, water resources etc, can also be directed to target populations, so the advantage may go either way.

I doubt whether developing countries would prefer to defer consumption increments to later generations, whether what is deferred comes out of their own resources or out of resources made available by wealthier countries. I would expect, if offered a choice of immediate development assistance or equivalent investments in carbon abatement, potential aid recipients would elect for the immediate. So if we, the developed, elect carbon abatement for their benefit, it is *our* choice of their descendants over themselves.

Whichever we choose, there is no reliable way we can constrain our own descendants' choices to continue or to discontinue what we began. We can invest in the consumption of future generations via carbon abatement or via direct investments; in 2050 they can discontinue the direct investments, or they can discontinue the carbon abatement. There may be institutional reasons for expecting discontinuance to be more likely with one approach or the other, but the preference is not self-evident.

A final word. This paper has been directed only at the valuation of future material consumption by individuals in GNP projections. That is what the optimization models assess. It has not examined the projections themselves, and it has omitted any reference to values other than those conventionally included in material consumption. It omits damages and benefits relating to nature, biodiversity, wildlife etc. It ignores environmental influences on public health, which are potentially important, for good or ill, in poor countries where the hazards remain more biological than chemical. It ignores the possibility that in developed countries the income elasticity of demand for environmental and recreational benefits may be high enough to bring developed-country benefits back into the picture. Any implications for greenhouse policy are limited accordingly.

References

Cline, W R (1992) *The Economics of Global Warming* Institute for International Economics, Washington, DC

Fankhauser, S (1993) *The Social Costs of Greenhouse Gas Emissions: An Expected Value Approach* Centre for Social and Economic Research on the Global Environment, University College and University of East Anglia, Norwich

Lerner, A P (1994) *The Economics of Control: Principles of Welfare Economics* Macmillan, New York

Nordhaus, W D (1992) 'An optimal transition path for controlling greenhouse gases' *Science* **258** 1315–1319

Okun, A (1975) *Equality and Efficiency: The Big Trade-off* The Brookings Institution, Washington

Rawls, J (1971) *A Theory of Justice* Harvard University Press, Cambridge, MA

[31]

Carbon Dioxide and Intergenerational Choice

By RALPH C. D'ARGE, WILLIAM D. SCHULZE, AND DAVID S. BROOKSHIRE*

Benefit-cost analysis has become an accepted *tool* for evaluation of government investment programs in natural resources during the past few decades (Otto Eckstein, 1958; John Krutilla, 1964). In recent years, it has been tested and applied to a few long-term environmental problems including chlorofluoromethanes and ozone, climatic change caused by SST emissions, and to CO_2 climate changes (d'Arge et al., 1975, 1976; J. A. Laurman, 1979; and Martin Bailey, 1980).

These studies have worked with very uncertain estimates of both costs and benefits, however, each has had a similar set of underlying problems in assessing results. First, in each case the process of economic discounting of the future has led to small present values for even almost catastrophic future economic losses. For example, a complete loss of the world's *GNP* in 100 years would be worth about one million dollars today if discounted at the current prime rate. If world *GNP* in 100 years were the same as today's *GNP*, discounting would reduce the value to only approximately $70,000 presently. Thus, catastrophic losses in the distant future are almost valueless to the present generation *if* benefit-cost analysis, as it is generally applied, is used in valuing the future. Second, given changing lifestyles, substantial future shifts in technologies, and probabilities of drastic world political-social events, any quantitative measures of benefits or costs in 100 years are not subject to better than 2–4 orders of magnitude accuracy, and may even switch sign (d'Arge, 1979).[1] Given these uncertainties, perhaps the best current measure of benefits or costs to future generations is how much current generations would pay to prevent future risks (Krutilla, 1967).[2] That is, current decisions on regulation affecting future generations should be predicated on the preferences and values of those now populating the earth's surface. Third, the ethical basis of benefit-cost analysis has not been explored to see what forms of bias are introduced when the current values are imposed on future generations. It is argued in this paper that the appropriate social rate of discount varies substantially depending on the underlying ethical beliefs of society. Some ethical beliefs lead to discounting future effects to zero (an implied rate of discount of infinity), while others would require valuing future effects at more than present costs or benefits.

I. CO_2 Effects

The CO_2 problem, from the standpoint of an economic perspective, has its most interesting elements in terms of intergenerational choice. The reason for this is the anticipated positive economic benefits accruing for several generations and the potential disastrous economic costs that may accrue to generations after that. In the NAS Geophysics Study Committee report, *Energy and Climate*, several potential physical and environmental effects of increasing CO_2 are rather dramatically proposed over a period of 170 years (see Panel on Energy and Climate; George Woodwell et al. 1979). They include the "melting of glaciers" causing a rise in the sea level of three to five meters, a potential shift in agricultural zones upward in latitude in the northern hemisphere with an expansion of frost-free days, rates of photosynthesis, and production of biomass in

*University of Wyoming.

[1]For example, current estimates of the net costs of a gradual warming in the United States depend crucially on whether future increased air-conditioning costs will be greater than benefits from reduced heating costs.

[2]There is substantial indirect evidence that individuals are willing to pay some proportion of their current income to endow future generations with certain types of assets.

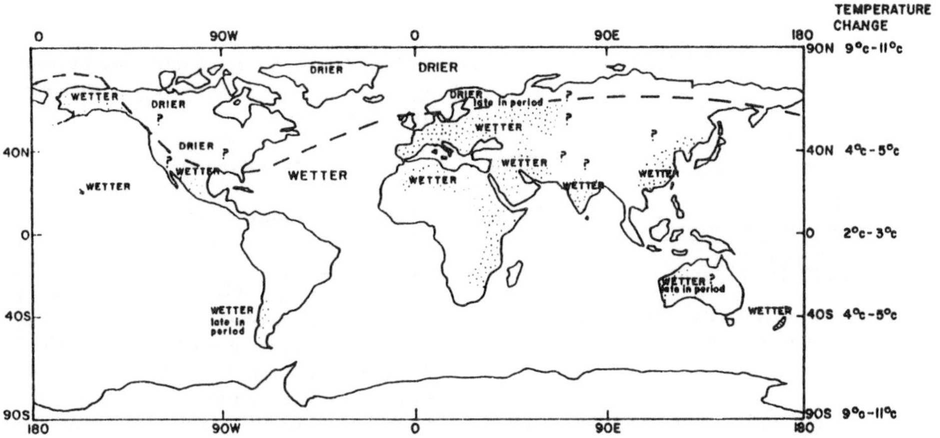

FIGURE 1. POTENTIAL CLIMATIC CHANGES AND THE ALTITHERMAL PERIOD (4,000–8,000 YEARS AGO)

Source: Adapted from Kellogg. Temperature changes were formulated in discussions at the Aspen Institute Workshop on "The Consequences of a Hypothetical World Climate Change Produced by Atmospheric Carbon Dioxide," Maryland, February 25–29, 1980.

most regions of the world, along with expansion of arid and semi-arid regions. Thus, from a qualitative perspective there may well be substantial benefits and costs to agriculture, forestry, and other natural resources.

Figure 1 is a world map adapted from William Kellogg (1979) indicating potential temperature changes by latitudinal zone given a doubling of atmospheric CO_2. A rather arbitrary line divides the world into drier and wetter zones based on the altithermal period occurring four to eight thousand years ago. If the line separating drier and wetter is accurate, a whole sequence of economic difficulties in assessing benefits and costs becomes inevitable. For example, there is a lot of uncertainty as to where the line should pass through the Continental United States and if, for example, the Northern Great Plains and Midwest were significantly drier and warmer, this would suggest a wholly different set of adaptive policies and thereby benefits and costs as contrasted to a warmer and wetter area. The same difficulty arises for the Soviet Union in terms of wheat yields. The line bisects the Soviet Union where only a slight error would make the winter wheat crop more or less vulnerable. If there was more variation in the hydrologic cycle in the wetter zones identified in Figure 1, this could cause greater variations in crop yields and thereby cause substantial impacts to agrarian communities on the food threshold. It would also be dependent on the possibilities of new varieties of crops which are either more drought resistant or water tolerant (S. Wittwer, 1979).

A drier climate in the Northern Great Plains of the United States would cause an increase in the probability of crop failure in any one year. Over a long time, the Midwest, through development of substantial irrigation systems, could possibly even increase yields above those existing currently but at a substantial capital cost (i.e., \$30–\$50 per acre). In addition to the direct agricultural effects and impact on growing seasons, a substantially warmer climate would undoubtedly mean a reduction in heating costs, a potential increase in air conditioning costs (I. Hoch, 1979), a movement towards or away from an optimal urban environment depending on the initial location (Bailey, 1980; K. Attaran, 1973; and d'Arge et al., 1975), a potential increase in diseases (particularly in the tropical zones) such as elephantiasis and

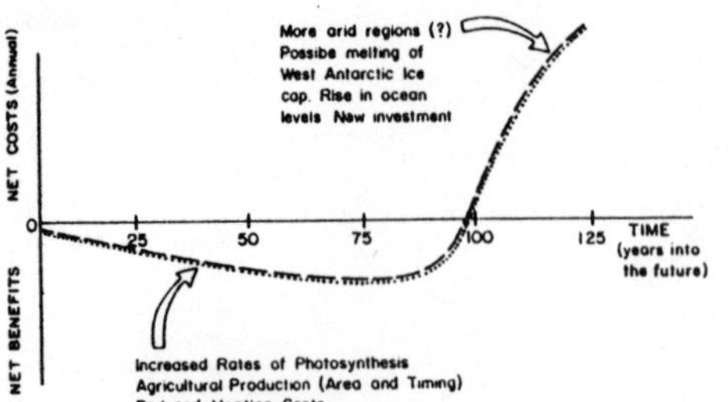

FIGURE 2. LONG-TERM NET COSTS OF CLIMATIC CHANGE: A QUALITATIVE PICTURE

dengue fever, and greater capital requirements for reducing growth of molds in relatively high humidity locations. In addition to changes in location of agriculture, there would be an implied movement of population and thereby location of urban areas. Whether such a movement would be rapid and substantial enough to induce large adjustment costs is not now known (J. Neidercorn, 1976). The cost of adjustment would depend substantially on the depreciation rate for existing capital assets and if the adjustment time is slow enough, such costs might be minimal (Laurman).[3] It appears reasonable to presume, that given the existing quantitative evidence for these sectors, that very small positive changes in global temperature in the range of 0–2° centigrade, would be beneficial to at least the next several generations.

Given the highly incomplete and tentative findings on costs and benefits, the CO_2 problem might be visualized as is depicted in Figure 2. That is, for a period of perhaps up to 100 years, there is likely to be a net annual benefit from increasing levels of CO_2 in the atmosphere. Beyond that time, however, as the climate becomes distinctly warmer with greater potential variability in precipitation patterns and of melting the antarctic ice caps, the likelihood of very large future economic costs increases. Thus, we visualize the CO_2 problem as a problem of tradeoffs between succeeding human generations where the first few benefit substantially by imposing uncertain but possibly very large costs on future generations.

[3]Laurman suggests the lead time necessary to convert to nonfossil fuels may be the important consideration for CO_2 induced climate change.

II. Ethical Systems

Ethical systems attempt to provide a mechanism for answering the question: "Is a contemplated action right or wrong?" An ethical system can take the form of a list of rules. Examples of the first class of ethical systems include such specific lists as the Ten Commandments and Kant's Categorical Imperative, which states: "Act only on the maxim whereby thou canst at the same time will that it should become a universal law." Note that the Ten Commandments provide a list of specific behavioral rules while Kant provides a mechanism for generating such a list. The difficulties with lists, however, is first that some rules may well come into conflict (be inconsistent) under some circumstances, which would require a hierarchical ordering of rules to resolve conflicts. Second, such lists, if explicit, may fail to cover certain eventualities.

An alternative specification of an ethical system can take the form of a criterion for evaluation. Thus, for example, "do unto others as you would have them do unto you," can be applied to nearly all ethical behavior decisions. Similarly, the statements "turn the other cheek" and "individuals should have freedom of choice where no one else is bothered" imply that ethical behavior involves not harming others inclusive of future generations under any circumstances, and yield a general criterion or decision rule.

Here we explore the issue of discounting using a two-period model of climate change resulting from combustion of fossil fuels. The strategy is to model the welfare of utility of the present generation, U_1, to compare to the expected utility of future generations, $E(U_2)$. We try to capture two explicit features of the decision problem. First, current depletion of fossil fuels by the present generation, D_1, reduces the availability of fossil fuels for future consumption. Thus, if the initial stock of fossil fuels is S, the remaining quantity available for future generations, is $D_2 = S - D_1$. Second, depletion by the first generation results in the possibility of losing a fraction $f(D_1)$, $0 \leq f \leq 1$, of the capital stock, K_2, of future generations. This might occur because of flooding of coastal areas or more variable future climates. We assume that climate change is uncertain so there is a probability, P_f, of the reduction of the future's capital stock from K_2 to $(1-f)\cdot K_2$.

Utility of the initial generation then depends on available consumption. Initial output, F_1, is taken to be a function of the capital stock K_1 (given for our analysis) and fossil fuel depletion D_1. Thus, output in period 1 is

(1) $$F_1 = F(K_1, D_1),$$

where F is a strictly concave production function. Where the first generation's investment in future generations is given as I_1, consumption of the first generation is $F_1 - I_1$ so utility of the first generation is

(2) $$U_1(F_1 - I_1).$$

Expected utility of the future depends on available consumption in two possible states of the world. If climate change does not occur, output is a function of capital, $K_2 = K_1 + I_1$, and fossil fuel availability, $D_2 = S - D_1$. Thus, given the same production function defined in (1), output is

(3) $$F_2 = F(K_1 + I_1, S - D_1),$$

with probability $(1 - P_f)$. If climate change does occur then output is

(4) $$\tilde{F}_2 = F([K_1 + I_1]\cdot[1 - f(D_1)], S - D_1),$$

with probability P_f. Given that we represent the utility of the entire future as U_2, no investment in any succeeding generation occurs, so consumption is equal to output for "generation two." Thus, expected utility is

(5) $$E(U_2) = (1 - P_f)\cdot U_2(F_2) + P_f \cdot U_2(\tilde{F}_2),$$

or the probability weighted sum of future utility if climate change does not occur and future utility if climate change does occur.

We assume that no individuals overlap between the present and succeeding generations and view the decision on I_1, investment in the future by the first generation, and depletion of fossil fuels by generation 1, D_1, in a purely ethical context. Ethical solutions to the decision problem over investment and fossil fuel depletion are derived from the following simple optimization problems over different ethical systems.

(6) Utilitarian (Benthamite)

$$max\, U_1 + E(U_2),$$

(7) Total Egalitarian (Rawlsian)

$$max\, min\{U_1, E(U_2)\},$$

(8) Total Elitist (Nietzschean)

$$max\, max\{U_1, E(U_2)\},$$

(9) Libertarian (Paretian)

$$max\, U_1 \text{ such that } E(U_2) \geq U_2^0,$$

where U_1 and $E(U_2)$ are defined by (1)–(5) above and U_2^0 in (9) is the initial position (in expected utility) for the future. We analyze outcomes for each of the four criteria below. However, each of the criteria yields as a condition for use of fossil fuels in the present ($D_1 > 0$) that

(10) $$F_{D_1} > \left(\frac{1}{1+d}\right)(F_{D_2} + P_f \cdot F_{K_2} \cdot f' K_2),$$

when the terms in (10) are evaluated at $D_1 = 0$. Expression (10) simply states that the marginal benefits of fuel use in period 1 (increase in output or *GNP* for an increase in D_1, F_{D1}) must exceed the discounted (d is defined as the discount rate) sum of marginal benefits of future fossil fuel use (F_{D2}) plus marginal expected damages to the future combustion in period one ($P_f F_{K2} f' K_2$). If the inequality in (10) is reversed, then optimally $D_1 = 0$. The ethical criteria are entirely embodied in the discount rate, d, which weights relative benefits and costs to present vs. future generations. Thus, the analysis of ethical criteria here takes the form of an analysis of the discount rate.

TABLE 1—DISCOUNT RATES AND ETHICAL SYSTEMS

Ethical System	Compensation to Future (Investment) is Possible ($I \geq 0$)	Compensation to Future (Investment) is not Possible ($I \equiv 0$)
Utilitarian:		
$max\ U_1 + E(U_2)$	$d = r$	$+\infty > d > -1$
Rawlsian:		
Case a		
$max\ U_1 < E(U_2)$	$d = +\infty$	$d = +\infty$
Case b		
$U_1 = E(U_2)$	$d \geq r$	$+\infty > d > -1$
Case c		
$U_2^0 < U_1^0$ or $E(U_2) < U_1$	$d = r$	$d = -1$
Nietzschean:		
Case a		
$max\ U_1 > max\ E(U_2)$	$d = +\infty$	$d = +\infty$
Case b		
$max\ E(U_2) > max\ U_1$	$d = r$	$d = -1$
Libertarian:		
$max\ U_1$ such that $E(U_2) \geq U_2^0$	$d = r$	$+\infty > d > -1$

As shown in Table 1, the discount rate in cases where investment in the future by the present, I_1, *is* possible and occurs is generally equal to the expected market rate of return of future investment, r, which is defined, in our model as

$$(11) \qquad r = (1 - P_f f) \cdot F_{K_2} - 1.$$

If compensation is not possible, the discount rate tends to take on values of -1 or $+\infty$ depending on the ethical criterion as shown in Table 1. Given the analysis of discount rates presented, the implications for present use of fossil fuels or regulating CO_2 emissions generally can be obtained by examining expression (10) (Schulze, Brookshire, and Todd Sandler, forthcoming). Clearly, if $d = \infty$ then $D_1 > 0$. If $d = -1$ then $D_1 = 0$. In other words, an infinite discount rate implies fossil fuels should be used to the maximum desirable level in the present, a discount rate of -1 implies additional fossil fuel use is "unethical."

Perhaps the most important result of the preceeding analysis is that ethics do matter for economic analysis. The traditional economic criterion of discounting future benefits and costs at the market rate of return only holds in special ethical cases and where *actual* compensation or investment in the future is possible. Economists often use the notion of "hypothetical" compensation to justify discounting. In an ethical context such arguments play no role whatsoever. Rather, if no actual compensation occurs, the market rate of return has no relevance for discount rates. Discount rates are then determined solely by the ethical criteria employed in the analysis.

III. Conclusions

Depending on ethical beliefs, different decisions emerge for resolving the CO_2 issue. It is doubtful that an international consensus can be reached on a "correct" ethical criterion. Perhaps the best strategy would be to delay acceptance of either a particular set of beliefs or the existing scientific evidence and wait for more accurate and conclusive research to emerge. If the scientific evidence is accepted as valid, and all future generations that will exist are evaluated equally, then the optimal current regulatory strategy is to restrict, as much as possible, current emissions of CO_2 (d'Arge and K. C. Kogiku, 1973).

REFERENCES

Attaran, K., "Inter-Metropolitan Budget Cost Determinants: An Econometric Analysis," unpublished doctoral dissertation, University of Southern California, 1973.

Bailey, Martin, "Uncertainties and Benefit-Cost Analysis of CFC Control," University of Maryland, February 1, 1980.

d'Arge, Ralph C., "Climate and Economic Activity," paper presented to the World Climate Conference, WMO, Geneva, February 1979.

______ **and Kogiku, K. C.,** "Economic Growth and the Environment," *Review of Economic Studies*, January 1973, *40*, 61–77.

______ **et al.,** *Economic and Social Measures of Biologic and Climatic Change*, U.S. Department of Transportation, 1975.

______, *Benefits and Costs of Regulating Fluorocarbons 11–12*, Final Report, University of Wyoming to U.S. Environmental Protection Agency, 1976.

Eckstein, Otto, *Water Resource Development: The Economics of Project Evaluation*, Cambridge: Harvard University Press, 1958.

Hoch, I., "Climate, Energy Use, and Wages," in J. H. Cumberland, ed., *Economic Aspects of Effects Upon Health and Climate from the Management and Control of Ozone Depletion*, Washington: U.S. Environmental Protection Agency, 1979.

Kellogg, William W., "Influences of Mankind on Climate," *Annual Review of Earth and Planetary Sciences*, 1979.

Krutilla, John V., "Welfare Aspects of Benefit-Cost Analysis," in S. C. Smith and E. N. Castle, eds., *Economics and Public Policy in Water Resource Development*, Ames: University of Iowa Press, 1964.

______, "Conservation Reconsidered," *American Economic Review*, September 1967, *57*, 777–86.

Laurman, J. A., "Economic Impact of CO_2-Induced Climatic Change," Division of Applied Mechanics, Stanford University, July 1979.

Neidercorn, J., "The Capital Costs of Climatically Induced Shifts: The Example of the American Corn Belt," in T. Ferrar, ed., *The Value Costs of Climate Modification*, New York: John Wiley and Sons, 1976.

Schulze, William D., Brookshire, David S. and Sandler, Todd, "Economics and Ethics?: Evaluating the Risks of Storing Nuclear Waste," *Natural Resources Journal*, forthcoming.

Wittwer, S., Remarks to the National Academy of Sciences Workshop on *Critical Future Environmental Issues*, Washington, November 1979.

Woodwell, George et al., "The Carbon Dioxide Problem: Implications for Policy in the Management of Energy and Other Resources," Report to the Council on Environmental Quality, Washington, July 1979.

Panel on Energy and Climate, *Energy and Climate*, National Academy of Sciences, Washington, 1977.

Name Index

The International Library of Critical Writings in Economics

1. Multinational Corporations
   Mark Casson
2. The Economics of Innovation
   Christopher Freeman
3. Entrepreneurship
   Mark Casson
4. International Investment
   Peter J. Buckley
5. Game Theory in Economics
   Ariel Rubinstein
6. The History of Economic Thought
   Mark Blaug
7. Monetary Theory
   Thomas Mayer
8. Joint Production of Commodities
   Neri Salvadori and Ian Steedman
9. Industrial Organization
   Oliver E. Williamson
10. Growth Theory (Volumes I, II and III)
    R. Becker and E. Burmeister
11. Microeconomics: Theoretical and Applied (Volumes I, II and III)
    Robert E. Kuenne
12. The Economics of Health (Volumes I and II)
    A.J. Culyer
13. Recent Developments in Macroeconomics (Volumes I, II and III)
    Edmund S. Phelps
14. Urban and Regional Economics
    Paul C. Cheshire and Alan W. Evans
15. Modern Public Finance (Volumes I and II)
    A.B. Atkinson
16. Exchange Rate Economics (Volumes I and II)
    Ronald MacDonald and Mark P. Taylor

17. The Economic Value of Education: Studies in the Economics of Education
Mark Blaug

18. Development Economics (Volumes I, II, III and IV)
Deepak Lal

19. The New Classical Macroeconomics (Volumes I, II and III)
Kevin D. Hoover

20. The Economics of the Environment
Wallace E. Oates

21. Post Keynesian Theory of Growth and Distribution
Carlo Panico and Neri Salvadori

22. Dynamic Labor Demand and Adjustment Costs
Giorgio Galeazzi and Daniel S. Hamermesh

23. The Philosophy and Methodology of Economics (Volumes I, II and III)
Bruce J. Caldwell

24. Public Choice Theory (Volumes I, II and III)
Charles K. Rowley

25. Evolutionary Economics
Ulrich Witt

26. Economics and Psychology
Shlomo Maital and Sharone L. Maital

27. Social Choice Theory (Volumes I, II and III)
Charles K. Rowley

28. Non-Linear Dynamics in Economic Theory
Marc Jarsulic

29. Recent Developments in Experimental Economics (Volumes I and II)
John D. Hey and Graham Loomes

30. Monopoly and Competition Policy (Volumes I and II)
F.M. Scherer

31. The Economics of Technical Change
Edwin Mansfield and Elizabeth Mansfield

32. The Economics of Exhaustible Resources
Geoffrey Heal

33. The Economics of Institutions
Geoffrey M. Hodgson

34. The Economics of Transport (Volumes I and II)
Herbert Mohring

35. Implicit Contract Theory
Sherwin Rosen

36. Foundations of Analytical Marxism (Volumes I and II)
John E. Roemer

37. The Economics of Product Differentiation (Volumes I and II)
Jacques-François Thisse and George Norman

38. Economic Growth in Theory and Practice: A Kaldorian Perspective
John E. King

39. Markets and Socialism
Alec Nove and Ian D. Thatcher

40. Recent Developments in the Economics of Education
Elchanan Cohn and Geraint Johnes

41. The Theory of Inflation
Michael Parkin

42. The Economics of Location (Volumes I, II and III)
Melvin L. Greenhut and George Norman

43. Financial Intermediaries
Mervyn K. Lewis

44. The Political Economy of Privatization and Deregulation
Elizabeth E. Bailey and Janet Rothenberg Pack

45. Gender and Economics
Jane Humphries

46. Macroeconomics and Imperfect Competition
Jean-Pascal Bénassy

47. Labor Economics (Volumes I, II, III and IV)
Orley C. Ashenfelter and Kevin F. Hallock

48. The Economics of Altruism
Stefano Zamagni

49. The Economic Analysis of Rent Seeking
Robert D. Tollison and Roger D. Congleton

50. Economics and Biology
Geoffrey M. Hodgson

51. The Economics of Ageing
John Creedy

52. Fiscal and Monetary Policy (Volumes I and II)
Thomas Mayer and Steven M. Sheffrin

53. The Economics of Information (Volumes I and II)
David K. Levine and Steven A. Lippman

54. Transaction Cost Economics (Volumes I and II)
Oliver E. Williamson and Scott E. Masten

55. Agricultural Economics
George H. Peters

56. International Debt (Volumes I and II)
Graham Bird and P. Nicholas Snowden

57. Economics and Discrimination (Volumes I and II)
William A. Darity, Jr

58. Business Cycle Theory
Finn E. Kydland

59. International Trade (Volumes I and II)
J. Peter Neary

60. The Money Supply in the Economic Process
Marco Musella and Carlo Panico

61. Small Firms and Economic Growth (Volumes I and II)
Zoltan J. Acs

62. Producer Cooperatives and Labor-Managed Systems (Volumes I and II)
David L. Prychitko and Jaroslav Vanek

63. Ethics and Economics (Volumes I and II)
Alan P. Hamlin

64. The Economics of the Family
Nancy Folbre

65. The Economics of Training (Volumes I and II)
Orley C. Ashenfelter and Robert J. LaLonde

66. Chaos Theory in Economics: Methods, Models and Evidence
W. Davis Dechert

67. General Equilibrium Theory (Volumes I, II and III)
Gérard Debreu

68. Economic Growth: Theory and Evidence (Volumes I and II)
Gene M. Grossman

69. Long Wave Theory
Christopher Freeman

70. The Economics of Communication and Information
Donald M. Lamberton

71. The Foundations of Public Finance (Volumes I and II)
Peter Jackson

72. The Theory of the Firm
Mark Casson

73. The Economics of Uncertainty (Volumes I and II)
John D. Hey

74. The Economics of Global Warming
Tom Tietenberg

Future titles will include:

The Development of Ecological Economics
Robert Costanza, Charles Perrings and Cutler Cleveland

Law and Economics
Richard A. Posner and Francesco Parisi

Economic Growth in the Long Run
Bart van Ark

Political Business Cycles
Bruno Frey

Cultural Economics: The Arts, the Heritage and the Media Industries
Ruth Towse

International Finance
Robert Z. Aliber

Welfare Economics
William J. Baumol and Janusz A. Ordover

The Economics of Inequality and Poverty
A.B. Atkinson

The Economics of Housing
John M. Quigley

Population Economics
Julian L. Simon

The Economics of Crime
Isaac Ehrlich

The Economics of Integration
Willem Molle

The Rhetoric of Economics
Deirdre McCloskey

Migration
Oded Stark

The Economics of Defence
Keith Hartley and Nicholas Hooper

Consumer Theory
Kelvin Lancaster

The Economics of Business Policy
John Kay

Microeconomic Theories of Imperfect Competition
Jacques Thisse and Jean Gabszewicz

The Economics of Increasing Returns
Geoffrey Heal

The Balance of Payments
Michael J. Artis

Cost-Benefit Analysis
Arnold Harberger and Glenn P. Jenkins

The Economics of Unemployment
P.N. Junankar

Mathematical Economics
Graciela Chichilnisky

Gender in Economic and Social History
K.J. Humphries and J. Lewis

The Economics of Local Finance and Fiscal Federalism
Wallace Oates

Privatization in Developing and Transitional Economies
Colin Kirkpatrick and Paul Cook

Input-Output Analysis
Heinz Kurz, Erik Dietzenbacher and Christian Lager

The Economics of Energy
Paul Stevens

The Economics of Intellectual Property
Ruth Towse

The Economics of Productivity
Edward Wolff

The Economics of Tourism
Clem Tisdell

The Economics of Organization and Bureaucracy
Peter Jackson

Independent Central Banks and Economic Performance
Sylvester Eijffinger

The Economics of Commodity Markets
David Greenaway and Wyn Morgan

Realism and Economics: Studies in Ontology
Tony Lawson

Women in the Labor Market
Marianne A. Ferber

New Developments in Game Theory
Eric S. Maskin

Economic Demography
T. Paul Schultz

The International Economic Institutions of the Twentieth Century
David Greenaway

The Economics of Structural Change
Harald Hagemann, Michael Landesmann and Roberto Scazzieri

Market Process Theories
Peter Boettke and David Prychitko